OXFORD PAPERBACK REFERENCE

A Dictionary of Chemistry

:NGTON

Oxford Paperback Reference

The most authoritative and up-to-date reference books for both students and the general reader.

A Dictionary of Chemistry

FIFTH EDITION

Edited by
JOHN DAINTITH

OXFORD
UNIVERSITY PRESS
Great Clarendon Street, Oxford OX2 6DP

Oxford University Press is a department of the University of Oxford.
It furthers the University's objective of excellence in research, scholarship,
and education by publishing worldwide in

Oxford New York

Auckland Bangkok Buenos Aires Cape Town Chennai
Dar es Salaam Delhi Hong Kong Istanbul Karachi Kolkata
Kuala Lumpur Madrid Melbourne Mexico City Mumbai Nairobi
São Paulo Shanghai Singapore Taipei Tokyo Toronto

Oxford is a registered trade mark of Oxford University Press
in the UK and in certain other countries

First published 1985 as *A Concise Dictionary of Chemistry*
Second edition 1990
Third edition 1996
Fourth edition 2000
Fifth edition 2004

British Library Cataloguing in Publication Data
Data available

Library of Congress Cataloging in Publication Data
Data available

ISBN-13: 978-0-19-860918-6
ISBN-10: 0-19-860918-3

3

Typeset in Swift by Market House Books Ltd.

Printed in Great Britain by Clays Ltd, St Ives plc

Preface

This dictionary was originally derived from the *Concise Science Dictionary*, first published by Oxford University Press in 1984 (fourth edition, retitled *Dictionary of Science*, 1999). It consisted of all the entries relating to chemistry in this dictionary, including physical chemistry, as well as many of the terms used in biochemistry. Subsequent editions included special feature articles on important topics as well as several chronologies tracing the history of some topics and short biographical entries on the chemists and other scientists who have been responsible for the development of the subject. For this fifth edition the text has been revised, some entries have been substantially expanded, and over 160 new entries have been added covering all branches of the subject.

An asterisk placed before a word used in an entry indicates that this word can be looked up in the dictionary and will provide further explanation or clarification. However, not every word that appears in the dictionary has an asterisk placed before it. Some entries simply refer the reader to another entry, indicating either that they are synonyms or abbreviations or that they are most conveniently explained in one of the dictionary's longer articles or features. Synonyms and abbreviations are usually placed within brackets immediately after the headword. Terms that are explained within an entry are highlighted by being printed in italic type.

The more physical aspects of physical chemistry and the physics itself will be found in *A Dictionary of Physics*, which is a companion volume to this dictionary. *A Dictionary of Biology* contains a more thorough coverage of the biophysical and biochemical entries from the *Dictionary of Science* together with the entries relating to biology.

SI units are used throughout this book and its companion volumes.

J.D.

2004

Editor

John Daintith BSc, PhD

Advisers

B. S. Beckett BSc, BPhil, MA (Ed)
R. A. Hands BSc
Michael Lewis MA

Contributors

John Clark BSc
H. M. Clarke MA, MSc
R. Cutler BSc
Derek Cooper PhD, FRIC
D. E. Edwards BSc, MSc
Richard Rennie PhD, MSc, BSc
David Eric Ward BSc, MSc, PhD

AAS *See* **atomic absorption spectroscopy.**

abherent *See* **release agent.**

ab-initio calculation A method of calculating atomic and molecular structure directly from the first principles of quantum mechanics, without using quantities derived from experiment (such as ionization energies found by spectroscopy) as parameters. Ab-initio calculations require a large amount of numerical computation; the amount of computing time required increases rapidly as the size of the atom or molecule increases. The development of computing power has enabled the properties of both small and large molecules to be calculated accurately, so that this form of calculation can now replace *semi-empirical calculations. Ab initio calculations can, for example, be used to determine the bond lengths and bond angles of molecules by calculating the total energy of the molecule for a variety of molecular geometries and finding which conformation has the lowest energy.

absolute 1. Not dependent on or relative to anything else, e.g. *absolute zero. **2.** Denoting a temperature measured on an *absolute scale*, a scale of temperature based on absolute zero. The usual absolute scale now is that of thermodynamic *temperature; its unit, the kelvin, was formerly called the *degree absolute* (°A) and is the same size as the degree Celsius. In British engineering practice an absolute scale with Fahrenheit size degrees has been used: this is the Rankine scale.

absolute alcohol *See* **ethanol.**

absolute configuration A way of denoting the absolute structure of an optical isomer (*see* **optical activity**). Two conventions are in use: The D–L convention relates the structure of the molecule to some reference molecule. In the case of sugars and similar compounds, the dextrorotatory form of glyceraldehyde ($HOCH_2CH(OH)CHO$), 2,3-dihydroxypropanal) was used. The rule is as follows. Write the structure of this molecule down with the asymmetric carbon in the centre, the –CHO group at the top, the –OH on the right, the $–CH_2OH$ at the bottom, and the –H on the left. Now imagine that the central carbon atom is at the centre of a tetrahedron with the four groups at the corners and that the –H and –OH come out of the paper and the –CHO and $–CH_2OH$ groups go into the paper. The resulting three-dimensional structure was taken to be that of *d*-glyceraldehyde and called D-glyceraldehyde. Any compound that contains an asymmetric carbon atom having this configuration belongs to the D-series. One having the opposite configuration belongs to the L-series. It is important to note that the prefixes D- and L- do not stand for dextrorotatory and laevorotatory (i.e. they are not the same as *d*- and *l*-). In

CHO / H—C—OH / CH_2OH — planar formula

CHO / C / H, OH / CH_2OH — structure in 3 dimensions

CHO / HCOH / CH_2OH — Fischer projection

D-(+)-glyceraldehyde (2,3-dihydroxypropanal)

H / C / CH_3, COOH / NH_2

COOH / H / CH_3, NH_2

D-alanine (R is CH_3 in the CORN rule). The molecule is viewed with H on top

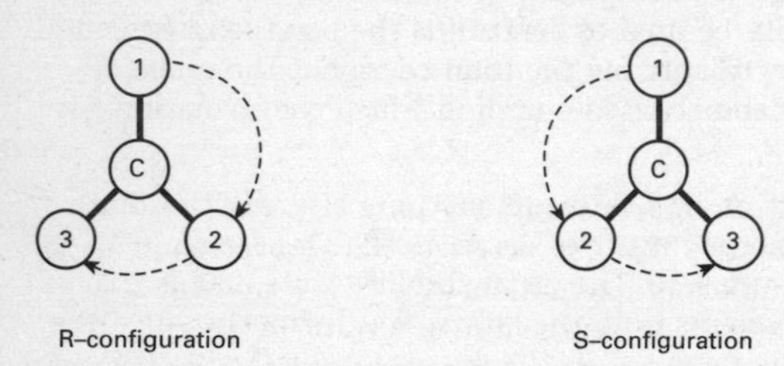

R–S system. The lowest priority group is behind the chiral carbon atom

fact the arbitrary configuration assigned to D-glyceraldehyde is now known to be the correct one for the dextrorotatory form, although this was not known at the time. However, all D-compounds are not dextrorotatory. For instance, the acid obtained by oxidizing the –CHO group of glyceraldehyde is glyceric acid (1,2-dihydroxypropanoic acid). By convention, this belongs to the D-series, but it is in fact laevorotatory; i.e. its name can be written as D-glyceric acid or *l*-glyceric acid. To avoid confusion it is better to use + (for dextrorotatory) and – (for laevorotatory), as in D-(+)-glyceraldehyde and D-(–)-glyceric acid.

The D–L convention can also be used with alpha amino acids (compounds with the $-NH_2$ group on the same carbon as the –COOH group). In this case the molecule is imagined as being viewed along the H–C bond between the hydrogen and the asymmetric carbon atom. If the clockwise order of the other three groups is –**CO**OH, –**R**, $-\mathbf{N}H_2$, the amino acid belongs to the D-series; otherwise it belongs to the L-series. This is known as the *CORN rule.*

The R–S convention is a convention based on priority of groups attached to the chiral carbon atom. The order of priority is I, Br, Cl, SO_3H, $OCOCH_3$, OCH_3, OH, NO_2, NH_2, $COOCH_3$, $CONH_2$, $COCH_3$, CHO, CH_2OH, C_6H_5, C_2H_5, CH_3, H, with hydrogen lowest. The molecule is viewed with the group of lowest priority behind the chiral atom. If the clockwise arrangement of the other three groups is in descending priority, the compound belongs to the R-series; if the descending order is anticlockwise it is in the S-series. D-(+)-glyceraldehyde is R-(+)-glyceraldehyde. See illustration.

absolute temperature *See* **absolute; temperature.**

absolute zero Zero of thermodynamic *temperature (0 kelvin) and the lowest temperature theoretically attainable. It is the temperature at which the kinetic energy of atoms and molecules is minimal. It is equivalent to −273.15°C or −459.67°F. *See also* **zero-point energy**.

absorption 1. ***(in chemistry)*** The take up of a gas by a solid or liquid, or the take up of a liquid by a solid. Absorption differs from *adsorption in that the absorbed substance permeates the bulk of the absorbing substance. **2.** ***(in physics)*** The conversion of the energy of electromagnetic radiation, sound, streams of particles, etc., into other forms of energy on passing through a medium. A beam of light, for instance, passing through a medium, may lose intensity because of two effects: scattering of light out of the beam, and absorption of photons by atoms or molecules in the medium. When a photon is absorbed, there is a transition to an excited state.

absorption coefficient 1. ***(in spectroscopy)*** The *molar absorption coefficient* (symbol ε) is a quantity that characterizes the absorption of light (or any other type of electromagnetic radiation) as it passes through a sample of the absorbing material. It has the dimensions of 1/(concentration × length). ε is dependent on the frequency of the incident light; its highest value occurs where the absorption is most intense. Since absorption bands usually spread over a range of values of the frequency ν it is useful to define a quantity called the *integrated absorption coefficient*, *A*, which is the integral of all the absorption coefficients in the band, i.e. $A = \int\varepsilon(\nu)d\nu$. This quantity characterizes the intensity of a transition. *See also* **Beer–Lambert law. 2.** The volume of a given gas, measured at standard temperature and pressure, that will dissolve in unit volume of a given liquid.

absorption indicator *See* **adsorption indicator.**

absorption spectrum *See* **spectrum.**

absorption tower A long vertical column used in industry for absorbing gases. The gas is introduced at the bottom of the column and the absorbing liquid, often water, passes in at the top and falls down against the countercurrent of gas. The towers are also known as *scrubbers*.

ABS plastic Any of a class of plastics based on acrylonitrile–butadiene–styrene copolymers.

abstraction A chemical reaction that involves bimolecular removal of an atom or ion from a molecule. An example is the abstraction of hydrogen from methane by reaction with a radical:

$CH_4 + X\cdot \rightarrow H_3C\cdot + HX.$

abundance 1. The ratio of the total mass of a specified element in the earth's crust to the total mass of the earth's crust, often expressed as a percentage. For example, the abundance of aluminium in the earth's crust is about 8%. **2.** The ratio of the number of atoms of a particular isotope of an element to the total number of atoms of all the isotopes present, often expressed as a percentage. For example, the abundance of uranium-235 in natural uranium is 0.71%. This is the *natural abundance*, i.e. the abundance as found in nature before any enrichment has taken place.

ac Anticlinal. *See* **torsion angle.**

accelerator A substance that increases the rate of a chemical reaction, i.e. a catalyst.

acceptor 1. ***(in chemistry and biochemistry)*** A compound, molecule, ion, etc., to which electrons are donated in the formation of a coordinate bond. **2.** ***(in physics)*** A substance that is added as an impurity to a *semiconductor because of its ability to accept electrons from the valence bands, causing *p*-type conduction by the mobile positive holes left. *Compare* **donor.**

accessory pigment A *photosynthetic pigment that traps light energy and channels it to chlorophyll *a*, the primary pigment, which initiates the reactions of photosynthesis. Accessory pigments include the carotenes and chlorophylls *b*, *c*, and *d*.

accumulator (secondary cell; storage battery) A type of *voltaic cell or battery that can be recharged by passing a current through it from an external d.c. supply. The charging current, which is passed in the opposite direction to that in which the cell supplies current, reverses the chemical reactions in the cell. The common types are the *lead–acid accumulator, the *nickel–iron accumulator, and the *nickel–cadmium cell. *See also* **sodium–sulphur cell.**

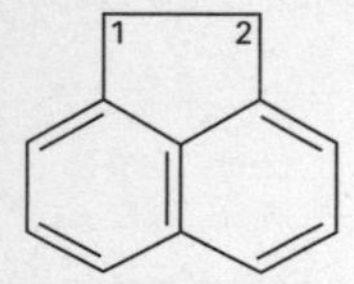

Acenaphthene

acenaphthene A colourless crystalline aromatic compound, $C_{12}H_{10}$; m.p. 95°C; b.p. 278°C. It is an intermediate in the production of some dyes.

acetaldehyde *See* **ethanal.**

acetaldol *See* **aldol reaction.**

acetals Organic compounds formed by addition of alcohol molecules to aldehyde molecules. If one molecule of aldehyde (RCHO) reacts with one

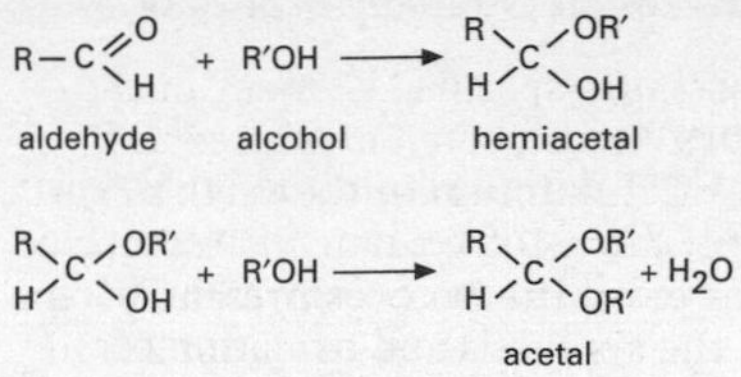

$$R{-}C(=O)H + R'OH \longrightarrow RC(H)(OR')(OH)$$

aldehyde alcohol hemiacetal

$$RC(H)(OR')(OH) + R'OH \longrightarrow RC(H)(OR')(OR') + H_2O$$

acetal

Formation of acetals

molecule of alcohol (R′OH) a *hemiacetal* is formed (RCH(OH)OR′). The rings of aldose sugars are hemiacetals. Further reaction with a second alcohol molecule produces a full acetal ($RCH(OR')_2$). It is common to refer to both types of compound simply as 'acetals'. The formation of acetals is reversible; acetals can be hydrolysed back to aldehydes in acidic solutions. In synthetic organic chemistry aldehyde groups are often converted into acetal groups to protect them before performing other reactions on different groups in the molecule. *See also* **ketals.**

acetamide *See* **ethanamide.**

acetanilide A white crystalline primary amide of ethanoic acid, $CH_3CONHC_6H_5$; r.d. 1.2; m.p. 114.3°C; b.p. 304°C. It is made by reacting phenylamine (aniline) with excess ethanoic acid or ethanoic anhydride and is used in the manufacture of dyestuffs and rubber. The full systematic name is N-*phenylethanamide.*

acetate *See* **ethanoate.**

acetate process *See* **rayon.**

acetic acid *See* **ethanoic acid.**

acetic anhydride *See* **ethanoic anhydride.**

acetoacetic acid *See* **3-oxobutanoic acid.**

acetoacetic ester *See* **ethyl 3-oxobutanoate.**

acetone *See* **propanone.**

acetonitrile *See* **ethanenitrile.**

acetophenone *See* **phenyl methyl ketone.**

acetylating agent *See* **ethanoylating agent.**

acetylation *See* **acylation.**

acetyl chloride *See* **ethanoyl chloride.**

acetylcholine A substance that is released at some (*cholinergic*) nerve endings. Its function is to pass on a nerve impulse to the next nerve (i.e. at a synapse) or to initiate muscular contraction. Once acetylcholine has been released, it has only a transitory effect because it is rapidly broken down by the enzyme *cholinesterase*.

acetyl coenzyme A (acetyl CoA) A compound formed in the mitochondria when an acetyl group (CH_3CO–), derived from the breakdown of fats, proteins, or carbohydrates (via *glycolysis), combines with the thiol group (–SH) of *coenzyme A. Acetyl CoA feeds into the energy generating the *Krebs cycle and also plays a role in the synthesis and oxidation of fatty acids.

acetylene *See* **ethyne.**

acetylenes *See* **alkynes.**

acetyl group *See* **ethanoyl group.**

acetylide *See* **carbide.**

Acheson process An industrial process for the manufacture of graphite by heating coke mixed with clay. The reaction involves the production of silicon carbide, which loses silicon at 4150°C to leave graphite. The process was patented in 1896 by the US inventor Edward Goodrich Acheson (1856–1931).

acid 1. A type of compound that contains hydrogen and dissociates in water to produce positive hydrogen ions. The reaction, for an acid HX, is commonly written:

$$HX \rightleftharpoons H^+ + X^-$$

In fact, the hydrogen ion (the proton) is solvated, and the complete reaction is:

$$HX + H_2O \rightleftharpoons H_3O^+ + X^-$$

The ion H_3O^+ is the *oxonium ion* (or *hydroxonium ion* or *hydronium ion*). This definition of acids comes from the *Arrhenius theory*. Such acids tend to be corrosive substances with a sharp taste, which turn litmus red and give colour changes with other *indicators. They are referred to as *protonic acids* and are classified into *strong acids*, which are almost completely dissociated in water (e.g. sulphuric acid and hydrochloric acid), and *weak acids*, which are only partially dissociated (e.g. ethanoic acid and hydrogen sulphide). The strength of an acid depends on the extent to which it dissociates, and is measured by its *dissociation constant. *See also* **base.**
2. In the *Lowry–Brønsted theory* of acids and bases (1923), the definition was extended to one in which an acid is a proton donor (a *Brønsted acid*), and a base is a proton acceptor (a *Brønsted base*). For example, in

$$HCN + H_2O \rightleftharpoons H_3O^+ + CN^-$$

the HCN is an acid, in that it donates a proton to H_2O. The H_2O is acting as a base in accepting a proton. Similarly, in the reverse reaction H_3O^+ is an acid and CN^- a base. In such reactions, two species related by loss or gain of a proton are said to be *conjugate*. Thus, in the reaction above HCN is the *conjugate acid* of the base CN^-, and CN^- is the *conjugate base* of the acid

HCN. Similarly, H_3O^+ is the conjugate acid of the base H_2O. An equilibrium, such as that above, is a competition for protons between an acid and its conjugate base. A strong acid has a weak conjugate base, and vice versa. Under this definition water can act as both acid and base. Thus in

$$NH_3 + H_2O \rightleftharpoons NH_4^+ + OH^-$$

the H_2O is the conjugate acid of OH^-. The definition also extends the idea of acid–base reaction to solvents other than water. For instance, liquid ammonia, like water, has a high dielectric constant and is a good ionizing solvent. Equilibria of the type

$$NH_3 + Na^+Cl^- \rightleftharpoons Na^+NH_2^- + HCl$$

can be studied, in which NH_3 and HCl are acids and NH_2^- and Cl^- are their conjugate bases.

3. A further extension of the idea of acids and bases was made in the *Lewis theory* (G. N. Lewis, 1923). In this, a *Lewis acid* is a compound or atom that can accept a pair of electrons and a *Lewis base* is one that can donate an electron pair. This definition encompasses 'traditional' acid–base reactions. In

$$HCl + NaOH \rightarrow NaCl + H_2O$$

the reaction is essentially

$$H^+ + :OH^- \rightarrow H:OH$$

i.e. donation of an electron pair by OH^-. But it also includes reactions that do not involve ions, e.g.

$$H_3N: + BCl_3 \rightarrow H_3NBCl_3$$

in which NH_3 is the base (donor) and BCl_3 the acid (acceptor). The Lewis theory establishes a relationship between acid–base reactions and *oxidation–reduction reactions.

See also **aqua acid; hydroxoacid; oxoacid.**

R.CO.OH + HO.CO.R′ $\xrightarrow{-H_2O}$ R.CO.O.CO.R′

carboxylic acids — acid anhydride

Formation of a carboxylic acid anhydride

acid anhydrides (acyl anhydrides) Compounds that react with water to form an acid. For example, carbon dioxide reacts with water to give carbonic acid:

$$CO_2(g) + H_2O(aq) \rightleftharpoons H_2CO_3(aq)$$

A particular group of acid anhydrides are anhydrides of carboxylic acids. They have a general formula of the type R.CO.O.CO.R′, where R and R′ are alkyl or aryl groups. For example, the compound ethanoic anhydride (CH_3.CO.O.CO.CH_3) is the acid anhydride of ethanoic (acetic) acid. Organic acid anhydrides can be produced by dehydrating acids (or mixtures of acids). They are usually made by reacting an acyl halide with the sodium salt of the acid. They react readily with water, alcohols, phenols, and amines and are used in *acylation reactions.

acid–base indicator *See* indicator.

acid dissociation constant *See* dissociation.

acid dye *See* dyes.

acid halides *See* acyl halides.

acidic **1.** Describing a compound that is an acid. **2.** Describing a solution that has an excess of hydrogen ions. **3.** Describing a compound that forms an acid when dissolved in water. Carbon dioxide, for example, is an acidic oxide.

acidic hydrogen (acid hydrogen) A hydrogen atom in an *acid that forms a positive ion when the acid dissociates. For instance, in methanoic acid

$$HCOOH \rightleftharpoons H^+ + HCOO^-$$

the hydrogen atom on the carboxylate group is the acidic hydrogen (the one bound directly to the carbon atom does not dissociate).

acidimetry Volumetric analysis using standard solutions of acids to determine the amount of base present.

acidity constant *See* dissociation.

acid rain Precipitation having a pH value of less than about 5.0, which has adverse effects on the fauna and flora on which it falls. Rainwater typically has a pH value of 5.6, due to the presence of dissolved carbon dioxide (forming carbonic acid). Acid rain results from the emission into the atmosphere of various pollutant gases, in particular sulphur dioxide and various oxides of nitrogen, which originate from the burning of fossil fuels and from car exhaust fumes, respectively. These gases dissolve in atmospheric water to form sulphuric and nitric acids in rain, snow, or hail (*wet deposition*). Alternatively, the pollutants are deposited as gases or minute particles (*dry deposition*). Both types of acid deposition affect plant growth – by damaging the leaves and impairing photosynthesis and by increasing the acidity of the soil, which results in the leaching of essential nutrients. This acid pollution of the soil also leads to acidification of water draining from the soil into lakes and rivers, which become unable to support fish life. Lichens are particularly sensitive to changes in pH and can be used as indicators of acid pollution.

acid salt A salt of a polybasic acid (i.e. an acid having two or more acidic hydrogens) in which not all the hydrogen atoms have been replaced by positive ions. For example, the dibasic acid carbonic acid (H_2CO_3) forms acid salts (hydrogencarbonates) containing the ion HCO_3^-. Some salts of monobasic acids are also known as acid salts. For instance, the compound potassium hydrogendifluoride, KHF_2, contains the ion $[F...H–F]^-$, in which there is hydrogen bonding between the fluoride ion F^- and a hydrogen fluoride molecule.

acid value A measure of the amount of free acid present in a fat, equal to the number of milligrams of potassium hydroxide needed to neutralize this acid. Fresh fats contain glycerides of fatty acids and very little free acid, but the glycerides decompose slowly with time and the acid value increases.

Acridine

acridine A colourless crystalline heterocyclic compound, $C_{12}H_9N$; m.p. 110°C. The ring structure is similar to that of anthracene, with three fused rings, the centre ring containing a nitrogen heteroatom. Several derivatives of acridine (such as acridine orange) are used as dyes or biological stains.

Acrilan A tradename for a synthetic fibre. *See* **acrylic resins.**

acrolein *See* **propenal.**

acrylamide An inert gel (polyacrylamide) employed as a medium in *electrophoresis. It is used particularly in the separation of macromolecules, such as nucleic acids and proteins.

acrylate *See* **propenoate.**

acrylic acid *See* **propenoic acid.**

acrylic resins Synthetic resins made by polymerizing esters or other derivatives of acrylic acid (propenoic acid). Examples are poly(propenonitrile) (e.g. *Acrilan*), and poly(methyl 2-methylpropenoate) (polymethyl methacrylate, e.g. *Perspex*).

acrylonitrile *See* **propenonitrile.**

ACT *See* **activated-complex theory.**

actinic radiation Electromagnetic radiation that is capable of initiating a chemical reaction. The term is used especially of ultraviolet radiation and also to denote radiation that will affect a photographic emulsion.

actinides *See* **actinoids.**

actinium Symbol Ac. A silvery radioactive metallic element belonging to group 3 (formerly IIIA) of the periodic table; a.n. 89; mass number of most stable isotope 227 (half-life 21.7 years); m.p. 1050 ± 50°C; b.p. 3200°C (estimated). Actinium–227 occurs in natural uranium to an extent of about 0.715%. Actinium–228 (half-life 6.13 hours) also occurs in nature. There are 22 other artificial isotopes, all radioactive and all with very short half-lives.

Its chemistry is similar to that of lanthanum. Its main use is as a source of alpha particles. The element was discovered by A. Debierne in 1899.

actinium series *See* **radioactive series.**

actinoid contraction A smooth decrease in atomic or ionic radius with increasing proton number found in the *actinoids.

actinoids (actinides) A series of elements in the *periodic table, generally considered to range in atomic number from thorium (90) to lawrencium (103) inclusive. The actinoids all have two outer *s*-electrons (a $7s^2$ configuration), follow actinium, and are classified together by the fact that increasing proton number corresponds to filling of the $5f$ level. In fact, because the $5f$ and $6d$ levels are close in energy the filling of the $5f$ orbitals is not smooth. The outer electron configurations are as follows:
89 actinium (Ac) $6d^1 7s^2$
90 thorium (Th) $6d^2 7s^2$
91 protactinium (Pa) $5f^2 6d^1 7s^2$
92 uranium (Ur) $5f^3 6d 7s^2$
93 neptunium (Np) $5f^5 7s^2$ (or $5f^4 6d^1 7s^2$)
94 plutonium (Pu) $5f^6 7s^2$
95 americium (Am) $5f^7 7s^2$
96 curium (Cm) $5f^7 6d^1 s^2$
97 berkelium (Bk) $5f^8 6d 7s^2$ (or $5f^9 7s^2$)
98 californium (Cf) $5f^{10} 7s^2$
99 einsteinium (Es) $5f^{11} 7s^2$
100 fermium (Fm) $5f^{12} 7s^2$
101 mendelevium (Md) $5f^{13} 7s^2$
102 nobelium (Nb) $5f^{14} 7s^2$
103 lawrencium (Lw) $5f^{14} 6d^1 s^2$

The first four members (Ac to Ur) occur naturally. All are radioactive and this makes investigation difficult because of self-heating, short lifetimes, safety precautions, etc. Like the *lanthanoids, the actinoids show a smooth decrease in atomic and ionic radius with increasing proton number. The lighter members of the series (up to americium) have *f*-electrons that can participate in bonding, unlike the lanthanoids. Consequently, these elements resemble the transition metals in forming coordination complexes and displaying variable valency. As a result of increased nuclear charge, the heavier members (curium to lawrencium) tend not to use their inner *f*-electrons in forming bonds and resemble the lanthanoids in forming compounds containing the M^{3+} ion. The reason for this is pulling of these inner electrons towards the centre of the atom by the increased nuclear charge. Note that actinium itself does not have a $5f$ electron, but it is usually classified with the actinoids because of its chemical similarities. *See also* **transition elements.**

actinometer *See* **actinometry.**

actinometry The measurement of the intensity of electromagnetic radiation. An instrument that measures this quantity is called an *actinometer*. Recent actinometers use the *photoelectric effect but earlier instruments depended either on the fluorescence produced by the radiation on a screen or on the amount of chemical change induced in some suitable substance. Different types of actinometer have different names according to the type of radiation they measure. A *pyroheliometer* measures the intensity of radiation from the sun. A *pyranometer* measures the intensity of radiation that reaches the surface of the earth after being scattered by molecules or objects suspended in the atmosphere. A *pyrogeometer* measures the difference between the outgoing infrared radiation from the earth and the incoming radiation from the sun that penetrates the earth's atmosphere.

action potential The change in electrical potential that occurs across a cell membrane during the passage of a nerve impulse. As an impulse travels in a wavelike manner along the axon of a nerve, it causes a localized and transient switch in electric potential across the cell membrane from −60 mV (the resting potential) to +45 mV. The change in electric potential is caused by an influx of sodium ions. Nervous stimulation of a muscle fibre has a similar effect.

action spectrum A graphical plot of the efficiency of electromagnetic radiation in producing a photochemical reaction against the wavelength of the radiation used. For example, the action spectrum for photosynthesis using light shows a peak in the region 670–700 nm. This corresponds to a maximum absorption in the absorption spectrum of chlorophylls in this region.

activated adsorption *Adsorption that involves an activation energy. This occurs in certain cases of chemisorption.

activated alumina *See* **aluminium hydroxide**.

activated charcoal *See* **charcoal**.

activated complex *See* **activated-complex theory**.

activated-complex theory (ACT) A theory enabling the rate constants in chemical reactions to be calculated using statistical thermodynamics. The events assumed to be taking place can be shown in a diagram with the potential energy as the vertical axis, while the horizontal axis, called the *reaction coordinate*, represents the course of the reaction. As two reactants A and B approach each other, the potential energy rises to a maximum. The collection of atoms near the maximum is called the *activated complex*. After the atoms have rearranged in the chemical reaction, the value of the potential energy falls as the products of the reaction are formed. The point of maximum potential energy is called the *transition state* of the reaction, as reactants passing through this state become products. In ACT, it is assumed that the reactants are in equilibrium with the activated complex, and that this decomposes along the reaction coordinate to give the products. ACT was developed by the US chemist Henry Eyring and colleagues in the 1930s. *See also* **Eyring equation**.

activated sludge process A sewage and waste-water treatment. The sludge produced after primary treatment is pumped into aeration tanks, where it is continuously stirred and aerated, resulting in the formation of small aggregates of suspended colloidal organic matter called *floc*. Floc contains numerous slime-forming and nitrifying bacteria, as well as protozoans, which decompose organic substances in the sludge. Agitation or air injection maintains high levels of dissolved oxygen, which helps to reduce the *biochemical oxygen demand. Roughly half the sewage in Britain is treated using this method.

activation analysis An analytical technique that can be used to detect most elements when present in a sample in milligram quantities (or less). In *neutron activation analysis* the sample is exposed to a flux of thermal neutrons in a nuclear reactor. Some of these neutrons are captured by nuclides in the sample to form nuclides of the same atomic number but a higher mass number. These newly formed nuclides emit gamma radiation, which can be used to identify the element present by means of a gamma-ray spectrometer. Activation analysis has also been employed using charged particles, such as protons or alpha particles.

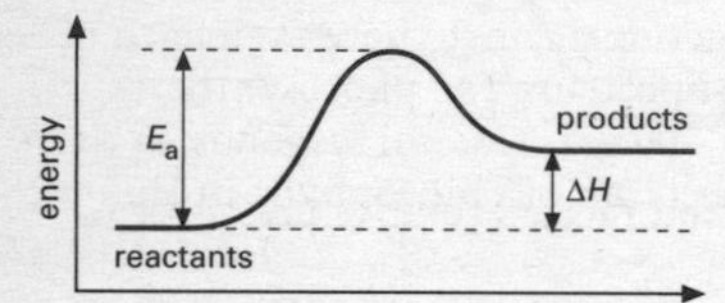

Reaction profile (for an endothermic reaction)

activation energy Symbol E_a. The minimum energy required for a chemical reaction to take place. In a reaction, the reactant molecules come together and chemical bonds are stretched, broken, and formed in producing the products. During this process the energy of the system increases to a maximum, then decreases to the energy of the products (see illustration). The activation energy is the difference between the maximum energy and the energy of the reactants; i.e. it is the energy barrier that has to be overcome for the reaction to proceed. The activation energy determines the way in which the rate of the reaction varies with temperature (*see* **Arrhenius equation**). It is usual to express activation energies in joules per mole of reactants. *See also* **activated-complex theory**.

activator 1. A substance that increases the activity of a catalyst; for example, a substance that – by binding to an *allosteric site on an enzyme – enables the active site of the enzyme to bind to the substrate. **2.** Any compound that potentiates the activity of a drug or other foreign substance in the body.

active mass *See* **mass action**.

active site (active centre) 1. A site on the surface of a catalyst at which activity occurs. **2.** The site on the surface of an *enzyme molecule that binds the substrate molecule. The properties of an active site are determined by the three-dimensional arrangement of the polypeptide chains of the enzyme and their constituent amino acids. These govern the nature of the interaction that takes place and hence the degree of substrate specificity and susceptibility to *inhibition.

activity 1. Symbol a. A thermodynamic function used in place of concentration in equilibrium constants for reactions involving nonideal gases and solutions. For example, in a reaction

$$A \rightleftharpoons B + C$$

the true equilibrium constant is given by

$$K = a_B a_C / a_A$$

where a_A, a_B, and a_C are the activities of the components, which function as concentrations (or pressures) corrected for nonideal behaviour. *Activity coefficients* (symbol γ) are defined for gases by $\gamma = a/p$ (where p is pressure) and for solutions by $\gamma = aX$ (where X is the mole fraction). Thus, the equilibrium constant of a gas reaction has the form

$$K_p = \gamma_B p_B \gamma_C p_C / \gamma_A p_A$$

The equilibrium constant of a reaction in solution is

$$K_c = \gamma_B X_B \gamma_C X_C / \gamma_A X_A$$

The activity coefficients thus act as correction factors for the pressures or concentrations. *See also* **fugacity**.
2. Symbol A. The number of atoms of a radioactive substance that disintegrate per unit time. The *specific activity* (a) is the activity per unit mass of a pure radioisotope. *See* **radiation units**.

activity series *See* **electromotive series**.

acyclic Describing a compound that does not have a ring in its molecules.

acyl anhydrides *See* **acid anhydrides**.

acylation The process of introducing an acyl group (RCO–) into a compound. The usual method is to react an alcohol with an acyl halide or a carboxylic acid anhydride; e.g.

$$RCOCl + R'OH \rightarrow RCOOR' + HCl$$

The introduction of an acetyl group (CH_3CO–) is *acetylation*, a process used for protecting –OH groups in organic synthesis.

acyl fission The breaking of the carbon–oxygen bond in an acyl group. It occurs in the hydrolysis of an *ester to produce an alcohol and a carboxylic acid.

acylglycerol *See* **glyceride**.

acyl group A group of the type RCO–, where R is an organic group. An example is the acetyl group CH_3CO–.

$$R\!\!>\!C=O \quad (R, X \text{ bonded to } C)$$

Acyl halide: X is a halogen atom

acyl halides (acid halides) Organic compounds containing the group –CO.X, where X is a halogen atom (see formula). Acyl chlorides, for instance, have the general formula RCOCl. The group RCO– is the *acyl group*. In systematic chemical nomenclature acyl-halide names end in the suffix *-oyl*; for example, ethanoyl chloride, CH_3COCl. Acyl halides react readily with water, alcohols, phenols, and amines and are used in *acylation reactions. They are made by replacing the –OH group in a carboxylic acid by a halogen using a halogenating agent such as PCl_5.

addition polymerization *See* **polymerization.**

addition reaction A chemical reaction in which one molecule adds to another. Addition reactions occur with unsaturated compounds containing double or triple bonds, and may be *electrophilic or *nucleophilic. An example of electrophilic addition is the reaction of hydrogen chloride with an alkene, e.g.

$$HCl + CH_2{:}CH_2 \rightarrow CH_3CH_2Cl$$

An example of nucleophilic addition is the addition of hydrogen cyanide across the carbonyl bond in aldehydes to form *cyanohydrins. *Addition–elimination* reactions are ones in which the addition is followed by elimination of another molecule (*see* **condensation reaction**).

additive A substance added to another substance or material to improve its properties in some way. Additives are often present in small amounts and are used for a variety of purposes, as in preventing corrosion, stabilizing polymers, and preserving and improving foods (*see* **food additive**).

adduct A compound formed by an addition reaction. The term is used particularly for compounds formed by coordination between a Lewis acid (acceptor) and a Lewis base (donor). *See* **acid.**

adenine A *purine derivative. It is one of the major component bases of *nucleotides and the nucleic acids *DNA and *RNA.

adenosine A nucleoside comprising one adenine molecule linked to a D-ribose sugar molecule. The phosphate-ester derivatives of adenosine, AMP, ADP, and *ATP, are of fundamental biological importance as carriers of chemical energy.

adenosine diphosphate (ADP) *See* ATP.

adenosine monophosphate (AMP) *See* ATP.

adenosine triphosphate *See* ATP.

adhesive A substance used for joining surfaces together. Adhesives are generally colloidal solutions, which set to gels. There are many types including animal glues (based on collagen), vegetable mucilages, and synthetic resins (e.g. *epoxy resins).

adiabatic approximation An approximation used in *quantum mechanics when the time dependence of parameters, such as the internuclear distance between atoms in a molecule, is slowly varying. This approximation means that the solution of the *Schrödinger equation at one time goes continuously over to the solution at a later time. It was formulated by Max *Born and the Soviet physicist Vladimir Alexandrovich Fock (1898–1974) in 1928. The *Born–Oppenheimer approximation is an example of the adiabatic approximation.

adiabatic demagnetization A technique for cooling a paramagnetic salt, such as potassium chrome alum, to a temperature near *absolute zero. The salt is placed between the poles of an electromagnet and the heat produced during magnetization is removed by liquid helium. The salt is then isolated thermally from the surroundings and the field is switched off; the salt is demagnetized adiabatically and its temperature falls. This is because the demagnetized state, being less ordered, involves more energy than the magnetized state. The extra energy can come only from the internal, or thermal, energy of the substance.

adiabatic process Any process that occurs without heat entering or leaving a system. In general, an adiabatic change involves a fall or rise in temperature of the system. For example, if a gas expands under adiabatic conditions, its temperature falls (work is done against the retreating walls of the container). The *adiabatic equation* describes the relationship between the pressure (p) of an ideal gas and its volume (V), i.e. $pV^{\gamma} = K$, where γ is the ratio of the principal specific *heat capacities of the gas and K is a constant.

adipic acid *See* **hexanedioic acid.**

ADP *See* ATP.

Adrenaline

adrenaline (epinephrine) A hormone, produced by the medulla of the adrenal glands, that increases heart activity, improves the power and prolongs the action of muscles, and increases the rate and depth of breathing to prepare the body for 'fright, flight, or fight'. At the same time it inhibits digestion and excretion.

adsorbate A substance that is adsorbed on a surface.

adsorbent A substance on the surface of which a substance is adsorbed.

adsorption The formation of a layer of gas, liquid, or solid on the surface of a solid or, less frequently, of a liquid. There are two types depending on the nature of the forces involved. In *chemisorption* a single layer of molecules, atoms, or ions is attached to the adsorbent surface by chemical bonds. In *physisorption* adsorbed molecules are held by the weaker *van der Waals' forces. Adsorption is an important feature of surface reactions, such as corrosion, and heterogeneous catalysis. The property is also utilized in adsorption *chromatography.

adsorption indicator (absorption indicator) A type of indicator used in reactions that involve precipitation. The yellow dye fluorescein is a common example, used for the reaction

$$NaCl(aq) + AgNO_3(aq) \rightarrow AgCl(s) + NaNO_3(aq)$$

As silver nitrate solution is added to the sodium chloride, silver chloride precipitates. As long as Cl^- ions are in excess, they adsorb on the precipitate particles. At the end point, no Cl^- ions are left in solution and negative fluorescein ions are then adsorbed, giving a pink colour to the precipitate.

adsorption isotherm An equation that describes how the amount of a substance adsorbed onto a surface depends on its pressure (if a gas) or its concentration (if in a solution), at a constant temperature. Several adsorption isotherms are used in surface chemistry including the *BET isotherm and the *Langmuir adsorption isotherm. The different isotherms correspond to different assumptions about the surface and the adsorbed molecules.

aerogel A low-density porous transparent material that consists of more than 90% air. Usually based on metal oxides or silica, aerogels are used as drying agents and insulators.

aerosol A colloidal dispersion of a solid or liquid in a gas. The commonly used aerosol sprays contain an inert propellant liquefied under pressure. *Chlorofluorocarbons, such as dichlorodifluoromethane, are commonly used in aerosol cans. This use has been criticized on the grounds that these compounds persist in the atmosphere and may lead to depletion of the *ozone layer.

AES *See* **atomic emission spectroscopy.**

A-factor *See* **Arrhenius equation.**

affinity chromatography A biochemical technique for purifying natural polymers, especially proteins. It functions by attaching a specific ligand by covalent bonding to an insoluble inert support. The ligand has to have a specific affinity for the polymer, so that when a solution containing the ligand is passed down a column of the material it is specifically retarded and thus separated from any contaminating molecules. An example of a suitable ligand is the substrate of an enzyme, provided that it does not change irreversibly during the chromatography.

aflatoxin A poisonous compound, $C_{15}H_{12}O_6$, produced by the fungus

Aspergillus flavus. It is extremely toxic to farm animals and can cause liver cancer in humans. It may occur as a contaminant of stored cereal crops, cotton seed, and, especially, peanuts. There are four isomeric forms.

AFM *See* **atomic force microscope**.

agar An extract of certain species of red seaweeds that is used as a gelling agent in microbiological culture media, foodstuffs, medicines, and cosmetic creams and jellies. *Nutrient agar* consists of a broth made from beef extract or blood that is gelled with agar and used for the cultivation of bacteria, fungi, and some algae.

agarose A carbohydrate polymer that is a component of agar. It is used in chromatography and electrophoresis.

agate A variety of *chalcedony that forms in rock cavities and has a pattern of concentrically arranged bands or layers that lie parallel to the cavity walls. These layers are frequently alternating tones of brownish-red. *Moss agate* does not show the same banding and is a milky chalcedony containing mosslike or dendritic patterns formed by inclusions of manganese and iron oxides. Agates are used in jewellery and for ornamental purposes.

agitator A bladelike instrument used in fermenters and *bioreactors to mix the medium continuously in order to maintain the rate of oxygen transfer and to help keep the cells in suspension.

air *See* **earth's atmosphere**.

air pollution (atmospheric pollution) The release into the atmosphere of substances that cause a variety of harmful effects to the natural environment. Most air pollutants are gases that are released into the troposphere, which extends about 8 km above the surface of the earth. The burning of fossil fuels, for example in power stations, is a major source of air pollution as this process produces such gases as sulphur dioxide and carbon dioxide. Released into the atmosphere, both these gases are thought to contribute to the greenhouse effect. Sulphur dioxide and nitrogen oxides, released in car exhaust fumes, are air pollutants that are responsible for the formation of *acid rain; nitrogen oxides also contribute to the formation of *photochemical smog. *See also* **ozone layer**; **pollution**.

alabaster *See* **gypsum**.

alanine *See* **amino acid**.

albumin (albumen) One of a group of globular proteins that are soluble in water but form insoluble coagulates when heated. Albumins occur in egg white, blood, milk, and plants. Serum albumins, which constitute about 55% of blood plasma protein, help regulate the osmotic pressure and hence plasma volume. They also bind and transport fatty acids. α-lactalbumin is one of the proteins in milk.

alcoholic fermentation *See* **fermentation**.

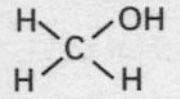

H OH
C
H H
primary alcohol (methanol)

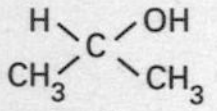

H OH
C
CH_3 CH_3
secondary alcohol (propan-2-ol)

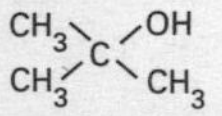

CH_3 OH
C
CH_3 CH_3
tertiary alcohol (2-methylpropan-2-ol)

Examples of alcohols

alcohols Organic compounds that contain the –OH group. In systematic chemical nomenclature alcohol names end in the suffix *-ol*. Examples are methanol, CH_3OH, and ethanol, C_2H_5OH. *Primary alcohols* have two hydrogen atoms on the carbon joined to the –OH group (i.e. they contain the group $-CH_2-OH$); *secondary alcohols* have one hydrogen on this carbon (the other two bonds being to carbon atoms, as in $(CH_3)_2CHOH$); *tertiary alcohols* have no hydrogen on this carbon (as in $(CH_3)_3COH$): see formulae. The different types of alcohols may differ in the way they react chemically. For example, with potassium dichromate(VI) in sulphuric acid the following reactions occur:
primary alcohol → aldehyde → carboxylic acid
secondary alcohol → ketone
tertiary alcohol – no reaction

Other characteristics of alcohols are reaction with acids to give *esters and dehydration to give *alkenes or *ethers. Alcohols that have two –OH groups in their molecules are *diols* (or *dihydric alcohols*), those with three are *triols* (or *trihydric alcohols*), etc.

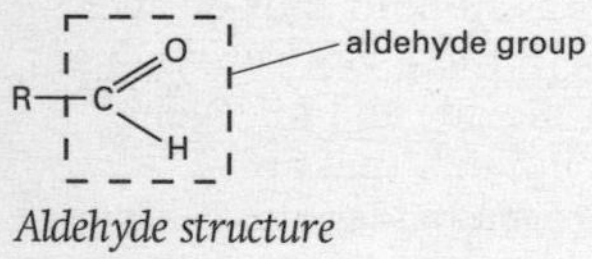

O
R–C
H
aldehyde group

Aldehyde structure

aldehydes Organic compounds that contain the group –CHO (the *aldehyde group*; i.e. a carbonyl group (C=O) with a hydrogen atom bound to the carbon atom). In systematic chemical nomenclature, aldehyde names end with the suffix *-al*. Examples of aldehydes are methanal (formaldehyde), HCOH, and ethanal (acetaldehyde), CH_3CHO. Aldehydes are formed by oxidation of primary *alcohols; further oxidation yields carboxylic acids. They are reducing agents and tests for aldehydes include *Fehling's test and *Tollens reagent. Aldehydes have certain characteristic addition and condensation reactions. With sodium hydrogensulphate(IV) they form addition compounds of the type $[RCOH(SO_3)H]^- Na^+$. Formerly these were known as *bisulphite addition compounds*. They also form addition compounds with hydrogen cyanide to give *cyanohydrins and with alcohols to give *acetals and undergo condensation reactions to yield *oximes, *hydrazones, and *semicarbazones. Aldehydes readily polymerize. *See also* **ketones**.

aldohexose *See* **monosaccharide**.

aldol *See* **aldol reaction**.

aldol reaction A reaction of aldehydes of the type

$2RCH_2CHO \rightleftharpoons RCH_2CH(OH)CHRCHO$

where R is a hydrocarbon group. The resulting compound is a hydroxy-aldehyde, i.e. an aldehyde–alcohol or *aldol*, containing alcohol (–OH) and aldehyde (–CHO) groups on adjacent carbon atoms. The reaction is base-catalysed, the first step being the formation of a carbanion of the type $RHC^{-}CHO$, which adds to the carbonyl group of the other aldehyde molecule. For the carbanion to form, the aldehyde must have a hydrogen atom on the carbon next to the carbonyl group.

Aldols can be further converted to other products; in particular, they are a source of unsaturated aldehydes. For example, the reaction of ethanal gives 3-hydroxybutenal (*acetaldol*):

$2CH_3CHO \rightleftharpoons CH_3CH(OH)CH_2CHO$

This can be further dehydrated to 2-butenal (*crotonaldehyde*):

$CH_3CH(OH)CH_2CHO \rightarrow H_2O + CH_3CH{:}CHCHO$

aldose *See* **monosaccharide**.

aldosterone A hormone produced by the adrenal glands that controls excretion of sodium by the kidneys and thereby maintains the balance of salt and water in the body fluids.

algin (alginic acid) A complex polysaccharide occurring in the cell walls of the brown algae (Phaeophyta). Algin strongly absorbs water to form a viscous gel. It is produced commercially from a variety of species of *Laminaria* and from *Macrocystis pyrifera* in the form of *alginates*, which are used mainly as a stabilizer and texturing agent in the food industry.

alicyclic compound A compound that contains a ring of atoms and is aliphatic. Cyclohexane, C_6H_{12}, is an example.

aliphatic compounds Organic compounds that are *alkanes, *alkenes, or *alkynes or their derivatives. The term is used to denote compounds that do not have the special stability of *aromatic compounds. All noncyclic organic compounds are aliphatic. Cyclic aliphatic compounds are said to be *alicyclic*.

alizarin An orange-red compound, $C_{14}H_8O_4$. The compound is a derivative of *anthraquinone, with hydroxyl groups substituted at the 1 and 2 positions. It is an important dyestuff producing red or violet *lakes with metal hydroxide. Alizarin occurs naturally as the glucoside in madder. It can be synthesized by heating anthraquinone with sodium hydroxide.

alkali A *base that dissolves in water to give hydroxide ions.

alkali metals (group 1 elements) The elements of group 1 (formerly IA) of the *periodic table: lithium (Li), sodium (Na), potassium (K), rubidium (Rb), caesium (Cs), and francium (Fr). All have a characteristic electron configuration that is a noble gas structure with one outer *s*-electron. They are typical metals (in the chemical sense) and readily lose their outer electron to form stable M^+ ions with noble-gas configurations. All are highly reactive, with the reactivity (i.e. metallic character) increasing down the group. There is a decrease in ionization energy from lithium (520 $kJ\,mol^{-1}$) to caesium (380 $kJ\,mol^{-1}$). The second ionization energies are much higher and divalent ions are not formed. Other properties also change down the group. Thus, there is an increase in atomic and ionic radius, an increase in density, and a decrease in melting and boiling point. The standard electrode potentials are low and negative, although they do not show a regular trend because they depend both on ionization energy (which decreases down the group) and the hydration energy of the ions (which increases).

All the elements react with water (lithium slowly; the others violently) and tarnish rapidly in air. They can all be made to react with chlorine, bromine, sulphur, and hydrogen. The hydroxides of the alkali metals are strongly alkaline (hence the name) and do not decompose on heating. The salts are generally soluble. The carbonates do not decompose on heating, except at very high temperatures. The nitrates (except for lithium) decompose to give the nitrite and oxygen:

$$2MNO_3(s) \rightarrow 2MNO_2(s) + O_2(g)$$

Lithium nitrate decomposes to the oxide. In fact lithium shows a number of dissimilarities to the other members of group 1 and in many ways resembles magnesium (*see* **diagonal relationship**). In general, the stability of salts of oxo acids increases down the group (i.e. with increasing size of the M^+ ion). This trend occurs because the smaller cations (at the top of the group) tend to polarize the oxo anion more effectively than the larger cations at the bottom of the group.

alkalimetry Volumetric analysis using standard solutions of alkali to determine the amount of acid present.

alkaline 1. Describing an alkali. **2.** Describing a solution that has an excess of hydroxide ions (i.e. a pH greater than 7).

alkaline-earth metals (group 2 elements) The elements of group 2 (formerly IIA) of the *periodic table: beryllium (Be), magnesium (Mg), calcium (Ca), strontium (Sr), and barium (Ba). The elements are sometimes referred to as the 'alkaline earths', although strictly the 'earths' are the oxides of the elements. All have a characteristic electron configuration that is a noble-gas structure with two outer *s*-electrons. They are typical metals (in the chemical sense) and readily lose both outer electrons to form stable M^{2+} ions; i.e. they are strong reducing agents. All are reactive, with the reactivity increasing down the group. There is a decrease in both first and second ionization energies down the group. Although there is a significant difference between the first and second ionization energies of each element, compounds containing univalent ions are not known. This is because the divalent ions have a smaller size and larger charge, leading to higher hydration energies (in solution) or lattice energies (in solids). Consequently, the overall energy change favours the formation of divalent compounds. The third ionization energies are much higher than the second ionization energies, and trivalent compounds (containing M^{3+}) are unknown.

Beryllium, the first member of the group, has anomalous properties because of the small size of the ion; its atomic radius (0.112 nm) is much less than that of magnesium (0.16 nm). From magnesium to radium there is a fairly regular increase in atomic and ionic radius. Other regular changes take place in moving down the group from magnesium. Thus, the density and melting and boiling points all increase. Beryllium, on the other hand, has higher boiling and melting points than calcium and its density lies between those of calcium and strontium. The standard electrode potentials are negative and show a regular small decrease from magnesium to barium. In some ways beryllium resembles aluminium (*see* **diagonal relationship**).

All the metals are rather less reactive than the alkali metals. They react with water and oxygen (beryllium and magnesium form a protective surface film) and can be made to react with chlorine, bromine, sulphur, and hydrogen. The oxides and hydroxides of the metals show the increasing ionic character in moving down the group: beryllium hydroxide is amphoteric, magnesium hydroxide is only very slightly soluble in water and is weakly basic, calcium hydroxide is sparingly soluble and distinctly basic, strontium and barium hydroxides are quite soluble and basic. The hydroxides decompose on heating to give the oxide and water:

$$M(OH)_2(s) \rightarrow MO(s) + H_2O(g)$$

The carbonates also decompose on heating to the oxide and carbon dioxide:

$$MCO_3(s) \rightarrow MO(s) + CO_2(g)$$

The nitrates decompose to give the oxide:

$$2M(NO_3)_2(s) \rightarrow 2MO(s) + 4NO_2(g) + O_2(g)$$

As with the *alkali metals, the stability of salts of oxo acids increases down the group. In general, salts of the alkaline-earth elements are soluble if the anion has a single charge (e.g. nitrates, chlorides). Most salts with a doubly charged anion (e.g. carbonates, sulphates) are insoluble. The solubilities of salts of a particular acid tend to decrease down the group. (Solubilities of hydroxides increase for larger cations.)

alkaloid One of a group of nitrogenous organic compounds derived from plants and having diverse pharmacological properties. Alkaloids include morphine, cocaine, atropine, quinine, and caffeine, most of which are used in medicine as analgesics (pain relievers) or anaesthetics. Some alkaloids are poisonous, e.g. strychnine and coniine, and colchicine inhibits cell division.

alkanal An aliphatic aldehyde.

alkanes (paraffins) Saturated hydrocarbons with the general formula C_nH_{2n+2}. In systematic chemical nomenclature alkane names end in the suffix *-ane*. They form a *homologous series (the *alkane series*) methane (CH_4), ethane (C_2H_6), propane (C_3H_8), butane (C_4H_{10}), pentane (C_5H_{12}), etc. The lower members of the series are gases; the high-molecular weight alkanes are waxy solids. Alkanes are present in natural gas and petroleum. They can be made by heating the sodium salt of a carboxylic acid with soda lime:

$$RCOO^-Na^+ + Na^+OH^- \rightarrow Na_2CO_3 + RH$$

Other methods include the *Wurtz reaction and *Kolbe's method. Generally the alkanes are fairly unreactive. They form haloalkanes with halogens when irradiated with ultraviolet radiation.

alkanol An aliphatic alcohol.

```
       H  H  H
H\     |  |  |
  >C = C--C--C--H        but-1-ene
H/        |  |
          H  H

  H     H  H  H
  |     |  |  |
H-C-  C=C--C--H          but-2-ene
  |        |
  H        H
```

Butene isomers

alkenes (olefines; olefins) Unsaturated hydrocarbons that contain one or more double carbon–carbon bonds in their molecules. In systematic chemical nomenclature alkene names end in the suffix *-ene*. Alkenes that have only one double bond form a homologous series (the *alkene series*) starting ethene (ethylene), $CH_2{:}CH_2$, propene, $CH_3CH{:}CH_2$, etc. The general formula is C_nH_{2n}. Higher members of the series show isomerism depending on position of the double bond; for example, butene (C_4H_8) has two isomers, which are (1) but-1-ene ($C_2H_5CH{:}CH_2$) and (2) but-2-ene ($CH_3CH{:}CHCH_3$): see formulae. Alkenes can be made by dehydration of alcohols (passing the vapour over hot pumice):

$$RCH_2CH_2OH - H_2O \rightarrow RCH{:}CH_2$$

An alternative method is the removal of a hydrogen atom and halogen atom from a haloalkane by potassium hydroxide in hot alcoholic solution:

$$RCH_2CH_2Cl + KOH \rightarrow KCl + H_2O + RCH{:}CH_2$$

Alkenes typically undergo *addition reactions to the double bond. *See also* **hydrogenation; oxo process; ozonolysis; Ziegler process.**

alkoxides Compounds formed by reaction of alcohols with sodium or potassium metal. Alkoxides are saltlike compounds containing the ion $R{-}O^-$.

alkyd resin A type of *polyester resin used in paints and other surface coating. The original alkyd resins were made by copolymerizing phthalic anhydride with glycerol, to give a brittle cross-linked polymer. The properties of such resins can be modified by adding monobasic acids or alcohols during the polymerization.

alkylation A chemical reaction that introduces an *alkyl group into an organic molecule. The *Friedel–Crafts reaction results in alkylation of aromatic compounds.

alkylbenzenes Organic compounds that have an alkyl group bound to a benzene ring. The simplest example is methylbenzene (toluene), $CH_3C_6H_5$. Alkyl benzenes can be made by the *Friedel–Crafts reaction.

alkyl group A group obtained by removing a hydrogen atom from an alkane, e.g. methyl group, CH_3–, derived from methane.

alkyl halides *See* **haloalkanes.**

alkynes (acetylenes) Unsaturated hydrocarbons that contain one or more triple carbon–carbon bonds in their molecules. In systematic chemical nomenclature alkyne names end in the suffix *-yne*. Alkynes that have only one triple bond form a *homologous series: ethyne (acetylene), $CH\equiv CH$, propyne, $CH_3CH\equiv CH$, etc. They can be made by the action of potassium hydroxide in alcohol solution on haloalkanes containing halogen atoms on adjacent carbon atoms; for example:

$$RCHClCH_2Cl + 2KOH \rightarrow 2KCl + 2H_2O + RCH\equiv CH$$

Like *alkenes, alkynes undergo addition reactions.

allenes Compounds that contain the group $>C=C=C<$, in which three carbon atoms are linked by two adjacent double bonds. The outer carbon atoms are each linked to two other atoms or groups by single bonds. The simplest example is 1,2-propadiene, CH_2CCH_2. Allenes are *dienes with typical reactions of alkenes. Under basic conditions, they often convert to alkynes. In an allene, the two double bonds lie in planes that are perpendicular to each other. Consequently, in an allene of the type $R_1R_2C{:}C{:}CR_3R_4$, the groups R_1 and R_2 lie in a plane perpendicular to the plane containing R_3 and R_4. Under these circumstances, the molecule is chiral and can show optical activity.

allosteric enzyme An enzyme that has two structurally distinct forms, one of which is active and the other inactive. In the active form, the quaternary structure (*see* **protein**) of the enzyme is such that a substrate can interact with the enzyme at the active site (*see* **enzyme–substrate complex**). The conformation of the substrate-binding site becomes altered in the inactive form and interaction with the substrate is not possible. Allosteric enzymes tend to catalyse the initial step in a pathway leading to the synthesis of molecules. The end product of this synthesis can act as a feedback inhibitor (*see* **inhibition**) and the enzyme is converted to the inactive form, thereby controlling the amount of product synthesized.

allosteric site A binding site on the surface of an enzyme other than the *active site. In noncompetitive *inhibition, binding of the inhibitor to an allosteric site inhibits the activity of the enzyme. In an *allosteric enzyme, the binding of a regulatory molecule to the allosteric site changes the overall shape of the enzyme, either enabling the substrate to bind to the active site or preventing the binding of the substrate.

allotropy The existence of elements in two or more different forms (*allotropes*). In the case of oxygen, there are two forms: 'normal' dioxygen (O_2) and ozone, or trioxygen (O_3). These two allotropes have different molecular configurations. More commonly, allotropy occurs because of different crystal structures in the solid, and is particularly prevalent in groups 14, 15, and 16 of the periodic table. In some cases, the allotropes are stable over a temperature range, with a definite transition point at which one changes into the other. For instance, tin has two allotropes: white (metallic) tin stable above 13.2°C and grey (nonmetallic) tin stable below 13.2°C. This form of allotropy is called *enantiotropy*. Carbon also has two allotropes – diamond and graphite – although graphite is the stable form at all temperatures. This form of allotropy, in which there is no transition temperature at which the two are in equilibrium, is called *monotropy*. *See also* **polymorphism**.

allowed bands *See* **energy bands**.

allowed transition A transition between two electronic states allowed according to *selection rules associated with group theory. The probability of a transition between states m and n produced by the interaction of electromagnetic radiation with an atomic system is proportional to the square of the magnitude of the matrix elements of the electric dipole moment. If this quantity is not zero, the transition is an allowed transition; if it is zero the transition is a *forbidden transition as a dipole transition. It may, however, be an allowed transition for magnetic dipole or quadrupole-moment transitions, which have much smaller transition probabilities and consequently give much weaker lines in the spectrum.

alloy A material consisting of two or more metals (e.g. brass is an alloy of copper and zinc) or a metal and a nonmetal (e.g. steel is an alloy of iron and carbon, sometimes with other metals included). Alloys may be compounds, *solid solutions, or mixtures of the components.

alloy steels *See* **steel**.

allyl alcohol *See* **propenol**.

allyl group *See* **propenyl group**.

Alnico A tradename for a series of alloys, containing iron, aluminium, nickel, cobalt, and copper, used to make permanent magnets.

alpha helix The most common form of secondary structure in *proteins, in which the polypeptide chain is coiled into a helix. The helical structure is held in place by weak hydrogen bonds between the N–H and C=O groups in successive turns of the helix (see illustration). *Compare* **beta sheet**.

alpha-iron *See* iron.

alpha-naphthol test A biochemical test to detect the presence of carbohydrates in solution, also known as *Molisch's test* (after the Austrian chemist H. Molisch (1856–1937), who devised it). A small amount of alcoholic alpha-naphthol is mixed with the test solution and concentrated sulphuric acid is poured slowly down the side of the test tube. A positive reaction is indicated by the formation of a violet ring at the junction of the liquids.

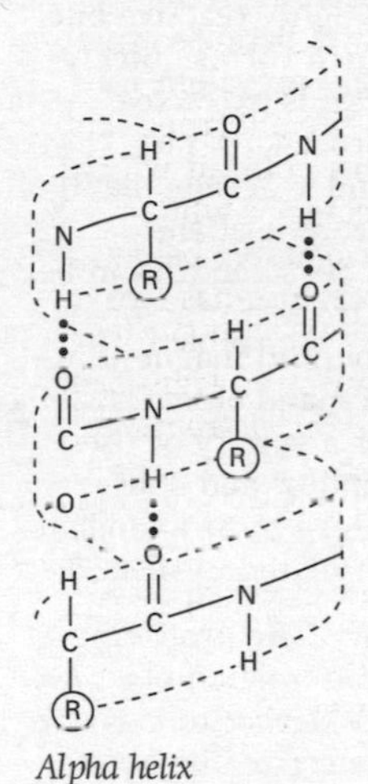

Alpha helix

alpha particle A helium nucleus emitted by a larger nucleus during the course of the type of radioactive decay known as *alpha decay*. As a helium nucleus consists of two protons and two neutrons bound together as a stable entity the loss of an alpha particle involves a decrease in *nucleon number of 4 and decrease of 2 in the *atomic number, e.g. the decay of a uranium–238 nucleus into a thorium–234 nucleus. A stream of alpha particles is known as an *alpha-ray* or *alpha-radiation*.

alternant Describing a conjugated molecule in which the atoms can be divided into two sets of alternate atoms such that no atom has a direct link to another atom in the same set. Naphthalene, for example, has an alternant conjugated system.

alum *See* **aluminium potassium sulphate; alums.**

alumina *See* **aluminium oxide; aluminium hydroxide**.

aluminate A salt formed when aluminium hydroxide or γ-alumina is dissolved in solutions of strong bases, such as sodium hydroxide. Aluminates exist in solutions containing the aluminate ion, commonly written $[Al(OH)_4]^-$. In fact the ion probably is a complex hydrated ion and can be regarded as formed from a hydrated Al^{3+} ion by removal of four hydrogen ions:

$$[Al(H_2O)_6]^{3+} + 4OH^- \rightarrow 4H_2O + [Al(OH)_4(H_2O)_2]^-$$

Other aluminates and polyaluminates, such as $[Al(OH)_6]^{3-}$ and $[(HO)_3AlOAl(OH)_3]^{2-}$, are also present. *See also* **aluminium hydroxide**.

aluminium Symbol Al. A silvery-white lustrous metallic element belonging to *group 3 (formerly IIIB) of the periodic table; a.n. 13; r.a.m. 26.98; r.d. 2.7; m.p. 660°C; b.p. 2467°C. The metal itself is highly reactive but is protected by a thin transparent layer of the oxide, which forms quickly in air. Aluminium and its oxide are amphoteric. The metal is extracted from purified bauxite (Al_2O_3) by electrolysis; the main process uses a *Hall–Heroult cell but other electrolytic methods are under development, including conversion of bauxite with chlorine and electrolysis of the molten chloride. Pure aluminium is soft and ductile but its strength can be increased by work-hardening. A large number of alloys are manufactured; alloying elements include copper, manganese, silicon, zinc, and magnesium. Its lightness, strength (when alloyed), corrosion resistance, and electrical conductivity (62% of that of copper) make it suitable for a variety of uses, including vehicle and aircraft construction, building (window and door frames), and overhead power cables. Although it is the third most abundant element in the earth's crust (8.1% by weight) it was not isolated until 1825 by H. C. Oersted.

aluminium acetate *See* **aluminium ethanoate**.

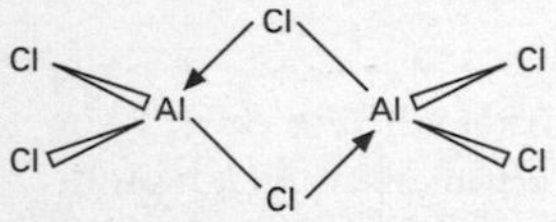

Structure of aluminium trichloride dimer

aluminium chloride A whitish solid, $AlCl_3$, which fumes in moist air and reacts violently with water (to give hydrogen chloride). It is known as the anhydrous salt (hexagonal; r.d. 2.44 (fused solid); m.p. 190°C (2.5 atm.); sublimes at 178°C) or the hexahydrate $AlCl_3.6H_2O$ (rhombic; r.d. 2.398; loses water at 100°C), both of which are deliquescent. Aluminium chloride may be prepared by passing hydrogen chloride or chlorine over hot aluminium or (industrially) by passing chlorine over heated aluminium oxide and carbon. The chloride ion is polarized by the small positive aluminium ion and the bonding in the solid is intermediate between covalent and ionic. In the liquid and vapour phases dimer molecules exist, Al_2Cl_6, in which there are chlorine bridges making coordinate bonds to aluminium atoms (see formula). The $AlCl_3$ molecule can also form compounds with other molecules that donate pairs of electrons (e.g. amines or hydrogen sulphide); i.e. it acts as a Lewis *acid. At high temperatures the Al_2Cl_6 molecules in the vapour dissociate to (planar) $AlCl_3$ molecules. Aluminium chloride is used commercially as a catalyst in the cracking of oils. It is also a catalyst in certain other organic reactions, especially the Friedel–Crafts reaction.

aluminium ethanoate **(aluminium acetate)** A white solid, $Al(OOCCH_3)_3$, which decomposes on heating, is very slightly soluble in cold water, and decomposes in warm water. The normal salt, $Al(OOCCH_3)_3$, can only be made in the absence of water (e.g. ethanoic anhydride and aluminium chloride at 180°C); in water it forms the basic salts $Al(OH)(OOCCH_3)_2$ and $Al_2(OH)_2(OOCCH_3)_4$. The reaction of aluminium hydroxide with ethanoic acid gives these basic salts directly. The compound is used extensively in dyeing as a mordant, particularly in combination with aluminium sulphate (known as *red liquor*); in the paper and board industry for sizing and hardening; and in tanning. It was previously used as an antiseptic and astringent.

aluminium hydroxide A white crystalline compound, $Al(OH)_3$; r.d. 2.42–2.52. The compound occurs naturally as the mineral *gibbsite* (monoclinic). In the laboratory it can be prepared by precipitation from solutions of aluminium salts. Such solutions contain the hexaquoaluminium(III) ion with six water molecules coordinated, $[Al(H_2O)_6]^{3+}$. In neutral solution this ionizes:

$$[Al(H_2O)_6]^{3+} \rightleftharpoons H^+ + [Al(H_2O)_5OH]^{2+}$$

The presence of a weak base such as S^{2-} or CO_3^{2-} (by bubbling hydrogen sulphide or carbon dioxide through the solution) causes further ionization with precipitation of aluminium hydroxide

$$[Al(H_2O)_6]^{3+}(aq) \rightarrow Al(H_2O)_3(OH)_3(s) + 3H^+(aq)$$

The substance contains coordinated water molecules and is more correctly termed *hydrated aluminium hydroxide*. In addition, the precipitate has water molecules trapped in it and has a characteristic gelatinous form. The substance is amphoteric. In strong bases the *aluminate ion is produced by loss of a further proton:

$$Al(H_2O)_3(OH)_3(s) + OH^-(aq) \rightleftharpoons [Al(H_2O)_2(OH)_4]^-(aq) + H_2O(l)$$

On heating, the hydroxide transforms to a mixed oxide hydroxide, AlO.OH (rhombic; r.d. 3.01). This substance occurs naturally as *diaspore* and *boehmite*. Above 450°C it transforms to γ-alumina.

In practice various substances can be produced that are mixed crystalline forms of $Al(OH)_3$, AlO.OH, and aluminium oxide (Al_2O_3) with water molecules. These are known as *hydrated alumina*. Heating the hydrated hydroxide causes loss of water, and produces various *activated aluminas*, which differ in porosity, number of remaining –OH groups, and particle size. These are used as catalysts (particularly for organic dehydration reactions), as catalyst supports, and in chromatography. Gelatinous freshly precipitated aluminium hydroxide was formerly widely used as a mordant for dyeing and calico printing because of its ability to form insoluble coloured *lakes with vegetable dyes. *See also* **aluminium oxide**.

aluminium oxide **(alumina)** A white or colourless oxide of aluminium occurring in two main forms. The stable form α-alumina (r.d. 3.97; m.p. 2015°C; b.p. 2980 ± 60°C) has colourless hexagonal or rhombic crystals; γ-alumina (r.d. 3.5–3.9) transforms to the α-form on heating and is a white microcrystalline solid. The compound occurs naturally as *corundum* or *emery* in the α-form with a hexagonal-close-packed structure of oxide ions with aluminium ions in the octahedral interstices. The gemstones ruby and sapphire are aluminium oxide coloured by minute traces of chromium and cobalt respectively. A number of other forms of aluminium oxide have been described (β-, δ-, and ζ-alumina) but these contain alkali-metal ions. There is also a short-lived spectroscopic suboxide AlO. The highly protective film of oxide formed on the surface of aluminium metal is yet another structural variation, being a defective rock-salt form (every third Al missing).

Pure aluminium oxide is obtained by dissolving the ore bauxite in sodium hydroxide solution; impurities such as iron oxides remain insoluble because they are not amphoteric. The hydrated oxide is precipitated by seeding with material from a previous batch and this is then roasted at 1150–1200°C to give pure α-alumina, or at 500–800°C to give γ-alumina. The bonding in aluminium hydroxide is not purely ionic due to polarization of the oxide ion. Although the compound might be expected to be amphoteric, α-alumina is weakly acidic, dissolving in alkalis to give solutions containing aluminate ions; it is resistant to acid attack. In contrast γ-alumina is typically amphoteric dissolving both in acids to give aluminium salts and in bases to give aluminates. α-alumina is one of the hardest materials known (silicon carbide and diamond are harder) and is widely used as an abrasive in both natural (corundum) and synthetic forms. Its refractory nature makes alumina brick an ideal material for furnace linings and alumina is also used in cements for high-temperature conditions. *See also* **aluminium hydroxide**.

aluminium potassium sulphate (potash alum; alum) A white or colourless crystalline compound, $Al_2(SO_4)_3.K_2SO_4.24H_2O$; r.d. 1.757; loses $18H_2O$ at 92.5°C; becomes anhydrous at 200°C. It forms cubic or octahedral crystals that are soluble in cold water, very soluble in hot water, and insoluble in ethanol and acetone. The compound occurs naturally as the mineral *kalinite*. It is a double salt and can be prepared by recrystallization from a solution containing equimolar quantities of potassium sulphate and aluminium sulphate. It is used as a mordant for dyeing and in the tanning and finishing of leather goods (for white leather). *See also* **alums**.

aluminium sulphate A white or colourless crystalline compound, $Al_2(SO_4)_3$, known as the anhydrous compound (r.d. 2.71; decomposes at 770°C) or as the hydrate $Al_2(SO)_3.18H_2O$ (monoclinic; r.d. 1.69; loses water at 86.5°C). The anhydrous salt is soluble in water and slightly soluble in ethanol; the hydrate is very soluble in water and insoluble in ethanol. The compound occurs naturally in the rare mineral *alunogenite* ($Al_2(SO)_3.18H_2O$). It may be prepared by dissolving aluminium hydroxide or china clays (aluminosilicates) in sulphuric acid. It decomposes on heating to sulphur dioxide, sulphur trioxide, and aluminium oxide. Its solutions are acidic because of hydrolysis.

Aluminium sulphate is commercially one of the most important aluminium compounds; it is used in sewage treatment (as a flocculating agent) and in the purification of drinking water, the paper industry, and in the preparation of mordants. It is also a fire-proofing agent. Aluminium sulphate is often wrongly called *alum* in these industries.

aluminium trimethyl *See* **trimethylaluminium**.

alums A group of double salts with the formula $A_2SO_4.B_2(SO_4)_3.24H_2O$, where A is a monovalent metal and B a trivalent metal. The original example contains potassium and aluminium (called *potash alum* or simply *alum*); its formula is often written $AlK(SO_4)_2.12H_2O$ (aluminium potassium sulphate-12-water). *Ammonium alum* is $AlNH_4(SO_4)_2.12H_2O$, *chrome alum* is $KCr(SO_4)_2.12H_2O$ (*see* **potassium chromium sulphate**), etc. The alums are isomorphous and can be made by dissolving equivalent amounts of the two salts in water and recrystallizing. *See also* **aluminium sulphate**.

alunogenite A mineral form of hydrated *aluminium sulphate, $Al_2(SO_4)_3.18H_2O$.

amalgam An alloy of mercury with one or more other metals. Most metals form amalgams (iron and platinum are exceptions), which may be liquid or solid. Some contain definite intermetallic compounds, such as $NaHg_2$.

amatol A high explosive consisting of a mixture of ammonium nitrate and trinitrotoluene.

ambident Describing a chemical species that has two alternative reactive centres such that reaction at one centre stops or inhibits reaction at the other. An example is the *enolate ion in which electrophilic attack can occur at either the oxygen atom or at the beta-carbon atom.

ambidentate Describing a ligand that can coordinate at two different sites. For example, the NO_2 molecule can coordinate through the N atom (the *nitro ligand*) or through an O atom (the *nitrido ligand*). Complexes that differ only in the way the ligand coordinates display *linkage isomerism*.

ambo- A prefix used to indicate that a substance is present as a mixture of racemic diastereoisomers in unspecified proportions. For example, if L-alanine is reacted with DL-leucine, the resulting dipeptide can be described as L-alanyl-*ambo*-leucine, to indicate the mixture.

americium Symbol Am. A radioactive metallic transuranic element belonging to the *actinoids; a.n. 95; mass number of most stable isotope 243 (half-life 7.95×10^3 years); r.d. 13.67 (20°C); m.p. 994 ± 4°C; b.p. 2607°C. Ten isotopes are known. The element was discovered by G. T. Seaborg and associates in 1945, who obtained it by bombarding uranium–238 with alpha particles.

amethyst The purple variety of the mineral *quartz. It is found chiefly in Brazil, the Urals (Russia), Arizona (USA), and Uruguay. The colour is due to impurities, especially iron oxide. It is used as a gemstone.

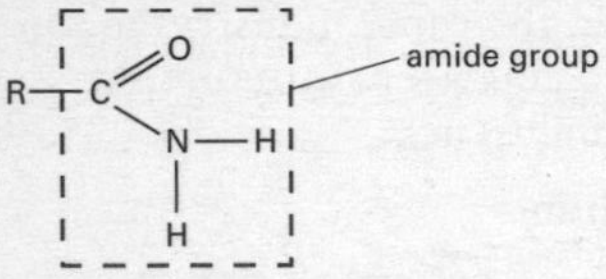

Amide structure

amides 1. Organic compounds containing the group $-CO.NH_2$ (the *amide group*). Compounds containing this group are *primary amides. Secondary* and *tertiary amides* can also exist, in which the hydrogen atoms on the nitrogen are replaced by one or two other organic groups respectively. Simple examples of primary amides are ethanamide, CH_3CONH_2, and propanamide, $C_2H_5CONH_2$. They are made by heating the ammonium salt of the corresponding carboxylic acid. Amides can also be made by reaction of ammonia (or an amine) with an acyl halide. *See also* **Hofmann's reaction. 2.** Inorganic compounds containing the ion NH_2^-, e.g. KNH_2 and $Cd(NH_2)_2$. They are formed by the reaction of ammonia with electropositive metals.

amidol *See* **aminophenol.**

amination A chemical reaction in which an amino group ($-NH_2$) is introduced into a molecule. Examples of amination reaction include the reaction of halogenated hydrocarbons with ammonia (high pressure and temperature) and the reduction of nitro compounds and nitriles.

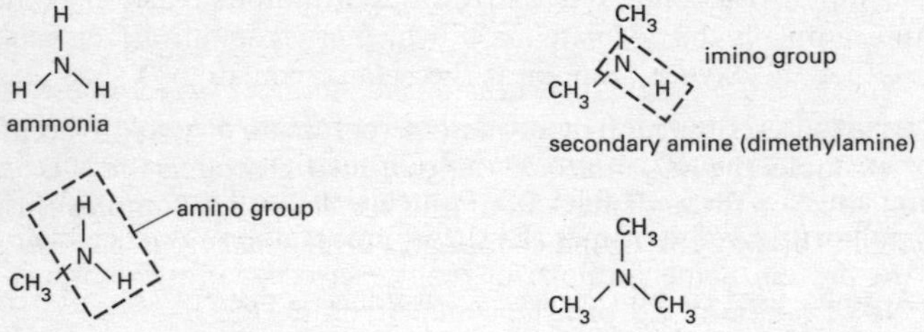

Examples of amines

amines Organic compounds derived by replacing one or more of the hydrogen atoms in ammonia by organic groups (see illustration). *Primary amines* have one hydrogen replaced, e.g. methylamine, CH_3NH_2. They contain the functional group $-NH_2$ (the *amino group*). *Secondary amines* have two hydrogens replaced, e.g. methylethylamine, $CH_3(C_2H_5)NH$. *Tertiary amines* have all three hydrogens replaced, e.g. trimethylamine, $(CH_3)_3N$.

Amines are produced by the decomposition of organic matter. They can be made by reducing nitro compounds or amides. *See also* **imines**.

amine salts Salts similar to ammonium salts in which the hydrogen atoms attached to the nitrogen are replaced by one or more organic groups. Amines readily form salts by reaction with acids, gaining a proton to form a positive ammonium ion, They are named as if they were substituted derivatives of ammonium compounds; for example, dimethylamine ($(CH_3)_2NH$) will react with hydrogen chloride to give dimethylammonium chloride, which is an ionic compound $[(CH_3)_2NH_2]^+Cl^-$. When the amine has a common nonsystematic name the suffix *-ium* can be used; for example, phenylamine (aniline) would give $[C_6H_5NH_3]^+Cl^-$, known as anilinium chloride. Formerly, such compounds were sometimes called *hydrochlorides*, e.g. aniline hydrochloride with the formula $C_6H_5NH_2.HCl$.

Salts formed by amines are crystalline substances that are readily soluble in water. Many insoluble *alkaloids (e.g. quinine and atropine) are used medicinally in the form of soluble salts ('hydrochlorides'). If alkali (sodium hydroxide) is added to solutions of such salts the free amine is liberated.

If all four hydrogen atoms of an ammonium salt are replaced by organic groups a *quaternary ammonium compound* is formed. Such compounds are made by reacting tertiary amines with halogen compounds; for example, trimethylamine ($(CH_3)_3N$) with chloromethane (CH_3Cl) gives tetramethylammonium chloride, $(CH_3)_4N^+Cl^-$. Salts of this type do not liberate the free amine when alkali is added, and quaternary hydroxides (such as $(CH_3)_4N^+OH^-$) can be isolated. Such compounds are strong alkalis, comparable to sodium hydroxide.

amino acid Any of a group of water-soluble organic compounds that possess both a carboxyl (–COOH) and an amino (–NH_2) group attached to the same carbon atom, called the α-carbon atom. Amino acids can be represented by the general formula R–CH(NH_2)COOH. R may be hydrogen or an organic group and determines the properties of any particular amino acid. Through the formation of peptide bonds, amino acids join together to form short chains (*peptides) or much longer chains (*polypeptides). Proteins are composed of various proportions of about 20 commonly occurring amino acids (see table). The sequence of these amino acids in the protein polypeptides determines the shape, properties, and hence biological role of the protein. Some amino acids that never occur in proteins are nevertheless important, e.g. ornithine and citrulline, which are intermediates in the urea cycle.

Plants and many microorganisms can synthesize amino acids from simple inorganic compounds, but animals rely on adequate supplies in their diet. The *essential amino acids must be present in the diet whereas others can be manufactured from them.

aminobenzene *See* **phenylamine**.

amino group *See* **amines**.

aminophenol Any of various organic compounds used as reducing agents,

amino acid	abbreviation	formula
alanine	Ala	$CH_3-\overset{H}{\underset{NH_2}{C}}-COOH$
*arginine	Arg	$H_2N-\underset{\overset{\|}{NH}}{C}-NH-CH_2-CH_2-CH_2-\overset{H}{\underset{NH_2}{C}}-COOH$
asparagine	Asn	$H_2N-\underset{\overset{\|}{O}}{C}-CH_2-\overset{H}{\underset{NH_2}{C}}-COOH$
aspartic acid	Asp	$HOOC-CH_2-\overset{H}{\underset{NH_2}{C}}-COOH$
cysteine	Cys	$HS-CH_2-\overset{H}{\underset{NH_2}{C}}-COOH$
glutamic acid	Glu	$HOOC-CH_2-CH_2-\overset{H}{\underset{NH_2}{C}}-COOH$
glutamine	Gln	$\genfrac{}{}{0pt}{}{H_2N}{O=}\!\!>C-CH_2-CH_2-\overset{H}{\underset{NH_2}{C}}-COOH$
glycine	Gly	$H-\overset{H}{\underset{NH_2}{C}}-COOH$
*histidine	His	$\underset{\underset{\underset{H}{C}}{N\quad NH}}{HC=C}-CH_2-\overset{H}{\underset{NH_2}{C}}-COOH$
*isoleucine	Ile	$CH_3-CH_2-\underset{CH_3}{CH}-\overset{H}{\underset{NH_2}{C}}-COOH$
*leucine	Leu	$\genfrac{}{}{0pt}{}{H_3C}{H_3C}\!\!>CH-CH_2-\overset{H}{\underset{NH_2}{C}}-COOH$
*lysine	Lys	$H_2N-CH_2-CH_2-CH_2-CH_2-\overset{H}{\underset{NH_2}{C}}-COOH$

*methionine	Met	$CH_3-S-CH_2-CH_2-C(H)(NH_2)-COOH$
*phenylalanine	Phe	$C_6H_5-CH_2-C(H)(NH_2)-COOH$
proline	Pro	H_2C-CH_2, H_2C, $CH-COOH$, N, H → H, $C-CH_2$, OH, H_2C, $CH-COOH$, N, H 4–hydroxyproline
serine	Ser	$HO-CH_2-C(H)(NH_2)-COOH$
*threonine	Thr	$CH_3-CH(OH)-C(H)(NH_2)-COOH$
*tryptophan	Trp	$C-CH_2-C(H)(NH_2)-COOH$, CH, N, H
*tyrosine	Tyr	$HO-C_6H_4-CH_2-C(H)(NH_2)-COOH$
*valine	Val	$(H_3C)(H_3C)CH-C(H)(NH_2)-COOH$

*an essential amino acid

The amino acids occurring in proteins

especially as photographic developers, and for making dyes. Examples include *amidol* (the dihydrochloride of 2,4-diaminophenol), *metol* (the hemisulphate of 4-methylaminophenol) and *rhodinol* (4-methylaminophenol).

α-aminotoluene *See* **benzylamine**.

ammine A coordination *complex in which the ligands are ammonia molecules. An example of an ammine is the tetraamminecopper(II) ion $[Cu(NH_3)_4]^{2+}$.

ammonia A colourless gas, NH_3, with a strong pungent odour; r.d. 0.59 (relative to air); m.p. −77.7°C; b.p. −33.35°C. It is very soluble in water and soluble in alcohol. The compound may be prepared in the laboratory by reacting ammonium salts with bases such as calcium hydroxide, or by the hydrolysis of a nitride. Industrially it is made by the *Haber process and over 80 million tonnes per year are used either directly or in combination. Major uses are the manufacture of nitric acid, ammonium nitrate, ammonium phosphate, and urea (the last three as fertilizers), explosives, dyestuffs and resins.

Liquid ammonia has some similarity to water as it is hydrogen bonded and has a moderate dielectric constant, which permits it to act as an ionizing solvent. It is weakly self-ionized to give ammonium ions, NH_4^+ and amide ions, NH_2^-. It also dissolves electropositive metals to give blue solutions, which are believed to contain solvated electrons. Ammonia is extremely soluble in water giving basic solutions that contain solvated NH_3 molecules and small amounts of the ions NH_4^+ and OH^-. The combustion of ammonia in air yields nitrogen and water. In the presence of catalysts NO, NO_2, and water are formed; this last reaction is the basis for the industrial production of nitric acid. Ammonia is a good proton acceptor (i.e. it is a base) and gives rise to a series of ammonium salts, e.g.

$$NH_3 + HCl \rightarrow NH_4^+ + Cl^-.$$

It is also a reducing agent.

The participation of ammonia in the *nitrogen cycle is a most important natural process. Nitrogen-fixing bacteria are able to achieve similar reactions to those of the Haber process, but under normal conditions of temperature and pressure. These release ammonium ions, which are converted by nitrifying bacteria into nitrite and nitrate ions.

ammoniacal Describing a solution in which the solvent is aqueous ammonia.

ammonia clock A form of atomic clock in which the frequency of a quartz oscillator is controlled by the vibrations of excited ammonia molecules. The ammonia molecule (NH_3) consists of a pyramid with a nitrogen atom at the apex and one hydrogen atom at each corner of the triangular base. When the molecule is excited, once every 20.9 microseconds the nitrogen atom passes through the base and forms a pyramid the other side: 20.9 microseconds later it returns to its original position. This vibration back and forth has a frequency of 23 870 hertz and ammonia gas will only absorb excitation energy at exactly this frequency. By using a crystal oscillator to feed energy to the gas and a suitable feedback mechanism, the oscillator can be locked to exactly this frequency.

ammonia–soda process *See* **Solvay process.**

ammonium alum *See* **alums.**

ammonium bicarbonate *See* **ammonium hydrogencarbonate.**

ammonium carbonate A colourless or white crystalline solid, $(NH_4)_2CO_3$, usually encountered as the monohydrate. It is very soluble in cold water. The compound decomposes slowly to give ammonia, water, and carbon dioxide. Commercial 'ammonium carbonate' is a double salt of ammonium hydrogencarbonate and ammonium aminomethanoate (carbamate), $NH_4HCO_3.NH_2COONH_4$. This material is manufactured by heating a mixture of ammonium chloride and calcium carbonate and recovering the product as a sublimed solid. It readily releases ammonia and is the basis of sal volatile. It is also used in dyeing and wool preparation and in baking powders.

ammonium chloride (sal ammoniac) A white or colourless cubic solid, NH_4Cl; r.d. 1.53; sublimes at 340°C. It is very soluble in water and slightly soluble in ethanol but insoluble in ether. It may be prepared by fractional crystallization from a solution containing ammonium sulphate and sodium chloride or ammonium carbonate and calcium chloride. Pure samples may be made directly by the gas-phase reaction of ammonia and hydrogen chloride. Because of its ease of preparation it can be manufactured industrially alongside any plant that uses or produces ammonia. The compound is used in dry cells, metal finishing, and in the preparation of cotton for dyeing and printing.

ammonium hydrogencarbonate (ammonium bicarbonate) A white crystalline compound, NH_4HCO_3. It is formed naturally as a decay product of nitrogenous matter and is made commercially by various methods: the action of carbon dioxide and steam on a solution of ammonium carbonate; heating commercial ammonium carbonate (which always contains some hydrogencarbonate); and the interaction of ammonia, carbon dioxide, and water vapour. It is used in some *baking powders and medicines.

ammonium ion The monovalent cation NH_4^+. It may be regarded as the product of the reaction of ammonia (a Lewis base) with a hydrogen ion. The ion has tetrahedral symmetry. The chemical properties of ammonium salts are frequently very similar to those of equivalent alkali-metal salts.

ammonium nitrate A colourless crystalline solid, NH_4NO_3; r.d. 1.72; m.p. 169.6°C; b.p. 210°C. It is very soluble in water and soluble in ethanol. The crystals are rhombic when obtained below 32°C and monoclinic above 32°C. It may be readily prepared in the laboratory by the reaction of nitric acid with aqueous ammonia. Industrially, it is manufactured by the same reaction using ammonia gas. Vast quantities of ammonium nitrate are used as fertilizers (over 20 million tonnes per year) and it is also a component of some explosives.

ammonium sulphate A white rhombic solid, $(NH_4)_2SO_4$; r.d. 1.77; decomposes at 235°C. It is very soluble in water and insoluble in ethanol. It occurs naturally as the mineral *mascagnite*. Ammonium sulphate was formerly manufactured from the 'ammoniacal liquors' produced during coal-gas manufacture but is now produced by the direct reaction between ammonia gas and sulphuric acid. It is decomposed by heating to release ammonia (and ammonium hydrogensulphate) and eventually water, sulphur dioxide, and ammonia. Vast quantities of ammonium sulphate are used as fertilizers.

ammonium thiocyanate A colourless, soluble crystalline compound, NH_4NCS. It is made by the action of hydrogen cyanide on ammonium sulphide or from ammonia and carbon disulphide in ethanol. On heating, it turns into its isomer thiourea, $SC(NH_2)_2$. Its solutions give a characteristic blood-red colour with iron(III) compounds and so are employed as a test for ferric iron. Ammonium thiocyanate is used as a rapid fixative in photography and as an ingredient in making explosives.

amorphous Describing a solid that is not crystalline; i.e. one that has no long-range order in its lattice. Many powders that are described as 'amorphous' in fact are composed of microscopic crystals, as can be demonstrated by X-ray diffraction. *Glasses are examples of true amorphous solids.

amount of substance Symbol n. A measure of the number of entities present in a substance. The specified entity may be an atom, molecule, ion, electron, photon, etc., or any specified group of such entities. The amount of substance of an element, for example, is proportional to the number of atoms present. For all entities, the constant of proportionality is the *Avogadro constant. The SI unit of amount of substance is the *mole.

AMP *See* ATP; cyclic AMP.

ampere Symbol A. The SI unit of electric current. The constant current that, maintained in two straight parallel infinite conductors of negligible cross section placed one metre apart in a vacuum, would produce a force between the conductors of 2×10^{-7} N m^{-1}. This definition replaced the earlier international ampere defined as the current required to deposit 0.001 118 00 gram of silver from a solution of silver nitrate in one second. The unit is named after the French physicist André Marie Ampère (1775–1836).

ampere-hour A practical unit of electric charge equal to the charge flowing in one hour through a conductor passing one ampere. It is equal to 3600 coulombs.

ampere-turn The SI unit of magnetomotive force equal to the magnetomotive force produced when a current of one ampere flows through one turn of a magnetizing coil.

amperometric titration A method of determining the chemical composition of a solution by measuring the current passing through a cell containing the solution; the potential is held constant during the titration for both the indicator and reference electrodes, with changes in the current being measured. The current flowing through the cell is measured as a function of the amount of substance being titrated.

amphetamine A drug, 1-phenyl-2-aminopropane (or a derivative of this compound), that stimulates the central nervous system by causing the release of the transmitters noradrenaline and dopamine from nerve endings. It inhibits sleep, suppresses the appetite, and has variable effects on mood; prolonged use can lead to addiction.

amphiboles A large group of rock-forming metasilicate minerals. They have a structure of silicate tetrahedra linked to form double endless chains, in contrast to the single chains of the *pyroxenes, to which they are closely related. They are present in many igneous and metamorphic rocks. The amphiboles show a wide range of compositional variation but conform to the general formula: $X_{2-3}Y_5Z_8O_{22}(OH)_2$, where X = Ca, Na, K, Mg, or Fe^{2+}; Y = Mg, Fe^{2+}, Fe^{3+}, Al, Ti, or Mn; and Z = Si or Al. The hydroxyl ions may be replaced by F, Cl, or O. Most amphiboles are monoclinic, including: cummingtonite, $(Mg,Fe^{2+})_7(Si_8O_{22})(OH)_2$; tremolite, $Ca_2Mg_5(Si_8O_{22})(OH,F)_2$; actinolite, $Ca_2(Mg,Fe^{2+})_5(Si_8O_{22})(OH,F)_2$; *hornblende, $NaCa_2(Mg,Fe^{2+},Fe^{3+},Al)_5((Si,Al)_8O_{22})(OH,F)_2$; edenite, $NaCa_2(Mg,Fe^{2+})_5(Si_7AlO_{22})(OH,F)_2$; and riebeckite, $Na_2,Fe_3^{2+}(Si_8O_{22})(OH,F)_2$. Anthophyllite, $(Mg,Fe^{2+})_7(Si_8O_{22})(OH,F)_2$, and gedrite, $(Mg,Fe^{2+})_6Al(Si,Al)_8O_{22})(OH,F)_2$, are orthorhombic amphiboles.

amphibolic pathway A biochemical pathway that serves both anabolic and catabolic processes. An important example of an amphibolic pathway is the *Krebs cycle, which involves both the catabolism of carbohydrates and fatty acids and the synthesis of anabolic precursors for amino-acid synthesis.

amphiphilic Describing a molecule that has both hydrophilic and hydrophobic parts, as in *detergents.

amphiprotic *See* **amphoteric; solvent.**

ampholyte A substance that can act as either an acid, in the presence of a strong base, or a base, when in the presence of a strong acid.

ampholyte ion *See* **zwitterion.**

amphoteric Describing a compound that can act as both an acid and a base (in the traditional sense of the term). For instance, aluminium hydroxide is amphoteric: as a base $Al(OH)_3$ it reacts with acids to form aluminium salts; as an acid H_3AlO_3 it reacts with alkalis to give *aluminates. Oxides of metals are typically basic and oxides of nonmetals tend to be acidic. The existence of amphoteric oxides is sometimes regarded as evidence that an element is a *metalloid. Compounds such as the amino acids, which contain both acidic and basic groups in their molecules, can also be described as amphoteric. Solvents, such as water, that can both donate and accept protons are usually described as *amphiprotic* (*see* **solvent**).

a.m.u. *See* **atomic mass unit.**

amylase Any of a group of closely related enzymes that degrade starch, glycogen, and other polysaccharides. Plants contain both α- and β-amylases; the name *diastase* is given to the component of malt containing β-amylase, important in the brewing industry. Animals possess only α-amylases, found in pancreatic juice (as *pancreatic amylase*) and also (in humans and some other species) in saliva (as *salivary amylase* or *ptyalin*). Amylases cleave the long polysaccharide chains, producing a mixture of glucose and maltose.

amyl group Formerly, any of several isomeric groups with the formula C_5H_{11}–.

amylopectin A *polysaccharide comprising highly branched chains of glucose molecules. It is one of the constituents (the other being amylose) of *starch.

amylose A *polysaccharide consisting of linear chains of between 100 and 1000 linked glucose molecules. Amylose is a constituent of *starch. In water, amylose reacts with iodine to give a characteristic blue colour.

anabolism The metabolic synthesis of proteins, fats, and other constituents of living organisms from molecules or simple precursors. This process requires energy in the form of ATP. Drugs that promote such metabolic activity are described as *anabolic*. *See* **metabolism**. *Compare* **catabolism**.

Analar reagent A chemical reagent of high purity with known contaminants for use in chemical analyses.

analyser A device, used in the *polarization of light, that is placed in the eyepiece of a *polarimeter to observe plane-polarized light. The analyser, which may be a *Nicol prism or *Polaroid, can be oriented in different directions to investigate in which plane an incoming wave is polarized or if the light is plane polarized. If there is one direction from which light does not emerge from the analyser when it is rotated, the incoming wave is plane polarized. If the analyser is horizontal when extinction of light takes place, the polarization of light must have been in the vertical plane. The intensity of a beam of light transmitted through an analyser is proportional to $\cos^2\theta$, where θ is the angle between the plane of polarization and the plane of the analyser. Extinction is said to be produced by 'crossing' the *polarizer and analyser.

analysis The determination of the components in a chemical sample. *Qualitative analysis* involves determining the nature of a pure unknown compound or the compounds present in a mixture. Various chemical tests exist for different elements or types of compound, and systematic analytical procedures can be used for mixtures. *Quantitative analysis* involves measuring the proportions of known components in a mixture. Chemical techniques for this fall into two main classes: *volumetric analysis and *gravimetric analysis. In addition, there are numerous physical methods of qualitative and quantitative analysis, including spectroscopic techniques, mass spectrometry, polarography, chromatography, activation analysis, etc.

Andrews titration A titration used to estimate amounts of reducing agents. The reducing agent being estimated is dissolved in concentrated hydrochloric acid and titrated with a solution of potassium iodate. A drop of tetrachloromethane is added to the solution. The end point of the titration is reached when the iodine colour disappears from this layer. This is due to the reducing agent being oxidized and the iodate being reduced to ICl. This reaction involves a four-electron change.

angle-resolved photoelectron spectroscopy **(ARPES)** A technique for studying the composition and structure of surfaces by measuring both the kinetic energy and angular distribution of photoelectrons ejected from a surface by electromagnetic radiation. *See also* **photoelectron spectroscopy**.

anglesite A mineral form of *lead(II) sulphate, $PbSO_4$.

angle strain The departure of a bond angle from its normal value. The effects of angle strain are often apparent in aliphatic ring compounds. For example, in cyclopentane, C_5H_{10}, the angles between the bonds of the ring differ from the normal tetrahedral angle (109° 28′), which is found in cyclohexane. This ring form of angle strain is often called *Baeyer strain.*

angstrom Symbol Å. A unit of length equal to 10^{-10} metre. It was formerly used to measure wavelengths and intermolecular distances but has now been replaced by the nanometre. 1 Å = 0.1 nanometre. The unit is named after A. J. *Ångström.

Ångström, Anders Jonas (1814–74) Swedish astronomer and physicist who became professor of physics at the University of Uppsala from 1858 until his death. He worked mainly with emission *spectra, demonstrating the presence of hydrogen in the sun. He also worked out the wavelengths of *Fraunhofer lines. Wavelengths and intermolecular distances were formerly expressed in *angstroms.

angular momentum A property of a rotating body. In the case of a rigid rotating body the angular momentum is given by $I\omega$, where I is the *moment of inertia of the body and ω is its angular velocity. The quantum theory of angular momentum is closely associated with the *rotation group and has important applications in the electronic structure of atoms and diatomic molecules and *rotational spectroscopy of molecules. Electron spin is a type of angular momentum.

anharmonicity The extent to which the oscillation of an oscillator differs from simple harmonic motion. In molecular vibrations the anharmonicity is very small near the equilibrium position, becomes large as the vibration moves away from the equilibrium position, and is very large as dissociation is approached. Anharmonicity is taken into account in molecular vibrations by adding an anharmonicity term to the potential energy function of the molecule. For a harmonic oscillator the potential energy function U is given by $U = f(r - r_e)^2$ where r is the interatomic distance, r_e is the equilibrium interatomic distance, and f is a constant. Anharmonicity is taken into account by adding a cubic term $g(r - r_e)^3$ to the quadratic term, where g is much smaller than f. Higher terms in $(r - r_e)$ can be added to improve the description of anharmonicity.

anharmonic oscillator An oscillating system (in either classical mechanics or *quantum mechanics) that is not oscillating in simple harmonic motion. In general, the problem of an anharmonic oscillator is not exactly soluble, although many systems approximate to harmonic oscillators and for such systems the *anharmonicity can be calculated using *perturbation theory.

anhydride A compound that produces a given compound on reaction with water. For instance, sulphur trioxide is the (acid) anhydride of sulphuric acid

$$SO_3 + H_2O \rightarrow H_2SO_4$$

See also **acid anhydrides**.

anhydrite An important rock-forming anhydrous mineral form of calcium sulphate, $CaSO_4$. It is chemically similar to *gypsum but is harder and heavier and crystallizes in the rhombic form (gypsum is monoclinic). Under natural conditions anhydrite slowly hydrates to form gypsum. It occurs chiefly in white and greyish granular masses and is often found in the caprock of certain salt domes. It is used as a raw material in the chemical industry and in the manufacture of cement and fertilizers.

anhydrous Denoting a chemical compound lacking water: applied particularly to salts lacking their water of crystallization.

aniline *See* **phenylamine**.

anilinium ion The ion $C_6H_5NH_3^+$, derived from *phenylamine.

animal charcoal *See* **charcoal**.

animal starch *See* **glycogen**.

anion A negatively charged *ion, i.e. an ion that is attracted to the *anode in *electrolysis. *Compare* **cation**.

anionic detergent *See* **detergent**.

anionic resin *See* **ion exchange**.

anisotropic Denoting a medium in which certain physical properties are different in different directions. Wood, for instance, is an anisotropic material: its strength along the grain differs from that perpendicular to the grain. Single crystals that are not cubic are anisotropic with respect to some physical properties, such as the transmission of electromagnetic radiation. *Compare* **isotropic**.

annealing A form of heat treatment applied to a metal to soften it, relieve internal stresses and instabilities, and make it easier to work or machine. It consists of heating the metal to a specified temperature for a specified time, both of which depend on the metal involved, and then allowing it to cool slowly. It is applied to both ferrous and nonferrous metals and a similar process can be applied to other materials, such as glass.

annelation *See* annulation.

annulation A type of chemical reaction in which a ring is fused to a molecule by formation of two new bonds. Sometimes the term *annelation* is used. *See also* **cyclization**.

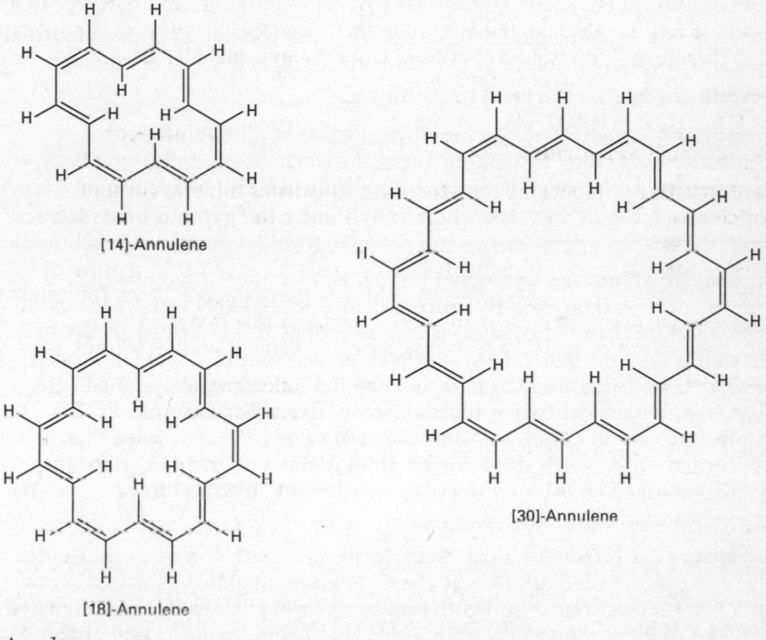

[14]-Annulene

[30]-Annulene

[18]-Annulene

Annulenes

annulenes Organic hydrocarbons that have molecules containing simple single rings of carbon atoms linked by alternating single and double bonds. Such compounds have even numbers of carbon atoms. *Cyclo-octatetraene, C_8H_8, is the next in the series following benzene. Higher annulenes are usually referred to by the number of carbon atoms in the ring, as in [10]-annulene, $C_{10}H_{10}$, [12]-annulene, $C_{12}H_{12}$, etc. The lower members are not stable as a result of the interactions between hydrogen atoms inside the ring. This is true even for molecules that have the necessary number of pi electrons to be *aromatic compounds. Thus, [10]-annulene has $4n + 2$ pi electrons with $n = 2$, but is not aromatic because it is not planar. [14]-annulene also has a suitable number of pi electrons to be aromatic ($n = 3$) but is not planar because of interaction between the inner hydrogens.

The compound [18]-annulene is large enough to be planar and obeys the

Hückel rule ($4n + 2 = 18$, with $n = 4$). It is a brownish red fairly stable reactive solid. NMR evidence shows that it has aromatic character. The annulene with $n = 7$, [30]-annulene, can also exist in a planar form but is highly unstable. *See also* **pseudoaromatic**.

anode A positive electrode. In *electrolysis anions are attracted to the anode. In an electronic vacuum tube it attracts electrons from the *cathode and it is therefore from the anode that electrons flow out of the device. In these instances the anode is made positive by external means; however in a *voltaic cell the anode is the electrode that spontaneously becomes positive and therefore attracts electrons to it from the external circuit.

anode sludge *See* **electrolytic refining**.

anodizing A method of coating objects made of aluminium with a protective oxide film, by making them the anode in an electrolytic bath containing an oxidizing electrolyte. Anodizing can also be used to produce a decorative finish by formation of an oxide layer that can absorb a coloured dye.

anomeric effect If a pyranose ring has an electronegative substituent at the anomeric carbon then this substituent is more likely to take the axial configuration than the equatorial configuration. This is a consequence of a general effect (the generalized anomeric effect) that in a chain of atoms X–C–Y–C, in which Y and X are atoms with nonbonding electron pairs (e.g. F, O, N), the synclinal conformation is more likely. For example, in the compound CH_3–O–CH_2Cl, rotation can occur along the O–C bond. The conformation in which the chlorine atom is closer to the methyl group is preferred (the lone pairs on the oxygen atom act as groups in determining the conformation).

anomers Diastereoisomers of cyclic forms of sugars or similar molecules differing in the configuration at the C1 atom of an aldose or the C2 atom of a ketose. This atom is called the *anomeric carbon*. Anomers are designated α if the configuration at the anomeric carbon is the same as that at the reference asymmetric carbon in a Fischer projection. If the configuration differs the anomer is designated β. The α- and β-forms of glucose are examples of anomers (*see* **monosaccharide**).

anoxic reactor A *bioreactor in which the organisms being cultured are anaerobes or in which the reaction being exploited does not require oxygen.

anthocyanin One of a group of *flavonoid pigments. Anthocyanins occur in various plant organs and are responsible for many of the blue, red, and purple colours in plants (particularly in flowers).

anthracene A white crystalline solid, $C_{14}H_{10}$; r.d. 1.28; m.p. 215.8°C; b.p. 341.4°C. It is an aromatic hydrocarbon with three fused rings (see formula), and is obtained by the distillation of crude oils. The main use is in the manufacture of dyes.

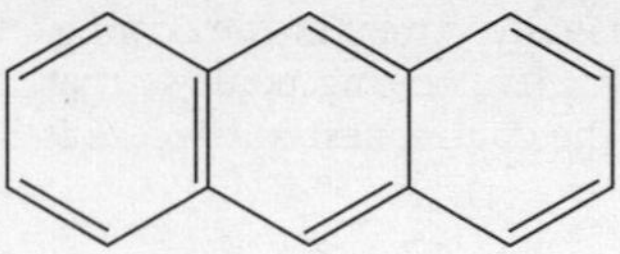

Anthracene

anthracite *See* coal.

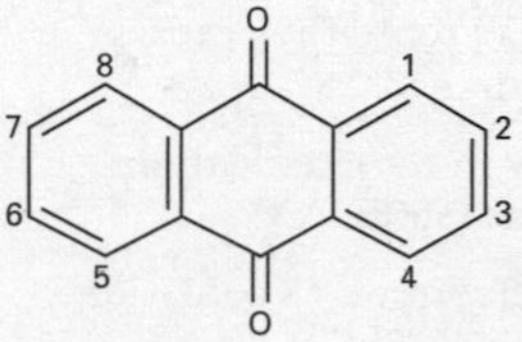

8 1 7 2 6 3 5 4 O O

Anthraquinone

anthraquinone A colourless crystalling *quinone; m.p. 154°C. It may be prepared by reacting benzene with phthalic anhydride. The compound is the basis of a range of dyestuffs. *Alizarin is an example.

anti 1. *See* **torsion angle. 2.** *See* **E–Z convention. 3.** *See* **conformation.**

antiaromatic *See* **pseudoaromatic.**

antibiotics Substances that destroy or inhibit the growth of microorganisms, particularly disease-producing bacteria. Antibiotics are obtained from microorganisms (especially moulds) or synthesized. Common antibiotics include the penicillins, streptomycin, and the tetracyclines. They are used to treat various infections but tend to weaken the body's natural defence mechanisms and can cause allergies. Overuse of antibiotics can lead to the development of resistant strains of microorganisms.

antibonding orbital *See* **orbital.**

anticlinal *See* **torsion angle.**

antiferromagnetism *See* **magnetism.**

antifluorite structure A crystal structure for ionic compounds of the type M_2X, which is the same as the fluorite (CaF_2) structure except that the positions of the anions and cations are reversed, i.e. the cations occupy the F^- sites and the anions occupy the Ca^{2+} sites. Examples of the antifluorite structure are given by K_2O and K_2S.

antifoaming agent A substance that prevents a foam from forming, employed in such processes as electroplating and paper-making, and in water for boilers. They are compounds that are absorbed strongly by the liquid (usually water) but lack the properties that allow foaming. Substances used include polyamides, polysiloxanes and silicones.

antifreeze A substance added to the liquid (usually water) in the cooling systems of internal-combustion engines to lower its freezing point so that it does not solidify at sub-zero temperatures. The commonest antifreeze is *ethane-1,2-diol (ethylene glycol).

antigorite *See* **serpentine.**

***anti*-isomer** *See* **isomerism.**

antiknock agent A petrol additive that inhibits preignition ('knocking') in internal-combustion engines. Antiknock agents work by retarding combustion chain reactions. The commonest, lead(IV) tetraethyl, causes environmental pollution, and its use is being discouraged.

antimonic compounds Compounds of antimony in its +5 oxidation state; e.g. antimonic chloride is antimony(V) chloride ($SbCl_5$).

antimonous compounds Compounds of antimony in its +3 oxidation state; e.g. antimonous chloride is antimony(III) chloride ($SbCl_3$).

antimony Symbol Sb. An element belonging to *group 15 (formerly VB) of the periodic table; a.n. 51; r.a.m. 121.75; r.d. 6.68; m.p. 630.5°C; b.p. 1750°C. Antimony has several allotropes. The stable form is a bluish-white metal. Yellow antimony and black antimony are unstable nonmetallic allotropes made at low temperatures. The main source is stibnite (Sb_2S_3), from which antimony is extracted by reduction with iron metal or by roasting (to give the oxide) followed by reduction with carbon and sodium carbonate. The main use of the metal is as an alloying agent in lead-accumulator plates, type metals, bearing alloys, solders, Britannia metal, and pewter. It is also an agent for producing pearlitic cast iron. Its compounds are used in flame-proofing, paints, ceramics, enamels, glass dyestuffs, and rubber technology. The element will burn in air but is unaffected by water or dilute acids. It is attacked by oxidizing acids and by halogens. It was first reported by Tholden in 1450.

antioxidants Substances that slow the rate of oxidation reactions. Various antioxidants are used to preserve foodstuffs and to prevent the deterioration of rubber, synthetic plastics, and many other materials. Some antioxidants act as chelating agents to sequester the metal ions that catalyse oxidation reactions. Others inhibit the oxidation reaction by removing oxygen free radicals. Naturally occurring antioxidants include *vitamin E and β-carotene; they limit the cell and tissue damage caused by foreign substances, such as toxins and pollutants, in the body.

antiparallel spins Neighbouring spinning electrons in which the *spins, and hence the magnetic moments, of the electrons are aligned in the opposite direction. Under some circumstances the interactions between magnetic moments in atoms favour *parallel spins, while under other conditions they favour antiparallel spins. The case of antiferromagnetism (*see* **magnetism**) is an example of a system with antiparallel spins.

antiperiplanar *See* **torsion angle.**

anti-Stokes lines *See* **anti-Stokes radiation; Stokes radiation.**

anti-Stokes radiation Electromagnetic radiation occurring in the *Raman effect, which is of much lower intensity than the Rayleigh scattering in which the frequency of the radiation is higher than that of the incident light, i.e. displaced towards shorter wavelengths. If the frequency of the original light is ν, the frequency of the anti-Stokes radiation is $\nu + \nu_k$, where ν_k is the frequency of the rotation or vibration of the molecule. The spectral lines associated with anti-Stokes radiation are called *anti-Stokes lines.* The quantum theory of the Raman effect provides an explanation for the intensity of the anti-Stokes line being much less than the intensity of the Stokes line.

ap Antiperiplanar. *See* **torsion angle.**

apatite A complex mineral form of *calcium phosphate, $Ca_5(PO_4)_3(OH,F,Cl)$; the commonest of the phosphate minerals. It has a hexagonal structure and occurs widely as an accessory mineral in igneous rocks (e.g. pegmatite) and often in regional and contact metamorphic rocks, especially limestone. Large deposits occur in the Kola Peninsula, Russia. It is used in the production of fertilizers and is a major source of phosphorus. The enamel of teeth is composed chiefly of apatite.

apical Designating certain bonds and positions in a bipyramidal or pyramidal structure. For example, in a trigonal bipyramid the two bonds aligned through the central atom are the apical bonds and the groups attached to these bonds are in apical positions. The three other bonds and positions are *equatorial.* Similarly, in a square pyramidal structure the bond perpendicular to the base of the pyramid is the apical bond. The four other coplanar bonds (and positions) are said to be *basal.*

aprotic *See* **solvent.**

aqua acid A type of acid in which the acidic hydrogen is on a water molecule coordinated to a metal ion. For example,

$$Al(OH_2)_6^{3+} + H_2O \rightarrow Al(OH_2)_5(OH)^{2+} + H_3O^+$$

aqua regia A mixture of concentrated nitric acid and concentrated hydrochloric acid in the ratio 1:3 respectively. It is a very powerful oxidizing mixture and will dissolve all metals (except silver, which forms an insoluble chloride) including such noble metals as gold and platinum, hence its name ('royal water'). Nitrosyl chloride (NOCl) is believed to be one of the active constituents.

aquation The process in which water molecules solvate or form coordination complexes with ions.

aqueous Describing a solution in water.

aquo ion A hydrated positive ion present in a crystal or in solution.

arachidonic acid An unsaturated fatty acid,

$CH_3(CH_2)_3(CH_2CH{:}CH)_4(CH_2)_3COOH$, that is essential for growth in mammals. It can be synthesized from *linoleic acid. Arachidonic acid acts as a precursor to several biologically active compounds, including prostaglandins, and plays an important role in membrane production and fat metabolism. The release of arachidonic acid from membrane phospholipids is triggered by certain hormones.

arachno-structure *See* **borane.**

aragonite A rock-forming anhydrous mineral form of calcium carbonate, $CaCO_3$. It is much less stable than *calcite, the commoner form of calcium carbonate, from which it may be distinguished by its greater hardness and specific gravity. Over time aragonite undergoes recrystallization to calcite. Aragonite occurs in cavities in limestone, as a deposit in limestone caverns, as a precipitate around hot springs and geysers, and in high-pressure low-temperature metamorphic rocks; it is also found in the shells of a number of molluscs and corals and is the main constituent of pearls. It is white or colourless when pure but the presence of impurities may tint it grey, blue, green, or pink.

arenes Aromatic hydrocarbons, such as benzene, toluene, and naphthalene.

argentic compounds Compounds of silver in its higher (+2) oxidation state; e.g. argentic oxide is silver(II) oxide (AgO).

argentite A sulphide ore of silver, Ag_2S. It crystallizes in the cubic system but most commonly occurs in massive form. It is dull grey-black in colour but bright when first cut and occurs in veins associated with other silver minerals. Important deposits occur in Mexico, Peru, Chile, Bolivia, and Norway.

argentous compounds Compounds of silver in its lower (+1) oxidation state; e.g. argentous chloride is silver(I) chloride.

arginine *See* **amino acid.**

argon Symbol Ar. A monatomic noble gas present in air (0.93%); a.n. 18; r.a.m. 39.948; d. 0.00178 g cm^{-3}; m.p. −189°C; b.p. −185°C. Argon is separated from liquid air by fractional distillation. It is slightly soluble in water, colourless, and has no smell. Its uses include inert atmospheres in welding and special-metal manufacture (Ti and Zr), and (when mixed with 20% nitrogen) in gas-filled electric-light bulbs. The element is inert and has no true compounds. Lord *Rayleigh and Sir William *Ramsay identified argon in 1894.

aromatic compound An organic compound that contains a benzene ring in its molecules or that has chemical properties similar to benzene. Aromatic compounds are unsaturated compounds, yet they do not easily partake in addition reactions. Instead they undergo electrophilic substitution.

Benzene, the archetypal aromatic compound, has an hexagonal ring of carbon atoms and the classical formula (the Kekulé structure) would have alternating double and single bonds. In fact all the bonds in benzene are the same length intermediate between double and single C–C bonds. The properties arise because the electrons in the π-orbitals are delocalized over the ring, giving an extra stabilization energy of 150 kJ mol^{-1} over the energy of a Kekulé structure. The condition for such delocalization is that a compound should have a planar ring with $(4n + 2)$ pi electrons – this is known as the *Hückel rule*. Aromatic behaviour is also found in heterocyclic compounds such as pyridine. Aromatic character can be detected by the presence of a ring current using NMR. *See also* **annulenes**; **nonbenzenoid aromatics**; **pseudoaromatic**.

aromaticity The property characteristic of *aromatic compounds.

ARPES *See* **angle-resolved photoelectron spectroscopy**.

Arrhenius, Svante (August) (1859–1927) Swedish physical chemist who first demonstrated that electrolytes conduct because of the presence of ions. He also worked on kinetics and proposed the *Arrhenius equation. He was the first to predict the greenhouse effect resulting from carbon dioxide. Arrhenius was awarded the 1903 Nobel Prize for chemistry.

Arrhenius equation An equation of the form

$$k = A\exp(-E_a/RT)$$

where k is the rate constant of a given reaction and E_a the *activation energy. A is a constant for a given reaction, called the *pre-exponential factor* (or *A-factor*). Often the equation is written in logarithmic form

$$\ln k = \ln A - E_a/RT$$

A graph of $\ln k$ against $1/T$ is a straight line with a gradient $-E_a/R$ and an intercept on the $\ln k$ axis of $\ln A$. It is named after Svante *Arrhenius.

Arrhenius theory *See* **acid**.

arsenate(III) *See* **arsenic(III) oxide**.

arsenate(V) *See* **arsenic(V) oxide**.

arsenic Symbol As. A metalloid element of *group 15 (formerly VB) of the periodic table; a.n. 33; r.a.m. 74.92; r.d. 5.7; sublimes at 613°C. It has three allotropes – yellow, black, and grey. The grey metallic form is the stable and most common one. Over 150 minerals contain arsenic but the main sources are as impurities in sulphide ores and in the minerals orpiment (As_2S_3) and realgar (As_4S_4). Ores are roasted in air to form arsenic oxide and then reduced by hydrogen or carbon to metallic arsenic. Arsenic compounds are used in insecticides and as doping agents in semiconductors. The element is included in some lead-based alloys to promote hardening. Confusion can arise because As_4O_6 is often sold as white arsenic. Arsenic compounds are accumulative poisons. The element will react with halogens, concentrated oxidizing acids, and hot alkalis. Albertus Magnus is believed to have been the first to isolate the element in 1250.

arsenic acid *See* **arsenic(V) oxide**.

arsenic(III) acid *See* **arsenic(III) oxide.**

arsenic hydride *See* **arsine.**

arsenic(III) oxide (arsenic trioxide; arsenious oxide; white arsenic) A white or colourless compound, As_4O_6, existing in three solid forms. The commonest has cubic or octahedral crystals (r.d. 3.87; sublimes at 193°C) and is soluble in water, ethanol, and alkali solutions. It occurs naturally as *arsenolite*. A vitreous form can be prepared by slow condensation of the vapour (r.d. 3.74); its solubility in cold water is more than double that of the cubic form. The third modification, which occurs naturally as *claudetite*, has monoclinic crystals (r.d. 4.15). Arsenic(III) oxide is obtained commercially as a byproduct from the smelting of nonferrous sulphide ores; it may be produced in the laboratory by burning elemental arsenic in air. The structure of the molecule is similar to that of P_4O_6, with a tetrahedral arrangement of As atoms edge linked by oxygen bridges. Arsenic(III) oxide is acidic; its solutions were formerly called *arsenious acid* (technically, *arsenic(III) acid*). It forms *arsenate(III)* salts (formerly called *arsenites*). Arsenic(III) oxide is extremely toxic and is used as a poison for vermin; trace doses are used for a variety of medicinal purposes. It is also used for producing opalescent glasses and enamels.

arsenic(V) oxide (arsenic oxide) A white amorphous deliquescent solid, As_2O_5; r.d. 4.32; decomposes at 315°C. It is soluble in water and ethanol. Arsenic(V) oxide cannot be obtained by direct combination of arsenic and oxygen; it is usually prepared by the reaction of arsenic with nitric acid followed by dehydration of the arsenic acid thus formed. It readily loses oxygen on heating to give arsenic(III) oxide. Arsenic(V) oxide is acidic, dissolving in water to give arsenic(V) acid (formerly called *arsenic acid*), H_3AsO_4; the acid is tribasic and slightly weaker than phosphoric acid and should be visualized as $(HO)_3AsO$. It gives *arsenate(V)* salts (formerly called *arsenates*).

arsenic trioxide *See* **arsenic(III) oxide.**

arsenious acid *See* **arsenic(III) oxide.**

arsenious oxide *See* **arsenic(III) oxide.**

arsenite *See* **arsenic(III) oxide.**

arsenolite A mineral form of *arsenic(III) oxide, As_4O_6.

arsine (arsenic hydride) A colourless gas, AsH_3; m.p. −116.3°C; b.p. −55°C. It is soluble in water, chloroform, and benzene. Liquid arsine has a relative density of 1.69. Arsine is produced by the reaction of mineral acids with arsenides of electropositive metals or by the reduction of many arsenic compounds using nascent hydrogen. It is extremely poisonous and, like the hydrides of the heavier members of group 15 (formerly VB), is readily decomposed at elevated temperatures (around 260–300°C). Like ammonia and phosphine, arsine has a pyramidal structure.

Arsine gas has a very important commercial application in the production of modern microelectronic components. It is used in a dilute gas mixture with an inert gas and its ready thermal decomposition is exploited to enable other growing crystals to be doped with minute traces of arsenic to give *n*-type semiconductors.

artinite A mineral form of basic *magnesium carbonate, $MgCO_3.Mg(OH)_2.3H_2O$.

aryl group A group obtained by removing a hydrogen atom from an aromatic compound, e.g. phenyl group, C_6H_5-, derived from benzene.

aryne A compound that can be regarded as formed from an arene by removing two adjacent hydrogen atoms to convert a double bond into a triple bond. Arynes are transient intermediates in a number of reactions. The simplest example is *benzyne.

asbestos Any one of a group of fibrous amphibole minerals (amosite, crocidolite (blue asbestos), tremolite, anthophyllite, and actinolite) or the fibrous serpentine mineral chrysotile. Asbestos has widespread commercial uses because of its resistance to heat, chemical inertness, and high electrical resistance. The fibres may be spun and woven into fireproof cloth for use in protective clothing, curtains, brake linings, etc., or moulded into blocks. Since the 1970s short asbestos fibres have been recognized as a cause of asbestosis, a serious lung disorder, and mesothelioma, a fatal form of lung cancer. These concerns have limited its use and imposed many safety procedures when it is used. Canada is the largest producer of asbestos; others include Russia, South Africa, Zimbabwe, and China.

ascorbic acid *See* **vitamin C**.

asparagine *See* **amino acid**.

aspartic acid *See* **amino acid**.

asphalt Bitumen. *See* **petroleum**.

associated liquid *See* **association**.

association The combination of molecules of one substance with those of another to form chemical species that are held together by forces weaker than normal chemical bonds. For example, ethanol and water form a mixture (an *associated liquid*) in which hydrogen bonding holds the different molecules together.

astatine Symbol At. A radioactive *halogen element; a.n. 85; r.a.m. 211; m.p. 302°C; b.p. 337°C. It occurs naturally by radioactive decay from uranium and thorium isotopes. Astatine forms at least 20 isotopes, the most stable astatine–210 has a half-life of 8.3 hours. It can also be produced by alpha bombardment of bismuth–200. Astatine is stated to be more metallic than iodine; at least 5 oxidation states are known in aqueous solutions. It will form interhalogen compounds, such as AtI and AtCl. The existence of At_2 has not yet been established. The element was synthesized by nuclear bombardment in 1940 by D. R. Corson, K. R. MacKenzie, and E. Segré at the University of California.

Aston, Francis William (1877–1945) British chemist and physicist, who until 1910 worked at Mason College (later Birmingham University) and then with J. J. *Thomson at Cambridge University. In 1919 Aston designed the mass spectrograph (*see* **mass spectroscopy**), for which he was awarded the Nobel Prize for chemistry in 1922. With it he discovered the {isotopes} of neon, and was thus able to explain nonintegral atomic weights.

asymmetric atom *See* **optical activity**.

asymmetric induction The preferential formation of one particular enantiomer or diastereoisomer in a reaction as a result of a chiral element in one of the reactants or in the catalyst used.

asymmetric top *See* **moment of inertia**.

atactic polymer *See* **polymer**.

ATLEED (automated tensor low-energy electron diffraction) A form of *LEED (low-energy electron diffraction) in which the information obtained can readily be stored and analysed using a computer.

atmolysis The separation of a mixture of gases by means of their different rates of diffusion. Usually, separation is effected by allowing the gases to diffuse through the walls of a porous partition or membrane.

atmosphere 1. (atm.) A unit of pressure equal to 101 325 pascals. This is equal to 760.0 mmHg. The actual *atmospheric pressure fluctuates around this value. The unit is usually used for expressing pressures well in excess of standard atmospheric pressure, e.g. in high-pressure chemical processes. **2.** *See* **earth's atmosphere**.

atmospheric pressure The pressure exerted by the weight of the air above it at any point on the earth's surface. At sea level the atmosphere will support a column of mercury about 760 mm high. This decreases with increasing altitude. The standard value for the atmospheric pressure at sea level in SI units is 101 325 pascals.

atom The smallest part of an element that can exist chemically. Atoms consist of a small dense nucleus of protons and neutrons surrounded by moving electrons. The number of electrons equals the number of protons so the overall charge is zero. The electrons are considered to move in circular or elliptical orbits (*see* **Bohr theory**) or, more accurately, in regions of space around the nucleus (*see* **orbital**).

The *electronic structure* of an atom refers to the way in which the electrons are arranged about the nucleus, and in particular the *energy levels that they occupy. Each electron can be characterized by a set of four quantum numbers, as follows:

(1) The *principal quantum number n* gives the main energy level and has values 1, 2, 3, etc. (the higher the number, the further the electron from

the nucleus). Traditionally, these levels, or the orbits corresponding to them, are referred to as *shells* and given letters K, L, M, etc. The K-shell is the one nearest the nucleus. The maximum number of electrons in a given shell is $2n^2$.
(2) The *orbital quantum number* l, which governs the angular momentum of the electron. The possible values of l are $(n-1), (n-2), \ldots 1, 0$. Thus, in the first shell ($n = 1$) the electrons can only have angular momentum zero ($l = 0$). In the second shell ($n = 2$), the values of l can be 1 or 0, giving rise to two *subshells* of slightly different energy. In the third shell ($n = 3$) there are three subshells, with $l = 2$, 1, or 0. The subshells are denoted by letters $s(l = 0)$, $p(l = 1)$, $d(l = 2)$, $f(l = 3)$. The number of electrons in each subshell is written as a superscript numeral to the subshell symbol, and the maximum number of electrons in each subshell is s^2, p^6, d^{10}, and f^{14}. The orbital quantum number is sometimes called the *azimuthal quantum number*.
(3) The *magnetic quantum number* m, which governs the energies of electrons in an external magnetic field. This can take values of $+l, +(l-1), \ldots 1, 0, -1, \ldots -(l-1), -l$. In an s-subshell (i.e. $l = 0$) the value of $m = 0$. In a p-subshell ($l = 1$), m can have values +1, 0, and −1; i.e. there are three p-orbitals in the p-subshell, usually designated p_x, p_y, and p_z. Under normal circumstances, these all have the same energy level.
(4) The *spin quantum number* m_s, which gives the spin of the individual electrons and can have the values +½ or −½.

According to the *Pauli exclusion principle, no two electrons in the atom can have the same set of quantum numbers. The numbers define the *quantum state* of the electron, and explain how the electronic structures of atoms occur. See Chronology: Atomic Theory

atomic absorption spectroscopy (AAS) An analytical technique in which a sample is vaporized and the nonexcited atoms absorb electromagnetic radiation at characteristic wavelengths.

atomic clock An apparatus for standardizing time based on periodic phenomena within atoms or molecules. *See* **ammonia clock; caesium clock.**

atomic emission spectroscopy (AES) An analytical technique in which a sample is vaporized and the atoms present are detected by their emission of electromagnetic radiation at characteristic wavelengths.

atomic force microscope (AFM) A type of microscope in which a small probe, consisting of a tiny chip of diamond, is held on a spring-loaded cantilever in contact with the surface of the sample. The probe is moved slowly across the surface and the tracking force between the tip and the surface is monitored. The probe is raised and lowered so as to keep this force constant, and a profile of the surface is produced. Scanning the probe over the sample gives a computer-generated contour map of the surface. The instrument is similar to the scanning tunnelling microscope, but uses mechanical forces rather than electrical signals. It can resolve individual

ATOMIC THEORY

c.430 BC	Greek natural philosopher Empedocles (d. c. 430 BC) proposes that all matter consists of four elements: earth, air, fire, and water.
c.400 BC	Greek natural philosopher Democritus of Abdera (c. 460–370 BC) proposes that all matter consists of atoms.
306 BC	Greek philosopher Epicurus (c. 342–270 BC) champions Democritus' atomic theory.
1649	French philosopher Pierre Gassendi (1592–1655) proposes an atomic theory (having read Epicurus).
1803	John Dalton proposes Dalton's atomic theory.
1897	J. J. Thomson discovers the electron.
1904	J. J. Thomson proposes his 'plum pudding' model of the atom, with electrons embedded in a nucleus of positive charges. Japanese physicist Hantaro Nagaoka (1865–1950) proposes a 'Saturn' model of the atom with a central nucleus having a ring of many electrons.
1911	New Zealand-born physicist Ernest Rutherford (1871–1937) discovers the atomic nucleus.
1913	Niels Bohr proposes model of the atom with a central nucleus surrounded by orbiting electrons. British physicist Henry Moseley (1887–1915) equates the positive charge on the nucleus with its atomic number. Frederick Soddy discovers isotopes.
1916	German physicist Arnold Sommerfield (1868–1951) modifies Bohr's model of the atom specifying elliptical orbits for the electrons.
1919	Ernest Rutherford discovers the proton.
1920	Ernest Rutherford postulates the existence of the neutron.
1926	Erwin Schrödinger proposes a wave-mechanical model of the atom (with electrons represented as wave trains).
1932	British physicist James Chadwick (1891–1974) discovers the neutron. Werner Heisenberg proposes a model of the atomic nucleus in which protons and neutrons exchange electrons to achieve stability.
1939	Niels Bohr proposes a 'liquid drop' model of the atomic nucleus.
1948	German-born US physicist Marie Goeppert-Meyer (1906–72) and German physicist Hans Jensen (1907–73) independently propose the 'shell' structure of the nucleus.
1950	US physicist Leo Rainwater (1917–86) combines the 'liquid-drop' and 'shell' models of the nucleus into a single theory.

molecules and, unlike the scanning tunnelling microscope, can be used with nonconducting samples, such as biological specimens.

atomicity The number of atoms in a given molecule. For example, oxygen (O_2) has an atomicity of 2, ozone (O_3) an atomicity of 3, benzene (C_6H_6) an atomicity of 12, etc.

atomic mass unit (a.m.u.) A unit of mass used to express *relative atomic masses. It is equal to 1/12 of the mass of an atom of the isotope carbon–12 and is equal to $1.660\,33 \times 10^{-27}$ kg. This unit superseded both the physical and chemical mass units based on oxygen–16 and is sometimes called the *unified mass unit* or the *dalton.*

atomic number (proton number) Symbol Z. The number of protons in the nucleus of an atom. The atomic number is equal to the number of electrons orbiting the nucleus in a neutral atom.

atomic orbital *See* orbital.

atomic volume The relative atomic mass of an element divided by its density.

atomic weight *See* relative atomic mass.

atom-probe field-ion microscopy A technique for identifying individual atoms on surfaces. In the atom-probe *field-ionization microscope (FIM) there is a hole in the fluorescent screen, with which the FIM image of an adsorbed atom is brought into coincidence. The gas causing the imaging is removed. The adsorbed atom is removed as an ion by a pulse of potential difference. The atom then passes in the same direction as the gas ions, through the hole in the screen. This enables the atom to be identified by a mass spectrometer behind the screen. Atom-probe FIM identifies both the type and the position of the atom and can be used to observe atomic processes, such as evaporation, with the pulse used for analysis lasting about 2 nanoseconds.

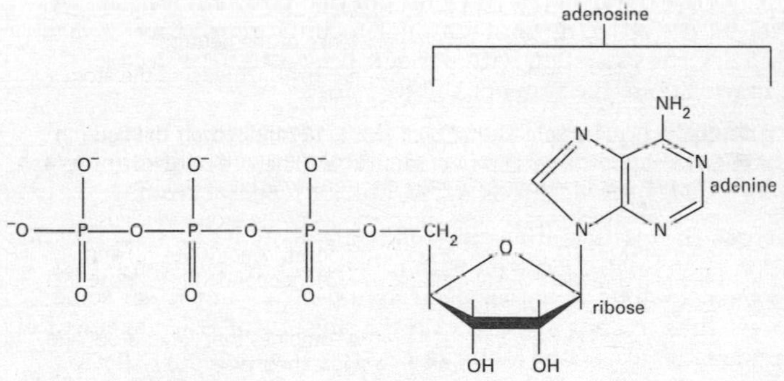

adenosine
adenine
ribose

ATP

ATP (adenosine triphosphate) A nucleotide that is of fundamental importance as a carrier of chemical energy in all living organisms. It consists of adenine linked to D-ribose (i.e. adenosine); the D-ribose component bears three phosphate groups, linearly linked together by covalent bonds (see formula). These bonds can undergo hydrolysis to yield either a molecule of *ADP* (*adenosine diphosphate*) and inorganic phosphate or a molecule of *AMP* (*adenosine monophosphate*) and pyrophosphate. Both these reactions yield a large amount of energy (about 30.6 kJ mol^{-1}) that is used to bring about such biological processes as muscle contraction, the active transport of ions and molecules across cell membranes, and the synthesis of biomolecules. The reactions bringing about these processes often involve the enzyme-catalysed transfer of the phosphate group to intermediate substrates. Most ATP-mediated reactions require Mg^{2+} ions as *cofactors.

ATP is regenerated by the rephosphorylation of AMP and ADP using the chemical energy obtained from the oxidation of food. This takes place during *glycolysis and the *Krebs cycle but, most significantly, is also a result of the reduction–oxidation reactions of the *electron transport chain, which ultimately reduces molecular oxygen to water (oxidative phosphorylation).

atropine A poisonous crystalline alkaloid, $C_{17}H_{23}NO_3$; m.p. 118–119°C. It can be extracted from deadly nightshade and other solanaceous plants and is used in medicine to treat colic, to reduce secretions, and to dilate the pupil of the eye.

atropisomers Conformers that have highly restricted rotation about a single bond and can consequently be separated as distinct species. This can occur, for example, in the case of substituted biphenyls with large groups at the ortho positions of the rings.

atto- Symbol a. A prefix used in the metric system to denote 10^{-18}. For example, 10^{-18} second = 1 attosecond (as).

attractor The set of points in phase space to which the representative point of a dissipative system (i.e. one with internal friction) tends as the system evolves. The attractor can be: a single point; a closed curve (a *limit cycle*), which describes a system with periodic behaviour; or a *strange attractor*, in which case the system exhibits *chaos.

Aufbau principle A principle that gives the order in which orbitals are filled in successive elements in the periodic table. The order of filling is 1*s*, 2*s*, 2*p*, 3*s*, 3*p*, 4*s*, 3*d*, 4*p*, 5*s*, 4*d*, 5*p*, 6*s*, 4*f*, 5*d*, 6*p*, 7*s*, 5*f*, 6*d*. *See* **atom**.

Auger effect The ejection of an electron from an atom as a result of the de-excitation of an excited electron within the atom. An electron is first ejected from an atom by a photon, electron impact, ion impact, or some other process, thus creating a vacancy. In the subsequent rearrangement of the electronic structure of the atom, an electron from a higher energy level falls into the vacancy. This process is associated with excess energy, which is released by the ejection of a second electron (rather than by emission of a photon). This second electron is called the *Auger electron*. *Auger spectroscopy* is a form of electron spectroscopy using this effect to study the energy levels of ions. It is also a form of analysis and can be used to identify the presence of elements in surface layers of solids. The effect was discovered by the French physicist Pierre Auger (1899–1994) in 1925.

Auger electron *See* **Auger effect**.

auric compounds Compounds of gold in its higher (+3) oxidation state; e.g. auric chloride is gold(III) chloride ($AuCl_3$).

aurous compounds Compounds of gold in its lower (+1) oxidation state; e.g. aurous chloride is gold(I) chloride (AuCl).

austenite *See* steel.

autocatalysis *Catalysis in which one of the products of the reaction is a catalyst for the reaction. Reactions in which autocatalysis occurs have a characteristic S-shaped curve for reaction rate against time – the reaction starts slowly and increases as the amount of catalyst builds up, falling off again as the products are used up.

autoclave A strong steel vessel used for carrying out chemical reactions, sterilizations, etc., at high temperature and pressure.

automated tensor low-energy electron diffraction *See* ATLEED.

autoprotolysis A transfer of a hydrogen ion (H^+) between molecules of an amphiprotic *solvent, one molecule acting as a Brønsted acid and the other as a Brønsted base. It occurs in the autoionization of water.

autoprotolysis constant *See* **ionic product**.

autoradiography An experimental technique in which a radioactive specimen is placed in contact with (or close to) a photographic plate, so as to produce a record of the distribution of radioactivity in the specimen. The film is darkened by the ionizing radiation from radioactive parts of the sample. Autoradiography has a number of applications, particularly in the study of living tissues and cells.

auxochrome A group in a dye molecule that influences the colour due to the *chromophore. Auxochromes are groups, such as $-OH$ and $-NH_2$, containing lone pairs of electrons that can be delocalized along with the delocalized electrons of the chromophore. The auxochrome intensifies the colour of the dye. Formerly, the term was also used of such groups as $-SO_2O^-$, which make the molecule soluble and affect its application.

Avogadro, Amedeo (1776–1856) Italian chemist and physicist. In 1811 he published his hypothesis (*see* **Avogadro's law**), which provided a method of calculating molecular weights from vapour densities. The importance of the work remained unrecognized, however, until championed by Stanislao *Cannizzaro in 1860.

Avogadro constant Symbol N_A or L. The number of atoms or molecules in one *mole of substance. It has the value $6.022\,1367(36) \times 10^{23}$. Formerly it was called *Avogadro's number*.

Avogadro's law Equal volumes of all gases contain equal numbers of molecules at the same pressure and temperature. The law, often called *Avogadro's hypothesis*, is true only for ideal gases. It was first proposed in 1811 by Amedeo *Avogadro.

axial *See* **ring conformations.**

Axilrod–Teller formula An expression for three-body interactions in intermolecular forces. These dipole dispersion terms, called *Axilrod–Teller terms* (named after B. M. Axilrod and E. Teller, who discovered them in 1943) are of importance in the third virial coefficient. At low temperatures, the Axilrod–Teller terms can have a comparable size to two-body interactions, making them of importance to such properties of liquids as pressure and surface tension. The Axilrod–Teller formula for the dispersion energy V of three closed-shell atoms A, B, and C is given by

$$V = -C_6/r_{AB}^6 - C_6/r_{BC}^6 - C_6/r_{CA}^6 + C'/(r_{AB}r_{BC}r_{CA})^3,$$

where C_6 is a coefficient depending on the identity of the atoms, C' is a coefficient depending on another constant and the angles between the atoms, and the r terms give distances between the indicated atoms.

AX spectrum A general pattern of the nuclear magnetic resonance spectrum of a molecule AX, where A and X are both spin-½ nuclei. If X is spin α, the spin–spin interaction between the nuclei of A and X results in one line in the spectrum of A being shifted by J/2 from the frequency of precession it has in the absence of coupling. If X is spin β, the change in frequency of precession of A is −J/2. Thus, single lines in the resonances of both A and X are split into doublets, with splittings J. The A resonance in molecules of the type A_nX is also a doublet with the splitting J.

azeotrope (azeotropic mixture; constant-boiling mixture) A mixture of two liquids that boils at constant composition; i.e. the composition of the vapour is the same as that of the liquid. Azeotropes occur because of deviations in Raoult's law leading to a maximum or minimum in the *boiling-point–composition diagram. When the mixture is boiled, the vapour initially has a higher proportion of one component than is present in the liquid, so the proportion of this in the liquid falls with time. Eventually, the maximum and minimum point is reached, at which the two liquids distil together without change in composition. The composition of an azeotrope depends on the pressure.

azeotropic distillation A technique for separating components of an azeotrope by adding a third liquid to form a new azeotrope with one of the original components. It is most commonly used to separate ethanol from water, adding benzene to associate with the ethanol.

azides Compounds containing the ion N_3^- or the group $-N_3$.

azimuthal quantum number *See* **atom.**

azine An organic heterocyclic compound containing a six-membered ring formed from carbon and nitrogen atoms. Pyridine is an example containing one nitrogen atom (C_5H_5N). *Diazines* have two nitrogen atoms in the ring (e.g. $C_4H_4N_2$), and isomers exist depending on the relative positions of the nitrogen atoms. *Triazines* contain three nitrogen atoms.

azo compounds Organic compounds containing the group −N=N− linking two other groups. They can be formed by reaction of a diazonium ion with a benzene ring.

azo dye *See* dyes.

azoimide *See* hydrogen azide.

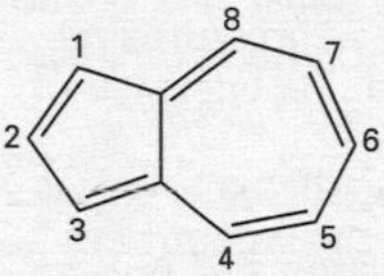

Azulene

azulene 1. A blue crystalline compound, $C_{10}H_8$; m.p. 99°C. It contains a five-membered ring fused to a seven-membered ring and has aromatic properties. When heated it is converted into naphthalene. **2.** Any of a number of blue oils that can be produced by distilling or heating essential oils from plants.

azurite A secondary mineral consisting of hydrated basic copper carbonate, $Cu_3(OH)_2(CO_3)_2$, in monoclinic crystalline form. It is generally formed in the upper zone of copper ore deposits and often occurs with *malachite. Its intense azure-blue colour made it formerly important as a pigment. It is a minor ore of copper and is used as a gemstone.

Babbit metal Any of a group of related alloys used for making bearings. They consist of tin containing antimony (about 10%) and copper (1–2%), and often lead. The original alloy was invented in 1839 by the US inventor Isaac Babbit (1799–1862).

Babo's law The vapour pressure of a liquid is decreased when a solute is added, the amount of the decrease being proportional to the amount of solute dissolved. The law was discovered in 1847 by the German chemist Lambert Babo (1818–99). *See also* **Raoult's law**.

backbiting A rearrangement that can occur in some *polymerization reactions involving free radicals. A radical that has an unpaired electron at the end of the chain changes into a radical with the unpaired electron elsewhere along the chain, the new radical being more stable than the one from which it originates. For example, the radical

$RCH_2CH_2CH_2CH_2CH_2CH_2\cdot$

may change into

$RCH_2CH\cdot CH_2CH_2CH_2CH_3$

The rearrangement is equivalent to a hydrogen atom being transferred within the molecule. The new unpaired electron initiates further polymerization, with the production of polymers with butyl ($CH_3CH_2CH_2CH_2$–) side chains.

back donation A form of chemical bonding in which a *ligand forms a sigma bond to an atom or ion by donating a pair of electrons, and the central atom donates electrons back by overlap of its *d*-orbitals with empty *p*- or *d*-orbitals on the ligand. It occurs, for example, in metal carbonyls.

back e.m.f. An electromotive force that opposes the main current flow in a circuit. For example, in an electric cell, *polarization causes a back e.m.f. to be set up by chemical means.

background radiation Low intensity *ionizing radiation present on the surface of the earth and in the atmosphere as a result of cosmic radiation and the presence of radioisotopes in the earth's rocks, soil, and atmosphere. The radioisotopes are either natural or the result of nuclear fallout or waste gas from power stations. Background counts must be taken into account when measuring the radiation produced by a specified source.

back titration A technique in *volumetric analysis in which a known excess amount of a reagent is added to the solution to be estimated. The unreacted amount of the added reagent is then determined by titration, allowing the amount of substance in the original test solution to be calculated.

bacteriocidal Capable of killing bacteria. Common bacteriocides are some antibiotics, antiseptics, and disinfectants.

bacteriorhodopsin A membrane-bound protein of the halophilic (salt-resistant) bacterium *Halobacterium halobium*. When activated by light, it pumps protons out of the cell; this creates a concentration gradient, which enables ATP to be synthesized. Bacteriorhodopsin is composed of seven α-helix segments, which span the membrane and are joined together by short amino-acid chains. It contains the prosthetic group retinal, which is also found in the pigment rhodopsin in the rod cells of vertebrates.

Baeyer, Adolf von (1835–1917) German organic chemist noted for his work on organic synthesis. He synthesized and determined the structure of indigo and also synthesized some of the first barbiturates. In his work on carbon rings he formulated his strain theory. Baeyer was awarded the 1905 Nobel Prize for chemistry.

Baeyer strain *See* **angle strain**.

Baeyer–Villiger reaction A rearrangement reaction, sometimes known as the *Dakin reaction*, commonly used in organic synthesis in which a ketone reacts with a peroxy acid to form an ester. For example,

R–CO–R → R–CO–O–R

The reaction is equivalent to the insertion of an oxygen atom next to the ketone's carbonyl (>C=O) group. *Meta*-chloroperbenzoic acid (*m*-CPBA; $ClC_6H_4.CO.O.OH$) and trifluoroperethanoic acid ($CF_3.CO.O.OH$) are typical peroxy acids employed in the reaction. It was discovered by Adolf von *Baeyer and the German chemist V. Villiger in 1899.

Bakelite A trade name for certain phenol–formaldehyde resins, first introduced in 1909 by the Belgian–US chemist Leo Hendrik Baekeland (1863–1944).

baking powder A mixture of powdered compounds added to dough or cake mixture to make it rise in cooking. It is used as a substitute for yeast in bread-making. Baking powders consist of a source of carbon dioxide, such as sodium hydrogencarbonate or ammonium hydrogencarbonate, and an acidic substance such as calcium hydrogenphosphate, potassium hydrogentartrate (cream of tartar), or sodium hydrogenphosphate. In the hot wet mixture, the acid releases bubbles of carbon dioxide gas from the hydrogencarbonate, which make the mixture rise.

baking soda *See* **sodium hydrogencarbonate**.

balance An accurate weighing device. The simple *beam balance* consists of two pans suspended from a centrally pivoted beam. Known masses are placed on one pan and the substance or body to be weighed is placed in the other. When the beam is exactly horizontal the two masses are equal. An accurate laboratory balance weighs to the nearest hundredth of a milligram. Specially designed balances can be accurate to a millionth of a milligram. More modern *substitution balances* use the substitution principle.

In this calibrated weights are removed from the single lever arm to bring the single pan suspended from it into equilibrium with a fixed counter weight. The substitution balance is more accurate than the two-pan device and enables weighing to be carried out more rapidly. In automatic electronic balances, mass is determined not by mechanical deflection but by electronically controlled compensation of an electric force. A scanner monitors the displacement of the pan support generating a current proportional to the displacement. This current flows through a coil forcing the pan support to return to its original position by means of a magnetic force. The signal generated enables the mass to be read from a digital display. The mass of the empty container can be stored in the balance's computer memory and automatically deducted from the mass of the container plus its contents.

Balmer series *See* **hydrogen spectrum**.

banana bond Informal name for the type of electron-deficient three-centre bond holding the B–H–B bridges in *boranes and similar compounds.

band spectrum *See* **spectrum**.

band theory *See* **energy bands**.

bar A c.g.s. unit of pressure equal to 10^6 dynes per square centimetre or 10^5 pascals (approximately 750 mmHg or 0.987 atmosphere). The *millibar* (100 Pa) is commonly used in meteorology.

barbiturate Any one of a group of drugs derived from barbituric acid that have a depressant effect on the central nervous system. Barbiturates were originally used as sedatives and sleeping pills but their clinical use is now limited due to their toxic side-effects; prolonged use can lead to addiction. Specific barbiturates in clinical use include phenobarbitone, for treating epilepsy, and methohexitane sodium, used as an anaesthetic.

Barfoed's test A biochemical test to detect monosaccharide (reducing) sugars in solution, devised by the Swedish physician Christen Barfoed (1815–99). *Barfoed's reagent*, a mixture of ethanoic (acetic) acid and copper(II) acetate, is added to the test solution and boiled. If any reducing sugars are present a red precipitate of copper(II) oxide is formed. The reaction will be negative in the presence of disaccharide sugars as they are weaker reducing agents.

barite *See* **barytes**.

barium Symbol Ba. A silvery-white reactive element belonging to *group 2 (formerly IIA) of the periodic table; a.n. 56; r.a.m. 137.34; r.d. 3.51; m.p. 725°C; b.p. 1640°C. It occurs as the minerals barytes ($BaSO_4$) and witherite ($BaCO_3$). Extraction is by high-temperature reduction of barium oxide with aluminium or silicon in a vacuum, or by electrolysis of fused barium chloride. The metal is used as a getter in vacuum systems. It oxidizes readily in air and reacts with ethanol and water. Soluble barium compounds are extremely poisonous. It was first identified in 1774 by Karl *Scheele, and was extracted by Humphry *Davy in 1808.

barium bicarbonate *See* **barium hydrogencarbonate**.

barium carbonate A white insoluble compound, $BaCO_3$; r.d. 4.43. It decomposes on heating to give barium oxide and carbon dioxide:

$$BaCO_3(s) \rightarrow BaO(s) + CO_2(g)$$

The compound occurs naturally as the mineral *witherite* and can be prepared by adding an alkaline solution of a carbonate to a solution of a barium salt. It is used as a raw material for making other barium salts, as a flux for ceramics, and as a raw material in the manufacture of certain types of optical glass.

barium chloride A white compound, $BaCl_2$. The anhydrous compound has two crystalline forms: an α form (monoclinic; r.d. 3.856), which transforms at 962°C to a β form (cubic; r.d. 3.917; m.p. 963°C; b.p. 1560°C). There is also a dihydrate, $BaCl_2.2H_2O$ (cubic; r.d. 3.1), which loses water at 113°C. It is prepared by dissolving barium carbonate (witherite) in hydrochloric acid and crystallizing out the dihydrate. The compound is used in the extraction of barium by electrolysis.

barium hydrogencarbonate (barium bicarbonate) A compound, $Ba(HCO_3)_2$, which is only stable in solution. It can be formed by the action of carbon dioxide on a suspension of barium carbonate in cold water:

$$BaCO_3(s) + CO_2(g) + H_2O(l) \rightarrow Ba(HCO_3)_2(aq)$$

On heating, this reaction is reversed.

barium hydroxide (baryta) A white solid, $Ba(OH)_2$, sparingly soluble in water. The common form is the octahydrate, $Ba(OH)_2.8H_2O$; monoclinic; r.d. 2.18; m.p. 78°C. It can be produced by adding water to barium monoxide or by the action of sodium hydroxide on soluble barium compounds and is used as a weak alkali in volumetric analysis.

barium oxide A white or yellowish solid, BaO, obtained by heating barium in oxygen or by the thermal decomposition of barium carbonate or nitrate; cubic; r.d. 5.72; m.p. 1923°C; b.p. 2000°C. When barium oxide is heated in oxygen the peroxide, BaO_2, is formed in a reversible reaction that was once used as a method for obtaining oxygen (the *Brin process*). Barium oxide is now used in the manufacture of lubricating-oil additives.

barium peroxide A dense off-white solid, BaO_2, prepared by carefully heating *barium oxide in oxygen; r.d. 4.96; m.p. 450°C. It is used as a bleaching agent. With acids, hydrogen peroxide is formed and the reaction is used in the laboratory preparation of hydrogen peroxide.

barium sulphate An insoluble white solid, $BaSO_4$, that occurs naturally as the mineral *barytes (or *heavy spar*) and can be prepared as a precipitate by adding sulphuric acid to barium chloride solution; r.d. 4.50; m.p. 1580°C. The rhombic form changes to a monoclinic form at 1149°C. It is used as a raw material for making other barium salts, as a pigment extender in surface coating materials (called *blanc fixe*), and in the glass and rubber industries. Barium compounds are opaque to X-rays, and a suspension of the sulphate in water is used in medicine to provide a contrast medium for X-rays of the stomach and intestine. Although barium compounds are extremely poisonous, the sulphate is safe to use because it is very insoluble.

barrel A measurement of volume, widely used in the chemical industry, equal to 35 UK gallons (approximately 159 litres).

Bartlett, Neil (1932–) British inorganic chemist who prepared oxygen hexachloroplatinate in 1961 and, in 1962, xenon hexachloroplatinate – the first compound of a noble gas.

Barton, Sir Derek (Harold Richard) (1918–98) British organic chemist who worked on steroids and other natural products. He is noted for his investigations into the conformation of organic compounds and its effect on chemical properties. Barton was awarded the 1969 Nobel Prize for chemistry (with O. Hassell (1897–1981)).

baryta *See* **barium hydroxide**.

barytes (barite) An orthorhombic mineral form of *barium sulphate, $BaSO_4$; the chief ore of barium. It is usually white but may also be yellow, grey, or brown. Large deposits occur in Andalusia, Spain, and in the USA.

basal *See* **apical**.

basalt A fine-grained basic igneous rock. It is composed chiefly of calcium-rich plagioclase feldspar and pyroxene; other minerals present may be olivine, magnetite, and apatite. Basalt is the commonest type of lava.

base A compound that reacts with a protonic acid to give water (and a salt). The definition comes from the Arrhenius theory of acids and bases. Typically, bases are metal oxides, hydroxides, or compounds (such as ammonia) that give hydroxide ions in aqueous solution. Thus, a base may be either: (1) An insoluble oxide or hydroxide that reacts with an acid, e.g.

$$CuO(s) + 2HCl(aq) \rightarrow CuCl_2(aq) + H_2O(l)$$

Here the reaction involves hydrogen ions from the acid

$$CuO(s) + 2H^+(aq) \rightarrow H_2O(l) + Cu^{2+}(aq)$$

(2) A soluble hydroxide, in which case the solution contains hydroxide ions. The reaction with acids is a reaction between hydrogen ions and hydroxide ions:

$$H^+ + OH^- \rightarrow H_2O$$

(3) A compound that dissolves in water to produce hydroxide ions. For example, ammonia reacts as follows:

$$NH_3(g) + H_2O(l) \rightleftharpoons NH_4^+(aq) + {}^-OH$$

Similar reactions occur with organic *amines (*see also* **nitrogenous base**; **amine salts**). A base that dissolves in water to give hydroxide ions is called an *alkali*. Ammonia and sodium hydroxide are common examples.

The original Arrhenius definition of a base has been extended by the Lowry–Brønsted theory and by the Lewis theory. *See* **acid.**

base dissociation constant *See* **dissociation.**

base metal A common relatively inexpensive metal, such as iron or lead, that corrodes, oxidizes, or tarnishes on exposure to air, moisture, or heat, as distinguished from precious metals, such as gold and silver.

base unit A unit that is defined arbitrarily rather than being defined by simple combinations of other units. For example, the ampere is a base unit in the SI system defined in terms of the force produced between two current-carrying conductors, whereas the coulomb is a *derived unit*, defined as the quantity of charge transferred by one ampere in one second.

basic 1. Describing a compound that is a base. **2.** Describing a solution containing an excess of hydroxide ions; alkaline.

basic dye *See* **dyes.**

basicity constant *See* **dissociation.**

basic-oxygen process (BOP process) A high-speed method of making high-grade steel. It originated in the *Linz–Donawitz (L–D) process.* Molten pig iron and scrap are charged into a tilting furnace, similar to the Bessemer furnace except that it has no tuyeres. The charge is converted to steel by blowing high-pressure oxygen onto the surface of the metal through a water-cooled lance. The excess heat produced enables up to 30% of scrap to be incorporated into the charge. The process has largely replaced the Bessemer and open-hearth processes.

basic salt A compound that can be regarded as being formed by replacing some of the oxide or hydroxide ions in a base by other negative ions. Basic salts are thus mixed salt–oxides (e.g. bismuth(III) chloride oxide, BiOCl) or salt–hydroxides (e.g. lead(II) chloride hydroxide, Pb(OH)Cl).

basic slag *Slag formed from a basic flux (e.g. calcium oxide) in a blast furnace. The basic flux is used to remove acid impurities in the ore and contains calcium silicate, phosphate, and sulphide. If the phosphorus content is high the slag can be used as a fertilizer.

bathochromic shift A shift of a spectral band to longer wavelengths as a result of substitution in a molecule or a change in the conditions. *Compare* **hypsochromic shift.**

battery A number of electric cells joined together. The common car battery, or *accumulator, usually consists of six secondary cells connected in series to give a total e.m.f. of 12 volts. A torch battery is usually a dry version of the *Leclanché cell, two of which are often connected in series. Batteries may also have cells connected in parallel, in which case they have the same e.m.f. as a single cell, but their capacity is increased, i.e. they will provide more total charge. The capacity of a battery is usually specified in ampere-hours, the ability to supply 1 A for 1 hr, or the equivalent.

bauxite The chief ore of aluminium, consisting of hydrous aluminium oxides and aluminous laterite. It is a claylike amorphous material formed by the weathering of silicate rocks under tropical conditions. The chief producers are Australia, Guinea, Jamaica, Russia, Brazil, and Surinam.

b.c.c. Body-centred cubic. *See* **cubic crystal**.

beam balance *See* **balance**.

Beattie–Bridgman equation An *equation of state that relates the pressure, volume, and temperature of a gas and the gas constant. The Beattie–Bridgman equation uses empirical constants to take into account the reduction in the effective number of molecules due to various types of molecular aggregation. It is given by

$$P = RT(1 - \varepsilon)(V + B)/V^2 - A/V^2,$$

where P is the pressure, T is the thermodynamic temperature, V is the volume, R is the gas constant, and A, B, and ε are constants related to five empirical constants A_0, B_0, a, b, and c by: $A = A_0(1 - a/V)$, $B = B_0(1 - b/V)$, and $\varepsilon = c/VT^3$.

Beckmann rearrangement The chemical conversion of a ketone *oxime into an *amide, usually using sulphuric acid as a catalyst. The reaction, used in the manufacture of nylon and other polyamides, is named after the German chemist Ernst Beckmann (1853–1923).

Beckmann thermometer A thermometer for measuring small changes of temperature (see illustration). It consists of a mercury-in-glass thermometer with a scale covering only 5 or 6°C calibrated in hundredths of a degree. It has two mercury bulbs, the range of temperature to be measured is varied by running mercury from the upper bulb into the larger lower bulb. It is used particularly for measuring *depression of freezing point or *elevation of boiling point of liquids when solute is added, in order to find relative molecular masses.

becquerel Symbol Bq. The SI unit of activity (*see* **radiation units**). The unit is named after the discoverer of radioactivity A. H. *Becquerel.

Becquerel, Antoine Henri (1852–1908) French physicist. His early researches were in optics, then in 1896 he accidentally discovered *radioactivity in fluorescent salts of uranium. Three years later he showed that it consists of charged particles that are deflected by a magnetic field. For this work he was awarded the 1903 Nobel Prize, which he shared with Pierre and Marie *Curie.

Beer–Lambert law A law relating the reduction in luminous intensity of light passing through a material to the length of the light's path through the material: i.e.

$$\log(I/I_0) = -\varepsilon[J]l,$$

where ε is the molar *absorption coefficient, I is the intensity after passing through a sample of length l, I_0 is the incident intensity, and $[J]$ is the molar concentration of species *J*. The Beer–Lambert law was formed empirically; however, it can be derived on the basis that the loss in intensity dI is proportional to the thickness dl of the sample, the concentration $[J]$, and the intensity I (since the rate of absorption is proportional to the intensity). The Beer–Lambert law means that the intensity of light (or any other form of electromagnetic radiation) passing through a sample diminishes exponentially with the concentration and the thickness of the sample (for a given wave number).

beet sugar *See* sucrose.

Beilstein's test A test for the presence of a halogen (chlorine, bromine, or iodine) in an organic compound. A piece of copper wire or gauze is preheated strongly in the oxidizing flame of a Bunsen burner (until the flame is no longer green) and the test substance placed on the wire or gauze, which is re-heated. A green flame indicates the presence of a halogen.

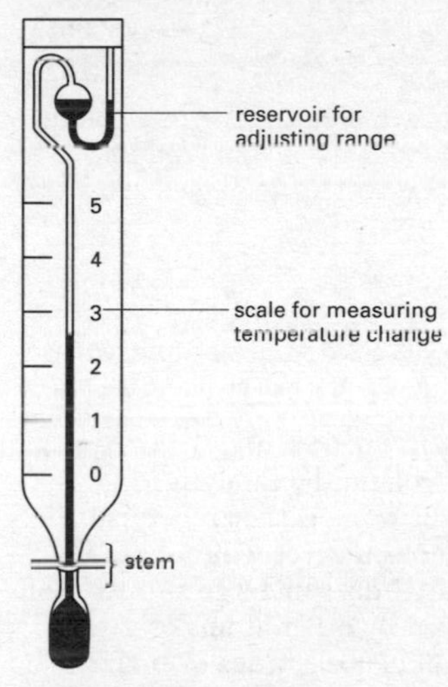

Beckmann thermometer

bell metal A type of *bronze used in casting bells. It consists of 60–85% copper alloyed with tin, often with some zinc and lead included.

Belousov–Zhabotinskii reaction *See* B–Z reaction.

Bénard cell A structure associated with a layer of liquid that is confined by two horizontal parallel plates, in which the lateral dimensions are much larger than the width of the layer. Before heating the liquid is homogeneous. However, if after heating from below the temperatures of the plates are T_1 and T_2, at a critical value of the temperature gradient $\Delta T = T_1 - T_2$ the liquid abruptly starts to convect. The liquid spontaneously organizes itself into a set of convection rolls, i.e. the liquid goes round in a series of 'cells', called Bénard cells.

Benedict's test A biochemical test to detect reducing sugars in solution, devised by the US chemist S. R. Benedict (1884–1936). *Benedict's reagent* – a mixture of copper(II) sulphate and a filtered mixture of hydrated sodium citrate and hydrated sodium carbonate – is added to the test solution and boiled. A high concentration of reducing sugars induces the formation of a red precipitate; a lower concentration produces a yellow precipitate. Benedict's test is a more sensitive alternative to *Fehling's test.

beneficiation (ore dressing) The separation of an ore into the valuable components and the waste material (gangue). This may be achieved by a number of processes, including crushing, grinding, magnetic separation, froth flotation, etc. The dressed ore, consisting of a high proportion of valuable components, is then ready for smelting or some other refining process.

bent sandwich *See* **sandwich compound**.

benzaldehyde *See* **benzenecarbaldehyde**.

Kekulé structures

Dewar structures (3 in all)

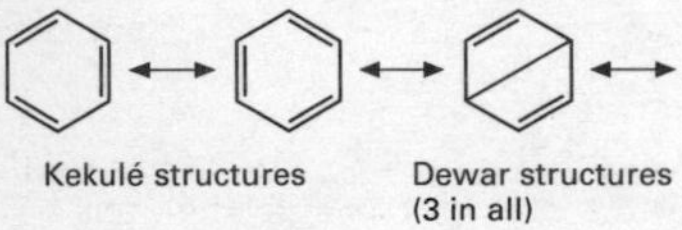

Benzene

benzene A colourless liquid hydrocarbon, C_6H_6; r.d. 0.88; m.p. 5.5°C; b.p. 80.1°C. It is now made from gasoline from petroleum by catalytic reforming (formerly obtained from coal tar). Benzene is the archetypal *aromatic compound. It has an unsaturated molecule, yet will not readily undergo addition reactions. On the other hand, it does undergo substitution reactions in which hydrogen atoms are replaced by other atoms or groups. This behaviour occurs because of delocalization of *p*-electrons over the benzene ring, and all the C–C bonds in benzene are equivalent and intermediate in length between single and double bonds. It can be regarded as a resonance hybrid of Kekulé and Dewar structures (see formulae). In formulae it can be represented by a hexagon with a ring inside it.

benzene-1,4-diol (hydroquinone; quinol) A white crystalline solid, $C_6H_4(OH)_2$; r.d. 1.33; m.p. 173–174°C; b.p. 285°C. It is used in making dyes. *See also* **quinhydrone electrode**.

benzenecarbaldehyde (benzaldehyde) A yellowish volatile oily liquid, C_6H_5CHO; r.d. 1.04; m.p. −26°C; b.p. 178.1°C. The compound occurs in almond kernels and has an almond-like smell. It is made from methylbenzene (by conversion to dichloromethyl benzene, $C_6H_5CHCl_2$, followed by hydrolysis).

Benzenecarbaldehyde is used in flavourings, perfumery, and the dyestuffs industry.

benzenecarbonyl chloride **(benzoyl chloride)** A colourless liquid, C_6H_5COCl; r.d. 1.21; m.p. 0°C; b.p. 197.2°C. It is an *acyl halide, used to introduce benzenecarbonyl groups into molecules. *See* **acylation**.

benzenecarbonyl group **(benzoyl group)** The organic group $C_6H_5CO–$.

benzenecarboxylate **(benzoate)** A salt or ester of benzenecarboxylic acid.

benzenecarboxylic acid **(benzoic acid)** A white crystalline compound, C_6H_5COOH; r.d. 1.27; m.p. 122.4°C; b.p. 249°C. It occurs naturally in some plants and is used as a food preservative. Benzenecarboxylic acid has a carboxyl group bound directly to a benzene ring. It is a weak carboxylic acid ($K_a = 6.4 \times 10^{-5}$ at 25°C), which is slightly soluble in water. It also undergoes substitution reactions on the benzene ring.

benzene-1,4-diol **(hydroquinone; quinol)** A white crystalline solid, $C_6H_4(OH)_2$; r.d. 1.33; m.p. 170°C; b.p. 285°C. It is used in making dyes. *See also* **quinhydrone electrode**.

benzene hexacarboxylic acid *See* **mellitic acid**.

benzene hexachloride **(BHC)** A crystalline substance, $C_6H_6Cl_6$, made by adding chlorine to benzene. It is used as a pesticide and, like *DDT, concern has been expressed at its environmental effects.

benzenesulphonic acid A colourless deliquescent solid, $C_6H_5SO_2OH$, m.p. 43–44°C, usually found as an oily liquid. It is made by treating benzene with concentrated sulphuric acid. Its alkyl derivatives are used as *detergents.

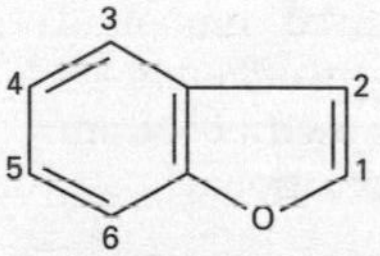

Benzfuran

benzfuran **(coumarone)** A crystalline aromatic compound, C_8H_6O. It is a heterocyclic compound having a benzene ring fused to a five-membered *furan ring.

benzil 1,2-diphenylethan-1,2-dione. *See* **benzilic acid rearrangement**.

benzilic acid rearrangement An organic rearrangement reaction in which *benzil* (1,2-diphenylethan-1,2-dione) is treated with hydroxide and then acid to give *benzilic acid* (2-hydroxy-2,2-diphenylethanoic acid):

$$C_6H_5.CO.CO.C_6H_5 \rightarrow (C_6H_5)_2C(OH).COOH$$

In the reaction a phenyl group ($C_6H_5–$) migrates from one carbon atom to another. The reaction was discovered in 1828 by Justus von *Liebig; it was the first rearrangement reaction to be described.

benzoate *See* **benzenecarboxylate.**

benzoic acid *See* **benzenecarboxylic acid.**

benzoin A colourless crystalline compound, $C_6H_5CHOHCOC_6H_5$; m.p. 137°C. It is a condensation product of *benzenecarbaldehyde (benzaldehyde), made by the action of sodium cyanide on benzenecarbaldehyde in alcoholic solution. It also occurs naturally as the resin of a tropical tree. It is both a secondary alcohol and a ketone, and gives reactions characteristic of both types of compound.

benzoquinone *See* **cyclohexadiene-1,4-dione.**

benzoylation A chemical reaction in which a benzoyl group (benzenecarbonyl group, C_6H_5CO) is introduced into a molecule. *See* **acylation.**

benzoyl chloride *See* **benzenecarbonyl chloride.**

benzoyl group *See* **benzenecarbonyl group.**

benzpyrene A pale yellow solid, $C_{20}H_{12}$, m.p. 179°C, whose molecules consist of five fused benzene rings. It occurs in tars from coal and tobacco smoke and is a *carcinogen.

benzyl alcohol *See* **phenylmethanol.**

benzylamine (α-aminotoluene; phenylmethylamine) A colourless liquid, $C_6H_5CH_2NH_2$; r.d. 0.981; b.p. 185°C. It behaves in the same way as primary aliphatic amines.

benzyne (1,2-didehydrobenzene) A highly reactive short-lived compound, C_6H_4, having a hexagonal ring of carbon atoms containing two double bonds and one triple bond. Benzyne, which is the simplest example of an *aryne, is thought to be an intermediate in a number of reactions.

Bergius, Friedrich Karl Rudolf (1884–1949) German organic chemist. While working with Fritz *Haber in Karlsruhe, he become interested in reactions at high pressures. In 1912 he devised an industrial process for making light hydrocarbons by the high-pressure hydrogenation of coal or heavy oil. The work earned him a share of the 1931 Nobel Prize for chemistry with Carl Bosch (1874–1940). The Bergius process proved important for supplying petrol for the German war effort in World War II.

Bergius process A process for making hydrocarbon mixtures (for fuels) from coal by heating powdered coal mixed with tar and iron(III) oxide catalyst at 450°C under hydrogen at a pressure of about 200 atmospheres. In later developments of the process, the coal was suspended in liquid hydrocarbons and other catalysts were used. The process was developed by Friedrich *Bergius during World War I as a source of motor fuel.

berkelium Symbol Bk. A radioactive metallic transuranic element belonging to the *actinoids; a.n. 97; mass number of the most stable isotope 247 (half-life 1.4×10^3 years); r.d. (calculated) 14. There are eight known isotopes. It was first produced by G. T. Seborg and associates in 1949 by bombarding americium–241 with alpha particles.

Berthelot, Marcellin (Pierre Eugène) (1827–1907) French chemist who pioneered organic synthesis, producing many organic compounds from simple starting materials. Berthelot was also one of the first to investigate thermochemistry.

Berthelot equation An *equation of state that relates the pressure, volume, and temperature of a gas and the gas constant. It is given by:

$$PV = RT[1 + 9PT_c(1 - 6T_c^2/T^2)/128P_cT],$$

where P is the pressure, V is the volume, R is the gas constant, T is the thermodynamic temperature, and T_c and P_c are the critical temperature and pressure of the gas. The Berthelot equation can be derived from the *Clapeyron–Clausius equation.

Berthollide compound A solid compound with slight variations in chemical composition (*see* **nonstoichiometric compound**). Berthollide compounds are named after the French inorganic chemist Claude Louis Berthollet (1748–1822), who attacked the law of constant composition.

beryl A hexagonal mineral form of beryllium aluminium silicate, $Be_3Al_2Si_6O_{18}$; the chief ore of beryllium. It may be green, blue, yellow, or white and has long been used as a gemstone. Beryl occurs throughout the world in granite and pegmatites. *Emerald, the green gem variety, occurs more rarely and is of great value. Important sources of beryl are found in Brazil, Madagascar, and the USA.

beryllate A compound formed in solution when beryllium metal, or the oxide or hydroxide, dissolves in strong alkali. The reaction (for the metal) is often written

$$Be + 2OH^-(aq) \rightarrow BeO_2^{2-}(aq) + H_2(g)$$

The ion BeO_2^{2-} is the beryllate ion. In fact, as with the *aluminates, the ions present are probably hydroxy ions of the type $Be(OH)_4^{2-}$ (the *tetrahydroxoberyllate(II) ion*) together with polymeric ions.

beryllia *See* **beryllium oxide**.

beryllium Symbol Be. A grey metallic element of *group 2 (formerly IIA) of the periodic table; a.n. 4; r.a.m. 9.012; r.d. 1.85; m.p. 1278°C; b.p. 2970°C. Beryllium occurs as beryl ($3BeO.Al_2O_3.6SiO_2$) and chrysoberyl ($BeO.Al_2O_3$). The metal is extracted from a fused mixture of BeF_2/NaF by electrolysis or by magnesium reduction of BeF_2. It is used to manufacture Be–Cu alloys, which are used in nuclear reactors as reflectors and moderators because of their low absorption cross section. Beryllium oxide is used in ceramics and in nuclear reactors. Beryllium and its compounds are toxic and can cause serious lung diseases and dermatitis. The metal is resistant to oxidation by air because of the formation of an oxide layer, but will react with dilute hydrochloric and sulphuric acids. Beryllium compounds show high covalent character. The element was isolated independently by F. *Wöhler and A. A. Bussy in 1828.

beryllium bronze A hard, strong type of *bronze containing about 2% beryllium, in addition to copper and tin.

beryllium hydroxide A white crystalline compound, $Be(OH)_2$, precipitated from solutions of beryllium salts by adding alkali. Like the oxide, it is amphoteric and dissolves in excess alkali to give *beryllates.

beryllium oxide (beryllia) An insoluble solid compound, BeO; hexagonal; r.d. 3.01; m.p. 2530°C; b.p. 3900°C. It occurs naturally as *bromellite*, and can be made by burning beryllium in oxygen or by the decomposition of beryllium carbonate or hydroxide. It is an important amphoteric oxide, reacting with acids to form salts and with alkalis to form compounds known as *beryllates. Beryllium oxide is used in the production of beryllium and beryllium–copper refractories, transistors, and integrated circuits.

Berzelius, Jöns Jacob (1779–1848) Swedish chemist. After moving to Stockholm he worked with mining chemists and, with them, discovered several elements, including cerium (1803), selenium (1817), lithium (1818), thorium (1828), and vanadium (1830). He produced the first accurate table of atomic weights and was extremely influential in the general development of 19th-century chemistry.

Bessemer process A process for converting *pig iron from a *blast furnace into *steel. The molten pig iron is loaded into a refractory-lined tilting furnace (*Bessemer converter*) at about 1250°C. Air is blown into the furnace from the base and *spiegel is added to introduce the correct amount of carbon. Impurities (especially silicon, phosphorus, and manganese) are removed by the converter lining to form a slag. Finally the furnace is tilted so that the molten steel can be poured off. In the modern VLN (very low nitrogen) version of this process, oxygen and steam are blown into the furnace in place of air to minimize the absorption of nitrogen from the air by the steel. The process is named after the British engineer Sir Henry Bessemer (1813–98), who announced it in 1856. *See also* **basic-oxygen process.**

beta decay A type of radioactive decay in which an unstable atomic nucleus changes into a nucleus of the same mass number but different proton number. The change involves the conversion of a neutron into a proton with the emission of an electron and an antineutrino ($n \rightarrow p + e^- + \bar{\nu}$) or of a proton into a neutron with the emission of a positron and a neutrino ($p \rightarrow n + e^+ + \nu$). An example is the decay of carbon–14:

$$^{14}_{6}C \rightarrow {}^{14}_{7}N + e^- + \bar{\nu}$$

The electrons or positrons emitted are called *beta particles* and streams of beta particles are known as *beta radiation*.

beta-iron A nonmagnetic allotrope of iron that exists between 768°C and 900°C.

beta particle *See* **beta decay**.

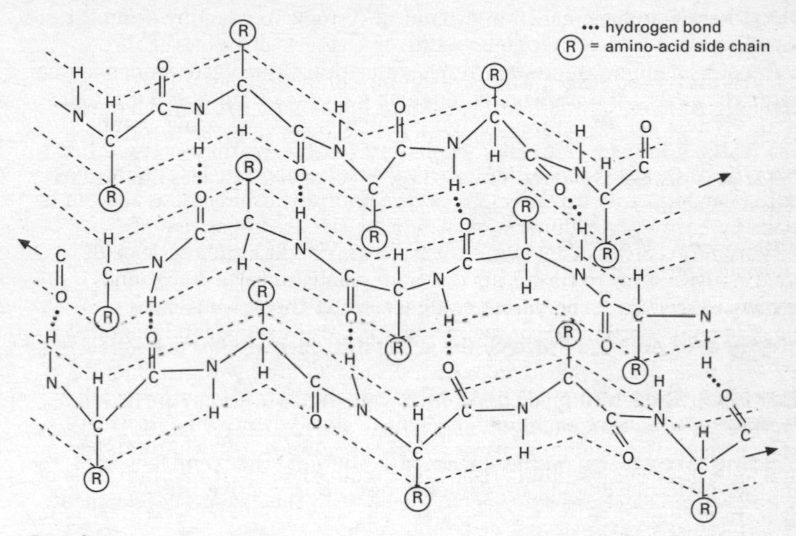

Beta sheet

beta sheet (β-pleated sheet) A form of secondary structure in *proteins in which extended polypeptide chains lie parallel to each other and are linked by hydrogen bonds between the N–H and C=O groups (see illustration). Beta sheets occur in many globular proteins and link polypeptides of the same type in certain fibrous proteins, including fibroin (the protein of silk).

BET isotherm An isotherm that takes account of the possibility that the monolayer in the *Langmuir adsorption isotherm can act as a substrate for further adsorption. The BET isotherm (named after S. Brunauer, P. Emmett, and E. Teller) has the form:

$$V/V_{mon} = cz/\{(1-z)[1-(1-c)z]\},$$

where $z = p/p^*$ (p^* is the vapour pressure above a macroscopically thick layer of liquid on the surface), V_{mon} is the volume that corresponds to the surface being covered by a monolayer, V and p are the volume and pressure of the gas respectively, and c is a constant. In the BET isotherm, the isotherm rises indefinitely at high pressures (in contrast to the Langmuir isotherm). It provides a useful approximation over some ranges of pressure but underestimates adsorption for low pressures and overestimates adsorption for high pressures.

BHC *See* **benzene hexachloride.**

bicarbonate *See* **hydrogencarbonate.**

bicarbonate of soda *See* **sodium hydrogencarbonate.**

bifurcation A phenomenon in dynamical systems in which the number of solutions for a type of behaviour suddenly changes when one of the parameters defining the system reaches a critical value. Bifurcations can be depicted in a *bifurcation diagram* in which one axis of a graph is a variable of the system and the other axis is a parameter that can bring about bifurcations. In many systems a whole series of bifurcations occur, called *bifurcation cascades.* In certain cases these bifurcation cascades can lead to chaotic reactions.

bimolecular reaction A step in a chemical reaction that involves two molecules. *See* **molecularity.**

binary Describing a compound or alloy formed from two elements.

binary acid An *acid in which the acidic hydrogen atom(s) are bound directly to an atom other than oxygen. Examples are hydrogen chloride (HCl) and hydrogen sulphide (H_2S). Such compounds are sometimes called *hydracids. Compare* **oxo acid.**

binding site An area on the surface of a molecule that combines with another molecule. Binding sites on enzymes can be *active sites or *allosteric sites.

bioaccumulation An increase in the concentration of chemicals, such as pesticides, in organisms that live in environments contaminated by a wide variety of organic compounds. These compounds are not usually decomposed in the environment (i.e. they are not biodegradable) or metabolized by the organisms, so that their rate of absorption and storage is greater than their rate of excretion. The chemicals are normally stored in fatty tissues. *DDT is known as a *persistent pesticide,* as it is not easily broken down and bioaccumulates along food chains, so that increasing concentrations occur in individual organisms at each trophic level.

bioactivation A metabolic process in which a product that is chemically reactive is produced from a relatively inactive precursor.

biochemical fuel cell A system that exploits biological reactions for the conversion of biomass (chemical energy) to electricity (electrical energy). One potential application is the generation of electricity from industrial waste and sewage. *Methyltrophic* organisms (i.e. organisms that use methane or methanol as their sole carbon sources) are being investigated for their potential use in biochemical fuel cells.

biochemical oxygen demand (BOD) The amount of oxygen taken up by microorganisms that decompose organic waste matter in water. It is therefore used as a measure of the amount of certain types of organic pollutant in water. BOD is calculated by keeping a sample of water containing a known amount of oxygen for five days at 20°C. The oxygen content is measured again after this time. A high BOD indicates the presence of a large number of microorganisms, which suggests a high level of pollution.

biochemistry The study of the chemistry of living organisms, especially the structure and function of their chemical components (principally proteins, carbohydrates, lipids, and nucleic acids). Biochemistry has advanced rapidly with the development, from the mid-20th century, of such techniques as chromatography, X-ray diffraction, radioisotopic labelling, and electron microscopy. Using these techniques to separate and analyse biologically important molecules, the steps of the metabolic pathways in which they are involved (e.g. glycolysis) have been determined. This has provided some knowledge of how organisms obtain and store energy, how they manufacture and degrade their biomolecules, and how they sense and respond to their environment. See Chronology.

biodiesel *See* biofuel.

bioelement Any chemical element that is found in the molecules and compounds that make up a living organism. In the human body the most common bioelements (in decreasing order of occurrence) are oxygen, carbon, hydrogen, nitrogen, calcium, and phosphorus. Other bioelements include sodium, potassium, magnesium, and copper. *See* **essential element.**

bioenergetics The study of the flow and the transformations of energy that occur in living organisms. Typically, the amount of energy that an organism takes in (from food or sunlight) is measured and divided into the amount used for growth of new tissues; that lost through death, wastes, and (in plants) transpiration; and that lost to the environment as heat (through respiration).

biofuel A gaseous, liquid, or solid fuel that contains an energy content derived from a biological source. The organic matter that makes up living organisms provides a potential source of trapped energy that is beginning to be exploited to supply the ever-increasing energy demand around the world. An example of a biofuel is rapeseed oil, which can be used in place of diesel fuel in modified engines. The methyl ester of this oil, *rapeseed methyl ester* (*RME*), can be used in unmodified diesel engines and is sometimes known as *biodiesel*. Other biofuels include *biogas and *gasohol.

biogas A mixture of methane and carbon dioxide resulting from the anaerobic decomposition of such waste materials as domestic, industrial, and agricultural sewage. The decomposition is carried out by methanogenic bacteria; these obligate anaerobes produce methane, the main component of biogas, which can be collected and used as an energy source for domestic processes, such as heating, cooking, and lighting. The production of biogas is carried out in special *digesters*, which are widely used in China and India.

BIOCHEMISTRY

1833	French chemist Anselme Payen (1795–1871) discovers diastase (the first enzyme to be discovered).
1836	Theodor Schwann discovers the digestive enzyme pepsin.
c.1860	Louis Pasteur demonstrates fermentation is caused by 'ferments' in yeasts and bacteria.
1869	German biochemist Johann Friedrich Miescher (1844–95) discovers nucleic acid.
1877	Pasteur's 'ferments' are designated as enzymes.
1890	German chemist Emil Fischer (1852–1919) proposes the 'lock-and-key' mechanism to explain enzyme action.
1901	Japanese chemist Jokichi Takamine (1854–1922) isolates adrenaline (the first hormone to be isolated).
1903	German biologist Eduard Buchner (1860–1917) discovers the enzyme zymase (causing fermentation).
1904	British biologist Arthur Harden (1865–1940) discovers coenzymes.
1909	Russian-born US biochemist Phoebus Levene (1869–1940) identifies ribose in RNA.
1921	Canadian physiologist Frederick Banting (1891–1941) and US physiologist Charles Best (1899–1978) isolate insulin.
1922	Alexander Fleming discovers the enzyme lysozyme.
1925	Russian-born British biologist David Keilin (1887–1963) discovers cytochrome.
1926	US biochemist James Sumner (1877–1955) crystallizes urease (the first enzyme to be isolated).
1929	German chemist Hans Fischer (1881–1945) determines the structure of haem (in haemoglobin). K. Lohman isolates ATP from muscle.
1930	US biochemist John Northrop (1891–1987) isolates the enzyme pepsin.
1932	Swedish biochemist Hugo Theorell (1903–82) isolates the muscle protein myoglobin.
1937	Hans Krebs discovers the Krebs cycle.
1940	German-born US biochemist Fritz Lipmann (1899–1986) proposes that ATP is the carrier of chemical energy in many cells.
1943	US biochemist Britton Chance (1913–) discovers how enzymes work (by forming an enzyme–substrate complex).
1952	US biologist Alfred Hershey (1908–) proves that DNA carries genetic information.

As well as providing a source of fuel, these systems also enable sewage, which contains pathogenic bacteria, to be digested, thereby removing the danger to humans that could otherwise result from untreated domestic and agricultural waste.

1953	Francis Crick and James Watson discover the structure of DNA.
1955	Frederick Sanger discovers the amino acid sequence of insulin.
1956	US biochemist Arthur Kornberg (1918–) discovers DNA polymerase. US molecular biologist Paul Berg (1926–) identifies the nucleic acid later known as transfer RNA.
1957	British biologist Alick Isaacs (1921–67) discovers interferon.
1959	Austrian-born British biochemist Max Perutz (1914–) determines the structure of haemoglobin.
1960	South African-born British molecular biologist Sydney Brenner (1927–) and French biochemist François Jacob (1920–) discover messenger RNA.
1961	British biochemist Peter Mitchell (1920–92) proposes the chemiosmotic theory. Brenner and Crick discover that the genetic code consists of a series of base triplets.
1969	US biochemist Gerald Edelman (1929–) discovers the amino acid sequence of immunoglobulin G.
1970	US virologists Howard Temin (1934–94) and David Baltimore (1938–) discover the enzyme reverse transcriptase.
1970	US molecular biologist Hamilton Smith (1931–) discovers restriction enzymes.
1973	US biochemists Stanley Cohen (1935–) and Herbert Boyer (1936–) use restriction enzymes to produce recombinant DNA.
1977	Sanger determines the complete base sequence of DNA in bacteriophage ϕX174.
1984	British biochemist Alec Jeffreys (1950–) devises DNA fingerprinting.
1985	US biochemist Kary Mullis (1944–) invents the polymerase chain reaction.
1986	US pharmacologists Robert Furchgott (1916–) and Louis Ignarro (1941–) demonstrate the importance of nitric oxide as a signal molecule in the blood vascular system.
1988	US biochemist Peter Agre (1949–) identifies a water-channel protein (aquaporin) in the plasma membrane of cells.
1994	Beginnings of DNA chip technology.
1998	US biochemist Roderick MacKinnon (1956–) reveals detailed three-dimensional structure of potassium-ion channel in brain cells.
2001	US molecular biologist Harry Noller and colleagues produce first detailed X-ray crystallographic image of a complete ribosome.

bioinorganic chemistry Biochemistry involving compounds that contain metal atoms or ions. Two common examples of bioinorganic compounds are haemoglobin (which contains iron) and chlorophyll (which contains magnesium). Many enzymes contain metal atoms and bioinorganics are important in a number of biochemical processes, including oxygen transport, electron transfer, and protein folding.

bioluminescence The emission of light without heat (*see* **luminescence**) by living organisms. The phenomenon occurs in glow-worms and fireflies, bacteria and fungi, and in many deep-sea fish (among others); in animals it may serve as a means of protection (e.g. by disguising the shape of a fish) or species recognition or it may provide mating signals. The light is produced during the oxidation of a compound called *luciferin* (the composition of which varies according to the species), the reaction being catalysed by an enzyme, *luciferase*. Bioluminescence may be continuous (e.g. in bacteria) or intermittent (e.g. in fireflies).

biomarker A normal metabolite that, when present in abnormal concentrations in certain body fluids, can indicate the presence of a particular disease or toxicological condition. For example, abnormal concentrations of glucose in the blood can be indicative of diabetes mellitus (*see* **insulin**).

biomolecule Any molecule that is involved in the maintenance and metabolic processes of living organisms (*see* **metabolism**). Biomolecules include carbohydrate, lipid, protein, nucleic acid, and water molecules; some biomolecules are macromolecules.

bioreactor (industrial fermenter) A large stainless steel tank used to grow producer microorganisms in the industrial production of enzymes and other chemicals. After the tank is steam-sterilized, an inoculum of the producer cells is introduced into a medium that is maintained by probes at optimum conditions of temperature, pressure, pH, and oxygen levels for enzyme production. An *agitator* (stirrer) mixes the medium, which is constantly aerated. It is essential that the culture medium is sterile and contains the appropriate nutritional requirements for the microorganism. When the nutrients have been utilized the product is separated; if the product is an extracellular compound the medium can be removed during the growth phase of the microorganisms, but an intracellular product must be harvested when the batch culture growth stops. Some bioreactors are designed for continuous culture.

biosensor A device that uses an immobilized agent to detect or measure a chemical compound. The agents include enzymes, antibiotics, organelles, or whole cells. A reaction between the immobilized agent and the molecule being analysed is transduced into an electronic signal. This signal may be produced in response to the presence of a reaction product, the movement of electrons, or the appearance of some other factor (e.g. light). Biosensors are being used increasingly in diagnostic tests: these allow quick, sensitive, and specific analysis of a wide range of biological products, including antibiotics, vitamins, and other important biomolecules (such as glucose), as well as the determination of certain *xenobiotics, such as synthetic organic compounds.

biosynthesis The production of molecules by a living cell, which is the essential feature of *anabolism.

biotechnology The development of techniques for the application of biological processes to the production of materials of use in medicine and industry. For example, the production of antibiotics, cheese, and wine rely on the activity of various fungi and bacteria. Genetic engineering can modify bacterial cells to synthesize completely new substances, e.g. hormones, vaccines, monoclonal antibodies, etc., or introduce novel traits into plants or animals.

biotin A vitamin in the *vitamin B complex. It is the *coenzyme for various enzymes that catalyse the incorporation of carbon dioxide into various compounds. Adequate amounts are normally produced by the intestinal bacteria; other sources include cereals, vegetables, milk, and liver.

biotite An important rock-forming silicate mineral, a member of the *mica group of minerals, in common with which it has a sheetlike crystal structure. It is usually black, dark brown, or green in colour.

bipy *See* **dipyridyl.**

bipyramid *See* **complex.**

bipyridyl *See* **dipyridyl.**

biradical (diradical) A radical that has two unpaired electrons at different points in the molecule, so that the radical centres are independent of each other.

birefringence *See* **double refraction.**

Birge–Sponer extrapolation A method used to calculate the heat of dissociation of a molecule by extrapolation from observed band spectra. The dissociation energy D_0 is equal to the sum of the vibrational quanta ΔG, where ΔG represents the energy between two successive vibrational states. This means that D_0 is approximately equal to the area under the curve of ΔG plotted against the vibrational quantum number v. If the first few vibrational quanta are observed, an approximate value of D_0 can be obtained by a linear extrapolation of this curve. If a sufficient number of vibrational quanta have been observed, a considerable improvement in the value of D_0 can be obtained by taking the curvature of the ΔG curve into account by quadratic (and/or) higher terms. The technique was put forward by R. T. Birge and H. Sponer in 1926.

Birkeland–Eyde process A process for the fixation of nitrogen by passing air through an electric arc to produce nitrogen oxides. It was introduced in 1903 by the Norwegian chemists Kristian Birkeland (1867–1913) and Samuel Eyde (1866–1940). The process is economic only if cheap hydroelectricity is available.

bisecting *See* **conformation.**

bismuth Symbol Bi. A white crystalline metal with a pinkish tinge belonging to *group 15 (formerly VB) of the periodic table; a.n. 83; r.a.m. 208.98; r.d. 9.78; m.p. 271.3°C; b.p. 1560°C. The most important ores are bismuthinite (Bi_2S_3) and bismite (Bi_2O_3). Peru, Japan, Mexico, Bolivia, and Canada are major producers. The metal is extracted by carbon reduction of its oxide. Bismuth is the most diamagnetic of all metals and its thermal conductivity is lower than any metal except mercury. The metal has a high electrical resistance and a high Hall effect when placed in magnetic fields. It is used to make low-melting-point casting alloys with tin and cadmium. These alloys expand on solidification to give clear replication of intricate features. It is also used to make thermally activated safety devices for fire-detection and sprinkler systems. More recent applications include its use as a catalyst for making acrylic fibres, as a constituent of malleable iron, as a carrier of uranium–235 fuel in nuclear reactors, and as a specialized thermocouple material. Bismuth compounds (when lead-free) are used for cosmetics and medical preparations. It is attacked by oxidizing acids, steam (at high temperatures), and by moist halogens. It burns in air with a blue flame to produce yellow oxide fumes. C. G. Junine first demonstrated that it was different from lead in 1753.

bistability The ability of a system to exist in two steady states. Bistability is a necessary condition for oscillations to occur in chemical reactions. The two steady states of a bistable system are not states of thermodynamic equilibrium, as they are associated with conditions far from equilibrium. Chemical reactions involving bistability jump suddenly from one state to the other state when a certain concentration of one of the participants in the reaction is reached. The effect is analogous to the phenomenon of supercooling, i.e. the cooling of a liquid below its freezing point without freezing. *See also* **oscillating reaction**.

bisulphate *See* **hydrogensulphate**.

bisulphite *See* **hydrogensulphite; aldehydes**.

bite angle *See* **chelate**.

bittern The solution of salts remaining when sodium chloride is crystallized from sea water.

bitumen *See* **petroleum**.

bituminous coal *See* **coal**.

bituminous sand *See* **oil sand**.

biuret test A biochemical test to detect proteins in solution, named after the substance *biuret* ($H_2NCONHCONH_2$), which is formed when urea is heated. Sodium hydroxide is mixed with the test solution and drops of 1% copper(II) sulphate solution are then added slowly. A positive result is indicated by a violet ring, caused by the reaction of *peptide bonds in the proteins or peptides. Such a result will not occur in the presence of free amino acids.

bivalent (divalent) Having a valency of two.

Black, Joseph (1728–99) British chemist and physician, born in France. He studied at Glasgow and Edinburgh, where his thesis (1754) contained the first accurate description of the chemistry of carbon dioxide. In 1757 he discovered latent heat, and was the first to distinguish between heat and temperature.

blackdamp (choke damp) Air left depleted in oxygen following the explosion of firedamp in a mine.

black lead *See* **carbon**.

blanc fixe *See* **barium sulphate**.

blast furnace A furnace for smelting iron ores, such as haematite (Fe_2O_3) or magnetite (Fe_3O_4), to make *pig iron. The furnace is a tall refractory-lined cylindrical structure that is charged at the top with the dressed ore (*see* **beneficiation**), coke, and a flux, usually limestone. The conversion of the iron oxides to metallic iron is a reduction process in which carbon monoxide and hydrogen are the reducing agents. The overall reaction can be summarized thus:

$$Fe_3O_4 + 2CO + 2H_2 \rightarrow 3Fe + 2CO_2 + 2H_2O$$

The CO is obtained within the furnace by blasting the coke with hot air from a ring of tuyeres about two-thirds of the way down the furnace. The reaction producing the CO is:

$$2C + O_2 \rightarrow 2CO$$

In most blast furnaces hydrocarbons (oil, gas, tar, etc.) are added to the blast to provide a source of hydrogen. In the modern *direct-reduction process* the CO and H_2 may be produced separately so that the reduction process can proceed at a lower temperature. The pig iron produced by a blast furnace contains about 4% carbon and further refining is usually required to produce steel or cast iron.

blasting gelatin A high explosive made from nitroglycerine and gun cotton (cellulose nitrate).

bleaching powder A white solid regarded as a mixture of calcium chlorate(I), calcium chloride, and calcium hydroxide. It is prepared on a large scale by passing chlorine gas through a solution of calcium hydroxide. Bleaching powder is sold on the basis of available chlorine, which is liberated when it is treated with a dilute acid. It is used for bleaching paper pulps and fabrics and for sterilizing water.

blende A naturally occurring metal sulphide, e.g. zinc blende ZnS.

Bloch's theorem A theorem relating to the *quantum mechanics of crystals stating that the wave function ψ for an electron in a periodic potential has the form $\psi(\boldsymbol{r}) = \exp(i\boldsymbol{k}\cdot\boldsymbol{r})U(\boldsymbol{r})$, where $\boldsymbol{k}$ is the wave vector, $\boldsymbol{r}$ is a position vector, and $U(\boldsymbol{r})$ is a periodic function that satisfies $U(\boldsymbol{r} + \boldsymbol{R}) = U(\boldsymbol{r})$, for all vectors $\boldsymbol{R}$ of the Bravais lattice of the crystal. Block's theorem is interpreted to mean that the wave function for an electron in a periodic potential is a plane wave modulated by a periodic function. This explains why a free-electron model has some success in describing the properties of certain metals although it is inadequate to give a quantitative description of the properties of most metals. Block's theorem was formulated by the German-born US physicist Felix Bloch (1905–83) in 1928. *See also* **energy band.**

block *See* **periodic table.**

block copolymer *See* **polymer.**

blood pigment Any one of a group of metal-containing coloured protein compounds whose function is to increase the oxygen-carrying capacity of blood.

blue vitriol *See* **copper(II) sulphate.**

boat *See* **ring conformations.**

BOD *See* **biochemical oxygen demand.**

body-centred cubic (b.c.c.) *See* **cubic crystal.**

boehmite A mineral form of a mixed aluminium oxide and hydroxide, AlO.OH. It is named after the German scientist J. Böhm. *See* **aluminium hydroxide.**

Bohr, Niels Henrik David (1885–1962) Danish physicist. In 1913 he published his explanation of how atoms, with electrons orbiting a central nucleus, achieve stability by assuming that their angular momentum is quantized. Movement of electrons from one orbit to another is accompanied by the absorption or emission of energy in the form of light, thus accounting for the series of lines in the emission *spectrum of hydrogen. For this work Bohr was awarded the 1922 Nobel Prize. *See* **Bohr theory.**

bohrium Symbol Bh. A radioactive *transactinide element; a.n. 107. It was first made in 1981 by Peter Armbruster and a team in Darmstadt, Germany, by bombarding bismuth-209 nuclei with chromium-54 nuclei. Only a few atoms of bohrium have ever been detected.

Bohr magneton *See* **magneton.**

Bohr theory The theory published in 1913 by Niels *Bohr to explain the line spectrum of hydrogen. He assumed that a single electron of mass m travelled in a circular orbit of radius r, at a velocity v, around a positively charged nucleus. The angular momentum of the electron would then be mvr. Bohr proposed that electrons could only occupy orbits in which this angular momentum had certain fixed values, $h/2\pi$, $2h/2\pi$, $3h/2\pi$,...$nh/2\pi$, where h is the Planck constant. This means that the angular momentum is quantized, i.e. can only have certain values, each of which is a multiple of n. Each permitted value of n is associated with an orbit of different radius and Bohr assumed that when the atom emitted or absorbed radiation of frequency ν, the electron jumped from one orbit to another; the energy emitted or absorbed by each jump is equal to $h\nu$. This theory gave good results in predicting the lines observed in the spectrum of hydrogen and simple ions such as He^+, Li^{2+}, etc. The idea of quantized values of angular momentum was later explained by the wave nature of the electron. Each orbit has to have a whole number of wavelengths around it; i.e. $n\lambda = 2\pi r$, where λ is the wavelength and n a whole number. The wavelength of a particle is given by h/mv, so $nh/mv = 2\pi r$, which leads to $mvr = nh/2\pi$. Modern atomic theory does not allow subatomic particles to be treated in the same way as large objects, and Bohr's reasoning is somewhat discredited. However, the idea of quantized angular momentum has been retained.

boiling point (b.p.) The temperature at which the saturated vapour pressure of a liquid equals the external atmospheric pressure. As a consequence, bubbles form in the liquid and the temperature remains constant until all the liquid has evaporated. As the boiling point of a liquid depends on the external atmospheric pressure, boiling points are usually quoted for standard atmospheric pressure (760 mmHg = 101 325 Pa).

boiling-point–composition diagram A graph showing how the boiling point and vapour composition of a mixture of two liquids depends on the composition of the mixture. The abscissa shows the range of compositions from 100% A at one end to 100% B at the other. The diagram has two curves: the lower one gives the boiling points (at a fixed pressure) for the different compositions. The upper one is plotted by taking the composition of vapour at each temperature on the boiling-point curve. The two curves would coincide for an ideal mixture, but generally they are different because of deviations from *Raoult's law. In some cases, they may show a maximum or minimum and coincide at some intermediate composition, explaining the formation of *azeotropes.

boiling-point elevation *See* **elevation of boiling point**.

Boltzmann, Ludwig Eduard (1844–1906) Austrian physicist. He held professorships in Graz, Vienna, Munich, and Leipzig, where he worked on the kinetic theory of gases (*see* **Maxwell–Boltzmann distribution**) and on thermodynamics (*see* **Boltzmann equation**). He suffered from depression and committed suicide.

Boltzmann constant Symbol k. The ratio of the universal gas constant (R) to the Avogadro constant (N_A). It may be thought of therefore as the gas constant per molecule:

$$k = R/N_A = 1.380\,658(12) \times 10^{-23}\ \mathrm{J\,K^{-1}}$$

It is named after Ludwig *Boltzmann.

Boltzmann equation An equation used in the study of a collection of particles in *nonequilibrium statistical mechanics, particularly their transport properties. The Boltzmann equation describes a quantity called the *distribution function*, f, which gives a mathematical description of the state and how it is changing. The distribution function depends on a position vector $\boldsymbol{r}$, a velocity vector $\boldsymbol{v}$, and the time t; it thus provides a statistical statement about the positions and velocities of the particles at any time. In the case of one species of particle being present, Boltzmann's equation can be written

$$\partial f/\partial t + \boldsymbol{a}.(\partial f/\partial \boldsymbol{v}) + \boldsymbol{v}.(\partial f/\partial \boldsymbol{r}) = (\partial f/\partial t)_{\text{coll}},$$

where $\boldsymbol{a}$ is the acceleration of bodies between collisions and $(\partial f/\partial t)_{\text{coll}}$ is the rate of change of $f(\boldsymbol{r},\boldsymbol{v},t)$ due to collisions. The Boltzmann equation can be used to calculate *transport coefficients, such as conductivity. The equation was proposed by Ludwig *Boltzmann in 1872.

Boltzmann formula An equation concerning the entropy S of a system that derives from statistical mechanics. It states that entropy is related to the number W of distinguishable ways in which the equation $S = k \ln W$, where k is the Boltzmann constant, can describe the system. It expresses in quantitative terms the concept that entropy is a measure of the disorder of a system. It was discovered in the late 19th century by Ludwig *Boltzmann while he was studying statistical mechanics.

bomb calorimeter An apparatus used for measuring heats of combustion (e.g. calorific values of fuels and foods). It consists of a strong container in which the sample is sealed with excess oxygen and ignited electrically. The heat of combustion at constant volume can be calculated from the resulting rise in temperature.

bond *See* **chemical bond.**

bond dissociation energy *See* **bond energy.**

bond energy An amount of energy associated with a bond in a chemical compound. It is obtained from the heat of atomization. For instance, in methane the bond energy of the C–H bond is one quarter of the enthalpy of the process

$$CH_4(g) \rightarrow C(g) + 4H(g)$$

Bond energies (or *bond enthalpies*) can be calculated from the standard enthalpy of formation of the compound and from the enthalpies of atomization of the elements. Energies calculated in this way are called *average bond energies* or *bond–energy terms.* They depend to some extent on the molecule chosen; the C–H bond energy in methane will differ slightly from that in ethane. The *bond dissociation energy* is a different measurement, being the energy required to break a particular bond; e.g. the energy for the process:

$$CH_4(g) \rightarrow CH_3{\cdot}(g) + H{\cdot}(g)$$

bond enthalpy *See* **bond energy.**

bonding orbital *See* **orbital.**

bond order A value indicating the degree of bonding between two atoms in a molecule relative to a single bond. Bond orders are theoretical values depending on the way the calculation is done. For example, in ethane the bond order of the carbon–carbon bond is 1. In ethene, the bond order is 2. In benzene the bond order as calculated by molecular orbital theory is 1.67.

bone black *See* **charcoal.**

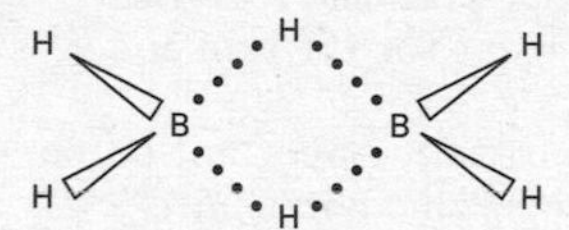

Diborane, the simplest of the boranes

borane (boron hydride) Any of a group of compounds of boron and hydrogen, many of which can be prepared by the action of acid on magnesium boride (MgB_2). Others are made by pyrolysis of the products of this reaction in the presence of hydrogen and other reagents. They are all volatile, reactive, and oxidize readily in air, some explosively so. The boranes are a remarkable group of compounds in that their structures cannot be described using the conventional two-electron covalent bond model (*see* **electron deficient compound**). The simplest example is *diborane* (B_2H_6): see formula. Other boranes include B_4H_{10}, B_5H_9, B_5H_{11}, B_6H_{10}, and $B_{10}H_4$. The larger borane molecules have open or closed polyhedra of boron atoms. In addition, there is a wide range of borane derivatives containing atoms of other elements, such as carbon and phosphorus. *Borohydride ions* of the type $B_6H_6^{2-}$ also exist. Boranes and borohydride ions are classified according to their structure. Those with a complete polyhedron are said to have a *closo-structure*. Those in which the polyhedron is incomplete by loss of one vertex have a *nido-structure* (from the Greek for 'nest'). Those with open structures by removal of two or more vertices have an *arachno-structure* (from the Greek for 'spider').

borate Any of a wide range of ionic compounds that have negative ions containing boron and oxygen (see formulae). Lithium borate, for example, contains the simple anion $B(OH)_4^-$. Most borates, however, are inorganic polymers with rings, chains, or other networks based on the planar BO_3 group or the tetrahedral $BO_3(OH)$ group. 'Hydrated' borates are ones containing –OH groups; many examples occur naturally. Anhydrous borates, which contain BO_3 groups, can be made by melting together boric acid and metal oxides.

borax (disodium tetraborate-10-water) A colourless monoclinic solid, $Na_2B_4O_7.10H_2O$, soluble in water and very slightly soluble in ethanol; monoclinic; r.d. 1.73; loses $8H_2O$ at 75°C; loses $10H_2O$ at 320°C. The formula gives a misleading impression of the structure. The compound contains the ion $[B_4O_5(OH)_4]^{2-}$ (*see* **borate**). Attempts to recrystallize this compound above 60.8°C yield the pentahydrate. The main sources are the borate minerals *kernite* ($Na_2B_4O_7.4H_2O$) and *tincal* ($Na_2B_4O_7.10H_2O$). The ores are purified by carefully controlled dissolution and recrystallization. On treatment with mineral acids borax gives boric acid.

Borax is a very important substance in the glass and ceramics industries as a raw material for making borosilicates. It is also important as a metallurgical flux because of the ability of molten borates to dissolve metal oxides. In solution it partially hydrolyses to boric acid and can thus act as a buffer. For this reason it is used as a laundry pre-soak. It is used medicinally as a mild alkaline antiseptic and astringent for the skin and mucous membranes.

Disodium tetraborate is the source of many industrially important boron compounds, such as barium borate (fungicidal paints), zinc borate (fire-retardant additive in plastics), and boron phosphate (heterogeneous acid catalyst in the petrochemicals industry).

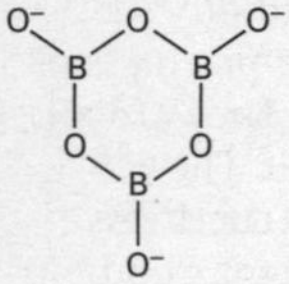

$B_3O_6^{3-}$ as in $Na_3B_3O_6$

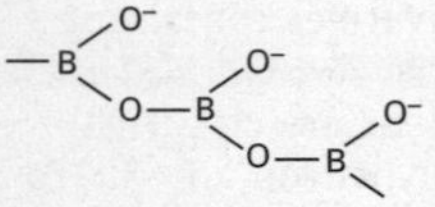

$(BO_2)_n^{n-}$ as in CaB_2O_4

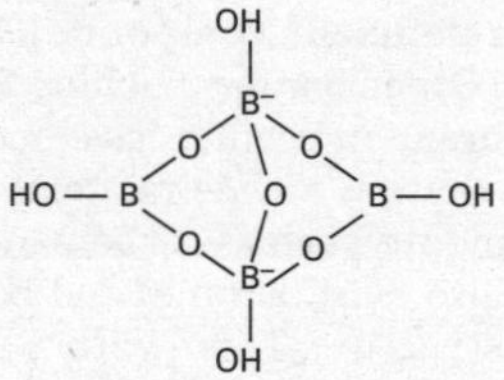

$[B_4O_5(OH)_4]^{2-}$ as in borax $Na_2B_4O_7.10H_2O$

Structure of some typical borate ions

borax-bead test A simple laboratory test for certain metal ions in salts. A small amount of the salt is mixed with borax and a molten bead formed on the end of a piece of platinum wire. Certain metals can be identified by the colour of the bead produced in the oxidizing and reducing parts of a Bunsen flame. For example, iron gives a bead that is red when hot and yellow when cold in the oxidizing flame and a green bead in the reducing flame.

borazon *See* **boron nitride**.

Bordeaux mixture A mixture of copper(II) sulphate and calcium hydroxide in water, used as a fungicide.

boric acid Any of a number of acids containing boron and oxygen. Used without qualification the term applies to the compound H_3BO_3 (which is also called *orthoboric acid* or, technically, *trioxoboric(III) acid*). This is a white or colourless solid that is soluble in water and ethanol; triclinic; r.d. 1.435; m.p. 169°C. It occurs naturally in the condensate from volcanic steam vents (suffioni). Commercially, it is made by treating borate minerals (e.g. kernite, $Na_2B_4O_7.4H_2O$) with sulphuric acid followed by recrystallization.

In the solid there is considerable hydrogen bonding between H_3BO_3 molecules resulting in a layer structure, which accounts for the easy cleavage of the crystals. H_3BO_3 molecules also exist in dilute solutions but in more concentrated solutions polymeric acids and ions are formed (e.g. $H_4B_2O_7$; *pyroboric acid* or *tetrahydroxomonoxodiboric(III) acid*). The compound is a very weak acid but also acts as a Lewis *acid in accepting hydroxide ions:

$$B(OH)_3 + H_2O \rightleftharpoons B(OH)_4^- + H^+$$

If solid boric acid is heated it loses water and transforms to another acid at 300°C. This is given the formula HBO_2 but is in fact a polymer $(HBO_2)_n$. It is called *metaboric acid* or, technically, *polydioxoboric(III)* acid.

Boric acid is used in the manufacture of glass (borosilicate glass), glazes and enamels, leather, paper, adhesives, and explosives. It is widely used (particularly in the USA) in detergents, and because of the ability of fused boric acid to dissolve other metal oxides it is used as a flux in brazing and welding. Because of its mild antiseptic properties it is used in the pharmaceutical industry and as a food preservative.

boric anhydride *See* **boron(III) oxide.**

boric oxide *See* **boron(III) oxide.**

boride A compound of boron with a metal. Most metals form at least one boride of the type MB, MB_2, MB_4, MB_6, or MB_{12}. The compounds have a variety of structures; in particular, the hexaborides contain clusters of B_6 atoms. The borides are all hard high-melting materials with metal-like conductivity. They can be made by direct combination of the elements at high temperatures (over 2000°C) or, more usually, by high-temperature reduction of a mixture of the metal oxide and boron oxide using carbon or aluminium. Chemically, they are stable to nonoxidizing acids but are attacked by strong oxidizing agents and by strong alkalis. Magnesium boride (MgB_2) is unusual in that it can be hydrolysed to boranes. Industrially, metal borides are used as refractory materials. The most important are CrB, CrB_2, TiB_2, and ZnB_2. Generally, they are fabricated using high-temperature powder metallurgy, in which the article is produced in a graphite die at over 2000°C and at very high pressure. Items are pressed as near to final shape as possible as machining requires diamond cutters and is extremely expensive.

Born, Max (1882–1970) German-born British physicist who was awarded the 1954 Nobel Prize for physics (with W. Bothe) for his work on statistical mechanics. With *Heisenberg he also developed matrix mechanics.

Born–Haber cycle A cycle of reactions used for calculating the lattice energies of ionic crystalline solids. For a compound MX, the lattice energy is the enthalpy of the reaction

$M^+(g) + X^-(g) \rightarrow M^+X^-(s)\ \Delta H_L$

The standard enthalpy of formation of the ionic solid is the enthalpy of the reaction

$M(s) + ½X_2(g) \rightarrow M^+X^-(s)\ \Delta H_f$

The cycle involves equating this enthalpy (which can be measured) to the sum of the enthalpies of a number of steps proceeding from the elements to the ionic solid. The steps are:

(1) Atomization of the metal:

$M(s) \rightarrow M(g)\ \Delta H_1$

(2) Atomization of the nonmetal:

$½X_2(g) \rightarrow X(g)\ \Delta H_2$

(3) Ionization of the metal:

$M(g) \rightarrow M^+(g) + e\ \Delta H_3$

This is obtained from the ionization potential.

(4) Ionization of the nonmetal:

$X(g) + e \rightarrow X^-(g)\ \Delta H_4$

This is the electron affinity.

(5) Formation of the ionic solids:

$M^+(g) + X^-(g) \rightarrow M^+X^-(s)\ \Delta H_L$

Equating the enthalpies gives:

$\Delta H_f = \Delta H_1 + \Delta H_2 + \Delta H_3 + \Delta H_4 + \Delta H_L$

from which ΔH_L can be found. It is named after Max *Born and Fritz *Haber.

bornite An important ore of copper composed of a mixed copper–iron sulphide, Cu_5FeS_4. Freshly exposed surfaces of the mineral are a metallic reddish-brown but a purplish iridescent tarnish soon develops – hence it is popularly known as *peacock ore*. Bornite is mined in Chile, Peru, Bolivia, Mexico, and the USA.

Born–Oppenheimer approximation An *adiabatic approximation used in molecular and solid-state physics in which the motion of atomic nuclei is taken to be so much slower than the motion of electrons that, when calculating the motions of electrons, the nuclei can be taken to be in fixed positions. This approximation was justified using perturbation theory by Max Born and the US physicist Julius Robert Oppenheimer (1904–67) in 1927.

borohydride ions *See* borane.

boron Symbol B. An element of *group 13 (formerly IIIB) of the periodic table; a.n. 5; r.a.m. 10.81; r.d. 2.34–2.37 (amorphous); m.p. 2300°C; b.p. 2550°C. It forms two allotropes; amorphous boron is a brown powder but metallic boron is black. The metallic form is very hard (9.3 on Mohs' scale) and is a poor electrical conductor at room temperature. At least three crystalline forms are possible; two are rhombohedral and the other tetragonal. The element is never found free in nature. It occurs as orthoboric acid in volcanic springs in Tuscany, as borates in kernite ($Na_2B_4O_7.4H_2O$), and as colemanite ($Ca_2B_6O_{11}.5H_2O$) in California. Samples usually contain isotopes in the ratio of 19.78% boron–10 to 80.22% boron–11. Extraction is achieved by vapour-phase reduction of boron trichloride with hydrogen on electrically heated filaments. Amorphous boron can be obtained by reducing the trioxide with magnesium powder. Boron when heated reacts with oxygen, halogens, oxidizing acids, and hot alkalis. It is used in semiconductors and in filaments for specialized aerospace applications. Amorphous boron is used in flares, giving a green coloration. The isotope boron–10 is used in nuclear reactor control rods and shields. The element was discovered in 1808 by Sir Humphry *Davy and by J. L. *Gay-Lussac and L. J. Thenard.

boron carbide A black solid, B_4C, soluble only in fused alkali; it is extremely hard, over 9½ on Mohs' scale; rhombohedral; r.d. 2.52; m.p. 2350°C; b.p. >3500°C. Boron carbide is manufactured by the reduction of boric oxide with petroleum coke in an electric furnace. It is used largely as an abrasive, but objects can also be fabricated using high-temperature powder metallurgy. Boron nitride is also used as a neutron absorber because of its high proportion of boron–10.

boron hydride *See* **borane.**

boron nitride A solid, BN, insoluble in cold water and slowly decomposed by hot water; r.d. 2.25 (hexagonal); sublimes above 3000°C. Boron nitride is manufactured by heating boron oxide to 800°C on an acid-soluble carrier, such as calcium phosphate, in the presence of nitrogen or ammonia. It is isoelectronic with carbon and, like carbon, it has a very hard cubic form (*borazon*) and a softer hexagonal form, unlike graphite this is a nonconductor. It is used in the electrical industries where its high thermal conductivity and high resistance are of especial value.

boron(III) oxide (boric anhydride; boric oxide; diboron trioxide) A glassy solid, B_2O_3, that gradually absorbs water to form boric acid. It has some amphoteric characteristics and forms various salts.

boron trichloride A colourless fuming liquid, BCl_3, which reacts with water to give hydrogen chloride and boric acid; r.d. 1.349; m.p. −107°C; b.p. 12.5°C. Boron trichloride is prepared industrially by the exothermic chlorination of boron carbide at above 700°C, followed by fractional distillation. An alternative, but more expensive, laboratory method is the reaction of dry chlorine with boron at high temperature. Boron trichloride is a Lewis *acid, forming stable addition compounds with such donors as ammonia and the amines and is used in the laboratory to promote reactions that liberate these donors. The compound is important industrially as a source of pure boron (reduction with hydrogen) for the electronics industry. It is also used for the preparation of boranes by reaction with metal hydrides.

borosilicate Any of a large number of substances in which BO_3 and SiO_4 units are linked to form networks with a wide range of structures. Borosilicate glasses are particularly important; the addition of boron to the silicate network enables the glass to be fused at lower temperatures than pure silica and also extends the plastic range of the glass. Thus such glasses as Pyrex have a wider range of applications than soda glasses (narrow plastic range, higher thermal expansion) or silica (much higher melting point). Borosilicates are also used in glazes and enamels and in the production of glass wools.

Bosch process *See* **Haber process.**

Bose–Einstein statistics *See* **quantum statistics.**

boson A particle or system that obeys Bose–Einstein statistics (*see* **quantum statistics**). Combining quantum mechanics with special relativity theory gives a result that a boson has to have an integer spin. A *photon is an example of a boson.

bound state A system in which two (or more) parts are bound together in such a way that energy is required to split them. An example of a bound state is a *molecule formed from two (or more) *atoms.

bowsprit *See* **ring conformations.**

Boyle, Robert (1627–91) English chemist and physicist, born in Ireland. After moving to Oxford in 1654 he worked on gases, using an air pump made by Robert Hooke. In 1662 he discovered *Boyle's law. In chemistry he worked on *flame tests and acid-base *indicators. Boyle is generally regarded as the person who established chemistry as a modern subject, distinct from alchemy. He was the first to give a definition of a chemical element.

Boyle's law The volume (V) of a given mass of gas at a constant temperature is inversely proportional to its pressure (p), i.e. pV = constant. This is true only for an *ideal gas. This law was discovered in 1662 by Robert *Boyle. On the continent of Europe it is known as *Mariotte's law* after E. Mariotte (1620–84), who discovered it independently in 1676. *See also* **gas laws.**

Brackett series *See* **hydrogen spectrum.**

Bragg, Sir William Henry (1862–1942) British physicist, who with his son *Sir (William) Lawrence Bragg* (1890–1971), was awarded the 1915 Nobel Prize for physics for their pioneering work on *X-ray crystallography. He also constructed an X-ray spectrometer for measuring the wavelengths of X-rays. In the 1920s, while director of the Royal Institution in London, he initiated X-ray diffraction studies of organic molecules.

Bragg peak A peak in the scattering pattern in X-ray diffraction of a crystal. The intensity of Bragg peaks is proportional to the square of the number of the scatterers. If X-ray scattering from a solid produces Bragg peaks this indicates that the solid has long-range order. Bragg peaks are named after Sir Lawrence *Bragg, who discovered them in 1912. The scattering of X-rays from a set of planes in a crystal that gives rise to Bragg peaks is called *Bragg scattering*.

Bragg's law When a beam of X-rays (wavelength λ) strikes a crystal surface in which the layers of atoms or ions are separated by a distance d, the maximum intensity of the reflected ray occurs when $\sin\theta = n\lambda/2d$, where θ (known as the *Bragg angle*) is the complement of the angle of incidence and n is an integer. The law enables the structure of many crystals to be determined. It was discovered in 1912 by Sir Lawrence *Bragg.

branched chain *See* **chain**.

Bravais lattice An infinite array of lattice points. A Bravais lattice can have only 14 space groups divided into 7 crystal systems. It is named after Auguste Bravais (1811–63).

brass A group of alloys consisting of copper and zinc. A typical yellow brass might contain about 67% copper and 33% zinc.

Bremsstrahlung (German: braking radiation) The *X-rays emitted when a charged particle, especially a fast electron, is rapidly slowed down, as when it passes through the electric field around an atomic nucleus. The X-rays cover a whole continuous range of wavelengths down to a minimum value, which depends on the energy of the incident particles. Bremsstrahlung are produced by a metal target when it is bombarded by electrons.

brewing The process by which beer is made. Fermentation of sugars from barley grain by the yeasts *Saccharomyces cerevisiae* and *S. uvarum* (or *S. carlsbergenesis*) produces alcohol. In the first stage the barley grain is soaked in water, a process known as *malting*. The grain is then allowed to germinate and the natural enzymes of the grain (the amylases and the maltases) convert the starch to maltose and then to glucose. The next stage is *kilning* or *roasting*, in which the grains are dried and crushed. The colour of a beer depends on the temperature used for this process: the higher the temperature, the darker the beer. In the next stage, *mashing*, the crushed grain is added to water at a specific temperature and any remaining starch is converted to sugar; the resultant liquid is the raw material of brewing, called *wort*. The yeast is then added to the wort to convert the sugar to alcohol, followed by hops, which give beer its characteristic flavour. Hops are the female flowers of the vine *Humulus lupulus*; they contain resins (humulones, cohumulones, and adhumulones) that give beer its distinctive bitter taste.

bridge An atom or group joining two other atoms in a molecule. *See* **aluminium chloride**; **borane**.

brighteners Substances added to detergents or used to treat textiles or paper in order to brighten the colours or, particularly, to enhance whiteness. Blueing agents are used in laundries to give a slight blue cast to white material in order to counteract yellowing. Fluorescent brighteners are compounds that absorb visible or ultraviolet radiation and fluoresce in the blue region of the optical spectrum.

Brillouin zone A cell in a reciprocal lattice. The first Brillouin zone is the cell of smallest volume enclosed by those planes that are perpendicular bisectors of reciprocal lattice vectors. Higher Brillouin zones also exist. Brillouin zones are used in the theory of energy levels in a periodic potential, as in a crystal. They are named after the French physicist Léon Brillouin (1889–1969), who introduced them into the theory of crystals in 1930.

Brinell hardness A scale for measuring the hardness of metals introduced around 1900 by the Swedish metallurgist Johann Brinell (1849–1925). A small chromium-steel ball is pressed into the surface of the metal by a load of known weight. The ratio of the mass of the load in kilograms to the area of the depression formed in square millimetres is the *Brinell number*.

Brin process A process formerly used for making oxygen by heating barium oxide in air to form the peroxide and then heating the peroxide at higher temperature (>800°C) to produce oxygen

$$2BaO_2 \rightarrow 2BaO + O_2$$

Britannia metal A silvery alloy consisting of 80–90% tin, 5–15% antimony, and sometimes small percentages of copper, lead, and zinc. It is used in bearings and some domestic articles.

British thermal unit (Btu) The Imperial unit of heat, being originally the heat required to raise the temperature of 1lb of water by 1°F. 1 Btu is now defined as 1055.06 joules.

bromate A salt or ester of a bromic acid.

bromic(I) acid (hypobromous acid) A yellow liquid, HBrO. It is a weak acid but a strong oxidizing agent.

bromic(V) acid A colourless liquid, $HBrO_3$, made by adding sulphuric acid to barium bromate. It is a strong acid.

bromide *See* **halide**.

bromination A chemical reaction in which a bromine atom is introduced into a molecule. *See also* **halogenation**.

bromine Symbol Br. A *halogen element; a.n. 35; r.a.m. 79.909; r.d. 3.13; m.p. −7.2°C; b.p. 58.78°C. It is a red volatile liquid at room temperature, having a red-brown vapour. Bromine is obtained from brines in the USA (displacement with chlorine); a small amount is obtained from sea water in Anglesey. Large quantities are used to make 1,2-dibromoethane as a petrol additive. It is also used in the manufacture of many other compounds. Chemically, it is intermediate in reactivity between chlorine and iodine. It forms compounds in which it has oxidation states of 1, 3, 5, or 7. The liquid is harmful to human tissue and the vapour irritates the eyes and throat. The element was discovered in 1826 by Antoine Balard.

bromoethane (ethyl bromide) A colourless flammable liquid, C_2H_5Br; r.d. 1.46; m.p. −119°C; b.p. 38.4°C. It is a typical *haloalkane, which can be prepared from ethene and hydrogen bromide. Bromoethane is used as a refrigerant.

bromoform *See* **tribromomethane; haloforms.**

bromomethane (methyl bromide) A colourless volatile nonflammable liquid, CH_3Br; r.d. 1.68; m.p. −93°C; b.p. 3.56°C. It is a typical *haloalkane.

bromothymol blue An acid–base *indicator that is yellow in acid solutions and blue in alkaline solutions. It changes colour over the pH range 6–8.

Brønsted, Johannes Nicolaus (1879–1947) Danish physical chemist. He worked on thermochemistry and electrochemistry and is best known for the Lowry–Brønsted theory of *acids and bases, which he proposed (independently of Lowry) in 1923.

Brønsted acid *See* **acid.**

Brønsted base *See* **acid.**

bronze Any of a group of alloys of copper and tin, sometimes with lead and zinc present. The amount of tin varies from 1% to 30%. The alloy is hard and easily cast and extensively used in bearings, valves, and other machine parts. Various improved bronzes are produced by adding other elements; for instance, *phosphor bronzes* contain up to 1% phosphorus. In addition certain alloys of copper and metals other than tin are called bronzes – *aluminium bronze* is a mixture of copper and aluminium. Other special bronzes include *bell metal, *gun metal, and *beryllium bronze.

Brownian movement The continuous random movement of microscopic solid particles (of about 1 micrometre in diameter) when suspended in a fluid medium. First observed by the British botanist Robert Brown (1773–1858) in 1827 when studying pollen particles, it was originally thought to be the manifestation of some vital force. It was later recognized to be a consequence of bombardment of the particles by the continually moving molecules of the liquid. The smaller the particles the more extensive is the motion. The effect is also visible in particles of smoke suspended in a still gas.

brown-ring test A test for ionic nitrates. The sample is dissolved and iron(II) sulphate solution added in a test tube. Concentrated sulphuric acid is then added slowly so that it forms a separate layer. A brown ring (of $Fe(NO)SO_4$) at the junction of the liquids indicates a positive result.

brucite A mineral form of *magnesium hydroxide, $Mg(OH)_2$.

brusselator A type of chemical reaction mechanism that leads to an *oscillating reaction. It involves the conversion of reactants A and B into products C and B by a series of four steps:

$A \rightarrow X$

$2X + Y \rightarrow 3Y$

$B + X \rightarrow Y + C$

$X \rightarrow D$

Autocatalysis occurs as in the *Lotka–Volterra mechanism and the *oregonator. If the concentrations of A and B are maintained constant, the concentrations of X and Y oscillate with time. A graph of the concentration of X against that of Y is a closed loop (the *limit cycle* of the reaction). The reaction settles down to this limit cycle whatever the initial concentrations of X and Y, i.e. the limit cycle is an *attractor for the system. The reaction mechanism is named after the city of Brussels, where the research group that discovered it is based.

Buchner funnel A type of funnel with an internal perforated tray on which a flat circular filter paper can be placed, used for filtering by suction. It is named after the German chemist Eduard Buchner (1860–1917).

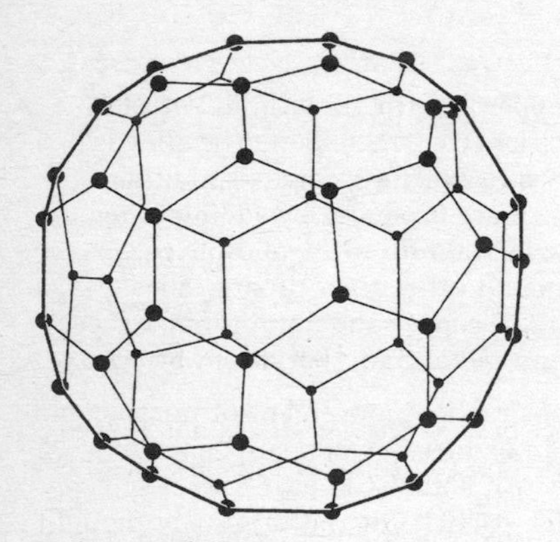

Buckminsterfullerene

buckminsterfullerene A form of carbon composed of clusters of 60 carbon atoms bonded together in a polyhedral structure composed of pentagons and hexagons (see illustration). Originally it was identified in 1985 in products obtained by firing a high-power laser at a graphite target. It can be made by an electric arc struck between graphite electrodes in an inert atmosphere. The molecule, C_{60}, was named after the US architect Richard Buckminster Fuller (1895–1983) because of the resemblance of the structure to the geodesic dome, which Fuller invented. The molecules are informally called *buckyballs*; more formally, the substance itself is also called *fullerene*. The substance is a yellow crystalline solid (*fullerite*), soluble in benzene.

Various fullerene derivatives are known in which organic groups are attached to carbon atoms on the sphere. In addition, it is possible to produce novel enclosure compounds by trapping metal ions within the C_{60} cage. Some of these have semiconducting properties. The electric-arc method of producing C_{60} also leads to a smaller number of fullerenes such as C_{70}, which have less symmetrical molecular structures. It is also possible to produce forms of carbon in which the atoms are linked in a cylindrical, rather than spherical, framework with a diameter of a few nanometres. They are known as *buckytubes* (or *nanotubes*).

buckyball *See* **buckminsterfullerene**.

buckytube *See* **buckminsterfullerene**.

buffer A solution that resists change in pH when an acid or alkali is added or when the solution is diluted. Acidic buffers consist of a weak acid with a salt of the acid. The salt provides the negative ion A^-, which is the conjugate base of the acid HA. An example is carbonic acid and sodium hydrogencarbonate. Basic buffers have a weak base and a salt of the base (to provide the conjugate acid). An example is ammonia solution with ammonium chloride.

In an acidic buffer, for example, molecules HA and ions A^- are present. When acid is added most of the extra protons are removed by the base:

$$A^- + H^+ \rightarrow HA$$

When base is added, most of the extra hydroxide ions are removed by reaction with undissociated acid:

$$OH^- + HA \rightarrow A^- + H_2O$$

Thus, the addition of acid or base changes the pH very little. The hydrogen-ion concentration in a buffer is given by the expression

$$K_a = [H^+] = [A^-]/[HA]$$

i.e. it depends on the ratio of conjugate base to acid. As this is not altered by dilution, the hydrogen-ion concentration for a buffer does not change much during dilution.

In the laboratory, buffers are used to prepare solutions of known stable pH. Natural buffers occur in living organisms, where the biochemical reactions are very sensitive to change in pH. The main natural buffers are H_2CO_3/HCO_3^- and $H_2PO_4^-/HPO_4^{2-}$. Buffer solutions are also used in medicine (e.g. in intravenous injections), in agriculture, and in many industrial processes (e.g. dyeing, fermentation processes, and the food industry).

bumping Violent boiling of a liquid caused by superheating so that bubbles form at a pressure above atmospheric pressure. It can be prevented by putting pieces of porous pot in the liquid to enable bubbles of vapour to form at the normal boiling point.

Bunsen, Robert Wilhelm (1811–99) German chemist, who held professorships at Kassel, Marburg, and Heidelberg. His early researches on arsenic-containing compounds cost him an eye in an explosion. He then turned to gas analysis and spectroscopy, enabling him and *Kirchhoff to discover the elements caesium (1860) and rubidium (1861). He also popularized the use of the *Bunsen burner and developed the *Bunsen cell.

Bunsen burner A laboratory gas burner having a vertical metal tube into which the gas is led, with a hole in the side of the base of the tube to admit air. The amount of air can be regulated by a sleeve on the tube. When no air is admitted the flame is luminous and smoky. With air, it has a faintly visible hot outer part (the oxidizing part) and an inner blue cone where combustion is incomplete (the cooler reducing part of the flame). The device is named after Robert *Bunsen, who used a similar device (without a regulating sleeve) in 1855.

Bunsen cell A *primary cell devised by Robert *Bunsen consisting of a zinc cathode immersed in dilute sulphuric acid and a carbon anode immersed in concentrated nitric acid. The electrolytes are separated by a porous pot. The cell gives an e.m.f. of about 1.9 volts.

burette A graduated glass tube with a tap at one end leading to a fine outlet tube, used for delivering known volumes of a liquid (e.g. in titration).

buta-1,3-diene (butadiene) A colourless gaseous hydrocarbon, $CH_2{:}CHCH{:}CH_2$; m.p. −109°C; b.p. −4.5°C. It is made by catalytic dehydrogenation of butane (from petroleum or natural gas) and polymerized in the production of synthetic rubbers. The compound is a conjugated *diene in which the electrons in the pi orbitals are partially delocalized over the whole molecule. It can have trans and cis forms, the latter taking part in *Diels–Alder reactions.

butanal (butyraldehyde) A colourless flammable liquid aldehyde, C_3H_7CHO; r.d. 0.8; m.p. −99°C; b.p. 75.7°C.

butane A gaseous hydrocarbon, C_4H_{10}; d. 0.58 g cm^{-3}; m.p. −138°C; b.p. 0°C. Butane is obtained from petroleum (from refinery gas or by cracking higher hydrocarbons). The fourth member of the *alkane series, it has a straight chain of carbon atoms and is isomeric with 2-methylpropane ($CH_3CH(CH_3)CH_3$, formerly called *isobutane*). It can easily be liquefied under pressure and is supplied in cylinders for use as a fuel gas. It is also a raw material for making buta-1,3-diene (for synthetic rubber).

butanedioic acid (succinic acid) A colourless crystalline fatty acid, $(CH_2)_2(COOH)_2$; r.d. 1.6; m.p. 185°C; b.p. 235°C. A weak carboxylic acid, it is produced by fermentation of sugar or ammonium tartrate and used as a sequestrant and in making dyes. It occurs in living organisms as an intermediate in metabolism, especially in the *Krebs cycle.

butanoic acid (butyric acid) A colourless liquid water-soluble acid, C_3H_7COOH; r.d. 0.96; b.p. 163°C. It is a weak acid ($K_a = 1.5 \times 10^{-5}$ mol dm^{-3} at

25°C) with a rancid odour. Its esters are present in butter and in human perspiration. The acid is used to make esters for flavourings and perfumery.

butanol Either of two aliphatic alcohols with the formula C_4H_9OH. *Butan-1-ol*, $CH_3(CH_2)_3OH$, is a primary alcohol; r.d. 0.81; m.p. −89.5°C; b.p. 117.3°C. *Butan-2-ol*, $CH_3CH(OH)C_2H_5$, is a secondary alcohol; r.d. 0.81; m.p. −114.7°C; b.p. 100°C. Both are colourless volatile liquids obtained from butane and are used as solvents.

butanone (methyl ethyl ketone) A colourless flammable water-soluble liquid, $CH_3COC_2H_5$; r.d. 0.8; m.p. −86.4°C; b.p. 79.6°C. It can be made by the catalytic oxidation of butane and is used as a solvent.

butene (butylene) Either of two isomers with the formula C_4H_8: *1-butene* ($CH_3CH_2CH{:}CH_2$), which is made by passing 1-butanol vapour over heated alumina, and *2-butene* ($CH_3CH{:}CHCH_3$), which is made by heating 2-butanol with sulphuric acid. They are unpleasant-smelling gases used in the manufacture of polymers. The isomer *2-methylpentene* ($(CH_3)_2C{:}CH_2$) was formerly called *isobutene* or *isobutylene*.

butenedioic acid Either of two isomers with the formula HCOOHC:CHCOOH. Both compounds can be regarded as derivatives of ethene in which a hydrogen atom on each carbon has been replaced by a −COOH group. The compounds show cis−trans isomerism. The trans form is *fumaric acid* (r.d. 1.64; sublimes at 165°C) and the cis form is *maleic acid* (r.d. 1.59; m.p. 139−140°C). Both are colourless crystalline compounds used in making synthetic resins. The cis form is rather less stable than the trans form and converts to the trans form at 120°C. Unlike the trans form it can eliminate water on heating to form a cyclic anhydride containing a −CO.O.CO− group (*maleic anhydride*). Fumaric acid is an intermediate in the *Krebs cycle.

Butler–Volmer equation An equation for the rate of an electrochemical reaction; it describes the current density at an electrode in terms of the overpotential. The Butler−Volmer equation is given by:

$$j = j_a - j_c - j_e[\exp(1 - \alpha)F\eta/RT - \exp(-\alpha F/RT)],$$

where j_a and j_c are the individual cathode and anode currents respectively, and j_e is the equilibrium current, called the *exchange current density*. By definition $j_e = j_{ce} = j_{ae}$, where j_{ce} is the equilibrium cathode current and j_{ae} is the equilibrium anode current. F is the Faraday constant, η is the overpotential, R is the gas constant, T is the thermodynamic temperature, and α is a quantity called the *transfer coefficient.

butterfly effect *See* chaos.

butylene *See* butene.

butyl group The organic group $CH_3(CH_2)_3-$.

butyl rubber A type of synthetic rubber obtained by copolymerizing

2-methylpropene (CH_2:C(CH_3)CH_3; isobutylene) and methylbuta-1,3-diene (CH_2:C(CH_3)CH:CH_2, isoprene). Only small amounts of isoprene (about 2 mole %) are used. The rubber can be vulcanized. Large amounts were once used for tyre inner tubes.

butyraldehyde *See* **butanal.**

butyric acid *See* **butanoic acid.**

by-product A compound formed during a chemical reaction at the same time as the main product. Commercially useful by-products are obtained from a number of industrial processes. For example, calcium chloride is a by-product of the *Solvay process for making sodium carbonate. Propanone is a by-product in the manufacture of *phenol.

B–Z reaction (Belousov–Zhabotinskii reaction) A chemical reaction that shows a periodic colour change between magenta and blue with a period of about one minute. It occurs with a mixture of sulphuric acid, potassium bromate(V), cerium sulphate, and propanedioic acid. The colour change is caused by alternating oxidation–reductions in which cerium changes its oxidation state (Ce^{3+} gives a magenta solution while Ce^{4+} gives a blue solution). The B–Z reaction is an example of a chemical *oscillating reaction – a reaction in which there is a regular periodic change in the concentration of one or more reactants. The mechanism is highly complicated, involving a large number of steps. *See* **brusselator**.

C

cadmium Symbol Cd. A soft bluish metal belonging to *group 12 (formerly IIB) of the periodic table; a.n. 48; r.a.m. 112.41; r.d. 8.65; m.p. 320.9°C; b.p. 765°C. The element's name is derived from the ancient name for calamine, zinc carbonate $ZnCO_3$, and it is usually found associated with zinc ores, such as sphalerite (ZnS), but does occur as the mineral greenockite (CdS). Cadmium is usually produced as an associate product when zinc, copper, and lead ores are reduced. Cadmium is used in low-melting-point alloys to make solders, in Ni–Cd batteries, in bearing alloys, and in electroplating (over 50%). Cadmium compounds are used as phosphorescent coatings in TV tubes. Cadmium and its compounds are extremely toxic at low concentrations; great care is essential where solders are used or where fumes are emitted. It has similar chemical properties to zinc but shows a greater tendency towards complex formation. The element was discovered in 1817 by F. Stromeyer.

cadmium cell *See* Weston cell.

cadmium sulphide A water-insoluble compound, CdS; r.d. 4.82. It occurs naturally as the mineral *greenockite* and is used as a pigment and in semiconductors and fluorescent materials.

caesium Symbol Cs. A soft silvery-white metallic element belonging to *group 1 (formerly IA) of the periodic table; a.n. 55; r.a.m. 132.905; r.d. 1.88; m.p. 28.4°C; b.p. 678°C. It occurs in small amounts in a number of minerals, the main source being carnallite ($KCl.MgCl_2.6H_2O$). It is obtained by electrolysis of molten caesium cyanide. The natural isotope is caesium–133. There are 15 other radioactive isotopes. Caesium–137 (half-life 33 years) is used as a gamma source. As the heaviest alkali metal, caesium has the lowest ionization potential of all elements, hence its use in photoelectric cells, etc.

caesium clock An *atomic clock that depends on the energy difference between two states of the caesium–133 nucleus when it is in a magnetic field. In one type, atoms of caesium–133 are irradiated with radio-frequency radiation, whose frequency is chosen to correspond to the energy difference between the two states. Some caesium nuclei absorb this radiation and are excited to the higher state. These atoms are deflected by a further magnetic field, which causes them to hit a detector. A signal from this detector is fed back to the radio-frequency oscillator to prevent it drifting from the resonant frequency of 9 192 631 770 hertz. In this way the device is locked to this frequency with an accuracy better than 1 part in 10^{13}. The caesium clock is used in the *SI unit definition of the second.

cage compound *See* clathrate.

cage effect An effect occurring in certain condensed-phase reactions in which fragments are formed and their diffusion is hindered by a surrounding 'cage' of molecules. The initial fragments are consequently more likely to recombine or to react together to form new products.

Cahn–Ingold–Prelog system *See* **CIP system**.

calamine A mineral form of zinc carbonate, $ZnCO_3$ (smithsonite), although in the USA the same name is given to a hydrated zinc silicate (hemimorphite). The calamine used medicinally in lotions for treating sunburn and other skin conditions is basic zinc carbonate coloured pink with a trace of iron(III) oxide.

calcination The formation of a calcium carbonate deposit from hard water. *See* **hardness of water**.

calcinite A mineral form of *potassium hydrogencarbonate, $KHCO_3$.

calcite One of the most common and widespread minerals, consisting of crystalline calcium carbonate, $CaCO_3$. Calcite crystallizes in the rhombohedral system; it is usually colourless or white and has a hardness of 3 on the Mohs' scale. It has the property of double refraction, which is apparent in Iceland spar – the transparent variety of calcite. It is an important rock-forming mineral and is a major constituent in limestones, marbles, and carbonatites.

calcium Symbol Ca. A soft grey metallic element belonging to *group 2 (formerly IIA) of the periodic table; a.n. 20; r.a.m. 40.08; r.d. 1.54; m.p. 839°C; b.p. 1484°C. Calcium compounds are common in the earth's crust; e.g. limestone and marble ($CaCO_3$), gypsum ($CaSO_4.2H_2O$), and fluorite (CaF_2). The element is extracted by electrolysis of fused calcium chloride and is used as a getter in vacuum systems and a deoxidizer in producing nonferrous alloys. It is also used as a reducing agent in the extraction of such metals as thorium, zirconium, and uranium.

Calcium is an essential element for living organisms, being required for normal growth and development. In animals it is an important constituent of bones and teeth and is present in the blood, being required for muscle contraction and other metabolic processes. In plants it is a constituent (in the form of calcium pectate) of the middle lamella.

calcium acetylide *See* **calcium dicarbide**.

calcium bicarbonate *See* **calcium hydrogencarbonate**.

calcium carbide *See* **calcium dicarbide**.

calcium carbonate A white solid, $CaCO_3$, which is only sparingly soluble in water. Calcium carbonate decomposes on heating to give *calcium oxide (quicklime) and carbon dioxide. It occurs naturally as the minerals *calcite (rhombohedral; r.d. 2.71) and *aragonite (rhombic; r.d. 2.93). Rocks containing calcium carbonate dissolve slowly in acidified rainwater (containing dissolved CO_2) to cause temporary hardness. In the laboratory, calcium carbonate is precipitated from *limewater by carbon dioxide. Calcium carbonate is used in making lime (calcium oxide) and is the main raw material for the *Solvay process.

calcium chloride A white deliquescent compound, $CaCl_2$, which is soluble in water; r.d. 2.15; m.p. 782°C; b.p. >1600°C. There are a number of hydrated forms, including the monohydrate, $CaCl_2.H_2O$, the dihydrate, $CaCl_2.2H_2O$ (r.d. 0.84), and the hexahydrate, $CaCl_2.6H_2O$ (trigonal; r.d. 1.71; the hexahydrate loses $4H_2O$ at 30°C and the remaining $2H_2O$ at 200°C). Large quantities of it are formed as a byproduct of the *Solvay process and it can be prepared by dissolving calcium carbonate or calcium oxide in hydrochloric acid. Crystals of the anhydrous salt can only be obtained if the hydrated salt is heated in a stream of hydrogen chloride. Solid calcium chloride is used in mines and on roads to reduce dust problems, whilst the molten salt is the electrolyte in the extraction of calcium. An aqueous solution of calcium chloride is used in refrigeration plants.

calcium cyanamide A colourless solid, $CaCN_2$, which sublimes at 1300°C. It is prepared by heating calcium dicarbide at 800°C in a stream of nitrogen:

$$CaC_2(s) + N_2(g) \rightarrow CaCN_2(s) + C(s)$$

The reaction has been used as a method of fixing nitrogen in countries in which cheap electricity is available to make the calcium dicarbide (the *cyanamide process*). Calcium cyanamide can be used as a fertilizer because it reacts with water to give ammonia and calcium carbonate:

$$CaCN_2(s) + 3H_2O(l) \rightarrow CaCO_3(s) + 2NH_3(g)$$

It is also used in the production of melamine, urea, and certain cyanide salts.

calcium dicarbide (calcium acetylide; calcium carbide; carbide) A colourless solid compound, CaC_2; tetragonal; r.d. 2.22; m.p. 450°C; b.p. 2300°C. In countries in which electricity is cheap it is manufactured by heating calcium oxide with either coke or ethyne at temperatures above 2000°C in an electric arc furnace. The crystals consist of Ca^{2+} and C_2^- ions arranged in a similar way to the ions in sodium chloride. When water is added to calcium dicarbide, the important organic raw material ethyne (acetylene) is produced:

$$CaC_2(s) + 2H_2O(l) \rightarrow Ca(OH)_2(s) + C_2H_2(g)$$

calcium fluoride A white crystalline solid, CaF_2; r.d. 3.2; m.p. 1360°C; b.p. 2500°C. It occurs naturally as the mineral *fluorite (or fluorspar) and is the main source of fluorine. The *calcium fluoride structure* (*fluorite structure*) is a crystal structure in which the calcium ions are each surrounded by eight fluoride ions arranged at the corners of a cube. Each fluoride ion is surrounded by four calcium ions at the corners of a tetrahedron.

calcium hydrogencarbonate (calcium bicarbonate) A compound,

$Ca(HCO_3)_2$, that is stable only in solution and is formed when water containing carbon dioxide dissolves calcium carbonate:

$$CaCO_3(s) + H_2O(l) + CO_2(g) \rightarrow Ca(HCO_3)_2(aq)$$

It is the cause of temporary *hardness in water, because the calcium ions react with soap to give scum. Calcium hydrogencarbonate is unstable when heated and decomposes to give solid calcium carbonate. This explains why temporary hardness is removed by boiling and the formation of 'scale' in kettles and boilers.

calcium hydroxide **(slaked lime)** A white solid, $Ca(OH)_2$, which dissolves sparingly in water (*see* **limewater**); hexagonal; r.d. 2.24. It is manufactured by adding water to calcium oxide, a process that evolves much heat and is known as slaking. It is used as a cheap alkali to neutralize the acidity in certain soils and in the manufacture of mortar, whitewash, bleaching powder, and glass.

calcium nitrate A white deliquescent compound, $Ca(NO_3)_2$, that is very soluble in water; cubic; r.d. 2.50; m.p. 561°C. It can be prepared by neutralizing nitric acid with calcium carbonate and crystallizing it from solution as the tetrahydrate $Ca(NO_3)_2.4H_2O$, which exists in two monoclinic crystalline forms (α, r.d. 1.9; β, r.d. 1.82). There is also a trihydrate, $Ca(NO_3)_2.3H_2O$. The anhydrous salt can be obtained from the hydrate by heating but it decomposes on strong heating to give the oxide, nitrogen dioxide, and oxygen. Calcium nitrate is sometimes used as a nitrogenous fertilizer.

calcium octadecanoate **(calcium stearate)** An insoluble white salt, $Ca(CH_3(CH_2)_{16}COO)_2$, which is formed when soap is mixed with water containing calcium ions and is the scum produced in hard-water regions.

calcium oxide **(quicklime)** A white solid compound, CaO, formed by heating calcium in oxygen or by the thermal decomposition of calcium carbonate; cubic; r.d. 3.35; m.p. 2580°C; b.p. 2850°C. On a large scale, calcium carbonate in the form of limestone is heated in a tall tower (lime kiln) to a temperature above 550°C:

$$CaCO_3(s) \rightleftharpoons CaO(s) + CO_2(g)$$

Although the reaction is reversible, the carbon dioxide is carried away by the upward current through the kiln and all the limestone decomposes. Calcium oxide is used to make calcium hydroxide, as a cheap alkali for treating acid soil, and in extractive metallurgy to produce a slag with the impurities (especially sand) present in metal ores.

calcium phosphate(V) A white insoluble powder, $Ca_3(PO_4)_2$; r.d. 3.14. It is found naturally in the mineral *apatite, $Ca_5(PO_4)_3(OH,F,Cl)$, and as rock phosphate. It is also the main constituent of animal bones. Calcium phosphate can be prepared by mixing solutions containing calcium ions and hydrogenphosphate ions in the presence of an alkali:

$$HPO_4^{2-} + OH^- \rightarrow PO_4^{3-} + H_2O$$

$$3Ca^{2+} + 2PO_4^{3-} \rightarrow Ca_3(PO_4)_2$$

It is used extensively as a fertilizer. The compound was formerly called *calcium orthophosphate* (*see* **phosphates**).

calcium stearate *See* **calcium octadecanoate**.

calcium sulphate A white solid compound, $CaSO_4$; r.d. 2.96; 1450°C. It occurs naturally as the mineral *anhydrite, which has a rhombic structure, transforming to a monoclinic form at 200°C. More commonly, it is found as the dihydrate, *gypsum, $CaSO_4.2H_2O$ (monoclinic; r.d. 2.32). When heated, gypsum loses water at 128°C to give the hemihydrate, $2CaSO_4.H_2O$, better known as *plaster of Paris. Calcium sulphate is sparingly soluble in water and is a cause of permanent *hardness of water. It is used in the manufacture of certain paints, ceramics, and paper. The naturally occurring forms are used in the manufacture of sulphuric acid.

Calgon Tradename for a water-softening agent. *See* **hardness of water**.

caliche A mixture of salts found in deposits between gravel beds in the Atacama and Tarapaca regions of Chile. They vary from 4 m to 15 cm thick and were formed by periodic leaching of soluble salts during wet geological epochs, followed by drying out of inland seas in dry periods. They are economically important as a source of nitrates. A typical composition is $NaNO_3$ 17.6%, NaCl 16.1%, Na_2SO_4 6.5%, $CaSO_4$ 5.5%, $MgSO_4$ 3.0%, KNO_3 1.3%, $Na_2B_4O_7$ 0.94%, $KClO_3$ 0.23%, $NaIO_3$ 0.11%, sand and gravel to 100%.

californium Symbol Cf. A radioactive metallic transuranic element belonging to the *actinoids; a.n. 98; mass number of the most stable isotope 251 (half-life about 700 years). Nine isotopes are known; californium-252 is an intense neutron source, which makes it useful in neutron *activation analysis and potentially useful as a radiation source in medicine. The element was first produced by Glenn Seaborg (1912–99) and associates in 1950.

calixarenes Compounds that have molecules with a cuplike structure (the name comes from the Greek *calix*, cup). The simplest, has four phenol molecules joined by four $-CH_2-$ groups into a ring (forming the base of the 'cup'). The four phenol hexagons point in the same direction to form a cavity that can bind substrate molecules. Interest has been shown in the potential ability of calixarene molecules to mimic enzyme action.

calmodulin A protein, consisting of 148 amino-acid residues, that acts as a receptor for calcium ions in many calcium-regulated processes in both animal and plant cells. Calmodulin mediates reactions catalysed by many enzymes.

calomel *See* **mercury(I) chloride**.

calomel half cell (calomel electrode) A type of half cell in which the electrode is mercury coated with calomel (HgCl) and the electrolyte is a solution of potassium chloride and saturated calomel. The standard electrode potential is −0.2415 volt (25°C). In the calomel half cell the reactions are

$$HgCl(s) \rightleftharpoons Hg^+(aq) + Cl^-(aq)$$

$$Hg^+(aq) + e \rightleftharpoons Hg(s)$$

The overall reaction is

$$HgCl(s) + e \rightleftharpoons Hg(s) + Cl^-(aq)$$

This is equivalent to a $Cl_2(g)|Cl^-(aq)$ half cell.

calorie The quantity of heat required to raise the temperature of 1 gram of water by 1°C (1K). The calorie, a c.g.s. unit, is now largely replaced by the *joule, an *SI unit. 1 calorie = 4.1868 joules.

Calorie (kilogram calorie; kilocalorie) 1000 calories. This unit is still in limited use in estimating the energy value of foods, but is obsolescent.

calorific value The heat per unit mass produced by complete combustion of a given substance. Calorific values are used to express the energy values of fuels; usually these are expressed in megajoules per kilogram ($MJ\,kg^{-1}$). They are also used to measure the energy content of foodstuffs; i.e. the energy produced when the food is oxidized in the body. The units here are kilojoules per gram ($kJ\,g^{-1}$), although Calories (kilocalories) are often still used in nontechnical contexts. Calorific values are measured using a *bomb calorimeter.

calorimeter Any of various devices used to measure thermal properties, such as calorific values or heats of chemical reactions. *See also* **bomb calorimeter**.

Calvin, Melvin (1911–97) US biochemist. After World War II, at the Lawrence Radiation Laboratory, Berkeley, he investigated the light-independent reactions of *photosynthesis. Using radioactive carbon-14 to label carbon dioxide he discovered the *Calvin cycle, for which he was awarded the 1961 Nobel Prize for chemistry.

Calvin cycle The metabolic pathway of the light-independent stage of *photosynthesis, which occurs in the stroma of the chloroplasts. The pathway was elucidated by Melvin *Calvin and his co-workers and involves the fixation of carbon dioxide and its subsequent reduction to carbohydrate. During the cycle, carbon dioxide combines with *ribulose bisphosphate, through the mediation of the enzyme ribulose bisphosphate carboxylase, to form an unstable six-carbon compound that breaks down to form two molecules of the three-carbon compound glycerate 3-phosphate. This is converted to glyceraldehyde 3-phosphate, which is used to regenerate ribulose bisphosphate and to produce glucose and fructose.

calx A metal oxide formed by heating an ore in air.

camphor A white crystalline cyclic ketone, $C_{10}H_{16}O$; r.d. 0.99; m.p. 179°C; b.p. 204°C. It was formerly obtained from the wood of the Formosan

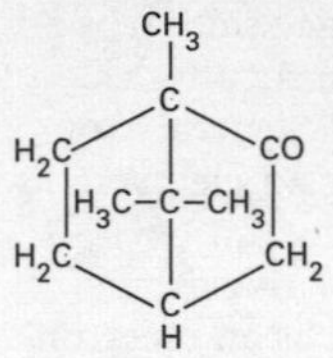

Camphor

camphor tree, but can now be synthesized. The compound has a characteristic odour associated with its use in mothballs. It is a plasticizer in celluloid.

Canada balsam A yellow-tinted resin used for mounting specimens in optical microscopy. It has similar optical properties to glass.

candela Symbol Cd. The *SI unit of luminous intensity equal to the luminous intensity in a given direction of a source that emits monochromatic radiation of frequency 540×10^{12} Hz and has a radiant intensity in that direction of 1/683 watt per steradian.

cane sugar *See* **sucrose.**

Cannizzaro, Stanislao (1826–1910) Italian chemist. He discovered the *Cannizzaro reaction in 1853. His main contribution to chemistry was his recognition of the validity of *Avogadro's hypothesis and his deduction that common gases such as H_2 and O_2 are diatomic molecules. This led to the establishment of a reliable system of atomic and molecular weights.

Cannizzaro reaction A reaction of aldehydes to give carboxylic acids and alcohols. It occurs in the presence of strong bases with aldehydes that do not have alpha hydrogen atoms. For example, benzenecarbaldehyde gives benzenecarboxylic acid and benzyl alcohol:

$$2C_6H_5CHO \rightarrow C_6H_5COOH + C_6H_5CH_2OH$$

Aldehydes that have alpha hydrogen atoms undergo the *aldol reaction instead. The Cannizzaro reaction is an example of a *disproportionation. It was discovered by Stanislao *Cannizzaro.

canonical form One of the possible structures of a molecule that together form a *resonance hybrid.

capillary A tube of small diameter.

capric acid *See* **decanoic acid.**

caproic acid *See* **hexanoic acid.**

caprolactam **(6-hexanelactam)** A white crystalline substance, $C_6H_{11}NO$; r.d. 1.02; m.p. 69–71°C; b.p. 139°C. It is a *lactam containing the –NH.CO– group with five CH_2 groups making up the rest of the seven-membered ring. Caprolactam is used in making *nylon.

caprylic acid *See* **octanoic acid.**

capture Any of various processes in which a system of particles absorbs an extra particle. There are several examples in atomic and nuclear physics. For instance, a positive ion may capture an electron to give a neutral atom or molecule. Similarly, a neutral atom or molecule capturing an electron becomes a negative ion. An atomic nucleus may capture a neutron to produce a different (often unstable) nucleus. Another type of nuclear capture is the process in which the nucleus of an atom absorbs an electron from the innermost orbit (the K shell) to transform into a different nucleus. In this process (called *K capture*) the atom is left in an excited state and generally decays by emission of an X-ray photon.

Radiative capture is any such process in which the capture results in an excited state that decays by emission of photons. A common example is neutron capture to yield an excited nucleus, which decays by emission of a gamma ray.

carat 1. A measure of fineness (purity) of gold. Pure gold is described as 24-carat gold. 14-carat gold contains 14 parts in 24 of gold, the remainder usually being copper. **2.** A unit of mass equal to 0.200 gram, used to measure the masses of diamonds and other gemstones.

carbamide *See* **urea**.

carbanion An organic ion with a negative charge on a carbon atom; i.e. an ion of the type R_3C^-. Carbanions are intermediates in certain types of organic reaction (e.g. the *aldol reaction).

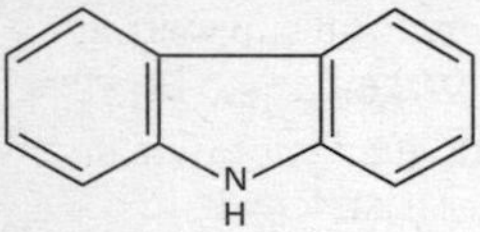

Carbazole

carbazole A white crystalline compound found with anthracene, $C_{12}H_9N$; m.p. 238°C; b.p. 335°C. It is used in the manufacture of dyestuffs.

carbene A species of the type R_2C:, in which the carbon atom has two electrons that do not form bonds. *Methylene*, $:CH_2$, is the simplest example. Carbenes are highly reactive and exist only as transient intermediates in certain organic reactions. They attack double bonds to give cyclopropane derivatives. They also cause insertion reactions, in which the carbene group is inserted between the carbon and hydrogen atoms of a C–H bond:

$$C{-}H + :CR_2 \rightarrow C{-}CR_2{-}H$$

carbenium ion *See* **carbonium ion**.

carbide Any of various compounds of carbon with metals or other more electropositive elements. True carbides contain the ion C^{4-} as in Al_4C_3. These are saltlike compounds giving methane on hydrolysis, and were formerly called *methanides*. Compounds containing the ion C_2^{2-} are also saltlike and are known as *dicarbides*. They yield ethyne (acetylene) on hydrolysis and were formerly called *acetylides*. The above types of compound are ionic but have partially covalent bond character, but boron and silicon form true covalent carbides, with giant molecular structures. In addition, the transition metals form a range of interstitial carbides in which the carbon atoms occupy interstitial positions in the metal lattice. These substances are generally hard materials with metallic conductivity. Some transition metals (e.g. Cr, Mn, Fe, Co, and Ni) have atomic radii that are too small to allow individual carbon atoms in the interstitial holes. These form carbides in which the metal lattice is distorted and chains of carbon atoms exist (e.g. Cr_3C_2, Fe_3C). Such compounds are intermediate in character between interstitial carbides and ionic carbides. They give mixtures of hydrocarbons on hydrolysis with water or acids.

carbocation *See* **carbonium ion**.

carbocyclic *See* **cyclic**.

carbohydrate One of a group of organic compounds based on the general formula $C_x(H_2O)_y$. The simplest carbohydrates are the *sugars (saccharides), including glucose and sucrose. *Polysaccharides are carbohydrates of much greater molecular weight and complexity; examples are starch, glycogen, and cellulose. Carbohydrates perform many vital roles in living organisms. Sugars, notably glucose, and their derivatives are essential intermediates in the conversion of food to energy. Starch and other polysaccharides serve as energy stores in plants. Cellulose, lignin, and others form the supporting cell walls and woody tissue of plants. Chitin is a structural polysaccharide found in the body shells of many invertebrate animals.

carbolic acid *See* **phenol**.

carbon Symbol C. A nonmetallic element belonging to *group 14 (formerly IVB) of the periodic table; a.n. 6; r.a.m. 12.011; m.p. ~3550°C; b.p. ~4827°C. Carbon has three main allotropic forms (*see* **allotropy**).

*Diamond (r.d. 3.52) occurs naturally and can be produced synthetically. It is extremely hard and has highly refractive crystals. The hardness of diamond results from the covalent crystal structure, in which each carbon atom is linked by covalent bonds to four others situated at the corners of a tetrahedron. The C–C bond length is 0.154 nm and the bond angle is 109.5°.

Graphite (r.d. 2.25) is a soft black slippery substance (sometimes called *black lead* or *plumbago*). It occurs naturally and can also be made by the *Acheson process. In graphite the carbon atoms are arranged in layers, in which each carbon atom is surrounded by three others to which it is bound by single or double bonds. The layers are held together by much weaker van der Waals' forces. The carbon–carbon bond length in the layers is 0.142 nm and the layers are 0.34 nm apart. Graphite is a good conductor of heat and electricity. It has a variety of uses including electrical contacts, high-temperature equipment, and as a solid lubricant. Graphite mixed with clay is the 'lead' in pencils (hence its alternative name). The third crystalline allotrope is fullerite (*see* **buckminsterfullerene**). There are also several amorphous forms of carbon, such as *carbon black and *charcoal.

There are two stable isotopes of carbon (proton numbers 12 and 13) and four radioactive ones (10, 11, 14, 15). Carbon-14 is used in *carbon dating.

Carbon forms a large number of compounds because of its unique ability to form stable covalent bonds with other carbon atoms and also with hydrogen, oxygen, nitrogen, and sulphur atoms, resulting in the formation of a variety of compounds containing chains and rings of carbon atoms.

carbon assimilation The incorporation of carbon from atmospheric carbon dioxide into organic molecules. This process occurs during *photosynthesis. *See* **carbon cycle**.

carbonate A salt of carbonic acid containing the carbonate ion, CO_3^{2-}. The free ion has a plane triangular structure. Metal carbonates may be ionic or may contain covalent metal–carbonate bonds (complex carbonates) via one or two oxygen atoms. The carbonates of the alkali metals are all soluble but other carbonates are insoluble; they all react with mineral acids to release carbon dioxide.

carbonate minerals A group of common rock-forming minerals containing the anion CO_3^{2-} as the fundamental unit in their structure. The most important carbonate minerals are *calcite, *dolomite, and *magnesite. *See also* **aragonite**.

carbonation The solution of carbon dioxide in a liquid under pressure.

carbon bisulphide *See* **carbon disulphide**.

carbon black A fine carbon powder made by burning hydrocarbons in insufficient air. It is used as a pigment and a filler (e.g. for rubber).

carbon cycle 1. One of the major cycles of chemical elements in the environment. Carbon (as carbon dioxide) is taken up from the atmosphere and incorporated into the tissues of plants in *photosynthesis. It may then pass into the bodies of animals as the plants are eaten. During the respiration of plants, animals, and organisms that bring about decomposition, carbon dioxide is returned to the atmosphere. The combustion of fossil fuels (e.g. coal and peat) also releases carbon dioxide into the atmosphere. See illustration.

2. ***(in physics)*** A series of nuclear reactions in which four hydrogen nuclei combine to form a helium nucleus with the liberation of energy, two positrons, and two neutrinos. The process is believed to be the source of energy in many stars and to take place in six stages. In this series carbon-12 acts as if it were a catalyst, being reformed at the end of the series:

$${}^{12}_{6}C + {}^{1}_{1}H \rightarrow {}^{13}_{7}N + \gamma$$

$${}^{13}_{7}N \rightarrow {}^{13}_{6}C + e^{+} + \nu_e$$

$${}^{13}_{6}C + {}^{1}_{1}H \rightarrow {}^{14}_{7}N + \gamma$$

$$^{14}_{7}N + ^{1}_{1}H \rightarrow ^{15}_{8}O + \gamma$$

$$^{15}_{8}O \rightarrow ^{15}_{7}N + e^{+} + \nu_e$$

$$^{15}_{7}N + ^{1}_{1}H \rightarrow ^{12}_{6}C + ^{4}_{2}He.$$

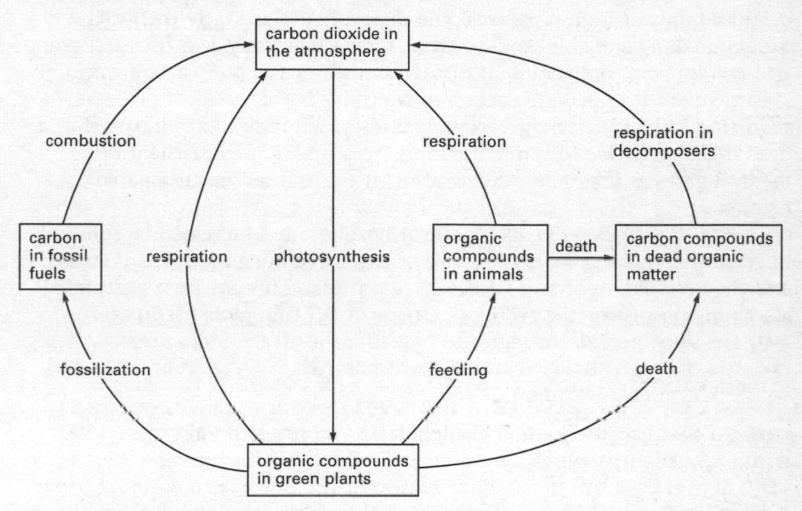

The carbon cycle in nature

carbon dating **(radiocarbon dating)** A method of estimating the ages of archaeological specimens of biological origin. As a result of cosmic radiation a small number of atmospheric nitrogen nuclei are continuously being transformed by neutron bombardment into radioactive nuclei of carbon–14:

$$^{14}_{7}N + n \rightarrow ^{14}_{6}C + p$$

Some of these radiocarbon atoms find their way into living trees and other plants in the form of carbon dioxide, as a result of photosynthesis. When the tree is cut down photosynthesis stops and the ratio of radiocarbon atoms to stable carbon atoms begins to fall as the radiocarbon decays. The ratio $^{14}C/^{12}C$ in the specimen can be measured and enables the time that has elapsed since the tree was cut down to be calculated. The method has been shown to give consistent results for specimens up to some 40 000 years old, though its accuracy depends upon assumptions concerning the past intensity of the cosmic radiation. The technique was developed by Willard F. Libby (1908–80) and his coworkers in 1946–47.

carbon dioxide A colourless odourless gas, CO_2, soluble in water, ethanol, and acetone; d. 1.977 g dm^{-3} (0°C); m.p. −56.6°C; b.p. −78.5°C. It occurs in the atmosphere (0.04% by volume) but has a short residence time in this phase as it is both consumed by plants during *photosynthesis and produced by respiration and by combustion. It is readily prepared in the laboratory by the action of dilute acids on metal carbonates or of heat on heavy-metal carbonates. Carbon dioxide is a by-product from the manufacture of lime and from fermentation processes.

Carbon dioxide has a small liquid range and liquid carbon dioxide is produced only at high pressures. The molecule CO_2 is linear with each oxygen making a double bond to the carbon. Chemically, it is unreactive and will not support combustion. It dissolves in water to give *carbonic acid.

Large quantities of solid carbon dioxide (*dry ice*) are used in processes requiring large-scale refrigeration. It is also used in fire extinguishers as a desirable alternative to water for most fires, and as a constituent of medical gases as it promotes exhalation. It is also used in carbonated drinks.

The level of carbon dioxide in the atmosphere has increased by some 12% in the last 100 years, mainly because of extensive burning of fossil fuels and the destruction of large areas of rain forest. This has been postulated as the main cause of the average increase of 0.5°C in global temperatures over the same period, through the *greenhouse effect. Steps are now being taken to prevent further increases in atmospheric CO_2 concentration and subsequent global warming.

carbon disulphide **(carbon bisulphide)** A colourless highly refractive liquid, CS_2, slightly soluble in water and soluble in ethanol and ether; r.d. 1.261; m.p. −110°C; b.p. 46.3°C. Pure carbon disulphide has an ethereal odour but the commercial product is contaminated with a variety of other sulphur compounds and has a most unpleasant smell. It was previously manufactured by heating a mixture of wood, sulphur, and charcoal; modern processes use natural gas and sulphur. Carbon disulphide is an excellent solvent for oils, waxes, rubber, sulphur, and phosphorus, but its use is decreasing because of its high toxicity and its flammability. It is used for the preparation of xanthates in the manufacture of viscose yarns.

carbon fibres Fibres of carbon in which the carbon has an oriented crystal structure. Carbon fibres are made by heating textile fibres and are used in strong composite materials for use at high temperatures.

carbonic acid A dibasic acid, H_2CO_3, formed in solution when carbon dioxide is dissolved in water:

$$CO_2(aq) + H_2O(l) \rightleftharpoons H_2CO_3(aq)$$

The acid is in equilibrium with dissolved carbon dioxide, and also dissociates as follows:

$$H_2CO_3 \rightleftharpoons H^+ + HCO_3^-$$

$$K_a = 4.5 \times 10^{-7}\ \text{mol dm}^{-3}$$

$$HCO_3^- \rightleftharpoons CO_3^{2-} + H^+$$

$$K_a = 4.8 \times 10^{-11}\ \text{mol dm}^{-3}$$

The pure acid cannot be isolated, although it can be produced in ether solution at −30°C. Carbonic acid gives rise to two series of salts: the *carbonates and the *hydrogencarbonates.

carbonium ion **(carbenium ion)** An organic ion with a positive charge on a carbon atom; i.e. an ion of the type R_3C^+. Carbonium ions are intermediates in certain types of organic reaction (e.g. *Williamson's synthesis). Certain fairly stable carbonium ions can be formed (*carbocations*). Carbonium ions can be prepared from an alkyl fluoride and a *superacid, such as antimony pentafluoride (SbF_5), at low temperature. Carbonium atoms always have a strong affinity for such nucleophiles as water.

carbonize **(carburize)** To change an organic compound into carbon by heating, or to coat something with carbon in this way.

C≡O	carbon monoxide
O=C=O	carbon dioxide
O=C=C=C=O	tricarbon dioxide (carbon suboxide)

Oxides of carbon

carbon monoxide A colourless odourless gas, CO, sparingly soluble in water and soluble in ethanol and benzene; d. 1.25 g dm^{-3} (0°C); m.p. −199°C; b.p. −191.5°C. It is flammable and highly toxic. In the laboratory it can be made by the dehydration of methanoic acid (formic acid) using concentrated sulphuric acid. Industrially it is produced by the oxidation of natural gas (methane) or (formerly) by the water-gas reaction. It is formed by the incomplete combustion of carbon and is present in car-exhaust gases.

It is a neutral oxide, which burns in air to give carbon dioxide, and is a good reducing agent, used in a number of metallurgical processes. It has the interesting chemical property of forming a range of transition metal carbonyls, e.g. $Ni(CO)_4$. Carbon monoxide is able to use vacant *p*-orbitals in bonding with metals; the stabilization of low oxidation states, including the zero state, is a consequence of this. This also accounts for its toxicity, which is due to the binding of the CO to the iron in haemoglobin, thereby blocking the uptake of oxygen.

carbon suboxide *See* **tricarbon dioxide.**

carbon tetrachloride *See* **tetrachloromethane.**

carbonyl chloride **(phosgene)** A colourless gas, $COCl_2$, with an odour of freshly cut hay. It is used in organic chemistry as a chlorinating agent, and was formerly used as a war gas.

carbonyl compound A compound containing the carbonyl group >C=O. Aldehydes, ketones, and carboxylic acids are examples of organic carbonyl compounds. Inorganic carbonyls are complexes in which carbon monoxide has coordinated to a metal atom or ion, as in *nickel carbonyl, $Ni(CO)_4$. *See also* **ligand.**

carbonyl group The group >C=O, found in aldehydes, ketones, carboxylic acids, amides, etc., and in inorganic carbonyl complexes (*see* **carbonyl compound**).

carborundum *See* **silicon carbide.**

carboxyhaemoglobin The highly stable product formed when *haemoglobin combines with carbon monoxide. Carbon monoxide competes with oxygen for haemoglobin, with which it binds strongly: the affinity of haemoglobin for carbon monoxide is 250 times greater than that for oxygen. This reduces the availability of haemoglobin for combination with (and transport of) oxygen and accounts for the toxic effects of carbon monoxide on the respiratory system.

carboxylate An anion formed from a *carboxylic acid. For example, ethanoic acid gives rise to the ethanoate ion, CH_3COO^-.

carboxyl group The organic group –COOH, present in *carboxylic acids.

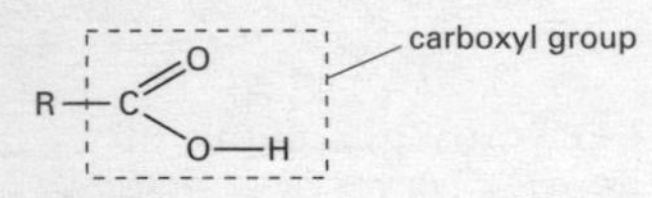

Carboxylic acid structure

carboxylic acids Organic compounds containing the group –CO.OH (the *carboxyl group*; i.e. a carbonyl group attached to a hydroxyl group). In systematic chemical nomenclature carboxylic-acid names end in the suffix *-oic*, e.g. ethanoic acid, CH_3COOH. They are generally weak acids. Many long-chain carboxylic acids occur naturally as esters in fats and oils and are therefore also known as *fatty acids. *See also* **glyceride.**

carburize *See* **carbonize.**

carbylamine reaction *See* **isocyanide test.**

carbyne A transient species of the type R–C≡, with three nonbonding electrons on the carbon atom. Formerly, carbynes were called *methylidynes.*

carcinogen Any agent that produces cancer, e.g. tobacco smoke, certain industrial chemicals, and ionizing radiation (such as X-rays and ultraviolet rays).

Carius method A method of determining the amount of sulphur and halogens in an organic compound, by heating the compound in a sealed tube with silver nitrate in concentrated nitric acid. The compound is decomposed and silver sulphide and halides are precipitated, separated, and weighed.

carnallite A mineral consisting of a hydrated mixed chloride of potassium and magnesium, $KCl.MgCl_2.6H_2O$.

carnauba wax A natural wax obtained from the leaves of the copaiba palm of South America. It is extremely hard and brittle and is used in varnishes and to add hardness and lustre to other waxes in polishes.

Carnot, Nicolas Léonard Sadi (1796–1832) French physicist, who first worked as a military engineer. He then turned to scientific research and in 1824 published his analysis of the efficiency of heat engines. The key to this analysis is the thermodynamic *Carnot cycle and the eventual introduction of the idea of *entropy in thermodynamics. He died at an early age from cholera.

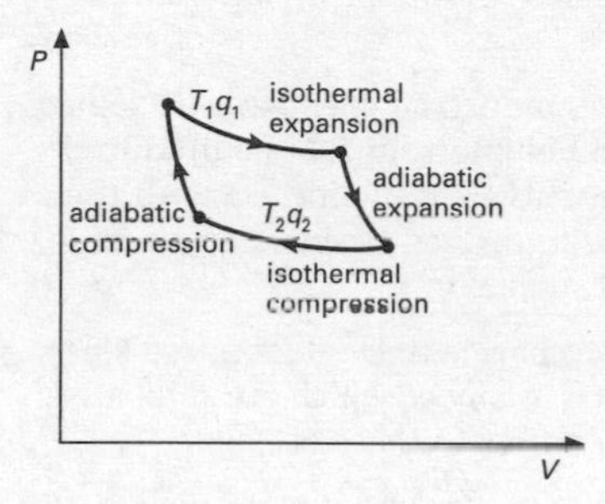

Carnot cycle

Carnot cycle The most efficient cycle of operations for a reversible *heat engine. Published in 1824 by Nicolas *Carnot, it consists of four operations on the working substance in the engine (see illustration):
a. Isothermal expansion at thermodynamic temperature T_1 with heat q_1 taken in.
b. Adiabatic expansion with a fall of temperature to T_2.
c. Isothermal compression at temperature T_2 with heat q_2 given out.
d. Adiabatic compression with a rise of temperature back to T_1.
According to the *Carnot principle*, the efficiency of any reversible heat engine depends only on the temperature range through which it works, rather than the properties of the working substances. In any reversible engine, the efficiency (η) is the ratio of the work done (W) to the heat input (q_1), i.e. $\eta = W/q_1$. As, according to the first law of *thermodynamics, $W = q_1 - q_2$, it follows that $\eta = (q_1 - q_2)/q_1$. For the Kelvin temperature scale, $q_1/q_2 = T_1/T_2$ and $\eta = (T_1 - T_2)/T_1$. For maximum efficiency T_1 should be as high as possible and T_2 as low as possible.

carnotite A radioactive mineral consisting of hydrated uranium potassium vanadate, $K_2(UO_2)_2(VO_4)_2.nH_2O$. It varies in colour from bright yellow to lemon- or greenish-yellow. It is a source of uranium, radium, and vanadium. The chief occurrences are in the Colorado Plateau, USA; Radium Hill, Australia; and Katanga, Congo.

Carnot principle *See* **Carnot cycle.**

Caro's acid *See* **peroxosulphuric(VI) acid.**

carotene A member of a class of *carotenoid pigments. Examples are β-carotene and lycopene, which colour carrot roots and ripe tomato fruits respectively. α- and β-carotene yield vitamin A when they are broken down during animal digestion.

carotenoid Any of a group of yellow, orange, red, or brown plant pigments chemically related to terpenes. Carotenoids are responsible for the characteristic colour of many plant organs, such as ripe tomatoes, carrots, and autumn leaves. They also absorb light energy and pass this on to chlorophyll molecules in the light-dependent reactions of photosynthesis.

Carothers, Wallace Hume (1896–1937) US industrial chemist, who joined the Du Pont company where he worked on polymers. In 1931 he produced *neoprene, a synthetic rubber. His greatest success came in 1935 with the discovery of the polyamide that came to be known as *nylon. Carothers, who suffered from depression, committed suicide.

carrageenan A naturally occurring polysaccharide isolated from red algae. The polymer is composed of D-galactose units, many of which are sulphated. *K-carrageenan* is a gelling agent with uses similar to those of *agar.

carrier gas The gas that carries the sample in *gas chromatography.

carrier molecule 1. A molecule that plays a role in transporting electrons through the *electron transport chain. Carrier molecules are usually proteins bound to a nonprotein group; they can undergo oxidation and reduction relatively easily, thus allowing electrons to flow through the system. **2.** A lipid-soluble molecule that can bind to lipid-insoluble molecules and transport them across membranes. Carrier molecules have specific sites that interact with the molecules they transport. Several different molecules may compete for transport by the same carrier.

Carr–Purcell–Meiboom–Gill sequence *See* **CPMG sequence**.

CARS (coherent anti-Stokes Raman spectroscopy) A form of Raman spectroscopy (*see* **Raman effect**) enabling the intensity of Raman transitions to be increased. In this technique two laser beams with different frequencies pass through a sample producing electromagnetic radiation of several frequencies. It is possible to adjust the frequency of the lasers so that one of the frequencies corresponds to that of a Stokes line from the sample and the coherent emission has the frequency of the anti-Stokes line with a high intensity. CARS enables Raman spectra to be obtained even in the presence of other radiation. One application of CARS is to obtain Raman spectra from bodies in flames. Using this technique temperatures in different parts of the flame can be estimated from the intensity of the transitions.

cascade liquefier An apparatus for liquefying a gas of low *critical temperature. Another gas, already below its critical temperature, is liquified and evaporated at a reduced pressure in order to cool the first gas to below its critical temperature. In practice a series of steps is often used, each step enabling the critical temperature of the next gas to be reached.

cascade process Any process that takes place in a number of steps, usually because the single step is too inefficient to produce the desired result. For example, in various uranium-enrichment processes the separation of the desired isotope is only poorly achieved in a single stage; to achieve better separation the process has to be repeated a number of times, in a series, with the enriched fraction of one stage being fed to the succeeding stage for further enrichment. Another example of cascade process is that operating in a *cascade liquefier.

case hardening The hardening of the surface layer of steel, used for tools and certain mechanical components. The commonest method is to carburize the surface layer by heating the metal in a hydrocarbon or by dipping the red hot metal into molten sodium cyanide. Diffusion of nitrogen into the surface layer to form nitrides is also used.

casein One of a group of phosphate-containing proteins (phosphoproteins) found in milk. Caseins are easily digested by the enzymes of young mammals and represent a major source of phosphorus.

cassiterite A yellow, brown, or black form of tin(IV) oxide, SnO_2, that forms tetragonal, often twinned, crystals; the principal ore of tin. It occurs in hydrothermal veins and metasomatic deposits associated with acid igneous rocks and in alluvial (placer) deposits. The chief producers are Malaysia, Indonesia, Congo, and Nigeria.

cast iron A group of iron alloys containing 1.8 to 4.5% of carbon. It is usually cast into specific shapes ready for machining, heat treatment, or assembly. It is sometimes produced direct from the *blast furnace or it may be made from remelted *pig iron.

Castner–Kellner cell An electrolytic cell used industrially for the production of *sodium hydroxide. The usually iron cell is filled with brine (sodium chloride solution) and employs liquid mercury as the cathode. The sodium liberated there forms an amalgam with the mercury, which is run off and reacted with water to give sodium hydroxide (and hydrogen); the mercury is then re-used. Chlorine gas produced at the anode is another valuable by-product.

castor oil A pale-coloured oil extracted from the castor-oil plant. It contains a mixture of glycerides of fatty acids, the predominant acid being ricinoleic acid, $C_{17}H_{32}(OH)COOH$. It is used as a *drying oil in paints and varnishes and medically as a laxative.

catabolism The metabolic breakdown of large molecules in living organisms to smaller ones, with the release of energy. Respiration is an example of a catabolic series of reactions. *See* **metabolism**. *Compare* **anabolism**.

catalysis The process of changing the rate of a chemical reaction by use of a *catalyst.

catalyst A substance that increases the rate of a chemical reaction without itself undergoing any permanent chemical change. Catalysts that have the same phase as the reactants are *homogeneous catalysts* (e.g. *enzymes in biochemical reactions or transition-metal complexes used in the liquid phase for catalysing organic reactions). Those that have a different phase are *hetereogeneous catalysts* (e.g. metals or oxides used in many industrial gas reactions). The catalyst provides an alternative pathway by which the reaction can proceed, in which the activation energy is lower. It thus increases the rate at which the reaction comes to equilibrium, although it does not alter the position of the equilibrium. The catalyst itself takes part in the reaction and consequently may undergo physical change (e.g. conversion into powder). In certain circumstances, very small quantities of catalyst can speed up reactions. Most catalysts are also highly specific in the type of reaction they catalyse, particularly enzymes in biochemical reactions. Generally, the term is used for a substance that increases reaction rate (a *positive catalyst*). Some reactions can be slowed down by *negative catalysts* (*see* **inhibition**).

catalytic converter A device used in the exhaust systems of motor vehicles to reduce atmospheric pollution. The three main pollutants produced by petrol engines are: unburnt hydrocarbons, carbon monoxide produced by incomplete combustion of hydrocarbons, and nitrogen oxides produced by nitrogen in the air reacting with oxygen at high engine temperatures. Hydrocarbons and carbon monoxide can be controlled by a higher combustion temperature and a weaker mixture. However, the higher temperature and greater availability of oxygen arising from these measures encourage formation of nitrogen oxides. The use of three-way catalytic converters solves this problem by using platinum and palladium catalysts to oxidize the hydrocarbons and the CO and rhodium catalysts to reduce the nitrogen oxides back to nitrogen. These three-way catalysts require that the air–fuel ratio is strictly stochiometric. Some catalytic converters promote oxidation reactions only, leaving the nitrogen oxides unchanged. Three-way converters can reduce hydrocarbons and CO emissions by some 85%, at the same time reducing nitrogen oxides by 62%.

catalytic cracking *See* **cracking**.

cataphoresis *See* **electrophoresis**.

catechol *See* **1,2-dihydroxybenzene**.

catecholamine Any of a class of amines that possess a catechol ($C_6H_4(OH)_2$) ring. Including *dopamine, *adrenaline, and *noradrenaline, they function as neurotransmitters and/or hormones.

catenation The formation of chains of atoms in chemical compounds.

cathetometer A telescope or microscope fitted with crosswires in the eyepiece and mounted so that it can slide along a graduated scale. Cathetometers are used for accurate measurement of lengths without mechanical contact. The microscope type is often called a *travelling microscope.*

cathode A negative electrode. In *electrolysis cations are attracted to the cathode. In vacuum electronic devices electrons are emitted by the cathode and flow to the *anode. It is therefore from the cathode that electrons flow into these devices. However, in a primary or secondary cell the cathode is the electrode that spontaneously becomes negative during discharge, and from which therefore electrons emerge.

cathodic protection *See* **sacrificial protection**.

cation A positively charged ion, i.e. an ion that is attracted to the cathode in *electrolysis. *Compare* **anion**.

cationic detergent *See* **detergent**.

cationic dye *See* **dyes**.

cationic resin *See* **ion exchange**.

caustic Describing a substance that is strongly alkaline (e.g. caustic soda).

caustic potash *See* **potassium hydroxide**.

caustic soda *See* **sodium hydroxide**.

Cavendish, Henry (1731–1810) British chemist and physicist, born in France. Although untrained, his inheritance from his grandfather, the Duke of Devonshire, enabled him to live as a recluse and study science. In his experiments with gases (1766), he correctly distinguished between hydrogen and carbon dioxide. In 1784 he showed that sparking mixtures of hydrogen and oxygen produced pure water, showing that water is not an element. Also in sparking experiments with air, Cavendish recognized a small percentage unchanged (later identified as noble gases). Cavendish also did significant work in physics.

c.c.p. Cubic close packing. *See* **close packing**.

CD spectrum **(circular dichroism spectrum)** The spectrum obtained by plotting the variable $I_R - I_L$ against frequency of the incident electromagnetic radiation, where I_R and I_L are the absorption intensities for right- and left-circularly polarized light, respectively. One application of CD spectroscopy is to determine the configurations of complexes of transition metals. If the CD spectra of similar transition metal complexes (including similarity of geometry) are taken, the features of their CD spectra are also similar.

celestine A mineral form of strontium sulphate, $SrSO_4$.

cell 1. A system in which two electrodes are in contact with an electrolyte. The electrodes are metal or carbon plates or rods or, in some cases, liquid metals (e.g. mercury). In an *electrolytic cell a current from an outside source is passed through the electrolyte to produce chemical change (*see* **electrolysis**). In a *voltaic cell, spontaneous reactions between the electrodes and electrolyte(s) produce a potential difference between the two electrodes.

Voltaic cells can be regarded as made up of two *half cells, each composed of an electrode in contact with an electrolyte. For instance, a zinc rod dipped in zinc sulphate solution is a $Zn|Zn^{2+}$ half cell. In such a system zinc atoms dissolve as zinc ions, leaving a negative charge on the electrode

$$Zn(s) \rightarrow Zn^{2+}(aq) + 2e$$

The solution of zinc continues until the charge build-up is sufficient to prevent further ionization. There is then a potential difference between the zinc rod and its solution. This cannot be measured directly, since measurement would involve making contact with the electrolyte, thereby introducing another half cell (*see* **electrode potential**). A rod of copper in copper sulphate solution comprises another half cell. In this case the spontaneous reaction is one in which copper ions in solution take electrons from the electrode and are deposited on the electrode as copper atoms. In this case, the copper acquires a positive charge.

The two half cells can be connected by using a porous pot for the liquid junction (as in the *Daniell cell) or by using a salt bridge. The resulting cell can then supply current if the electrodes are connected through an external circuit. The cell is written

$$Zn(s)|Zn^{2+}(aq)|Cu^{2+}(aq)|Cu$$

$$E = 1.10\,V$$

Here, E is the e.m.f. of the cell equal to the potential of the right-hand electrode minus that of the left-hand electrode for zero current. Note that 'right' and 'left' refer to the cell as written. Thus, the cell could be written

$$Cu(s)|Cu^{2+}(aq)|Zn^{2+}(aq)|Zn(s)$$

$$E = -1.10\,V$$

The overall reaction for the cell is

$$Zn(s) + Cu^{2+}(aq) \rightarrow Cu(s) + Zn^{2+}(aq)$$

This is the direction in which the cell reaction occurs for a positive e.m.f.

The cell above is a simple example of a *chemical cell*; i.e. one in which the e.m.f. is produced by a chemical difference. *Concentration cells* are cells in which the e.m.f. is caused by a difference of concentration. This may be a difference in concentration of the electrolyte in the two half cells. Alternatively, it may be an electrode concentration difference (e.g. different concentrations of metal in an amalgam, or different pressures of gas in two gas electrodes). Cells are also classified into cells *without transport* (having a single electrolyte) and *with transport* (having a liquid junction across which ions are transferred). Various types of voltaic cell exist, used as sources of current, standards of potential, and experimental set-ups for

studying electrochemical reactions. *See also* **dry cell**; **primary cell**; **secondary cell**.
2. *See* **Kerr effect** (for Kerr cell).

cellophane Reconstituted cellulose in the form of a thin transparent sheet, made by extruding a viscous cellulose xanthate solution through a fine slit into a bath of acid (*see* **rayon**). It is commonly used as a wrapping material, especially for foodstuffs, but is being replaced by polypropene because of its flammability.

cellular plastics Solid synthetic materials having an open structure. A common example is rigid *polystyrene foam used in insulation and packaging.

celluloid A transparent highly flammable substance made from cellulose nitrate with a camphor plasticizer. It was formerly widely used as a thermoplastic material, especially for film (a use now discontinued owing to the flammability of celluloid).

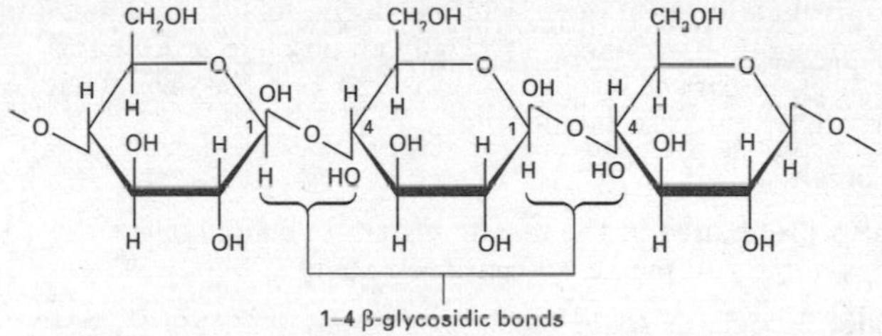

1–4 β-glycosidic bonds

Cellulose

cellulose A polysaccharide that consists of a long unbranched chain of glucose units. It is the main constituent of the cell walls of all plants, many algae, and some fungi and is responsible for providing the rigidity of the cell wall. It is an important constituent of dietary fibre. The fibrous nature of extracted cellulose has led to its use in the textile industry for the production of cotton, artificial silk, etc.

cellulose acetate *See* **cellulose ethanoate**.

cellulose ethanoate (cellulose acetate) A compound prepared by treating cellulose (cotton linters or wood pulp) with a mixture of ethanoic anhydride, ethanoic acid, and concentrated sulphuric acid. Cellulose in the cotton is ethanoylated and when the resulting solution is treated with water, cellulose ethanoate forms as a flocculent white mass. It is used in lacquers, nonshatterable glass, varnishes, and as a fibre (*see also* **rayon**).

cellulose nitrate A highly flammable material made by treating cellulose (wood pulp) with concentrated nitric acid. Despite the alternative name *nitrocellulose*, the compound is in fact an ester (containing $CONO_2$ groups), not a nitro compound (which would contain C–NO_2). It is used in explosives (as *guncotton*) and celluloid.

Celsius scale A *temperature scale in which the fixed points are the temperatures at standard pressure of ice in equilibrium with water (0°C) and water in equilibrium with steam (100°C). The scale, between these two temperatures, is divided in 100 degrees. The degree Celsius (°C) is equal in magnitude to the *kelvin. This scale was formerly known as the *centigrade scale*; the name was officially changed in 1948 to avoid confusion with a hundredth part of a grade. It is named after the Swedish astronomer Anders Celsius (1701–44), who devised the inverted form of this scale (ice point 100°, steam point 0°) in 1742.

cement Any of various substances used for bonding or setting to a hard material. Portland cement is a mixture of calcium silicates and aluminates made by heating limestone ($CaCO_3$) with clay (containing aluminosilicates) in a kiln. The product is ground to a fine powder. When mixed with water it sets in a few hours and then hardens over a longer period of time due to the formation of hydrated aluminates and silicates.

cementation Any metallurgical process in which the surface of a metal is impregnated by some other substance, especially an obsolete process for making steel by heating bars of wrought iron to red heat for several days in a bed of charcoal. *See also* **case hardening**.

cementite *See* **steel**.

centi- Symbol c. A prefix used in the metric system to denote one hundredth. For example, 0.01 metre = 1 centimetre (cm).

centigrade scale *See* **Celsius scale**.

centrifugal pump *See* **pump**.

centrifuge A device in which solid or liquid particles of different densities are separated by rotating them in a tube in a horizontal circle. The denser particles tend to move along the length of the tube to a greater radius of rotation, displacing the lighter particles to the other end.

ceramics Inorganic materials, such as pottery, enamels, and refractories. Ceramics are metal silicates, oxides, nitrides, etc.

cerebroside Any one of a class of *glycolipids in which a single sugar unit is bound to a sphingolipid (*see* **phospholipid**). The most common cerebrosides are *galactocerebrosides*, containing the sugar group galactose; they are found in the plasma membranes of neural tissue and are abundant in the myelin sheaths of neurones.

cerium Symbol Ce. A silvery metallic element belonging to the *lanthanoids; a.n. 58; r.a.m. 140.12; r.d. 6.77 (20°C); m.p. 799°C; b.p. 3426°C. It occurs in allanite, bastnasite, cerite, and monazite. Four isotopes occur naturally: cerium-136, -138, -140, and -142; fifteen radioisotopes have been identified. Cerium is used in mischmetal, a rare-earth metal containing 25%

cerium, for use in lighter flints. The oxide is used in the glass industry. It was discovered by Martin Klaproth (1743–1817) in 1803.

cermet A composite material consisting of a ceramic in combination with a sintered metal, used when a high resistance to temperature, corrosion, and abrasion is needed.

cerussite An ore of lead consisting of lead carbonate, $PbCO_3$. It is usually of secondary origin, formed by the weathering of *galena. Pure cerussite is white but the mineral may be grey due to the presence of impurities. It forms well-shaped orthorhombic crystals. It occurs in the USA, Spain, and SW Africa.

cetane *See* **hexadecane**.

cetane number A number that provides a measure of the ignition characteristics of a Diesel fuel when it is burnt in a standard Diesel engine. It is the percentage of cetane (hexadecane) in a mixture of cetane and 1-methylnaphthalene that has the same ignition characteristics as the fuel being tested. *Compare* **octane number**.

CFC *See* **chlorofluorocarbon**.

c.g.s. units A system of *units based on the centimetre, gram, and second. Derived from the metric system, it was badly adapted to use with thermal quantities (based on the inconsistently defined *calorie) and with electrical quantities (in which two systems, based respectively on unit permittivity and unit permeability of free space, were used). For scientific purposes c.g.s. units have now been replaced by *SI units.

chain A line of atoms of the same type in a molecule. In a *straight chain* the atoms are attached only to single atoms, not to groups. Propane, for instance, is a straight-chain alkane, $CH_3CH_2CH_3$, with a chain of three carbon atoms. A *branched chain* is one in which there are side groups attached to the chain. Thus, 3-ethyloctane, $CH_3CH_2CH(C_2H_5)C_5H_{11}$, is a branched-chain alkane in which there is a *side chain* (C_2H_5) attached to the third carbon atom. A *closed chain* is a *ring of atoms in a molecule; otherwise the molecule has an *open chain*.

Chain, Sir Ernst Boris (1906–79) German-born British biochemist, who began his research career at Cambridge University in 1933. Two years later he joined *Florey at Oxford, where they isolated and purified *penicillin. They also developed a method of producing the drug in large quantities and carried out its first clinical trials. The two men shared the 1945 Nobel Prize for physiology or medicine with penicillin's discoverer, Alexander *Fleming.

chain reaction A reaction that is self-sustaining as a result of the products of one step initiating a subsequent step. Chemical chain reactions usually involve free radicals as intermediates. An example is the reaction of chlorine with hydrogen initiated by ultraviolet radiation. A chlorine molecule is first split into atoms:

$Cl_2 \rightarrow Cl\cdot + Cl\cdot$

These react with hydrogen as follows

$Cl\cdot + H_2 \rightarrow HCl + H\cdot$

$H\cdot + Cl_2 \rightarrow HCl + Cl\cdot$ etc.

Combustion and explosion reactions involve similar free-radical chain reactions.

chair *See* **ring conformations**.

chalcedony A mineral consisting of a microcrystalline variety of *quartz. It occurs in several forms, including a large number of semiprecious gemstones; for example, sard, carnelian, jasper, onyx, chrysoprase, agate, and tiger's-eye.

chalcogens *See* **group 16 elements**.

chalconides Binary compounds formed between metals and group 16 elements; i.e. oxides, sulphides, selenides, and tellurides.

chalcopyrite (copper pyrites) A brassy yellow mineral consisting of a mixed copper–iron sulphide, $CuFeS_2$, crystallizing in the tetragonal system; the principal ore of copper. It is similar in appearance to pyrite and gold. It crystallizes in igneous rocks and hydrothermal veins associated with the upper parts of acid igneous intrusions. Chalcopyrite is the most widespread of the copper ores, occurring, for example, in Cornwall (UK), Sudbury (Canada), Chile, Tasmania (Australia), and Rio Tinto (Spain).

chalk A very fine-grained white rock composed of the skeletal remains of microscopic sea creatures, such as plankton, and consisting largely of *calcium carbonate ($CaCO_3$). It is used in toothpaste and cosmetics. It should not be confused with blackboard 'chalk', which is made from calcium sulphate.

change of phase (change of state) A change of matter in one physical *phase (solid, liquid, or gas) into another. The change is invariably accompanied by the evolution or absorption of energy.

chaos Unpredictable and apparently random behaviour arising in a system that is governed by deterministic laws (i.e. laws that, given the state of the system at a given time, uniquely determine the state at any other time). *Chaotic dynamics* are widespread in nature: examples are the turbulent flow of fluids, the dynamics of planetary moons, oscillations in electrical circuits, and the long-term unpredictability of the weather. In such situations, chaos arises because the mathematical equations expressing the (deterministic) laws in question are nonlinear and extremely sensitive to the initial conditions. It is impossible to specify data at a given time to a sufficient degree of precision to be able to predict the future behaviour of the system because two pieces of initial data differing by even a very small amount will give widely different results at a later time. Originally, the theory was introduced to describe unpredictability in meteorology, as exemplified by the *butterfly effect*. It has been suggested that the dynamical equations governing the weather are so sensitive to the initial data that whether or not a butterfly flaps its wings in one part of the world may make the difference between a tornado occurring or not occurring in some other part of the world. In chemistry, chaos theory has been used to explain oscillatory (clock) reactions, such as the *B–Z reaction, and chaotic reactions.

chaotic dynamics *See* **chaos**.

chaotic reaction A type of chemical reaction in which the concentrations of reactants show chaotic behaviour. This may occur when the reaction involves a large number of complex interlinked steps. Under such conditions, it is possible for the reaction to display unpredictable changes with time. *See also* **bistability**; **oscillating reaction**.

charcoal A porous form of carbon produced by the destructive distillation of organic material. Charcoal from wood is used as a fuel. All forms of charcoal are porous and are used for adsorbing gases and purifying and clarifying liquids. There are several types depending on the source. Charcoal from coconut shells is a particularly good gas adsorbent. *Animal charcoal* (or *bone black*) is made by heating bones and dissolving out the calcium phosphates and other mineral salts with acid. It is used in sugar refining. *Activated charcoal* is charcoal that has been activated for adsorption by steaming or by heating in a vacuum.

charge A property of some *elementary particles that gives rise to an interaction between them and consequently to the host of material phenomena described as electrical. Charge occurs in nature in two forms, conventionally described as *positive* and *negative* in order to distinguish between the two kinds of interaction between particles. Two particles that have similar charges (both negative or both positive) interact by repelling each other; two particles that have dissimilar charges (one positive, one negative) interact by attracting each other.

The natural unit of negative charge is the charge on an *electron, which is equal but opposite in effect to the positive charge on the proton. Large-scale matter that consists of equal numbers of electrons and protons is electrically neutral. If there is an excess of electrons the body is negatively charged; an excess of protons results in a positive charge. A flow of charged particles, especially a flow of electrons, constitutes an electric current. Charge is measured in coulombs, the charge on an electron being 1.602×10^{-19} coulombs.

charge carrier The entity that transports electric charge in an electric current. The nature of the carrier depends on the type of conductor: in metals, the charge carriers are electrons; in *semiconductors the carriers are electrons (*n*-type) or positive *holes (*p*-type); in gases the carriers are positive ions and electrons; in electrolytes they are positive and negative ions.

charge density 1. The electric charge per unit volume of a medium or body (*volume charge density*). **2.** The electric charge per unit surface area of a body (*surface charge density*).

charge-transfer complex A chemical compound in which there is weak coordination involving the transfer of charge between two molecules. An example is phenoquinone, in which the phenol and quinone molecules are not held together by formal chemical bonds but are associated by transfer of charge between the compounds' aromatic ring systems.

Charles, Jacques Alexandre César (1746–1823) French chemist and physicist, who became professor of physics at the Paris Conservatoire des Arts et Métiers. He is best remembered for discovering *Charles' law (1787), relating to the volume and temperature of a gas. In 1783 he became the first person to make an ascent in a hydrogen balloon.

Charles' law The volume of a fixed mass of gas at constant pressure expands by a constant fraction of its volume at 0°C for each Celsius degree or kelvin its temperature is raised. For any *ideal gas the fraction is approximately 1/273. This can be expressed by the equation $V = V_0(1 + t/273)$, where V_0 is the volume at 0°C and V is its volume at t°C. This is equivalent to the statement that the volume of a fixed mass of gas at constant pressure is proportional to its thermodynamic temperature, $V = kT$, where k is a constant. The law resulted from experiments begun around 1787 by Jacques *Charles but was properly established only by the more accurate results published in 1802 by Joseph *Gay-Lussac. Thus the law is also known as *Gay-Lussac's law*. An equation similar to that given above applies to pressures for ideal gases: $p = p_0(1 + t/273)$, a relationship known as *Charles' law of pressures*. *See also* **gas laws**.

cheddite Any of a group of high explosives made from nitro compounds mixed with sodium or potassium chlorate.

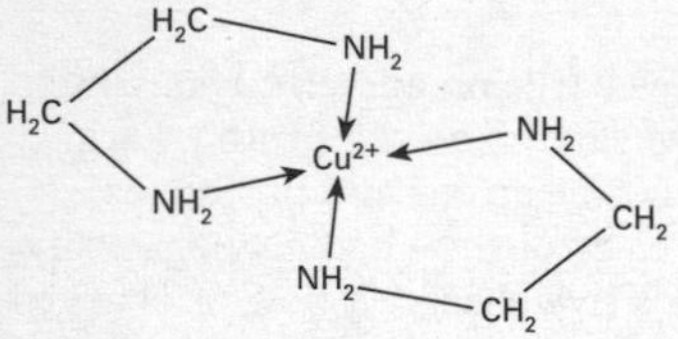

Chelate formed by coordination of two molecules of $H_2N(CH_2)_2NH_2$

chelate An inorganic complex in which a *ligand is coordinated to a metal ion at two (or more) points, so that there is a ring of atoms including the metal. The process is known as *chelation*. Chelating agents are classified according to the number of points at which they can coordinate. A ligand such as diaminoethane ($H_2N(CH_2)_2NH_2$), which can coordinate at two points, is said to be *bidentate*. The angle made between two bonds coordinating to a metal atom is the *bite angle* of the ligand. *See also* **sequestration**.

cheletropic reaction A type of addition reaction in which a conjugated molecule forms two single bonds from terminal atoms of the conjugated system to a single atom on another molecule to give a cyclic adduct. Cheletropic reactions are types of *pericyclic reaction.

chemical bond A strong force of attraction holding atoms together in a molecule or crystal. Typically chemical bonds have energies of about 1000 kJ mol^{-1} and are distinguished from the much weaker forces between molecules (*see* **van der Waals' force**). There are various types. *Ionic* (or *electrovalent*) bonds can be formed by transfer of electrons. For instance, the calcium atom has an electron configuration of [Ar]$4s^2$, i.e. it has two electrons in its outer shell. The chlorine atom is [Ne]$3s^2 3p^5$, with seven outer electrons. If the calcium atom transfers two electrons, one to each chlorine atom, it becomes a Ca^{2+} ion with the stable configuration of an inert gas [Ar]. At the same time each chlorine, having gained one electron, becomes a Cl^- ion, also with an inert-gas configuration [Ar]. The bonding in calcium chloride is the electrostatic attraction between the ions. *Covalent* bonds are formed by sharing of valence electrons rather than by transfer. For instance, hydrogen atoms have one outer electron ($1s^1$). In the hydrogen molecule, H_2, each atom contributes 1 electron to the bond. Consequently, each hydrogen atom has control of 2 electrons – one of its own and the second from the other atom – giving it the electron configuration of an inert gas [He]. In the water molecule, H_2O, the oxygen atom, with six outer electrons, gains control of an extra two electrons supplied by the two hydrogen atoms. This gives it the configuration [Ne]. Similarly, each hydrogen atom gains control of an extra electron from the oxygen, and has the [He] electron configuration.

A particular type of covalent bond is one in which one of the atoms supplies both the electrons. These are known as *dipolar bonds* (or *coordinate*, *semipolar*, or *dative bonds*), and written A→B, where the direction of the arrow denotes the direction in which electrons are donated.

Covalent or coordinate bonds in which one pair of electrons is shared are *electron-pair bonds* and are known as *single bonds*. Atoms can also share two pairs of electrons to form *double bonds* or three pairs in *triple bonds*. *See* **orbital**.

In a compound such as sodium chloride, Na^+Cl^-, there is probably complete transfer of electrons in forming the ionic bond (the bond is said to be *heteropolar*). Alternatively, in the hydrogen molecule H–H, the pair of electrons is equally shared between the two atoms (the bond is *homopolar*). Between these two extremes, there is a whole range of *intermediate bonds*, which have both ionic and covalent contributions. Thus, in hydrogen chloride, H–Cl, the bonding is predominantly covalent with one pair of electrons shared between the two atoms. However, the chlorine atom is more electronegative than the hydrogen and has more control over the electron pair; i.e. the molecule is polarized with a positive charge on the hydrogen and a negative charge on the chlorine. *See also* **banana bond**; **hydrogen bond**; **metallic bond**; **multicentre bond**; **multiple bond**.

chemical cell *See* cell.

chemical combination The combination of elements to give compounds. There are three laws of chemical combination.
(1) The *law of constant composition* states that the proportions of the elements in a compound are always the same, no matter how the compound is made. It is also called the *law of constant proportions* or *definite proportions.*
(2) The *law of multiple proportions* states that when two elements A and B combine to form more than one compound, then the masses of B that combine with a fixed mass of A are in simple ratio to one another. For example, carbon forms two oxides. In one, 12 grams of carbon is combined with 16 grams of oxygen (CO); in the other 12 g of carbon is combined with 32 grams of oxygen (CO_2). The oxygen masses combining with a fixed mass of carbon are in the ratio 16:32, i.e. 1:2.
(3) The *law of equivalent proportions* states that if two elements A and B each form a compound with a third element C, then a compound of A and B will contain A and B in the relative proportions in which they react with C. For example, sulphur and carbon both form compounds with hydrogen. In methane 12 g of carbon react with 4 g of hydrogen. In hydrogen sulphide, 32 g of sulphur react with 2 g of hydrogen (i.e. 64 g of S for 4 g of hydrogen). Sulphur and carbon form a compound in which the C:S ratio is 12:64 (i.e. CS_2). The law is sometimes called the law of *reciprocal proportions.*

chemical dating An absolute *dating technique that depends on measuring the chemical composition of a specimen. Chemical dating can be used when the specimen is known to undergo slow chemical change at a known rate. For instance, phosphate in buried bones is slowly replaced by fluoride ions from the ground water. Measurement of the proportion of fluorine present gives a rough estimate of the time that the bones have been in the ground. Another, more accurate, method depends on the fact that amino acids in living organisms are L-optical isomers. After death, these racemize and the age of bones can be estimated by measuring the relative amounts of D- and L-amino acids present.

chemical engineering The study of the design, manufacture, and operation of plant and machinery in industrial chemical processes.

chemical equation A way of denoting a chemical reaction using the symbols for the participating particles (atoms, molecules, ions, etc.); for example,

$$xA + yB \rightarrow zC + wD$$

The single arrow is used for an irreversible reaction; double arrows ($\rightleftharpoons$) are used for reversible reactions. When reactions involve different phases it is usual to put the phase in brackets after the symbol (s = solid; l = liquid; g = gas; aq = aqueous). The numbers x, y, z, and w, showing the relative numbers of molecules reacting, are called the *stoichiometric coefficients.* The sum of the coefficients of the reactants minus the sum of the coefficients of the products ($x + y - z - w$ in the example) is the *stoichiometric sum.* If this is zero the equation is balanced. Sometimes a generalized chemical equation is considered

$\nu_1 A_1 + \nu_2 A_2 + \ldots \rightarrow \ldots \nu_n A_n + \nu_{n+1} A_{n+1} \ldots$

In this case the reaction can be written $\Sigma \nu_i A_i = 0$, where the convention is that stoichiometric coefficients are positive for reactants and negative for products. The stoichiometric sum is $\Sigma \nu_i$.

chemical equilibrium A reversible chemical reaction in which the concentrations of reactants and products are not changing with time because the system is in thermodynamic equilibrium. For example, the reversible reaction

$3H_2 + N_2 \rightleftharpoons 2NH_3$

is in chemical equilibrium when the rate of the *forward reaction*

$3H_2 + N_2 \rightarrow 2NH_3$

is equal to the rate of the *back reaction*

$2NH_3 \rightarrow 3H_2 + N_2$

See also **equilibrium constant.**

chemical equivalent *See* **equivalent weight.**

chemical fossil Any of various organic compounds found in ancient geological strata that appear to be biological in origin and are assumed to indicate that life existed when the rocks were formed. The presence of chemical fossils in Archaean strata indicates that life existed over 3000 million years ago.

chemically induced dynamic nuclear polarization *See* **CIDNP.**

chemical potential Symbol: μ. For a given component in a mixture, the coefficient $\partial G/\partial n$, where G is the Gibbs free energy and n the amount of substance of the component. The chemical potential is the change in Gibbs free energy with respect to change in amount of the component, with pressure, temperature, and amounts of other components being constant. Components are in equilibrium if their chemical potentials are equal.

chemical reaction A change in which one or more chemical elements or compounds (the *reactants*) form new compounds (the *products*). All reactions are to some extent *reversible*; i.e. the products can also react to give the original reactants. However, in many cases the extent of this back reaction is negligibly small, and the reaction is regarded as *irreversible*.

chemical shift A change in the normal wavelength of absorption or emission of electromagnetic wavelength in a process in which there is a nuclear energy change (as in the *Mössbauer effect and *nuclear magnetic resonance) or a change in electron energy levels in the inner shells of an atom (as in X-ray *photoelectron spectroscopy).

chemiluminescence *See* **luminescence.**

chemiosmotic theory A theory postulated by the British biochemist

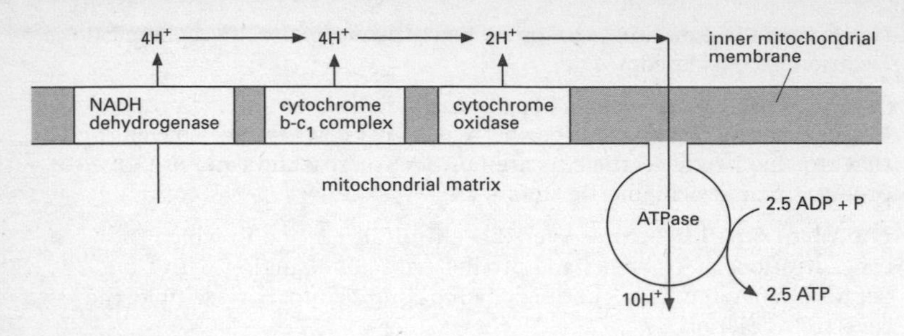

Chemiosmotic theory

Peter Mitchell (1920–92) to explain the formation of ATP in the mitochondrial *electron transport chain. As electrons are transferred along the electron carrier system in the inner mitochondrial membrane, hydrogen ions (protons) are actively transported (via *hydrogen carriers) into the space between the inner and outer mitochondrial membranes, which thus contains a higher concentration of protons than the matrix. This creates an electrochemical gradient across the inner membrane, down which protons move back into the matrix. This movement occurs through special channels associated with ATP synthetase, the enzyme that catalyses the conversion of ADP to ATP, and is coupled with the phosphorylation of ADP. A similar gradient is created across the thylakoid membranes of chloroplasts during the light-dependent reactions of *photosynthesis.

chemisorption *See* **adsorption.**

chemistry The study of the elements and the compounds they form. Chemistry is mainly concerned with effects that depend on the outer electrons in atoms. *See* **biochemistry**; **geochemistry**; **inorganic chemistry**; **organic chemistry**; **physical chemistry**.

chert *See* **flint.**

Chile saltpetre A commercial mineral largely composed of *sodium nitrate from the caliche deposits in Chile. Before the ammonia-oxidation process for nitrates most imported Chilean saltpetre was used by the chemical industry; its principal use today is as an agricultural source of nitrogen.

china clay *See* **kaolin.**

Chinese white *See* **zinc oxide.**

chirality The property of existing in left- and right-handed structural forms. *See* **optical activity.**

chirality element The part of a molecule that makes it exist in left- and right-handed forms. In most cases this is a *chirality centre* (i.e. an asymmetric atom). In certain cases the element is a *chirality axis.* For example, in allenenes of the type $R_1R_2C{=}C{=}CR_3R_4$ the C=C=C chain is a chirality axis. Certain ring compounds may display chirality as a result of a *chirality plane* in the molecule.

chirooptical spectroscopy Spectroscopy making use of the properties of chiral substances when they interact with polarized light of various wavelengths. *Optical rotatory dispersion* (the change of optical rotation with wavelength) and *circular dichroism are old examples of chirooptical spectroscopy. More recent types of chirooptical spectroscopy are *vibrational optical rotatory dispersion, vibrational circular dichroism. (VCD)*, and *Raman optical activity* (*ROA*), which are all the result of the interaction between chiral substances and polarized infrared electromagnetic radiation; these techniques are known as *vibrational optical activity* (VOA), as they are associated with transitions in the vibrational energy levels in the electronic ground state of a molecule. Chirooptical spectroscopy is used in the analysis of the structure of molecules.

chitin A *polysaccharide comprising chains of N-acetyl-D-glucosamine, a derivative of glucose. Chitin is structurally very similar to cellulose and serves to strengthen the supporting structures of various invertebrates. It also occurs in fungi.

chloral *See* **trichloroethanal.**

chloral hydrate *See* **2,2,2-trichloroethanediol.**

chlorates Salts of the chloric acids; i.e. salts containing the ions ClO^- (chlorate(I) or *hypochlorite*), ClO_2^- (chlorate(III) or *chlorite*), ClO_3^- (chlorate(V)), or ClO_4^- (chlorate(VII) or *perchlorate*). When used without specification of an oxidation state the term 'chlorate' refers to a chlorate(V) salt.

chloric acid Any of the oxoacids of chlorine: *chloric(I) acid, *chloric(III) acid, *chloric(V) acid, and *chloric(VII) acid. The term is commonly used without specification of the oxidation state of chlorine to mean chloric(V) acid, $HClO_3$.

chloric(I) acid **(hypochlorous acid)** A liquid acid that is stable only in solution, HOCl. It may be prepared by the reaction of chlorine with an agitated suspension of mercury(II) oxide. Because the disproportionation of the ion ClO^- is slow at low temperatures chloric(I) acid may be produced, along with chloride ions by the reaction of chlorine with water at 0°C. At higher temperatures disproportionation to the chlorate(V) ion, ClO_3^-, takes place. Chloric(I) acid is a very weak acid but is a mild oxidizing agent and is widely used as a bleaching agent.

chloric(III) acid **(chlorous acid)** A pale-yellow acid known only in solution, $HClO_2$. It is formed by the reaction of chlorine dioxide and water and is a weak acid and an oxidizing agent.

chloric(V) acid **(chloric acid)** A colourless unstable liquid, $HClO_3$; r.d. 1.2; m.p. <-20°C; decomposes at 40°C. It is best prepared by the reaction of barium chlorate with sulphuric acid although chloric(V) acid is also formed by the disproportionation of chloric(I) acid in hot solutions. It is both a strong acid and a powerful oxidizing agent; hot solutions of the acid or its salts have been known to detonate in contact with readily oxidized organic material.

chloric(VII) acid **(perchloric acid)** An unstable liquid acid, $HClO_4$; r.d. 1.76; m.p. −112°C; b.p. 39°C (50 mmHg); explodes at about 90°C at atmospheric pressure. There is also a monohydrate (r.d. 1.88 (solid), 1.77 (liquid); m.p. 48°C; explodes at about 110°C) and a dihydrate (r.d. 1.65; m.p. −17.8°C; b.p. 200°C). Commercial chloric(VII) acid is a water azeotrope, which is 72.5% $HClO_4$, boiling at 203°C. The anhydrous acid may be prepared by vacuum distillation of the concentrated acid in the presence of magnesium perchlorate as a dehydrating agent. Chloric(VII) acid is both a strong acid and a strong oxidizing agent. It is widely used to decompose organic materials prior to analysis, e.g. samples of animal or vegetable matter requiring heavy-metal analysis.

chloride *See* **halide**.

chlorination **1.** A chemical reaction in which a chlorine atom is introduced into a compound. *See* **halogenation**. **2.** The treatment of water with chlorine to disinfect it.

chlorine Symbol Cl. A *halogen element; a.n. 17; r.a.m. 35.453; d. 3.214 g dm^{-3}; m.p. −100.98°C; b.p. −34.6°C. It is a poisonous greenish-yellow gas and occurs widely in nature as sodium chloride in seawater and as halite (NaCl), carnallite ($KCl.MgCl_2.6H_2O$), and sylvite (KCl). It is manufactured by the electrolysis of brine and also obtained in the *Downs process for making sodium. It has many applications, including the chlorination of drinking water, bleaching, and the manufacture of a large number of organic chemicals.

It reacts directly with many elements and compounds and is a strong oxidizing agent. Chlorine compounds contain the element in the 1, 3, 5, and 7 oxidation states. It was discovered by Karl Scheele in 1774 and Humphry Davy confirmed it as an element in 1810.

chlorine dioxide A yellowish-red explosive gas, ClO_2; d. 3.09 g dm^{-3}; m.p. −59.5°C; b.p. 9.9°C. It is soluble in cold water but decomposed by hot water to give chloric(VII) acid, chlorine, and oxygen. Because of its high reactivity, chlorine dioxide is best prepared by the reaction of sodium chlorate and moist oxalic acid at 90°–100°C, as the product is then diluted by liberated carbon dioxide. Commercially the gas is produced by the reaction of sulphuric acid containing chloride ions with sulphur dioxide. Chlorine dioxide is widely used as a bleach in flour milling and in wood pulping and also finds application in water purification.

chlorine monoxide *See* **dichlorine oxide**.

chlorite **1.** *See* **chlorates**. **2.** A group of layered silicate minerals, usually green or white in colour, that are similar to the micas in structure and crystallize in the monoclinic system. Chlorites are composed of complex silicates of aluminium, magnesium, and iron in combination with water, with the formula $(Mg,Al,Fe)_{12}(Si,Al)_8O_{20}(OH)_{16}$. They are most common in low-grade metamorphic rocks and also occur as secondary minerals in igneous rocks as alteration products of pyroxenes, amphiboles, and micas. The term is derived from *chloros*, the Greek word for green.

chloroacetic acids *See* **chloroethanoic acids.**

chlorobenzene A colourless highly flammable liquid, C_6H_5Cl; r.d. 1.106; m.p. −45.43°C; b.p. 131.85°C. It is prepared by the direct chlorination of benzene using a halogen carrier (*see* **Friedel–Crafts reaction**), or manufactured by the *Raschig process. It is used mainly as an industrial solvent.

2-chlorobuta-1,3-diene **(chloroprene)** A colourless liquid chlorinated diene, $CH_2{:}CClCH{:}CH_2$; r.d. 0.96; b.p. 59°C. It is polymerized to make synthetic rubbers (e.g. neoprene).

chlorocruorin A greenish iron-containing respiratory pigment that occurs in the blood of polychaete worms. It closely resembles *haemoglobin.

chloroethane **(ethyl chloride)** A colourless flammable gas, C_2H_5Cl; m.p. −136.4°C; b.p. 12.3°C. It is made by reaction of ethene and hydrogen chloride and used in making lead tetraethyl for petrol.

chloroethanoic acids **(chloroacetic acids)** Three acids in which hydrogen atoms in the methyl group of ethanoic acid have been replaced by chlorine atoms. They are: *monochloroethanoic acid* ($CH_2ClCOOH$); *dichloroethanoic acid* ($CHCl_2COOH$); *trichloroethanoic acid* (CCl_3COOH). The presence of chlorine atoms in the methyl group causes electron withdrawal from the COOH group and makes the chloroethanoic acids stronger acids than ethanoic acid itself. The K_a values (in moles dm^{-3} at 25°C) are

CH_3COOH 1.7×10^{-5}

$CH_2ClCOOH$ 1.3×10^{-3}

$CHCl_2COOH$ 5.0×10^{-2}

CCl_3COOH 2.3×10^{-1}

chloroethene **(vinyl chloride)** A gaseous compound, $CH_2{:}CHCl$; r.d. 0.911; m.p. −153.8°C; b.p. −13.37°C. It is made by chlorinating ethene to give dichloroethane, then removing HCl:

$$C_2H_4 + Cl_2 \rightarrow CH_2ClCH_2Cl \rightarrow CH_2CHCl$$

The compound is used in making PVC.

chlorofluorocarbon **(CFC)** A type of compound in which some or all of the hydrogen atoms of a hydrocarbon (usually an alkane) have been replaced by chlorine and fluorine atoms. Most chlorofluorocarbons are chemically unreactive and are stable at high temperatures. They are used as aerosol propellants, refrigerants, and solvents, and in the manufacture of rigid packaging foam. A commonly encountered commercial name for these compounds is *freon*, e.g. freon 12 is dichlorodifluoromethane (CCl_2F_2).

Chlorofluorocarbons, because of their chemical inertness, can diffuse unchanged into the upper atmosphere. Here, photochemical reactions cause them to break down and react with ozone (*see* **ozone layer**). For this reason, their use has been discouraged.

chloroform *See* **trichloromethane.**

chloromethane (methyl chloride) A colourless flammable gas, CH_3Cl; r.d. 0.916; m.p. −97.1°C; b.p. −24.2°C. It is a *haloalkane, made by direct chlorination of methane and used as a local anaesthetic and refrigerant.

chlorophenol Any of the various compounds produced by chlorinating a *phenol. Chlorophenols are fairly acidic and have many uses, including antiseptics, disinfectants, herbicides, insecticides, and wood preservatives, and in making dyes. They form condensation polymers of the bakelite type with methanal.

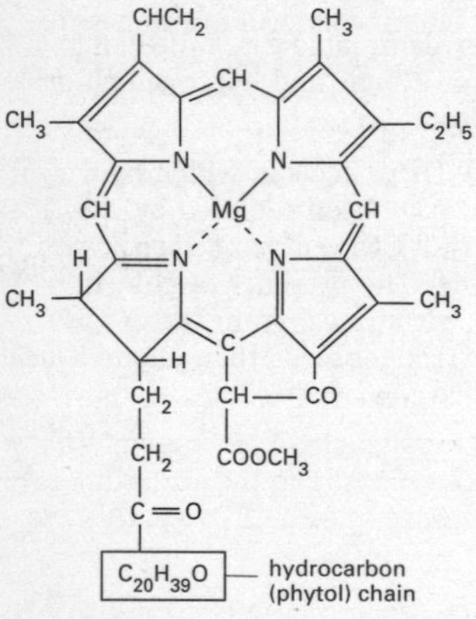

Chlorophyll a

chlorophyll One of a number of pigments, including (*chlorophyll a* and *chlorophyll b*), that are responsible for the green colour of most plants. Chlorophyll molecules are the principal sites of light absorption in the light-dependent reactions of *photosynthesis. They are magnesium-containing *porphyrins, chemically related to *cytochrome and *haemoglobin.

chloroplatinic acid A reddish crystalline compound, H_2PtCl_6, made by dissolving platinum in aqua regia.

chloroprene *See* **2-chlorobuta-1,3-diene.**

chlorosulphanes *See* **disulphur dichloride.**

chlorous acid *See* chloric(III) acid.

choke damp *See* blackdamp.

cholecalciferol *See* vitamin D.

cholesteric crystal A type of *liquid crystal in which molecules lie in sheets at angles that change by a small amount between each sheet. The pitch of the helix so formed is similar to the wavelength of visible light. Cholesteric liquid crystals therefore diffract light and have colours that depend on the temperature. (The word cholesteric is derived from the Greek for bile solid.)

cholesterol A *sterol occurring widely in animal tissues and also in some higher plants and algae. It can exist as a free sterol or esterified with a long-chain fatty acid. Cholesterol is absorbed through the intestine or manufactured in the liver. It serves principally as a constituent of blood plasma lipoproteins and of the lipid–protein complexes that form cell membranes. It is also important as a precursor of various steroids, especially the bile acids, sex hormones, and adrenocorticoid hormones. The derivative 7-dehydrocholesterol is converted to vitamin D_3 by the action of sunlight on skin. Increased levels of dietary and blood cholesterol have been associated with *atherosclerosis*, a condition in which lipids accumulate on the inner walls of arteries and eventually obstruct blood flow.

choline An amino alcohol, $CH_2OHCH_2N(CH_3)_3OH$. It occurs widely in living organisms as a constituent of certain types of phospholipids – the lecithins and sphingomyelins – and in the neurotransmitter *acetylcholine. It is sometimes classified as a member of the vitamin B complex.

chromate A salt containing the ion CrO_4^{2-}.

chromatogram A record obtained by *chromatography. The term is applied to the developed records of *paper chromatography and *thin-layer chromatography and also to the graphical record produced in *gas chromatography.

chromatography A technique for analysing or separating mixtures of gases, liquids, or dissolved substances. The original technique (invented by the Russian botanist Mikhail Tsvet in 1906) is a good example of *column chromatography*. A vertical glass tube is packed with an adsorbing material, such as alumina. The sample is poured into the column and continuously washed through with a solvent (a process known as *elution*). Different components of the sample are adsorbed to different extents and move down the column at different rates. In Tsvet's original application, plant pigments were used and these separated into coloured bands in passing down the column (hence the name chromatography). The usual method is to collect the liquid (the *eluate*) as it passes out from the column in fractions.

In general, all types of chromatography involve two distinct phases – the *stationary phase* (the adsorbent material in the column in the example above) and the *moving phase* (the solution in the example). The separation depends on competition for molecules of sample between the moving phase and the stationary phase. The form of column chromatography above is an example of *adsorption chromatography*, in which the sample molecules are adsorbed on the alumina. In *partition chromatography*, a liquid (e.g. water) is first absorbed by the stationary phase and the moving phase is an immiscible liquid. The separation is then by *partition between the two liquids. In ion-exchange chromatography (*see* **ion exchange**), the process involves competition between different ions for ionic sites on the stationary phase. *Gel filtration is another chromatographic technique in which the size of the sample molecules is important.

See also **affinity chromatography; gas chromatography; high-performance liquid chromatography; paper chromatography; R_F value; thin-layer chromatography**.

chrome alum *See* **potassium chromium sulphate**.

chrome iron ore A mixed iron–chromium oxide, $FeO.Cr_2O_3$, used to make ferrochromium for chromium steels.

chrome red A basic lead chromate, $PbO.PbCrO_4$, used as a red pigment.

chrome yellow Lead chromate, $PbCrO_4$, used as a pigment.

chromic acid A hypothetical acid, H_2CrO_4, known only in chromate salts.

chromic anhydride *See* **chromium(VI) oxide**.

chromic compounds Compounds containing chromium in a higher (+3 or +6) oxidation state; e.g. chromic oxide is chromium(VI) oxide (CrO_3).

chromite A spinel mineral, $FeCr_2O_4$; the principal ore of chromium. It is black with a metallic lustre and usually occurs in massive form. It is a common constituent of peridotites and serpentines. The chief producing countries are Turkey, South Africa, Russia, the Philippines, and Zimbabwe.

chromium Symbol Cr. A hard silvery *transition element; a.n. 24; r.a.m. 52.00; r.d. 7.19; m.p. 1857°C; b.p. 2672°C. The main ore is chromite ($FeCr_2O_4$). The metal has a body-centred-cubic structure. It is extracted by heating chromite with sodium chromate, from which chromium can be obtained by electrolysis. Alternatively, chromite can be heated with carbon in an electric furnace to give ferrochrome, which is used in making alloy steels. The metal is also used as a shiny decorative electroplated coating and in the manufacture of certain chromium compounds.

At normal temperatures the metal is corrosion-resistant. It reacts with dilute hydrochloric and sulphuric acids to give chromium(II) salts. These readily oxidize to the more stable chromium(III) salts. Chromium also forms compounds with the +6 oxidation state, as in chromates, which contain the CrO_4^{2-} ion. The element was discovered in 1797 by Vauquelin.

chromium dioxide *See* **chromium(IV) oxide**.

chromium(II) oxide A black insoluble powder, CrO. Chromium(II) oxide is prepared by oxidizing chromium amalgam with air. At high temperatures hydrogen reduces it to the metal.

chromium(III) oxide **(chromium sesquioxide)** A green crystalline water-insoluble salt, Cr_2O_3; r.d. 5.21; m.p. 2435°C; b.p. 4000°C. It is obtained by heating chromium in a stream of oxygen or by heating ammonium dichromate. The industrial preparation is by reduction of sodium dichromate with carbon. Chromium(III) oxide is amphoteric, dissolving in acids to give chromium(III) ions and in concentrated solutions of alkalis to give *chromites*. It is used as a green pigment in glass, porcelain, and oil paint.

chromium(IV) oxide **(chromium dioxide)** A black insoluble powder, CrO_2; m.p. 300°C. It is prepared by the action of oxygen on chromium(VI) oxide or chromium(III) oxide at 420–450°C and 200–300 atmospheres. The compound is unstable.

chromium(VI) oxide **(chromium trioxide; chromic anhydride)** A red compound, CrO_3; rhombic; r.d. 2.70; m.p. 196°C. It can be made by careful addition of concentrated sulphuric acid to an ice-cooled concentrated aqueous solution of sodium dichromate with stirring. The mixture is then filtered through sintered glass, washed with nitric acid, then dried at 120°C in a desiccator.

Chromium(VI) oxide is an extremely powerful oxidizing agent, especially to organic matter; it immediately inflames ethanol. It is an acidic oxide and dissolves in water to form 'chromic acid', a powerful oxidizing agent and cleansing fluid for glassware. At 400°C, chromium(VI) oxide loses oxygen to give chromium(III) oxide.

chromium potassium sulphate A red crystalline solid, $K_2SO_4.Cr_2(SO_4)_3.24H_2O$; r.d. 1.91. It is used as a mordant *See also* **alums**.

chromium sesquioxide *See* **chromium(III) oxide**.

chromium steel Any of a group of *stainless steels containing 8–25% of chromium. A typical chromium steel might contain 18% of chromium, 8% of nickel, and 0.15% of carbon. Chromium steels are highly resistant to corrosion and are used for cutlery, chemical plant, ball bearings, etc.

chromium trioxide *See* **chromium(VI) oxide**.

chromophore A group causing coloration in a *dye. Chromophores are generally groups of atoms having delocalized electrons.

chromous compounds Compounds containing chromium in its lower (+2) oxidation state; e.g. chromous chloride is chromium(II) chloride ($CrCl_2$).

chromyl chloride **(chromium oxychloride)** A dark red liquid, CrO_2Cl_2; r.d. 1.911; m.p. −96.5°C; b.p. 117°C. It is evolved as a dark-red vapour on addition of concentrated sulphuric acid to a mixture of solid potassium dichromate and sodium chloride; it condenses to a dark-red covalent liquid, which is immediately hydrolysed by solutions of alkalis to give the yellow chromate.

Since bromides and iodides do not give analogous compounds this is a specific test for chloride ions. The compound is a powerful oxidizing agent, exploding on contact with phosphorus and inflaming sulphur and many organic compounds.

chrysotile *See* **serpentine.**

CI *See* **colour index.**

CIDNP (chemically induced dynamic nuclear polarization) A mechanism enabling nuclear spin to influence the direction of a chemical reaction. This can occur in certain cases, even though the gap between energy levels of nuclear spin states in a magnetic field is very much smaller than the dissociation energies of chemical bonds. Two radicals, the electrons of which have parallel spins, can only combine if the conversion of a *triplet to a *singlet can take place. In a magnetic field, a triplet has three nondegenerate states called T_0, T_+, and T_-. For a triplet-to-singlet conversion to take place, one electron must precess faster than the other for a sufficient time to enable a 180° phase difference to develop. This difference in precession can arise when the nuclear spin interacts with the electron on the radical, by means of hyperfine coupling.

cine-substitution A type of substitution reaction in which the entering group takes a position on an atom adjacent to the atom to which the leaving group is attached. *See also* **tele-substitution**.

cinnabar A bright red mineral form of mercury(II) sulphide, HgS, crystallizing in the hexagonal system; the principal ore of mercury. It is deposited in veins and impregnations near recent volcanic rocks and hot springs. The chief sources include Spain, Italy, and Yugoslavia.

cinnamic acid (3-phenylpropenoic acid) A white crystalline aromatic *carboxylic acid, $C_6H_5CH{:}CHCOOH$; r.d. 1.248 (trans isomer); m.p. 135–136°C; b.p. 300°C. Esters of cinnamic acid occur in some essential oils.

CIP system (Cahn–Ingold–Prelog system) A system for the unambiguous description of stereoisomers used in the *R–S* convention (*see* **absolute configuration**) and in the **E–Z* convention. The system involves a *sequence rule* for determining a conventional order of ligands. The rule is that the atom bonded directly to the chiral centre or double bond is considered and the ligand in which this atom has the highest proton number takes precedence. So, for example, I takes precedence over Cl. If two ligands have bonding atoms with the same proton number, then substituents are taken into account (with the substituent of highest proton number taking precedence). Thus, $-C_2H_5$ has a higher precedence than $-CH_3$. If a double (or triple) bond occurs to a substituent, then the substituent is counted twice (or three times). An isotope of high nucleon number takes precedence over one of lower nucleon number. Hydrogen always has lowest priority in this system. For example, the sequence for some common ligands is I, Br, Cl, SO_3H, $OCOCH_3$, OCH_3, OH, NO_2, NH_2, $COOCH_3$, $CONH_2$, $COCH_3$, CHO, CH_2OH, C_6H_5, C_2H_5, CH_3, H. The system was jointly developed by the British chemists Robert Sidney Cahn (1899–1981) and Sir Christopher Kelk Ingold (1893–1970) and the Bosnian–Swiss chemist Vladimir Prelog (1906–).

circular birefringence A phenomenon in which there is a difference between the refractive indices of the molecules of a substance for right- and left-circularly polarized light. Circular birefringence depends on the way in which the electromagnetic field interacts with the molecule, and is affected by the handedness of the molecule, and hence its polarizability. If a molecule has the shape of a helix, the polarizability is dependent on whether or not the electric field of the electromagnetic field rotates in the same sense as the helix, thus giving rise to circular birefringence.

circular dichroism (CD) The production of an elliptically polarized wave when a linearly polarized light wave passes through a substance that has differences in the extinction coefficients for left- and right-handed polarized light. The size of this effect is given by $\phi = \pi/\lambda(\eta_l - \eta_r)$, where ϕ is the ellipticity of the beam that emerges (in radians), λ is the wavelength of the light, and η_l and η_r are the absorption indices of the left- and right-handed circularly polarized light, respectively. Circular dichroism is a property of optically active molecules and is used to obtain information about the solution environment of proteins. *See also* **CD spectrum**.

circular polarization *See* **polarization of light**.

cis *See* **isomerism; torsion angle**.

cisplatin A platinum complex, *cis*-$[PtCl_2(NH_3)_2]$, used in cancer treatment to inhibit the growth of tumours. It acts by binding between strands of DNA.

citrate A salt or ester of citric acid.

citric acid A white crystalline hydroxycarboxylic acid, $HOOCCH_2C(OH)(COOH)CH_2COOH$; r.d. 1.54; m.p. 153°C. It is present in citrus fruits and is an intermediate in the *Krebs cycle in plant and animal cells.

citric-acid cycle *See* **Krebs cycle**.

Claisen condensation A reaction of esters in which two molecules of the ester react to give a keto ester, e.g.

$$2CH_3COOR \rightarrow CH_3COCH_2COOR + ROH$$

The reaction is catalysed by sodium ethoxide, the mechanism being similar to that of the *aldol reaction. It is named after Ludwig Claisen (1851–1930).

Clapeyron–Clausius equation A differential equation that describes the relationship between variables when there is a change in the state of a system. In a system that has two *phases of the same substance, for example solid and liquid, heat is added or taken away very slowly so that one phase can change reversibly into the other while the system remains at equilibrium. If the two phases are denoted A and B, the Clapeyron–Clausius equation is:

$$dp/dT = L/T(V_B - V_A),$$

where p is the pressure, T is the thermodynamic temperature, L is the heat absorbed per mole in the change from A to B, and V_B and V_A are the volumes of B and A respectively. In the case of a transition from liquid to vapour, the volume of the liquid can be ignored. Taking the vapour to be an *ideal gas, the Clapeyron–Clausius equation can be written:

$$\mathrm{d}\log_e p/\mathrm{d}T = L/RT^2$$

The Clapeyron–Clausius equation is named after the French engineer Benoit-Pierre-Émile Clapeyron (1799–1864) and Rudolf *Clausius.

Clark cell A type of *voltaic cell consisting of an anode made of zinc amalgam and a cathode of mercury both immersed in a saturated solution of zinc sulphate. The Clark cell was formerly used as a standard of e.m.f.; the e.m.f. at 15°C is 1.4345 volts. It is named after the British scientist Hosiah Clark (*d.* 1898).

Clark process *See* **hardness of water.**

clathrate A solid mixture in which small molecules of one compound or element are trapped in holes in the crystal lattice of another substance. Clathrates are sometimes called *enclosure compounds* or *cage compounds*, but they are not true compounds (the molecules are not held by chemical bonds). Quinol and ice both form clathrates with such substances as sulphur dioxide and xenon.

Claude process A process for liquefying air on a commercial basis. Air under pressure is used as the working substance in a piston engine, where it does external work and cools adiabatically. This cool air is fed to a counter-current heat exchanger, where it reduces the temperature of the next intake of high-pressure air. The same air is re-compressed and used again, and after several cycles eventually liquefies. The process was perfected in 1902 by the French scientist Georges Claude (1870–1960).

claudetite A mineral form of *arsenic(III) oxide, As_4O_6.

Clausius, Rudolf Julius Emmanuel (1822–88) German physicist, who held teaching posts in Berlin and Zurich, before going to Würzburg in 1869. He is best known for formulating the second law of *thermodynamics in 1850, independently of William Thomson (Lord *Kelvin). In 1865 he introduced the concept of *entropy, and later contributed to electrochemistry and electrodynamics (*see* *Clausius–Mossotti equation).

Clausius inequality An inequality that relates the change in entropy $\mathrm{d}S$ in an irreversible process to the heat supplied to the system $\mathrm{d}Q$ and the thermodynamic temperature T, i.e. $\mathrm{d}S \geq \mathrm{d}Q/T$. In the case of an irreversible adiabatic change, where $\mathrm{d}Q = 0$, the Clausius inequality has the form $\mathrm{d}S > 0$, which means that for this type of change the entropy of the system must increase. The inequality is named after Rudolf Julius Emmanuel *Clausius. The Clausius inequality can be used to demonstrate that entropy increases in such processes as the free expansion of a gas and the cooling of an initially hot substance.

Clausius–Mossotti equation A relation between the *polarizability α of a molecule and the dielectric constant ε of a dielectric substance made up of molecules with this polarizability. The Clausius–Mossotti equation can be written in the form

$$\alpha = (3/4\pi N)/[(\varepsilon - 1)/(\varepsilon - 2)],$$

where N is the number of molecules per unit volume. The equation provides a link between a microscopic quantity (the polarizability) and a macroscopic quantity (the dielectric constant); it was derived using macroscopic electrostatics by the Italian physicist Ottaviano Fabrizio Mossotti (1791–1863) in 1850 and independently by Rudolf *Clausius in 1879. It works best for gases and is only approximately true for liquids or solids, particularly if the dielectric constant is large. *Compare* **Lorentz–Lorenz equation.**

Claus process A process for obtaining sulphur from hydrogen sulphide (from natural gas or crude oil). It involves two stages. First, part of the hydrogen sulphide is oxidized to sulphur dioxide:

$$2H_2S + 3O_2 \rightarrow 2SO_2 + 2H_2O$$

Subsequently, the sulphur dioxide reacts with hydrogen sulphide to produce sulphur:

$$SO_2 + 2H_2S \rightarrow 3S + 2H_2O$$

The second stage occurs at 300°C and needs an iron or aluminium oxide catalyst.

clay A fine-grained deposit consisting chiefly of *clay minerals. It is characteristically plastic and virtually impermeable when wet and cracks when it dries out. In geology the size of the constituent particles is usually taken to be less than 1/256 mm. In soil science clay is regarded as a soil with particles less than 0.002 mm in size.

clay minerals Very small particles, chiefly hydrous silicates of aluminium, sometimes with magnesium and/or iron substituting for all or part of the aluminium, that are the major constituents of clay materials. The particles are essentially crystalline (either platy or fibrous) with a layered structure, but may be amorphous or metalloidal. The clay minerals are responsible for the plastic properties of clay; the particles have the property of being able to hold water. The chief groups of clay minerals are: *kaolinite*, $Al_4Si_4O_{10}(OH)_8$, the chief constituent of *kaolin; *halloysite*, $Al_4Si_4(OH)_8O_{10}.4H_2O$; *illite*, $KAl_4(Si,Al)_8O_{18}.2H_2O$; *montmorillonite*, $(Na,Ca)_{0.33}(Al,Mg)_2Si_4O_{10}(OH)_2.nH_2O$, formed chiefly through alteration of volcanic ash; and *vermiculite*, $(Mg,Fe,Al)_3(Al,Si)_4O_{10}(OH)_2.4H_2O$, used as an insulating material and potting soil.

cleavage The splitting of a crystal along planes of atoms in the lattice.

Clemmensen reduction A method of reducing a *carbonyl group (C=O) to CH_2, using zinc amalgam and concentrated hydrochloric acid. It is used as a method of ascending a homologous series. The reaction is named after Erik Clemmensen (1876–1941).

clinal *See* **torsion angle.**

clock reaction *See* **B–Z reaction**; **oscillating reaction.**

closed chain *See* **chain**; **ring.**

close packing The packing of spheres so as to occupy the minimum amount of space. In a single plane, each sphere is surrounded by six close neighbours in a hexagonal arrangement. The spheres in the second plane fit into depressions in the first layer, and so on. Each sphere has 12 other touching spheres. There are two types of close packing. In *hexagonal close packing* the spheres in the third layer are directly over those in the first, etc., and the arrangement of planes is ABAB.... In *cubic close packing* the spheres in the third layer occupy a different set of depressions than those in the first. The arrangement is ABCABC.... *See also* **cubic crystal.**

closo-structure *See* **borane.**

Clusius column A device for separating isotopes by thermal diffusion. One form consists of a vertical column some 30 metres high with a heated electric wire running along its axis. The lighter isotopes in a gaseous mixture of isotopes diffuse faster than the heavier isotopes. Heated by the axial wire, and assisted by natural convection, the lighter atoms are carried to the top of the column, where a fraction rich in lighter isotopes can be removed for further enrichment.

cluster compound A compound in which groups of metal atoms are joined together by metal–metal bonds. The formation of such compounds is a feature of the chemistry of certain transition elements, particularly molybdenum and tungsten, but also vanadium, tantalum, niobium, and uranium. *Isopoly compounds* are ones in which the cluster contains atoms of the same element; *heteropoly compounds* contain a mixture of different elements.

coacervate An aggregate of macromolecules, such as proteins, lipids, and nucleic acids, that form a stable *colloid unit with properties that resemble living matter. Many are coated with a lipid membrane and contain enzymes that are capable of converting such substances as glucose into more complex molecules, such as starch. Coacervate droplets arise spontaneously under appropriate conditions and may have been the prebiological systems from which living organisms originated.

coagulation The process in which colloidal particles come together irreversibly to form larger masses. Coagulation can be brought about by adding ions to change the ionic strength of the solution and thus destabilize the colloid (*see* **flocculation**). Ions with a high charge are particularly effective (e.g. alum, containing Al^{3+}, is used in styptics to coagulate blood). Another example of ionic coagulation is in the formation of river deltas, which occurs when colloidal silt particles in rivers are coagulated by ions in sea water. Alum and iron (III) sulphate are also used for coagulation in sewage treatment. Heating is another way of coagulating certain colloids (e.g. boiling an egg coagulates the albumin).

coal A brown or black carbonaceous deposit derived from the accumulation and alteration of ancient vegetation, which originated largely in swamps or other moist environments. As the vegetation decomposed it formed layers of peat, which were subsequently buried (for example, by marine sediments following a rise in sea level or subsidence of the land). Under the increased pressure and resulting higher temperatures the peat was transformed into coal. Two types of coal are recognized: *humic* (or *woody*) *coals*, derived from plant remains; and *sapropelic coals*, which are derived from algae, spores, and finely divided plant material.

As the processes of coalification (i.e. the transformation resulting from the high temperatures and pressures) continue, there is a progressive transformation of the deposit: the proportion of carbon relative to oxygen rises and volatile substances and water are driven out. The various stages in this process are referred to as the *ranks* of the coal. In ascending order, the main ranks of coal are: *lignite* (or *brown coal*), which is soft, brown, and has a high moisture content; *subbituminous coal*, which is used chiefly by generating stations; *bituminous coal*, which is the most abundant rank of coal; *semibituminous coal*; *semianthracite coal*, which has a fixed carbon content of between 86% and 92%; and *anthracite coal*, which is hard and black with a fixed carbon content of between 92% and 98%.

Most deposits of coal were formed during the Carboniferous and Permian periods. More recent periods of coal formation occurred during the early Jurassic and Tertiary periods. Coal deposits occur in all the major continents; the leading producers include the USA, China, Ukraine, Poland, UK, South Africa, India, Australia, and Germany. Coal is used as a fuel and in the chemical industry; by-products include coke and coal tar.

coal gas A fuel gas produced by the destructive distillation of coal. In the late-19th and early-20th centuries coal gas was a major source of energy and was made by heating coal in the absence of air in local gas works. Typically, it contained hydrogen (50%), methane (35%), and carbon monoxide (8%). By-products of the process were *coal tar and coke. The use of this type of gas declined with the increasing availability of natural gas, although since the early 1970s interest has developed in using coal in making *SNG.

coal tar A tar obtained from the destructive distillation of coal. Formerly, coal tar was obtained as a by-product in manufacturing *coal gas. Now it is produced in making coke for steel making. The crude tar contains a large number of organic compounds, such as benzene, naphthalene, methylbenzene, phenols, etc., which can be obtained by distillation. The residue is *pitch*. At one time coal tar was the major source of organic chemicals, most of which are now derived from petroleum and natural gas.

cobalamin (vitamin B_{12}) *See* **vitamin B complex**.

cobalt Symbol Co. A light-grey *transition element; a.n. 27; r.a.m. 58.933; r.d. 8.9; m.p. 1495°C; b.p. 2870°C. Cobalt is ferromagnetic below its Curie point of 1150°C. Small amounts of metallic cobalt are present in meteorites but it is usually extracted from ore deposits worked in Canada, Morocco, and Zaïre. It is present in the minerals cobaltite, smaltite, and erythrite but also associated with copper and nickel as sulphides and arsenides. Cobalt ores are usually roasted to the oxide and then reduced with carbon or water gas. Cobalt is usually alloyed for use. Alnico is a well-known magnetic alloy and cobalt is also used to make stainless steels and in high-strength alloys that are resistant to oxidation at high temperatures (for turbine blades and cutting tools).

The metal is oxidized by hot air and also reacts with carbon, phosphorus, sulphur, and dilute mineral acids. Cobalt salts, usual oxidation states II and III, are used to give a brilliant blue colour in glass, tiles, and pottery. Anhydrous cobalt(II) chloride paper is used as a qualitative test for water and as a heat-sensitive ink. Small amounts of cobalt salts are essential in a balanced diet for mammals (*see* **essential element**). Artificially produced cobalt–60 is an important radioactive tracer and cancer-treatment agent. The element was discovered by Georg Brandt (1694–1768) in 1737.

cobalt(II) oxide A pink solid, CoO; cubic; r.d. 6.45; m.p. 1935°C. The addition of potassium hydroxide to a solution of cobalt(II) nitrate gives a bluish-violet precipitate, which on boiling is converted to pink impure cobalt(II) hydroxide. On heating this in the absence of air, cobalt(II) oxide is formed. The compound is readily oxidized in air to form tricobalt tetroxide, Co_3O_4, and is readily reduced by hydrogen to the metal.

cobalt(III) oxide (cobalt sesquioxide) A black grey insoluble solid, Co_2O_3; hexagonal or rhombic; r.d. 5.18; decomposes at 895°C. It is produced by the ignition of cobalt nitrate; the product however never has the composition corresponding exactly to cobalt(III) oxide. On heating it readily forms Co_3O_4, which contains both Co(II) and Co(III), and is easily reduced to the metal by hydrogen. Cobalt(III) oxide dissolves in strong acid to give unstable brown solutions of trivalent cobalt salts. With dilute acids cobalt(II) salts are formed.

cobalt steel Any of a group of alloy *steels containing 5–12% of cobalt, 14–20% of tungsten, usually with 4% of chromium and 1–2% of vanadium. They are very hard but somewhat brittle. Their main use is in high-speed tools.

codeine A pain-relieving drug that is derived from the plant *Papaver somniferum*.

coenzyme An organic nonprotein molecule that associates with an enzyme molecule in catalysing biochemical reactions. Coenzymes usually participate in the substrate–enzyme interaction by donating or accepting certain chemical groups. Many vitamins are precursors of coenzymes. *See also* **cofactor**.

coenzyme A (CoA) A complex organic compound that acts in conjunction with enzymes involved in various biochemical reactions, notably the oxidation of pyruvate via the *Krebs cycle and fatty-acid oxidation and synthesis. It comprises principally the B vitamin pantothenic acid, the nucleotide adenine, and a ribose-phosphate group.

coenzyme Q *See* **ubiquinone**.

cofactor A nonprotein component essential for the normal catalytic activity of an enzyme. Cofactors may be organic molecules (*coenzymes) or inorganic ions. They may activate the enzyme by altering its shape or they may actually participate in the chemical reaction.

coherent anti-Stokes Raman spectroscopy *See* CARS.

coherent units A system of *units of measurement in which derived units are obtained by multiplying or dividing base units without the use of numerical factors. *SI units form a coherent system; for example the unit of force is the newton, which is equal to 1 kilogram metre per second squared ($kg\,m\,s^{-2}$), the kilogram, metre, and second all being base units of the system.

cohesion The force of attraction between like molecules.

coinage metals A group of three malleable ductile transition metals forming group 11 (formerly IB) of the *periodic table: copper (Cu), silver (Ag), and gold (Au). Their outer electronic configurations have the form $nd^{10}(n+1)s^1$. Although this is similar to that of alkali metals, the coinage metals all have much higher ionization energies and higher (and positive) standard electrode potentials. Thus, they are much more difficult to oxidize and are more resistant to corrosion. In addition, the fact that they have *d*-electrons makes them show variable valency (Cu^I, Cu^{II}, and Cu^{III}; Ag^I and Ag^{II}; Au^I and Au^{III}) and form a wide range of coordination compounds. They are generally classified with the *transition elements.

coke A form of carbon made by the destructive distillation of coal. Coke is used for blast-furnaces and other metallurgical and chemical processes requiring a source of carbon. Lower-grade cokes, made by heating the coal to a lower temperature, are used as smokeless fuels for domestic heating.

colchicine An *alkaloid derived from the autumn crocus, *Colchicum autumnale*. It inhibits cell division. Colchicine is used in genetics, cytology, and plant breeding research and also in cancer therapy to inhibit cell division.

collagen An insoluble fibrous protein found extensively in the connective tissue of skin, tendons, and bone. The polypeptide chains of collagen (containing the amino acids glycine and proline predominantly) form triple-stranded helical coils that are bound together to form fibrils, which have great strength and limited elasticity. Collagen accounts for over 30% of the total body protein of mammals.

collective oscillation An oscillation in a many-body system in which all the particles in the system take part in a cooperative. Plasma oscillations provide an example of collective oscillations. In systems described by quantum mechanics, collective oscillations are quantized to give collective excitations.

colligation The combination of two free radicals to form a covalent bond, as in $H_3C\cdot + Cl\cdot \rightarrow CH_3Cl$. It is the reverse of *homolytic fission.

colligative properties Properties that depend on the concentration of particles (molecules, ions, etc.) present in a solution, and not on the nature of the particles. Examples of colligative properties are osmotic pressure (*see* **osmosis**), *lowering of vapour pressure, *depression of freezing point, and *elevation of boiling point.

collision density The number of collisions that occur in unit volume in unit time when a given particle flux passes through matter.

collision quenching *See* **external conversion**.

collodion A thin film of cellulose nitrate made by dissolving the cellulose nitrate in ethanol or ethoxyethane, coating the surface, and evaporating the solvent.

colloid mills Machines used to grind aggregates into very fine particles or to apply very high shearing forces within a fluid to produce colloid suspensions or emulsions in which the particle sizes are less than 1 micrometer. One type of colloid mill is called a *disc mill*, in which a mixture of a solid and liquid (or two liquids) is passed between two discs a small distance apart, which rotate very rapidly relative to each other. Other types of colloid mills are the *valve* and *orifice* types, in which the mixture is forced through an orifice at a very high speed and then strikes a breaker ring. Applications of colloid mills occur in food processing, in paint manufacture, and in the pharmaceutical industry.

colloids Colloids were originally defined by Thomas *Graham in 1861 as substances, such as starch or gelatin, which will not diffuse through a membrane. He distinguished them from *crystalloids* (e.g. inorganic salts), which would pass through membranes. Later it was recognized that colloids were distinguished from true solutions by the presence of particles that were too small to be observed with a normal microscope yet were much larger than normal molecules. Colloids are now regarded as systems in which there are two or more phases, with one (the *dispersed phase*) distributed in the other (the *continuous phase*). Moreover, at least one of the phases has small dimensions (in the range 10^{-9}–10^{-6} m). Colloids are classified in various ways.

Sols are dispersions of small solid particles in a liquid. The particles may be macromolecules or may be clusters of small molecules. *Lyophobic sols* are those in which there is no affinity between the dispersed phase and the liquid. An example is silver chloride dispersed in water. In such colloids the solid particles have a surface charge, which tends to stop them coming together. Lyophobic sols are inherently unstable and in time the particles aggregate and form a precipitate. *Lyophilic sols*, on the other hand, are more like true solutions in which the solute molecules are large and have an affinity for the solvent. Starch in water is an example of such a system. *Association colloids* are systems in which the dispersed phase consists of clusters of molecules that have lyophobic and lyophilic parts. Soap in water is an association colloid (*see* **micelle**).

Emulsions are colloidal systems in which the dispersed and continuous phases are both liquids, e.g. oil-in-water or water-in-oil. Such systems require an emulsifying agent to stabilize the dispersed particles.

Gels are colloids in which both dispersed and continuous phases have a three-dimensional network throughout the material, so that it forms a jelly-like mass. Gelatin is a common example. One component may sometimes be removed (e.g. by heating) to leave a rigid gel (e.g. silica gel).

Other types of colloid include *aerosols* (dispersions of liquid or solid particles in a gas, as in a mist or smoke) and foams (dispersions of gases in liquids or solids).

colophony *See* **rosin**.

colorimetric analysis Quantitative analysis of solutions by estimating their colour, e.g. by comparing it with the colours of standard solutions.

colour index (CI) A list, regarded as definitive, of dyes and pigments, which includes information on their commercial names, method of application, colour fastness, etc.

colour photography Any of various methods of forming coloured images on film or paper by photographic means. One common process is a subtractive reversal system that utilizes a film with three layers of light-sensitive emulsion, one responding to each of the three primary colours. On development a black image is formed where the scene is blue. The white areas are dyed yellow, the complementary colour of blue, and the blackened areas are bleached clean. A yellow filter between this emulsion layer and the next keeps blue light from the second emulsion, which is green-sensitive. This is dyed magenta where no green light has fallen. The final emulsion is red-sensitive and is given a cyan (blue-green) image on the negative after dying. When white light shines through the three dye layers the cyan dye subtracts red where it does not occur in the scene, the magenta subtracts green, and the yellow subtracts blue. The light projected by the negative therefore reconstructs the original scene either as a transparency or for use with printing paper.

columbium A former name for the element *niobium.

column chromatography *See* **chromatography**.

combinatorial chemistry A technique in which very large numbers of related compounds are formed in small quantities in cells on a plate, and properties are investigated by some automated technique. It is particularly used in the development of new drugs.

combustion A chemical reaction in which a substance reacts rapidly with oxygen with the production of heat and light. Such reactions are often free-radical chain reactions, which can usually be summarized as the oxidation of carbon to form its oxides and the oxidation of hydrogen to form water. *See also* **flame**.

commensurate lattice A lattice that can be divided into two or more sublattices, with the basis vectors of the lattice being a rational multiple of the basis vectors of the sublattice. The phase transition between a commensurate lattice and an incommensurate lattice can be analysed using the Frenkel–Kontorowa model.

common ion An ion that is common to two or more components in a mixture. In a solution of XCl, for example, the addition of another chloride, YCl, may precipitate XCl because the solubility product is exceeded as a result of the extra concentration of Cl^- ions. This is an example of a *common-ion effect*.

common salt *See* **sodium chloride**.

competitive inhibition *See* **inhibition**.

complementarity The concept that a single model may not be adequate to explain all the observations made of atomic or subatomic systems in different experiments. For example, *electron diffraction is best explained by assuming that the electron is a wave (*see* **de Broglie wavelength**), whereas the *photoelectric effect is described by assuming that it is a particle. The idea of two different but complementary concepts to treat quantum phenomena was first put forward by the Danish physicist Niels Bohr (1855–1962) in 1927.

complex A compound in which molecules or ions form coordinate bonds to a metal atom or ion (see illustration). Often complexes occur as *complex ions*, such as $[Cu(H_2O)_6]^{2+}$ or $Fe[(CN)_6]^{3-}$. A complex may also be a neutral molecule (e.g. $PtCl_2(NH_3)_2$). The formation of such coordination complexes is typical behaviour of transition metals. The complexes formed are often coloured and have unpaired electrons (i.e. are paramagnetic). *See also* **ligand**; **chelate**.

complex ion *See* **complex**.

complexometric analysis A type of volumetric analysis in which the reaction involves the formation of an inorganic *complex.

component A distinct chemical species in a mixture. If there are no reactions taking place, the number of components is the number of separate chemical species. A mixture of water and ethanol, for instance, has two components (but is a single phase). A mixture of ice and water has two phases but one component (H_2O). If an equilibrium reaction occurs, the number of components is taken to be the number of chemical species minus the number of reactions. Thus, in

$H_2 + I_2 \rightleftharpoons 2HI$

there are two components. *See also* **phase rule.**

compound A substance formed by the combination of elements in fixed proportions. The formation of a compound involves a chemical reaction; i.e. there is a change in the configuration of the valence electrons of the atoms. Compounds, unlike mixtures, cannot be separated by physical means. *See also* **molecule.**

comproportionation A reaction in which an element in a higher oxidation state reacts with the same element in a lower oxidation state to give the element in an intermediate oxidation state. For example

$Ag^{2+}(aq) + Ag(s) \rightarrow 2Ag^{+}(aq)$

It is the reverse of *disproportionation.

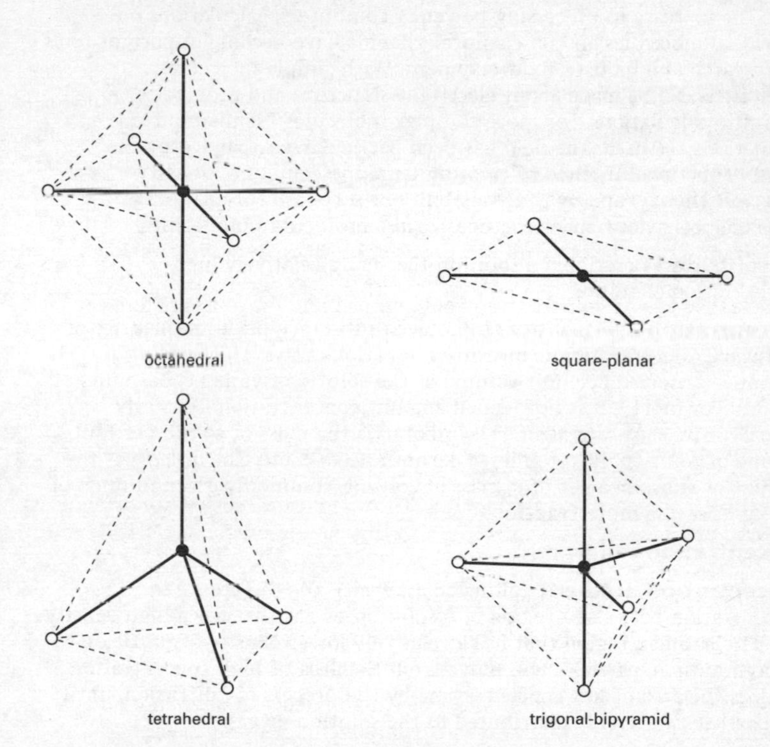

Some common shapes of coordination complexes

Compton, Arthur Holly (1892–1962) US physicist, who became professor of physics at the University of Chicago in 1923. He is best known for his discovery (1923) of the *Compton effect, for which he shared the 1927 Nobel Prize with C. T. R. Wilson.

Compton effect The reduction in the energy of high-energy (X-ray or gamma-ray) photons when they are scattered by free electrons, which thereby gain energy. The phenomenon, first observed in 1923 by A. H. *Compton, occurs when the photon collides with an electron; some of the photon's energy is transferred to the electron and consequently the photon loses energy $h(\nu_1 - \nu_2)$, where h is the *Planck constant and ν_1 and ν_2 are the frequencies before and after collision. As $\nu_1 > \nu_2$, the wavelength of the radiation increases after the collision. This type of inelastic scattering is known as *Compton scattering* and is similar to the *Raman effect. *See also* **inverse Compton effect**.

computational chemistry The use of computers in chemical research. With the increase in processing power of computers, calculations on individual molecules and on chemical systems have become important tools for research and industrial development. With simple molecules, predictions can be made about electronic structure and properties using *ab-initio calculations. For more complex molecules *semi-empirical calculations are used. The field has been particularly expanded by the density-functional method of treating large molecules (*see* **density-function theory**) and by the availability of software for analysing molecular behaviour and structure. *See also* **molecular modelling**.

concentrated Describing a solution that has a relatively high concentration of solute.

concentration The quantity of dissolved substance per unit quantity of a solution. Concentration is measured in various ways. The amount of substance dissolved per unit volume of the solution (symbol c) has units of $\mathrm{mol\,dm^{-3}}$ or $\mathrm{mol\,l^{-1}}$. It is now called amount concentration (formerly *molarity*). The *mass concentration* (symbol ρ) is the mass of solute per unit volume of solution. It has units of $\mathrm{kg\,dm^{-3}}$, $\mathrm{g\,cm^{-3}}$, etc. The *molality* is the amount of substance per unit mass of solvent, commonly given in units of $\mathrm{mol\,kg^{-1}}$. *See also* **mole fraction**.

concentration cell *See* **cell**.

concentration gradient (diffusion gradient) The difference in concentration between a region of a solution or gas that has a high density of particles and a region that has a relatively lower density of particles. By random motion, particles will move from the area of high concentration towards the area of low concentration, by the process of *diffusion, until the particles are evenly distributed in the solution or gas.

condensation The change of a vapour or gas into a liquid. The change of phase is accompanied by the evolution of heat (*see* **latent heat**).

condensation polymerization *See* **polymer**.

condensation pump *See* **diffusion pump**.

condensation reaction A chemical reaction in which two molecules combine to form a larger molecule with elimination of a small molecule (e.g. H_2O). *See* **aldehydes; ketones.**

condenser A device used to cool a vapour to cause it to condense to a liquid. *See* **Liebig condenser.**

conducting polymer An organic polymer that conducts electricity. Conducting polymers have a crystalline structure in which chains of conjugated unsaturated carbon–carbon bonds are aligned. Examples are polyacetylene and polypyrrole. There has been considerable interest in the development of such materials because they would be cheaper and lighter than metallic conductors. They do, however, tend to be chemically unstable and, so far, no commercial conducting polymers have been developed.

conductiometric titration A type of titration in which the electrical conductivity of the reaction mixture is continuously monitored as one reactant is added. The equivalence point is the point at which this undergoes a sudden change. The method is used for titrating coloured solutions, which cannot be used with normal indicators.

conduction band *See* **energy bands.**

conductivity water *See* **distilled water.**

Condy's fluid A mixture of calcium and potassium permanganates (manganate(VII)) used as an antiseptic.

configuration 1. The arrangement of atoms or groups in a molecule. **2.** The arrangement of electrons about the nucleus of an *atom.

configuration space The *n*-dimensional space with coordinates $(q_1,q_2,...,q_n)$ associated with a system that has *n* degrees of freedom, where the values *q* describe the degrees of freedom. For example, in a gas of *N* atoms each atom has three positional coordinates, so the configuration space is 3*N*-dimensional. If the particles also have internal degrees of freedom, such as those caused by vibration and rotation in a molecule, then these must be included in the configuration space, which is consequently of a higher dimension. *See also* **statistical mechanics.**

conformation One of the very large number of possible spatial arrangements of atoms that can be interconverted by rotation about a single bond in a molecule. In the case of ethane, $H_3C–CH_3$, one methyl group can rotate relative to the other. There are two extreme cases. In one, the C–H bonds on one group align with the C–H bonds on the other (as viewed along the C–C bond). This is an *eclipsed* conformation (or *eclipsing* conformation) and corresponds to a maximum in a graph of potential energy against rotation angle. In the other the C–H bonds on one group bisect the angle between two C–H bonds on the other. This is a *staggered* conformation (or *bisecting* conformation) and corresponds to a minimum in the potential-energy curve. In the case of ethane, the energy difference

eclipsed conformation anti conformation gauche conformation

● = methyl group

Conformations of butane (sawhorse projection)

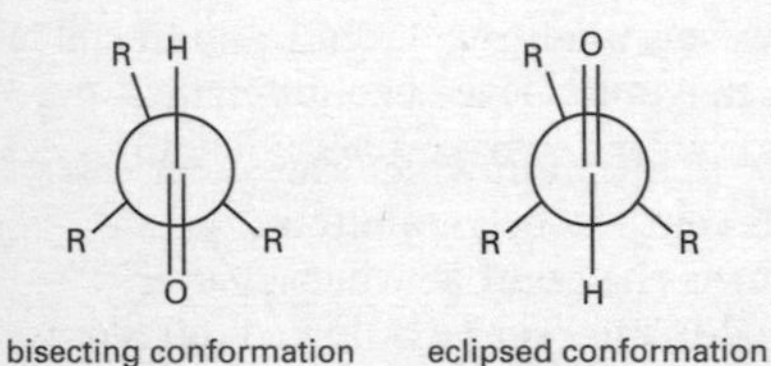

Conformations of R_3CHO (Newman projection)

between the two conformations is small (2.8 kcal/mole) and effective free rotation occurs under normal conditions.

In the case of butane, rotation can occur about the bond between the second and third carbon atoms, $MeH_2C–CH_2Me$. The highest maximum occurs when the methyl groups are eclipsed (they are at their closest). Maxima of lower energies occur when the methyl groups on one carbon atom eclipse the hydrogen atoms on the other. The lowest potential energy occurs when a C–Me bond bisects the angle between the C–H bonds (i.e. the two methyl groups are furthest apart). This is called the *anti* conformation. A lesser minimum occurs when the methyl group bisects the angle between a C–H bond and a C–Me bond. This arrangement is called the *gauche* conformation.

Similar rotational arrangements occur in other types of molecule. For instance, in a compound such as $R_3C–CHO$, with a C=O double bond, one conformation has the C=O bond bisecting the angle between C–R bonds. This is called the *bisecting* conformation (note that the C–H bond eclipses one of the C–R bonds in this case). The other has the C=O bond eclipsing a C–R bond. This is the *eclipsed* or *eclipsing* conformation. (In this case the C–H bond bisects the angle between the C–R bonds.)

See also **torsion angle**; **ring conformations.**

conformational analysis The determination or estimation of the relative energies of possible conformations of a molecule and the effect of these on the molecule's properties.

conformational isomer (conformer) One of a set of stereoisomers that differ from each other by torsion angle or angles (where only structures that are minima of potential energy are considered).

congeners Elements that belong to the same group in the periodic table.

conjugate acid *or* **base** *See* acid.

conjugated Describing double or triple bonds in a molecule that are separated by one single bond. For example, the organic compound buta-1,3-diene, $H_2C{=}CH{-}CH{=}CH_2$, has conjugated double bonds. In such molecules, there is some delocalization of electrons in the pi orbitals between the carbon atoms linked by the single bond.

conjugate solutions Solutions formed between two liquids that are partially miscible with one another; such solutions are in equilibrium at a particular temperature. Examples of conjugate solutions are phenol in water and water in phenol.

conjugation Delocalization of pi electrons as occurs in *conjugated systems. Conjugation can also involve *d*-orbitals and lone pairs of electrons.

conservation The sensible use of the earth's natural resources in order to avoid excessive degradation and impoverishment of the environment. It should include the search for alternative food and fuel supplies when these are endangered (as by deforestation and overfishing); an awareness of the dangers of *pollution; and the maintenance and preservation of natural habitats and the creation of new ones, e.g. nature reserves, national parks, and sites of special scientific intrest (SSSIs).

conservation law A law stating that the total magnitude of a certain physical property of a system, such as its mass, energy, or charge, remain unchanged even though there may be exchanges of that property between components of the system. For example, imagine a table with a bottle of salt solution (NaCl), a bottle of silver nitrate solution ($AgNO_3$), and a beaker standing on it. The mass of this table and its contents will not change even when some of the contents of the bottles are poured into the beaker. As a result of the reaction between the chemicals two new substances (silver chloride and sodium nitrate) will appear in the beaker:

$$NaCl + AgNO_3 \rightarrow AgCl + NaNO_3,$$

but the total mass of the table and its contents will not change. This *conservation of mass* is a law of wide and general applicability, which is true for the universe as a whole, provided that the universe can be considered a closed system (nothing escaping from it, nothing being added to it). According to Einstein's mass–energy relationship, every quantity of energy (E) has a mass (m), which is given by E/c^2, where c is the speed of light. Therefore if mass is conserved, the law of *conservation of energy* must be of equally wide application.

consolute temperature The temperature at which two partially miscible liquids become fully miscible as the temperature is increased.

constantan An alloy having an electrical resistance that varies only very slightly with temperature (over a limited range around normal room temperatures). It consists of copper (50–60%) and nickel (40–50%) and is used in resistance wire, thermocouples, etc.

constant-boiling mixture *See* **azeotrope**.

constant proportions *See* **chemical combination**.

constitutional isomerism *See* **isomerism**.

contact insecticide Any insecticide (*see* **pesticide**) that kills its target insect by being absorbed through the cuticle or by blocking the spiracles, rather than by being ingested.

contact process A process for making sulphuric acid from sulphur dioxide (SO_2), which is made by burning sulphur or by roasting sulphide ores. A mixture of sulphur dioxide and air is passed over a hot catalyst

$$2SO_2 + O_2 \rightarrow 2SO_3$$

The reaction is exothermic and the conditions are controlled to keep the temperature at an optimum 450°C. Formerly, platinum catalysts were used but vanadium–vanadium oxide catalysts are now mainly employed (although less efficient, they are less susceptible to poisoning). The sulphur trioxide is dissolved in sulphuric acid

$$H_2SO_4 + SO_3 \rightarrow H_2S_2O_7$$

and the oleum is then diluted.

continuous phase *See* **colloids**.

continuous spectrum *See* **spectrum**.

convection A process by which heat is transferred from one part of a fluid to another by movement of the fluid itself. In *natural convection* the movement occurs as a result of gravity; the hot part of the fluid expands, becomes less dense, and is displaced by the colder denser part of the fluid as this drops below it. This is the process that occurs in most domestic hot-water systems between the boiler and the hot-water cylinder. A natural convection current is set up transferring the hot water from the boiler up to the cylinder (always placed above the boiler) so that the cold water from the cylinder can move down into the boiler to be heated. In some modern systems, where small-bore pipes are used or it is inconvenient to place the cylinder above the boiler, the circulation between boiler and hot-water cylinder relies upon a pump. This is an example of *forced convection*, where hot fluid is transferred from one region to another by a pump or fan.

converter The reaction vessel in the *Bessemer process or some similar steel-making process.

coomassie blue A biological dye used for staining proteins.

coordinate bond *See* **chemical bond**.

coordination compound A compound in which coordinate bonds are formed (*see* **chemical bond**). The term is used especially for inorganic *complexes.

coordination number The number of groups, molecules, atoms, or ions surrounding a given atom or ion in a complex or crystal. For instance, in a square-planar complex the central ion has a coordination number of four. In a close-packed crystal (*see* **close packing**) the coordination number is twelve.

copolymer *See* **polymer.**

copper Symbol Cu. A red-brown *transition element; a.n. 29; r.a.m. 63.546; r.d. 8.92; m.p. 1083.4°C; b.p. 2567°C. Copper has been extracted for thousands of years; it was known to the Romans as cuprum, a name linked with the island of Cyprus. The metal is malleable and ductile and an excellent conductor of heat and electricity. Copper-containing minerals include cuprite (Cu_2O) as well as azurite ($2CuCO_3.Cu(OH)_2$), chalcopyrite ($CuFeS_2$), and malachite ($CuCO_3.Cu(OH)_2$). Native copper appears in isolated pockets in some parts of the world. The large mines in the USA, Chile, Canada, Zambia, Congo, and Peru extract ores containing sulphides, oxides, and carbonates. They are usually worked by smelting, leaching, and electrolysis. Copper metal is used to make electric cables and wires. Its alloys, brass (copper–zinc) and bronze (copper–tin), are used extensively.

Water does not attack copper but in moist atmospheres it slowly forms a characteristic green surface layer (patina). The metal will not react with dilute sulphuric or hydrochloric acids, but with nitric acid oxides of nitrogen are formed. Copper compounds contain the element in the +1 and +2 oxidation states. Copper(I) compounds are mostly white (the oxide is red). Copper(II) salts are blue in solution. The metal also forms a large number of coordination complexes.

copperas *See* **iron(II) sulphate.**

copper(I) chloride A white solid compound, CuCl; cubic; r.d. 4.14; m.p. 430°C; b.p. 1490°C. It is obtained by boiling a solution containing copper(II) chloride, excess copper turnings, and hydrochloric acid. Copper(I) is present as the $[CuCl_2]^-$ complex ion. On pouring the solution into air-free distilled water copper(I) chloride precipitates. It must be kept free of air and moisture since it oxidizes to copper(II) chloride under those conditions.

Copper(I) chloride is essentially covalent and its structure is similar to that of diamond; i.e. each copper atom is surrounded tetrahedrally by four chlorine atoms and vice versa. In the vapour phase, dimeric and trimeric species are present. Copper(I) chloride is used in conjunction with ammonium chloride as a catalyst in the dimerization of ethyne to but-1-ene-3-yne (vinyl acetylene), which is used in the production of synthetic rubber. In the laboratory a mixture of copper(I) chloride and hydrochloric acid is used for converting benzene diazonium chloride to chlorobenzene – the Sandmeyer reaction.

copper(II) chloride A brown-yellow powder, $CuCl_2$; r.d. 3.386; m.p. 620°C.

It exists as a blue-green dihydrate (rhombic; r.d. 2.54; loses H_2O at 100°C). The anhydrous solid is obtained by passing chlorine over heated copper. It is predominantly covalent and adopts a layer structure in which each copper atom is surrounded by four chlorine atoms at a distance of 0.23 and two more at a distance of 0.295. A concentrated aqueous solution is dark brown in colour due to the presence of complex ions such as $[CuCl_4]^{2-}$. On dilution the colour changes to green and then blue because of successive replacement of chloride ions by water molecules, the final colour being that of the $[Cu(H_2O)_6]^{2+}$ ion. The dihydrate can be obtained by crystallizing the solution.

copper glance A mineral form of copper(I) sulphide, Cu_2S.

copper(II) nitrate A blue deliquescent solid, $Cu(NO_3)_2.3H_2O$; r.d. 2.32; m.p. 114.5°C. It may be obtained by reacting either copper(II) oxide or copper(II) carbonate with dilute nitric acid and crystallizing the resulting solution. Other hydrates containing 6 or 9 molecules of water are known. On heating it readily decomposes to give copper(II) oxide, nitrogen dioxide, and oxygen. The anhydrous form can be obtained by reacting copper with a solution of nitrogen dioxide in ethyl ethanoate. It sublimes on heating suggesting that it is appreciably covalent.

copper(I) oxide A red insoluble solid, Cu_2O; r.d. 6.0; m.p. 1235°C. It is obtained by reduction of an alkaline solution of copper(II) sulphate. Since the addition of alkalis to a solution of copper(II) salt results in the precipitation of copper(II) hydroxide the copper(II) ions are complexed with tartrate ions; under such conditions the concentration of copper(II) ions is so low that the solubility product of copper(II) hydroxide is not exceeded.

When copper(I) oxide reacts with dilute sulphuric acid a solution of copper(II) sulphate and a deposit of copper results, i.e. disproportionation occurs.

$$Cu_2O + 2H^+ \rightarrow Cu^{2+} + Cu + H_2O$$

When dissolved in concentrated hydrochloric acid the $[CuCl_2]^-$ complex ion is formed. Copper(I) oxide is used in the manufacture of rectifiers and the production of red glass.

copper(II) oxide A black insoluble solid, CuO; monoclinic; r.d. 6.3; m.p. 1326°C. It is obtained by heating either copper(II) carbonate or copper(II) nitrate. It decomposes on heating above 800°C to copper(I) oxide and oxygen. Copper(II) oxide reacts readily with mineral acids on warming, with the formation of copper(II) salts; it is also readily reduced to copper on heating in a stream of hydrogen. Copper(II) oxide is soluble in dilute acids forming blue solutions of cupric salts.

copper pyrites *See* **chalcopyrite**.

copper(II) sulphate A blue crystalline solid, $CuSO_4.5H_2O$; triclinic; r.d. 2.284. The pentahydrate loses $4H_2O$ at 110°C and the fifth H_2O at 150°C to form the white anhydrous compound (rhombic; r.d. 3.6; decomposes above 200°C). The pentahydrate is prepared either by reacting copper(II) oxide or copper(II) carbonate with dilute sulphuric acid; the solution is heated to saturation and the blue pentahydrate crystallizes out on cooling (a few drops of dilute sulphuric acid are generally added to prevent hydrolysis). It is obtained on an industrial scale by forcing air through a hot mixture of copper and dilute sulphuric acid. In the pentahydrate each copper(II) ion is surrounded by four water molecules at the corner of a square, the fifth and sixth octahedral positions are occupied by oxygen atoms from the sulphate anions, and the fifth water molecule is held in place by hydrogen bonding. Copper(II) sulphate has many industrial uses, including the preparation of the Bordeaux mixture (a fungicide) and the preparation of other copper compounds. It is also used in electroplating and textile dying and as a timber preservative. The anhydrous form is used in the detection of traces of moisture.

Copper(II) sulphate pentahydrate is also known as *blue vitriol.*

cordite An explosive mixture of cellulose nitrate and nitroglycerin, with added plasticizers and stabilizers, used as a propellant for guns.

core orbital An atomic orbital that is part of the inner closed shell of an atom. Unlike the *valence orbitals*, which are the atomic orbitals of the valence shell of an atom, core orbitals on one atom have a small overlap with core orbitals on another atom. They are therefore usually ignored in simple calculations of chemical bonding.

Corey–Pauling rules A set of rules, formulated by Robert Corey and Linus *Pauling, that govern the secondary nature of proteins. The Corey–Pauling rules are concerned with the stability of structures provided by *hydrogen bonds associated with the –CO–NH– peptide link. The Corey–Pauling rules state that:
(1) All the atoms in the peptide link lie in the same plane.
(2) The N, H, and O atoms in a hydrogen bond are approximately on a straight line.
(3) All the CO and NH groups are involved in bonding.
Two important structures in which the Corey–Pauling rules are obeyed are the *alpha helix and the *beta sheet.

CORN rule *See* **absolute configuration.**

correlation diagram A diagram that relates the energy levels of separate atoms to the energy levels of diatomic molecules and united atoms, the two limiting states being correlated with each other. Using this diagram, the types of molecular orbital possible for the molecules and their order as a function of increasing energy can be seen as a function of interatomic distance. Lines are drawn linking each united-atom orbital to a separated atom orbital. An important rule for determining which energy levels correlate is the *noncrossing rule*. This states that two curves of energy plotted against interatomic distance never cross if the invariant properties (e.g. parity, spin, etc.) are the same.

correlation functions Quantities used in condensed-matter physics that are *ensemble averages of products of such quantities as *density at different points in space. Each particle affects the behaviour of its neighbouring particles, so creating *correlations*, the range of which is at least as long as the range of the intermolecular forces. Thus, correlation functions contain a great deal of information about systems in condensed-matter physics. Correlation functions can be measured in considerable detail, frequently by experiments involving scattering of such particles as neutrons or of *electromagnetic radiation (e.g. light or X-rays) and can be calculated theoretically using *statistical mechanics.

correlation spectroscopy (COSY) A type of spectroscopy used in *nuclear magnetic resonance (NMR) in which the pulse sequence used is: $90°x - t_1 - 90°x$ – acquire (t_2). The delay t_1 is variable and a series of acquisitions is made. Fourier transforms are then performed on both the delay t_1 and the real time t_2. The information thus gained can be shown on a diagram that plots contours of signal intensity. This representation of the information enables NMR spectra to be interpreted more readily than in one-dimensional NMR, particularly for complex spectra.

corrosion Chemical or electrochemical attack on the surface of a metal. *See also* **electrolytic corrosion**; **rusting**.

corundum A mineral form of aluminium oxide, Al_2O_3. It crystallizes in the trigonal system and occurs as well-developed hexagonal crystals. It is colourless and transparent when pure but the presence of other elements gives rise to a variety of colours. *Ruby is a red variety containing chromium; *sapphire is a blue variety containing iron and titanium. Corundum occurs as a rock-forming mineral in both metamorphic and igneous rocks. It is chemically resistant to weathering processes and so also occurs in alluvial (placer) deposits. The second hardest mineral after diamond (it has a hardness of 9 on the Mohs' scale), it is used as an abrasive.

COSY *See* **correlation spectroscopy**.

COT *See* **cyclo-octatetraene**.

Cotton effect The wavelength dependence of the optical rotary dispersion curve or the *circular dichroism curve in the neighbourhood of an absorption band, both having characteristic shapes. If the wavelength is decreased, the rotation angle increases until it reaches a maximum and then decreases, passing through zero at the wavelength at which the maximum of absorption occurs. As the wavelength is decreased further the angle becomes negative, until it reaches a minimum after which it rises again. This pattern is called the *positive Cotton effect*. A mirror image of this pattern can also occur about the λ-axis, where λ is the wavelength; this is called the *negative Cotton effect*. These effects occur for coloured substances and for colourless substances with bands in the ultraviolet. It is named after the French physicist Aimé Cotton (1859–1951).

coulomb Symbol C. The *SI unit of electric charge. It is equal to the charge transferred by a current of one ampere in one second. The unit is named after the French physicist Charles de Coulomb (1736–1806).

Coulomb explosion The sudden disruption of a molecule from which the electrons have been stripped to leave only the nuclei, which repel each other because of their electric charge. The technique of *Coulomb explosion imaging* uses this effect to investigate the shape of molecules. A beam of high-energy neutral molecules is produced by first adding electrons, accelerating the ions in an electric field, and then removing the electrons. The beam collides with a thin metal foil having a thickness of about 30 atoms. As the molecules pass through this foil their electrons are scattered and only the nuclei of the molecules emerge. The process occurs within a very short period of time, shorter than the time required for a complete molecular vibration, and consequently the nuclei retain the molecular shape until they are suddenly repulsed by the like charges. The nuclei then impinge on a detector that records their velocity and direction, thus enabling the spatial arrangement of the original molecule to be derived.

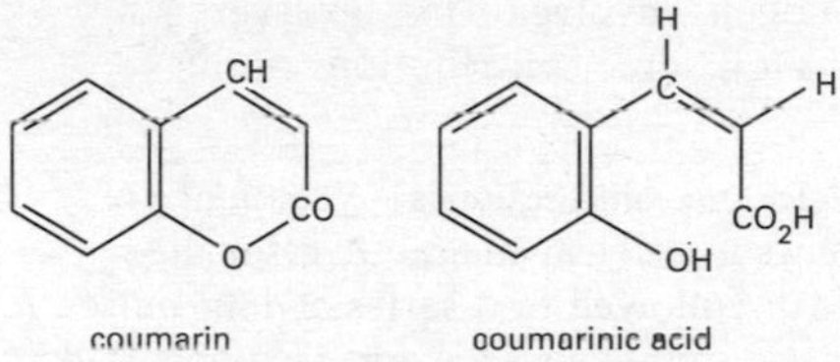

coumarin coumarinic acid

Coumarin and coumarinic acid

coumarin (1,2-benzopyrone) A pleasant-smelling colourless crystalline compound, $C_9H_6O_2$, m.p. 70°C. It occurs naturally in tonka (or tonquin) beans, and is synthesized from salicylaldehyde. It forms *coumarinic acid* on hydrolysis with sodium hydroxide. (See formulae.) Coumarin is used in making perfumes, to scent tobacco, and as an anticoagulant in medicine; *warfarin is derived from it.

coumarone *See* **benzfuran**.

coupling 1. An interaction between two different parts of a system or between two or more systems. Examples of coupling in the *spectra of atoms and nuclei are *Russell–Saunders coupling, *j-j coupling, and spin–orbit coupling. In the spectra of molecules there are five idealized ways (called the *Hund coupling cases*) in which the different types of angular momentum in a molecule (the electron orbital angular momentum *L*, the electron spin angular momentum *S*, and the angular momentum of nuclear rotation *N*) couple to form a resultant angular momentum *J*. (In practice, the coupling for many molecules is intermediate between Hund's cases due to interactions, which are ignored in the idealized cases.) **2.** A type of chemical reaction in which two molecules join together; for example, the formation of an *azo compound by coupling of a diazonium ion with a benzene ring.

covalent bond *See* **chemical bond.**

covalent crystal A crystal in which the atoms are held together by covalent bonds. Covalent crystals are sometimes called *macromolecular* or *giant-molecular crystals.* They are hard high-melting substances. Examples are diamond and boron nitride.

covalent radius An effective radius assigned to an atom in a covalent compound. In the case of a simple diatomic molecule, the covalent radius is half the distance between the nuclei. Thus, in Cl_2 the internuclear distance is 0.198 nm so the covalent radius is taken to be 0.099 nm. Covalent radii can also be calculated for multiple bonds; for instance, in the case of carbon the values are 0.077 nm for single bonds, 0.0665 nm for double bonds, and 0.0605 nm for triple bonds. The values of different covalent radii can sometimes be added to give internuclear distances. For example, the length of the bond in interhalogens (e.g. ClBr) is nearly equal to the sum of the covalent radii of the halogens involved. This, however, is not always true because of other effects (e.g. ionic contributions to the bonding).

CPMG sequence (Carr–Purcell–Meiboom–Gill sequence) A sequence of pulses used for spin-echo experiments in nuclear magnetic resonance (NMR) in which the initial pulse is 90° followed by a series of 180° pulses. A CPMG sequence is designed so that the spin echoes die away exponentially with time. Spin–spin relaxation occurs characterized by a time constant T_2, which can be determined from the decay signal.

cracking The process of breaking down chemical compounds by heat. The term is applied particularly to the cracking of hydrocarbons in the kerosine fraction obtained from *petroleum refining to give smaller hydrocarbon molecules and alkenes. It is an important process, both as a source of branched-chain hydrocarbons suitable for gasoline (for motor fuel) and as a source of ethene and other alkenes. *Catalytic cracking* is a similar process in which a catalyst is used to lower the temperature required and to modify the products obtained.

cream of tartar *See* **potassium hydrogentartrate.**

creosote 1. (wood creosote) An almost colourless liquid mixture of phenols obtained by distilling tar obtained by the destructive distillation of wood. It is used medically as an antiseptic and expectorant. **2. (coal-tar creosote)** A dark liquid mixture of phenols and cresols obtained by distilling coal tar. It is used for preserving timber.

cresols *See* **methylphenols.**

cristobalite A mineral form of *silicon(IV) oxide, SiO_2.

critical pressure The pressure of a fluid in its *critical state; i.e. when it is at its critical temperature and critical volume.

critical state The state of a fluid in which the liquid and gas phases both have the same density. The fluid is then at its critical temperature, critical pressure, and critical volume.

critical temperature **1.** The temperature above which a gas cannot be liquefied by an increase of pressure. *See also* **critical state.** **2.** *See* **transition point.**

critical volume The volume of a fixed mass of a fluid in its *critical state; i.e. when it is at its critical temperature and critical pressure. The *critical specific volume* is its volume per unit mass in this state: in the past this has often been called the critical volume.

Crookes, Sir William (1832–1919) British chemist and physicist, who in 1861 used *spectroscopy to discover *thallium and in 1875 invented the radiometer. He also developed an improved vacuum tube (*Crookes' tube*) for studying gas discharges. Crookes was also involved in industrial chemistry and realized the importance of nitrogen fixation for fertilizers.

crossed-beam reaction A chemical reaction in which two molecular beams are crossed; one beam is regarded as the incident beam of gas and the other as the target gas. This technique enables a great deal of information to be gained about the chemical reaction since the states of both the target and projectile molecules can be controlled. The incident beam is characterized by its *incident beam flux*, I, which is the number of particles per unit area per unit time. The scattered molecules can be detected first by ionizing them and then detecting the ions electrically or using spectroscopy if changes in the vibrational or rotational states of molecules in a reaction are of interest.

cross linkage A short side chain of atoms linking two longer chains in a polymeric material.

crown *See* **ring conformations.**

crown ethers Organic compounds with molecules containing large rings of carbon and oxygen atoms. The crown ethers are macrocyclic polyethers. The first to be synthesized was the compound 18-crown-6, which consists of a ring of six $-CH_2-CH_2-O-$ units (i.e. $C_{12}H_{24}O_6$). The general method of naming crown ethers is to use the form n-crown-m, where n is the number of atoms in the ring and m is the number of oxygen atoms. Substituted crown ethers can also be made. The crown ethers are able to form strongly bound complexes with metal ions by coordination through the oxygen atoms. The stability of these complexes depends on the size of the ion relative to the cavity available in the ring of the particular crown ether. Crown ethers also form complexes with ammonium ions (NH_4^+) and alkyl ammonium ions (RNH_3^+). They can be used for increasing the solubility of ionic salts in nonpolar solvents. For example, dicyclohexyl-18-crown-6

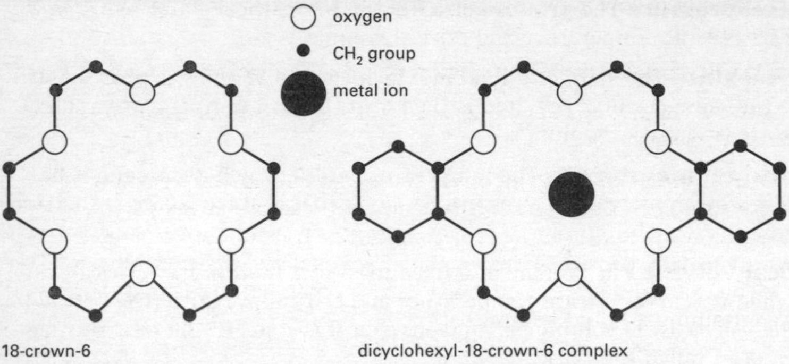

oxygen
CH_2 group
metal ion

18-crown-6 dicyclohexyl-18-crown-6 complex

Crown ethers

complexes with the potassium ion of potassium permanganate and allows it to dissolve in benzene, giving a purple neutral solution that can oxidize many organic compounds. They also act as catalysts in certain reactions involving organic salts by complexing with the positive metal cation and thereby increasing its separation from the organic anion, which shows a consequent increase in activity. Some of the uses of crown ethers depend on their selectivity for specific sizes of anions. Thus they can be used to extract specific ions from mixtures and enrich isotope mixtures. Their selectivity also makes them useful analytical reagents. *See also* **cryptands**.

crucible A dish or other vessel in which substances can be heated to a high temperature. *See also* **Gooch crucible**.

crude oil *See* **petroleum**.

Crum Brown's rule A rule that predicts how substituents will enter the benzene ring. If C_6H_5X is a compound with one substituent in the benzene ring, it will produce the 1,3 (meta) disubstituted derivative if the substance HX can be directly oxidized to HOX. If not, a mixture of 1,2 (ortho) and 1,4 (para) compounds will be formed. The rule was proposed in 1892 by the British chemist Alexander Crum Brown (1838–1922).

cryogenic pump A *vacuum pump in which pressure is reduced by condensing gases on surfaces maintained at about 20 K by means of liquid hydrogen or at 4 K by means of liquid helium. Pressures down to 10^{-8} mmHg (10^{-6} Pa) can be maintained; if they are used in conjunction with a *diffusion pump, pressures as low as 10^{-15} mmHg (10^{-13} Pa) can be reached.

cryohydrate A eutectic mixture of ice and some other substance (e.g. an ionic salt) obtained by freezing a solution.

cryolite A rare mineral form of sodium aluminofluoride, Na_3AlF_6, which crystallizes in the monoclinic system. It is usually white but may also be colourless. The only important occurrence of the mineral is in Greenland. It is used chiefly to lower the melting point of alumina in the production of aluminium.

cryoscopic constant *See* **depression of freezing point.**

cryoscopy The use of *depression of freezing point to determine relative molecular masses.

cryostat A vessel enabling a sample to be maintained at a very low temperature. The *Dewar flask is the most satisfactory vessel for controlling heat leaking in by radiation, conduction, or convection. Cryostats usually consist of two or more Dewar flasks nesting in each other. For example, a liquid nitrogen bath is often used to cool a Dewar flask containing a liquid helium bath.

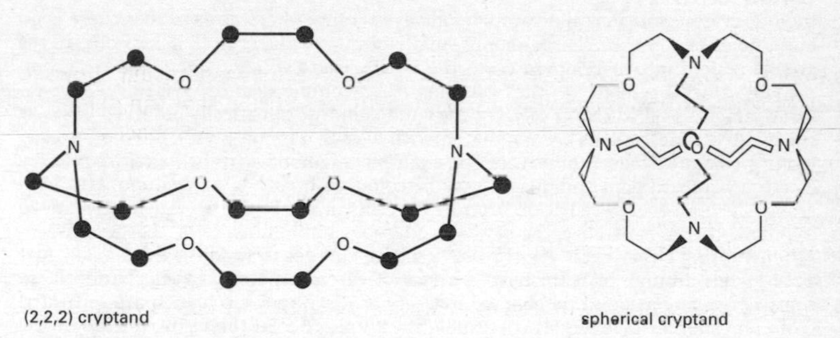

(2,2,2) cryptand

spherical cryptand

Cryptands

cryptands Compounds with large three-dimensional molecular structures containing ether chains linked by three-coordinate nitrogen atoms. Thus cryptands are macropolycyclic polyaza-polyethers. For example, the compound (2,2,2)-cryptand has three chains of the form

$-CH_2CH_2OCH_2CH_2OCH_2CH_2-$.

These chains are linked at each end by a nitrogen atom. Cryptands, like the *crown ethers, can form coordination complexes with ions that can fit into the cavity formed by the open three-dimensional structure, i.e. they can 'cryptate' the ion. Various types of cryptand have been produced having both spherical and cylindrical cavities. The cryptands have the same kind of properties as the crown ethers and the same uses. In general, they form much more strongly bound complexes and can be used to stabilize unusual ionic species. For example, it is possible to produce the negative Na^- ion in the compound $[(2,2,2)\text{-cryptand-Na}]^+Na^-$, which is a gold-coloured crystalline substance stable at room temperature. Cluster ions, such as Pb_5^{2-}, can be similarly stabilized.

crystal A solid with a regular polyhedral shape. All crystals of the same substance grow so that they have the same angles between their faces.

A *crystal lattice is formed by a repeated arrangement of atoms, ions, or molecules. Within one cubic centimetre of material one can expect to find up to 10^{22} atoms and it is extremely unlikely that all of these will be arranged in perfect order. Some atoms will not be exactly in the right place, with the result that the lattice will contain defects. The presence of defects within the crystal structure has profound consequences for certain bulk properties of the solid, such as the electrical resistance and the mechanical strength. It can also affect the chemical behaviour.

Point defects

Local crystal defects, called *point defects*, appear as either impurity atoms or gaps in the lattice. Impurity atoms can occur in the lattice either at *interstitial sites* (between atoms in a non-lattice site) or at *substitutional sites* (replacing an atom in the host lattice). Lattice gaps are called *vacancies* and arise when an atom is missing from its site in the lattice. Vacancies are sometimes called *Schottky defects*. A vacancy in which the missing atom has moved to an interstitial position is known as a *Frenkel defect*.

Colour centres

In ionic crystals, the ions and vacancies always arrange themselves so that there is no build-up of one type of charge in any small volume of the crystal. If ions or charges are introduced into or removed from the lattice, there will, in general, be an accompanying rearrangement of the ions and their outer valence electrons. This rearrangement is called *charge compensation* and is most dramatically observed in *colour centres*. If certain crystals are irradiated with X-rays, gamma rays, neutrons, or electrons a colour change is observed. For example, diamond may be coloured blue by electron bombardment and quartz may be coloured brown by irradiation with neutrons. The high-energy radiation produces defects in the lattice and, in an attempt to maintain charge neutrality, the crystal undergoes some measure of charge compensation, causing either extra electrons or positive charges to reside at the defects. Just as electrons around an atom have a series of discrete permitted energy levels, so charges residing at point defects exhibit sets of discrete levels, which are separated from one another by energies corresponding to photons in the visible region of the spectrum. Thus light of certain wavelengths can be absorbed at the defect sites, and the material appears to be coloured. Heating the irradiated crystal can, in many cases, repair the irradiation damage and the crystal loses its coloration.

Dislocations

Non-local defects may involve entire planes of atoms. The most important of these is called a *dislocation*. Dislocations are essentially *line-defects*; that is, there is an incomplete plane of atoms in the crystal lattice. In 1934, Taylor, Orowan, and Polanyi independently proposed the concept of the dislocation to account for the mechanical strength of metal crystals. Their microscopic studies revealed that when a metal crystal is plastically deformed, the deformation does not occur by a separation of individual atoms but rather by a slip of one plane of atoms over another plane. Dislo-

Point defects in a two-dimensional crystal

cations provide a mechanism for this slipping of planes that does not require the bulk movement of crystal material. The passage of a dislocation in a crystal is similar to the movement of a ruck in a carpet. A relatively large force is required to slide the carpet as a whole. However, moving a ruck over the carpet can inch it forward without needing such large forces. This movement of dislocations is called *plastic flow*.

Strength of materials

In practice most metal samples are *polycrystalline*; that is, they consist of many small crystals or grains at different angles to each other. The boundary between two such grains is called a *grain boundary*. The plastic flow of dislocations may be hindered by the presence of grain boundaries, impurity atoms, and other dislocations. Pure metals produced commercially are generally too weak to be of much mechanical use. A rod of pure copper the thickness of a pencil is easily bent in the hand. The weakness of these samples can be attributed to the ease with which the dislocations are able to move within the sample. Slip, and therefore deformation, can then occur under relatively low stresses. Impurity atoms, other dislocations, and grain boundaries can all act as obstructions to the slip of atomic planes. Traditionally, methods of making metals stronger involved introducing defects that provide regions of disorder in the material. For example, in an alloy, such as steel, impurity atoms (e.g. carbon) are introduced into the lattice during the forging process. The perfection of the iron lattice structure is disturbed and the impurities oppose the dislocation motion. This makes for greater strength and stiffness.

The complete elimination of dislocations may seem an obvious way to strengthen materials. However, this has only proved possible for hairlike single crystal specimens called *whiskers*. These whiskers are only a few micrometres thick and are seldom more than a few millimetres long; nevertheless their strength approaches the theoretical value.

Thermal vibrations

However carefully one may prepare a specimen, using extremely high purity materials and specially controlled methods of crystal growth, defects will occur due to the random thermal vibrations of atoms. Atoms are always vibrating about their fixed lattice sites. At room temperature they are vibrating at a frequency of about 10^{13} Hz. Even at very low temperatures the *zero-point motions of the atoms occurs. As the temperature rises, the amplitude of these atomic vibrations increases; as a result, virtually every property of a material determined by the positions of the atoms within it, or by its thermal energy, changes with temperature.

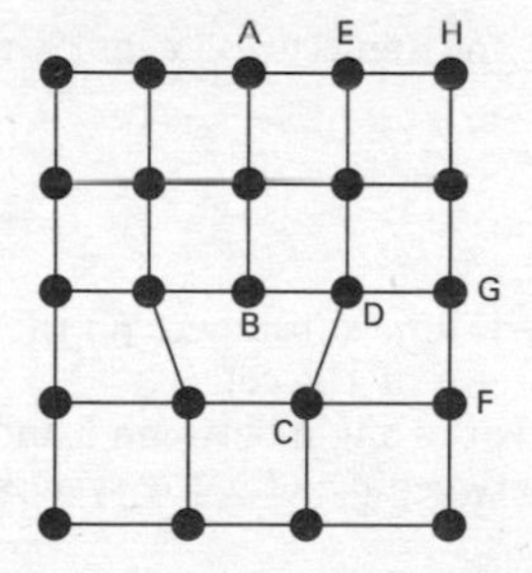

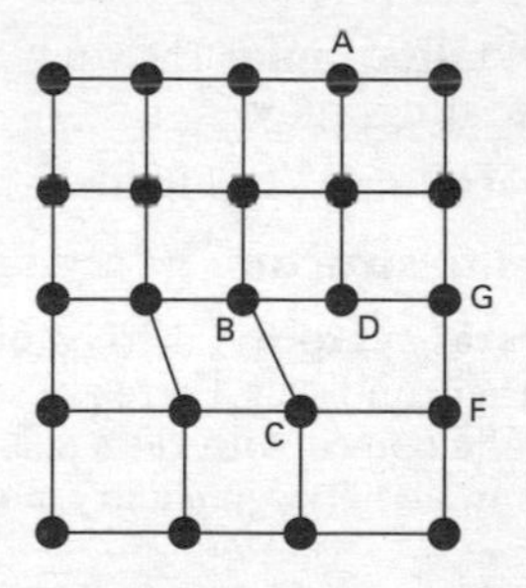

Dislocation in a two-dimensional crystal: the extra plane of atoms AB causes strain at bond CD. On breaking, the bond flips across to form CB. This incremental movement shifts the dislocation across so that the overall effect is to slide the two planes BDG and CF over each other

However, they may not have the same external appearance because different faces can grow at different rates, depending on the conditions. The external form of the crystal is referred to as the *crystal habit*. The atoms, ions, or molecules forming the crystal have a regular arrangement and this is the *crystal structure*.

crystal defect An imperfection in the regular lattice of a crystal. See Feature (pp 156–157).

crystal-field theory A theory of the electronic structures of inorganic *complexes, in which the complex is assumed to consist of a central metal atom or ion surrounded by ligands that are ions. For example, the complex $[PtCl_4]^{2-}$ is thought of as a Pt^{2+} ion surrounded by four Cl^- ions at the corners of a square. The presence of these ions affects the energies of the *d*-orbitals, causing a splitting of energy levels. The theory can be used to explain the spectra of complexes and their magnetic properties. *Ligand-field theory* is a development of crystal-field theory in which the overlap of orbitals is taken into account. Crystal-field theory was initiated in 1929 by the German-born US physicist Hans Albrecht Bethe (1906–) and extensively developed in the 1930s.

crystal habit *See* **crystal**.

crystal lattice The regular pattern of atoms, ions, or molecules in a crystalline substance. A crystal lattice can be regarded as produced by repeated translations of a *unit cell* of the lattice. *See also* **crystal system**.

crystalline Having the regular internal arrangement of atoms, ions, or molecules characteristic of crystals. Crystalline materials need not necessarily exist as crystals; all metals, for example, are crystalline although they are not usually seen as regular geometric crystals.

crystallite A small crystal, e.g. one of the small crystals forming part of a microcrystalline substance.

crystallization The process of forming crystals from a liquid or gas.

crystallography The study of crystal form and structure. *See also* **X-ray crystallography**.

crystalloids *See* **colloids**.

crystal structure *See* **crystal**.

crystal system A method of classifying crystalline substances on the basis of their unit cell. There are seven crystal systems. If the cell is a parallelopiped with sides a, b, and c and if α is the angle between b and c, β the angle between a and c, and γ the angle between a and b, the systems are:

(1) *cubic* $a=b=c$ and $\alpha=\beta=\gamma=90°$
(2) *tetragonal* $a=b\neq c$ and $\alpha=\beta=\gamma=90°$
(3) *rhombic* (or *orthorhombic*) $a\neq b\neq c$ and $\alpha=\beta=\gamma=90°$
(4) *hexagonal* $a=b\neq c$ and $\alpha=\beta=\gamma=90°$

(5) *trigonal* $a=b\neq c$ and $\alpha=\beta=\gamma\neq 90°$
(6) *monoclinic* $a\neq b\neq c$ and $\alpha=\gamma=90°\neq\beta$
(7) *triclinic* $a=b=c$ and $\alpha\neq\beta\neq\gamma$

CS gas The vapour from a white solid, $C_6H_4(Cl)CH:C(CN)_2$, causing tears and choking, used in 'crowd control'.

cubic close packing *See* **close packing**.

cubic crystal A crystal in which the unit cell is a cube (*see* **crystal system**). There are three possible packings for cubic crystals: *simple cubic*, *face-centred cubic*, and *body-centred cubic*. See illustration.

cumene process An industrial process for making phenol from benzene. A mixture of benzene vapour and propene is passed over a phosphoric acid catalyst at 250°C and high pressure

$C_6H_6 + CH_3CH{:}CH_2 \rightarrow C_6H_5CH(CH_3)_2$

The product is called *cumene*, and it can be oxidized in air to a peroxide, $C_6H_5C(CH_3)_2O_2H$. This reacts with dilute acid to give phenol (C_6H_5OH) and propanone (acetone, CH_3OCH_3), which is a valuable by-product.

cupellation A method of separating noble metals (e.g. gold or silver) from base metals (e.g. lead) by melting the mixture with a blast of hot air in a shallow porous dish (the *cupel*). The base metals are oxidized, the oxide being carried away by the blast of air or absorbed by the porous container.

cuprammonium ion The tetraamminecopper(II) ion $[Cu(NH_3)_4]^{2+}$. *See* **ammine**.

cupric compounds Compounds containing copper in its higher (+2) oxidation state; e.g. cupric chloride is copper(II) chloride ($CuCl_2$).

cuprite A red mineral cubic form of copper(I) oxide, Cu_2O; an important ore of copper. It occurs where deposits of copper have been subjected to oxidation. The mineral has been mined as a copper ore in Chile, Congo, Bolivia, Australia, Russia, and the USA.

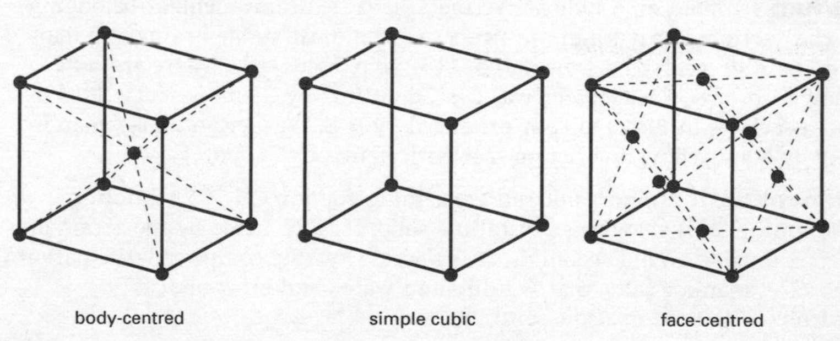

Cubic crystal structures

cupronickel A type of corrosion-resistant alloy of copper and nickel containing up to 45% nickel.

cuprous compounds Compounds containing copper in its lower (+1) oxidation state; e.g. cuprous chloride is copper(I) chloride (CuCl).

curare A resin obtained from the bark of South American trees of the genera *Strychnos* and *Chondrodendron* that causes paralysis of voluntary muscle. It acts by blocking the action of the neurotransmitter *acetylcholine at neuromuscular junctions. Curare is used as an arrow poison by South American Indians and was formerly used as a muscle relaxant in surgery.

curie The former unit of *activity (*see* **radiation units**). It is named after Marie *Curie.

Curie, Marie (Marya Sklodowska; 1867–1934) Polish-born French chemist, who went to Paris in 1891. She married the physicist *Pierre Curie* (1859–1906) in 1895 and soon began work on seeking radioactive elements other than uranium in pitchblende (to account for its unexpectedly high radioactivity). By 1898 she had discovered *radium and *polonium, although it took her four years to purify them. In 1903 the Curies shared the Nobel Prize for physics with Henri *Becquerel, who had discovered radioactivity.

Curie point **(Curie temperature)** The temperature at which a ferromagnetic substance loses its ferromagnetism and becomes only paramagnetic. For iron the Curie point is 760°C and for nickel 356°C. It is named after Pierre *Curie.

Curie's law The susceptibility (χ) of a paramagnetic substance is proportional to the thermodynamic temperature (T), i.e. $\chi = C/T$, where C is the Curie constant. A modification of this law, the *Curie–Weiss law*, is more generally applicable. It states that $\chi = C/(T - \theta)$, where θ is the Weiss constant, a characteristic of the material. The law was first proposed by Pierre *Curie and modified by another French physicist, Pierre-Ernest Weiss (1865–1940).

curium Symbol Cm. A radioactive metallic transuranic element belonging to the *actinoids; a.n. 96; mass number of the most stable isotope 247 (half-life 1.64×10^7 years); r.d. (calculated) 13.51; m.p. 1340±40°C. There are nine known isotopes. The element was first identified by Glenn Seaborg (1912–99) and associates in 1944 and first produced by L. B. Werner and I. Perlman in 1947 by bombarding americium-241 with neutrons.

cyanamide 1. An inorganic salt containing the ion CN_2^{2-}. *See* **calcium cyanamide**. **2.** A colourless crystalline solid, H_2NCN, made by the action of carbon dioxide on hot sodamide. It is a weakly acidic compound (the parent acid of cyanamide salts) that is soluble in water and ethanol. It is hydrolysed to urea in acidic solutions.

cyanamide process *See* **calcium cyanamide**.

cyanate *See* **cyanic acid.**

cyanic acid An unstable explosive acid, HOCN. The compound has the structure H–O–C≡N, and is also called *fulminic acid.* Its salts and esters are *cyanates* (or *fulminates*). The compound is a volatile liquid, which readily polymerizes. In water it hydrolyses to ammonia and carbon dioxide. It is isomeric with another acid, H–N=C=O, which is known as *isocyanic acid.* Its salts and esters are *isocyanates.*

cyanide 1. An inorganic salt containing the cyanide ion CN^-. Cyanides are extremely poisonous because of the ability of the CN^- ion to coordinate with the iron in haemoglobin, thereby blocking the uptake of oxygen by the blood. **2.** A metal coordination complex formed with cyanide ions.

cyanide process A method of extracting gold by dissolving it in potassium cyanide (to form the complex ion $[Au(CN)_2]^-$). The ion can be reduced back to gold with zinc.

cyanine dyes A class of dyes that contain a –CH= group linking two nitrogen containing heterocyclic rings. They are used as sensitizers in photography.

cyanoacrylate A compound formed by the condensation of an alkyl cyanoethanoate and methanal. Cyanoacrylates have the general formula CH_2:C(CN)COOR. The methyl and ethyl esters are the basis of various 'superglues', which rapidly polymerize in air (under the action of moisture) to form strong adhesives.

cyanocobalamin *See* **vitamin B complex.**

cyanogen A colourless gas, $(CN)_2$, with a pungent odour; soluble in water, ethanol, and ether; d. 2.335 g dm^{-3}; m.p. –27.9°C; b.p. –20.7°C. The compound is very toxic. It may be prepared in the laboratory by heating mercury(II) cyanide; industrially it is made by gas-phase oxidation of hydrogen cyanide using air over a silver catalyst, chlorine over activated silicon(IV) oxide, or nitrogen dioxide over a copper(II) salt. Cyanogen is an important intermediate in the preparation of various fertilizers and is also used as a stabilizer in making nitrocellulose. It is an example of a *pseudohalogen.

cyano group The group –CN in a chemical compound. *See* **nitriles.**

cyanohydrins Organic compounds formed by the addition of hydrogen cyanide to aldehydes or ketones (in the presence of a base). The first step is attack by a CN^- ion on the carbonyl carbon atom. The final product is a compound in which a –CN and –OH group are attached to the same carbon atom. For example, ethanal reacts as follows

$$CH_3CHO + HCN \rightarrow CH_3CH(OH)(CN)$$

The product is 2-hydroxypropanonitrile. Cyanohydrins of this type can be oxidized to α-hydroxy carboxylic acids.

cyanuric acid A white crystalline water-soluble trimer of cyanic acid,

$(HNCO)_3$. It is a cyclic compound having a six-membered ring made of alternating imide (NH) and carbonyl (CO) groups.

cyclamates Salts of the acid, $C_6H_{11}.NH.SO_3H$, where $C_6H_{11}-$ is a cyclohexyl group. Sodium and calcium cyclamates were formerly used as sweetening agents in soft drinks, etc, until their use was banned when they were suspected of causing cancer.

cyclic Describing a compound that has a ring of atoms in its molecules. In *homocyclic* compounds all the atoms in the ring are the same type, e.g. benzene (C_6H_6) and cyclohexane (C_6H_{12}). These two examples are also examples of *carbocyclic* compounds; i.e. the rings are of carbon atoms. If different atoms occur in the ring, as in pyridine (C_5H_5N), the compound is said to be *heterocyclic*.

cyclic AMP A derivative of *ATP that is widespread in animal cells as a second messenger in many biochemical reactions induced by hormones. Upon reaching their target cells, the hormones activate adenylate cyclase, the enzyme that catalyses cyclic AMP production. Cyclic AMP ultimately activates the enzymes of the reaction induced by the hormone concerned. Cyclic AMP is also involved in controlling gene expression and cell division, in immune responses, and in nervous transmission.

cyclization The formation of a cyclic compound from an open-chain compound. *See* **ring**.

cyclo- Prefix designating a cyclic compound, e.g. a cycloalkane or a cyclosilicate.

cycloaddition A reaction in which two or more unsaturated compounds form a cyclic adduct or in which a cyclic compound is formed by addition between unsaturated parts of the same molecule. In cycloaddition, there is no net reduction in bond multiplicity. The *Diels–Alder reaction is an example. Cycloadditions may be stepwise reactions or may be *pericyclic reactions.

cycloalkanes Cyclic saturated hydrocarbons containing a ring of carbon atoms joined by single bonds. They have the general formula C_nH_{2n}, for example cyclohexane, C_6H_{12}, etc. In general they behave like the *alkanes but are rather less reactive.

cyclobutadiene A short-lived cyclic diene hydrocarbon, C_4H_4, with its four carbon atoms in a square ring. It is made by degradation of its metal complexes, and has a short lifetime of a few seconds. It undergoes addition reactions with alkynes.

cyclohexadiene-1,4-dione **(benzoquinone; quinone)** A yellow solid, $C_6H_4O_2$; r.d. 1.3; m.p. 116°C. It has a six-membered ring of carbon atoms with two opposite carbon atoms linked to oxygen atoms (C=O) and the other two pairs of carbon atoms linked by double bonds (HC=CH). The compound is used in making dyes. *See also* **quinhydrone electrode**.

cyclohexane A colourless liquid *cycloalkane, C_6H_{12}; r.d. 0.78; m.p. 6.5°C; b.p. 81°C. It occurs in petroleum and is made by passing benzene and hydrogen under pressure over a heated Raney nickel catalyst at 150°C, or by the reduction of cyclohexanone. It is used as a solvent and paint remover and can be oxidized using hot concentrated nitric acid to hexanedioic acid (adipic acid). The cyclohexane ring is not planar and can adopt different *ring conformations; in formulae it is represented by a single hexagon.

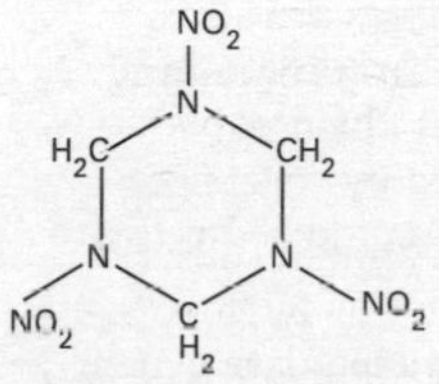

Cyclonite

cyclonite (RDX) A highly explosive nitro compound, $(CH_2N.NO_2)_3$. It has a cyclic structure with a six-membered ring of alternating CH_2 groups and nitrogen atoms, with each nitrogen being attached to a NO_2 group. It is made by nitrating hexamine, $C_6H_{12}N_4$, which is obtained from ammonia and methanal. Cyclonite is a very powerful explosive used mainly for military purposes.

cyclo-octatetraene (COT) A yellow liquid cyclic compound, C_8H_8, with eight carbon atoms and four double bonds in its ring; b.p. 142°C. It is made by polymerizing ethyne (acetylene) in the presence of nickel salts. It forms addition compounds and certain metal complexes, such as uranocene, $(C_8H_8)_2U$.

cyclopentadiene A colourless liquid cyclic *alkene, C_5H_6; r.d. 0.8021; m.p. −97.2°C; b.p. 40.0°C. It is prepared as a by-product during the fractional distillation of crude benzene from coal tar. It undergoes condensation reactions with ketones to give highly coloured compounds (fulvenes) and readily undergoes polymerization at room temperature to give the dimer, dicyclopentadiene. The compound itself is not aromatic because it does not have the required number of pi electrons (*see* **aromatic compound**). However, removal of a hydrogen atom produces the stable *cyclopentadienyl ion*, $C_5H_5^-$, which does have aromatic properties. In particular, the ring can coordinate to positive ions in such compounds as *ferrocene.

cyclopentadienyl ion *See* **cyclopentadiene**.

cyclopropane A colourless gas, C_3H_6, b.p. −34.5°C, whose molecules contain a triangular ring of carbon atoms. It is made by treating 1,3-dibromopropane with zinc metal, and is used as a general anaesthetic.

cysteine *See* **amino acid**.

cystine A molecule resulting from the oxidation reaction between the sulphydryl (–SH) groups of two cysteine molecules (*see* **amino acid**). This often occurs between adjacent cysteine residues in polypeptides. The resultant *disulphide bridges* (–S–S–) are important in stabilizing the structure of protein molecules.

cytidine A nucleoside comprising one cytosine molecule linked to a D-ribose sugar molecule. The derived nucleotides, cytidine mono-, di-, and triphosphate (CMP, CDP, and CTP respectively) participate in various biochemical reactions, notably in phospholipid synthesis.

cytochrome Any of a group of proteins, each with an iron-containing *haem group, that form part of the *electron transport chain in mitochondria and chloroplasts. Electrons are transferred by reversible changes in the iron atom between the reduced Fe(II) and oxidized Fe(III) states.

cytosine A *pyrimidine derivative. It is one of the principal component bases of *nucleotides and the nucleic acids *DNA and *RNA.

2,4-D 2,4-dichlorophenoxyacetic acid (2,4-dichlorophenoxyethanoic acid): a synthetic auxin frequently used as a weedkiller of broad-leaved weeds. *See also* **pesticide.**

Dakin reaction *See* **Baeyer–Villiger reaction.**

dalton *See* **atomic mass unit.**

Dalton, John (1766–1844) British chemist and physicist. In 1801 he formulated his law of partial pressures (*see* **Dalton's law**), but he is best remembered for *Dalton's atomic theory, which he announced in 1803. Dalton also studied colour blindness (a condition, once called Daltonism, that he shared with his brother).

Dalton's atomic theory A theory of *chemical combination, first stated by John *Dalton in 1803. It involves the following postulates:
(1) Elements consist of indivisible small particles (atoms).
(2) All atoms of the same element are identical; different elements have different types of atom.
(3) Atoms can neither be created nor destroyed.
(4) 'Compound elements' (i.e. compounds) are formed when atoms of different elements join in simple ratios to form 'compound atoms' (i.e. molecules).

Dalton also proposed symbols for atoms of different elements (later replaced by the present notation using letters).

Dalton's law The total pressure of a mixture of gases or vapours is equal to the sum of the partial pressures of its components, i.e. the sum of the pressures that each component would exert if it were present alone and occupied the same volume as the mixture of gases. Strictly speaking, the principle is true only for ideal gases. The law was discovered by John *Dalton.

Daniell cell A type of primary *voltaic cell with a copper positive electrode and a negative electrode of a zinc amalgam. The zinc-amalgam electrode is placed in an electrolyte of dilute sulphuric acid or zinc sulphate solution in a porous pot, which stands in a solution of copper sulphate in which the copper electrode is immersed. While the reaction takes place ions move through the porous pot, but when it is not in use the cell should be dismantled to prevent the diffusion of one electrolyte into the other. The e.m.f. of the cell is 1.08 volts with sulphuric acid and 1.10 volts with zinc sulphate. It was invented in 1836 by the British chemist John Daniell (1790–1845).

dark reaction *See* **photosynthesis.**

darmstadtium Symbol Ds. A radioactive transactinide; a.n. 110. It has several isotopes; the most stable being ^{281}Ds, with a half-life of about 1.6 minutes. It can be produced by bombarding a plutonium target with sulphur nuclei or by bombarding a lead target with nickel nuclei. Its chemical properties probably resemble those of platinum. Darmstadtium was named after the German city of Darmstadt, the location of the Institute for Heavy Ion Research where it was first produced.

dating techniques Methods of estimating the age of rocks, palaeontological specimens, archaeological sites, etc. *Relative dating techniques* date specimens in relation to one another; for example, stratigraphy is used to establish the succession of fossils. *Absolute* (or *chronometric*) *techniques* give an absolute estimate of the age and fall into two main groups. The first depends on the existence of something that develops at a seasonally varying rate, as in dendrochronology and varve dating. The other uses some measurable change that occurs at a known rate, as in *chemical dating, *radioactive* (or *radiometric*) *dating* (*see* **carbon dating; fission-track dating; potassium–argon dating; rubidium–strontium dating; uranium–lead dating**), and *thermoluminescence.

dative bond *See* **chemical bond.**

daughter 1. A nuclide produced by radioactive *decay of some other nuclide (the *parent*). **2.** An ion or free radical produced by dissociation or reaction of some other (*parent*) ion or radical.

Davisson–Germer experiment *See* **electron diffraction.**

Davy, Sir Humphry (1778–1829) British chemist, who studied gases at the Pneumatic Institute in Bristol, where he discovered the anaesthetic properties of *dinitrogen oxide (nitrous oxide). He moved to the Royal Institution, London, in 1801 and five years later isolated potassium and sodium by electrolysis. He also prepared barium, boron, calcium, and strontium as well as proving that chlorine and iodine are elements. In 1816 he invented the *Davy lamp.

Davy lamp An oil-burning miner's safety lamp invented by Sir Humphry Davy in 1816 when investigating firedamp (methane) explosions in coal mines. The lamp has a metal gauze surrounding the flame, which cools the hot gases by conduction and prevents ignition of gas outside the gauze. If firedamp is present it burns within the gauze cage, and lamps of this type are still used for testing for gas.

***d*-block elements** The block of elements in the *periodic table consisting of scandium, yttrium, and lanthanum together with the three periods of transition elements: titanium to zinc, zirconium to cadmium, and hafnium to mercury. These elements all have two outer *s*-electrons and have *d*-electrons in their penultimate shell; i.e. an outer electron configuration of the form $(n-1)d^x ns^2$, where x is 1 to 10. *See also* **transition elements.**

DDT Dichlorodiphenyltrichloroethane; a colourless organic crystalline compound, $(ClC_6H_4)_2CH(CCl_3)$, made by the reaction of trichloromethanal with chlorobenzene. DDT is the best known of a number of chlorine-containing *pesticides used extensively in agriculture in the 1940s and 1950s. The compound is stable, accumulates in the soil, and concentrates in fatty tissue, reaching dangerous levels in carnivores high in the food chain. Restrictions are now placed on the use of DDT and similar pesticides.

deacetylation The removal of an acetyl group ($-COCH_3$) from a molecule. Deacetylation is an important reaction in several biochemical pathways, including the *Krebs cycle.

Deacon process A former process for making chlorine by oxidizing hydrogen chloride in air at 450°C using a copper chloride catalyst. It was patented in 1870 by Henry Deacon (1822–76).

deactivation A partial or complete reduction in the reactivity of a substance, as in the poisoning of a catalyst.

deamination The removal of an amino group ($-NH_2$) from a compound. Enzymatic deamination occurs in the liver and is important in amino-acid metabolism, especially in their degradation and subsequent oxidation. The amino group is removed as ammonia and excreted, either unchanged or as urea or uric acid.

de Broglie, Louis-Victor Pierre Raymond (1892–1987) French physicist, who taught at the Sorbonne in Paris for 34 years. He is best known for his 1924 theory of wave–particle duality (*see* **de Broglie wavelength**), which reconciled the corpuscular and wave theories of light and proved important in quantum theory. For this work he was awarded the 1929 Nobel Prize.

de Broglie wavelength The wavelength of the wave associated with a moving particle. The wavelength (λ) is given by $\lambda = h/mv$, where h is the Planck constant, m is the mass of the particle, and v its velocity. The *de Broglie wave* was first suggested by Louis de Broglie in 1924 on the grounds that electromagnetic waves can be treated as particles (photons) and one could therefore expect particles to behave in some circumstances like waves. The subsequent observation of *electron diffraction substantiated this argument and the de Broglie wave became the basis of *wave mechanics.

debye A unit of electric dipole moment in the electrostatic system, used to express dipole moments of molecules. It is the dipole moment produced by two charges of opposite sign, each of 1 statcoulomb and placed 10^{-18} cm apart, and has the value $3.335\,64 \times 10^{-30}$ coulomb metre. It is named after Peter Debye.

Debye, Peter Joseph William (1884–1966) Dutch-born physical chemist who worked on a number of topics. He introduced the idea of electric dipole moments in molecules and, in 1923, working with Erich Hückel, he published the *Debye–Hückel theory of electrolytes. Debye was awarded the 1936 Nobel Prize for chemistry.

Debye–Hückel–Onsager theory A theory providing quantitative results for the conductivity of ions in dilute solutions of strong electrolytes, which enables the *Kohlrausch equation to be derived. This theory can be stated as: $K = A + B\Lambda^{\ominus}_{m}$, where $\Lambda^{\ominus}_{m}$ is the *limiting molar conductivity.*

$A = z^2eF^2/3\pi\eta(2/\varepsilon RT)^{1/2}$.

$B = qz^3eF/24\pi\varepsilon RT(2/\varepsilon RT)^{1/2}$.

where z is the charge of an ion, e is the charge of an electron, F is Faraday's constant, η is the viscosity of the liquid, R is the gas constant, T is the thermodynamic temperature, and $q = 0.586$ in the case of a 1,1 electrolyte. The Debye–Hückel–Onsager theory uses the same assumptions and approximations as the *Debye–Hückel theory and is also limited to very dilute solutions (usually less than 10^{-3} M) for which there is good agreement between theory and experiment. The modifications were made by the Norwegian-born US chemist Lars Onsager (1903–76).

Debye–Hückel theory A theory to explain the nonideal behaviour of electrolytes, published in 1923 by Peter *Debye and Erich Hückel (1896–1980). It assumes that electrolytes in solution are fully dissociated and that nonideal behaviour arises because of electrostatic interactions between the ions. The theory shows how to calculate the extra free energy per ion resulting from such interactions, and consequently the activity coefficient. It gives a good description of nonideal electrolyte behaviour for very dilute solutions, but cannot be used for more concentrated electrolytes.

Debye–Scherrer method A technique used in *X-ray diffraction in which a crystal in powder form is fixed to a thin fibre or thin silica tube, which is then rotated in the path of a beam of monochromatic X-rays. A circular diffraction ring, called the *Debye–Scherrer ring*, concentric with the undeflected beam is formed. The diffraction pattern is recorded on a cylindrical film, which has its axis parallel to the axis of rotation of the material. The Debye–Scherrer method is used to obtain information about the material. The grains of the powdered crystal must be much larger than the atomic dimensions in order for them to diffract X-rays.

Debye temperature *See* **Debye theory of specific heat**.

Debye theory of specific heat A theory of the specific heat capacity of solids put forward by Peter *Debye in 1912, in which it was assumed that the specific heat is a consequence of the vibrations of the atoms of the lattice of the solid. In contrast to the *Einstein theory of specific heat, which assumes that each atom has the same vibrational frequency, Debye postulated that there is a continuous range of frequencies that cuts off at a maximum frequency ν_D, which is characteristic of a particular solid. The theory leads to the conclusion that the specific heat capacity of solids is proportional to T^3, where T is the thermodynamic temperature. This result is in very good agreement with experiment at low temperatures. A key quantity in this theory is the *Debye temperature*, θ_D, defined by $\theta_D = h\nu_D k$, where *h* is the Planck constant and *k* is the Boltzmann constant. The Debye temperature is characteristic of a particular solid. For example, the Debye temperature of sodium is 150 K and the Debye temperature of copper is 315 K.

Debye–Waller factor A quantity that characterizes the effect of lattice vibrations on the scattering intensity in X-ray diffraction by crystals. The existence of the Debye–Waller factor was pointed out and calculated by Peter *Debye in 1913–1914 and Ivar Waller in 1923–1925. Because the amplitude of lattice vibrations increases with temperature, it was thought in the very early days of X-ray diffraction studies that the diffraction pattern would disappear at high temperatures. The work of Debye and Waller showed that the lattice vibrations at higher temperatures reduce the intensities of the diffracted radiation but do not destroy the diffraction pattern altogether.

deca- Symbol da. A prefix used in the metric system to denote ten times. For example, 10 coulombs = 1 decacoulomb (daC).

decahydrate A crystalline hydrate containing ten moles of water per mole of compound.

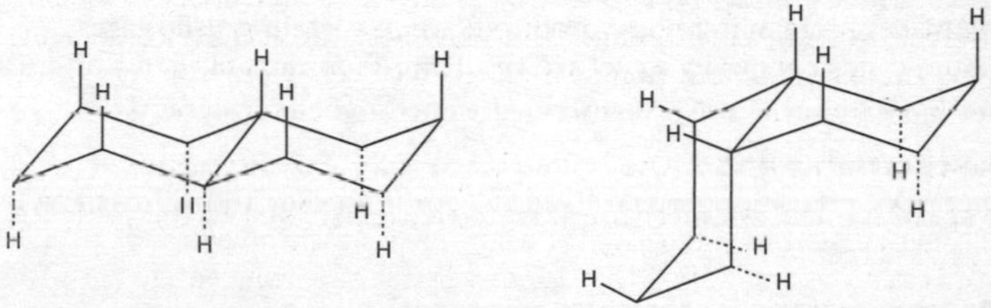

trans-decalin

cis-decalin

Isomers of decalin

decalin (decahydronaphthalene) A liquid bicyclic hydrocarbon, $C_{10}H_{18}$, used as a solvent. There are two stereoisomers, *cis* (b.p. 198°C) and *trans* (b.p. 185°C), made by the catalytic hydrogenation of naphthalene at high temperatures and pressures.

decanoic acid (capric acid) A white crystalline straight-chain saturated *carboxylic acid, $CH_3(CH_2)_8COOH$; m.p. 31.5°C. Its esters are used in perfumes and flavourings.

decantation The process of separating a liquid from a settled solid suspension or from a heavier immiscible liquid by carefully pouring it into a different container.

decarboxylation The removal of carbon dioxide from a molecule.

Decarboxylation is an important reaction in many biochemical processes, such as the *Krebs cycle and the synthesis of fatty acids.

decay 1. The spontaneous transformation of one radioactive nuclide into a daughter nuclide, which may be radioactive or may not, with the emission of one or more particles or photons. The decay of N_0 nuclides to give N nuclides after time t is given by $N = N_0\exp(-\gamma t)$, where γ is called the *decay constant* or the *disintegration constant*. The reciprocal of the decay constant is the *mean life*. The time required for half the original nuclides to decay (i.e. $N = \frac{1}{2}N_0$) is called the *half-life* of the nuclide. The same terms are applied to elementary particles that spontaneously transform into other particles. For example, a free neutron decays into a proton and an electron. **2.** The reversion of excited states of atoms or molecules to the ground state.

deci- Symbol d. A prefix used in the metric system to denote one tenth. For example, 0.1 coulomb = 1 decicoulomb (dC); 0.1 metre = 1 decimetre (dm).

decomposition 1. The chemical breakdown of organic matter into its constituents by the action of bacteria and other organisms. **2.** A chemical reaction in which a compound breaks down into simpler compounds or into elements.

decrepitation A crackling noise produced when certain crystals are heated, caused by changes in structure resulting from loss of water of crystallization.

defect *See* **crystal defect**.

defect state A quantum mechanical state that exists due to the presence of a *crystal defect.

definite proportions *See* **chemical combination**.

deflagration A type of explosion in which the shock wave arrives before the reaction is complete (because the reaction front moves more slowly than the speed of sound in the medium).

degassing The removal of dissolved, absorbed, or adsorbed gases from a liquid or solid. Degassing is important in vacuum systems, where gas absorbed in the walls of the vacuum vessel starts to desorb as the pressure is lowered.

degeneracy The state of being *degenerate.

degenerate Having quantum states with the same energy. For example, the five *d*-orbitals in an isolated transition-metal atom have the same energy (although they have different spatial arrangements) and are thus degenerate. The application of a magnetic or electric field may cause the quantum states to have different energies (*see* **crystal-field theory**). In this case, the degeneracy is said to be 'lifted'.

degenerate rearrangement A rearrangement of a molecule in which the product is chemically indistinguishable from the reactant. Degenerate rearrangements can be detected by using isotopic labelling.

degradation A type of organic chemical reaction in which a compound is converted into a simpler compound in stages.

degree A division on a *temperature scale.

degrees absolute *See* **absolute.**

degrees of freedom 1. The number of independent parameters required to specify the configuration of a system. This concept is applied in the *kinetic theory to specify the number of independent ways in which an atom or molecule can take up energy. There are however various sets of parameters that may be chosen, and the details of the consequent theory vary with the choice. For example, in a monatomic gas each atom may be allotted three degrees of freedom, corresponding to the three coordinates in space required to specify its position. The mean energy per atom for each degree of freedom is the same, according to the principle of the *equipartition of energy, and is equal to $kT/2$ for each degree of freedom (where k is the *Boltzmann constant and T is the thermodynamic temperature). Thus for a monatomic gas the total molar energy is $3LkT/2$, where L is the Avogadro constant (the number of atoms per mole). As $k = R/L$, where R is the molar gas constant, the total molar energy is $3RT/2$.

In a diatomic gas the two atoms require six coordinates between them, giving six degrees of freedom. Commonly these are interpreted as six independent ways of storing energy: on this basis the molecule has three degrees of freedom for different directions of translational motion, and in addition there are two degrees of freedom for rotation of the molecular axis and one vibrational degree of freedom along the bond between the atoms. The rotational degrees of freedom each contribute their share, $kT/2$, to the total energy; similarly the vibrational degree of freedom has an equal share of kinetic energy and must on average have as much potential energy. The total energy per molecule for a diatomic gas is therefore $3kT/2$ (for translational energy of the whole molecule) plus $2kT/2$ (for rotational energy) plus $2kT/2$ (for vibrational energy), i.e. a total of $7kT/2$.
2. The least number of independent variables required to define the state of a system in the *phase rule. In this sense a gas has two degrees of freedom (e.g. temperature and pressure).

dehydration 1. Removal of water from a substance. **2.** A chemical reaction in which a compound loses hydrogen and oxygen in the ratio 2:1. For instance, ethanol passed over hot pumice undergoes dehydration to ethene:

$$C_2H_5OH - H_2O \rightarrow CH_2{:}CH_2$$

Substances such as concentrated sulphuric acid, which can remove H_2O in this way, are known as *dehydrating agents*. For example, with sulphuric acid, methanoic acid gives carbon monoxide:

$HCOOH - H_2O \rightarrow CO$

dehydrogenase Any enzyme that catalyses the removal of hydrogen atoms (*dehydrogenation) in biological reactions. Dehydrogenases occur in many biochemical pathways but are particularly important in driving the *electron-transport-chain reactions of cell respiration. They work in conjunction with the hydrogen-accepting coenzymes *NAD and *FAD.

dehydrogenation A chemical reaction in which hydrogen is removed from a compound. Dehydrogenation of organic compounds converts single carbon–carbon bonds into double bonds. It is usually effected by means of a metal catalyst or – in biological systems – by *dehydrogenases.

deionized water Water from which ionic salts have been removed by ion-exchange. It is used for many purposes as an alternative to distilled water.

deliquescence The absorption of water from the atmosphere by a hygroscopic solid to such an extent that a concentrated solution of the solid eventually forms.

delocalization The spreading of valence electrons over two or more bonds in a chemical compound. In certain compounds, the valence electrons cannot be regarded as restricted to definite bonds between the atoms but move over several atoms in the molecule. Such electrons are said to be *delocalized*. Delocalization occurs particularly when the compound contains alternating (conjugated) double or triple bonds, the delocalized electrons being those in the pi *orbitals. The molecule is then more stable than it would be if the electrons were localized, an effect accounting for the properties of benzene and other aromatic compounds. The energy difference between the actual delocalized state and a localized state is the *delocalization energy*. Another example is in the ions of carboxylic acids, containing the carboxylate group $-COO^-$. In terms of a simple model of chemical bonding, this group would have the carbon joined to one oxygen by a double bond (i.e. C=O) and the other joined to O^- by a single bond ($C-O^-$). In fact, the two C–O bonds are identical because the extra electron on the O^- and the electrons in the pi bond of C=O are delocalized over the three atoms. Delocalization of electrons is a feature of metallic bonding. The delocalization energy of molecules can be calculated approximately using the *Hückel approximation, as was done originally by Hückel. However, modern computing power enables delocalization energy to be calculated using *ab-initio calculations, even for large molecules. *See also* **localization**.

delta bonding Chemical bonding involving *delta* (δ) *orbitals*. A δ orbital is so called because it resembles a d-orbital when viewed along the axis of a molecule and has two units of orbital angular momentum around the internuclear axis. The formation of δ bonds originates in the overlap of d-orbitals on different atoms. Delta orbitals contribute to the bonding of cluster compounds of transition metals.

delta-brass A strong hard type of *brass that contains, in addition to copper and zinc, a small percentage of iron. It is mainly used for making cartridge cases.

delta-iron *See* **iron**.

delta orbital *See* **delta bonding**.

delta value A quantity that measures the shift in *nuclear magnetic resonance (NMR).

denature 1. To add a poisonous or unpleasant substance to ethanol to make it unsuitable for human consumption (*see* **methylated spirits**). **2.** To produce a structural change in a protein or nucleic acid that results in the reduction or loss of its biological properties. Denaturation is caused by heat, chemicals, and extremes of pH. The differences between raw and boiled eggs are largely a result of denaturation. **3.** To add another isotope to a fissile material to make it unsuitable for use in a nuclear weapon.

dendrimer (dendritic polymer) A type of macromolecule in which a number of chains radiate out from a central atom or cluster of atoms. Dendritic polymers have a number of possible applications. *See also* **supramolecular chemistry**.

dendrite A crystal that has branched in growth into two parts. Crystals that grow in this way (*dendritic growth*) have a branching treelike appearance.

dendrochronology An absolute *dating technique using the growth rings of trees. It depends on the fact that trees in the same locality show a characteristic pattern of growth rings resulting from climatic conditions. Thus it is possible to assign a definite date for each growth ring in living trees, and to use the ring patterns to date fossil trees or specimens of wood (e.g. used for buildings or objects on archaeological sites) with lifespans that overlap those of living trees. The bristlecone pine (*Pinus aristata*), which lives for up to 5000 years, has been used to date specimens over 8000 years old. Fossil specimens accurately dated by dendrochronology have been used to make corrections to the *carbon-dating technique. Dendrochronology is also helpful in studying past climatic conditions. Analysis of trace elements in sections of rings can also provide information on past atmospheric pollution.

denitrification A chemical process in which nitrates in the soil are reduced to molecular nitrogen, which is released into the atmosphere. This process is effected by the bacterium *Pseudomonas denitrificans*, which uses nitrates as a source of energy for other chemical reactions in a manner similar to respiration in other organisms. *Compare* **nitrification**. *See* **nitrogen cycle**.

de novo pathway Any metabolic pathway in which a *biomolecule is synthesized from simple precursor molecules. Nucleotide synthesis is an example.

density The mass of a substance per unit of volume. In *SI units it is measured in kg m^{-3}. *See also* **relative density**; **vapour density**.

density-function theory A method for the theoretical treatment of molecules, in which the electron density is considered rather than the interactions of individual electrons. It is successful for large molecules.

deoxyribonucleic acid *See* DNA.

depolarization The prevention of *polarization in a *primary cell. For example, maganese(IV) oxide (the *depolarizer*) is placed around the positive electrode of a *Leclanché cell to oxidize the hydrogen released at this electrode.

depression of freezing point The reduction in the freezing point of a pure liquid when another substance is dissolved in it. It is a *colligative property – i.e. the lowering of the freezing point is proportional to the number of dissolved particles (molecules or ions), and does not depend on their nature. It is given by $\Delta t = K_f C_m$, where C_m is the molar concentration of dissolved solute and K_f is a constant (the *cryoscopic constant*) for the solvent used. Measurements of freezing-point depression (using a Beckmann thermometer) can be used for finding relative molecular masses of unknown substances.

derivative A compound that is derived from some other compound and usually maintains its general structure, e.g. trichloromethane (chloroform) is a derivative of methane.

derived unit *See* **base unit**.

desalination The removal of salt from sea water for irrigation of the land or to provide drinking water. The process is normally only economic if a cheap source of energy, such as the waste heat from a nuclear power station, can be used. Desalination using solar energy has the greatest economic potential since shortage of fresh water is most acute in hot regions. The methods employed include evaporation, often under reduced pressure (flash evaporation); freezing (pure ice forms from freezing brine); *reverse osmosis; *electrodialysis; and *ion exchange.

desiccant A drying agent. There are many types, including anhydrous calcium chloride, anhydrous calcium sulphate, concentrated sulphuric acid, phosphorus(V) oxide, solid sodium hydroxide, lime, and *silica gel.

desiccation A method of preserving organic material by the removal of its water content. Cells and tissues can be preserved by desiccation after lowering the samples to freezing temperatures; thereafter they can be stored at room temperature.

desiccator A container for drying substances or for keeping them free from moisture. Simple laboratory desiccators are glass vessels containing a drying agent, such as silica gel. They can be evacuated through a tap in the lid.

desorption The removal of adsorbed atoms, molecules, or ions from a surface.

destructive distillation The process of heating complex organic substances in the absence of air so that they break down into a mixture of volatile products, which are condensed and collected. At one time the destructive distillation of coal (to give coke, coal tar, and coal gas) was the principal source of industrial organic chemicals.

desulphuration The removal of sulphur from a compound. Desulphuration has been implicated in the toxicity of a number of sulphur-containing organic compounds: it is postulated that the atomic sulphur released into a cell by desulphuration, which is highly electrophilic, can bind to proteins and thereby alter their function.

detailed balance The cancellation of the effect of one process by another process that operates at the same time with the opposite effect. An example of detailed balance is provided by a chemical reaction between two molecular species A and B, which results in the formation of the molecular species C and D. Detailed balance for this chemical reaction occurs if the rate at which the reaction $A + B \rightarrow C + D$ occurs is equal to the rate at which the reaction $C + D \rightarrow A + B$ occurs. The equilibrium state in thermodynamics is characterized by detailed balance.

detergent A substance added to water to improve its cleaning properties. Although water is a powerful solvent for many compounds, it will not dissolve grease and natural oils. Detergents are compounds that cause such nonpolar substances to go into solution in water. *Soap is the original example, owing its action to the presence of ions formed from long-chain fatty acids (e.g. the octadecanoate (stearate) ion, $CH_3(CH_2)_{16}COO^-$). These have two parts: a nonpolar part (the hydrocarbon chain), which attaches to the grease; and a polar part (the $-COO^-$ group), which is attracted to the water. A disadvantage of soap is that it forms a scum with hard water (*see* **hardness of water**) and is relatively expensive to make. Various synthetic ('soapless') detergents have been developed from petrochemicals. The commonest, used in washing powders, is sodium dodecylbenzenesulphonate, which contains $CH_3(CH_2)_{11}C_6H_4SO_2O^-$ ions. This, like soap, is an example of an *anionic detergent*, i.e. one in which the active part is a negative ion. *Cationic detergents* have a long hydrocarbon chain connected to a positive ion. Usually they are amine salts, as in $CH_3(CH_2)_{15}N(CH_3)_3{}^+Br^-$, in which the polar part is the $-N(CH_3)_3{}^+$ group. *Nonionic detergents* have nonionic polar groups of the type $-C_2H_4-O-C_2H_4-OH$, which form hydrogen bonds with the water. Synthetic detergents are also used as wetting agents, emulsifiers, and stabilizers for foam.

detonating gas *See* **electrolytic gas.**

deuterated compound A compound in which some or all of the hydrogen-1 atoms have been replaced by deuterium atoms.

deuterium (heavy hydrogen) Symbol D. The isotope of hydrogen that has a mass number 2 (r.a.m. 2.0144). Its nucleus contains one proton and one neutron. The abundance of deuterium in natural hydrogen is about 0.015%. It is present in water as the oxide HDO (*see also* **heavy water**), from which it is usually obtained by electrolysis or fractional distillation. Its chemical behaviour is almost identical to hydrogen although deuterium compounds tend to react rather more slowly than the corresponding hydrogen compounds. Its physical properties are slightly different from those of hydrogen, e.g. b.p. 23.6 K (hydrogen 20.4 K).

deuterium oxide *See* **heavy water**.

deuteron A nucleus of a deuterium atom, consisting of a proton and a neutron bound together; the ion D^+ formed by ionization of a deuterium atom. *See also* **hydron**.

Devarda's alloy An alloy of copper (50%), aluminium (45%) and zinc (5%), used in chemical tests for the nitrate ion (in alkaline solutions it reduces a nitrate to ammonia).

devitrification Loss of the amorphous nature of glass as a result of crystallization.

Dewar, Sir James (1842–1923) British chemist and physicist, born in Scotland. In 1875 he became a professor at Cambridge University, while carrying out much of his experimental work at the Royal Institution in London. He began studying gases at low temperatures and in 1872 invented the *Dewar flask. In 1891, together with Frederick Abel (1827–1902), he developed the smokeless propellant explosive *cordite, and in 1898 was the first to liquefy hydrogen.

Dewar flask A vessel for storing hot or cold liquids so that they maintain their temperature independently of the surroundings. Heat transfer to the surroundings is reduced to a minimum: the walls of the vessel consist of two thin layers of glass (or, in large vessels, steel) separated by a vacuum to reduce conduction and convection; the inner surface of a glass vessel is silvered to reduce radiation; and the vessel is stoppered to prevent evaporation. It was devised around 1872 by Sir James *Dewar and is also known by its first trade name *Thermos flask*. *See also* **cryostat**.

Dewar structure A structure of *benzene proposed by Sir James *Dewar, having a hexagonal ring of six carbon atoms with two opposite atoms joined by a long single bond across the ring and with two double C–C bonds, one on each side of the hexagon. Dewar structures contribute to the resonance hybrid of benzene.

dextran A glutinous glucose polymer produced by certain bacteria. It can be made by fermenting sucrose (cane sugar) and is used as a thickening agent, as a stabilizer in ice cream, and as a substitute for plasma in blood transfusions. Esters with sulphuric acid yield sodium salts that are employed as anticoagulant drugs.

dextrin An intermediate polysaccharide compound resulting from the hydrolysis of starch to maltose by amylase enzymes.

dextro- form *See* **optical activity**.

dextrorotatory Denoting a chemical compound that rotates the plane of polarization of plane-polarized light to the right (clockwise as observed by someone facing the oncoming radiation). *See* **optical activity**.

dextrose *See* **glucose**.

diagonal relationship A relationship within the periodic table by which certain elements in the second period have a close chemical similarity to their diagonal neighbours in the next group of the third period. This is particularly noticeable with the following pairs.

Lithium and magnesium:

(1) both form chlorides and bromides that hydrolyse slowly and are soluble in ethanol;
(2) both form colourless or slightly coloured crystalline nitrides by direct reaction with nitrogen at high temperatures;
(3) both burn in air to give the normal oxide only;
(4) both form carbonates that decompose on heating.

Beryllium and aluminium:

(1) both form highly refractory oxides with polymorphs;
(2) both form crystalline nitrides that are hydrolysed in water;
(3) addition of hydroxide ion to solutions of the salts gives an amphoteric hydroxide, which is soluble in excess hydroxide giving beryllate or aluminate ions $[Be(OH)_4]^{2-}$ and $[Al(OH)_4]^-$;
(4) both form covalent halides and covalent alkyl compounds that display bridging structures;
(5) both metals dissolve in alkalis.

Boron and silicon:

(1) both display semiconductor properties;
(2) both form hydrides that are unstable in air and chlorides that hydrolyse in moist air;
(3) both form acidic oxides with covalent crystal structures, which are readily incorporated along with other oxides into a wide range of glassy materials.

The reason for this relationship is a combination of the trends to increase size down a group and to decrease size along a period, and a similar, but reversed, effect in electronegativity, i.e. decrease down a group and increase along a period.

dialysis A method by which large molecules (such as starch or protein) and small molecules (such as glucose or amino acids) in solution may be separated by selective diffusion through a semipermeable membrane. For example, if a mixed solution of starch and glucose is placed in a closed container made of a semipermeable substance (such as Cellophane), which is then immersed in a beaker of water, the smaller glucose molecules will pass through the membrane into the water while the starch molecules remain behind. The cell membranes of living organisms are semipermeable, and dialysis takes place naturally in the kidneys for the excretion of nitrogenous waste. An artificial kidney (*dialyser*) utilizes the principle of dialysis by taking over the functions of diseased kidneys.

diamagnetism *See* magnetism.

diaminobenzene **(phenylenediamine)** Any one of three isomeric aromatic compounds, $C_6H_4(NH_2)_2$, with strong basic properties. *1,2-Diaminobenzene* is a yellow-brown crystalline solid, m.p. 104°C, used as a photographic developer and in making dyes. *1,3-Diaminobenzene* is a colourless crystalline solid, m.p. 63°C, which turns brown on standing in air. It is made by reducing 1,3-dinitrobenzene with iron and hydrochloric acid, and is used for making dyes. *1,4-Diaminobenzene* is a white crystalline solid, m.p. 147°C, which rapidly darkens on standing in air. It is made by the reduction of 1,4-nitrophenylamine and is used as a photographic developer and hair dye.

1,2-diaminoethane **(ethylenediamine)** A colourless fuming liquid that smells of ammonia, $H_2NCH_2CH_2NH_2$; b.p. 116°C. It is made from ammonia and 1,2-dichloroethane, which are heated under pressure with a copper(I) chloride catalyst. In air it absorbs water and carbon dioxide to form diaminoethane carbamate. With fatty acids diaminoethane produces soaps, employed as emulsifying agents and detergents. It is also used as a solvent (for resins) and in making coatings, paper, and textiles. Crystals of its tartrate derivative are employed in piezoelectric devices.

1,6-diaminohexane **(hexamethylenediamine)** A solid colourless amine, $H_2N(CH_2)_6NH_2$; m.p. 41°C; b.p. 204°C. It is made by oxidizing cyclohexane to hexanedioic acid, reacting this with ammonia to give the ammonium salt, and dehydrating the salt to give hexanedionitrile ($NC(CH_2)_6CN$). This is reduced with hydrogen to the diamine. The compound is used, with hexanedioic acid, for producing *nylon 6,6.

diamond The hardest known mineral (with a hardness of 10 on Mohs' scale). It is an allotropic form of pure *carbon that has crystallized in the cubic system, usually as octahedra or cubes, under great pressure. Diamond crystals may be colourless and transparent or yellow, brown, or black. They are highly prized as gemstones but also have extensive uses in industry, mainly for cutting and grinding tools. Diamonds occur in ancient volcanic pipes of kimberlite; the most important deposits are in South Africa but others are found in Tanzania, the USA, Russia, and Australia. Diamonds also occur in river deposits that have been derived from weathered kimberlite, notably in Brazil, the Congo, Sierra Leone, and India. Industrial diamonds are increasingly being produced synthetically.

diamond-anvil cell A device for producing very high pressures. A sample of material to be subject to the high pressure is placed in a cavity between two high-quality diamonds. The diamond-anvil cell operates like a nutcracker, with pressures up to 1 megabar (10^{11} Pa) being exerted by turning a screw. The pressure exerted can be determined by spectroscopy for small samples of ruby in the material being compressed, while the sample itself is observed optically. A use of the diamond-anvil cell is to study the insulator–metal transition in such substances as iodine as the pressure is increased. This type of study is the nearest laboratory approach to the structure of matter in the conditions obtaining in the interior of the earth.

diaspore A mineral form of a mixed aluminium oxide and hydroxide, AlO.OH. *See* **aluminium hydroxide.**

diastase *See* amylase.

diastereoisomers Stereoisomers that are not identical and yet not mirror images. For instance, the *d*-form of tartaric acid and the meso form constitute a pair of diastereoisomers. *See* **optical activity**.

diatomaceous earth *See* **kieselguhr.**

diatomic molecule A molecule formed from two atoms (e.g. H_2 or HCl).

diatomite *See* **kieselguhr.**

diazine *See* **azine**.

diazo compounds Organic compounds containing two linked nitrogen compounds. The term includes *azo compounds, diazonium compounds, and also such compounds as diazomethane, CH_2N_2.

Structure of diazonium ion $C_6H_5N_2^+$

diazonium salts Unstable salts containing the ion $C_6H_5N_2^+$ (the *diazonium ion*: see formula). They are formed by *diazotization reactions.

diazotization The formation of a *diazonium salt by reaction of an aromatic amine with nitrous acid at low temperature (below 5°C). The nitrous acid is produced in the reaction mixture from sodium nitrite and hydrochloric acid:

$$ArNH_2 + NaNO_2 + HCl \rightarrow ArN^+N + Cl^- + Na^+ + OH^- + H_2O$$

dibasic acid An *acid that has two acidic hydrogen atoms in its molecules. Sulphuric (H_2SO_4) and carbonic (H_2CO_3) acids are common examples.

diboron trioxide *See* **boron(III) oxide.**

1,2-dibromoethane A colourless liquid *haloalkane, $BrCH_2CH_2Br$; r.d. 2.2; m.p. 9.79°C; b.p. 131.36°C. It is made by addition of bromine to ethene and used as an additive in petrol to remove lead during combustion as the volatile lead bromide.

dicarbide *See* **carbide.**

dicarboxylic acid A *carboxylic acid having two carboxyl groups in its molecules. In systematic chemical nomenclature, dicarboxylic acids are denoted by the suffix *-dioic*; e.g. hexanedioic acid, $HOOC(CH_2)_4COOH$.

dichlorine oxide (chlorine monoxide) A strongly oxidizing orange gas, Cl_2O, made by oxidation of chlorine using mercury(II) oxide. It is the acid anhydride of chloric(I) acid.

dichlorobenzene Any one of three isomeric liquid aromatic compounds, $C_6H_4Cl_2$. *1,2-Dichlorobenzene* (b.p. 179°C) and *1,4-dichlorobenzene* (b.p. 174°C) are made by chlorinating benzene with an iron catalyst and separating the mixed isomers by fractional distillation; *1,3-dichlorobenzene* (b.p. 172°C) is made from one of the other two by catalytic isomerization. The 1,2 isomer is used as an insecticide and in making dyes; the 1,4 compound is employed as a deodorant and moth repellent.

dichloroethanoic acid *See* **chloroethanoic acids.**

dichloromethane (methylene chloride) A colourless, slightly toxic liquid, CH_2Cl_2, b.p. 41°C. It has a characteristic odour similar to that of trichloromethane (chloroform), from which it is made by heating with zinc and hydrochloric acid. It is used as a refrigerant and solvent (for paint stripping and degreasing).

2,4-dichlorophenoxyacetic acid *See* **2,4-D.**

dichroism The property of some crystals, such as tourmaline, of selectively absorbing light vibrations in one plane while allowing light vibrations at right angles to this plane to pass through. Polaroid is a synthetic dichroic material. *See* **polarization.**

dichromate(VI) A salt containing the ion $Cr_2O_7^-$. Solutions containing dichromate(VI) ions are strongly oxidizing.

1,2-didehydrobenzene *See* **benzyne.**

dielectric constant *See* **permittivity.**

Diels, Otto Paul Hermann (1876–1954) German organic chemist who worked mostly at the University of Kiel. In 1906 he discovered tricarbon dioxide (C_3O_2). Diels also did important work on steroids but is remembered for his discovery in 1928 of the *Diels–Alder reaction, which he made with his assistant Kurt Alder (1902–58). Diels and Alder shared the 1950 Nobel Prize for chemistry.

Diels–Alder reaction A type of chemical reaction in which a compound containing two double bonds separated by a single bond (i.e. a conjugated *diene) adds to a suitable compound containing one double bond (known as the *dienophile*) to give a ring compound. In the dienophile, the double bond

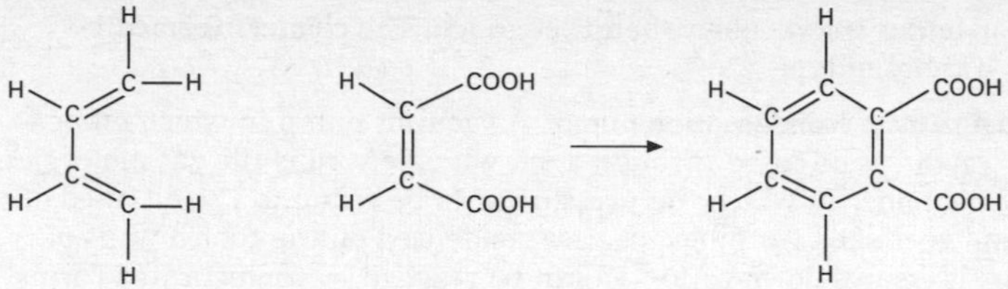

Example of a Diels–Alder reaction

must have a carbonyl group on each side. It is named after the German chemists Otto *Diels and Kurt Alder.

diene An *alkene that has two double bonds in its molecule. If the two bonds are separated by one single bond, as in buta-1,3-diene CH_2:CHCH:CH_2, the compound is a *conjugated diene.*

dienophile *See* **Diels–Alder reaction.**

Dieterici equation An *equation of state for a gas of the form

$$P(V - b)[\exp(a/VRT)] = RT,$$

where P is the pressure, V is the volume, T is the thermodynamic temperature, R is the gas constant, and a and b are constants characteristic of the gas. The Dieterici equation is a modification of van der Waals' equation, which takes account of the pressure gradient at the boundary of the gas. At low pressures the Dieterici equation becomes identical to van der Waals' equation.

diethanolamine *See* **ethanolamine.**

diethyl ether *See* **ethoxyethane.**

differential scanning calorimetry (DSC) *See* **thermal analysis.**

differential thermal analysis (DTA) *See* **thermal analysis.**

diffusion 1. The process by which different substances mix as a result of the random motions of their component atoms, molecules, and ions. In gases, all the components are perfectly miscible with each other and mixing ultimately becomes nearly uniform, though slightly affected by gravity (*see also* **Graham's law**). The diffusion of a solute through a solvent to produce a solution of uniform concentration is slower, but otherwise very similar to the process of gaseous diffusion. In solids, diffusion occurs very slowly at normal temperatures. *See also* **Fick's law. 2.** The passage of elementary particles through matter when there is a high probability of scattering and a low probability of capture.

diffusion constant *See* **Fick's law.**

diffusion gradient *See* **concentration gradient.**

diffusion limited aggregation (DLA) A process of aggregation dominated by particles diffusing and having a nonzero probability of sticking together irreversibly when they touch. The clusters formed by DLA are *fractal in type.

diffusion pump **(condensation pump)** A vacuum pump in which oil or mercury vapour is diffused through a jet, which entrains the gas molecules from the container in which the pressure is to be reduced. The diffused vapour and entrained gas molecules are condensed on the cooled walls of the pump. Pressures down to 10^{-7} Pa can be reached by sophisticated forms of the diffusion pump.

diffusive flux *See* **Fick's law.**

dihedral **(dihedron)** An angle formed by the intersection of two planes (e.g. two faces of a polyhedron). The *dihedral angle* is the angle formed by taking a point on the line of intersection and drawing two lines from this point, one in each plane, perpendicular to the line of intersection.

dihydrate A crystalline hydrate containing two moles of water per mole of compound.

dihydric alcohol *See* **diol.**

1,2-dihydroxybenzene **(catechol)** A colourless crystalline phenol, $C_6H_4(OH)_2$; r.d. 1.15; m.p. 105°C; b.p. 245°C. It is used as a photographic developer.

1,3-dihydroxybenzene **(resorcinol)** A colourless crystalline aromatic compound, $C_6H_4(OH)_2$; m.p. 111°C. It is made by the fusion of benzenedisulphonic acid with sodium hydroxide and used mainly in the dyestuffs industry. On fusing with phthalic anhydride it forms fluorescein dyes. It is also used to make cold-setting adhesives (with methanal), plasticizers, and resins.

2,3-dihydroxybutanedioic acid *See* **tartaric acid.**

diketones Organic compounds that have two carbonyl (>C=O) groups (*see* **ketones**). There are three kinds, depending on the location of the carbonyl groups. 1,2-Diketones (also called α-diketones), R.CO.CO.R′, have the carbonyl groups on adjacent carbon atoms. The aliphatic 1,2-diketones are pungent-smelling yellow oils, whereas the aromatic compounds are crystalline solids. 1,3-Diketones (or β-diketones), $R.CO.CH_2.CO.R'$, are more acidic and exist in both keto and enol forms (*see* **keto–enol tautomerism**); they form stable compounds with metals. 1,4-Diketones (or γ-diketones), $R.CO.CH_2.CH_2.CO.R'$, also exist and readily rearrange to form cyclic compounds.

dilatancy *See* **Newtonian fluid.**

dilation **(dilatation)** An increase in volume.

dilead(II) lead(IV) oxide A red powder, Pb_3O_4; r.d. 9.1; decomposes at 500°C to lead(II) oxide. It is prepared by heating lead(II) oxide to 400°C and has the unusual property of being black when hot and red-orange when cold. The compound is nonstoichiometric, generally containing less oxygen than implied by the formula. It is largely covalent and has $Pb(IV)O_6$ octahedral groups linked together by Pb(II) atoms, each joined to three oxygen atoms. It is used in glass making but its use in the paint industry has largely been discontinued because of the toxicity of lead. Dilead(II) lead(IV) oxide is commonly called *red lead* or, more accurately, *red lead oxide*.

diluent A substance added to dilute a solution or mixture (e.g. a *filler).

dilute Describing a solution that has a relatively low concentration of solute.

dilute spin species A type of nucleus in which it is statistically unlikely that there will be more than one such nucleus in a molecule (unless the substance has been artificially enriched with isotopes having that nucleus). An example of a dilute spin species is ^{13}C since it has a natural abundance of 1.1%. This means that for dilute spin systems it is usually not necessary to consider spin–spin interactions between two nuclei of the dilute spin species, e.g. ^{13}C–^{13}C in the same molecule. The opposite of a dilute spin species is an *abundant spin species*, an example being the proton.

dilution The volume of solvent in which a given amount of solute is dissolved.

dilution law *See* **Ostwald's dilution law**.

dimer An association of two identical molecules linked together. The molecules may react to form a larger molecule, as in the formation of dinitrogen tetroxide (N_2O_4) from nitrogen dioxide (NO_2), or the formation of an *aluminium chloride dimer (Al_2Cl_6) in the vapour. Alternatively, they may be held by hydrogen bonds. For example, carboxylic acids form dimers in organic solvents, in which hydrogen bonds exist between the O of the C=O group and the H of the –O–H group.

dimethylbenzenes (xylenes) Three compounds with the formula $(CH_3)_2C_6H_4$, each having two methyl groups substituted on the benzene ring. 1,2-Dimethylbenzene is ortho-xylene, etc. A mixture of the isomers (b.p. 135–145°C) is obtained from petroleum and is used as a clearing agent in preparing specimens for optical microscopy.

dimethylformamide (DMF) A colourless liquid, $HCON(CH_3)_2$; m.p. −61°C; b.p. 153°C. It is widely used as a solvent for organic compounds.

dimethylglyoxime (DMG) A colourless solid, $(CH_3CNOH)_2$, m.p. 234°C. It sublimes at 215°C and slowly polymerizes if left to stand. It is used in chemical tests for nickel, with which it forms a dark-red complex.

dimethyl sulphoxide (DMSO) A colourless solid, $(CH_3)_2SO$; m.p. 18°C; b.p. 189°C. It is used as a solvent and as a reagent in organic synthesis.

dimorphism *See* **polymorphism**.

dinitrogen oxide (nitrous oxide) A colourless gas, N_2O, d. 1.97 g dm^{-3}; m.p. −90.8°C; b.p. −88.5°C. It is soluble in water, ethanol, and sulphuric acid. It may be prepared by the controlled heating of ammonium nitrate (chloride free) to 250°C and passing the gas produced through solutions of iron(II) sulphate to remove impurities of nitrogen monoxide. It is relatively unreactive, being inert to halogens, alkali metals, and ozone at normal temperatures. It is decomposed on heating above 520°C to nitrogen and oxygen and will support the combustion of many compounds. Dinitrogen oxide is used as an anaesthetic gas ('laughing gas') and as an aerosol propellant.

dinitrogen tetroxide A colourless to pale yellow liquid or a brown gas, N_2O_4; r.d. 1.45 (liquid); m.p. −11.2°C; b.p. 21.2°C. It dissolves in water with reaction to give a mixture of nitric acid and nitrous acid. It may be readily prepared in the laboratory by the reaction of copper with concentrated nitric acid; mixed nitrogen oxides containing dinitrogen oxide may also be produced by heating metal nitrates. The solid compound is wholly N_2O_4 and the liquid is about 99% N_2O_4 at the boiling point; N_2O_4 is diamagnetic. In the gas phase it dissociates to give *nitrogen dioxide*

$$N_2O_4 \rightleftharpoons 2NO_2$$

Because of the unpaired electron this is paramagnetic and brown. Liquid N_2O_4 has been widely studied as a nonaqueous solvent system (self-ionizes to NO^+ and NO_3^-). Dinitrogen tetroxide, along with other nitrogen oxides, is a product of combustion engines and is thought to be involved in the depletion of stratospheric ozone.

dinucleotide A compound consisting of two *nucleotides.

diol (dihydric alcohol) An *alcohol containing two hydroxyl groups per molecule.

dioxan A colourless toxic liquid, $C_4H_8O_2$; r.d. 1.03; m.p. 11°C; b.p. 101.5°C. The molecule has a six-membered ring containing four CH_2 groups and two oxygen atoms at opposite corners. It can be made from ethane-1,2-diol and is used as a solvent.

dioxin (2,4,7,8-tetrachlorodibenzo-*p*-dioxin) A toxic solid, formed in the manufacture of the herbicide *2,4,5-T and present as an impurity in Agent Orange. It can cause skin disfigurement (chloracne) and severe fetal defects.

dioxonitric(III) acid *See* **nitrous acid.**

dioxygenyl compounds Compounds containing the positive ion O_2^+, as in dioxygenyl hexafluoroplatinate O_2PtF_6 – an orange solid that sublimes in vacuum at 100°C. Other ionic compounds of the type $O_2^+[MF_6]^-$ can be prepared, where M is P, As, or Sb.

dipeptide A compound consisting of two amino acid units joined at the amino ($-NH_2$) end of one and the carboxyl ($-COOH$) end of the other. This peptide bond (*see* **peptide**) is formed by a condensation reaction that involves the removal of one molecule of water.

diphenylamine A colourless crystalline aromatic compound, $(C_6H_5)_2NH$;

m.p. 54°C. It is made by heating phenylamine (aniline) with phenylamine hydrochloride. It is a secondary amine and is both slightly acidic (forming an N-potassium salt) and slightly basic (forming salts with mineral acids). Its derivatives are employed as stabilizers for synthetic rubber and rocket fuels.

diphenylmethanone **(benzophenone)** A colourless solid, $C_6H_5COC_6H_5$, m.p. 49°C. It has a characteristic smell and is used in making perfumes. It is made from benzene and benzoyl chloride using the *Friedel–Crafts reaction with aluminium chloride as catalyst.

diphosphane **(diphosphine)** A yellow liquid, P_2H_4, which is spontaneously flammable in air. It is obtained by hydrolysis of calcium phosphide. Many of the references to the spontaneous flammability of phosphine (PH_3) are in fact due to traces of P_2H_4 as impurities.

diphosphine *See* **diphosphane**.

dipolar bond *See* **chemical bond**.

dipole A pair of separated opposite electric charges. The *dipole moment* (symbol μ) is the product of the positive charge and the distance between the charges. Dipole moments are often stated in *debyes; the SI unit is the coulomb metre. In a diatomic molecule, such as HCl, the dipole moment is a measure of the polar nature of the bond (*see* **polar molecule**); i.e. the extent to which the average electron charge is displaced towards one atom (in the case of HCl, the electrons are attracted towards the more electronegative chlorine atom). In a polyatomic molecule, the dipole moment is the vector sum of the dipole moments of the individual bonds. In a symmetrical molecule, such as tetrachloromethane (CCl_4), there is no overall dipole moment, although the individual C–Cl bonds are polar.

dipole–dipole interaction The interaction of two systems, such as atoms or molecules, by their *dipole moments. The energy of dipole–dipole interaction depends on the relative orientation and the strength of the dipoles and how far apart they are. A water molecule has a permanent dipole moment, thus causing a dipole–dipole interaction if two water molecules are near each other. Although isolated atoms do not have permanent dipole moments, a dipole moment can be induced by the presence of another atom near it, thus leading to *induced dipole–dipole interactions*. Dipole–dipole interactions are responsible for *van der Waals' forces and *surface tension in liquids.

dipole radiation *See* **forbidden transitions**.

dipyridyl **(bipyridyl)** A compound formed by linking two pyridine rings by a single C–C bond, $(C_5H_4N)_2$. Various isomers are possible depending on the relative positions of the nitrogen atoms. A mixture of isomers can be made by reacting pyridine with sodium metal and oxidizing the resulting sodium salt. The 22′-isomer is a powerful chelating agent denoted *bipy* in chemical formulae. Both the 22′- and 44′-isomers can form quaternary

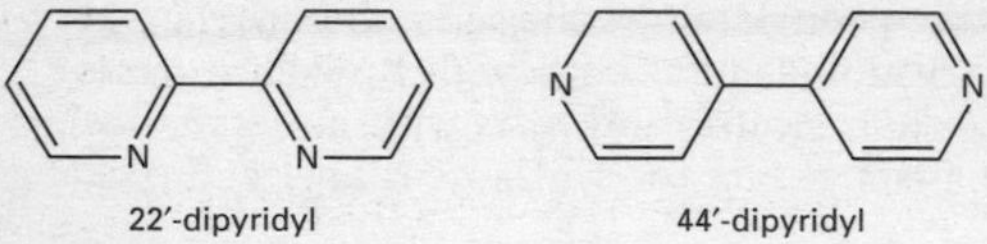

22′-dipyridyl 44′-dipyridyl

Dipyridyl

compounds that are used in herbicides such as *Paraquat* (44′) and *Diquat* (22′).

Diquat Tradename for a herbicide. *See* **dipyridyl.**

Dirac, Paul Adrien Maurice (1902–84) British physicist, who shared the 1933 Nobel Prize with Erwin *Schrödinger for developing Schrödinger's non-relativistic wave equations to take account of relativity. Dirac also invented, independently of Enrico Fermi, the form of *quantum statistics known as Fermi–Dirac statistics.

Dirac constant *See* **Planck constant.**

Dirac equation A version of the nonrelativistic *Schrödinger equation taking special relativity theory into account. The Dirac equation is needed to discuss the quantum mechanics of electrons in heavy atoms and, more generally, to discuss fine-structure features of atomic spectra, such as *spin–orbit coupling. The equation was put forward by Paul *Dirac in 1928. It can be solved exactly in the case of the hydrogen atom but can only be solved using approximation techniques for more complicated atoms.

diradical *See* **biradical.**

direct dye *See* **dyes.**

disaccharide A sugar consisting of two linked *monosaccharide molecules. For example, sucrose comprises one glucose molecule and one fructose molecule bonded together.

discharge 1. The conversion of the chemical energy stored in a *secondary cell into electrical energy. **2.** The release of electric charge from a capacitor in an external circuit. **3.** The passage of charge carriers through a gas at low pressure in a *discharge tube*. A potential difference applied between cathode and anode creates an electric field that accelerates any free electrons and ions to their appropriate electrodes. Collisions between electrons and gas molecules create more ions. Collisions also produce excited ions and molecules (*see* **excitation**), which decay with emission of light in certain parts of the tube.

disilane *See* **silane.**

dislocation *See* **crystal defect.**

disodium hydrogenphosphate(V) **(disodium orthophosphate)** A colourless crystalline solid, Na_2HPO_4, soluble in water and insoluble in ethanol. It is known as the dihydrate (r.d. 2.066), heptahydrate (r.d. 1.68), and dodecahydrate (r.d. 1.52). It may be prepared by titrating phosphoric acid with sodium hydroxide to an alkaline end point (phenolphthalein) and is used in treating boiler feed water and in the textile industry.

disodium orthophosphate *See* **disodium hydrogenphosphate(V).**

disodium tetraborate-10-water *See* **borax.**

***d*-isomer** *See* **optical activity.**

D-isomer *See* **absolute configuration.**

disordered solid A material that neither has the structure of a perfect *crystal lattice nor of a crystal lattice with isolated *crystal defects. In a *random alloy*, one type of disordered solid, the order of the different types of atom occurs at random. Another type of disordered solid is formed by introducing a high concentration of defects, with the defects distributed randomly throughout the solid. In an *amorphous solid, such as glass, there is a random network of atoms with no lattice.

disperse dye *See* **dyes.**

disperse phase *See* **colloids.**

dispersion forces *See* **van der Waals' force.**

displacement reaction *See* **substitution reaction.**

disproportionation A type of chemical reaction in which the same compound is simultaneously reduced and oxidized. For example, copper(I) chloride disproportionates thus:

$$2CuCl \rightarrow Cu + CuCl_2$$

The reaction involves oxidation of one molecule

$$Cu^{I} \rightarrow Cu^{II} + e$$

and reduction of the other

$$Cu^{I} + e \rightarrow Cu$$

The reaction of halogens with hydroxide ions is another example of a disproportionation reaction, for example

$$Cl_2(g) + 2OH^-(aq) \rightleftharpoons Cl^-(aq) + ClO^-(aq) + H_2O(l)$$

The reverse process is *comproportionation.

dissipative system A system that involves *irreversible processes. All real systems are dissipative (in contrast to such idealized systems as the frictionless pendulum, which is invariant under time reversal). In a dissipative system the system is moving towards a state of equilibrium, which can be regarded as moving toward a point attractor in phase space; this is equivalent to moving towards the minimum of the free energy, *F*.

dissociation The breakdown of a molecule, ion, etc., into smaller molecules, ions, etc. An example of dissociation is the reversible reaction of hydrogen iodide at high temperatures

$2HI(g) \rightleftharpoons H_2(g) + I_2(g)$

The *equilibrium constant of a reversible dissociation is called the *dissociation constant*. The term 'dissociation' is also applied to ionization reactions of *acids and *bases in water; for example

$HCN + H_2O \rightleftharpoons H_3O^+ + CN^-$

which is often regarded as a straightforward dissociation into ions

$HCN \rightleftharpoons H^+ + CN^-$

The equilibrium constant of such a dissociation is called the *acid dissociation constant* or *acidity constant*, given by

$K_a = [H^+][A^-]/[HA]$

for an acid HA (the concentration of water $[H_2O]$ can be taken as constant). K_a is a measure of the strength of the acid. Similarly, for a nitrogenous base B, the equilibrium

$B + H_2O \rightleftharpoons BH^+ + OH^-$

is also a dissociation; with the *base dissociation constant*, or *basicity constant*, given by

$K_b = [BH^+][OH^-]/[B]$

For a hydroxide MOH,

$K_b = [M^+][OH^-]/[MOH]$

dissociation pressure When a solid compound dissociates to give one or more gaseous products, the dissociation pressure is the pressure of gas in equilibrium with the solid at a given temperature. For example, when calcium carbonate is maintained at a constant high temperature in a closed container, the dissociation pressure at that temperature is the pressure of carbon dioxide from the equilibrium

$CaCO_3(s) \rightleftharpoons CaO(s) + CO_2(g)$

distillation The process of boiling a liquid and condensing and collecting the vapour. The liquid collected is the *distillate*. It is used to purify liquids and to separate liquid mixtures (*see* **fractional distillation**; **steam distillation**). *See also* **destructive distillation**.

distilled water Water purified by distillation so as to free it from dissolved salts and other compounds. Distilled water in equilibrium with the carbon dioxide in the air has a conductivity of about 0.8×10^{-6} siemens cm^{-1}. Repeated distillation in a vacuum can bring the conductivity down to 0.043×10^{-6} siemens cm^{-1} at 18°C (sometimes called *conductivity water*). The limiting conductivity is due to self ionization: $H_2O \rightleftharpoons H^+ + OH^-$. *See also* **deionized water**.

disulphur dichloride **(sulphur monochloride)** An orange–red liquid, S_2Cl_2, which is readily hydrolysed by water and is soluble in benzene and ether; r.d. 1.678; m.p. −80°C; b.p. 136°C. It may be prepared by passing chlorine over molten sulphur; in the presence of iodine or metal chlorides *sulphur dichloride*, SCl_2, is also formed. In the vapour phase S_2Cl_2 molecules have

Cl–S–S–Cl chains. The compound is used as a solvent for sulphur and can form higher *chlorosulphanes* of the type Cl–$(S)_n$–Cl ($n < 100$), which are of great value in *vulcanization processes.

disulphuric(VI) acid **(pyrosulphuric acid)** A colourless hygroscopic crystalline solid, $H_2S_2O_7$; r.d. 1.9; m.p. 35°C. It is commonly encountered mixed with sulphuric acid as it is formed by dissolving sulphur trioxide in concentrated sulphuric acid. The resulting fuming liquid, called *oleum* or *Nordhausen sulphuric acid*, is produced during the *contact process and is also widely used in the *sulphonation of organic compounds. *See also* **sulphuric acid.**

dithionate A salt of dithionic acid, containing the ion $S_2O_6^{2-}$, usually formed by the oxidation of a sulphite using manganese(IV) oxide. The ion has neither pronounced oxidizing nor reducing properties.

dithionic acid An acid, $H_2S_2O_6$, known in the form of its salts (dithionates).

dithionite *See* **sulphinate.**

dithionous acid *See* **sulphinic acid.**

divalent **(bivalent)** Having a valency of two.

DLA *See* **diffusion limited aggregation.**

D-lines Two close lines in the yellow region of the visible spectrum of sodium, having wavelengths 589.0 and 589.6 nm. As they are prominent and easily recognized they are used as a standard in spectroscopy.

***dl*-isomer** *See* **optical activity; racemic mixture.**

DLVO theory A theory of colloid stability proposed in the 1940s by the Soviet scientists Boris Derjaguin and Lev Landau and independently by the Dutch scientists Evert Verwey and Theo Overbeek. The DLVO theory takes account of two types of force in a stable colloid: the van der Waals' force, which is attractive and binds particles together, and electrostatic repulsion. The total interaction potential can be calculated as a function of distance, with colloid stability being attained when the two forces balance each other. The DLVO theory is the basis for understanding colloid stability and has a considerable amount of experimental support. However, it is inadequate for the properties of colloids in the aggregated state, which depend on short-range interactions taking into account the specific properties of ions, rather than regarding them as point particles.

DMF *See* **dimethylformamide.**

DMG *See* **dimethylglyoxime.**

DMSO *See* **dimethyl sulphoxide.**

DNA **(deoxyribonucleic acid)** The genetic material of most living organisms, which is a major constituent of the chromosomes within the

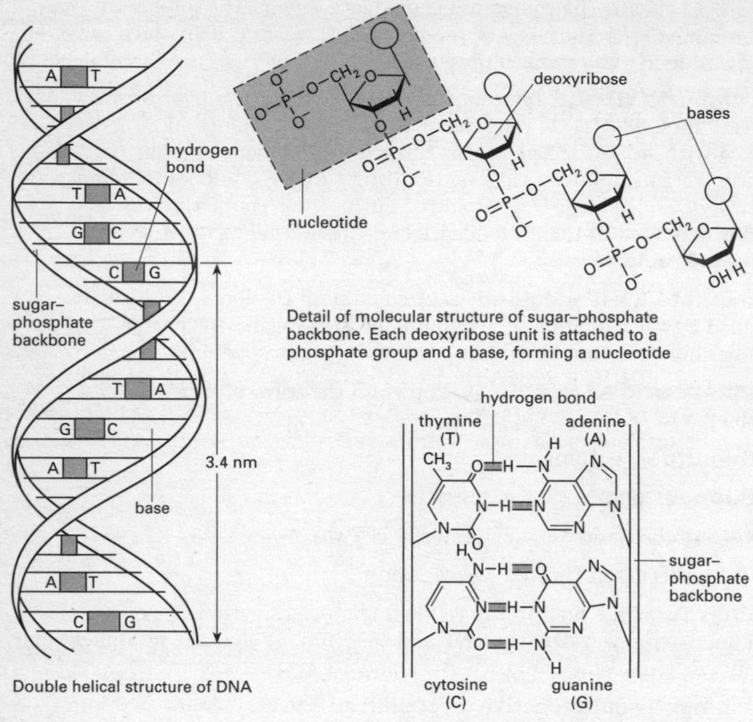

Double helical structure of DNA

Detail of molecular structure of sugar–phosphate backbone. Each deoxyribose unit is attached to a phosphate group and a base, forming a nucleotide

The four bases of DNA, showing the hydrogen bonding between base pairs

Molecular structure of DNA

cell nucleus and plays a central role in the determination of hereditary characteristics by controlling protein synthesis in cells. DNA is a nucleic acid composed of two chains of *nucleotides in which the sugar is *deoxyribose* and the bases are *adenine, *cytosine, *guanine, and *thymine (*compare* **RNA**). The two chains are wound round each other and linked together by hydrogen bonds between specific complementary bases to form a spiral ladder-shaped molecule (*double helix*: see illustration).

When the cell divides, its DNA also replicates in such a way that each of the two daughter molecules is identical to the parent molecule. The hydrogen bonds between the complementary bases on the two strands of the parent molecule break and the strands unwind. Using as building bricks nucleotides present in the nucleus, each strand directs the synthesis of a new one complementary to itself. Replication is initiated, controlled, and stopped by means of polymerase enzymes.

Döbereiner's triads A set of triads of chemically similar elements noted by Johann Döbereiner (1780–1849) in 1817. Even with the inaccurate atomic mass data of the day it was observed that when each triad was arranged in order of increasing atomic mass, then the mass of the central member was approximately the average of the values for the other two. The chemical and physical properties were similarly related. The triads are now recognized as consecutive members of the groups of the periodic table. Examples are: lithium, sodium, and potassium; calcium, strontium, and barium; and chlorine, bromine, and iodine.

dodecanoic acid **(lauric acid)** A white crystalline *fatty acid, $CH_3(CH_2)_{10}COOH$; r.d. 0.8; m.p. 44°C; b.p. 225°C. Glycerides of the acid are present in natural fats and oils (e.g. coconut and palm-kernel oil).

dodecene A straight-chain alkene, $CH_3(CH_2)_9CH{:}CH_2$, obtained from petroleum and used in making *dodecylbenzene.

dodecylbenzene A hydrocarbon, $CH_3(CH_2)_{11}C_6H_5$, manufactured by a Friedel–Crafts reaction between dodecene ($CH_3(CH_2)_9CH{:}CH_2$) and benzene. It can be sulphonated, and the sodium salt of the sulphonic acid is the basis of common *detergents.

dolomite A carbonate mineral consisting of a mixed calcium–magnesium carbonate, $CaCO_3.MgCO_3$, crystallizing in the rhombohedral system. It is usually white or colourless. The term is also used to denote a rock with a high ratio of magnesium to calcium carbonate. *See* **limestone**.

domain A functional unit of the tertiary structure of a *protein. It consists of chains of amino acids folded into alpha helices and *beta sheets to form a globular structure. Different domains are linked together by relatively straight sections of polypeptide chain to form the protein molecule. Domains allow a degree of movement in the protein structure.

Donnan equilibrium The equilibrium set up when two solutions are separated by a membrane permeable to some but not all of the ions in the solutions. In practice, the membrane is often permeable to the solvent and small ions but not to charged entities of colloidal size or such polyelectrolytes as proteins. An electrical potential develops between the two sides of the membrane with the two solutions having varying osmotic pressure. Donnan equilibrium is named after the British chemist Frederick George Donnan (1870–1956), who developed the theory of membrane equilibrium. Donnan equilibrium is important in biology.

donor An atom, ion, or molecule that provides a pair of electrons in forming a coordinate bond.

dopa **(dihydroxyphenylalanine)** A derivative of the amino acid tyrosine. It is found in particularly high levels in the adrenal glands and is a precursor in the synthesis of *dopamine, *noradrenaline, and *adrenaline. The laevorotatory form, *L-dopa*, is administered in the treatment of Parkinson's disease, in which brain levels of dopamine are reduced.

dopamine A *catecholamine that is a precursor in the synthesis of *noradrenaline and *adrenaline. It also functions as a neurotransmitter in the brain.

***d*-orbital** *See* **orbital.**

dose A measure of the extent to which matter has been exposed to *ionizing radiation. The *absorbed dose* is the energy per unit mass absorbed by matter as a result of such exposure. The SI unit is the gray, although it is often measured in rads (1 rad = 0.01 gray; *see* **radiation units**). The *maximum permissible dose* is the recommended upper limit of absorbed dose that a person or organ should receive in a specified period according to the International Commission on Radiological Protection.

dosimeter Any device used to measure absorbed *dose of ionizing radiation. Methods used include the ionization chamber, photographic film, or the rate at which certain chemical reactions occur in the presence of ionizing radiation.

double bond *See* **chemical bond.**

double decomposition (metathesis) A chemical reaction involving exchange of radicals, e.g.

$$AgNO_3(aq) + KCl(aq) \rightarrow KNO_3(aq) + AgCl(s)$$

double layer *See* **electrical double layer.**

double refraction The property, possessed by certain crystals (notably calcite), of forming two refracted rays from a single incident ray. The *ordinary ray* obeys the normal laws of refraction. The other refracted ray, called the *extraordinary ray*, follows different laws. The light in the ordinary ray is polarized at right angles to the light in the extraordinary ray. Along an optic axis the ordinary and extraordinary rays travel with the same speed. Some crystals, such as calcite, quartz, and tourmaline, have only one optic axis; they are *uniaxial crystals.* Others, such as mica and selenite, have two optic axes; they are *biaxial crystals.* The phenomenon is also known as *birefringence* and the double-refracting crystal as a *birefringent crystal. See also* **polarization.**

double salt A crystalline salt in which there are two different anions and/or cations. An example is the mineral dolomite, $CaCO_3.MgCO_3$, which contains a regular arrangement of Ca^{2+} and Mg^{2+} ions in its crystal lattice. *Alums are double sulphates. Double salts only exist in the solid; when dissolved they act as a mixture of the two separate salts. *Double oxides* are similar.

doublet A pair of associated lines in certain spectra, e.g. the two lines that make up the sodium D-lines.

Dow process A method of extracting magnesium from sea water by adding calcium hydroxide to precipitate magnesium hydroxide.

Downs process A process for extracting sodium by the electrolysis of molten sodium chloride. The *Downs cell* has a central graphite anode surrounded by a cylindrical steel cathode. Chlorine released is led away through a hood over the anode. Molten sodium is formed at the cathode and collected through another hood around the top of the cathode cylinder (it is less dense than the sodium chloride). The two hoods and electrodes are separated by a coaxial cylindrical steel gauze. A small amount of calcium chloride is added to the sodium chloride to lower its melting point. The sodium chloride is melted electrically and kept molten by the current through the cell. More sodium chloride is added as the electrolysis proceeds.

dropping-mercury electrode *See* **polarography**.

dry cell A primary or secondary cell in which the electrolytes are restrained from flowing in some way. Many torch, radio, and calculator batteries are *Leclanché cells in which the electrolyte is an ammonium chloride paste and the container is the negative zinc electrode (with an outer plastic wrapping). Various modifications of the Leclanché cell are used in dry cells. In the *zinc chloride cell*, the electrolyte is a paste of zinc chloride rather than ammonium chloride. The electrical characteristics are similar to those of the Leclanché cell but the cell works better at low temperatures and has more efficient depolarization characteristics. A number of alkaline secondary cells can be designed for use as dry cells. In these, the electrolyte is a liquid (sodium or potassium hydroxide) held in a porous material or in a gel. Alkaline dry cells typically have zinc–manganese dioxide, silver oxide–zinc, nickel–cadmium, or nickel–iron electrode systems (*see* **nickel–iron accumulator**). For specialized purposes, dry cells and batteries have been produced with solid electrolytes. These may contain a solid crystalline salt, such as silver iodide, an ion-exchange membrane, or an organic wax with a small amount of dissolved ionic material. Such cells deliver low currents. They are used in miniature cells for use in electronic equipment.

dry ice Solid carbon dioxide used as a refrigerant. It is convenient because it sublimes at −78°C (195 K) at standard pressure.

drying oil A natural oil, such as linseed oil, that hardens on exposure to the air. Drying oils contain unsaturated fatty acids, such as linoleic and linolenic acids, which polymerize on oxidation. They are used in paints, varnishes, etc.

DSC Differential scanning calorimetry. *See* **thermal analysis**.

D-series *See* **absolute configuration**.

DTA Differential thermal analysis. *See* **thermal analysis**.

dubnium Symbol Db. A radioactive *transactinide element; a.n. 105. It was first reported in 1967 by a group at Dubna near Moscow and was confirmed in 1970 at Dubna and at Berkeley, California. It can be made by bombarding californium–249 nuclei with nitrogen–15 nuclei. Only a few atoms have ever been made.

Dulong and Petit's law For a solid element the product of the relative atomic mass and the specific heat capacity is a constant equal to about 25 $J\,mol^{-1}\,K^{-1}$. Formulated in these terms in 1819 by the French scientists Pierre Dulong (1785–1838) and Alexis Petit (1791–1820), the law in modern terms states: the molar heat capacity of a solid element is approximately equal to $3R$, where R is the *gas constant. The law is only approximate but applies with fair accuracy at normal temperatures to elements with a simple crystal structure.

Dumas, Jean Baptiste André (1800–84) French chemist, who became an apothecary in Geneva, where in 1818 he investigated the use of iodine to treat goitre. He then took up chemistry and moved to Paris. In 1826 he devised a method of measuring *vapour density. He went on to discover various organic compounds, including anthracene (1832), urethane (1833), and methanol (1834), which led him in 1840 to propose the theory of types (functional groups).

Dumas' method 1. A method of finding the amount of nitrogen in an organic compound. The sample is weighed, mixed with copper(II) oxide, and heated in a tube. Any nitrogen present in the compound is converted into oxides of nitrogen, which are led over hot copper to reduce them to nitrogen gas. This is collected and the volume measured, from which the mass of nitrogen in a known mass of sample can be found. **2.** A method of finding the relative molecular masses of volatile liquids by weighing. A thin-glass bulb with a long narrow neck is used. This is weighed full of air at known temperature, then a small amount of sample is introduced and the bulb heated (in a bath) so that the liquid is vaporized and the air is driven out. The tip of the neck is sealed and the bulb cooled and weighed at known (room) temperature. The volume of the bulb is found by filling it with water and weighing again. If the density of air is known, the mass of vapour in a known volume can be calculated.

The techniques are named after Jean Baptiste André *Dumas.

duplet A pair of electrons in a covalent chemical bond.

Duralumin Tradename for a class of strong lightweight aluminium alloys containing copper, magnesium, manganese, and sometimes silicon. Duralumin alloys combine strength with lightness and are extensively used in aircraft, racing cars, etc.

Dutch metal An alloy of copper and zinc, which can be produced in very thin sheets and used as imitation gold leaf. It spontaneously inflames in chlorine.

dye laser A type of laser in which the active material is a dye dissolved in a suitable solvent (e.g. Rhodamine G in methanol). The dye is excited by an external source. The solvent broadens the states into bands and consequently laser action can be obtained over a range of wavelengths. This allows one to select a specific wavelength (using a grating) and to change the wavelength of the laser. Such a device is called a *tunable laser*. Dye lasers are also used in producing very short pulses of radiation. The technique is to use a dye that stops absorbing radiation when a high proportion of its molecules become excited. The cavity then becomes resonant and a pulse of radiation is produced. This technique can give pulses of about 10 nanoseconds duration and is used in *femtochemistry.

dyes Substances used to impart colour to textiles, leather, paper, etc. Compounds used for dyeing (*dyestuffs*) are generally organic compounds containing conjugated double bonds. The group producing the colour is the *chromophore; other noncoloured groups that influence or intensify the colour are called *auxochromes. Dyes can be classified according to the chemical structure of the dye molecule. For example, *azo dyes* contain the –N=N– group (*see* **azo compounds**). In practice, they are classified according to the way in which the dye is applied or is held on the substrate.

Acid dyes are compounds in which the chromophore is part of a negative ion (usually an organic sulphonate RSO_2O^-). They can be used for protein fibres (e.g. wool and silk) and for polyamide and acrylic fibres. Originally, they were applied from an acidic bath. *Metallized dyes* are forms of acid dyes in which the negative ion contains a chelated metal atom. *Basic dyes* have chromophores that are part of a positive ion (usually an amine salt or ionized imino group). They are used for acrylic fibres and also for wool and silk, although they have only moderate fastness with these materials.

Direct dyes are dyes that have a high affinity for cotton, rayon, and other cellulose fibres. They are applied directly from a neutral bath containing sodium chloride or sodium sulphate. Like acid dyes, they are usually sulphonic acid salts but are distinguished by their greater substantivity (affinity for the substrate), hence the alternative name *substantive dyes*.

Vat dyes are insoluble substances used for cotton dyeing. They usually contain keto groups, C=O, which are reduced to C–OH groups, rendering the dye soluble (the *leuco form* of the dye). The dye is applied in this form, then oxidized by air or oxidizing agents to precipitate the pigment in the fibres. Indigo and anthroquinone dyes are examples of vat dyes. *Sulphur dyes* are dyes applied by this technique using sodium sulphide solution to reduce and dissolve the dye. Sulphur dyes are used for cellulose fibres.

Disperse dyes are insoluble dyes applied in the form of a fine dispersion in water. They are used for cellulose acetate and other synthetic fibres.

Reactive dyes are compounds that contain groups capable of reacting with the substrate to form covalent bonds. They have high substantivity and are used particularly for cellulose fibres.

dynamical system A system governed by dynamics (either classical mechanics or quantum mechanics). The evolution of dynamical systems can be very complex, even for systems with only a few degrees of freedom, and may be studied using the phase space for the system. *Chaos is an example of the complex behaviour that can occur in a dynamical system.

dynamic equilibrium *See* **equilibrium**.

dynamite Any of a class of high explosives based on nitroglycerin. The original form, invented in 1867 by Alfred Nobel, consisted of nitroglycerin absorbed in kieselguhr. Modern dynamites, which are used for blasting, contain sodium or ammonium nitrate sensitized with nitroglycerin and use other absorbers (e.g. wood pulp).

dysprosium Symbol Dy. A soft silvery metallic element belonging to the *lanthanoids; a.n. 66; r.a.m. 162.50; r.d. 8.551 (20°C); m.p. 1412°C; b.p. 2562°C. It occurs in apatite, gadolinite, and xenotime, from which it is extracted by an ion-exchange process. There are seven natural isotopes. It finds limited use as a neutron absorber, particularly in nuclear technology. It was discovered by Paul Lecoq de Boisbaudran (1838–1912) in 1886.

dystectic mixture A mixture of substances that has a constant maximum melting point.

Earnshaw's theorem The principle that a system of particles interacting via an inverse square law, such as Coulomb's law of electrostatics, cannot exist in a state of static equilibrium. The Reverend Samuel Earnshaw (1805–88) proved this result in 1842. The theorem is of fundamental importance in chemistry since it shows that it is not possible to construct a correct model of atoms and molecules if the electrons are taken to be stationary.

earth The planet that orbits the sun between the planets Venus and Mars. The earth consists of three layers: the gaseous atmosphere (*see* **earth's atmosphere**), the liquid *hydrosphere*, and the solid *lithosphere*. The solid part of the earth also consists of three layers: the *crust* with a mean thickness of about 32 km under the land and 10 km under the seas; the *mantle*, which extends some 2900 km below the crust; and the *core*, part of which is believed to be liquid. The composition of the crust is: oxygen 47%, silicon 28%, aluminium 8%, iron 4.5%, calcium 3.5%, sodium and potassium 2.5% each, and magnesium 2.2%. Hydrogen, carbon, phosphorus, and sulphur are all present to an extent of less than 1%.

earth's atmosphere The gas that surrounds the earth. The composition of dry air at sea level is: nitrogen 78.08%, oxygen 20.95%, argon 0.93%, carbon dioxide 0.03%, neon 0.0018%, helium 0.0005%, krypton 0.0001%, and xenon 0.00001%. In addition to water vapour, air in some localities contains sulphur compounds, hydrogen peroxide, hydrocarbons, and dust particles.

ebonite *See* **vulcanite**.

ebullioscopic constant *See* **elevation of boiling point**.

ebullioscopy The use of *elevation of boiling point to determine relative molecular masses.

echelon A form of interferometer consisting of a stack of glass plates arranged stepwise with a constant offset. It gives a high resolution and is used in spectroscopy to study hyperfine line structure.

eclipsed *See* **conformation**.

eclipsing *See* **conformation**.

Edison cell *See* **nickel–iron accumulator**.

EDTA Ethylenediaminetetraacetic acid, $(HOOCCH_2)_2N(CH_2)_2N(CH_2COOH)_2$, a compound that acts as a chelating agent, reversibly binding with iron, magnesium, and other metal ions. It is used in certain culture media bound with iron, which it slowly releases into the medium, and also in some forms of quantitative analysis.

EELS (electron energy loss spectroscopy) A technique for studying adsorbates and their dissociation. A beam of electrons is reflected from a surface and the energy loss they suffer upon reflection is measured. This loss can be used to interpret the vibrational spectrum of the adsorbate. Very small amounts of adsorbate can be detected using EELS (as few as 50 atoms in a sample); this is particularly useful for light elements that are not readily detected using other techniques.

effervescence The formation of gas bubbles in a liquid by chemical reaction.

efficiency A measure of the performance of a machine, engine, etc., being the ratio of the energy or power it delivers to the energy or power fed to it. In general, the efficiency of a machine varies with the conditions under which it operates and there is usually a load at which it operates with the highest efficiency. The *thermal efficiency* of a heat engine is the ratio of the work done by the engine to the heat supplied by the fuel. For a reversible heat engine this efficiency equals $(T_1 - T_2)/T_1$, where T_1 is the thermodynamic temperature at which all the heat is taken up and T_2 is the thermodynamic temperature at which it is given out (*see* **Carnot cycle**). For real engines it is always less than this.

efflorescence The process in which a crystalline hydrate loses water, forming a powdery deposit on the crystals.

effusion The flow of a gas through a small aperture. The relative rates at which gases effuse, under the same conditions, is approximately inversely proportional to the square roots of their densities.

Ehrenfest classification A classification of phase transitions in terms of their thermodynamic properties put forward by the Dutch physicist Paul Ehrenfest (1880–1933). A *first-order phase transition* is a phase transition in which the first derivative of the chemical potential is discontinuous. In a first-order phase transition there is a nonzero change in the value of the enthalpy, entropy, and volume at the transition temperature. Melting and boiling are examples of first-order phase transitions. In a *second-order phase transition*, the first derivative of the chemical potential is continuous but its second derivative is not continuous. In a second-order phase transition there is no jump in the value of the enthalpy, entropy, and volume at the transition temperature. Examples of second-order phase transitions include the transition to ferromagnetism and order–disorder transitions in alloys.

eigenfunction In general, a solution of an *eigenvalue equation. In *quantum mechanics eigenfunctions occur as the allowed wave functions of a system and so satisfy the *Schrödinger equation.

eigenvalues The allowed set of values of the constants in an *eigenvalue equation*. In an eigenvalue equation the left-hand side consists of an operator and an *eigenfunction, upon which the operator operates; the right-hand side of the equation consists of the product of the same eigenfunction and a constant. Eigenvalue equations occur in quantum mechanics, with the eigenvalues being the values of the quantized quantities. In particular, the Schrödinger equation is an eigenvalue equation with the allowed quantized energy levels being the eigenvalues of this equation.

Einstein, Albert (1879–1955) German-born US physicist, who took Swiss nationality in 1901. A year later he went to work in the Bern patent office. In 1905 he published five enormously influential papers, one on *Brownian movement, one on the *photoelectric effect, one on the special theory of relativity, and one on energy and inertia (which included the famous expression $E = mc^2$). In 1915 he published the general theory of relativity, concerned mainly with gravitation. In 1921 he was awarded the Nobel Prize. In 1933, as a Jew, Einstein decided to remain in the USA (where he was lecturing), as Hitler had come to power. For the remainder of his life he sought a unified field theory. In 1939 he informed President Roosevelt that an atom bomb was feasible and that Germany might be able to make one.

Einstein coefficients Coefficients used in the quantum theory of radiation, related to the probability of a transition occurring between the ground state and an excited state (or vice versa) in the processes of *induced emission and *spontaneous emission. For an atom exposed to electromagnetic radiation, the rate of absorption R_a is given by $R_a = B\rho$, where ρ is the density of electromagnetic radiation and B is the *Einstein B coefficient* associated with absorption. The rate of induced emission is also given by $B\rho$, with the coefficient B of induced emission being equal to the coefficient of absorption. The rate of spontaneous emission is given by A, where A is the *Einstein A coefficient of spontaneous emission*. The A and B coefficients are related by $A = 8\pi h\nu^3 B/c^3$, where h is the Planck constant, ν is the frequency of electromagnetic radiation, and c is the speed of light. The coefficients were put forward by Albert *Einstein in 1916–17 in his analysis of the quantum theory of radiation.

Einstein equation 1. The mass–energy relationship announced by Albert *Einstein in 1905 in the form $E = mc^2$, where E is a quantity of energy, m its mass, and c is the speed of light. It presents the concept that energy possesses mass. **2.** The relationship $E_{max} = hf - W$, where E_{max} is the maximum kinetic energy of electrons emitted in the photoemissive effect, h is the Planck constant, f the frequency of the incident radiation, and W the *work function of the emitter. This is also written $E_{max} = hf - \phi e$, where e is the electronic charge and ϕ a potential difference, also called the work function. (Sometimes W and ϕ are distinguished as *work function energy* and *work function potential*.) The equation can also be applied to photoemission from gases, when it has the form: $E = hf - I$, where I is the ionization potential of the gas.

einsteinium Symbol Es. A radioactive metallic transuranic element belonging to the *actinoids; a.n. 99; mass number of the most stable isotope 254 (half-life 270 days). Eleven isotopes are known. The element was first identified by A. Ghiorso and associates in debris from the first hydrogen bomb explosion in 1952. Microgram quantities of the element did not become available until 1961. It is named after Albert *Einstein.

Einstein–Smoluchowski equation A relation between the diffusion coefficient D and the distance λ that a particle can jump when diffusing in a time τ. The Einstein–Smoluchowski equation, which is $D = \lambda^2/2\tau$, gives a connection between the microscopic details of particle diffusion and the macroscopic quantities associated with the diffusion, such as the viscosity. The equation is derived by assuming that the particles undergo a random walk. The quantities in the equation can be related to quantities in the *kinetic theory of gases, with λ/τ taken to be the mean speed of the particles and λ their mean free path. The Einstein–Smoluchowski equation was derived by Albert Einstein and the Polish physicist Marian Ritter von Smolan-Smoluchowski.

Einstein theory of specific heat A theory of the specific heat capacity of solids put forward by Albert *Einstein in 1906, in which it was assumed that the specific heat capacity is a consequence of the vibrations of the atoms of the lattice of the solid. Einstein assumed that each atom has the same frequency ν. The theory leads to the correct conclusion that the specific heat of solids tends to zero as the temperature goes to absolute zero, but does not give a correct quantitative description of the low-temperature behaviour of the specific heat capacity. In the *Debye theory of specific heat, and in other analyses of this problem, Einstein's simplifying approximation was improved on by taking account of the fact that the frequencies of lattice vibrations can have a range of values.

E-isomer *See* **E–Z convention**.

elastic collision A collision in which the total kinetic energy of the colliding bodies after collision is equal to their total kinetic energy before collision. Elastic collisions occur only if there is no conversion of kinetic energy into other forms, as in the collision of atoms. In the case of macroscopic bodies this will not be the case as some of the energy will become heat. In a collision between polyatomic molecules, some kinetic energy may be converted into vibrational and rotational energy of the molecules.

elastin A fibrous protein that is the major constituent of the yellow elastic fibres of connective tissue. It is rich in glycine, alanine, proline, and other nonpolar amino acids that are cross-linked, making the protein relatively insoluble. Elastic fibres can stretch to several times their length and then return to their original size. Elastin is particularly abundant in elastic cartilage, blood-vessel walls, ligaments, and the heart.

elastomer A natural or synthetic rubber or rubberoid material, which has the ability to undergo deformation under the influence of a force and regain its original shape once the force has been removed.

electret A permanently electrified substance or body that has opposite charges at its extremities. Electrets resemble permanent magnets in many ways. An electret can be made by cooling certain waxes in a strong electric field.

electrical double layer A model of the interface between an electrode and the solution close to it. In this model a sheet of one type of electrical charge surrounds the surface of the electrode and a sheet of the opposite charge surrounds the first sheet in the solution. In the *Helmholtz model* the double layer is regarded as consisting of two planes of charge, with the inner plane of ions from the solution being caused by the charge on the electrode and the outer plane being caused by oppositely charged ions in the solution responding to the first layer of ions. In the *Gouy–Chapman model (diffuse double layer) thermal motion of ions is taken into account. Neither model is completely successful since the Helmholtz model exaggerates the rigidity of the structure of the charges and the Gouy–Chapman model underestimates the rigidity of the structure. The *Stern model* improves on both models by assuming that the ions next to the electrode have a rigid structure, while taking the second layer to be as described by the Gouy–Chapman model.

electric-arc furnace A furnace used in melting metals to make alloys, especially in steel manufacture, in which the heat source is an electric arc. In the direct-arc furnace, such as the Héroult furnace, an arc is formed between the metal and an electrode. In the indirect-arc furnace, such as the Stassano furnace, the arc is formed between two electrodes and the heat is radiated onto the metal.

electrochemical cell *See* **cell.**

electrochemical equivalent Symbol *z*. The mass of a given element liberated from a solution of its ions in electrolysis by one coulomb of charge. *See* **Faraday's laws (of electrolysis).**

electrochemical series *See* **electromotive series.**

electrochemistry The study of chemical properties and reactions involving ions in solution, including electrolysis and electric cells.

electrochromatography *See* **electrophoresis.**

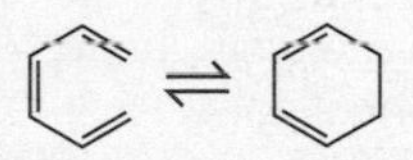

Electrocyclic reaction

electrocyclic reaction A type of cyclic rearrangement in which a sigma bond is formed between two terminal carbon atoms of a conjugated molecule, resulting in a decrease of one in the number of pi bonds present.

electrode 1. A conductor that emits or collects electrons in a cell, thermionic valve, semiconductor device, etc. The *anode* is the positive electrode and the *cathode* is the negative electrode. **2.** *See* **half cell.**

electrodeposition The process of depositing one metal on another by electrolysis, as in *electroforming and *electroplating.

electrode potential The potential difference produced between the electrode and the solution in a *half cell. It is not possible to measure this directly since any measurement involves completing the circuit with the electrolyte, thereby introducing another half cell. *Standard electrode potentials* $E^{\ominus}$ are defined by measuring the potential relative to a standard *hydrogen half cell using 1.0 molar solution at 25°C. The convention is to designate the cell so that the oxidized form is written first. For example,

$$Pt(s)|H_2(g)H^+(aq)|Zn^{2+}(aq)|Zn(s)$$

The e.m.f. of this cell is −0.76 volt (i.e. the zinc electrode is negative). Thus the standard electrode potential of the $Zn^{2+}|Zn$ half cell is −0.76 V. Electrode potentials are also called *reduction potentials*. *See also* **electromotive series**.

electrodialysis A method of obtaining pure water from water containing a salt, as in *desalination. The water to be purified is fed into a cell containing two electrodes. Between the electrodes is placed an array of *semipermeable membranes alternately semipermeable to positive ions and negative ions. The ions tend to segregate between alternate pairs of membranes, leaving pure water in the other gaps between membranes. In this way, the feed water is separated into two streams: one of pure water and the other of more concentrated solution.

electroforming A method of forming intricate metal articles or parts by *electrodeposition of the metal on a removable conductive mould.

electrokinetic potential (zeta potential) Symbol ζ. The electric potential associated with an *electrical double layer around a colloid at the *radius of shear*, relative to the value of the potential in the bulk of the solution far from the colloid, where the radius of shear is the radius of the entity made up of the colloid and the rigid layer of ions at the surface of the colloid.

electroluminescence *See* **luminescence**.

electrolysis The production of a chemical reaction by passing an electric current through an electrolyte. In electrolysis, positive ions migrate to the cathode and negative ions to the anode. The reactions occurring depend on electron transfer at the electrodes and are therefore redox reactions. At the anode, negative ions in solution may lose electrons to form neutral species. Alternatively, atoms of the electrode can lose electrons and go into solution as positive ions. In either case the reaction is an oxidation. At the cathode, positive ions in solution can gain electrons to form neutral species. Thus cathode reactions are reductions.

electrolyte A liquid that conducts electricity as a result of the presence of positive or negative ions. Electrolytes are molten ionic compounds or solutions containing ions, i.e. solutions of ionic salts or of compounds that ionize in solution. Liquid metals, in which the conduction is by free electrons, are not usually regarded as electrolytes. Solid conductors of ions, as in the sodium–sulphur cell, are also known as electrolytes.

electrolytic cell A cell in which electrolysis occurs; i.e. one in which current is passed through the electrolyte from an external source.

electrolytic corrosion Corrosion that occurs through an electrochemical reaction. *See* **rusting**.

electrolytic gas (detonating gas) The highly explosive gas formed by the electrolysis of water. It consists of two parts hydrogen and one part oxygen by volume.

electrolytic refining The purification of metals by electrolysis. It is commonly applied to copper. A large piece of impure copper is used as the anode with a thin strip of pure copper as the cathode. Copper(II) sulphate solution is the electrolyte. Copper dissolves at the anode: $Cu \rightarrow Cu^{2+} + 2e$, and is deposited at the cathode. The net result is transfer of pure copper from anode to cathode. Gold and silver in the impure copper form a so-called *anode sludge* at the bottom of the cell, which is recovered.

electrolytic separation A method of separating isotopes by exploiting the different rates at which they are released in electrolysis. It was formerly used for separating deuterium and hydrogen. On electrolysis of water, hydrogen is formed at the cathode more readily than deuterium, thus the water becomes enriched with deuterium oxide.

electromagnetic radiation Energy resulting from the acceleration of electric charge and the associated electric fields and magnetic fields. The energy can be regarded as waves propagated through space (requiring no supporting medium) involving oscillating electric and magnetic fields at right angles to each other and to the direction of propagation. In a vacuum the waves travel with a constant speed (the speed of light) of 2.9979×10^8 metres per second; if material is present they are slower. Alternatively, the energy can be regarded as a stream of *photons travelling at the speed of light, each photon having an energy hc/λ, where h is the Planck constant, c is the speed of light, and λ is the wavelength of the associated wave. A fusion of these apparently conflicting concepts is possible using the methods of *quantum mechanics. The characteristics of the radiation depend on its wavelength. *See* **electromagnetic spectrum**.

electromagnetic spectrum The range of wavelengths over which electromagnetic radiation extends. The longest waves (10^5–10^{-3} metres) are radio waves, the next longest (10^{-3}–10^{-6} m) are infrared waves, then comes the narrow band (4–7×10^{-7} m) of visible radiation, followed by ultraviolet waves (10^{-7}–10^{-9} m) and *X-rays and gamma radiation (10^{-9}–10^{-14} m).

electromeric effect *See* **electronic effects**.

electrometallurgy The uses of electrical processes in the separation of metals from their ores, the refining of metals, or the forming or plating of metals.

electromotive force **(e.m.f.)** The greatest potential difference that can be generated by a particular source of electric current. In practice this may be observable only when the source is not supplying current, because of its internal resistance.

electromotive series **(electrochemical series)** A series of chemical elements arranged in order of their *electrode potentials. The hydrogen electrode ($H^+ + e \rightarrow \frac{1}{2}H_2$) is taken as having zero electrode potential. Elements that have a greater tendency than hydrogen to lose electrons to their solution are taken as *electropositive*; those that gain electrons from their solution are below hydrogen in the series and are called *electronegative*. The series shows the order in which metals replace one another from their salts; electropositive metals will replace hydrogen from acids. The chief metals and hydrogen, placed in order in the series, are: potassium, calcium, sodium, magnesium, aluminium, zinc, cadmium, iron, nickel, tin, lead, hydrogen, copper, mercury, silver, platinum, gold. This type of series is sometimes referred to as an *activity series*.

electron An elementary particle with a rest mass of $9.109\,3897(54) \times 10^{-31}$ kg and a negative charge of $1.602\,177\,33(49) \times 10^{-19}$ coulomb. Electrons are present in all atoms in groupings called shells around the nucleus; when they are detached from the atom they are called *free electrons*. The antiparticle of the electron is the *positron*.

electron affinity Symbol *A*. The energy change occurring when an atom or molecule gains an electron to form a negative ion. For an atom or molecule X, it is the energy released for the electron-attachment reaction

$$X(g) + e \rightarrow X^-(g)$$

Often this is measured in electronvolts. Alternatively, the molar enthalpy change, ΔH, can be used.

electron capture **1.** The formation of a negative ion by an atom or molecule when it acquires an extra free electron. **2.** A radioactive transformation in which a nucleus acquires an electron from an inner orbit of the atom, thereby transforming, initially, into a nucleus with the same mass number but an atomic number one less than that of the original nucleus (capture of the electron transforms a proton into a neutron). This type of capture is accompanied by emission of an X-ray photon or Auger electron as the vacancy in the inner orbit is filled by an outer electron.

electron configuration *See* **configuration**.

electron-deficient compound A compound in which there are fewer electrons forming the chemical bonds than required in normal electron-pair bonds. Such compounds use *multicentre bonds. *See* **borane**.

electron diffraction Diffraction of a beam of electrons by atoms or molecules. The fact that electrons can be diffracted in a similar way to light and X-rays shows that particles can act as waves (*see* **de Broglie wavelength**). An electron (mass m, charge e) accelerated through a potential difference V acquires a kinetic energy $mv^2/2 = eV$, where v is the velocity of the electron (nonrelativistic). Thus, the momentum (p) of the electron is $\sqrt{(2eVm)}$. As the de Broglie wavelength (λ) of an electron is given by h/p, where h is the Planck constant, then $\lambda = h/\sqrt{(2eVm)}$. For an accelerating voltage of 3600 V, the wavelength of the electron beam is 0.02 nanometre, some 3×10^4 times shorter than visible radiation.

Electrons then, like X-rays, show diffraction effects with molecules and crystals in which the interatomic spacing is comparable to the wavelength of the beam. They have the advantage that their wavelength can be set by adjusting the voltage. Unlike X-rays they have very low penetrating power. The first observation of electron diffraction was by George Paget *Thomson in 1927, in an experiment in which he passed a beam of electrons in a vacuum through a very thin gold foil onto a photographic plate. Concentric circles were produced by diffraction of electrons by the lattice. The same year Clinton J. Davisson (1881–1958) and Lester Germer (1896–1971) performed a classic experiment in which they obtained diffraction patterns by glancing an electron beam off the surface of a nickel crystal. Both experiments were important verifications of de Broglie's theory and the new quantum theory.

Electron diffraction, because of the low penetration, cannot easily be used to investigate crystal structure. It is, however, employed to measure bond lengths and angles of molecules in gases. Moreover, it is extensively used in the study of solid surfaces and absorption. The main techniques are low-energy electron diffraction (*LEED*) in which the electron beam is reflected onto a fluorescent screen, and high-energy electron diffraction (*HEED*) used either with reflection or transmission in investigating thin films.

electronegative Describing elements that tend to gain electrons and form negative ions. The halogens are typical electronegative elements. For example, in hydrogen chloride, the chlorine atom is more electronegative than the hydrogen and the molecule is polar, with negative charge on the chlorine atom. There are various ways of assigning values for the *electronegativity* of an element. *Mulliken electronegativities* are calculated from $E = (I + A)/2$, where I is ionization potential and A is electron affinity. More commonly, *Pauling electronegativities* are used. These are based on bond dissociation energies using a scale in which fluorine, the most electronegative element, has a value 4. Some other values on this scale are B 2, C 2.5, N 3.0, O 3.5, Si 1.8, P 2.1, S 2.5, Cl 3.0, Br 2.8.

electron energy loss spectroscopy *See* **EELS**.

electron flow The transfer of electrons along a series of carrier molecules in the *electron transport chain.

electron gas A model of the electrons in a metal or a plasma in which they are regarded as forming a gas that interacts with a uniformly distributed background of positive charge to ensure that the system is electrically neutral. The electron gas is analysed theoretically using either classical or quantum statistical mechanics and the kinetic theory of gases. The electron-gas model accounts for many properties of metals and plasmas in a qualitative and approximately quantitative way but cannot give an accurate quantitative account of these systems, as this would require the motions of the positive ions to be taken into account.

electronic effects Effects by which the reactivity at one part of a molecule is affected by electron attraction or repulsion originating in another part of a molecule. Often this is called an *inductive effect (or *resonance effect*), although sometimes the term 'inductive effect' is reserved for an influence transmitted through chemical bonds and is distinguished from a *field effect*, which is transmitted through space. An inductive effect through chemical bonds was formerly called a *mesomeric effect* (or *mesomerism*) or an *electromeric effect*. It is common to refer to all effects (through bonds or through space) as *resonance effects*.

electronic spectra of molecules The spectra associated with transitions between the electronic states of molecules. These transitions correspond to the visible or ultraviolet regions of the electromagnetic spectrum. There are changes in vibrational and rotational energy when electronic transitions occur. Consequently there are spectral bands associated with changes in vibrational motion, with these bands having fine structure due to changes in rotational motion. Because electronic transitions are associated with changes in vibrational motion the corresponding spectra are sometimes called *vibrational spectra*. The electronic spectra of molecules are used to obtain information about energy levels in molecules, interatomic distances, dissociation energies of molecules, and force constants of chemical bonds.

electron microscope A form of microscope that uses a beam of electrons instead of a beam of light (as in the optical microscope) to form a large image of a very small object. In optical microscopes the resolution is limited by the wavelength of the light. High-energy electrons, however, can be associated with a considerably shorter wavelength than light; for example, electrons accelerated to an energy of 10^5 electronvolts have a wavelength of 0.004 nanometre (*see* **de Broglie wavelength**) enabling a resolution of 0.2–0.5 nm to be achieved. The *transmission electron microscope* has an electron beam, sharply focused by electron lenses, passing through a very thin metallized specimen (less than 50 nanometres thick) onto a fluorescent screen, where a visual image is formed. This image can be photographed. The *scanning electron microscope* can be used with thicker specimens and forms a perspective image, although the resolution and magnification are lower. In this type of instrument a beam of primary electrons scans the specimen and those that are reflected, together with any secondary electrons emitted, are collected. This current is used to modulate a separate electron beam in a TV monitor, which scans the screen at the same frequency, consequently building up a picture of the specimen. The resolution is limited to about 10–20 nm.

electron probe microanalysis **(EPM)** A method of analysing a very small quantity of a substance (as little as 10^{-13} gram). The method consists of directing a very finely focused beam of electrons on to the sample to produce the characteristic X-ray spectrum of the elements present. It can be used quantitatively for elements with atomic numbers in excess of 11.

electron-spin resonance **(ESR)** A spectroscopic method of locating electrons within the molecules of a paramagnetic substance (*see* **magnetism**) in order to provide information regarding its bonds and structure. The spin of an unpaired electron is associated with a magnetic moment that is able to align itself in one of two ways with an applied external magnetic field. These two alignments correspond to different energy levels, with a statistical probability, at normal temperatures, that there will be slightly more in the lower state than in the higher. By applying microwave radiation to the sample a transition to the higher state can be achieved. The precise energy difference between the two states of an electron depends on the surrounding electrons in the atom or molecule. In this way the position of unpaired electrons can be investigated. The technique is used particularly in studying free radicals and paramagnetic substances such as inorganic complexes. *See also* **nuclear magnetic resonance**.

electron-transfer reaction A chemical reaction that involves the transfer, addition, or removal of electrons. Electron-transfer reactions often involve complexes of transition metals. In such complexes one general mechanism for electron transfer is the *inner sphere mechanism*, in which two complexes form an intermediate, with ligand bridges enabling electrons to be transferred from one complex to another complex. The other main mechanism is the *outer-sphere mechanism*, in which two complexes retain all their ligands, with electrons passing from one complex to the other. The rates of electron-transfer reactions vary enormously. These rates can be explained in terms of the way in which molecules of the solvent solvating the reactants rearrange so as to solvate the products in the case of the outer-sphere mechanism. In the case of the inner-sphere (ligand-bridged) reactions the rate of the reaction depends on the intermediate and the way in which the electron is transferred.

electron transport chain **(electron transport system)** A sequence of biochemical reduction–oxidation reactions that effects the transfer of electrons through a series of carriers. An electron transport chain, also known as the *respiratory chain*, forms the final stage of aerobic respiration. It results in the transfer of electrons or hydrogen atoms derived from the *Krebs cycle to molecular oxygen, with the formation of water. At the same time it conserves energy from food or light in the form of *ATP. The chain comprises a series of carrier molecules that undergo reversible reduction–oxidation reactions, accepting electrons and then donating them

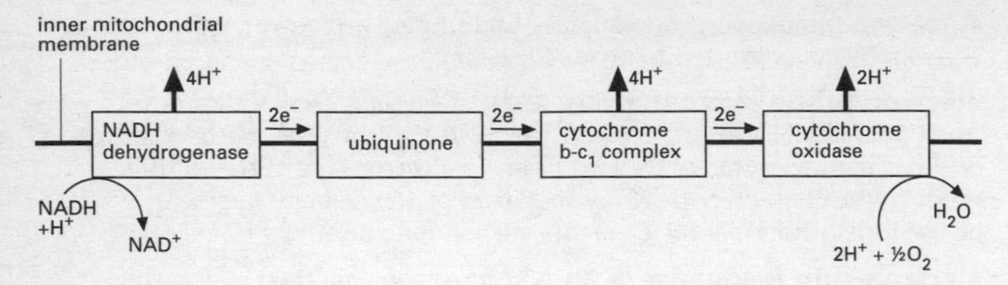

Electron transport chain

to the next carrier in the chain – a process known as *electron flow*. In the mitochondria, NADH and $FADH_2$, generated by the Krebs cycle, transfer their electrons to a chain including *ubiquinone and a series of *cytochromes. This process is coupled to the formation of ATP at three sites along the chain. The ATP is then carried across the mitochondrial membrane in exchange for ADP. An electron transport chain also occurs in *photosynthesis.

electronvolt Symbol eV. A unit of energy equal to the work done on an electron in moving it through a potential difference of one volt. It is used as a measure of particle energies although it is not an *SI unit. 1 eV = 1.602 $\times 10^{-19}$ joule.

electroorganic reaction An organic reaction produced in an electrolytic cell. Electroorganic reactions are used to synthesize compounds that are difficult to produce by conventional techniques. An example of an electroorganic reaction is *Kolbe's method of synthesizing alkanes.

electrophile An ion or molecule that is electron deficient and can accept electrons. Electrophiles are often reducing agents and Lewis *acids. They are either positive ions (e.g. NO_2^+) or molecules that have a positive charge on a particular atom (e.g. SO_3, which has an electron-deficient sulphur atom). In organic reactions they tend to attack negatively charged parts of a molecule. *Compare* **nucleophile**.

electrophilic addition An *addition reaction in which the first step is attack by an electrophile (e.g. a positive ion) on an electron-rich part of the molecule. An example is addition to the double bonds in alkenes.

electrophilic substitution A *substitution reaction in which the first step is attack by an electrophile. Electrophilic substitution is a feature of reactions of benzene (and its compounds) in which a positive ion approaches the delocalized pi electrons on the benzene ring.

electrophoresis (cataphoresis) A technique for the analysis and separation of colloids, based on the movement of charged colloidal particles in an electric field. There are various experimental methods. In one the sample is placed in a U-tube and a buffer solution added to each arm, so that there are sharp boundaries between buffer and sample. An electrode is placed in each arm, a voltage applied, and the motion of the boundaries under the influence of the field is observed. The rate of migration of the particles depends on the field, the charge on the particles, and on other factors, such as the size and shape of the particles. More simply, electrophoresis can be carried out using an adsorbent, such as a strip of filter paper, soaked in a buffer with two electrodes making contact. The sample is placed between the electrodes and a voltage applied. Different components of the mixture migrate at different rates, so the sample separates into zones. The components can be identified by the rate at which they move. In *gel electrophoresis* the medium is a gel, typically made of polyacrylamide (*see* **PAGE**), agarose, or starch.

Electrophoresis, which has also been called *electrochromatography*, is used extensively in studying mixtures of proteins, nucleic acids, carbohydrates, enzymes, etc. In clinical medicine it is used for determining the protein content of body fluids.

electroplating A method of plating one metal with another by *electrodeposition. The articles to be plated are made the cathode of an electrolytic cell and a rod or bar of the plating metal is made the anode. Electroplating is used for covering metal with a decorative, more expensive, or corrosion-resistant layer of another metal.

electropositive Describing elements that tend to lose electrons and form positive ions. The alkali metals are typical electropositive elements.

electrovalent bond *See* **chemical bond.**

electrum 1. An alloy of gold and silver containing 55–88% of gold. **2.** A *German silver alloy containing 52% copper, 26% nickel, and 22% zinc.

element A substance that cannot be decomposed into simpler substances. In an element, all the atoms have the same number of protons or electrons, although the number of neutrons may vary. There are 92 naturally occurring elements. *See also* **periodic table; transuranic elements; transactinide elements.**

elementary particle One of the fundamental particles of which matter is composed, such as the electron, proton, or neutron.

elementary reaction A reaction with no intermediates; i.e. one that takes place in a single step with a single transition state.

elevation of boiling point An increase in the boiling point of a liquid when a solid is dissolved in it. The elevation is proportional to the number of particles dissolved (molecules or ions) and is given by $\Delta t = k_B C$, where C is the molal concentration of solute. The constant k_B is the *ebullioscopic constant* of the solvent and if this is known, the molecular weight of the solute can be calculated from the measured value of Δt. The elevation is measured by a Beckmann thermometer. *See also* **colligative properties.**

elimination reaction A reaction in which one molecule decomposes into two, one much smaller than the other.

Elinvar Trade name for a nickel–chromium steel containing about 36% nickel, 12% chromium, and smaller proportions of tungsten and manganese. Its elasticity does not vary with temperature and it is therefore used to make hairsprings for watches.

Ellingham diagram A diagram used to predict whether or not a metal can be extracted from its oxide by reduction with carbon. The standard reaction Gibbs function $\Delta G^{\ominus}$ is plotted as a function of temperature (with $\Delta G^{\ominus}$ decreasing upwards) for the reactions of the oxidation of the metals and the oxidation reactions for carbon and carbon monoxide. A metal can only be extracted from its oxide using carbon if any carbon reaction lies above the line in the Ellingham diagram for the oxidation of the metal, since overall for the reduction process $\Delta G^{\ominus} < 0$. The temperatures at which this is possible can be seen from the diagram. The diagram was devised by the physical chemist H. J. T. Ellingham.

elliptical polarization *See* **polarization of light**.

elongation ***(in protein synthesis)*** The phase in which amino acids are linked together by sequentially formed peptide bonds to form a polypeptide chain. *Elongation factors* are proteins that – by binding to a tRNA–amino-acid complex – enable the correct positioning of this complex on the ribosome, so that translation can proceed.

eluate *See* **chromatography**; **elution**.

eluent *See* **chromatography**; **elution**.

elution The process of removing an adsorbed material (*adsorbate*) from an adsorbent by washing it in a liquid (*eluent*). The solution consisting of the adsorbate dissolved in the eluent is the *eluate*. Elution is the process used to wash components of a mixture through a *chromatography column.

elutriation The process of suspending finely divided particles in an upward flowing stream of air or water to wash and separate them into sized fractions.

emanation The former name for the gas radon, of which there are three isotopes: Rn–222 (radium emanation), Rn–220 (thoron emanation), and Rn–219 (actinium emanation).

emerald The green gem variety of *beryl: one of the most highly prized gemstones. The finest specimens occur in the Muzo mines, Colombia. Other occurrences include the Ural Mountains, the Transvaal in South Africa, and Kaligunan in India. Emeralds can also be successfully synthesized.

emery A rock composed of corundum (natural aluminium oxide, Al_2O_3) with magnetite, haematite, or spinel. It occurs on the island of Naxos (Greece) and in Turkey. Emery is used as an abrasive and polishing material and in the manufacture of certain concrete floors.

e.m.f. *See* **electromotive force**.

emission spectrum *See* **spectrum**.

empirical Denoting a result that is obtained by experiment or observation rather than from theory.

empirical formula *See* **formula**.

emulsification ***(in digestion)*** The breakdown of fat globules in the duodenum into tiny droplets, which provides a larger surface area on which the enzyme pancreatic *lipase can act to digest the fats into fatty acids and glycerol. Emulsification is assisted by the action of the bile salts in bile.

emulsion A *colloid in which small particles of one liquid are dispersed in another liquid. Usually emulsions involve a dispersion of water in an oil or a dispersion of oil in water, and are stabilized by an *emulsifier*. Commonly emulsifiers are substances, such as *detergents, that have lyophobic and lyophilic parts in their molecules.

enantiomeric pair A pair of molecules consisting of one chiral molecule and the mirror image of this molecule. The molecules making up an enantiomeric pair rotate the plane of polarized light in equal, but opposite, directions.

enantiomers *See* **optical activity**.

enantiomorphism *See* **optical activity**.

enantiotopic Denoting a ligand a, attached to a *prochiral centre, in which the replacement of this ligand with a ligand d (which is different from the ligands a, b, and c attached to the prochiral centre) gives rise to a pair of enantiomers C_{abca}.

enantiotropy *See* **allotropy**.

endothermic Denoting a chemical reaction that takes heat from its surroundings. *Compare* **exothermic**.

end point The point in a titration at which reaction is complete as shown by the *indicator.

energy A measure of a system's ability to do work. Like work itself, it is measured in joules. Energy is conveniently classified into two forms: *potential energy* is the energy stored in a body or system as a consequence of its position, shape, or state (this includes gravitational energy, electrical energy, nuclear energy, and chemical energy); *kinetic energy* is energy of motion and is usually defined as the work that will be done by the body possessing the energy when it is brought to rest. For a body of mass m having a speed v, the kinetic energy is $mv^2/2$ (classical) or $(m - m_0)c^2$ (relativistic). The rotational kinetic energy of a body having an angular velocity ω is $I\omega^2/2$, where I is its moment of inertia.

The *internal energy of a body is the sum of the potential energy and the kinetic energy of its component atoms and molecules.

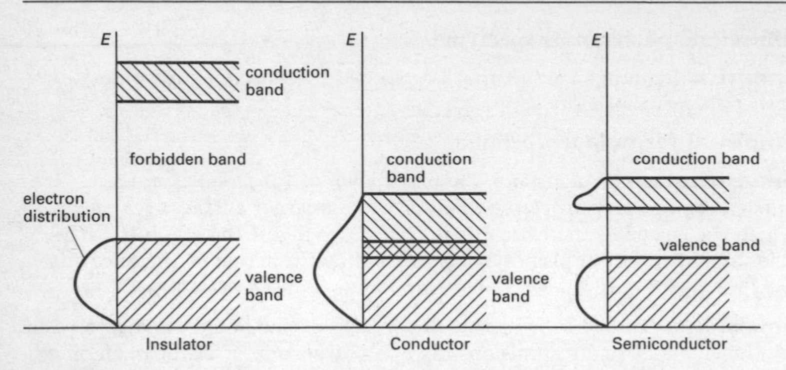

Energy bands

energy bands Ranges of energies that electrons can have in a solid. In a single atom, electrons exist in discrete *energy levels. In a crystal, in which large numbers of atoms are held closely together in a lattice, electrons are influenced by a number of adjacent nuclei and the sharply defined levels of the atoms become bands of allowed energy; this approach to energy levels in solids is often known as the *band theory*. Each band represents a large number of allowed quantum states. Between the bands are *forbidden bands*. The outermost electrons of the atoms (i.e. the ones responsible for chemical bonding) form the *valence band* of the solid. This is the band, of those occupied, that has the highest energy.

The band structure of solids accounts for their electrical properties. In order to move through the solid, the electrons have to change from one quantum state to another. This can only occur if there are empty quantum states with the same energy. In general, if the valence band is full, electrons cannot change to new quantum states in the same band. For conduction to occur, the electrons have to be in an unfilled band – the *conduction band*. Metals are good conductors either because the valence band and the conduction band are only half-filled or because the conduction band overlaps with the valence band; in either case vacant states are available. In insulators the conduction band and valence band are separated by a wide forbidden band and electrons do not have enough energy to 'jump' from one to the other. In intrinsic semiconductors the forbidden gap is narrow and, at normal temperatures, electrons at the top of the valence band can move by thermal agitation into the conduction band (at absolute zero, a semiconductor would act as an insulator). Doped semiconductors have extra bands in the forbidden gap. See illustration.

energy level A definite fixed energy that a molecule, atom, electron, or nucleus can have. In an atom, for example, the atom has a fixed energy corresponding to the *orbitals in which its electrons move around the nucleus. The atom can accept a quantum of energy to become an excited atom (*see* **excitation**) if that extra energy will raise an electron to a permitted orbital. Between the *ground state*, which is the lowest possible energy level for a particular system, and the first excited state there are no permissible energy levels. According to the *quantum theory, only certain energy levels are possible. An atom passes from one energy level to the next without passing through fractions of that energy transition. These levels are usually described by the energies associated with the individual electrons in the atoms, which are always lower than an arbitrary level for a free electron. The energy levels of molecules also involve quantized vibrational and rotational motion.

Engel's salt *See* **potassium carbonate.**

enolate ion A negative ion obtained from an enol, by removal of a hydrogen atom. Enolate ions can have two forms: one with a single C–C bond and the negative charge on the beta carbon atom and the other with a double C–C bond and the negative charge on the oxygen atom.

enols Compounds containing the group –CH=C(OH)– in their molecules. *See also* **keto–enol tautomerism**.

enrichment The process of increasing the abundance of a specified isotope in a mixture of isotopes. It is usually applied to an increase in the proportion of U–235, or the addition of Pu–239 to natural uranium for use in a nuclear reactor or weapon.

ensemble A set of systems of particles used in *statistical mechanics to describe a single system. The concept of an ensemble was put forward by Josiah Willard *Gibbs in 1902 as a way of calculating the time average of the single system, by averaging over the systems in the ensemble at a fixed time. An ensemble of systems is constructed from knowledge of the single system and can be represented as a set of points in phase space with each system of the ensemble represented by a point. Ensembles can be constructed both for isolated systems and for open systems.

enthalpy Symbol H. A thermodynamic property of a system defined by $H = U + pV$, where H is the enthalpy, U is the internal energy of the system, p its pressure, and V its volume. In a chemical reaction carried out in the atmosphere the pressure remains constant and the enthalpy of reaction, ΔH, is equal to $\Delta U + p\Delta V$. For an exothermic reaction ΔH is taken to be negative.

entropy Symbol S. A measure of the unavailability of a system's energy to do work; in a closed system, an increase in entropy is accompanied by a decrease in energy availability. When a system undergoes a reversible change the entropy (S) changes by an amount equal to the energy (Q) transferred to the system by heat divided by the thermodynamic temperature (T) at which this occurs, i.e. $\Delta S = \Delta Q/T$. However, all real processes are to a certain extent irreversible changes and in any closed system an irreversible change is always accompanied by an increase in entropy.

In a wider sense entropy can be interpreted as a measure of disorder; the higher the entropy the greater the disorder (*see* **Boltzmann formula**). As any real change to a closed system tends towards higher entropy, and therefore higher disorder, it follows that the entropy of the universe (if it can be considered a closed system) is increasing and its available energy is decreasing. This increase in the entropy of the universe is one way of stating the second law of *thermodynamics.

envelope *See* **ring conformations**.

enzyme A protein that acts as a *catalyst in biochemical reactions. Each enzyme is specific to a particular reaction or group of similar reactions. Many require the association of certain nonprotein *cofactors in order to function. The molecule undergoing reaction (the *substrate*) binds to a specific *active site on the enzyme molecule to form a short-lived intermediate (*see* **enzyme–substrate complex**): this greatly increases (by a factor of up to 10^{20}) the rate at which the reaction proceeds to form the product. Enzyme activity is influenced by substrate concentration and by temperature and pH, which must lie within a certain range. Other molecules may compete for the active site, causing *inhibition of the enzyme or even irreversible destruction of its catalytic properties.

Enzyme production is governed by a cell's genes. Enzyme activity is further controlled by pH changes, alterations in the concentrations of essential cofactors, feedback inhibition by the products of the reaction, and activation by another enzyme, either from a less active form or an inactive precursor (zymogen). Such changes may themselves be under the control of hormones or the nervous system. *See also* **enzyme kinetics**.

Enzymes are classified into six major groups, according to the type of reaction they catalyse: (1) oxidoreductases; (2) transferases; (3) hydrolases; (4) lyases; (5) isomerases; (6) ligases. The names of most individual enzymes also end in *-ase*, which is added to the names of the substrates on which they act. Thus lactase is the enzyme that acts to break down lactose; it is classified as a hydrolase.

enzyme inhibition *See* **inhibition**.

enzyme kinetics The study of the rates of enzyme-catalysed reactions. Rates of reaction are usually measured by using the purified enzyme *in vitro* with the substrate and then observing the formation of the product or disappearance of the substrate. As the concentration of the substrate is increased the rate of reaction increases proportionally up to a certain point, after which any further increase in substrate concentration no longer increases the reaction rate (*see* **Michaelis–Menten curve**). At this point, all active sites of the enzyme are saturated with substrate; any further increase in the rate of reaction will occur only if more enzyme is added. Reaction rates are also affected by the presence of inhibitors (*see* **inhibition**), temperature, and pH (*see* **enzyme**).

enzyme–substrate complex The intermediate formed when a substrate molecule interacts with the *active site of an enzyme. Following the formation of an enzyme–substrate complex, the substrate molecule undergoes a chemical reaction and is converted into a new product. Various mechanisms for the formation of enzyme–substrate complexes have been suggested, including the *induced-fit model and the *lock-and-key mechanism.

epimerism A type of optical isomerism in which a molecule has two chiral centres; two optical isomers (*epimers*) differ in the arrangement about one of these centres. *See also* **optical activity**.

epinephrine *See* **adrenaline**.

epitaxy (epitaxial growth) Growth of a layer of one substance on a single crystal of another, such that the crystal structure in the layer is the same as that in the substrate. It is used in making semiconductor devices.

EPM *See* **electron probe microanalysis**.

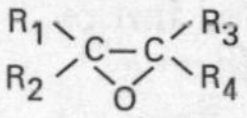

The functional group in epoxides

epoxides Compounds that contain oxygen atoms in their molecules as part of a three-membered ring (see formula). Epoxides are thus *cyclic ethers*.

epoxyethane (ethylene oxide) A colourless flammable gas, C_2H_4O; m.p. −111°C; b.p. 13.5°C. It is a cyclic ether (*see* **epoxides**), made by the catalytic oxidation of ethene. It can be hydrolysed to ethane-1,2-diol and also polymerizes to ...-OC_2H_4-O-C_2H_4-..., which is used for lowering the viscosity of water (e.g. in fire fighting).

epoxy resins Synthetic resins produced by copolymerizing epoxide compounds with phenols. They contain –O– linkages and epoxide groups and are usually viscous liquids. They can be hardened by addition of agents, such as polyamines, that form cross-linkages. Alternatively, catalysts may be used to induce further polymerization of the resin. Epoxy resins are used in electrical equipment and in the chemical industry (because of resistance to chemical attack). They are also used as adhesives.

epsomite A mineral form of *magnesium sulphate heptahydrate, $MgSO_4.7H_2O$.

Epsom salt *See* **magnesium sulphate**.

equation of state An equation that relates the pressure p, volume V, and thermodynamic temperature T of an amount of substance n. The simplest is the ideal *gas law:

$$pV = nRT,$$

where R is the universal gas constant. Applying only to ideal gases, this equation takes no account of the volume occupied by the gas molecules (according to this law if the pressure is infinitely great the volume becomes zero), nor does it take into account any forces between molecules. A more accurate equation of state would therefore be

$$(p + k)(V - nb) = nRT,$$

where k is a factor that reflects the decreased pressure on the walls of the container as a result of the attractive forces between particles, and nb is the volume occupied by the particles themselves when the pressure is infinitely high. In *van der Waals' equation*, proposed by the Dutch physicist J. D. van der Waals (1837–1923),

$$k = n^2a/V^2,$$

where a is a constant. This equation more accurately reflects the behaviour of real gases; several others have done better but are more complicated.

equatorial 1. *See* **ring conformations. 2.** *See* **apical.**

equilibrium A state in which a system has its energy distributed in the statistically most probable manner; a state of a system in which forces, influences, reactions, etc., balance each other out so that there is no net change. A body is said to be in *thermal equilibrium* if no net heat exchange is taking place within it or between it and its surroundings. A system is in *chemical equilibrium when a reaction and its reverse are proceeding at equal rates (*see also* **equilibrium constant**). These are examples of *dynamic equilibrium*, in which activity in one sense or direction is in aggregate balanced by comparable reverse activity.

equilibrium constant For a reversible reaction of the type

$$xA + yB \rightleftharpoons zC + wD$$

chemical equilibrium occurs when the rate of the forward reaction equals the rate of the back reaction, so that the concentrations of products and reactants reach steady-state values. It can be shown that at equilibrium the ratio of concentrations

$$[C]^z[D]^w/[A]^x[B]^y$$

is a constant for a given reaction and fixed temperature, called the equilibrium constant K_c (where the c indicates concentrations have been used). Note that, by convention, the products on the right-hand side of the reaction are used on the top line of the expression for equilibrium constant. This form of the equilibrium constant was originally introduced in 1863 by C. M. Guldberg and P. Waage using the law of *mass action. They derived the expression by taking the rate of the forward reaction

$$k_f[A]^x[B]^y$$

and that of the back reaction

$$k_b[C]^z[D]^w$$

Since the two rates are equal at equilibrium, the equilibrium constant K_c is the ratio of the rate constants k_f/k_b. The principle that the expression is a constant is known as the *equilibrium law* or *law of chemical equilibrium*.

The equilibrium constant shows the *position* of equilibrium. A low value of K_c indicates that [C] and [D] are small compared to [A] and [B]; i.e. that the back reaction predominates. It also indicates how the equilibrium shifts if concentration changes. For example, if [A] is increased (by adding A) the equilibrium shifts towards the right so that [C] and [D] increase, and K_c remains constant.

For gas reactions, partial pressures are used rather than concentrations. The symbol K_p is then used. Thus, in the example above

$$K_p = p_C^z p_D^w / p_A^x p_B^y$$

It can be shown that, for a given reaction $K_p = K_c(RT)^{\Delta\nu}$, where $\Delta\nu$ is the difference in stoichiometric coefficients for the reaction (i.e. $z + w - x - y$). Note that the units of K_p and K_c depend on the numbers of molecules appearing in the stoichiometric equation. The value of the equilibrium constant depends on the temperature. If the forward reaction is exothermic, the equilibrium constant decreases as the temperature rises; if endothermic it increases (*see also* **van't Hoff's isochore**).

The expression for the equilibrium constant can also be obtained by thermodynamics; it can be shown that the standard equilibrium constant $K^{\ominus}$ is given by $\exp(-\Delta G^{\ominus}/RT)$, where $\Delta G^{\ominus}$ is the standard Gibbs free energy change for the complete reaction. Strictly, the expressions above for equilibrium constants are true only for ideal gases (pressure) or infinite dilution (concentration). For accurate work *activities are used.

equilibrium law *See* **equilibrium constant.**

equipartition of energy The theory, proposed by Ludwig *Boltzmann and given some theoretical support by James Clerk *Maxwell, that the energy of gas molecules in a large sample under thermal *equilibrium is equally divided among their available *degrees of freedom, the average energy for each degree of freedom being $kT/2$, where k is the *Boltzmann constant and T is the thermodynamic temperature. The proposition is not generally true if quantum considerations are important, but is frequently a good approximation.

equivalence point The point in a titration at which reaction is complete. *See* **indicator.**

equivalent proportions *See* **chemical combination.**

equivalent weight The mass of an element or compound that could combine with or displace one gram of hydrogen (or eight grams of oxygen or 35.5 grams of chlorine) in a chemical reaction. The equivalent weight represents the 'combining power' of the substance. For an element it is the relative atomic mass divided by the valency. For a compound it depends on the reaction considered.

erbium Symbol Er. A soft silvery metallic element belonging to the *lanthanoids; a.n. 68; r.a.m. 167.26; r.d. 9.006 (20°C); m.p. 1529°C; b.p. 2863°C. It occurs in apatite, gadolinite, and xenotine from certain sources. There are six natural isotopes, which are stable, and twelve artificial isotopes are known. It has been used in alloys for nuclear technology as it is a neutron absorber; it is being investigated for other potential uses. It was discovered by Carl Mosander (1797–1858) in 1843.

ergocalciferol *See* **vitamin D.**

ergodic hypothesis A hypothesis in *statistical mechanics concerning phase space. If a system of N atoms or molecules is enclosed in a fixed volume, the state of this system is given by a point in $6N$-dimensional phase space with q_i representing coordinates and p_i representing momenta. Taking the energy E to be constant, a representative point in phase space describes an orbit on the surface $E(q_i,p_i) = c$, where c is a constant. The ergodic hypothesis states that the orbit of the representative point in phase space eventually goes through all points on the surface. The *quasi-ergodic hypothesis* states that the orbit of the representative point in phase space eventually comes close to all points on the surface. In general, it is very difficult to prove the ergodic or quasi-ergodic hypotheses for a given system. *See also* **ergodicity.**

ergodicity A property of a system that obeys the *ergodic hypothesis. The ergodicity of systems has been discussed extensively in the foundations of *statistical mechanics, although it is now thought by many physicists to be irrelevant to the problem. Considerations of ergodicity occur in dynamics, since the behaviour can be very complex even for simple *dynamical systems (*see* **attractor**). Systems, such as *spin glasses, in which ergodicity is thought not to hold are described as having *broken ergodicity.*

ergosterol A *sterol occurring in fungi, bacteria, algae, and plants. It is converted into vitamin D_2 by the action of ultraviolet light.

ESCA *See* **photoelectron spectroscopy.**

eserine (physostigmine) An alkaloid, derived from the calabar bean plant, that inhibits cholinesterase by covalently binding with it (*see* **inhibition**). Eserine is used to treat the eye condition glaucoma.

ESR *See* **electron-spin resonance.**

essential amino acid An *amino acid that an organism is unable to synthesize in sufficient quantities. It must therefore be present in the diet. In man the essential amino acids are arginine, histidine, lysine, threonine, methionine, isoleucine, leucine, valine, phenylalanine, and tryptophan. These are required for protein synthesis and deficiency leads to retarded growth and other symptoms. Most of the amino acids required by man are also essential for all other multicellular animals and for most protozoans.

essential element Any of a number of elements required by living organisms to ensure normal growth, development, and maintenance. Apart from the elements found in organic compounds (i.e. carbon, hydrogen, oxygen, and nitrogen), plants, animals, and microorganisms all require a range of elements in inorganic forms in varying amounts, depending on the type of organism. The *major elements*, present in tissues in relatively large amounts (greater than 0.005%), are calcium, phosphorus, potassium, sodium, chlorine, sulphur, and magnesium. The *trace elements* occur at much lower concentrations and thus requirements are much less. The most important are iron, manganese, zinc, copper, iodine, cobalt, selenium, molybdenum, chromium, and silicon. Each element may fulfil one or more of a variety of metabolic roles.

essential fatty acids *Fatty acids that must normally be present in the diet of certain animals, including man. Essential fatty acids, which include *linoleic and *linolenic acids, all possess double bonds at the same two positions along their hydrocarbon chain and so can act as precursors of *prostaglandins. Deficiency of essential fatty acids can cause dermatosis, weight loss, irregular oestrus, etc. An adult human requires 2–10 g linoleic acid or its equivalent per day.

essential oil A natural oil with a distinctive scent secreted by the glands of certain aromatic plants. *Terpenes are the main constituents. Essential oils are extracted from plants by steam distillation, extraction with cold neutral fats or solvents (e.g. alcohol), or pressing and used in perfumes, flavourings, and medicine. Examples are citrus oils, flower oils (e.g. rose, jasmine), and oil of cloves.

esterification A reaction of an alcohol with an acid to produce an ester and water; e.g.

$$CH_3OH + C_6H_5COOH \rightleftharpoons CH_3OOCC_6H_5 + H_2O$$

The reaction is an equilibrium and is slow under normal conditions, but can be speeded up by addition of a strong acid catalyst. The ester can often be distilled off so that the reaction can proceed to completion. The reverse reaction is ester hydrolysis or *saponification. *See also* **labelling**.

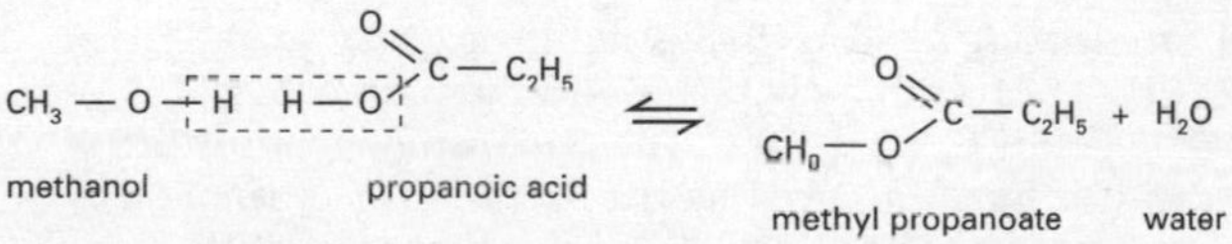

$CH_3-O-H \quad H-O-C(=O)-C_2H_5 \rightleftharpoons CH_3-O-C(=O)-C_2H_5 + H_2O$

methanol propanoic acid methyl propanoate water

Ester formation

esters Organic compounds formed by reaction between alcohols and acids (see illustration). Esters formed from carboxylic acids have the general formula RCOOR′. Examples are ethyl ethanoate, $CH_3COOC_2H_5$, and methyl propanoate, $C_2H_5COOCH_3$. Esters containing simple hydrocarbon groups are volatile fragrant substances used as flavourings in the food industry. Triesters, molecules containing three ester groups, occur in nature as oils and fats. *See also* **glycerides**.

ethanal (acetaldehyde) A colourless highly flammable liquid aldehyde, CH_3CHO; r.d. 0.78; m.p. −121°C; b.p. 20.8°C. It is made from ethene by the *Wacker process and used as a starting material for making many organic compounds. The compound polymerizes if dilute acid is added to give *ethanal trimer* (or *paraldehyde*), which contains a six-membered ring of alternating carbon and oxygen atoms with a hydrogen atom and a methyl group attached to each carbon atom. It is used as a drug for inducing sleep. Addition of dilute acid below 0°C gives *ethanal tetramer* (or *metaldehyde*), which has a similar structure to the trimer but with an eight-membered ring. It is used as a solid fuel in portable stoves and in slug pellets.

ethanamide **(acetamide)** A colourless solid crystallizing in the form of long white crystals with a characteristic smell of mice, CH_3CONH_2; r.d. 1.159; m.p. 82.3°C; b.p. 221.25°C. It is made by the dehydration of ammonium ethanoate or by the action of ammonia on ethanoyl chloride, ethanoic anhydride, or ethyl ethanoate.

ethane A colourless flammable gaseous hydrocarbon, C_2H_6; m.p. −183°C; b.p. −89°C. It is the second member of the *alkane series of hydrocarbons and occurs in natural gas.

ethanedial *See* **glyoxal.**

ethanedioic acid *See* **oxalic acid.**

ethane-1,2-diol **(ethylene glycol; glycol)** A colourless viscous hygroscopic liquid, CH_2OHCH_2OH; m.p. −11.5°C; b.p. 198°C. It is made by hydrolysis of epoxyethane (from ethene) and used as an antifreeze and a raw material for making *polyesters (e.g. Terylene).

ethanenitrile **(acetonitrile; methyl cyanide)** A poisonous liquid, CH_3CN; b.p. 82°C. It is made by dehydrating ethanamide or from ammonia and ethyne. It is a good polar solvent and is employed for dissolving ionic compounds when water cannot be used.

ethanoate **(acetate)** A salt or ester of ethanoic acid (acetic acid).

ethanoic acid **(acetic acid)** A clear viscous liquid or glassy solid *carboxylic acid, CH_3COOH, with a characteristically sharp odour of vinegar; r.d. 1.049; m.p. 16.6°C; b.p. 117.9°C. The pure compound is called *glacial ethanoic acid.* It is manufactured by the oxidation of ethanol or by the oxidation of butane in the presence of dissolved manganese(II) or cobalt(II) ethanoates at 200°C, and is used in making ethanoic anhydride for producing cellulose ethanoates. It is also used in making ethenyl ethanoate (for polyvinylacetate). The compound is formed by the fermentation of alcohol and is present in vinegar, which is made by fermenting beer or wine. 'Vinegar' made from ethanoic acid with added colouring matter is called 'nonbrewed condiment'. In living organisms it combines with *coenzyme A to form acetyl coenzyme A, which plays a crucial role in energy metabolism.

ethanoic anhydride **(acetic anhydride)** A pungent-smelling colourless liquid, $(CH_3CO)_2O$, b.p. 139.5°C. It is used in organic synthesis as an *ethanoylating agent (attacking an −OH or −NH group) and in the manufacture of aspirin and cellulose plastics. It hydrolyses in water to give ethanoic acid.

ethanol (ethyl alcohol) A colourless water-soluble *alcohol, C_2H_5OH; r.d. 0.789 (0°C); m.p. −117.3°C; b.p. −78.3°C. It is the active principle in intoxicating drinks, in which it is produced by fermentation of sugar using yeast

$$C_6H_{12}O_6 \rightarrow 2C_2H_5OH + 2CO_2$$

The ethanol produced kills the yeast and fermentation alone cannot produce ethanol solutions containing more than 15% ethanol by volume. Distillation can produce a constant-boiling mixture containing 95.6% ethanol and 4.4% water. Pure ethanol (*absolute alcohol*) is made by removing this water by means of drying agents.

The main industrial use of ethanol is as a solvent although at one time it was a major starting point for making other chemicals. For this it was produced by fermentation of molasses. Now ethene has replaced ethanol as a raw material and industrial ethanol is made by hydrolysis of ethene.

ethanolamine Any of three low-melting hygroscopic colourless solids. They are strong bases, smell of ammonia, and absorb water readily to form viscous liquids. *Monoethanolamine*, $HOCH_2CH_2NH_2$, is a primary *amine, m.p. 10.5°C; *diethanolamine*, $(HOCH_2CH_2)_2NH$, is a secondary amine, m.p. 28°C; and *triethanolamine*, $(HOCH_2CH_2)_3N$, is a tertiary amine, m.p. 21°C. All are made by heating epoxyethane with concentrated aqueous ammonia under pressure and separating the products by fractional distillation. With fatty acids they form neutral soaps, used as emulsifying agents and detergents, and in bactericides and cosmetics.

ethanoylating agent (acetylating agent) A chemical reagent used to introduce an ethanoyl group ($-COCH_3$) instead of hydrogen in an organic compound. Examples include *ethanoic anhydride and *ethanoyl chloride.

ethanoyl chloride (acetyl chloride) A colourless liquid acyl chloride (*see* **acyl halides**), CH_3COCl, with a pungent smell; r.d. 1.105; m.p. −112.15°C; b.p. 50.9°C. It is made by reacting ethanoic acid with a halogenating agent such as phosphorus(III) chloride, phosphorus(V) chloride, or sulphur dichloride oxide and is used to introduce ethanoyl groups into organic compounds containing $-OH$, $-NH_2$, and $-SH$ groups. *See* **acylation**.

ethanoyl group (acetyl group) The organic group CH_3CO-.

ethene (ethylene) A colourless flammable gaseous hydrocarbon, C_2H_4; m.p. −169°C; b.p. −103.7°C. It is the first member of the *alkene series of hydrocarbons. It is made by cracking hydrocarbons from petroleum and is now a major raw material for making other organic chemicals (e.g. ethanal, ethanol, ethane-1,2-diol). It can be polymerized to *polyethene. It occurs naturally in plants, in which it acts as a growth substance promoting the ripening of fruits.

ethenone *See* ketene.

ethenyl ethanoate (vinyl acetate) An unsaturated organic ester,

CH_2:$CHOOCCH_3$; r.d. 0.9; m.p. −100°C; b.p. 73°C. It is made by catalytic reaction of ethanoic acid and ethene and used to make polyvinylacetate.

ether *See* **ethoxyethane**; **ethers**.

ethers Organic compounds containing the group –O– in their molecules. Examples are dimethyl ether, CH_3OCH_3, and diethyl ether, $C_2H_5OC_2H_5$ (*see* **ethoxyethane**). They are volatile highly flammable compounds made by dehydrating alcohols using sulphuric acid.

ethoxyethane **(diethyl ether; ether)** A colourless flammable volatile *ether, $C_2H_5OC_2H_5$; r.d. 0.71; m.p. −116°C; b.p. 34.5°C. It can be made by *Williamson's synthesis. It is an anaesthetic and useful organic solvent.

ethyl 3-oxobutanoate **(ethyl acetoacetonate)** A colourless liquid ester with a pleasant odour, $CH_3COCH_2COOC_2H_5$; r.d. 1.03; m.p. <-80°C; b.p. 180.4°C. It can be prepared by reacting ethyl ethanoate ($CH_3COOC_2H_5$) with sodium or sodium ethoxide. The compound shows keto–enol *tautomerism and contains about 7% of the enol form, $CH_3C(OH)$:$CHCOOC_2H_5$, under normal conditions. Sometimes known as *acetoacetic ester*, it is used in organic synthesis.

ethyl acetate *See* **ethyl ethanoate**.

ethyl acetoacetonate *See* **ethyl 3-oxobutanoate**.

ethyl alcohol *See* **ethanol**.

ethylamine A colourless flammable volatile liquid, $C_2H_5NH_2$; r.d. 0.69; m.p. −81°C; b.p. 16.6°C. It is a primary amine made by reacting chloroethane with ammonia and used in making dyes.

ethylbenzene A colourless flammable liquid, $C_6H_5C_2H_5$; r.d. 0.867; m.p. −95°C; b.p. 136°C. It is made from ethene and ethybenzene by a *Friedel–Crafts reaction and is used in making phenylethene (for polystyrene).

ethyl bromide *See* **bromoethane**.

ethylene *See* **ethene**.

ethylenediamine *See* **1,2-diaminoethane**.

ethylene glycol *See* **ethane-1,2-diol**.

ethylene oxide *See* **epoxyethane**.

ethyl ethanoate **(ethyl acetate)** A colourless flammable liquid ester, $C_2H_5OOCCH_3$; r.d. 0.9; m.p. −83.6°C; b.p. 77.06°C. It is used as a solvent and in flavourings and perfumery.

ethyl group The organic group CH_3CH_2–.

ethyl iodide *See* **iodoethane**.

ethyne **(acetylene)** A colourless unstable gas, C_2H_2, with a characteristic sweet odour; r.d. 0.618; m.p. −80.8°C; b.p. −84.0°C. It is the simplest member of the *alkyne series of unsaturated hydrocarbons, and is prepared by the action of water on calcium dicarbide or by adding alcoholic potassium hydroxide to 1,2-dibromoethane. It can be manufactured by heating methane to 1500°C in the presence of a catalyst. It is used in oxyacetylene welding and in the manufacture of ethanal and ethanoic acid. Ethyne can be polymerized easily at high temperatures to give a range of products. The inorganic saltlike dicarbides contain the ion C_2^{2-}, although ethyne itself is a neutral compound (i.e. not a protonic acid).

eudiometer An apparatus for measuring changes in volume of gases during chemical reactions. A simple example is a graduated glass tube sealed at one end and inverted in mercury. Wires passing into the tube allow the gas mixture to be sparked to initiate the reaction between gases in the tube.

europium Symbol Eu. A soft silvery metallic element belonging to the *lanthanoids; a.n. 63; r.a.m. 151.96; r.d. 5.245 (20°C); m.p. 822°C; b.p. 1597°C. It occurs in small quantities in bastanite and monazite. Two stable isotopes occur naturally: europium–151 and europium–153, both of which are neutron absorbers. Experimental europium alloys have been tried for nuclear-reactor parts but until recently the metal has not been available in sufficient quantities. It is widely used in the form of the oxide in phosphors for television screens. It was discovered by Sir William Crookes in 1889.

eutectic mixture A solid solution consisting of two or more substances and having the lowest freezing point of any possible mixture of these components. The minimum freezing point for a set of components is called the *eutectic point*. Low-melting-point alloys are usually eutectic mixtures.

evaporation The change of state of a liquid into a vapour at a temperature below the boiling point of the liquid. Evaporation occurs at the surface of a liquid, some of those molecules with the highest kinetic energies escaping into the gas phase. The result is a fall in the average kinetic energy of the molecules of the liquid and consequently a fall in its temperature.

exa- Symbol E. A prefix used in the metric system to denote 10^{18} times. For example, 10^{18} metres = 1 exametre (Em).

EXAFS (extended X-ray absorption fine structure) Oscillations of the X-ray absorption coefficient beyond an absorption edge. The physical cause of EXAFS is the modification of the final state of a photoelectron *wave function caused by back-scattering from atoms surrounding the excited atom. EXAFS is used to determine structure in chemical, solid state, or biological systems; it is especially useful in those systems in which it is not possible to use diffraction techniques. EXAFS experiments are usually performed using synchrotron radiation. It is possible to interpret EXAFS experiments using single-scattering theory for short-range order.

excimer *See* **exciplex**.

exciplex A combination of two different atoms that exists only in an excited state. When an exciplex emits a photon of electromagnetic radiation, it immediately dissociates into the atoms, rather than reverting to the ground state. A similar transient excited association of two atoms of the same kind is an *excimer*. An example of an exciplex is the species XeCl* (the asterisk indicates an excited state), which can be formed by an electric discharge in xenon and chlorine. This is used in the *exciplex laser*, in which a population inversion is produced by an electrical discharge.

excitation A process in which a nucleus, electron, atom, ion, or molecule acquires energy that raises it to a quantum state (*excited state*) higher than that of its *ground state. The difference between the energy in the ground state and that in the excited state is called the *excitation energy*. *See* **energy level**.

exclusion principle *See* **Pauli exclusion principle**.

exothermic Denoting a chemical reaction that releases heat into its surroundings. *Compare* **endothermic**.

exotic atom A species in which some other charged particle replaces the electron or nucleon. An example is an atom in which an electron has been replaced by another negatively charged particle, such as a muon or meson. In this case the negative particle eventually collides with the nucleus with the emission of X-ray photons. Another system is one in which the nucleus of an atom has been replaced by a positively charged meson. An association of an electron and a positron is also regarded as an exotic atom (known as *positronium*). This has a mean life of about 10^{-7} s, decaying to give three photons.

explosive A compound or mixture that, when ignited or detonated, undergoes a rapid violent chemical reaction that produces large amounts of gas and heat, accompanied by light, sound and a high-pressure shock wave. *Low explosives* burn comparatively slowly when ignited, and are employed as propellants in firearms and guns; they are also used in blasting. Examples include *gunpowder and various smokeless propellants, such as *cordite. *High explosives* decompose very rapidly to produce an uncontrollable blast. Examples of this type include *dynamite, *nitroglycerine, and *trinitrotoluene (TNT); they are exploded using a detonator. Other high-power explosives include pentaerythritol tetranitrate (PETN) and ammonium nitride/fuel oil mixture (ANFO). Cyclonite (RDX) is a military high explosive; mixed with oils and waxes, it forms a plastic explosive (such as Semtex). See also Chronology.

extended X-ray absorption fine structure *See* **EXAFS**.

extender An inert substance added to a product (paint, rubber, washing powder, etc.) to dilute it (for economy) or to modify its physical properties.

extensive variable A quantity in a *macroscopic system that is proportional to the size of the system. Examples of extensive variables

EXPLOSIVES

900–1000	Gunpowder developed in China.
1242	English monk Roger Bacon (1220–92) describes the preparation of gunpowder (using an anagram).
c.1250	German alchemist Berthold Schwarz claims to have reinvented gunpowder.
1771	French chemist Pierre Woulfe discovers picric acid (originally used as a yellow dye).
1807	Scottish cleric Alexander Forsyth (1767–1843) discovers mercury fulminate.
1833	French chemist Henri Braconnot (1781–1855) nitrates starch, making a highly flammable compound (crude nitrocellulose).
1838	French chemist Théophile Pelouze (1807–67) nitrates paper, making crude nitrocellulose.
1845	German chemist Christian Schönbein (1799–1868) nitrates cotton, making nitrocellulose.
1846	Italian chemist Ascania Sobrero (1812–88) discovers nitroglycerine.
1863	Swedish chemist J. Wilbrand discovers trinitrotoluene (TNT). Swedish chemist Alfred Nobel (1833–96) invents a detonating cap based on mercury fulminate.
1867	Alfred Nobel invents dynamite by mixing nitroglycerine and kieselguhr.
1871	German chemist Hermann Sprengel shows that picric acid can be used as an explosive.
1875	Alfred Nobel invents blasting gelatin (nitroglycerine mixed with nitrocellulose).
1885	French chemist Eugène Turpin discovers ammonium picrate (Mélinite).
1888	Alfred Nobel invents a propellant from nitroglycerine and nitrocellulose (Ballistite).
1889	British scientists Frederick Abel (1826–1902) and James Dewar invent a propellant (Cordite) similar to Ballistite.
1891	German chemist Bernhard Tollens (1841–1918) discovers pentaerythritol tetranitrate (PETN).
1899	Henning discovers cyclotrimethylenetrinitramine (RDX or cyclonite).
1905	US army officer B. W. Dunn (1860–1936) invents ammonium picrate explosive (Dunnite).
1915	British scientists invent amatol (TNT + ammonium nitrate).
1955	US scientists develop ammonium nitrate–fuel oil mixtures (ANFO) as industrial explosives.

include the volume, mass, and total energy. If an extensive variable is divided by an arbitrary extensive variable, such as the volume, an *intensive variable results. A macroscopic system can be described by one extensive variable and a set of intensive variables.

external conversion A process in which molecules in electronically excited states pass to a lower electronic state (which is frequently the ground state) by colliding with other molecules. In this process the electronic energy is eventually converted into heat. Since this process involves collisions, the rate at which it occurs depends on how frequently collisons occur. As a result, this process occurs much faster in liquids than in gases. It is sometimes called *collision quenching*.

extraction The separation of a component from its mixture by selective solubility. *See* **partition**.

extraordinary ray *See* **double refraction**.

extrusion reaction *See* **insertion reaction**.

Eyring, Henry (1901–81) US chemist who worked at Princeton and Utah. His main work was on chemical kinetics and he is noted for the *Eyring equation for absolute reaction rates.

Eyring equation An equation used extensively to describe chemical reactions. For a rate constant κ, it is given by

$$\kappa = K(kT/h)\exp(-\Delta G^{\ddagger}/kT),$$

where k is the *Boltzmann constant, T is the thermodynamic temperature, h is the *Planck constant, $\Delta G^{\ddagger}$ is the free energy of activation, and K is a constant called the *transmission coefficient*, which is the probability that a chemical reaction takes place once the system has reached the activated state. A similar equation (without the K) has been used to describe transport processes, such as diffusion, thermal conductivity, and viscosity in dense gases and liquids. In these cases it is assumed that the main kinetic process is the motion of a molecule to a vacant site near it. The equation is derived by assuming that the reactants are in equilibrium with the excited state. This assumption of equilibrium is not necessarily correct for small activation energies. The Eyring equation is named after Henry *Eyring, who derived it and applied it widely in the theory of chemical reactions and transport processes.

***E–Z* convention** A convention for the description of a molecule showing cis-trans isomerism (*see* **isomerism**). In a molecule ABC=CDE, where A,B,D, and E are substituent groups, the sequence rule (*see* **CIP system**) is applied to the pair A and B to find which has priority and similarly to the pair C and D. If the two groups of highest priority are on the same side of the bond then the isomer is designated *Z* (from German *zusammen*, together). If they are on opposite sides the isomer is designated *E* (German *entgegen*, opposite). The letters are used in the names of compounds; for example (*E*)-butenedioic acid (fumaric acid) and (*Z*)-butenedioic acid (maleic acid). In compounds containing two (or more) double bonds numbers are used to designate the bonds (e.g. (2*E*, 4*Z*)-2,4-hexadienoic acid). The system is less ambiguous than the cis/trans system of describing isomers.

FAB mass spectroscopy *See* **fast-atom bombardment mass spectroscopy.**

face-centred cubic **(f.c.c.)** *See* **cubic crystal.**

fac-isomer *See* **isomerism.**

FAD **(flavin adenine dinucleotide)** A *coenzyme important in various biochemical reactions. It comprises a phosphorylated vitamin B_2 (riboflavin) molecule linked to the nucleotide adenine monophosphate (AMP). FAD is usually tightly bound to the enzyme forming a *flavoprotein*. It functions as a hydrogen acceptor in dehydrogenation reactions, being reduced to $FADH_2$. This in turn is oxidized to FAD by the *electron transport chain, thereby generating ATP (two molecules of ATP per molecule of $FADH_2$).

Fahrenheit, Gabriel Daniel (1686–1736) German physicist, who became an instrument maker in Amsterdam. In 1714 he developed the mercury-in-glass thermometer, and devised a temperature scale to go with it (*see* **Fahrenheit scale**).

Fahrenheit scale A temperature scale in which (by modern definition) the temperature of boiling water is taken as 212 degrees and the temperature of melting ice as 32 degrees. It was invented in 1714 by G. D. *Fahrenheit, who set the zero at the lowest temperature he knew how to obtain in the laboratory (by mixing ice and common salt) and took his own body temperature as 96°F. The scale is no longer in scientific use. To convert to the *Celsius scale the formula is $C = 5(F - 32)/9$.

Fajans' rules Rules indicating the extent to which an ionic bond has covalent character caused by polarization of the ions. Covalent character is more likely if:
(1) the charge of the ions is high;
(2) the positive ion is small or the negative ion is large;
(3) the positive ion has an outer electron configuration that is not a noble-gas configuration.

The rules were introduced by the Polish-born US chemist Kasimir Fajans (1887–1975).

fall-out **1.** **(radioactive fall-out)** Radioactive particles deposited from the atmosphere either from a nuclear explosion or from a nuclear accident. *Local fall-out*, within 250 km of an explosion, falls within a few hours of the explosion. *Tropospheric fall-out* consists of fine particles deposited all round the earth in the approximate latitude of the explosion within about one week. *Stratospheric fall-out* may fall anywhere on earth over a period of years. The most dangerous radioactive isotopes in fall-out are the fission fragments iodine–131 and strontium–90. Both can be taken up by grazing animals and passed on to human populations in milk, milk products, and meat. Iodine–131 accumulates in the thyroid gland and strontium–90 accumulates in bones. **2. (chemical fall-out)** Hazardous chemicals discharged into and subsequently released from the atmosphere, especially by factory chimneys.

farad Symbol F. The SI unit of capacitance, being the capacitance of a capacitor that, if charged with one coulomb, has a potential difference of one volt between its plates. $1\ F = 1\ C\ V^{-1}$. The farad itself is too large for most applications; the practical unit is the microfarad (10^{-6} F). The unit is named after Michael *Faraday.

Faraday, Michael (1791–1867) British chemist and physicist, who received little formal education. He started to experiment on electricity and in 1812 attended lectures by Sir Humphry *Davy at the Royal Institution; a year later he became Davy's assistant. He remained at the Institution until 1861. Faraday's chemical discoveries include the liquefaction of chlorine (1823) and benzene (1825) as well as the laws of electrolysis (*see* **Faraday's laws**). He is also remembered for his work in physics: in 1821 he demonstrated electromagnetic rotation (the principle of the electric motor) and in 1832 discovered electromagnetic induction (the principle of the dynamo).

Faraday constant Symbol *F*. The electric charge carried by one mole of electrons or singly ionized ions, i.e. the product of the *Avogadro constant and the charge on an electron (disregarding sign). It has the value $9.648\,5309(29) \times 10^4$ coulombs per mole. This number of coulombs is sometimes treated as a unit of electric charge called the *faraday*.

Faraday's laws Two laws describing electrolysis:
(1) The amount of chemical change during electrolysis is proportional to the charge passed.
(2) The charge required to deposit or liberate a mass m is given by $Q = Fmz/M$, where F is the Faraday constant, z the charge of the ion, and M the relative ionic mass.

These are the modern forms of the laws. Originally, they were stated by Michael *Faraday in a different form:
(1) The amount of chemical change produced is proportional to the quantity of electricity passed.
(2) The amount of chemical change produced in different substances by a fixed quantity of electricity is proportional to the electrochemical equivalent of the substance.

fast-atom bombardment mass spectroscopy (FAB mass spectroscopy) A technique in *mass spectroscopy in which ions are produced by bombardment with high-energy neutral atoms or molecules. It is used for samples that are nonvolatile or are thermally unstable.

fat A mixture of lipids, chiefly *triglycerides, that is solid at normal body temperatures. Fats occur widely in plants and animals as a means of storing food energy, having twice the calorific value of carbohydrates. In mammals, fat is deposited in a layer beneath the skin (subcutaneous fat) and deep within the body as a specialized adipose tissue.

Fats derived from plants and fish generally have a greater proportion of unsaturated *fatty acids than those from mammals. Their melting points thus tend to be lower, causing a softer consistency at room temperatures. Highly unsaturated fats are liquid at room temperatures and are therefore more properly called *oils.

fatty acid An organic compound consisting of a hydrocarbon chain and a terminal carboxyl group (*see* **carboxylic acids**). Chain length ranges from one hydrogen atom (methanoic, or formic, acid, HCOOH) to nearly 30 carbon atoms. Ethanoic (acetic), propanoic (propionic), and butanoic (butyric) acids are important in metabolism. Long-chain fatty acids (more than 8–10 carbon atoms) most commonly occur as constituents of certain lipids, notably glycerides, phospholipids, sterols, and waxes, in which they are esterified with alcohols. These long-chain fatty acids generally have an even number of carbon atoms; unbranched chains predominate over branched chains. They may be *saturated* (e.g. *palmitic (hexadecanoic) acid and *stearic (octadecanoic) acid) or *unsaturated*, with one double bond (e.g. *oleic (cis-octodec-9-enoic) acid) or two or more double bonds, in which case they are called *polyunsaturated fatty acids* (e.g. *linoleic acid and *linolenic acid). *See also* **essential fatty acids**.

The physical properties of fatty acids are determined by chain length, degree of unsaturation, and chain branching. Short-chain acids are pungent liquids, soluble in water. As chain length increases, melting points are raised and water-solubility decreases. Unsaturation and chain branching tend to lower melting points.

fatty-acid oxidation **(β-oxidation)** The metabolic pathway in which fats are metabolized to release energy. Fatty-acid oxidation occurs continually but does not become a major source of energy until the animal's carbohydrate resources are exhausted, for example during starvation. Fatty-acid oxidation occurs chiefly in the mitochondria. A series of reactions cleave off two carbon atoms at a time from the hydrocarbon chain of the fatty acid. These two-carbon fragments are combined with *coenzyme A to form acetyl coenzyme A (acetyl CoA), which then enters the *Krebs cycle. The formation of acetyl CoA occurs repeatedly until all the hydrocarbon chain has been used up.

***f*-block elements** The block of elements in the *periodic table consisting of the lanthanoid series (from cerium to lutetium) and the actinoid series (from thorium to lawrencium). They are characterized by having two *s*-electrons in their outer shell (n) and *f*-electrons in their inner (n–1) shell.

f.c.c. Face-centred cubic. *See* **cubic crystal**.

Fehling's test A chemical test to detect reducing sugars and aldehydes in solution, devised by the German chemist H. C. von Fehling (1812–85). *Fehling's solution* consists of Fehlings A (copper(II) sulphate solution) and Fehling's B (alkaline 2,3-dihydroxybutanedioate (sodium tartrate) solution), equal amounts of which are added to the test solution. After boiling, a positive result is indicated by the formation of a brick-red precipitate of copper(I) oxide. Methanal, being a strong reducing agent, also produces copper metal; ketones do not react. The test is now little used, having been replaced by *Benedict's test.

feldspars A group of silicate minerals, the most abundant minerals in the earth's crust. They have a structure in which $(Si,Al)O_4$ tetrahedra are linked together with potassium, sodium, and calcium and very occasionally barium ions occupying the large spaces in the framework. The chemical composition of feldspars may be expressed as combinations of the four components: *anorthite* (An), $CaAl_2Si_2O_8$; *albite* (Ab), $NaAlSi_3O_8$; *orthoclase* (Or), $KAlSi_3O_8$; *celsian* (Ce), $BaAl_2Si_2O_8$. The feldspars are subdivided into two groups: the *alkali feldspars* (including microcline, orthoclase, and sanidine), in which potassium is dominant with a smaller proportion of sodium and negligible calcium; and the *plagioclase feldspars*, which vary in composition in a series that ranges from pure sodium feldspar (albite) through to pure calcium feldspar (anorthite) with negligible potassium. Feldspars form colourless, white, or pink crystals with a hardness of 6 on the Mohs' scale.

feldspathoids A group of alkali aluminosilicate minerals that are similar in chemical composition to the *feldspars but are relatively deficient in silica and richer in alkalis. The structure consists of a framework of $(Si,Al)O_4$ tetrahedra with aluminium and silicon atoms at their centres. The feldspathoids occur chiefly with feldspars but do not coexist with free quartz (SiO_2) as they react with silica to yield feldspars. The chief varieties of feldspathoids are: nepheline, $KNa_3(AlSiO_4)_4$; leucite, $KAlSi_2O_6$; analcime, $NaAlSi_2O_6.H_2O$; cancrinite, $Na_8(AlSiO_4)_6(HCO_3)_2$; and the sodalite subgroup comprising: sodalite, $3(NaAlSiO_4).NaCl$; nosean, $3(NaAlSiO_4).Na_2SO_4$; haüyne, $3(NaAlSiO_4).CaSO_4$; lazurite $(Na,Ca)_8(Al,Si)_{12}O_{24}(S,SO_4)$ (*see* **lapis lazuli**).

FEM *See* **field-emission microscope**.

femto- Symbol f. A prefix used in the metric system to denote 10^{-15}. For example, 10^{-15} second = 1 femtosecond (fs).

femtochemistry The investigation of chemical processes that occur on the timescale of a femtosecond (10^{-15} s). Femtochemistry has become possible as a result of the development of *lasers capable of being pulsed in femtoseconds. This has enabled observations to be made on very short-lived species, such as activated complexes, which only exist for about a picosecond (10^{-12} s). In a femtochemical experiment a femtosecond pulse causes dissociation of a molecule. A series of femtosecond pulses is then released, the frequency of the pulses being that of an absorption of one of the products of the dissociation. The absorption can be used as a measure of the abundance of the product of the dissociation. This type of study enables the course of the mechanism of a chemical reaction to be studied in detail.

fermentation A form of anaerobic respiration occurring in certain microorganisms, e.g. yeasts. *Alcoholic fermentation* comprises a series of biochemical reactions by which pyruvate (the end product of *glycolysis) is converted to ethanol and carbon dioxide. Fermentation is the basis of the baking, wine, and beer industries.

fermi A unit of length formerly used in nuclear physics. It is equal to 10^{-15} metre. In SI units this is equal to 1 femtometre (fm). It was named after the Italian-born US physicist Enrico Fermi (1901–54).

Fermi–Dirac statistics *See* **quantum statistics**.

Fermi level The energy in a solid at which the average number of particles per quantum state is ½; i.e. one half of the quantum states are occupied. The Fermi level in conductors lies in the conduction band (*see* **energy bands**), in insulators it lies in the valence band, and in semiconductors it falls in the gap between the conduction band and the valence band. At absolute zero all the electrons would occupy energy levels up to the Fermi level and no higher levels would be occupied. It is named after Enrico Fermi.

fermion An *elementary particle (or bound state of an elementary particle, e.g. an atomic nucleus or an atom) with half-integral spin; i.e. a particle that conforms to Fermi–Dirac statistics (*see* **quantum statistics**).

fermium Symbol Fm. A radioactive metallic transuranic element belonging to the *actinoids; a.n. 100; mass number of the most stable isotope 257 (half-life 10 days). Ten isotopes are known. The element was first identified by A. Ghiorso and associates in debris from the first hydrogen-bomb explosion in 1952. It is named after Enrico Fermi.

ferrate An iron-containing anion, FeO_4^{2-}. It exists only in strong alkaline solutions, in which it forms purple solutions.

ferric alum One of the *alums, $K_2SO_4.Fe_2(SO_4)_3.24H_2O$, in which the aluminium ion Al^{3+} is replaced by the iron(III) (ferric) ion Fe^{3+}.

ferric compounds Compounds of iron in its +3 oxidation state; e.g. ferric chloride is iron(III) chloride, $FeCl_3$.

ferricyanide A compound containing the complex ion $[Fe(CN)_6]^{3-}$, i.e. the hexacyanoferrate(III) ion.

ferrimagnetism *See* **magnetism**.

ferrite 1. A member of a class of mixed oxides $MO.Fe_2O_3$, where M is a metal such as cobalt, manganese, nickel, or zinc. The ferrites are ceramic materials that show either ferrimagnetism or ferromagnetism, but are not electrical conductors. For this reason they are used in high-frequency circuits as magnetic cores. **2.** *See* **steel**.

ferroalloys Alloys of iron with other elements made by smelting mixtures of iron ore and the metal ore; e.g. ferrochromium, ferrovanadium, ferromanganese, ferrosilicon, etc. They are used in making alloy *steels.

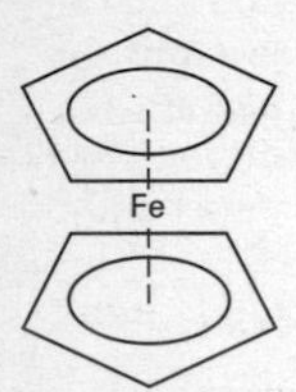

Ferrocene

ferrocene An orange-red crystalline solid, $Fe(C_5H_5)_2$; m.p. 173°C. It can be made by adding the ionic compound $Na^+C_5H_5^-$ (cyclopentadienyl sodium, made from sodium and cyclopentadiene) to iron(III) chloride. In ferrocene, the two rings are parallel, with the iron ion sandwiched between them (hence the name *sandwich compound*: see formula). The bonding is between pi orbitals on the rings and *d*-orbitals on the Fe^{2+} ion. The compound can undergo electrophilic substitution on the C_5H_5 rings (they have some aromatic character). It can also be oxidized to the blue ion $(C_5H_5)_2Fe^+$. Ferrocene is the first of a class of similar complexes called *sandwich compounds. Its systematic name is *di-π-cyclopentadienyl iron(II)*.

ferrocyanide A compound containing the complex ion $[Fe(CN)_6]^{4-}$, i.e. the hexacyanoferrate(II) ion.

ferroelectric materials Ceramic dielectrics, such as Rochelle salt and barium titanate, that have a domain structure making them analogous to ferromagnetic materials. They exhibit hysteresis and usually the piezoelectric effect.

ferromagnetism *See* magnetism.

ferrosoferric oxide *See* triiron tetroxide.

ferrous compounds Compounds of iron in its +2 oxidation state; e.g. ferrous chloride is iron(II) chloride, $FeCl_2$.

fertilizer Any substance that is added to soil in order to increase its productivity. Fertilizers can be of natural origin, such as composts, or they can be made up of synthetic chemicals, particularly nitrates and phosphates. Synthetic fertilizers can increase crop yields dramatically, but when leached from the soil by rain, which runs into lakes, they also increase the process of eutrophication. Bacteria that can fix nitrogen are sometimes added to the soil to increase its fertility; for example, in tropical countries the cyanobacterium *Anabaena* is added to rice paddies to increase soil fertility.

fibrous protein *See* **protein.**

Fick's law A law describing the diffusion that occurs when solutions of different concentrations come into contact, with molecules moving from regions of higher concentration to regions of lower concentration. Fick's law states that the rate of diffusion dn/dt, called the *diffusive flux* and denoted J, across an area A is given by: $dn/dt = J = -DA\partial c/\partial x$, where D is a constant called the *diffusion constant*, $\partial c/\partial x$ is the concentration gradient of the solute, and dn/dt is the amount of solute crossing the area A per unit time. D is constant for a specific solute and solvent at a specific temperature. Fick's law was formulated by the German physiologist Adolf Eugen Fick (1829–1901) in 1855.

field effect *See* **electronic effects.**

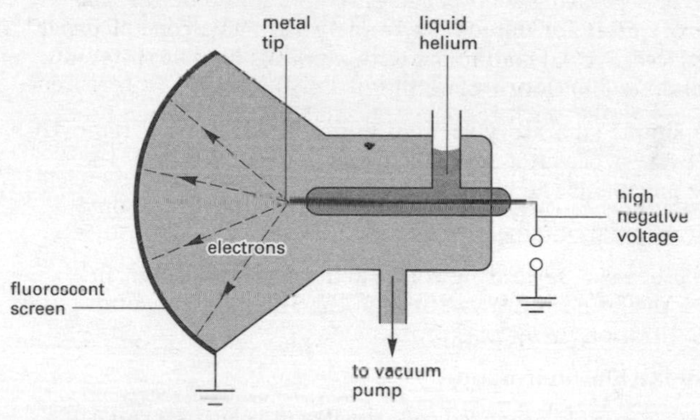

Field-emission microscope

field-emission microscope (FEM) A type of electron microscope in which a high negative voltage is applied to a metal tip placed in an evacuated vessel some distance from a glass screen with a fluorescent coating. The tip produces electrons by *field emission*, i.e. the emission of electrons from an unheated sharp metal part as a result of a high electric field. The emitted electrons form an enlarged pattern on the fluorescent screen, related to the individual exposed planes of atoms. As the resolution of the instrument is limited by the vibrations of the metal atoms, it is helpful to cool the tip in liquid helium. Although the individual atoms forming the point are not displayed, individual adsorbed atoms of other substances can be, and their activity is observable.

field-ionization microscope (field-ion microscope; FIM) A type of electron microscope that is similar in principle to the *field-emission microscope, except that a high positive voltage is applied to the metal tip, which is surrounded by low-pressure gas (usually helium) rather than a vacuum. The image is formed in this case by *field ionization*: ionization at the surface of an unheated solid as a result of a strong electric field creating positive ions by electron transfer from surrounding atoms or molecules. The image is formed by ions striking the fluorescent screen. Individual atoms on the surface of the tip can be resolved and, in certain cases, adsorbed atoms may be detected. *See also* **atom-probe field-ion microscopy**.

filler A solid inert material added to a synthetic resin or rubber, either to change its physical properties or simply to dilute it for economy.

film badge A lapel badge containing masked photographic film worn by personnel who could be exposed to ionizing radiation. The film is developed to indicate the extent that the wearer has been exposed to harmful radiation.

filter A device for separating solid particles from a liquid or gas. The simplest laboratory filter for liquids is a funnel in which a cone of paper (*filter paper*) is placed. Special containers with a porous base of sintered glass are also used. *See also* **Gooch crucible**.

filter pump A simple laboratory vacuum pump in which air is removed from a system by a jet of water forced through a narrow nozzle. The lowest pressure possible is the vapour pressure of water.

filtrate The clear liquid obtained by filtration.

filtration The process of separating solid particles using a filter. In *vacuum filtration*, the liquid is drawn through the filter by a vacuum pump. *Ultrafiltration* is filtration under pressure.

FIM *See* **field-ionization microscope**.

fine chemicals Chemicals produced industrially in relatively small quantities and with a high purity; e.g. dyes and drugs.

fineness of gold A measure of the purity of a gold alloy, defined as the parts of gold in 1000 parts of the alloy by mass. Gold with a fineness of 750 contains 75% gold, i.e. 18 *carat gold.

fine structure Closely spaced spectral lines arising from transitions between energy levels that are split by the vibrational or rotational motion of a molecule or by electron spin. They are visible only at high resolution. *Hyperfine structure*, visible only at very high resolution, results from the influence of the atomic nucleus on the allowed energy levels of the atom.

finger domain A finger-shaped structure produced in a protein when a series of the constituent amino acids combines with a metal atom.

firedamp Methane formed in coal mines.

fire extinguisher A substance that smothers flames or prevents their spreading. Many substances can be used. Liquids include water, sodium hydrogencarbonate solution, chlorinated organic compounds such as tetrachloromethane, and carbon dioxide-containing foams. Solid extinguishers, such as sodium or potassium hydrogencarbonate, produce carbon dioxide gas when strongly heated; solid carbon dioxide (dry ice) may also be used.

first-order reaction *See* **order**.

Fischer, Emil Hermann (1852–1919) German organic chemist who studied under *Kekulé in Bonn and later under *Baeyer in Strasbourg. He moved to Munich in 1875 and Würzburg in 1885. Fischer is noted for his extensive pioneering work on natural products and especially his work on sugar chemistry. In 1899 he began work on synthesizing peptides and proteins. Fischer was awarded the 1902 Nobel Prize for chemistry.

Fischer, Hans (1881–1945) German organic chemist who worked mainly in Munich. He worked on porphyrins, synthesizing haemin in 1927. Fischer worked also on chlorophyll, showing that it was a porphyrin containing magnesium. In 1944 he synthesized bilirubin. He was awarded the 1930 Nobel Prize for chemistry.

Fischer projection A type of *projection formula in which a molecule is drawn with horizontal bonds representing bonds coming out of the page and vertical bonds representing bonds in the plane of the page or behind the page. It is named after Emil Fischer. *See* **absolute configuration**.

Fischer–Tropsch process An industrial method of making hydrocarbon fuels from carbon monoxide and hydrogen. The process was invented in 1933 and used by Germany in World War II to produce motor fuel. Hydrogen and carbon monoxide are mixed in the ratio 2:1 (water gas was used with added hydrogen) and passed at 200°C over a nickel or cobalt catalyst. The resulting hydrocarbon mixture can be separated into a higher boiling fraction for Diesel engines and a lower-boiling gasoline fraction. The gasoline fraction contains a high proportion of straight-chain hydrocarbons and has to be reformed for use in motor fuel. Alcohols, aldehydes, and ketones are also present. The process is also used in the manufacture of SNG from coal. It is named after the German chemist Franz Fischer (1852–1932) and the Czech Hans Tropsch (1839–1935).

fission-track dating A method of estimating the age of glass and other mineral objects by observing the tracks made in them by the fission fragments of the uranium nuclei that they contain. By irradiating the objects with neutrons to induce fission and comparing the density and number of the tracks before and after irradiation it is possible to estimate the time that has elapsed since the object solidified.

Fittig reaction *See* **Wurtz reaction**.

fixation *See* **nitrogen fixation**.

fixed point A temperature that can be accurately reproduced to enable it to be used as the basis of a *temperature scale.

flagpole *See* **ring conformations**.

flame A hot luminous mixture of gases undergoing combustion. The chemical reactions in a flame are mainly free-radical chain reactions and the light comes from fluorescence of excited molecules or ions or from incandescence of small solid particles (e.g. carbon).

flame test A simple test for metals, in which a small amount of the sample (usually moistened with hydrochloric acid) is placed on the end of a platinum wire and held in a Bunsen flame. Certain metals can be detected by the colour produced: barium (green), calcium (brick red), lithium (crimson), potassium (lilac), sodium (yellow), strontium (red).

flash photolysis A technique for studying free-radical reactions in gases. The apparatus used typically consists of a long glass or quartz tube holding the gas, with a lamp outside the tube suitable for producing an intense flash of light. This dissociates molecules in the sample creating free radicals, which can be detected spectroscopically by a beam of light passed down the axis of the tube. It is possible to focus the spectrometer on an absorption line for a particular product and measure its change in intensity with time using an oscilloscope. In this way the kinetics of very fast free-radical gas reactions can be studied.

flash point The temperature at which the vapour above a volatile liquid forms a combustible mixture with air. At the flash point the application of a naked flame gives a momentary flash rather than sustained combustion, for which the temperature is too low.

flavin adenine dinucleotide *See* FAD.

flavonoid One of a group of naturally occurring phenolic compounds many of which are plant pigments. They include the anthocyanins, flavonols, and flavones. Patterns of flavonoid distribution have been used in taxonomic studies of plant species.

flavoprotein *See* FAD.

Fleming, Sir Alexander (1881–1955) British bacteriologist, born in Scotland. He studied medicine at St Mary's Hospital, London, where he remained all his life. In 1922 he identified lysozyme, an enzyme that destroys bacteria, and in 1928 discovered the antibiotic *penicillin. He shared the 1945 Nobel Prize for physiology or medicine with *Florey and *Chain, who first isolated the drug.

flint **(chert)** Very hard dense nodules of microcrystalline quartz and chalcedony found in chalk and limestone.

flip-flop The movement (*transverse diffusion*) of a lipid molecule from one surface of a *lipid bilayer membrane to the other, which occurs at a very slow rate. This contrasts with the much faster rate at which lipid molecules exchange places with neighbouring molecules on the same surface of the membrane (*lateral diffusion*).

flocculation The process in which particles in a colloid aggregate into larger clumps. Often, the term is used for a reversible aggregation of particles in which the forces holding the particles together are weak and the colloid can be redispersed by agitation. The stability of a lyophobic colloidal dispersion depends on the existence of a layer of electric charge on the surface of the particles. Around this are attracted electrolyte ions of opposite charge, which form a mobile ionic 'atmosphere'. The result is an electrical double layer on the particle, consisting of an inner shell of fixed charges with an outer mobile atmosphere. The potential energy between two particles depends on a repulsive interaction between double layers on adjacent particles and an attractive interaction due to *van der Waals' forces between the particles.

At large separations, the repulsive forces dominate, and this accounts for the overall stability of the colloid. As the particles become closer together, the potential energy increases to a maximum and then falls sharply at very close separations, where the van der Waals' forces dominate. This potential-energy minimum corresponds to *coagulation and is irreversible. If the *ionic strength of the solution is high, the ionic atmosphere around the particles is dense and the potential-energy curve shows a shallow minimum at larger separation of particles. This corresponds to flocculation of the particles. Ions with a high charge are particularly effective for causing flocculation and coagulation.

flocculent Aggregated in woolly masses; used to describe precipitates.

Florey, Howard Walter, Baron (1898–1968) Australian pathologist, who moved to Oxford in 1922. After working in Cambridge and Sheffield (studying lysozyme), he returned to Oxford in 1935. There he teamed up with Ernst *Chain and by 1939 they succeeded in isolating and purifying *penicillin. They also developed a method of producing the drug in large quantities and carried out its first clinical trials. The two men shared the 1945 Nobel Prize for physiology or medicine with penicillin's discoverer, Alexander *Fleming.

Flory temperature (theta temperature) Symbol θ. The unique temperature at which the attractions and repulsions of a polymer in a solution cancel each other. It is analogous to the Boyle temperature of a nonideal gas. A polymer solution at the Flory temperature is called a *theta* (θ) solution. At the Flory temperature the virial coefficient *B*, asociated with the excluded volume of the polymer, is zero, which results in the polymer chain behaving almost ideally. This enables the theory of polymer solutions at the Flory temperature to provide a more accurate description of events than for polymer solutions at other temperatures, even if the polymer solution is concentrated. It is not always possible to attain the Flory temperature experimentally. The Flory temperature is named after the US physicist Paul Flory (1910–85).

flotation *See* froth flotation.

fluctuation–dissipation theorem A theory relating quantities in equilibrium and *nonequilibrium statistical mechanics and microscopic and macroscopic quantities. The fluctuation–dissipation theorem was first derived for electrical circuits with noise in 1928 by H. Nyquist; a general theorem in statistical mechanics was derived by H. B. Callen and T. A. Welton in 1951. The underlying principle of the fluctuation–dissipation theorem is that a nonequilibrium state may have been reached either as a result of a random fluctuation or an external force (such as an electric or magnetic field) and that the evolution towards equilibrium is the same in both cases (for a sufficiently small fluctuation). The fluctuation–dissipation theorem enables *transport coefficients to be calculated in terms of response to external fields.

fluidization A technique used in some industrial processes in which solid particles suspended in a stream of gas are treated as if they were in the liquid state. Fluidization is useful for transporting powders, such as coal dust. *Fluidized beds*, in which solid particles are suspended in an upward stream, are extensively used in the chemical industry, particularly in catalytic reactions where the powdered catalyst has a high surface area. They are also used in furnaces, being formed by burning coal in a hot turbulent bed of sand or ash through which air is passed. The bed behaves like a fluid, enabling the combustion temperature to be reduced so that the production of polluting oxides of nitrogen is diminished. By adding limestone to the bed with the fuel, the emission of sulphur dioxide is reduced.

High-pressure fluidized beds are also used in power-station furnaces in a *combined cycle* in which the products of combustion from the fluidized bed are used to drive a gas turbine, while a steam-tube boiler in the fluid bed raises steam to drive a steam turbine. This system both increases the efficiency of the combustion process and reduces pollution.

fluorescein A yellowish-red dye that produces yellow solutions with a green fluorescence. It is used in tracing water flow and as an *adsorption indicator.

fluorescence *See* **luminescence**.

fluoridation The process of adding very small amounts of fluorine salts (e.g. sodium fluoride, NaF) to drinking water to prevent tooth decay. The fluoride becomes incorporated into the fluoroapatite (*see* **apatite**) of the growing teeth and reduces the incidence of dental caries.

fluoride *See* **halide**.

fluorination A chemical reaction in which a fluorine atom is introduced into a molecule. *See* **halogenation**.

fluorine Symbol F. A poisonous pale yellow gaseous element belonging to group 17 (formerly VIIB) of the periodic table (the *halogens); a.n. 9; r.a.m. 18.9984; d. 1.7 g dm^{-3}; m.p. −219.62°C; b.p. −188.1°C. The main mineral sources are *fluorite (CaF_2) and *cryolite (Na_3AlF). The element is obtained by electrolysis of a molten mixture of potassium fluoride and hydrogen fluoride. It is used in the synthesis of organic fluorine compounds.

Chemically, it is the most reactive and electronegative of all elements. It is a highly dangerous element, causing severe chemical burns on contact with the skin. The element was identified by Scheele in 1771 and first isolated by Moissan in 1886.

fluorite (fluorspar) A mineral form of calcium fluoride, CaF_2, crystallizing in the cubic system. It is variable in colour; the most common fluorites are green and purple (blue john), but other forms are white, yellow, or brown. Fluorite is used chiefly as a flux material in the smelting of iron and steel; it is also used as a source of fluorine and hydrofluoric acid and in the ceramic and optical-glass industries.

fluorite structure *See* **calcium fluoride.**

fluorocarbons Compounds obtained by replacing the hydrogen atoms of hydrocarbons by fluorine atoms. Their inertness and high stability to temperature make them suitable for a variety of uses as oils, polymers, etc. *See also* **chlorofluorocarbon**; **halon.**

flux 1. A substance applied to the surfaces of metals to be soldered to inhibit oxidation. **2.** A substance used in the smelting of metals to assist in the removal of impurities as slag.

fluxional molecule A molecule that undergoes alternate very rapid rearrangements of its atoms and thus only has a specific structure for a very short period of time. For example, the molecule ClF_3 has a T-shape at low temperatures (−60°C); at room temperature the fluorine atoms change position very rapidly and appear to have identical positions.

foam A dispersion of bubbles in a liquid. Foams can be stabilized by *surfactants. Solid foams (e.g. expanded polystyrene or foam rubber) are made by foaming the liquid and allowing it to set. *See also* **colloids.**

Fokker–Planck equation An equation in *nonequilibrium statistical mechanics that describes a superposition of a dynamic friction (slowing-down) process and a diffusion process for the evolution of variables in a system. The Fokker–Planck equation, which can be used to analyse such problems as *Brownian movement, is soluble using statistical methods and the theory of probability. It is named after the Dutch physicist Adriaan Fokker (1887–1968) and Max *Planck.

folacin *See* **folic acid.**

folic acid (folacin) A vitamin of the *vitamin B complex. In its active form, tetrahydrofolic acid, it is a *coenzyme in various reactions involved in the metabolism of amino acids, purines, and pyrimidines. It is synthesized by intestinal bacteria and is widespread in food, especially green leafy vegetables. Deficiency causes poor growth and nutritional anaemia.

food additive A substance added to a food during its manufacture or processing in order to improve its keeping qualities, texture, appearance, or stability or to enhance its taste or colour. Additives are usually present in minute quantities; they include colouring materials, sweeteners, preservatives, *antioxidants, emulsifiers, and stabilizers. In most countries the additives used must be selected from an approved list of such compounds, which have been tested for safety, and they must be listed on the food labels of individual products.

fool's gold *See* pyrite.

forbidden band *See* energy bands.

forbidden transitions Transitions between energy levels in a quantum-mechanical system that are not allowed to take place because of *selection rules. In practice, forbidden transitions can occur, but they do so with much lower probability than allowed transitions. There are three reasons why forbidden transitions may occur:
(1) the selection rule that is violated is only an approximate rule. An example is provided by those selection rules that are only exact in the absence of *spin–orbit coupling. When spin–orbit coupling is taken into account, the forbidden transitions become allowed – their strength increasing with the size of the spin–orbit coupling;
(2) the selection rule is valid for dipole radiation, i.e. in the interaction between a quantum-mechanical system, such as an atom, and an electromagnetic field, only the (variable) electric dipole moment is considered. Actual transitions may involve magnetic dipole radiation or quadrupole radiation;
(3) the selection rule only applies for an atom, molecule, etc., in isolation and does not necessarily apply if external fields, collisions, etc., are taken into account.

force constant A constant characterizing the strength of the bond in a diatomic molecule. Near the equilibrium position, R_e, of the potential energy curve of a diatomic molecule, the potential energy V is accurately represented by a parabola of the form $V = k/2(R - R_e)^2$, where R is the internuclear distance and k is the force constant. The greater the force constant, the stronger is the bond between the atoms, since the walls of the potential curve become steeper. Regarding the molecular vibrations as simple harmonic motion, the force constant occurs in the analysis of the vibrational energy levels.

forced convection *See* convection.

formaldehyde *See* methanal.

formalin A colourless solution of methanal (formaldehyde) in water with methanol as a stabilizer; r.d. 1.075–1.085. When kept at temperatures below 25°C a white polymer of methanal separates out. It is used as a disinfectant and preservative for biological specimens.

formate *See* methanoate.

formic acid *See* methanoic acid.

formula A way of representing a chemical compound using symbols for the atoms present. Subscripts are used for the numbers of atoms. The *molecular formula* simply gives the types and numbers of atoms present. For example, the molecular formula of ethanoic acid is $C_2H_4O_2$. The *empirical formula* gives the atoms in their simplest ratio; for ethanoic acid it is CH_2O. The *structural formula* gives an indication of the way the atoms are arranged. Commonly, this is done by dividing the formula into groups; ethanoic acid can be written $CH_3.CO.OH$ (or more usually simply CH_3COOH). Structural formulae can also show the arrangement of atoms or groups in space.

formula weight The relative molecular mass of a compound as calculated from its molecular formula.

formylation A chemical reaction that introduces a formyl group (methanoyl, –CHO) into an organic molecule.

formyl group The group HCO–.

fossil fuel Coal, oil, and natural gas, the fuels used by man as a source of energy. They are formed from the remains of living organisms and all have a high carbon or hydrogen content. Their value as fuels relies on the exothermic oxidation of carbon to form carbon dioxide ($C + O_2 \rightarrow CO_2$) and the oxidation of hydrogen to form water ($H_2 + \frac{1}{2}O_2 \rightarrow H_2O$).

four-circle diffractometer An instrument used in X-ray crystallography to automatically determine the shape and symmetry of the unit cell of a crystal. The crystal is placed in the goniometer head with an arbitrary orientation. If the dimensions of the unit cell have been determined, they can be used to calculate the settings of the four angles of the diffractometer needed to observe a specific (h k l) reflection, where (h k l) are Miller indices. A computer controls the settings so that each (h k l) is examined in turn and the diffraction intensity measured. This information enables the electron density to be calculated, as the intensity of the reflection for a set of (h k l) planes is proportional to the square of the modulus of a function called the *structure factor* F_{hkl} of the set, which is related to the electron density.

Fourier analysis The representation of a function f(x), which is periodic in x, as an infinite series of sine and cosine functions:

$$f(x) = a_0/2 + \sum_{n=1}^{\infty}(a_n\cos nx + b_n\sin nx)$$

A series of this type is called a *Fourier series. If the function is periodic with a period 2π, the coefficients a_0, a_n, b_n are:

$$a_0 = \int_{-\pi}^{+\pi} f(x)dx,$$

$$a_n = \int_{-\pi}^{+\pi} f(x)\cos nx dx \quad (n = 1,2,3,...),$$

$$b_n = 1/\pi \int_{-\pi}^{+\pi} f(x)\sin nx dx \quad (n = 1,2,3,...).$$

Fourier analysis and Fourier series are named after the French mathematician and engineer Joseph Fourier (1768–1830). Fourier series have many important applications in mathematics, science, and engineering, having been invented by Fourier in the first quarter of the 19th century in his analysis of the problem of heat conduction.

Fourier series An expansion of a periodic function as a series of trigonometric functions. Thus,

$$f(x) = a_0 + (a_1\cos x + b_1\sin x) + (a_2\cos 2x + b_2\sin 2x)+\ldots,$$

where a_0, a_1, b_1, b_2, etc., are constants, called *Fourier coefficients*. The series was first formulated by Joseph Fourier and is used in *Fourier analysis.

Fourier transform An integral transform of the type:

$$F(y) = \int_{-\infty}^{\infty} f(x)e^{-xy}dy.$$

The inverse is:

$$f(x) = (1/2\pi)\int_{-\infty}^{\infty} F(y)e^{ixy}dy.$$

Fourier transform techniques are used in obtaining information from spectra, especially in NMR and infrared spectroscopy (*see* **Fourier-transform infrared**).

Fourier-transform infrared (FT-IR) Infrared spectroscopy in which computers are part of the spectroscopic apparatus and use *Fourier transforms to enable the curve of intensity against wave number to be plotted with very high sensitivity. This has allowed spectra to be obtained in the far infrared region; previously it was difficult to attain spectra in this region as the resolution was obscured by the signal-to-noise ratio being too high to resolve the vibrational and/or rotational spectra of small molecules in their gas phase. FT-IR has been used in research on the atmosphere. Another application of this technique is the detection of impurities in samples of condensed matter.

four-level laser A laser in which four energy levels are involved. The disadvantage of a three-level laser is that it is difficult to attain population inversion because many molecules have to be raised from their ground state to an excited state by pumping. In a four-level laser, the laser transition finishes in an initially unoccupied state F, having started in a state I, which is not the ground state. As the state F is initially unoccupied, any population in I constitutes population inversion. Thus laser action is possible if I is sufficiently metastable. If transitions from F to the ground state G are rapid, population inversion is maintained since this lowers the population in F caused by the transition in the laser action.

fractal A curve or surface generated by a process involving successive subdivision. For example, a *snowflake curve* can be produced by starting with an equilateral triangle and dividing each side into three segments. The middle segments are then replaced by two equal segments, which would form the sides of a smaller equilateral triangle. This gives a 12-sided star-shaped figure. The next stage is to subdivide each of the sides of this figure in the same way, and so on. The result is a developing figure that resembles a snowflake. In the limit, this figure has 'fractional dimension' – i.e. a dimension between that of a line (1) and a surface (2); the dimension of the snowflake curve is 1.26. The study of this type of 'self-similar' figure is used in certain branches of chemistry – for example, crystal growth. Fractals are also important in *chaos theory and in computer graphics.

fraction *See* **fractional distillation.**

fractional crystallization A method of separating a mixture of soluble solids by dissolving them in a suitable hot solvent and then lowering the temperature slowly. The least soluble component will crystallize out first, leaving the other components in solution. By controlling the temperature, it is sometimes possible to remove each component in turn.

fractional distillation (fractionation) The separation of a mixture of liquids by distillation. Effective separation can be achieved by using a long vertical column (*fractionating column*) attached to the distillation vessel and filled with glass beads. Vapour from the liquid rises up the column until it condenses and runs back into the vessel. The rising vapour in the column flows over the descending liquid, and eventually a steady state is reached in which there is a decreasing temperature gradient up the column. The vapour in the column has more volatile components towards the top and less volatile components at the bottom. Various *fractions* of the mixture can be drawn off at points on the column. Industrially, fractional distillation is performed in large towers containing many perforated trays. It is used extensively in petroleum refining.

fractionating column *See* **fractional distillation.**

fractionation *See* **fractional distillation.**

francium Symbol Fr. A radioactive element belonging to *group 1 (formerly IA) of the periodic table; a.n. 87; r.d. 2.4; m.p. 27±1°C; b.p. 677±1°C. The element is found in uranium and thorium ores. All 22 known isotopes are radioactive, the most stable being francium–223. The existence of francium was confirmed by Marguerite Perey in 1939.

Franck–Condon principle A principle governing the intensity of transitions in the vibrational structure during an electronic transition in a molecule. The principle states that since nuclei are much heavier and move much more slowly than electrons (*see* **Born–Oppenheimer approximation**), an electronic transition occurs much more rapidly than the time required for the nuclei to respond to it. Therefore, in a diagram showing the electronic states of the molecule as a function of internuclear distance, the most intense electronic transition is represented by a vertical line. For this reason a transition obeying the Franck–Condon principle is called a *vertical transition*; when it occurs the relative positions of the nuclei remain unchanged. The Franck–Condon principle is named after James Franck (1882–1964), who stated it in 1925, and Edward Condon, who formulated it mathematically in terms of quantum mechanics in 1928.

Frankland, Sir Edward (1825–99) British organic chemist who was the first to produce organometallic compounds (zinc dialkyls). He is remembered as the originator of the theory of valency and introduced a method of writing structural formulas.

Frasch process A method of obtaining sulphur from underground deposits using a tube consisting of three concentric pipes. Superheated steam is passed down the outer pipe to melt the sulphur, which is forced up through the middle pipe by compressed air fed through the inner tube. The steam in the outer casing keeps the sulphur molten in the pipe. It was named after the German-born US chemist Hermann Frasch (1851–1914).

Fraunhofer, Josef von (1787–1826) German physicist, who trained as an optician. In 1814 he observed dark lines in the spectrum of the sun (*see* **Fraunhofer lines**). He also studied Fraunhofer diffraction.

Fraunhofer lines Dark lines in the solar spectrum, discovered by Josef von *Fraunhofer, that result from the absorption by elements in the solar chromosphere of some of the wavelengths of the visible radiation emitted by the hot interior of the sun.

free electron *See* **electron**.

free-electron approximation The approximation resulting from the assumption that electrons in *metals can be analysed using the *kinetic theory of gases, without taking the periodic potential of the metal into account. This approximation gives a good qualitative account of some properties of metals, such as their electrical conductivity. At very low temperatures it is necessary to use quantum statistical mechanics rather than classical statistical mechanics. The free-electron approximation does not, however, give an adequate quantitative description of the properties of metals. It can be improved by the *nearly free electron approximation*, in which the periodic potential is treated as a perturbation on the free electrons.

free energy A measure of a system's ability to do work. The *Gibbs free energy* (or *Gibbs function*), G, is defined by $G = H - TS$, where G is the energy liberated or absorbed in a reversible process at constant pressure and constant temperature (T), H is the *enthalpy and S the *entropy of the system. Changes in Gibbs free energy, ΔG, are useful in indicating the conditions under which a chemical reaction will occur. If ΔG is positive the reaction will only occur if energy is supplied to force it away from the equilibrium position (i.e. when $\Delta G = 0$). If ΔG is negative the reaction will proceed spontaneously to equilibrium.

The *Helmholtz free energy* (or *Helmholtz function*), F, is defined by $F = U - TS$, where U is the *internal energy. For a reversible isothermal process, ΔF represents the useful work available.

free radical An atom or group of atoms with an unpaired valence electron. Free radicals can be produced by photolysis or pyrolysis in which a bond is broken without forming ions (*see* **homolytic fission**). Because of their unpaired valence electron, most free radicals are extremely reactive. *See also* **chain reaction.**

freeze drying A process used in dehydrating food, blood plasma, and other heat-sensitive substances. The product is deep-frozen and the ice trapped in it is removed by reducing the pressure and causing it to sublime. The water vapour is then removed, leaving an undamaged dry product.

freezing mixture A mixture of components that produces a low temperature. For example, a mixture of ice and sodium chloride gives a temperature of −20°C.

freezing-point depression *See* **depression of freezing point.**

Frenkel defect *See* **crystal defect.**

Frenkel–Kontorowa model A one-dimensional model of atoms, such as xenon, adsorbed on a periodic substrate, such as graphite. This model, which can be used to investigate the nature of the lattice formed by the adsorbed gas, was invented in 1938 by Y. I. Frenkel and T. Kontorowa and independently in 1949 by F. C. Frank and J. H. Van der Merwe. The Frenkel–Kontorowa model can be used to investigate the phase transition between a *commensurate lattice and an *incommensurate lattice.

freon *See* **chlorofluorocarbon.**

Freundlich isotherm An isotherm for adsorption with the form $\theta = c_1P^{1/c_2}$, where the fractional coverage θ is the ratio of adsorption sites occupied to the number of adsorption sites available, c_1, and c_2 are constants, and P is the pressure of gas. The Freundlich isotherm is frequently used with adsorption from liquid solutions, when it has the form: $w = c_1 \times c^{1/c_2}$, where w is the mass of solute adsorbed per unit mass of adsorbent (a quantity called the mass fraction) and c is the concentration of the solution. Like other isotherms the Freundlich isotherm agrees with experiment for limited ranges of parameter. If the limits of applicability have been established these isotherms, such as the Freundlich isotherm, are useful in considering heterogeneous catalysis.

Friedel–Crafts reaction A type of reaction in which an alkyl group (from a haloalkane) or an acyl group (from an acyl halide) is substituted on a benzene ring (see illustration). The product is an alkylbenzene (for alkyl substitution) or an alkyl aryl ketone (for acyl substitution). The reactions occur at high temperature (about 100°C) with an aluminium chloride catalyst. The catalyst acts as an electron acceptor for a lone pair on the halide atom. This polarizes the haloalkane or acyl halide, producing a positive charge on the alkyl or acyl group. The mechanism is then electrophilic substitution. Alcohols and alkenes can also undergo Friedel–Crafts reactions. The reaction is named after the French chemist Charles Friedel (1832–99) and the US chemist James M. Craft (1839–1917).

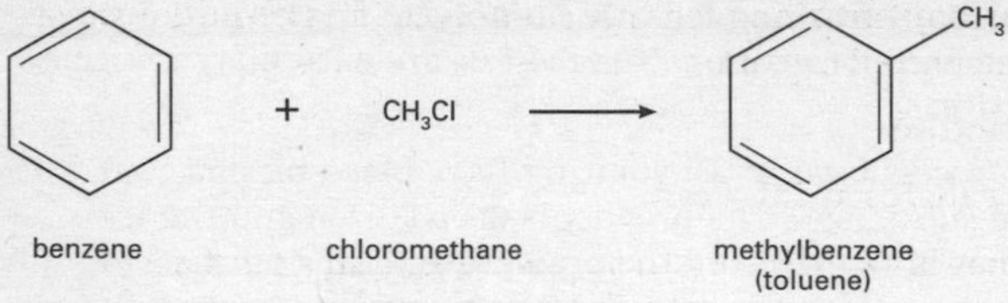

CH_3

+ CH_3Cl ⟶

benzene chloromethane methylbenzene (toluene)

Friedel–Crafts methylation

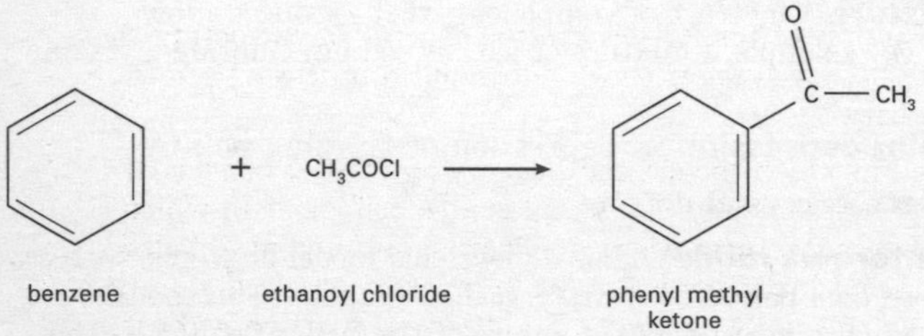

O

C — CH_3

+ CH_3COCl ⟶

benzene ethanoyl chloride phenyl methyl ketone

Friedel–Crafts acetylation

frontier orbital One of two orbitals in a molecule: the *highest occupied molecular orbital (HOMO) and the *lowest unoccupied molecular orbital (LUMO). These two molecular orbitals are usually the most important ones in determining chemical and spectroscopic properties of the molecule.

frontier-orbital theory A theory of the reactions of molecules that emphasizes the energies and symmetries of frontier orbitals. Frontier orbital theory was developed by the Japanese chemist Kenichi Fukui (1919–98) in the 1950s and is an alternative approach to the *Woodward–Hoffmann rules. It has been very successful in explaining such reactions as the *Diels–Alder reaction.

froth flotation A method of separating mixtures of solids, used industrially for separating ores from the unwanted gangue. The mixture is ground to a powder and water and a frothing agent added. Air is blown through the water. With a suitable frothing agent, the bubbles adhere only to particles of ore and carry them to the surface, leaving the gangue particles at the bottom.

fructose **(fruit sugar; laevulose)** A simple sugar, $C_6H_{12}O_6$, stereoisomeric with glucose (*see* **monosaccharide**). (Although natural fructose is the D-form, it is in fact laevorotatory.) Fructose occurs in green plants, fruits, and honey and tastes sweeter than sucrose (cane sugar), of which it is a constituent. Derivatives of fructose are important in the energy metabolism of living organisms. Some polysaccharide derivatives (fructans) are carbohydrate energy stores in certain plants.

fructose 1,6-bisphosphate An intermediate formed in the initial stage of *glycolysis by the phosphorylation of glucose using ATP.

fruit sugar *See* **fructose.**

FT-IR *See* **Fourier-transform infrared.**

fuel A substance that is oxidized or otherwise changed in a furnace or heat engine to release useful heat or energy. For this purpose wood, vegetable oil, and animal products have largely been replaced by *fossil fuels since the 18th century.

The limited supply of fossil fuels and the expense of extracting them from the earth has encouraged the development of nuclear fuels to produce electricity.

fuel cell A cell in which the chemical energy of a fuel is converted directly into electrical energy. The simplest fuel cell is one in which hydrogen is oxidized to form water over porous sintered nickel electrodes. A supply of gaseous hydrogen is fed to a compartment containing the porous anode and a supply of oxygen is fed to a compartment containing the porous cathode; the electrodes are separated by a third compartment containing a hot alkaline electrolyte, such as potassium hydroxide. The electrodes are porous to enable the gases to react with the electrolyte, with the nickel in the electrodes acting as a catalyst. At the anode the hydrogen reacts with the hydroxide ions in the electrolyte to form water, with the release of two electrons per hydrogen molecule:

$$H_2 + 2OH^- \rightarrow 2H_2O + 2e^-$$

At the cathode, the oxygen reacts with the water, taking up electrons, to form hydroxide ions:

$$\tfrac{1}{2}O_2 + H_2O + 2e^- \rightarrow 2OH^-$$

The electrons flow from the anode to the cathode through an external circuit as an electric current. The device is a more efficient converter of electric energy than a heat engine, but it is bulky and requires a continuous supply of gaseous fuels. Their use to power electric vehicles is being actively explored.

fugacity Symbol f. A thermodynamic function used in place of partial pressure in reactions involving real gases and mixtures. For a component of a mixture, it is defined by $d\mu = RTd(\ln f)$, where μ is the chemical potential. It has the same units as pressure and the fugacity of a gas is equal to the pressure if the gas is ideal. The fugacity of a liquid or solid is the fugacity of the vapour with which it is in equilibrium. The ratio of the fugacity to the fugacity in some standard state is the *activity. For a gas, the standard state is chosen to be the state at which the fugacity is 1. The activity then equals the fugacity.

fullerene *See* **buckminsterfullerene.**

fullerite *See* **buckminsterfullerene.**

fuller's earth A naturally occurring clay material (chiefly montmorillonite) that has the property of decolorizing oil and grease. In the past raw wool was cleaned of grease and whitened by kneading it in water with fuller's earth; a process known as *fulling*. Fuller's earth is now widely used to decolorize fats and oils and also as an insecticide carrier and drilling mud. The largest deposits occur in the USA, UK, and Japan.

fulminate *See* **cyanic acid.**

fulminic acid *See* **cyanic acid.**

fumaric acid *See* **butenedioic acid.**

functional group The group of atoms responsible for the characteristic reactions of a compound. The functional group is –OH for alcohols, –CHO for aldehydes, –COOH for carboxylic acids, etc.

fundamental *See* **harmonic.**

fundamental constants (universal constants) Those parameters that do not change throughout the universe. The charge on an electron, the speed of light in free space, the Planck constant, the gravitational constant, the electric constant, and the magnetic constant are all thought to be examples.

fundamental units A set of independently defined *units of measurement that forms the basis of a system of units. Such a set requires three mechanical units (usually of length, mass, and time) and one electrical unit; it has also been found convenient to treat certain other quantities as fundamental, even though they are not strictly independent. In the metric system the centimetre–gram–second (c.g.s.) system was replaced by the metre–kilogram–second (m.k.s.) system; the latter has now been adapted to provide the basis for *SI units. In British Imperial units the foot–pound–second (f.p.s.) system was formerly used.

fungicide *See* **pesticide.**

furan A colourless liquid compound, C_4H_4O; r.d. 0.94; m.p. –86°C; b.p. 31.4°C. It has a five-membered ring consisting of four CH_2 groups and one oxygen atom.

furanose A *sugar having a five-membered ring containing four carbon atoms and one oxygen atom.

furfural A colourless liquid, $C_5H_4O_2$, b.p. 162°C, which darkens on standing in air. It is the aldehyde derivative of *furan and occurs in various essential oils and in *fusel oil. It is used as a solvent for extracting mineral oils and natural resins and itself forms resins with some aromatic compounds.

fused ring *See* **ring.**

fusel oil A mixture of high-molecular weight *alcohols containing also esters and fatty acids, sometimes formed as a toxic impurity in the distillation products of alcoholic fermentation. It is used as a source of higher alcohols and in making paints and plastics.

fusible alloys Alloys that melt at low temperature (around 100°C). They have a number of uses, including constant-temperature baths, pipe bending, and automatic sprinklers to provide a spray of water to prevent fires from spreading. Fusible alloys are usually *eutectic mixtures of bismuth, lead, tin, and cadmium. *Wood's metal, *Rose's metal, and Lipowitz's alloy are examples of alloys that melt in the range 70–100°C.

fusion The process in which a solid turns into a liquid; melting.

GABA *See* **gamma-aminobutyric acid.**

Gabriel reaction A method of making a primary *amine (free from any secondary or tertiary amine impurities) from a haloalkane (alkyl halide) using potassium phthalimide. It is named after Siegmund Gabriel (1851–1924).

gadolinium Symbol Gd. A soft silvery metallic element belonging to the *lanthanoids; a.n. 64; r.a.m. 157.25; r.d. 7.901 (20°C); m.p. 1313°C; b.p. 3266°C. It occurs in gadolinite, xenotime, monazite, and residues from uranium ores. There are seven stable natural isotopes and eleven artificial isotopes are known. Two of the natural isotopes, gadolinium-155 and gadolinium-157, are the best neutron absorbers of all the elements. The metal has found limited applications in nuclear technology and in ferromagnetic alloys (with cobalt, copper, iron, and cerium). Gadolinium compounds are used in electronic components. The element was discovered by Jean de Marignac (1817–94) in 1880.

galactose A simple sugar, $C_6H_{12}O_6$, stereoisomeric with glucose, that occurs naturally as one of the products of the enzymic digestion of milk sugar (lactose) and as a constituent of gum arabic.

galena A mineral form of lead(II) sulphide, PbS, crystallizing in the cubic system; the chief ore of lead. It usually occurs as grey metallic cubes, frequently in association with silver, arsenic, copper, zinc, and antimony. Important deposits occur in Australia (at Broken Hill), Germany, the USA (especially in Missouri, Kansas, and Oklahoma), and the UK.

gallic acid (3,4,5-trihydroxybenzoic acid) A colourless crystalline aromatic compound, $C_6H_2(OH)_3COOH$; m.p. 253°C. It occurs in wood, oak galls, and tea, and is a component of tannins. It can be made from tannin by acid hydrolysis or fermentation. Gallic acid is used to make ink and various dyes. On heating it yields *pyrogallol.

gallium Symbol Ga. A soft silvery metallic element belonging to group 13 (formerly IIIB) of the periodic table; a.n. 31; r.a.m. 69.72; r.d. 5.90 (20°C); m.p. 29.78°C; b.p. 2403°C. It occurs in zinc blende, bauxite, and kaolin, from which it can be extracted by fractional electrolysis. It also occurs in gallite, $CuGaS_2$, to an extent of 1%; although bauxite only contains 0.01% this is the only commercial source. The two stable isotopes are gallium-69 and gallium-71; there are eight radioactive isotopes, all with short half-lives. The metal has only a few minor uses (e.g. as an activator in luminous paints), but gallium arsenide is extensively used as a semiconductor in many applications. Gallium corrodes most other metals because it rapidly diffuses into their lattices. Most gallium(I) and some gallium(II) compounds are unstable. The element was first identified by Paul Lecoq de Boisbaudran (1838–1912) in 1875.

GALP *See* **glyceraldehyde 3-phosphate**.

galvanic cell *See* **voltaic cell**.

galvanized iron Iron or steel that has been coated with a layer of zinc to protect it from corrosion in a process invented by Luigi Galvani. Corrugated mild-steel sheets for roofing and mild-steel sheets for dustbins, etc., are usually galvanized by dipping them in molten zinc. The formation of a brittle zinc–iron alloy is prevented by the addition of small quantities of aluminium or magnesium. Wire is often galvanized by a cold electrolytic process as no alloy forms in this process. Galvanizing is an effective method of protecting steel because even if the surface is scratched, the zinc still protects the underlying metal. *See* **sacrificial protection**.

gamma-aminobutyric acid (GABA) An inhibitory neurotransmitter in the central nervous system (principally the brain) that is capable of increasing the permeability of postsynaptic membranes. GABA is synthesized by *decarboxylation of the amino acid glutamate.

gamma-iron *See* **iron**.

gamma radiation Electromagnetic radiation emitted by excited atomic nuclei during the process of passing to a lower excitation state. Gamma radiation ranges in energy from about 10^{-15} to 10^{-10} joule (10 keV to 10 MeV) corresponding to a wavelength range of about 10^{-10} to 10^{-14} metre. A common source of gamma radiation is cobalt–60, the decay process of which is:

$${}^{60}_{27}Co \xrightarrow{\beta} {}^{60}_{28}Ni \xrightarrow{\gamma} {}^{60}_{28}Ni$$

The de-excitation of nickel–60 is accompanied by the emission of gamma-ray photons having energies 1.17 MeV and 1.33 MeV.

gangue Rock and other waste material present in an ore.

garnet Any of a group of silicate minerals that conform to the general formula $A_3B_2(SiO_4)_3$. The elements representing A may include magnesium, calcium, manganese, and iron(II); those representing B may include aluminium, iron(III), chromium, or titanium. Six varieties of garnet are generally recognized:
pyrope, $Mg_3Al_2Si_3O_{12}$;
almandine, $Fe_3^{2+}Al_2Si_3O_{12}$;
spessartite, $Mn_3Al_2Si_3O_{12}$;
grossularite, $Ca_3Al_2Si_3O_{12}$;
andradite, $Ca_3(Fe^{3+},Ti)_2Si_3O_{12}$;
uvarovite, $Ca_3Cr_2Si_3O_{12}$.
Varieties of garnet are used as gemstones and abrasives.

gas A state of matter in which the matter concerned occupies the whole of its container irrespective of its quantity. In an *ideal gas, which obeys the *gas laws exactly, the molecules themselves would have a negligible volume and negligible forces between them, and collisions between molecules would be perfectly elastic. In practice, however, the behaviour of real gases deviates from the gas laws because their molecules occupy a finite volume, there are small forces between molecules, and in polyatomic gases collisions are to a certain extent inelastic (*see* **equation of state**).

gas chromatography A technique for separating or analysing mixtures of gases by *chromatography. The apparatus consists of a very long tube containing the stationary phase. This may be a solid, such as kieselguhr (*gas–solid chromatography*, or *GSC*), or a nonvolatile liquid, such as a hydrocarbon oil coated on a solid support (*gas–liquid chromatography*, or *GLC*). The sample is often a volatile liquid mixture, which is vaporized and swept through the column by a carrier gas (e.g. hydrogen). The components of the mixture pass through the column at different rates because they adsorb to different extents on the stationary phase. They are detected as they leave, either by measuring the thermal conductivity of the gas or by a flame detector.

Gas chromatography is usually used for analysis; components can be identified by the time they take to pass through the column. It is sometimes also used for separating mixtures.

Gas chromatography is often used to separate a mixture into its components, which are then directly injected into a mass spectrometer. This technique is known as *gas chromatography–mass spectroscopy* or *GCMS*.

gas chromatography infrared (GC-IR) A form of *Fourier-transform infrared (FT-IR) used to identify small amounts of gas obtained by gas chromatography. Since functional groups in molecules are characteristic in infrared spectra this produces information to supplement that obtained by mass spectrometry. It is easier to perform FT-IR for gas chromatography than for liquid chromatography because carrier solvents in liquid chromatography absorb infrared radiation.

gas constant (universal molar gas constant) Symbol R. The constant that appears in the *universal gas equation* (*see* **gas laws**). It has the value 8.314 510(70) $J\,K^{-1}\,mol^{-1}$.

gas equation *See* **gas laws**.

gasification The conversion of solid or liquid hydrocarbons to fuel gas. Solid fuels such as coal or coke are converted into producer gas (carbon monoxide) or water gas (carbon monoxide and hydrogen) by the action of air (or oxygen) and steam. Solid fuels may also be hydrogenated to produce methane. Liquid fuels, from petroleum, are gasified to produce synthesis gas (carbon monoxide and hydrogen) or town gas (mostly hydrogen and methane), usually by *cracking or *hydrogenation.

gas laws Laws relating the temperature, pressure, and volume of an *ideal gas. *Boyle's law states that the pressure (p) of a specimen is inversely proportional to the volume (V) at constant temperature (pV =

constant). The modern equivalent of *Charles' law states that the volume is directly proportional to the thermodynamic temperature (T) at constant pressure (V/T = constant); originally this law stated the constant expansivity of a gas kept at constant pressure. The pressure law states that the pressure is directly proportional to the thermodynamic temperature for a specimen kept at constant volume. The three laws can be combined in the *universal gas equation*, $pV = nRT$, where n is the amount of gas in the specimen and R is the *gas constant. The gas laws were first established experimentally for real gases, although they are obeyed by real gases to only a limited extent; they are obeyed best at high temperatures and low pressures. *See also* **equation of state**.

gasohol A mixture of petrol (gasoline) and alcohol (i.e. typically ethanol at 10%, or methanol at 3%), used as an alternative fuel for cars and other vehicles in many countries. The ethanol is obtained as a *biofuel by fermentation of agricultural crops or crop residues, for example sugar cane waste. Many cars can also use a mixture of 85% ethanol and 15% petrol, called E85. Ethanol-based gasohol has a higher octane rating and burns more completely than conventional petrol, thus lowering some emissions. However, the ethanol can damage certain engine components, such as rubber seals. Methanol-based gasohol is more toxic and corrosive, and its emissions include formaldehyde, a known carcinogen.

gas oil A high-density petroleum fraction (between kerosene and lubricating oil), whose molecules have up to 25 carbon atoms. It is used as a domestic and industrial heating fuel.

gasoline *See* **petroleum**.

gas thermometer A device for measuring temperature in which the working fluid is a gas. It provides the most accurate method of measuring temperatures in the range 2.5 to 1337 K. Using a fixed mass of gas a *constant-volume thermometer* measures the pressure of a fixed volume of gas at relevant temperatures, usually by means of a mercury manometer and a barometer.

Gattermann reaction A variation of the *Sandmeyer reaction for preparing chloro- or bromoarenes by reaction of the diazonium compound. In the Gattermann reaction the aromatic amine is added to sodium nitrite and the halogen acid (10°C), then fresh copper powder (e.g. from Zn + $CuSO_4$) is added and the solution warmed. The diazonium salt then forms the haloarene, e.g.

$$C_6H_5N_2^+Cl^- \rightarrow C_6H_5Cl + N_2$$

The copper acts as a catalyst. The reaction is easier to perform than the Sandmeyer reaction and takes place at lower temperature, but generally gives lower yields. It was discovered in 1890 by the German chemist Ludwig Gattermann (1860–1920).

gauche *See* **conformation**; **torsion angle**.

Gay-Lussac, Joseph (1778–1850) French chemist and physicist whose discovery of the laws of chemical combination in gases helped to establish the atomic theory. It also led to *Avogadro's law. *See also* **Charles' law.**

Gay-Lussac's law 1. When gases combine chemically the volumes of the reactants and the volume of the product, if it is gaseous, bear simple relationships to each other when measured under the same conditions of temperature and pressure. The law was first stated in 1808 by J. L. *Gay-Lussac and led to *Avogadro's law. **2.** *See* **Charles' law.**

gaylussite A mineral consisting of a hydrated mixed carbonate of sodium and calcium, $Na_2CO_3.CaCO_3.5H_2O$.

GC-IR *See* **gas chromatography infrared.**

GCMS *See* **gas chromatography.**

Geiger counter **(Geiger–Müller counter)** A device used to detect and measure ionizing radiation. It consists of a tube containing a low-pressure gas (usually argon or neon with methane) and a cylindrical hollow cathode through the centre of which runs a fine-wire anode. A potential difference of about 1000 volts is maintained between the electrodes. An ionizing particle or photon passing through a window into the tube will cause an ion to be produced and the high p.d. will accelerate it towards its appropriate electrode, causing an avalanche of further ionizations by collision. The consequent current pulses can be counted in electronic circuits or simply amplified to work a small loudspeaker in the instrument. It was first devised in 1908 by the German physicist Hans Geiger (1882–1945). Geiger and W. Müller produced an improved design in 1928.

gel A lyophilic *olloid that has coagulated to a rigid or jelly-like solid. In a gel, the disperse medium has formed a loosely-held network of linked molecules through the dispersion medium. Examples of gels are silica gel and gelatin.

gelatin(e) A colourless or pale yellow water-soluble protein obtained by boiling collagen with water and evaporating the solution. It swells when water is added and dissolves in hot water to form a solution that sets to a gel on cooling. It is used in photographic emulsions and adhesives, and in jellies and other foodstuffs.

gel electrophoresis *See* **electrophoresis.**

gel filtration A type of column *chromatography in which a mixture of liquids is passed down a column containing a gel. Small molecules in the mixture can enter pores in the gel and move slowly down the column; large molecules, which cannot enter the pores, move more quickly. Thus, mixtures of molecules can be separated on the basis of their size. The technique is used particularly for separating proteins but it can also be applied to other polymers and to cell nuclei, viruses, etc.

gelignite A high explosive made from nitroglycerin, cellulose nitrate, sodium nitrate, and wood pulp.

gem Designating molecules in which two functional groups are attached to the same atom in a molecule. For example, 1,1-dichloroethane (CH_3CHCl_2) is a gem dihalide and can be named *gem*-dichloroethane. *Compare* **vicinal**.

geminate pair A pair of molecules, ions, etc., in close proximity surrounded by a solvent cage (*see* **cage effect**).

geochemistry The scientific study of the chemical composition of the earth. It includes the study of the abundance of the earth's elements and their isotopes and the distribution of the elements in environments of the earth (lithosphere, atmosphere, biosphere, and hydrosphere).

geometrical isomerism *See* **isomerism**.

gerade Symbol g. Describing a molecular orbital of a homonuclear diatomic molecule with *even* parity (*gerade* is the German word for even). This means that during the process of inversion, the sign of the orbital is unchanged upon going from any point in the molecule through the centre of inversion to the corresponding point on the other side. The symbol g is written as a subscript. The opposite of gerade is *ungerade*, symbol *u*. The symbols *g* and *u* are used to determine selection rules for diatomic molecules. These symbols are only applicable to homonuclear diatomic molecules, as heteronuclear diatomic molecules, such as CO, do not have a centre of inversion.

geraniol An alcohol, $C_9H_{15}CH_2OH$, present in a number of essential oils.

germanium Symbol Ge. A lustrous hard metalloid element belonging to group 14 (formerly IVB) of the periodic table; a.n. 32; r.a.m. 72.59; r.d. 5.36; m.p. 937°C; b.p. 2830°C. It is found in zinc sulphide and in certain other sulphide ores, and is mainly obtained as a by-product of zinc smelting. It is also present in some coal (up to 1.6%). Small amounts are used in specialized alloys but the main use depends on its semiconductor properties. Chemically, it forms compounds in the +2 and +4 oxidation states, the germanium(IV) compounds being the more stable. The element also forms a large number of organometallic compounds. Predicted in 1871 by *Mendeleev (eka-silicon), it was discovered by Winkler in 1886.

German silver (nickel silver) An alloy of copper, zinc, and nickel, often in the proportions 5:2:2. It resembles silver in appearance and is used in cheap jewellery and cutlery and as a base for silver-plated wire. *See also* **electrum**.

getter A substance used to remove small amounts of other substances from a system by chemical combination. For example, a metal such as magnesium may be used to remove the last traces of air when achieving a high vacuum. Various getters are also employed to remove impurities from semiconductors.

gibberellic acid (GA_3) A plant growth substance, abundant in young actively growing areas of the plant, that is involved in stem elongation. A *terpene, it was discovered in 1954. Gibberellic acid and related growth substances are called *gibberellins.*

Gibbs, Josiah Willard (1839–1903) US mathematician and physicist, who spent his entire academic career at Yale University. During the 1870s he developed the theory of chemical thermodynamics, devising functions such as Gibbs *free energy; he also derived the *phase rule. In mathematics he introduced vector notation.

Gibbs–Duhem equation An equation describing the relation between the chemical potentials of species in a mixture. If n_i is the amount of species i and μ_i is the chemical potential of species i, the Gibbs–Duhem equation states that

$$\sum_i n_i d\mu_i = 0$$

This equation implies that the chemical potentials for the species in a mixture do not change independently. Thus, in the case of a binary mixture, if the chemical potential of one species increases, the chemical potential of the other species must decrease. The equation was derived independently by J. W. *Gibbs and the French physicist P. Duhem (1861–1916).

Gibbs free energy (Gibbs function) *See* **free energy.**

Gibbs–Helmholtz equation An equation used in thermodynamics to show the temperature dependence of the *Gibbs free energy. It has the form:

$$(\partial G/\partial T)_p = (G - H)/T,$$

where G is the Gibbs free energy, H is the enthalpy, T is the thermodynamic temperature, and p is the pressure (which is held constant). The Gibbs–Helmholtz equation can be derived from $(\partial G/\partial T)_p = -S$ and $S = (H - G)T$ using the rules of differentiation. The equation was derived by J. W. *Gibbs and H. L. F. von *Helmholtz.

gibbsite A mineral form of hydrated *aluminium hydroxide ($Al(OH)_3$). It is named after the US mineralogist George Gibbs (d. 1833).

giga- Symbol G. A prefix used in the metric system to denote one thousand million times. For example, 10^9 joules = 1 gigajoule (GJ).

gilbert Symbol Gb. The c.g.s. unit of *magnetomotive force equal to $10/4\pi$ (= 0.795 77) ampere-turn. It is named after the English physician and physicist William Gilbert (1544–1603), who studied magnetism.

glacial ethanoic acid *See* **ethanoic acid.**

glass Any noncrystalline solid; i.e. a solid in which the atoms are random and have no long-range ordered pattern. Glasses are often regarded as supercooled liquids. Characteristically they have no definite melting point, but soften over a range of temperatures.

The common glass used in windows, bottles, etc., is *soda glass*, which is made by heating a mixture of lime (calcium oxide), soda (sodium carbonate), and sand (silicon(IV) oxide). It is a form of calcium silicate. Borosilicate glasses (e.g. *Pyrex*) are made by incorporating some boron oxide, so that silicon atoms are replaced by boron atoms. They are tougher than soda glass and more resistant to temperature changes, hence their use in cooking utensils and laboratory apparatus. Glasses for special purposes (e.g. optical glass) have other elements added (e.g. barium, lead). *See also* **spin glass.**

glass electrode A type of *half cell having a glass bulb containing an acidic solution of fixed pH, into which dips a platinum wire. The glass bulb is thin enough for hydrogen ions to diffuse through. If the bulb is placed in a solution containing hydrogen ions, the electrode potential depends on the hydrogen-ion concentration. Glass electrodes are used in pH measurement.

glass fibres Melted glass drawn into thin fibres some 0.005 mm–0.01 mm in diameter. The fibres may be spun into threads and woven into fabrics, which are then impregnated with resins to give a material that is both strong and corrosion resistant. It is used in car bodies, boat building, and similar applications.

glauberite A mineral consisting of a mixed sulphate of sodium and calcium, $Na_2SO_4.CaSO_4$.

Glauber's salt *Sodium sulphate decahydrate, $Na_2SO_4.10H_2O$, used as a laxative. It is named after Johann Glauber (1604–68).

GLC (gas–liquid chromatography) *See* **gas chromatography**.

global warming *See* **greenhouse effect**.

globin *See* **haemoglobin**.

globular protein *See* **protein**.

globulin Any of a group of globular proteins that are generally insoluble in water and present in blood, eggs, milk, and as a reserve protein in seeds. Blood serum globulins comprise four types: α_1-, α_2-, and β-globulins, which serve as carrier proteins; and γ-globulins, which include the immunoglobulins responsible for immune responses.

glove box A metal box that has gloves fitted to ports in its walls. It is used to manipulate mildly radioactive materials and in laboratory techniques in which an inert, sterile, dry, or dust-free atmosphere has to be maintained.

glow discharge An electrical discharge that passes through a gas at low pressure and causes the gas to become luminous. The glow is produced by the decay of excited atoms and molecules.

gluconic acid An optically active hydroxycarboxylic acid,

$CH_2(OH)(CHOH)_4COOH$. It is the carboxylic acid corresponding to the aldose sugar glucose, and can be made by the action of certain moulds.

glucosan Any one of a class of *polysaccharide compounds that can be converted to glucose by hydrolysis. Glucosans include dextrin, starch, and cellulose.

glucose (dextrose; grape sugar) A white crystalline sugar, $C_6H_{12}O_6$, occurring widely in nature. Like other *monosaccharides, glucose is optically active: most naturally occurring glucose is dextrorotatory. Glucose and its derivatives are crucially important in the energy metabolism of living organisms. Glucose is also a constituent of many polysaccharides, most notably starch and cellulose. These yield glucose when broken down, for example by enzymes during digestion.

glucuronic acid A compound, $OC_6H_9O_6$, derived from the oxidation of glucose. It is an important constituent of gums and mucilages. Glucuronic acid can combine with hydroxyl (–OH), carboxyl (–COOH), or amino ($-NH_2$) groups to form a *glucuronide*. The addition of a glucuronide group to a molecule (*glucuronidation*) generally increases the solubility of a compound; hence glucuronidation plays an important role in the excretion of foreign substances.

glucuronide *See* **glucuronic acid.**

glutamic acid *See* **amino acid.**

glutamine *See* **amino acid.**

glyceraldehyde 3-phosphate (GALP) A triose phosphate, $CHOCH(OH)CH_2OPO_3H_2$, that is an intermediate in the *Calvin cycle (*see also* **photosynthesis**) and glycolysis.

glycerate 3-phosphate A phosphorylated three-carbon monosaccharide that is an intermediate in the *Calvin cycle of photosynthesis and also in *glycolysis. It was formerly known as *3-phosphoglycerate* or *phosphoglyceric acid (PGA).*

glyceride (acylglycerol) A fatty-acid ester of glycerol. Esterification can occur at one, two, or all three hydroxyl groups of the glycerol molecule producing mono-, di-, and triglycerides respectively. *Triglycerides are the major constituent of fats and oils found in living organisms. Alternatively, one of the hydroxyl groups may be esterified with a phosphate group forming a phosphoglyceride (*see* **phospholipid**) or to a sugar forming a *glycolipid.

glycerine *See* **glycerol.**

glycerol (glycerine; propane-1,2,3,-triol) A trihydric alcohol, $HOCH_2CH(OH)CH_2OH$. Glycerol is a colourless sweet-tasting viscous liquid, miscible with water but insoluble in ether. It is widely distributed in all living organisms as a constituent of the *glycerides, which yield glycerol when hydrolysed.

glycerophospholipids *See* **phospholipid.**

glycine *See* **amino acid.**

glycobiology The study of carbohydrates and carbohydrate complexes, especially *glycoproteins.

glycogen (animal starch) A *polysaccharide consisting of a highly branched polymer of glucose occurring in animal tissues, especially in liver and muscle cells. It is the major store of carbohydrate energy in animal cells and is present as granular clusters of minute particles.

glycogenesis The conversion of glucose to glycogen, which is stimulated by insulin from the pancreas. Glycogenesis occurs in skeletal muscles and to a lesser extent in the liver. Glucose that is taken up by cells is phosphorylated to glucose 6-phosphate; this is converted successively to glucose 1-phosphate, uridine diphosphate glucose, and finally to glycogen. *Compare* **glycogenolysis.**

glycogenolysis The conversion of glycogen to glucose, which occurs in the liver and is stimulated by glucagon from the pancreas and adrenaline from the adrenal medulla. These hormones activate an enzyme that phosphorylates glucose molecules in the glycogen chain to form glucose 1-phosphate, which is converted to glucose 6-phosphate. This is then converted to glucose by a phosphatase enzyme. In skeletal muscle glycogen is degraded to glucose 6-phosphate, which is then converted into pyruvate and used in ATP production during glycolysis and the Krebs cycle. However, pyruvate can also be converted, in the liver, to glucose; thus muscle glycogen is indirectly a source of blood glucose. *Compare* **glycogenesis.**

glycol *See* **ethane-1,2-diol.**

glycolic acid (hydroxyethanoic acid) A colourless crystalline compound, $CH_2(OH)COOH$; m.p. 80°C. It occurs in sugar cane and sugar beet, and is made by the electrolytic reduction of oxalic acid or by boiling a solution of sodium monochloroethanoate. Glycolic acid is used in making textiles and leather and for cleaning metals.

glycolipid Any of a group of sugar-containing lipids, in which the lipid portion of the molecule is usually based on glycerol (*see* **glyceride**) or sphingosine and the sugar is typically galactose, glucose, or inositol. Glycolipids are components of biological membranes. In animal plasma membranes they are found in the outer layer of the lipid bilayer; the simplest animal glycolipids are the *cerebrosides. Plant glycolipids are glycerides in which the sugar group is most commonly galactose. They are the principal lipid constituents of chloroplasts.

glycolysis (Embden–Meyerhof pathway) The series of biochemical reactions in which glucose is broken down to pyruvate with the release of usable energy in the form of *ATP. One molecule of glucose undergoes two phosphorylation reactions and is then split to form two triose-phosphate molecules. Each of these is converted to pyruvate. The net energy yield is two ATP molecules per glucose molecule. In aerobic respiration pyruvate then enters the *Krebs cycle. Alternatively, when oxygen is in short supply or absent, the pyruvate is converted to various products by anaerobic respiration. Other simple sugars, e.g. fructose and galactose, and glycerol (from fats) enter the glycolysis pathway at intermediate stages.

glycoprotein A carbohydrate linked covalently to a protein. Formed in the Golgi apparatus in the process of *glycosylation, glycoproteins are important components of cell membranes. They are also constituents of body fluids, such as mucus, that are involved in lubrication. Many of the hormone receptors on the surfaces of cells have been identified as glycoproteins. Glycoproteins produced by viruses attach themselves to the surface of the host cell, where they act as markers for the receptors of leucocytes. Viral glycoproteins can also act as target molecules and help viruses to detect certain types of host cell; for example, a glycoprotein on the surface of HIV (the AIDS virus) enables the virus to find and infect white blood cells.

glycosaminoglycan Any one of a group of polysaccharides that contain amino sugars (such as glucosamine). Formerly known as *mucopolysaccharides*, they include *hyaluronic acid and chondroitin, which provide lubrication in joints and form part of the matrix of cartilage. The three-dimensional structure of these molecules enables them to trap water, which forms a gel and gives glycosaminoglycans their elastic properties.

glycoside Any one of a group of compounds consisting of a pyranose sugar residue, such as glucose, linked to a noncarbohydrate residue (R) by a *glycosidic bond: the hydroxyl group (–OH) on carbon-1 of the sugar is replaced by –OR. Glycosides are widely distributed in plants; examples are the *anthocyanin pigments and the *cardiac glycosides*, such as digoxin and ouabain, which are used medicinally for their stimulant effects on the heart.

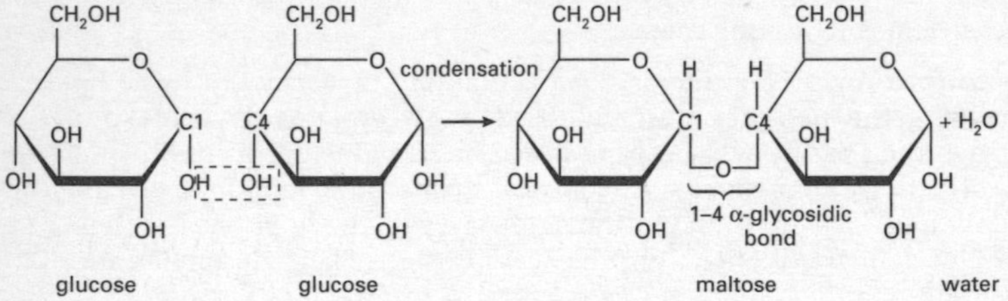

Formation of a glycosidic bond

glycosidic bond (glycosidic link) The type of chemical linkage between the monosaccharide units of disaccharides, oligosaccharides, and polysaccharides, which is formed by the removal of a molecule of water (i.e. a *condensation reaction). The bond is normally formed between the carbon-1 on one sugar and the carbon-4 on the other. An α-glycosidic bond is formed when the –OH group on carbon-1 is below the plane of the glucose ring and a β-glycosidic bond is formed when it is above the plane. Cellulose is formed of glucose molecules linked by 1-4 β-glycosidic bonds, whereas starch is composed of 1-4 α-glycosidic bonds.

glycosylation The process in which a carbohydrate is joined to another molecule, such as a protein to form a *glycoprotein or to a lipid to form a glycolipid (*see* **glyceride**). Glycosylation occurs in the rough endoplasmic reticulum and the Golgi apparatus of cells.

glyoxal (ethanedial) A low melting-point yellow crystalline solid, CHOCHO, m.p. 15°C. It is a dialdehyde. It gives off a green vapour when heated and burns with a violet-coloured flame. Glyoxal is made by the catalytic oxidation of ethanediol and is used to harden the gelatin employed for photographic emulsions.

goethite A yellow-brown mineral, FeO.OH, crystallizing in the orthorhombic system. It is formed as a result of the oxidation and hydration of iron minerals or as a direct precipitate from marine or fresh water (e.g. in swamps and bogs). Most *limonite is composed largely of cryptocrystalline goethite. Goethite is mined as an ore of iron

Golay cell A transparent cell containing gas, used to detect *infrared radiation. Incident radiation is absorbed within the cell, causing a rise in the gas temperature and pressure. The amount of incident radiation can be measured from the pressure rise in the tube.

gold Symbol Au. A soft yellow malleable metallic *transition element; a.n. 79; r.a.m. 196.967; r.d. 19.32; m.p. 1064.43°C; b.p. 2807±2°C. Gold has a face-centred-cubic crystal structure. It is found as the free metal in gravel or in quartz veins, and is also present in some lead and copper sulphide ores. It also occurs combined with silver in the telluride sylvanite, $(Ag,Au)Te_2$. It is used in jewellery, dentistry, and electronic devices. Chemically, it is unreactive, being unaffected by oxygen. It reacts with chlorine at 200°C to form gold(III) chloride. It forms a number of complexes with gold in the +1 and +3 oxidation states.

Goldschmidt process A method of extracting metals by reducing the oxide with aluminium powder, e.g.

$$Cr_2O_3 + 2Al \rightarrow 2Cr + Al_2O_3$$

The reaction can also be used to produce molten iron (*see* **thermite**). It was discovered by the German chemist Hans Goldschmidt (1861–1923).

Gooch crucible A porcelain dish with a perforated base over which a layer of asbestos is placed, used for filtration in gravimetric analysis. It is named after the US chemist Frank Gooch (1852–1929).

Gouy–Chapman model A model of the *electrical double layer in which thermal motion causing disordering of the layer is taken into account. This is very similar to the *Debye–Hückel theory of the ionic atmosphere around an ion, except that the concept of a central single ion is replaced by that of an infinite plane electrode. The Gouy–Chapman model understates the structure in a double layer, but can be improved by the *Stern model*, in which the ions closest to the electrode are ordered and the ions are described by the Gouy–Chapman model outside the first layer.

graft copolymer *See* **polymer**.

Graham, Thomas (1805–69) Scottish chemist, who became professor of chemistry at Glasgow University in 1830, moving to University College, London, in 1837. His 1829 paper on gaseous diffusion introduced *Graham's law. He went on to study diffusion in liquids, leading in 1861 to the definition of *colloids.

Graham's law The rates at which gases diffuse is inversely proportional to the square roots of their densities. This principle is made use of in the diffusion method of separating isotopes. The law was formulated in 1829 by Thomas *Graham.

gram Symbol g. One thousandth of a kilogram. The gram is the fundamental unit of mass in *c.g.s. units and was formerly used in such units as the *gram-atom*, *gram-molecule*, and *gram-equivalent*, which have now been replaced by the *mole.

grape sugar *See* **glucose**.

graphite *See* **carbon**.

graphitic compounds Substances in which atoms or molecules are trapped between the layers in graphite. *See* **lamellar solids**.

gravimetric analysis A type of quantitative analysis that depends on weighing. For instance, the amount of silver in a solution of silver salts could be measured by adding excess hydrochloric acid to precipitate silver chloride, filtering the precipitate, washing, drying, and weighing.

gray Symbol Gy. The derived SI unit of absorbed dose of ionizing radiation (*see* **radiation units**). It is named after the British radiobiologist L. H. Gray (1905–65).

greenhouse effect An effect occurring in the atmosphere because of the presence of certain gases (*greenhouse gases*) that absorb infrared radiation. Light and ultraviolet radiation from the sun are able to penetrate the atmosphere and warm the earth's surface. This energy is re-radiated as infrared radiation, which, because of its longer wavelength, is absorbed by such substances as carbon dioxide. Emissions of carbon dioxide from human activities have increased markedly in the last 150 years or so. The overall effect is that the average temperature of the earth and its atmosphere is increasing (so-called *global warming*). The effect is similar to that occurring in a greenhouse, where light and long-wavelength ultraviolet radiation can pass through the glass into the greenhouse but the infrared radiation is absorbed by the glass and part of it is re-radiated into the greenhouse.

The greenhouse effect is seen as a major environmental hazard. Average increases in temperature are likely to change weather patterns and agricultural output. It is already causing the polar ice caps to melt, with a corresponding rise in sea level. Carbon dioxide, from fossil-fuel power stations and car exhausts, is the main greenhouse gas. Other contributory pollutants are nitrogen oxides, ozone, methane, and chlorofluorocarbons.

greenhouse gas *See* **greenhouse effect.**

greenockite A mineral form of cadmium sulphide, CdS.

green vitriol *See* **iron(II) sulphate.**

Grignard reagents A class of organometallic compounds of magnesium, with the general formula RMgX, where R is an organic group and X a halogen atom (e.g. CH_3MgCl, C_2H_5MgBr, etc.). They actually have the structure $R_2Mg.MgCl_2$, and can be made by reacting a haloalkane with magnesium in ether; they are rarely isolated but are extensively used in organic synthesis, when they are made in one reaction mixture. Grignard reagents have a number of reactions that make them useful in organic synthesis. With methanal they give a primary alcohol

$$CH_3MgCl + HCHO \rightarrow CH_3CH_2OH$$

Other aldehydes give a secondary alcohol

$$CH_3CHO + CH_3MgCl \rightarrow (CH_3)_2CHOH$$

With alcohols, hydrocarbons are formed

$$CH_3MgCl + C_2H_5OH \rightarrow C_2H_5CH_3$$

Water also gives a hydrocarbon

$$CH_3MgCl + H_2O \rightarrow CH_4$$

The compounds are named after their discoverer, the French chemist Victor Grignard (1871–1935).

Grotrian diagram A diagram that summarizes the energy levels and the *allowed transitions between these energy levels in an atom. The energy is plotted vertically with a horizontal line for each energy level. The intensity of the transition can be represented on a Grotrian diagram by allowing the thickness of the line to represent the transition proportional to the intensity. Grotrian diagrams are named after the German spectroscopist W. Grotrian, who invented them in 1928.

Grottius–Draper law A law in photochemistry stating that only the light absorbed by a substance or substances is effective in bringing about chemical change. Not all the light falling on the substances will necessarily bring about chemical change, since some of it can be re-emitted in the form of heat or light. The light does not need to be absorbed directly by the reacting substances; it is possible, in photosensitization for example, that light can be absorbed by an inert substance, which subsequently transfers the absorbed energy (as thermal energy) to the reactants.

ground state The lowest stable energy state of a system, such as a molecule, atom, or nucleus. *See* **energy level**.

group 1. *See* **periodic table**. **2.** A mathematical structure consisting of a set of elements *A*, *B*, *C*, etc., for which there exists a law of composition, referred to as 'multiplication'. Any two elements can be combined to give a 'product' *AB*.
(1) Every product of two elements is an element of the set.
(2) The operation is associative, i.e. $A(BC) = (AB)C$.
(3) The set has an element *I*, called the *identity element*, such that $IA = AI = A$ for all *A* in the set.
(4) Each element of the set has an *inverse* A^{-1} belonging to the set such that $AA^{-1} = A^{-1}A = I$.

Although the law of combination is called 'multiplication' this does not necessarily have its usual meaning. For example, the set of integers forms a group if the law of composition is addition.

Two elements *A*, *B* of a group *commute* if $AB = BA$. If all the elements of a group commute with each other the group is said to be *Abelian*. If this is not the case the group is said to be *non-Abelian*.

The interest of group theory in physics and chemistry is in analysing symmetry. *Discrete groups* have a finite number of elements, such as the symmetries involved in rotations and reflections of molecules, which give rise to *point groups*. *Continuous groups* have an infinite number of elements where the elements are continuous. An example of a continuous group is the set of rotations about a fixed axis. The *rotation group* thus formed underlies the quantum theory of angular momentum, which has many applications to atoms and nuclei.

group 0 elements *See* **noble gases**.

group 1 elements A group of elements in the *periodic table: lithium (Li), sodium (Na), potassium (K), rubidium (Rb), caesium (Cs), and Francium (Fr). They are known as the *alkali metals. Formerly, they were classified in group I, which consisted of two subgroups: group IA (the main group) and group IB. Group IB consisted of the *coinage metals, copper, silver, and gold, which comprise group 11 and are usually considered with the *transition elements.

group 2 elements A group of elements in the *periodic table: beryllium (Be), magnesium (Hg), calcium (Ca), strontium (Sr), barium (Ba), and radium (Ra). They are known as the *alkaline-earth metals. Formerly, they were classified in group II, which consisted of two subgroups: group IIA (the main group, *see* **alkaline-earth metals**) and group IIB. Group IIB consisted of the three metals zinc (Zn), cadmium (Cd), and mercury (Hg), which have two *s*-electrons outside filled *d*-subshells. Moreover, none of their compounds have unfilled *d*-levels, and the metals are regarded as nontransition elements. They now form group 12 and are sometimes called the *zinc group*. Zinc and cadmium are relatively electropositive metals, forming compounds containing divalent ions Zn^{2+} or Cd^{2+}. Mercury is more unreactive and also unusual in forming mercury(I) compounds, which contain the ion Hg_2^{2+}.

groups 3–12 *See* **transition elements.**

group 13 elements A group of elements in the *periodic table: boron (B), aluminium (Al), gallium (Ga), indium (In), and thallium (Tl), which all have outer electronic configurations ns^2np^1 with no partly filled inner levels. They are the first members of the *p*-block. The group differs from the alkali metals and alkaline-earth metals in displaying a considerable variation in properties as the group is descended. Formerly, they were classified in group III, which consisted of two subgroups: group IIIB (the main group) and group IIIA. Group IIIA consisted of scandium (Sc), yttrium (Yt), and lanthanum (La), which are generally considered with the *lanthanoids, and actinium (Ac), which is classified with the *actinoids. Scandium and yttrium now belong to group 3 (along with lutetium and lawrencium).

Boron has a small atomic radius and a relatively high ionization energy. In consequence its chemistry is largely covalent and it is generally classed as a metalloid. It forms a large number of volatile hydrides, some of which have the uncommon bonding characteristic of *electron-deficient compounds. It also forms a weakly acidic oxide. In some ways, boron resembles silicon (*see* **diagonal relationship**).

As the group is descended, atomic radii increase and ionization energies are all lower than for boron. There is an increase in polar interactions and the formation of distinct M^{3+} ions. This increase in metallic character is clearly illustrated by the increasing basic character of the hydroxides: boron hydroxide is acidic, aluminium and gallium hydroxides are amphoteric, indium hydroxide is basic, and thallium forms only the oxide. As the elements of group 13 have a vacant *p*-orbital they display many electron-acceptor properties. For example, many boron compounds form adducts with donors such as ammonia and organic amines (acting as Lewis acids). A large number of complexes of the type $[BF_4]^-$, $[AlCl_4]^-$, $[InCl_4]^-$, $[TlI_4]^-$ are known and the heavier members can expand their coordination numbers to six as in $[AlF_6]^{3-}$ and $[TlCl_6]^{3-}$. This acceptor property is also seen in bridged dimers of the type Al_2Cl_6. Another feature of group 13 is the increasing stability of the monovalent state down the group. The electron configuration ns^2np^1 suggests that only one electron could be lost or shared in forming compounds. In fact, for the lighter members of the group the energy required to promote an electron from the *s*-subshell to a vacant *p*-subshell is small. It is more than compensated for by the resulting energy gain in forming three bonds rather than one. This energy gain is less important for the heavier members of the group. Thus, aluminium forms compounds of the type AlCl in the gas phase at high temperatures. Gallium similarly forms such compounds and gallium(I) oxide (Ga_2O) can be isolated. Indium has a number of known indium(I) compounds (e.g. InCl, In_2O, $In_3^{I}[In^{III}Cl_6]$). Thallium has stable monovalent compounds. In aqueous solution, thallium(I) compounds are more stable than the corresponding thallium(III) compounds. *See* **inert-pair effect**.

group 14 elements A group of elements in the *periodic table: carbon (C), silicon (Si), germanium (Ge), tin (Sn), and lead (Pb), which all have outer electronic configurations ns^2np^2 with no partly filled inner levels. Formerly, they were classified in group IV, which consisted of two subgroups: IVB (the main group) and group IVA. Group IVA consisted of titanium (Ti), zirconium (Zr), and hafnium (Hf), which now form group 4 and are generally considered with the *transition elements.

The main valency of the elements is 4, and the members of the group show a variation from nonmetallic to metallic behaviour in moving down the group. Thus, carbon is a nonmetal and forms an acidic oxide (CO_2) and a neutral oxide. Carbon compounds are mostly covalent. One allotrope (diamond) is an insulator, although graphite is a fairly good conductor. Silicon and germanium are metalloids, having semiconductor properties. Tin is a metal, but does have a nonmetallic allotrope (grey tin). Lead is definitely a metal. Another feature of the group is the tendency to form divalent compounds as the size of the atom increases. Thus carbon has only the highly reactive carbenes. Silicon forms analogous silylenes. Germanium has an unstable hydroxide ($Ge(OH)_2$), a sulphide (GeS), and halides. The sulphide and halides disproportionate to germanium and the germanium(IV) compound. Tin has a number of tin(II) compounds, which are moderately reducing, being oxidized to the tin(IV) compound. Lead has a stable lead(II) state. *See* **inert-pair effect**.

In general, the reactivity of the elements increases down the group from carbon to lead. All react with oxygen on heating. The first four form the dioxide; lead forms the monoxide (i.e. lead(II) oxide, PbO). Similarly, all will react with chlorine to form the tetrachloride (in the case of the first four) or the dichloride (for lead). Carbon is the only one capable of reacting directly with hydrogen. The hydrides all exist from the stable methane (CH_4) to the unstable plumbane (PbH_4).

group 15 elements A group of elements in the *periodic table: nitrogen (N), phosphorus (P), arsenic (As), antimony (Sb), and bismuth (Bi), which all have outer electronic configurations ns^2np^3 with no partly filled inner levels. Formerly, they were classified in group V, which consisted of two subgroups: group VB (the main group) and group VA. Group VA consisted of vanadium (V), niobium (Nb), and tantalum (Ta), which are generally considered with the *transition elements:

The lighter elements (N and P) are nonmetals; the heavier elements are metalloids. The lighter elements are electronegative in character and have fairly large ionization energies. Nitrogen has a valency of 3 and tends to form covalent compounds. The other elements have available *d*-sublevels and can promote an *s*-electron into one of these to form compounds with the V oxidation state. Thus, they have two oxides P_2O_3, P_2O_5, Sb_2O_3, Sb_2O_5, etc. In the case of bismuth, the pentoxide Bi_2O_5 is difficult to prepare and unstable – an example of the increasing stability of the III oxidation state in going from phosphorus to bismuth. The oxides also show how there is increasing metallic (electropositive) character down the group. Nitrogen and phosphorus have oxides that are either neutral (N_2O, NO) or acidic. Bismuth trioxide (Bi_2O_3) is basic. Bismuth is the only member of the group that forms a well-characterized positive ion Bi^{3+}.

group 16 elements A group of elements in the *periodic table: oxygen (O), sulphur (S), selenium (Se), tellurium (Te), and polonium (Po), which all have outer electronic configurations ns^2np^4 with no partly filled inner levels. They are also called the *chalcogens*. Formerly, they were classified in group VI, which consisted of two subgroups: group VIB (the main group) and group VIA. Group VIA consisted of chromium (Cr), molybdenum (Mo), and tungsten (W), which now form group 6 are generally classified with the *transition elements.

The configurations are just two electrons short of the configuration of a noble gas and the elements are characteristically electronegative and almost entirely nonmetallic. Ionization energies are high, (O 1314 to Po 813 kJ mol^{-1}) and monatomic cations are not known. Polyatomic cations do exist, e.g. O_2^+, S_8^{2+}, Se_8^{2+}, Te_4^{2+}. Electronegativity decreases down the group but the nearest approach to metallic character is the occurrence of 'metallic' allotropes of selenium, tellurium, and polonium along with some metalloid properties, in particular, marked photoconductivity. The elements of group 16 combine with a wide range of other elements and the bonding is largely covalent. The elements all form hydrides of the type XH_2. Apart from water, these materials are all toxic foul-smelling gases; they show decreasing thermal stability with increasing relative atomic mass of X. The hydrides dissolve in water to give very weak acids (acidity increases down the group). Oxygen forms the additional hydride H_2O_2 (hydrogen peroxide), but sulphur forms a range of sulphanes, such as H_2S_2, H_2S_4, H_2S_6.

Oxygen forms the fluorides O_2F_2 and OF_2, both powerful fluorinating agents; sulphur forms analogous fluorides along with some higher fluorides, S_2F_2, SF_2, SF_4, SF_6, S_2F_{10}. Selenium and tellurium form only the higher fluorides MF_4 and MF_6; this is in contrast to the formation of lower valence states by heavier elements observed in groups 13, 14, and 15. The chlorides are limited to M_2Cl_2 and MCl_4; the bromides are similar except that sulphur only forms S_2Br_2. All metallic elements form oxides and sulphides and many form selenides.

group 17 elements A group of elements in the *periodic table: fluorine (F), chlorine (Cl), bromine (Br), iodine (I), and astatine (At). They are known as the *halogens. Formerly, they were classified in group VII, which consisted of two subgroups: group VIIB (the main group) and group VIIA. Group VIIA consisted of the elements manganese (Mn), technetium (Te), and rhenium (Re), which now form group 7 and are usually considered with the transition elements.

group 18 elements A group of elements in the *periodic table: helium (He), neon (Ne), argon (Ar), krypton (Kr), xenon (Xe), and radon (Rn).

Formerly classified as *group 0 elements*, they are usually referred to as the *noble gases.

group representation A group of mathematical objects that is homomorphic to (i.e. has the same mathematical structure as) the original group. In particular, group representations made up of square matrices are of interest. The dimension of the representation is the number of rows or columns of the matrix. The *irreducible representations of a group, i.e. the representations that cannot be expressed in terms of lower-dimensional representations, are of great importance in quantum mechanics since the energy levels of a quantum mechanical system are labelled by the irreducible representations of the symmetry group of the system. This enables *selection rules for the system to be derived.

GSC (gas–solid chromatography) *See* **gas chromatography**.

guanidine A crystalline basic compound $HN{:}C(NH_2)_2$, related to urea.

guanine A *purine derivative. It is one of the major component bases of *nucleotides and the nucleic acids *DNA and *RNA.

gum Any of a variety of substances obtained from plants. Typically they are insoluble in organic solvents but form gelatinous or sticky solutions with water. Gum resins are mixtures of gums and natural resins. Gums are produced by the young xylem vessels of some plants (mainly trees) in response to wounding or pruning. The exudate hardens when it reaches the plant surface and thus provides a temporary protective seal while the cells below divide to form a permanent repair. Excessive gum formation is a symptom of some plant diseases.

guncotton *See* **cellulose nitrate**.

gun metal A type of bronze usually containing 88–90% copper, 8–10% tin, and 2–4% zinc. Formerly used for cannons, it is still used for bearings and other parts that require high resistance to wear and corrosion.

gunpowder An explosive consisting of a mixture of potassium nitrate, sulphur, and charcoal.

gypsum A monoclinic mineral form of hydrated *calcium sulphate, $CaSO_4.2H_2O$. It occurs in five varieties: *rock gypsum*, which is often red stained and granular; *gypsite*, an impure earthy form occurring as a surface deposit; *alabaster*, a pure fine-grained translucent form; *satin spar*, which is fibrous and silky; and *selenite*, which occurs as transparent crystals in muds and clays. It is used in the building industry and in the manufacture of cement, rubber, paper, and plaster of Paris.

gyromagnetic ratio Symbol γ. The ratio of the angular momentum of an atomic system to its magnetic moment. The inverse of the gyromagnetic ratio is called the *magnetomechanical ratio*.

Haber, Fritz (1868–1934) German chemist who worked at the Karlsruhe Technical Institute, where he perfected the *Haber process for making ammonia in 1908. As a Jew, he left Germany in 1933 to go into exile in Britain, working in Cambridge at the Cavendish Laboratory. For his Haber process, he was awarded the 1918 Nobel Prize for chemistry.

Haber process An industrial process for producing ammonia by reaction of nitrogen with hydrogen:

$$N_2 + 3H_2 \rightleftharpoons 2NH_3$$

The reaction is reversible and exothermic, so that a high yield of ammonia is favoured by low temperature (*see* **Le Chatelier's principle**). However, the rate of reaction would be too slow for equilibrium to be reached at normal temperatures, so an optimum temperature of about 450°C is used, with a catalyst of iron containing potassium and aluminium oxide promoters. The higher the pressure the greater the yield, although there are technical difficulties in using very high pressures. A pressure of about 250 atmospheres is commonly employed.

The process is of immense importance for the fixation of nitrogen for fertilizers. It was developed in 1908 by Fritz *Haber and was developed for industrial use by Carl Bosch (1874–1940), hence the alternative name *Haber–Bosch process.* The nitrogen is obtained from liquid air. Formerly, the hydrogen was from *water gas and the water–gas shift reaction (the *Bosch process*) but now the raw material (called *synthesis gas*) is obtained by steam *reforming natural gas.

habit *See* **crystal**.

haem (heme) An iron-containing molecule that binds with proteins as a *cofactor or *prosthetic group to form the *haemoproteins.* These are *haemoglobin, *myoglobin, and the *cytochromes. Essentially, haem comprises a *porphyrin with its four nitrogen atoms holding the iron(II) atom as a chelate. This iron can reversibly bind oxygen (as in haemoglobin and myoglobin) or (as in the cytochromes) conduct electrons by conversion between the iron(II) and iron(III) series.

haematite A mineral form of iron(III) oxide, Fe_2O_3. It is the most important ore of iron and usually occurs in two main forms: as a massive red kidney-shaped ore (*kidney ore*) and as grey to black metallic crystals known as *specular iron ore.* Haematite is the major red colouring agent in rocks; the largest deposits are of sedimentary origin. In industry haematite is also used as a polishing agent (jeweller's rouge) and in paints.

haemoerythrin A red iron-containing respiratory pigment that occurs in the blood of annelids and some other invertebrates. Its structure is essentially the same as that of *haemoglobin except the prosthetic group has a different chemical composition.

haemoglobin One of a group of globular proteins occurring widely in animals as oxygen carriers in blood. Vertebrate haemoglobin comprises two pairs of polypeptide chains, known as α-chains and β-chains (forming the *globin* protein), with each chain folded to provide a binding site for a *haem group. Each of the four haem groups binds one oxygen molecule to form *oxyhaemoglobin*. Dissociation occurs in oxygen-depleted tissues: oxygen is released and haemoglobin is reformed. The haem groups also bind other inorganic molecules, including carbon monoxide (to form carboxyhaemoglobin). In vertebrates, haemoglobin is contained in the red blood cells (erythrocytes).

haemoglobinic acid A very weak acid formed inside red blood cells when hydrogen ions combine with haemoglobin. The presence of the hydrogen ions, which are produced by the dissociation of carbonic acid, encourages oxyhaemoglobin to dissociate into haemoglobin and oxygen. The oxygen diffuses into the tissue cells and the haemoglobin acts as a *buffer for the excess hydrogen ions, which it takes up to form haemoglobinic acid.

hafnium Symbol Hf. A silvery lustrous metallic *transition element; a.n. 72; r.a.m. 178.49; r.d. 13.3; m.p. 2227±20°C; b.p. 4602°C. The element is found with zirconium and is extracted by formation of the chloride and reduction by the Kroll process. It is used in tungsten alloys in filaments and electrodes and as a neutron absorber. The metal forms a passive oxide layer in air. Most of its compounds are hafnium(IV) complexes; less stable hafnium(III) complexes also exist. The element was first reported by Urbain in 1911, and its existence was finally established by Dirk Coster (1889–1950) and George de Hevesey (1885–1966) in 1923.

Hahn, Otto (1879–1968) German chemist, who studied in London (with William *Ramsay) and Canada (with Ernest *Rutherford) before returning to Germany in 1907. In 1917, together with Lise *Meitner, he discovered protactinium. In the late 1930s he collaborated with Fritz Strassmann (1902–) and in 1938 bombarded uranium with slow neutrons. Among the products was barium, but it was Meitner (now in Sweden) who the next year interpreted the process as nuclear fission. In 1944 Hahn received the Nobel Prize for chemistry.

hahnium *See* **transactinide elements**.

half cell An electrode in contact with a solution of ions, forming part of a *cell. Various types of half cell exist, the simplest consisting of a metal electrode immersed in a solution of metal ions. Gas half cells have a gold or platinum plate in a solution with gas bubbled over the metal plate. The commonest is the *hydrogen half cell. Half cells can also be formed by a metal in contact with an insoluble salt or oxide and a solution. The *calomel half cell is an example of this. Half cells are commonly referred to as *electrodes.*

half chair *See* **ring conformations.**

half-life *See* **decay.**

half sandwich *See* **sandwich compound.**

half-thickness The thickness of a specified material that reduces the intensity of a beam of radiation to half its original value.

half-width Half the width of a spectrum line (or in some cases the full width) measured at half its height.

halide A compound of a halogen with another element or group. The halides of typical metals are ionic (e.g. sodium fluoride, Na^+F^-). Metals can also form halides in which the bonding is largely covalent (e.g. aluminium chloride, $AlCl_3$). Organic compounds are also sometimes referred to as halides; e.g. the alkyl halides (*see* **haloalkanes**) and the *acyl halides. Halides are named *fluorides, chlorides, bromides,* or *iodides.*

halite (rock salt) Naturally occurring *sodium chloride (common salt, NaCl), crystallizing in the cubic system. It is chiefly colourless or white (sometimes blue) when pure but the presence of impurities may colour it grey, pink, red, or brown. Halite often occurs in association with anhydrite and gypsum.

Hall–Heroult cell An electrolytic cell used industrially for the extraction of aluminium from bauxite. The bauxite is first purified by dissolving it in sodium hydroxide and filtering off insoluble constituents. Aluminium hydroxide is then precipitated (by adding CO_2) and this is decomposed by heating to obtain pure Al_2O_3. In the Hall–Heroult cell, the oxide is mixed with cryolite (to lower its melting point) and the molten mixture electrolysed using graphite anodes. The cathode is the lining of the cell, also of graphite. The electrolyte is kept in a molten state (about 850°C) by the current. Molten aluminium collects at the bottom of the cell and can be tapped off. Oxygen forms at the anode, and gradually oxidizes it away. The cell is named after the US chemist Charles Martin Hall (1863–1914), who discovered the process in 1886, and the French chemist Paul Heroult (1863–1914), who discovered it independently in the same year.

hallucinogen A drug or chemical that causes alterations in perception (usually visual), mood, and thought. Common hallucinogenic drugs include *lysergic acid diethylamide (LSD) and mescaline. There is no common mechanism of action for this class of compounds although many hallucinogens are structurally similar to neurotransmitters in the central nervous system, such as serotonin and the catecholamines.

halo A broad ring appearing in the electron diffraction, neutron diffraction, or X-ray diffraction patterns of materials that are not crystals.

Haloes of this type occur in gases and liquids as well as in noncrystalline solids.

haloalkanes **(alkyl halides)** Organic compounds in which one or more hydrogen atoms of an alkane have been substituted by halogen atoms. Examples are chloromethane, CH_3Cl, dibromoethane, CH_2BrCH_2Br, etc. Haloalkanes can be formed by direct reaction between alkanes and halogens using ultraviolet radiation. They are usually made by reaction of an alcohol with a halogen carrier.

halocarbons Compounds that contain carbon and halogen atoms and (sometimes) hydrogen. The simplest are compounds such as tetrachloromethane (CCl_4), tetrabromomethane (CBr_4), etc. The *haloforms are also simple halocarbons. The *chlorofluorocarbons (CFCs) contain carbon, chlorine, and fluorine. Similar to these are *hydrochlorofluorocarbons* (HCFCs), which contain carbon, chlorine, fluorine, and hydrogen, and the *hydrofluorocarbons* (HFCs), which contain carbon, fluorine, and hydrogen. The *halons are a class of halocarbons that contain bromine.

haloform reaction A reaction for producing *haloforms from methyl ketones. An example is the production of chloroform from propanone using sodium chlorate(I) (or bleaching powder):

$$CH_3COCH_3 + 3NaOCl \rightarrow CH_3COCCl_3 + 3NaOH$$

The substituted ketone then reacts to give chloroform (trichloromethane):

$$CH_3COCCl_3 + NaOH \rightarrow NaOCOCH_3 + CHCl_3$$

The reaction can also be used for making carboxylic acids, since $RCOCH_3$ gives the product NaOCOR. It is particularly useful for aromatic acids as the starting ketone can be made by a Friedel–Crafts acylation.

The reaction of methyl ketones with sodium iodate(I) gives iodoform (triiodomethane), which is a yellow solid with a characteristic smell. This reaction is used in the *iodoform test* to identify methyl ketones. It also gives a positive result with a secondary alcohol of the formula $RCH(OH)CH_3$ (which is first oxidized to a methylketone) or with ethanol (oxidized to ethanal, which also undergoes the reaction).

haloforms The four compounds with formula CHX_3, where X is a halogen atom. They are *chloroform* ($CHCl_3$), and, by analogy, *fluoroform* (CHF_3), *bromoform* ($CHBr_3$), and *iodoform* (CHI_3). The systematic names are trichloromethane, trifluoromethane, etc.

halogenating agent *See* **halogenation**.

halogenation A chemical reaction in which a halogen atom is introduced into a compound. Halogenations are described as *chlorination*, *fluorination*, *bromination*, etc., according to the halogen involved. Halogenation reactions may take place by direct reaction with the halogen. This occurs with alkanes, where the reaction involves free radicals and requires high temperature, ultraviolet radiation, or a chemical initiator; e.g.

$$C_2H_6 + Br_2 \rightarrow C_2H_5Br + HBr$$

The halogenation of aromatic compounds can be effected by electrophilic substitution using an aluminium chloride catalyst:

$$C_6H_6 + Cl_2 \rightarrow C_6H_5Cl + HCl$$

Halogenation can also be carried out using compounds, such as phosphorus halides (e.g. PCl_3) or sulphur dihalide oxides (e.g. $SOCl_2$), which react with –OH groups. Such compounds are called *halogenating agents*. Addition reactions are also referred to as halogenations; e.g.

$$C_2H_4 + Br_2 \rightarrow CH_2BrCH_2Br$$

halogens (group 17 elements) A group of elements in the *periodic table (formerly group VIIB): fluorine (F), chlorine (Cl), bromine (Br), iodine (I), and astatine (At). All have a characteristic electron configuration of noble gases but with outer ns^2np^5 electrons. The outer shell is thus one electron short of a noble-gas configuration. Consequently, the halogens are typical nonmetals; they have high electronegativities – high electron affinities and high ionization energies. They form compounds by gaining an electron to complete the stable configuration; i.e. they are good oxidizing agents. Alternatively, they share their outer electrons to form covalent compounds, with single bonds.

All are reactive elements with the reactivity decreasing down the group The electron affinity decreases down the group and other properties also show a change from fluorine to astatine. Thus, the melting and boiling points increase; at 20°C, fluorine and chlorine are gases, bromine a liquid, and iodine and astatine are solids. All exist as diatomic molecules.

The name 'halogen' comes from the Greek 'salt-producer', and the elements react with metals to form ionic halide salts. They also combine with nonmetals, the activity decreasing down the group: fluorine reacts with all nonmetals except nitrogen and the noble gases helium, neon, and argon; iodine does not react with any noble gas, nor with carbon, nitrogen, oxygen, or sulphur. The elements fluorine to iodine all react with hydrogen to give the acid, with the activity being greatest for fluorine, which reacts explosively. Chlorine and hydrogen react slowly at room temperature in the dark (sunlight causes a free-radical chain reaction). Bromine and hydrogen react if heated in the presence of a catalyst. Iodine and hydrogen react only slowly and the reaction is not complete. There is a decrease in oxidizing ability down the group from fluorine to iodine. As a consequence, each halogen will displace any halogen below it from a solution of its salt, for example:

$$Cl_2 + 2Br^- \rightarrow Br_2 + 2Cl^-$$

The halogens also form a wide variety of organic compounds in which the halogen atom is linked to carbon. In general, the aryl compounds are more stable than the alkyl compounds and there is decreasing resistance to chemical attack down the group from the fluoride to the iodide.

Fluorine has only a valency of 1, although the other halogens can have higher oxidation states using their vacant *d*-electron levels. There is also evidence for increasing metallic behaviour down the group. Chlorine and bromine form compounds with oxygen in which the halogen atom is assigned a positive oxidation state. Only iodine, however, forms positive ions, as in $I^+NO_3^-$.

halon A compound obtained by replacing the hydrogen atoms of a hydrocarbon by bromine along with other halogen atoms (*see* **halocarbons**), for instance halon 1211 is bromochlorodifluoromethane (CF_2BrCl) and halon 1301 is bromotrifluoromethane (CF_3Br). Halons are very stable and unreactive and are widely used in fire extinguishers. There is concern that they are being broken down in the atmosphere to bromine, which reacts with ozone, leading to depletion of *ozone layer, and their use is being curtailed. Although more *chlorofluorocarbons are present in the atmosphere, halons are between three and ten times more destructive of ozone.

halothane (1-chloro-1-bromo-2,2,2,-trifluoroethane) A colourless, non-flammable oily liquid, $CHBrClCF_3$; b.p. 51°C. It smells like trichloromethane and has a sickly burning taste. Halothane is widely used as a general anaesthetic, often administered also with oxygen and dinitrogen oxide.

Hamiltonian Symbol H. A function used to express the energy of a system in terms of its momentum and positional coordinates. In simple cases this is the sum of its kinetic and potential energies. In *Hamiltonian equations*, the usual equations used in mechanics (based on forces) are replaced by equations expressed in terms of momenta. This method of formulating mechanics (*Hamiltonian mechanics*) was first introduced by Sir William Rowan Hamilton (1805–65). The Hamiltonian operator is used in quantum mechanics in the *Schrödinger equation.

Hammett equation An equation relating the structure to the reactivity of side-chain derivatives of aromatic compounds. It arises from a comparison between rate constants for various reactions with the rate of hydrolysis of benzyl chloride on the one hand and a comparison between equilibrium constants (such as the dissociation constant of benzoic acid) on the other hand. The Hammett equation can be written in the form $\log(k/k_0) = \rho\log(K/K_0)$, where $\log(K/K_0)$ refers to comparing dissociation constants to the dissociation constant, K_0, of benzoic acid in water at 25°C, and $\log(k/k_0)$ refers to comparing rates of reaction to the rate, k_0, of hydrolysis of benzyl chloride. The term $\log(K/K_0) = \sigma$ is called the *substituent constant*, since the nature of the substituent affects the strength of the benzoic acid. If σ is positive, the substituent is electron attracting, while if σ is negative the substituent is electron donating. ρ is a reaction constant, which is determined for a given reaction by the slope of a graph of $\log(k/k_0)$ against σ. The numerical value of ρ depends on temperature and the type of solvent.

The Hammett equation applies to meta- and para- substituents (provided that resonance interaction from the substituents does not occur) but not to ortho- substituents.

hapticity Symbol η. The number of electrons in a ligand that are directly coordinated to a metal.

hardening of oils The process of converting unsaturated esters of *fatty acids into (more solid) saturated esters by hydrogenation using a nickel catalyst. It is used in the manufacture of margarine from vegetable oils.

hardness of water The presence in water of dissolved calcium or magnesium ions, which form a scum with soap and prevent the formation of a lather. The main cause of hard water is dissolved calcium hydrogencarbonate ($Ca(HCO_3)_2$), which is formed in limestone or chalk regions by the action of dissolved carbon dioxide on calcium carbonate. This type is known as *temporary hardness* because it is removed by boiling:

$$Ca(HCO_3)_2(aq) \rightarrow CaCO_3(s) + H_2O(l) + CO_2(g)$$

The precipitated calcium carbonate is the 'fur' (or 'scale') formed in kettles, boilers, pipes, etc. In some areas, hardness also results from dissolved calcium sulphate ($CaSO_4$), which cannot be removed by boiling (*permanent hardness*).

Hard water is a considerable problem in washing, reducing the efficiency of boilers, heating systems, etc., and in certain industrial processes. Various methods of *water softening* are used. In public supplies, the temporary hardness can be removed by adding lime (calcium hydroxide), which precipitates calcium carbonate

$$Ca(OH)_2(aq) + Ca(HCO_3)_2(aq) \rightarrow 2CaCO_3(s) + 2H_2O(l)$$

This is known as the *Clark process* (or as '*clarking*'). It does not remove permanent hardness. Both temporary and permanent hardness can be treated by precipitating calcium carbonate by added sodium carbonate – hence its use as a washing soda and in bath salts. Calcium (and other) ions can also be removed from water by ion-exchange using zeolites (e.g. *Permutit*). This method is used in small domestic water-softeners. Another technique is not to remove the Ca^{2+} ions but to complex them and prevent them reacting further. For domestic use polyphosphates (containing the ion $P_6O_{18}^{6-}$, e.g. *Calgon*) are added. Other sequestering agents are also used for industrial water. *See also* **sequestration**.

Hargreaves process *See* **potassium sulphate**.

harmonic An oscillation having a frequency that is a simple multiple of a *fundamental* sinusoidal oscillation. The fundamental frequency of a sinusoidal oscillation is usually called the *first harmonic*. The *second harmonic* has a frequency twice that of the fundamental, and so on.

harmonic oscillator A system that oscillates with simple harmonic motion. The harmonic oscillator is exactly soluble in both classical mechanics and quantum mechanics. Many systems exist for which harmonic oscillators provide very good approximations. Atoms vibrating about their mean positions in molecules or crystal lattices at low temperatures can be regarded as good approximations to harmonic oscillators in quantum mechanics. Even if a system is not exactly a harmonic oscillator the solution of the harmonic oscillator is frequently a useful starting point for treating such systems using *perturbation theory. *Compare* **anharmonic oscillator**.

harpoon mechanism A mechanism suggested to explain the reaction between such metal atoms as sodium or potassium and such halogen molecules as bromine, e.g. $K + Br_2 \rightarrow KBr + Br$. As the alkali atom approaches the bromine molecule its valence electron moves to the bromine molecule (thus providing a 'harpoon'). There are then two ions with a Coulomb attraction between them. As a result the ions move together and the reaction takes place. This mechanism, which has been worked out quantitatively, explains why the reaction occurs far more readily than might be expected taking into account only mechanical collisions between the alkali-metal atoms and halogen molecules.

Hartree–Fock procedure A self-consistent field (SCF) procedure used to find approximate *wave functions and energy levels in many-electron atoms. This procedure was introduced by the English mathematician and physicist Douglas Hartree in 1928 and improved by the Soviet physicist Vladimir Fock in 1930 (by taking into account the Pauli exclusion principle). The initial wave functions can be taken to be hydrogenic atomic orbitals. The resulting equations can be solved numerically using a computer. The results of the Hartree–Fock theory are sufficiently accurate to show that electron density occurs in shells around atoms and can be used quantitatively to show chemical periodicity.

hassium Symbol Hs. A radioactive *transactinide element; a.n. 108. It was first made in 1984 by Peter Armbruster and a team in Darmstadt, Germany. It can be produced by bombarding lead-208 nuclei with iron-58 nuclei. Only a few atoms have ever been produced. The name comes from the Latinized form of Hesse, the German state where it was first synthesized.

HCFC (hydrochlorofluorocarbon) *See* **halocarbons**.

h.c.p. Hexagonal close packing. *See* **close packing**.

heat capacity (thermal capacity) The ratio of the heat supplied to an object or specimen to its consequent rise in temperature. The *specific heat capacity* is the ratio of the heat supplied to unit mass of a substance to its consequent rise in temperature. The *molar heat capacity* is the ratio of the heat supplied to unit amount of a substance to its consequent rise in temperature. In practice, heat capacity (C) is measured in joules per kelvin, specific heat capacity (c) in $J\,K^{-1}\,kg^{-1}$, and molar heat capacity (C_m) in $J\,K^{-1}\,mol^{-1}$. For a gas, the values of c and C_m are commonly given either at *constant volume*, when only its *internal energy is increased, or at *constant pressure*, which requires a greater input of heat as the gas is allowed to expand and do work against the surroundings. The symbols for the specific and molar heat capacities at constant volume are c_v and C_v, respectively; those for the specific and molar heat capacities at constant pressure are c_p and C_p.

heat engine A device for converting heat into work. Engines usually work on cycles of operation, the most efficient of which would be the *Carnot cycle.

heat of atomization The energy required to dissociate one mole of a given substance into atoms.

heat of combustion The energy liberated when one mole of a given substance is completely oxidized.

heat of crystallization The energy liberated when one mole of a given substance crystallizes from a saturated solution of the same substance.

heat of dissociation The energy absorbed when one mole of a given substance is dissociated into its constituent elements.

heat of formation The energy liberated or absorbed when one mole of a compound is formed in their *standard states from its constituent elements.

heat of neutralization The energy liberated in neutralizing one mole of an acid or base.

heat of reaction The energy liberated or absorbed as a result of the complete chemical reaction of molar amounts of the reactants.

heat of solution The energy liberated or absorbed when one mole of a given substance is completely dissolved in a large volume of solvent (strictly, to infinite dilution).

heavy hydrogen *See* **deuterium**.

heavy metal A metal with a high relative atomic mass. The term is usually applied to common transition metals, such as copper, lead, and zinc. These metals are a cause of environmental *pollution (*heavy-metal pollution*) from a number of sources, including lead in petrol, industrial effluents, and leaching of metal ions from the soil into lakes and rivers by acid rain.

heavy spar A mineral form of *barium sulphate, $BaSO_4$

heavy water **(deuterium oxide)** Water in which hydrogen atoms, ^1H, are replaced by the heavier isotope deuterium, ^2H (symbol D). It is a colourless liquid, which forms hexagonal crystals on freezing. Its physical properties differ from those of 'normal' water; r.d. 1.105; m.p. 3.8°C; b.p. 101.4°C. Deuterium oxide, D_2O, occurs to a small extent (about 0.003% by weight) in natural water, from which it can be separated by fractional distillation or by electrolysis. It is useful in the nuclear industry because of its ability to reduce the energies of fast neutrons to thermal energies and because its absorption cross-section is lower than that of hydrogen and consequently it does not appreciably reduce the neutron flux. In the laboratory it is used for *labelling other molecules for studies of reaction mechanisms. Water also contains the compound HDO.

hecto- Symbol h. A prefix used in the metric system to denote 100 times. For example, 100 coulombs = 1 hectocoulomb (hC).

Heisenberg, Werner Karl (1901–76) German physicist, who became a professor at the University of Leipzig and, after World War II, at the Kaiser Wilhelm Institute in Göttingen. In 1923 he was awarded the Nobel Prize for his work on matrix mechanics. But he is best known for his 1927 discovery of the *uncertainty principle.

Heisenberg uncertainty principle *See* **uncertainty principle**.

helicate A type of inorganic molecule containing a double helix of bipyridyl-derived molecules formed around a chain of up to five copper(I) ions. *See also* **supramolecular chemistry**.

helium Symbol He. A colourless odourless gaseous nonmetallic element belonging to group 18 of the periodic table (*see* **noble gases**); a.n. 2; r.a.m. 4.0026; d. 0.178 g dm^{-3}; m.p. −272.2°C (at 20 atm.); b.p. −268.93°C. The element has the lowest boiling point of all substances and can be solidified only under pressure. Natural helium is mostly helium–4, with a small amount of helium–3. There are also two short-lived radioactive isotopes: helium–5 and –6. It occurs in ores of uranium and thorium and in some natural-gas deposits. It has a variety of uses, including the provision of inert atmospheres for welding and semiconductor manufacture, as a refrigerant for superconductors, and as a diluent in breathing apparatus. It is also used in filling balloons. Chemically it is totally inert and has no known compounds. It was discovered in the solar spectrum in 1868 by Joseph Lockyer (1836–1920).

helium–neon laser A laser in which the medium is a mixture of helium and neon in the mole ratio 1:5 respectively. An electric discharge is used to excite a He atom to the metastable $1s^1 2s^1$ configuration. Since the excitation energy coincides with an excitation energy of neon, transfer of energy between helium and neon atoms can readily occur during collisions. These collisions lead to highly excited neon atoms with unoccupied lower energy states. Thus, population inversion occurs giving rise to laser action with a wavelength of 633 nm. (Many other spectral lines are also produced in the process.)

Helmholtz, Hermann Ludwig Ferdinand von (1821–94) German physiologist and physicist. In 1850 he measured the speed of a nerve impulse and in 1851 invented the ophthalmoscope. Helmholtz discovered the conservation of energy (1847), giving many examples of its application, and also introduced the concept of *free energy.

Helmholtz free energy *See* **free energy**.

Helmholtz model *See* **electrical double layer**.

heme *See* **haem**.

hemiacetals *See* **acetals**.

hemicellulose A *polysaccharide found in the cell walls of plants. The branched chains of this molecule bind to cellulose microfibrils, forming a network of cross-linked fibres.

hemihydrate A crystalline hydrate containing two molecules of compound per molecule of water (e.g. $2CaSO_4.H_2O$).

hemiketals *See* **ketals.**

henry Symbol H. The *SI unit of inductance equal to the inductance of a closed circuit in which an e.m.f. of one volt is produced when the electric current in the circuit varies uniformly at a rate of one ampere per second. It is named after the US physicist Joseph Henry (1797–1878).

Henry's law At a constant temperature the mass of gas dissolved in a liquid at equilibrium is proportional to the partial pressure of the gas. The law, discovered in 1801 by the British chemist and physician William Henry (1775–1836), is a special case of the partition law. It applies only to gases that do not react with the solvent.

heparin A glycosaminoglycan (mucopolysaccharide) with anticoagulant properties, occurring in vertebrate tissues, especially the lungs and blood vessels.

heptahydrate A crystalline hydrate that has seven moles of water per mole of compound.

heptane A liquid straight-chain alkane obtained from petroleum, C_7H_{16}; r.d. 0.684; m.p. −90.6°C; b.p. 98.4°C. In standardizing *octane numbers, heptane is given a value zero.

heptaoxodiphosphoric(V) acid *See* **phosphoric(V) acid.**

heptavalent **(septivalent)** Having a valency of seven.

herbicide *See* **pesticide.**

Hermann–Mauguin system **(international system)** A notation used to describe the symmetry of point groups. In contrast to the *Schoenflies system, which is used for isolated molecules (e.g. in spectroscopy), the Hermann–Mauguin system is used in *crystallography. Some of the categories are the same as the Schoenflies system. n is the same group as C_n. nmm is the same group as C_{nv}. There are two ms because of two distinct types of mirror plane containing the n-fold axis. $n22$ is the same group as D_n. The other categories do not coincide with the Schoenflies system. $\bar{n}$ is a group with an n-fold rotation–inversion axis and includes C_{3h} as $\bar{6}$, S_4 as $\bar{4}$, S_6 as $\bar{3}$, and S_2 as $\bar{1}$. n/m is the same group as C_{nh} except that C_{3h} is regarded as $\bar{6}$. $n2m$ is the same group as D_{nd}, except that D_{3h} is regarded as $62m$. n/m $2/m$ $2/m$, abbreviated to n/mmm, is the same group as D_{nh}, except that D_{3h} is regarded as $\bar{6}2m$. (Unlike the Schoenflies system, the Hermann–Mauguin system regards the three-fold axis as a special case.) As regards the cubic groups, O_h is denoted $m3m$ (or $4/m\bar{3}2/m$), O is denoted 432, T_h is denoted $m3$ (or $2/m\ \bar{3}$), T_d is denoted $\bar{4}3m$, and T is denoted 23. In the

Hermann–Mauguin system all the cubic groups have 3 as the second number because of the three-fold axis that occurs in all cubic groups.

heroin (diacetylmorphine) A narcotic compound that is a synthetic derivative of morphine. The compound is easily absorbed by the brain, due to its lipid-like nature, and is used as a sedative and powerful analgesic. Highly addictive, it is abused by drug users.

hertz Symbol Hz. The *SI unit of frequency equal to one cycle per second. It is named after the German physicist Heinrich Hertz (1857–94).

Hess's law If reactants can be converted into products by a series of reactions, the sum of the heats of these reactions (with due regard to their sign) is equal to the heat of reaction for direct conversion from reactants to products. More generally, the overall energy change in going from reactants to products does not depend on the route taken. The law can be used to obtain thermodynamic data that cannot be measured directly. For example, the heat of formation of ethane can be found by considering the reactions:

$$2C(s) + 3H_2(g) + 3\tfrac{1}{2}O_2(g) \rightarrow 2CO_2(g) + 3H_2O(l)$$

The heat of this reaction is $2\Delta H_C + 3\Delta H_H$, where ΔH_C and ΔH_H are the heats of combustion of carbon and hydrogen respectively, which can be measured. By Hess's law, this is equal to the sum of the energies for two stages:

$$2C(s) + 3H_2(g) \rightarrow C_2H_6(g)$$

(the heat of formation of ethane, ΔH_f) and

$$C_2H_6(g) + 3\tfrac{1}{2}O_2 \rightarrow 2CO_2(g) + 3H_2O(l)$$

(the heat of combustion of ethane, ΔH_E). As ΔH_E can be measured and as

$$\Delta H_f + \Delta H_E = 2\Delta H_c + 3\Delta H_H$$

ΔH_f can be found. Another example is the use of the *Born–Haber cycle to obtain lattice energies. The law was first put forward in 1840 by the Swiss-born Russian chemist Germain Henri Hess (1802–50). It is sometimes called the *law of constant heat summation* and is a consequence of the law of conservation of energy.

hetero atom An odd atom in the ring of a heterocyclic compound. For instance, nitrogen is the hetero atom in pyridine.

heterocyclic *See* **cyclic**.

heterogeneous Relating to two or more phases, e.g. a heterogeneous *catalyst. *Compare* **homogeneous**.

heterolytic fission The breaking of a bond in a compound in which the two fragments are oppositely charged ions. For example, $HCl \rightarrow H^+ + Cl^-$. *Compare* **homolytic fission**.

heteronuclear Denoting a molecule in which the atoms are of different elements.

heteropolar bond *See* chemical bond.

heteropoly compound *See* cluster compound.

heteropolymer *See* polymer.

Heusler alloys Ferromagnetic alloys containing no ferromagnetic elements. The original alloys contained copper, manganese, and tin and were first made by Conrad Heusler (19th-century mining engineer).

hexachlorobenzene A colourless crystalline compound, C_6Cl_6; m.p. 227°C. It is made by the chlorination of benzene with an iron(III) chloride catalyst or by treating hexachlorocyclohexane with chlorine in hexachloroethane. It is used to preserve wood and dress seeds, and in the manufacture of hexafluorobenzene.

hexacyanoferrate(II) **(ferrocyanide)** A complex iron-containing anion, $[Fe(CN_6)]^{4-}$, used as a solution of its potassium salt as a test for ferric iron (iron(III)), with which it forms a dark blue precipitate of Prussian blue. The sodium salt is used as an anticaking agent in common salt.

hexacyanoferrate(III) **(ferricyanide)** A complex iron-containing anion, $[Fe(CN)_6)]^{3-}$, used as a solution of its potassium salt as a test for ferrous iron (iron(II)), with which it forms a dark blue precipitate of Prussian blue.

hexadecane **(cetane)** A colourless liquid straight-chain alkane hydrocarbon, $C_{16}H_{34}$, used in standardizing *cetane numbers of Diesel fuel.

hexadecanoate *See* palmitate.

hexadecanoic acid *See* palmitic acid.

hexagonal close packing *See* close packing.

hexagonal crystal *See* crystal system.

hexahydrate A crystalline compound that has six moles of water per mole of compound.

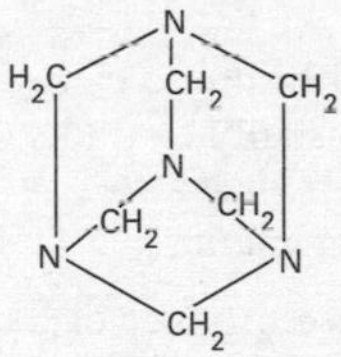

Hexamethylenetetramine

hexamethylenetetramine A bridged compound, $C_6H_{12}N_4$, formed by the condensation of methanal with ammonia. It can be reacted with nitric acid to produce *cyclonite.

hexamine **(hexamethylene tetramine)** A white crystalline compound, $C_6H_{12}N_4$, made by the condensation of methanal with ammonia. It has been used as a solid fuel for camping stoves and as an antiseptic for treating urinary infections in medicine. It is used in the production of *cyclonite.

hexanedioate **(adipate)** A salt or ester of hexanedioic acid.

hexanedioic acid **(adipic acid)** A carboxylic acid, $(CH_2)_4(COOH)_2$; r.d. 1.36; m.p. 153°C; b.p. 265°C (100 mmHg). It is used in the manufacture of *nylon 6,6. *See also* **polymerization**.

6-hexanelactam *See* **caprolactam**.

hexanoate **(caproate)** A salt or ester of hexanoic acid.

hexanoic acid **(caproic acid)** A liquid fatty acid, $CH_3(CH_2)_4COOH$; r.d. 0.93; m.p. −3.4°C; b.p. 205°C. Glycerides of the acid occur naturally in cow and goat milk and in some vegetable oils.

hexose A *monosaccharide that has six carbon atoms in its molecules.

hexyl group **(hexyl radical)** The organic group $CH_3CH_2CH_2CH_2CH_2CH_2-$, derived from hexane.

HFC **(hydrofluorocarbon)** *See* **halocarbons**.

highest occupied molecular orbital **(HOMO)** The orbital in a molecule that has the highest energy level occupied at the temperature of absolute zero. The highest occupied molecular orbital and the lowest unoccupied molecular orbital (LUMO) are the two *frontier orbitals of the molecule.

high frequency **(HF)** A radio frequency in the range 3–30 megahertz; i.e. having a wavelength in the range 10–100 metres.

high-performance liquid chromatography **(HPLC)** A sensitive technique for separating or analysing mixtures, in which the sample is forced through the chromatography column under pressure.

high-resolution electron-loss spectroscopy **(HRELS)** A technique for obtaining information about molecules absorbed on a surface by inelastic scattering of electrons.

high-speed steel A steel that will remain hard at dull red heat and can therefore be used in cutting tools for high-speed lathes. It usually contains 12–22% tungsten, up to 5% chromium, and 0.4–0.7% carbon. It may also contain small amounts of vanadium, molybdenum, and other metals.

Hilbert space A linear vector space that can have an infinite number of dimensions. The concept is of interest in physics because the state of a system in *quantum mechanics is represented by a vector in Hilbert space. The dimension of the Hilbert space has nothing to do with the physical dimension of the system. The Hilbert space formulation of quantum mechanics was put forward by the Hungarian-born US mathematician John von Neumann (1903–57) in 1927. Other formulations of quantum mechanics, such as *matrix mechanics and *wave mechanics, can be deduced from the Hilbert space formulation. It is named after the German mathematician

David Hilbert (1862–1943), who invented the concept early in the 20th century.

Hildebrand rule If the density of the vapour phase above a liquid is constant, the molar entropy of vaporization is constant. This law does not hold if molecular association takes place in the liquid or if the liquid is subject to quantum-mechanical effects, e.g. as in superfluidity. The rule is named after the US chemist Joel Henry Hildebrand (1881–1983).

Hill reaction The release of oxygen from isolated illuminated chloroplasts when suitable electron acceptors (e.g. potassium ferricyanide) are added to the surrounding water. The reaction was discovered by Robert Hill (1899–1991) in 1939; the electron acceptors substitute for $NADP^+$, the natural acceptor for the light-dependent reactions of *photosynthesis.

histidine *See* **amino acid**.

histochemistry The study of the distribution of the chemical constituents of tissues by means of their chemical reactions. It utilizes such techniques as staining, light and electron microscopy, autoradiography, and chromatography.

Hoff, Jacobus Henrikus van't (1852–1911) Dutch physical chemist who first recognized that a molecule could exist in two mirror-image forms. He proposed that these forms rotated the plane of polarization in opposite senses, and is generally regarded as the founder of stereochemistry. Van't Hoff also did important work on other branches of physical chemistry, especially chemical thermodynamics. He was awarded the 1901 Nobel Prize for chemistry. (*See* **van't Hoff factor**; **van't Hoff's isochore**.)

Hofmann's reaction (Hofmann rearrangement) A reaction for making primary *amines from *amides using bromine or chlorine and sodium hydroxide:

$$RCONH_2 \rightarrow RNH_2$$

The halogen replaces a hydrogen atom from the amido group to form a halo-amide. This then reacts with the alkali to produce an isocyanate, which decomposes into the amine and carbon dioxide. The amine has one carbon atom fewer than the amide from which it is produced. This technique is used to reduce the length of carbon chains in molecules (the *Hofmann degradation*). The reaction is named after the German chemist August Wilhelm von Hofmann (1818–92).

hole 1. A vacant electron position in the lattice structure of a solid that behaves like a mobile positive *charge carrier. **2.** A vacant electron position in one of the inner orbitals of an atom.

holmium Symbol Ho. A soft silvery metallic element belonging to the *lanthanoids; a.n. 67; r.a.m. 164.93; r.d. 8.795 (20°C); m.p. 1474°C; b.p. 2695°C. It occurs in apatite, xenotime, and some other rare-earth minerals. There is one natural isotope, holmium–165; eighteen artificial isotopes have been produced. There are no uses for the element, which was discovered by Per Cleve (1840–1905) and J. L. Soret in 1879.

HOMO *See* **highest occupied molecular orbital**.

homocyclic *See* **cyclic**.

homogeneous Relating to only one phase, e.g. a homogeneous mixture, a homogeneous *catalyst. *Compare* **heterogeneous**.

homologous series A series of related chemical compounds that have the same functional group(s) but differ in formula by a fixed group of atoms. For instance, the simple carboxylic acids: methanoic (HCOOH), ethanoic (CH_3COOH), propanoic (C_2H_5COOH), etc., form a homologous series in which each member differs from the next by CH_2. Successive members of such a series are called *homologues*.

homolytic fission The breaking of a bond in a compound in which the fragments are uncharged free radicals. For example, $Cl_2 \rightarrow Cl\cdot + Cl\cdot$. *Compare* **heterolytic fission**.

homonuclear Denoting a molecule in which the atoms are of the same element.

homopolar bond *See* **chemical bond**.

homopolymer *See* **polymer**.

hormone A substance that is manufactured and secreted in very small quantities into the bloodstream by an endocrine gland or a specialized nerve cell and regulates the growth or functioning of a specific tissue or organ in a distant part of the body. For example, the hormone insulin controls the rate and manner in which glucose is used by the body.

hornblende Any of a group of common rock-forming minerals of the amphibole group with the generalized formula $(Ca,Na)_2(Mg,Fe,Al)_5(Al,Si)_8O_{22}(OH,F)_2$. Hornblendes consist mainly of calcium, iron, and magnesium silicate.

host–guest chemistry *See* **supramolecular chemistry**.

HPLC *See* **high-performance liquid chromatography**.

HRELS *See* **high-resolution electron-loss spectroscopy**.

Hückel, Erich (1896–1980) German physicist and theoretical chemist who worked with Peter *Debye at Zürich on the theory of electrolytes. Later he moved to Copenhagen to work with Niels *Bohr and here he produced his work on bonding in aromatic molecules. *See* **Hückel approximation**; **aromatic compound**.

Hückel approximation A set of approximations used to simplify the molecular orbital analysis of conjugated molecules, suggested by Erich *Hückel in 1931. In the Hückel theory, the σ orbitals are treated separately from the π orbitals, the shape of the molecule being determined by the σ

orbitals. The Hückel theory makes the approximation that interactions between non-neighbouring atoms are taken to be zero. It enables calculations to be made for conjugated molecules. In particular, the *delocalization energy of such molecules can be estimated. Hückel explained the stability of benzene associated with aromaticity in this way. The theory can also be used to analyse delocalized bonding in solids.

Hückel rule *See* **aromatic compound.**

humectant A substance used to maintain moisture levels. Humectants are generally *hygroscopic. For example, glycerol is employed as a humectant in confectionery, foodstuffs, and tobacco. Other polyhydric alcohols, such as mannitol and sorbitol, are also used as humectant additives in the foodstuffs industry.

Humphreys series A series of lines in the *hydrogen spectrum with the form

$$1/\lambda = R(1/6^2 - 1/n^2),\ n = 7,8,9...,$$

where λ is the wavelength associated with the lines and R is the Rydberg constant. The Humphreys series was discovered by C. J. Humphreys in 1953 and lies in the far infrared.

Hund coupling cases *See* **coupling.**

Hund's rules Empirical rules for interpreting atomic *spectra used to determine the lowest energy level for a configuration of two equivalent electrons (i.e. electrons with the same n and l quantum numbers) in a many-electron *atom. (1) The lowest energy state has the maximum *multiplicity consistent with the *Pauli exclusion principle. (2) The lowest energy state has the maximum total electron orbital angular momentum quantum number, consistent with rule (1). These rules were put forward by the German physicist Friedrich Hund (1896–1993) in 1925. Hund's rules are explained by quantum theory involving the repulsion between two electrons.

hyaluronic acid A *glycosaminoglycan (mucopolysaccharide) that is part of the matrix of connective tissue. Hyaluronic acid binds cells together and helps to lubricate joints. It may play a role in the migration of cells at wounds; this activity ceases when hyaluronidase breaks down hyaluronic acid.

hybrid orbital *See* **orbital.**

hydracid *See* **binary acid.**

hydrate A substance formed by combination of a compound with water. *See* **water of crystallization.**

hydrated alumina *See* **aluminium hydroxide.**

hydrated aluminium hydroxide *See* **aluminium hydroxide.**

hydration *See* **solvation.**

hydrazine A colourless liquid or white crystalline solid, N_2H_4; r.d. 1.01 (liquid); m.p. 1.4°C; b.p. 113.5°C. It is very soluble in water and soluble in ethanol. Hydrazine is prepared by the *Raschig synthesis* in which ammonia reacts with sodium(I) chlorate (sodium hypochlorite) to give NH_2Cl, which then undergoes further reaction with ammonia to give N_2H_4. Industrial production must be carefully controlled to avoid a side reaction leading to NH_4Cl. The compound is a weak base giving rise to two series of salts, those based on $N_2H_5^+$, which are stable in water (sometimes written in the form $N_2H_4.HCl$ rather than $N_2H_5^+Cl^-$), and a less stable and extensively hydrolysed series based on $N_2H_6^{2+}$. Hydrazine is a powerful reducing agent and reacts violently with many oxidizing agents, hence its use as a rocket propellant. It reacts with aldehydes and ketones to give *hydrazones.

hydrazoic acid *See* **hydrogen azide.**

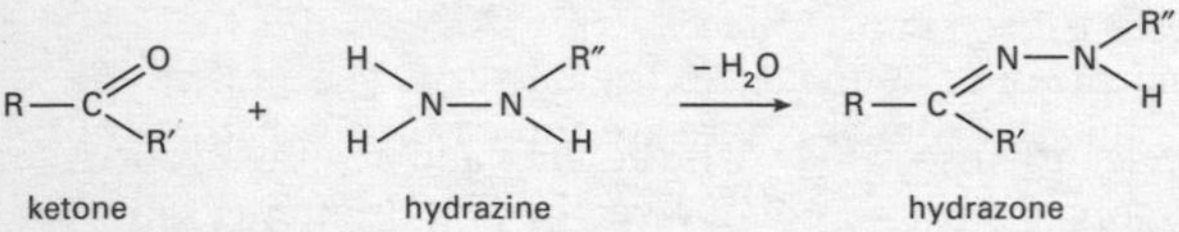

ketone + hydrazine —($-H_2O$)→ hydrazone

Formation of a hydrazone from a ketone. The same reaction occurs with an aldehyde (R′ = H). If R″ = C_6H_5, the product is phenylhydrazone

hydrazones Organic compounds containing the group $=C:NNH_2$, formed by condensation of substituted hydrazines with aldehydes or ketones (see illustration). *Phenylhydrazones* contain the group $=C:NNHC_6H_5$.

hydride A chemical compound of hydrogen and another element or elements. Non-metallic hydrides (e.g. ammonia, methane, water) are covalently bonded. The alkali metals and alkaline earths (*s-block elements) form salt-like hydrides containing the hydride ion H^-, which produce hydrogen on reacting with water. Hydride-forming *transition elements form interstitial hydrides, with the hydrogen atoms 'trapped' within the gaps in the lattice of metal atoms. Complex hydrides, such as *lithium tetrahydroaluminate(III), have hydride ions as *ligands; many are powerful reducing agents.

hydriodic acid *See* **hydrogen iodide.**

hydrobromic acid *See* **hydrogen bromide.**

hydrocarbons Chemical compounds that contain only carbon and hydrogen. A vast number of different hydrocarbon compounds exist, the main types being the *alkanes, *alkenes, *alkynes, and *arenes.

hydrochloric acid *See* **hydrogen chloride.**

hydrochloride *See* **amine salts.**

hydrochlorofluorocarbon **(HCFC)** *See* **halocarbons.**

hydrocyanic acid *See* **hydrogen cyanide**.

hydrodynamic radius The effective radius of an ion in a solution measured by assuming that it is a body moving through the solution and resisted by the solution's viscosity. If the solvent is water, the hydrodynamic radius includes all the water molecules attracted to the ion. As a result, it is possible for a small ion to have a larger hydrodynamic radius than a large ion – if it is surrounded by more solvent molecules. Experiments involving *nuclear magnetic resonance (NMR) and isotope tracers indicate that there is considerable movement between solvent molecules within the hydrodynamic radius and the rest of the solution.

hydrofluoric acid *See* **hydrogen fluoride**.

hydrofluorocarbon **(HFC)** *See* **halocarbons**.

hydrogen Symbol H. A colourless odourless gaseous chemical element; a.n. 1; r.a.m. 1.008; d. 0.0899 g dm^{-3}; m.p. −259.14°C; b.p. −252.87°C. It is the lightest element and the most abundant in the universe. It is present in water and in all organic compounds. There are three isotopes: naturally occurring hydrogen consists of the two stable isotopes hydrogen-1 (99.985%) and *deuterium. The radioactive *tritium is made artificially. The gas is diatomic and has two forms: *orthohydrogen*, in which the nuclear spins are parallel, and *parahydrogen*, in which they are antiparallel. At normal temperatures the gas is 25% parahydrogen. In the liquid it is 99.8% parahydrogen. The main source of hydrogen is steam *reforming of natural gas. It can also be made by the Bosch process (*see* **Haber process**) and by electrolysis of water. The main use is in the Haber process for making ammonia. Hydrogen is also used in various other industrial processes, such as the reduction of oxide ores, the refining of petroleum, the production of hydrocarbons from coal, and the hydrogenation of vegetable oils. Considerable interest has also been shown in its potential use in a 'hydrogen fuel economy' in which primary energy sources not based on fossil fuels (e.g. nuclear, solar, or geothermal energy) are used to produce electricity, which is employed in electrolysing water. The hydrogen formed is stored as liquid hydrogen or as metal hydrides. Chemically, hydrogen reacts with most elements. It was discovered by Henry *Cavendish in 1766.

hydrogen acceptor *See* **hydrogen carrier**.

hydrogenation **1.** A chemical reaction with hydrogen; in particular, an addition reaction in which hydrogen adds to an unsaturated compound. Nickel is a good catalyst for such reactions. **2.** The process of converting coal to oil by making the carbon in the coal combine with hydrogen to form hydrocarbons. *See* **Fischer–Tropsch process**; **Bergius process**.

hydrogen azide **(hydrazoic acid; azoimide)** A colourless liquid, HN_3; r.d. 1.09; m.p. −80°C; b.p. 37°C. It is highly toxic and a powerful reductant, which explodes in the presence of oxygen and other oxidizing agents. It may be prepared by the reaction of sodium amide and sodium nitrate at

175°C followed by distillation of a mixture of the resulting sodium azide and a dilute acid. *See also* **azides**.

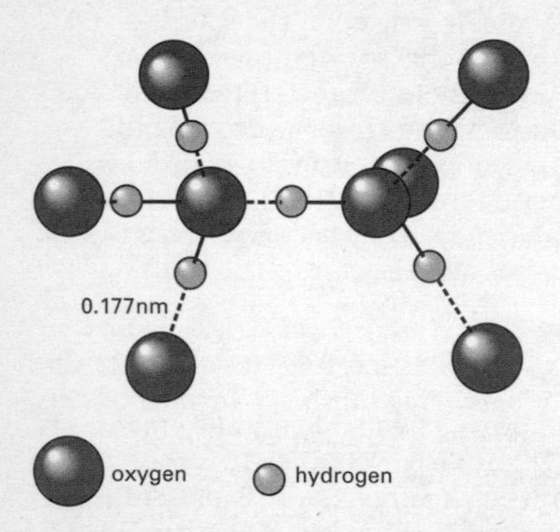

Hydrogen bonds (shown as dotted lines) between water molecules

hydrogen bond A type of electrostatic interaction between molecules occurring in molecules that have hydrogen atoms bound to electronegative atoms (F, N, O). It can be regarded as a strong dipole–dipole attraction caused by the electron-withdrawing properties of the electronegative atom. Thus, in the water molecule the oxygen atom attracts the electrons in the O–H bonds. The hydrogen atom has no inner shells of electrons to shield the nucleus, and there is an electrostatic interaction between the hydrogen proton and a lone pair of electrons on an oxygen atom in a neighbouring molecule. Each oxygen atom has two lone pairs and can make hydrogen bonds to two different hydrogen atoms. The strengths of hydrogen bonds are about one tenth of the strengths of normal covalent bonds. Hydrogen bonding does, however, have significant effects on physical properties. Thus it accounts for the unusual properties of *water and for the relatively high boiling points of H_2O, HF, and NH_3 (compared with H_2S, HCl, and PH_3). It is also of great importance in living organisms. Hydrogen bonding occurs between bases in the chains of DNA. It also occurs between the C=O and N–H groups in proteins, and is responsible for maintaining the secondary structure. Hydrogen bonds are not purely electrostatic and can be shown to have some covalent character.

hydrogen bromide A colourless gas, HBr; m.p. −88.5°C; b.p. −67°C. It can be made by direct combination of the elements using a platinum catalyst. It is a strong acid dissociating extensively in solution (*hydrobromic acid*).

hydrogencarbonate (bicarbonate) A salt of *carbonic acid in which one hydrogen atom has been replaced; it thus contains the hydrogencarbonate ion HCO_3^-.

hydrogen carrier (hydrogen acceptor) A molecule that accepts hydrogen atoms or ions, becoming reduced in the process (*see* **oxidation–reduction**).

The *electron transport chain, whose function is to generate energy in the form of ATP during respiration, involves a series of hydrogen carriers, including *NAD and *FAD, which pass on the hydrogen (derived from the breakdown of glucose) to the next carrier in the chain.

hydrogen chloride A colourless fuming gas, HCl; m.p. −114.8°C; b.p. −85°C. It can be prepared in the laboratory by heating sodium chloride with concentrated sulphuric acid (hence the former name *spirits of salt*). Industrially it is made directly from the elements at high temperature and used in the manufacture of PVC and other chloro compounds. It is a strong acid and dissociates fully in solution (*hydrochloric acid*).

hydrogen cyanide (hydrocyanic acid; prussic acid) A colourless liquid or gas, HCN, with a characteristic odour of almonds; r.d. 0.699 (liquid at 22°C); m.p. −14°C; b.p. 26°C. It is an extremely poisonous substance formed by the action of acids on metal cyanides. Industrially, it is made by catalytic oxidation of ammonia and methane with air and is used in producing acrylate plastics. Hydrogen cyanide is a weak acid ($K_a = 2.1 \times 10^{-9}$ mol dm^{-3}). With organic carbonyl compounds it forms *cyanohydrins.

hydrogen electrode *See* **hydrogen half cell.**

hydrogen fluoride A colourless liquid, HF; r.d. 0.99; m.p. −83°C; b.p. 19.5°C. It can be made by the action of sulphuric acid on calcium fluoride. The compound is an extremely corrosive fluorinating agent, which attacks glass. It is unlike the other hydrogen halides in being a liquid (a result of *hydrogen-bond formation). It is also a weaker acid than the others because the small size of the fluorine atom means that the H–F bond is shorter and stronger. Solutions of hydrogen fluoride in water are known as *hydrofluoric acid*.

hydrogen half cell (hydrogen electrode) A type of *half cell in which a metal foil is immersed in a solution of hydrogen ions and hydrogen gas is bubbled over the foil. The standard hydrogen electrode, used in measuring standard *electrode potentials, uses a platinum foil with a 1.0 M solution of hydrogen ions, the gas at 1 atmosphere pressure, and a temperature of 25°C. It is written $Pt(s)|H_2(g), H^+(aq)$, the effective reaction being

$$H_2 \rightarrow 2H^+ + 2e.$$

hydrogenic Describing an atom or ion that has only one electron; for example, H, He^+, Li^{2+}, C^{5+}. Hydrogenic atoms (or ions) do not involve electron–electron interactions and are easier to treat theoretically.

hydrogen iodide A colourless gas, HI; m.p. −51°C; b.p. −35.38°C. It can be made by direct combination of the elements using a platinum catalyst. It is a strong acid dissociating extensively in solution (*hydroiodic acid* or *hydriodic acid*). It is also a reducing agent.

hydrogen ion *See* **acid; pH scale.**

hydrogen molecule ion The simplest type of molecule (H_2^+), consisting of two hydrogen nuclei and one electron. In the *Born–Oppenheimer approximation the *Schrödinger equation for the hydrogen molecule ion can be solved exactly.

hydrogen peroxide A colourless or pale blue viscous unstable liquid, H_2O_2; r.d. 1.44; m.p. −0.41°C; b.p. 150.2°C. As with water, there is considerable hydrogen bonding in the liquid, which has a high dielectric constant. It can be made in the laboratory by adding dilute acid to barium peroxide at 0°C. Large quantities are made commercially by electrolysis of $KHSO_4.H_2SO_4$ solutions. Another industrial process involves catalytic oxidation (using nickel, palladium, or platinum with an anthraquinone) of hydrogen and water in the presence of oxygen. Hydrogen peroxide readily decomposes in light or in the presence of metal ions to give water and oxygen. It is usually supplied in solutions designated by volume strength. For example, 20-volume hydrogen peroxide would yield 20 volumes of oxygen per volume of solution. Although the *peroxides are formally salts of H_2O_2, the compound is essentially neutral. Thus, the acidity constant of the ionization

$$H_2O_2 + H_2O \rightleftharpoons H_3O^+ + HO_2^-$$

is 1.5×10^{-12} mol dm^{-3}. It is a strong oxidizing agent, hence its use as a mild antiseptic and as a bleaching agent for cloth, hair, etc. It has also been used as an oxidant in rocket fuels.

hydrogen spectrum The atomic spectrum of hydrogen is characterized by lines corresponding to radiation quanta of sharply defined energy. A graph of the frequencies at which these lines occur against the ordinal number that characterizes their position in the series of lines, produces a smooth curve indicating that they obey a formal law. In 1885 J. J. Balmer (1825–98) discovered the law having the form:

$$1/\lambda = R(1/n_1^2 + 1/n_2^2)$$

This law gives the so-called *Balmer series* of lines in the visible spectrum in which $n_1 = 2$ and $n_2 = 3,4,5...$, λ is the wavelength associated with the lines, and R is the *Rydberg constant.

In the *Lyman series*, discovered by Theodore Lyman (1874–1954), $n_1 = 1$ and the lines fall in the ultraviolet. The Lyman series is the strongest feature of the solar spectrum as observed by rockets and satellites above the earth's atmosphere. In the *Paschen series*, discovered by F. Paschen (1865–1947), $n_1 = 3$ and the lines occur in the far infrared. The *Brackett series* ($n_1 = 4$), *Pfund series* ($n_1 = 5$), and *Humphreys series* ($n_1 = 6$) also occur in the far infrared.

hydrogensulphate (bisulphate) A salt containing the ion HSO_4^- or an ester of the type $RHSO_4$, where R is an organic group. It was formerly called *hydrosulphate.*

hydrogen sulphide (sulphuretted hydrogen) A gas, H_2S, with an odour of rotten eggs; r.d. 1.54 (liquid); m.p. −85.5°C; b.p. −60.7°C. It is soluble in water and ethanol and may be prepared by the action of mineral acids on metal sulphides, typically hydrochloric acid on iron(II) sulphide (*see* **Kipp's apparatus**). Solutions in water (known as *hydrosulphuric acid*) contain the anions HS^- and minute traces of S^{2-} and are weakly acidic. Acid salts (those containing the HS^- ion) are known as *hydrogensulphides* (formerly *hydrosulphides*). In acid solution hydrogen sulphide is a mild reducing agent. Hydrogen sulphide has an important role in traditional qualitative chemical analysis, where it precipitates metals with insoluble sulphides. Hydrogen sulphide is exceedingly poisonous (more toxic than hydrogen cyanide). *See also* **Claus process.**

hydrogensulphite **(bisulphite)** A salt containing the ion $^-HSO_3$ or an ester of the type $RHSO_3$, where R is an organic group.

hydroiodic acid *See* **hydrogen iodide.**

hydrolysis A chemical reaction of a compound with water. For instance, salts of weak acids or bases hydrolyse in aqueous solution, as in

$$Na^+CH_3COO^- + H_2O \rightleftharpoons Na^+ + OH^- + CH_3COOH$$

The reverse reaction of *esterification is another example. *See also* **solvolysis.**

hydromagnesite A mineral form of basic *magnesium carbonate, $3MgCO_3.Mg(OH)_2.3H_2O$.

hydron The positive ion H^+. The name is used when it is not relevant to specify the isotope, as is usually the case in compounds in which the hydrogen is in natural abundance (i.e. about 0.015% deuterium). Thus in most cases it would be more correct to speak of 'hydron transfer' (for example) than 'proton transfer'. When the isotope is relevant then proton ($^1H^+$), deuteron ($^2H^+$), or triton ($^3H^+$) should be used.

hydronium ion *See* **oxonium ion**

hydrophilic Having an affinity for water. *See* **lyophilic.**

hydrophobic Lacking affinity for water. *See* **lyophobic.**

hydroquinone *See* **benzene-1,4-diol.**

hydrosol A sol in which the continuous phase is water. *See* **colloids.**

hydrosulphate *See* **hydrogensulphate.**

hydrosulphide *See* **hydrogen sulphide.**

hydrosulphuric acid *See* **hydrogen sulphide.**

hydroxide A metallic compound containing the ion OH^- (*hydroxide ion*) or containing the group –OH (hydroxyl group) bound to a metal atom. Hydroxides of typical metals are basic; those of *metalloids are amphoteric.

hydroxoacid A type of acid in which the acidic hydrogen is on a hydroxyl group attached to an atom that is not attached to an oxo (=O) group. An example is

$$Si(OH_4) + H_2O \rightarrow Si(OH)_3(O)^- + H_3O^+$$

Compare **oxoacid.**

hydroxonium ion *See* **oxonium ion**.

4-hydroxybutanoic acid lactone **(γ-butyrolactone)** A colourless liquid *lactone, $C_4H_6O_2$; b.p. 206°C. It is used as a solvent and in the production of certain synthetic resins.

hydroxycerussite *See* **lead(II) carbonate hydroxide**.

hydroxyethanoic acid *See* **glycolic acid**.

hydroxylamine A colourless solid, NH_2OH, m.p. 33°C. It explodes on heating and may be employed as an oxidizing agent or reducing agent. It is made by the reduction of nitrates or nitrites, and is used in making nylon. With aldehydes and ketones it forms *oximes.

hydroxylation The introduction of a hydroxyl group (–OH) into an organic compound. For example, alkenes can be hydroxylated using potassium permanganate or lead ethanoate to give alcohols. In biochemistry, various enzymes can bring about hydroxylation.

hydroxyl group The group –OH in a chemical compound.

2-hydroxypropanoic acid *See* **lactic acid**.

hygroscopic Describing a substance that can take up water from the atmosphere. *See also* **deliquescence**.

hyperconjugation The interaction of sigma bonds with pi bonds in a molecule, as in the interaction of a methyl group with the benzene ring in toluene. It is postulated to account for the stability of some carbonium ions. Hyperconjugation is often thought of as a contribution of resonance structures in which a sigma bond is broken to give ions; e.g. $C_6H_5CH_2^-H^+$ in toluene. It has been called *no-bond resonance*.

hyperfine structure *See* **fine structure**.

hypertonic solution A solution that has a higher osmotic pressure than some other solution. *Compare* **hypotonic solution**.

hypochlorite *See* **chlorates**.

hypochlorous acid *See* **chloric(I) acid**.

hypophosphorus acid *See* **phosphinic acid**.

hyposulphite *See* **sulphinate**.

hyposulphurous acid *See* **sulphinic acid**.

hypothesis *See* **laws, theories, and hypotheses**.

hypotonic solution A solution that has a lower osmotic pressure than some other solution. *Compare* **hypertonic solution**.

hypsochromic shift A shift of a spectral band to shorter wavelengths as a result of substitution in a molecule or as a result of a change in the physical conditions. *Compare* **bathochromic shift**.

ice *See* **water.**

ice point The temperature at which there is equilibrium between ice and water at standard atmospheric pressure (i.e. the freezing or melting point under standard conditions). It was used as a fixed point (0°) on the Celsius scale, but the kelvin and the International Practical Temperature Scale are based on the *triple point of water.

icosahedron A polyhedron having 20 triangular faces with five edges meeting at each vertex. The symmetry of an icosahedron (known as *icosahedral symmetry*) has fivefold rotation axes. It is impossible in *crystallography to have a periodic crystal with the point group symmetry of an icosahedron (icosahedral packing). However, it is possible for short-range order to occur with icosahedral symmetry in certain liquids and glasses because of the dense packing of icosahedra. Icosahedral symmetry also occurs in certain *quasicrystals, such as alloys of aluminium and manganese.

ideal crystal A single crystal with a perfectly regular lattice that contains no impurities, imperfections, or other defects.

ideal gas (perfect gas) A hypothetical gas that obeys the *gas laws exactly. An ideal gas would consist of molecules that occupy negligible space and have negligible forces between them. All collisions made between molecules and the walls of the container or between molecules and other molecules would be perfectly elastic, because the molecules would have no means of storing energy except as translational kinetic energy.

ideal solution *See* **Raoult's law.**

IE Ionization energy. *See* **ionization potential.**

ignition temperature The temperature to which a substance must be heated before it will burn in air.

Ilkovic equation A relation used in polarography relating the diffusion current i_a and the concentration c. The Ilkovic equation has the form $i_a = kc$, where k is a constant.

imides Organic compounds containing the group –CO.NH.CO.– (the *imido group*).

imido group *See* **imides.**

imines Compounds containing the group –NH– in which the nitrogen atom is part of a ring structure, or the group =NH, in which the nitrogen atom is linked to a carbon atom by a double bond. In either case, the group is referred to as an *imino group.*

imino group *See* imines.

Imperial units The British system of units based on the pound and the yard. The former f.p.s. system was used in engineering and was loosely based on Imperial units; for all scientific purposes *SI units are now used. Imperial units are also being replaced for general purposes by metric units.

implosion An inward collapse of a vessel, especially as a result of evacuation.

incandescence The emission of light by a substance as a result of raising it to a high temperature.

incommensurate lattice A lattice with long-range periodic order that has two or more periodicities in which an irrational number gives the ratio between the periodicities. An example of an incommensurate lattice occurs in certain magnetic systems in which the ratio of the magnetic period to the atomic lattice is an irrational number. The phase transition between a commensurate lattice and an incommensurate lattice can be analysed using the *Frenkel–Kontorowa model.

indene A colourless flammable hydrocarbon, C_9H_8; r.d. 0.996; m.p. −1.8°C; b.p. 182.6°C. Indene is an aromatic hydrocarbon with a five-membered ring fused to a benzene ring. It is present in coal tar and is used as a solvent and raw material for making other organic compounds.

independent-particle model A model for electrons in a many-electron system in which the correlation between electrons is either ignored or taken into account by regarding an electron as moving in an averaged-out potential that represents the interactions between the electron and all the other particles in the system. Although the independent-particle model cannot describe all aspects of many-electron systems, it has had some notable successes, such as explaining the shell structure of electrons in atoms.

indeterminacy *See* **uncertainty principle**.

indicator A substance used to show the presence of a chemical substance or ion by its colour. *Acid–base indicators* are compounds, such as phenolphthalein and methyl orange, that change colour reversibly, depending on whether the solution is acidic or basic. They are usually weak acids in which the un-ionized form HA has a different colour from the negative ion A^-. In solution the indicator dissociates slightly

$$HA \rightleftharpoons H^+ + A^-$$

In acid solution the concentration of H^+ is high, and the indicator is largely undissociated HA; in alkaline solutions the equilibrium is displaced to the right and A^- is formed. Useful acid–base indicators show a sharp colour change over a range of about 2 pH units. In titration, the point at which the reaction is complete is the *equivalence point* (i.e. the point at which equivalent quantities of acid and base are added). The *end point* is the point at which the indicator just changes colour. For accuracy, the two must be the same. During a titration the pH changes sharply close to the equivalence point, and the indicator used must change colour over the same range.

Other types of indicator can be used for other reactions. Starch, for example, is used in iodine titrations because of the deep blue complex it forms. *Oxidation–reduction indicators* are substances that show a reversible colour change between oxidized and reduced forms. *See also* **adsorption indicator**.

indigo A blue vat dye, $C_{16}H_{10}N_2O_2$. It occurs as the glucoside *indican* in the leaves of plants of the genus *Indigofera*, from which it was formerly extracted. It is now made synthetically.

indium Symbol In. A soft silvery element belonging to group 13 (formerly IIIB) of the periodic table; a.n. 49; r.a.m. 114.82; r.d. 7.31 (20°C); m.p. 156.6°C; b.p. 2080±2°C. It occurs in zinc blende and some iron ores and is obtained from zinc flue dust in total quantities of about 40 tonnes per annum. Naturally occurring indium consists of 4.23% indium–113 (stable) and 95.77% indium–115 (half-life 6×10^{14} years). There are a further five short-lived radioisotopes. The uses of the metal are small – some special-purpose electroplates and some special fusible alloys. Several semiconductor compounds are used, such as InAs, InP, and InSb. With only three electrons in its valency shell, indium is an electron acceptor and is used to dope pure germanium and silicon; it forms stable indium(I), indium(II), and indium(III) compounds. The element was discovered in 1863 by Ferdinand Reich (1799–1882) and Hieronymus Richter (1824–90).

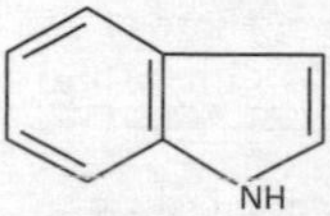

NH

Indole

indole A yellow solid, C_8H_7N, m.p. 52°C. Its molecules consist of a benzene ring fused to a nitrogen-containing five-membered ring. It occurs in some plants and in coal tar, and is produced in faeces by bacterial action. It is used in making perfumes.

induced emission (stimulated emission) The emission of a photon by an excited atom or molecule induced by an incident photon of suitable energy. The relation between induced emission and *spontaneous emission is given by the *Einstein coefficients. The process of induced emission is essential for the operation of lasers and masers.

induced-fit model A proposed mechanism of interaction between an enzyme and a substrate. It postulates that exposure of an enzyme to a substrate causes the *active site of the enzyme to change shape in order to allow the enzyme and substrate to bind (*see* **enzyme–substrate complex**). This hypothesis is generally preferred to the *lock-and-key mechanism.

inductive effect The effect of a group or atom of a compound in pulling electrons towards itself or in pushing them away. Inductive effects can be used to explain some aspects of organic reactions. For instance, electron-withdrawing groups, such as $-NO_2$, $-CN$, $-CHO$, $-COOH$, and the halogens substituted on a benzene ring, reduce the electron density on the ring and decrease its susceptibility to further (electrophilic) substitution. Electron-releasing groups, such as $-OH$, $-NH_2$, $-OCH_3$, and $-CH_3$, have the opposite effect. *See also* **electronic effects**.

industrial fermenter *See* **bioreactor**.

inelastic neutron scattering A technique for investigating the motion of molecules by scattering neutrons. The neutrons pick up or lose energy as they move through a sample of a liquid. The analysis of neutron scattering experiments enables information to be obtained about the liquid.

inert gases *See* **noble gases**.

inert-pair effect An effect seen especially in groups 13 and 14 of the periodic table, in which the heavier elements in the group tend to form compounds with a valency two lower than the expected group valency. It is used to account for the existence of thallium(I) compounds in group 13 and lead(II) in group 14. In forming compounds, elements in these groups promote an electron from a filled *s*-level state to an empty *p*-level. The energy required for this is more than compensated for by the extra energy gain in forming two more bonds. For the heavier elements, the bond strengths or lattice energies in the compounds are lower than those of the lighter elements. Consequently the energy compensation is less important and the lower valence states become favoured.

infrared chemiluminescence A technique for studying chemical reaction mechanisms by measuring and analysing weak infrared emissions from product molecules formed in certain chemical reactions. If products are formed with excess energy, this appears as excited vibrational states of the molecules, which decay with emission of infrared radiation. Spectroscopic investigation of this radiation gives information about the states in which the product molecules were formed.

infrared radiation (IR) Electromagnetic radiation with wavelengths longer than that of red light but shorter than radiowaves, i.e. radiation in the wavelength range 0.7 micrometre to 1 millimetre. It was discovered in 1800 by William Herschel (1738–1822) in the sun's spectrum. The natural vibrational frequencies of atoms and molecules and the rotational frequencies of some gaseous molecules fall in the infrared region of the electromagnetic spectrum. The infrared absorption spectrum of a molecule is highly characteristic of it and the spectrum can therefore be used for molecular identification. Glass is opaque to infrared radiation of wavelength greater than 2 micrometres and other materials, such as germanium, quartz, and polyethylene, have to be used to make lenses and prisms. Photographic film can be made sensitive to infrared up to about 1.2 μm.

infrared spectroscopy **(IR spectroscopy)** A technique for chemical analysis and the determination of structure. It is based on the principles that molecular vibrations occur in the infrared region of the electromagnetic spectrum and functional groups have characteristic absorption frequencies. The frequencies of most interest range from 2.5 to 16 μm; however, in IR spectroscopy it is common to use the reciprocal of the wavelength, and thus this range becomes 4000–625 cm^{-1}. Examples of typical vibrations are centred on 2900 cm^{-1} for C–H stretching in alkanes, 1600 cm^{-1} for N–H stretching in amino groups, and 2200 cm^{-1} for C≡C stretching in alkynes. In an IR spectrometer there is a source of IR light, covering the whole frequency range of the instrument, which is split into two beams of equal intensity. One beam is passed through the sample and the other is used as a reference against which the first is then compared. The spectrum is usually obtained as a chart showing absorption peaks, plotted against wavelength or frequency. The sample can be a gas, liquid, or solid. *See also* **Fourier-transform infrared.**

Ingold, Sir Christopher Kelk (1893–1970) British organic chemist who worked chiefly in London. His main research was on reaction mechanisms and, particularly, on *electronic effects in physical organic chemistry.

inhibition A reduction in the rate of a catalysed reaction by substances called *inhibitors*. Inhibitors may work by poisoning catalysts for the reaction or by removing free radicals in a chain reaction. *Enzyme inhibition* affects biochemical reactions, in which the catalysts are *enzymes. *Competitive inhibition* occurs when the inhibitor molecules resemble the substrate molecules and bind to the *active site of the enzyme, so preventing normal enzymatic activity. Competitive inhibition can be reversed by increasing the concentration of the substrate. In *noncompetitive inhibition* the inhibitor binds to a part of the enzyme or *enzyme substrate complex other than the active site, known as an *allosteric site*. This deforms the active site so that the enzyme cannot catalyse the reaction. Noncompetitive inhibition cannot be reversed by increasing the concentration of the substrate. The toxic effects of many substances are produced in this way. Inhibition by reaction products (*feedback inhibition*) is important in the control of enzyme activity. *See also* **allosteric enzyme.**

inner Describing a chemical compound formed by reaction of one part of a molecule with another part of the same molecule. Thus, a lactam is an inner amide; a lactone is an inner ester.

inner-sphere mechanism *See* **electron-transfer reaction.**

inner transition series *See* **transition elements.**

inorganic chemistry The branch of chemistry concerned with compounds of elements other than carbon. Certain simple carbon compounds, such as CO, CO_2, CS_2, and carbonates and cyanides, are usually treated in inorganic chemistry.

insecticide *See* **pesticide.**

insertion reaction A type of chemical reaction in which an atom or group is inserted between two atoms in a molecule. A common example is the insertion of carbene: $R_3C–X + CH_2: \rightarrow R_3CCH_2X$. The opposite of an insertion reaction is an *extrusion reaction.*

insulin A protein hormone, secreted by the β cells of the islets of Langerhans in the pancreas, that promotes the uptake of glucose by body cells, particularly in the liver and muscles, and thereby controls its concentration in the blood. Insulin was the first protein whose amino-acid sequence was fully determined (in 1955). Underproduction of insulin results in the accumulation of large amounts of glucose in the blood and its subsequent excretion in the urine. This condition, known as *diabetes mellitus*, can be treated successfully by insulin injections.

intensive variable A quantity in a macroscopic system that has a well defined value at every point inside the system and that remains (nearly) constant when the size of the system is increased. Examples of intensive variables are the pressure, temperature, density, specific heat capacity at constant volume, and viscosity. An intensive variable results when any *extensive variable is divided by an arbitrary extensive variable such as the volume. A macroscopic system can be described by one extensive variable and a set of intensive variables.

intercalation cell A type of secondary cell in which layered electrodes, usually made of metal oxides or graphite, store positive ions between the crystal layers of an electrode. In one type, lithium ions form an intercalation compound with a graphite electrode when the cell is charged. During discharge, the ions move through an electrolyte to the other electrode, made of manganese oxide, where they are more tightly bound. When the cell is being charged, the ions move back to their positions in the graphite. This backwards and forwards motion of the ions has led to the name *rocking-chair cell* for this type of system. Such cells have the advantage that only minor physical changes occur to the electrodes during the charging and discharging processes and the electrolyte is not decomposed but simply serves as a conductor of ions. Consequently, such cells can be recharged many more times than, say, a lead-acid accumulator, which eventually suffers from degeneration of the electrodes. *Lithium cells*, based on this principle, have been used in portable electronic equipment, such as camcorders. They have also been considered for use in electric vehicles.

intercalation compound A type of compound in which atoms, ions, or molecules are trapped between layers in a crystal lattice. There is no formal chemical bonding between the host crystal and the trapped molecules (*see also* **clathrate**). Such compounds are formed by *lamellar solids and are often nonstoichiometric; examples are graphitic oxide (graphite–oxygen) and the mineral *muscovite.

interhalogen A chemical compound formed between two *halogens. Interhalogens are highly reactive and volatile, made by direct combination of the elements. They include compounds with two atoms (ClF, IBr, etc.), four atoms (ClF_3, IF_3, etc.), six atoms (BrF_5, IF_5, etc.) and IF_7 with eight atoms.

intermediate bond *See* **chemical bond**.

intermediate coupling *See* **j-j coupling**.

intermetallic compound A compound consisting of two or more metallic elements present in definite proportions in an alloy.

intermolecular forces Weak forces occurring between molecules. *See* **van der Waals' force; hydrogen bond**.

internal conversion A process in which an excited atomic nucleus decays to the *ground state and the energy released is transferred by electromagnetic coupling to one of the bound electrons of that atom rather than being released as a photon. The coupling is usually with an electron in the K-, L-, or M-shell of the atom, and this *conversion electron* is ejected from the atom with a kinetic energy equal to the difference between the nuclear transition energy and the binding energy of the electron. The resulting ion is itself in an excited state and usually subsequently emits an Auger electron or an X-ray photon.

internal energy Symbol U. The total of the kinetic energies of the atoms and molecules of which a system consists and the potential energies associated with their mutual interactions. It does not include the kinetic and potential energies of the system as a whole nor their nuclear energies or other intra-atomic energies. The value of the absolute internal energy of a system in any particular state cannot be measured; the significant quantity is the change in internal energy, ΔU. For a closed system (i.e. one that is not being replenished from outside its boundaries) the change in internal energy is equal to the heat absorbed by the system (Q) from its surroundings, less the work done (W) by the system on its surroundings, i.e. $\Delta U = Q - W$. *See also* **energy; thermodynamics**.

internal resistance The resistance within a source of electric current, such as a cell or generator. It can be calculated as the difference between the e.m.f. (E) and the potential difference (V) between the terminals divided by the current being supplied (I), i.e. $r = (E - V)/I$, where r is the internal resistance.

interstitial *See* **crystal defect**.

interstitial compound A compound in which ions or atoms of a nonmetal occupy interstitial positions in a metal lattice. Such compounds often have metallic properties. Examples are found in the *carbides, *borides, and *silicides.

intersystem crossing A process in which a singlet excited electronic state makes a transition to a triplet excited state at the point where the potential energy curves for the excited singlet and triplet states cross. This transition is forbidden in the absence of *spin–orbit coupling but occurs in the presence of spin–orbit coupling. A triplet formed in this way is frequently in an excited vibrational state. This excited triplet state can reach its lowest vibrational state by collisions with other molecules. The transition from this state to the singlet state is forbidden in the absence of spin–orbit coupling but allowed when there is spin–orbit coupling. This gives rise to the slow emision of electromagnetic radiation known as *phosphorescence.

intrinsic factor *See* **vitamin B complex**.

Invar A tradename for an alloy of iron (63.8%), nickel (36%), and carbon (0.2%) that has a very low expansivity over a a restricted temperature range. It is used in watches and other instruments to reduce their sensitivity to changes in temperature.

inverse Compton effect The gain in energy of low-energy photons when they are scattered by free electrons of much higher energy. As a consequence, the electrons lose energy. *See also* **Compton effect**.

inversion A chemical reaction involving a change from one optically active configuration to the opposite configuration. The Walden inversion is an example. *See* **nucleophilic substitution**.

iodic acid Any of various oxoacids of iodine, such as iodic(V) acid and iodic(VII) acid. When used without an oxidation state specified, the term usually refers to iodic(V) acid (HIO_3).

iodic(V) acid A colourless or very pale yellow solid, HIO_3; r.d. 4.63; decomposes at 110°C. It is soluble in water but insoluble in pure ethanol and other organic solvents. The compound is obtained by oxidizing iodine with concentrated nitric acid, hydrogen peroxide, or ozone. It is a strong acid and a powerful oxidizing agent.

iodic(VII) acid **(periodic acid)** A hygroscopic white solid, H_5IO_6, which decomposes at 140°C and is very soluble in water, ethanol, and ethoxyethane. Iodic(VII) acid may be prepared by electrolytic oxidation of concentrated solutions of iodic(V) acid at low temperatures. It is a weak acid but a strong oxidizing agent.

iodide *See* **halide**.

iodine Symbol I. A dark violet nonmetallic element belonging to group 17 of the periodic table (*see* **halogens**); a.n. 53; r.a.m. 126.9045; r.d. 4.94; m.p. 113.5°C; b.p. 184.35°C. The element is insoluble in water but soluble in ethanol and other organic solvents. When heated it gives a violet vapour that sublimes. Iodine is required as a trace element (*see* **essential element**) by living organisms; in animals it is concentrated in the thyroid gland as a constituent of thyroid hormones. The element is present in sea water and was formerly extracted from seaweed. It is now obtained from oil-well brines (displacement by chlorine). There is one stable isotope, iodine-127, and fourteen radioactive isotopes. It is used in medicine as a mild antiseptic (dissolved in ethanol as *tincture of iodine*), and in the manufacture of iodine compounds. Chemically, it is less reactive than the other halogens and the most electropositive (metallic) halogen. In solution it can be determined by titration using thiosulphate solution:

$$I_2 + 2S_2O_3^{2-} \rightarrow 2I^- + S_4O_6^{2-}.$$

The molecule forms an intense blue complex with starch, which is consequently used as an indicator. It was discovered in 1812 by Courtois.

iodine(V) oxide **(iodine pentoxide)** A white solid, I_2O_5; r.d. 4.799; decomposes at 300–350°C. It dissolves in water to give iodic(V) acid and also acts as an oxidizing agent.

iodine value A measure of the amount of unsaturation in a fat or vegetable oil (i.e. the number of double bonds). It is obtained by finding the percentage of iodine by weight absorbed by the sample in a given time under standard conditions.

iodoethane **(ethyl iodide)** A colourless liquid *haloalkane, C_2H_5I; r.d. 1.9; m.p. −108°C; b.p. 72°C. It is made by reacting ethanol with a mixture of iodine and red phosphorus.

iodoform *See* **triiodomethane**.

iodoform test *See* **haloform reaction**.

iodomethane **(methyl iodide)** A colourless liquid haloalkane, CH_3I; r.d. 2.28; m.p. −66.45°C; b.p. 42.4°C. It can be made by reacting methanol with a mixture of iodine and red phosphorus.

ion An atom or group of atoms that has either lost one or more electrons, making it positively charged (a cation), or gained one or more electrons, making it negatively charged (an anion). *See also* **ionization**.

ion association The electrostatic attraction between ions in solutions of electrolytes that causes them to associate in pairs. As a result, complete dissociation does not take place and experimentally found electrical conductivities are lower than those predicted by the *Debye–Hückel theory. In 1926, Neils Bjerrum (1879–1958) proposed improving the Debye–Hückel theory by taking ion association into account. He found that the significance of ion association increases as the relative permittivity of the solvent decreases.

ion exchange The exchange of ions of the same charge between a solution (usually aqueous) and a solid in contact with it. The process occurs widely in nature, especially in the absorption and retention of water-soluble fertilizers by soil. For example, if a potassium salt is dissolved in water and applied to soil, potassium ions are absorbed by the soil and sodium and calcium ions are released from it.

The soil, in this case, is acting as an ion exchanger. Synthetic *ion-exchange resins* consist of various copolymers having a cross-linked three-dimensional structure to which ionic groups have been attached. An *anionic resin* has negative ions built into its structure and therefore exchanges positive ions. A *cationic resin* has positive ions built in and exchanges negative ions. Ion-exchange resins, which are used in sugar refining to remove salts, are synthetic organic polymers containing side groups that can be ionized. In anion exchange, the side groups are ionized basic groups, such as $-NH_3^+$ to which anions X^- are attached. The exchange reaction is one in which different anions in the solution displace the X^- from the solid. Similarly, cation exchange occurs with resins that have ionized acidic side groups such as $-COO^-$ or $-SO_2O^-$, with positive ions M^+ attached.

Ion exchange also occurs with inorganic polymers such as *zeolites, in which positive ions are held at sites in the silicate lattice. These are used for water-softening, in which Ca^{2+} ions in solution displace Na^+ ions in the zeolite. The zeolite can be regenerated with sodium chloride solution. *Ion-exchange membranes* are used as separators in electrolytic cells to remove salts from sea water and in producing deionized water. Ion-exchange resins are also used as the stationary phase in *ion-exchange chromatography*.

ion-exchange chromatography *See* **ion exchange**.

ionic bond *See* **chemical bond**.

ionic crystal *See* **crystal**.

ionic product The product of the concentrations of ions present in a given solution taking the stoichiometry into account. For a sodium chloride solution the ionic product is $[Na^+][Cl^-]$; for a calcium chloride solution it is $[Ca^{2+}][Cl^-]^2$. In pure water, there is an equilibrium with a small amount of self-ionization:

$$H_2O \rightleftharpoons H^+ + OH^-$$

The equilibrium constant of this dissociation is given by

$$K_W = [H^+][OH^-]$$

since the concentration $[H_2O]$ can be taken as constant. K_W is referred to as the ionic product of water. It has the value 10^{-14} mol^2 dm^{-6} at 25°C. In pure water (i.e. no added acid or added alkali) $[H^+] = [OH^-] = 10^{-7}$ mol dm^{-3}. In this type of self-ionization of a solvent the ionic product is also called the *autoprotolysis constant*. *See also* **solubility product**; **pH scale**.

ionic radius A value assigned to the radius of an ion in a crystalline solid, based on the assumption that the ions are spherical with a definite size. X-ray diffraction can be used to measure the internuclear distance in crystalline solids. For example, in NaF the Na – F distance is 0.231 nm, and this is assumed to be the sum of the Na^+ and F^- radii. By making certain assumptions about the shielding effect that the inner electrons have on the outer electrons, it is possible to assign individual values to the ionic radii – Na^+ 0.096 nm; F^- 0.135 nm. In general, negative ions have larger ionic radii than positive ions. The larger the negative charge, the larger the ion; the larger the positive charge, the smaller the ion. *See also* **hydrodynamic radius.**

ionic strength Symbol *I*. A function expressing the effect of the charge of the ions in a solution, equal to the sum of the molality of each type of ion present multiplied by the square of its charge. $I = \Sigma m_i z_i^2$.

ionization The process of producing *ions. Certain molecules (*see* **electrolyte**) ionize in solution; for example, *acids ionize when dissolved in water (*see also* **solvation**):

$$HCl \rightarrow H^+ + Cl^-$$

Electron transfer also causes ionization in certain reactions; for example, sodium and chlorine react by the transfer of a valence electron from the sodium atom to the chlorine atom to form the ions that constitute a sodium chloride crystal:

$$Na + Cl \rightarrow Na^+Cl^-$$

Ions may also be formed when an atom or molecule loses one or more electrons as a result of energy gained in a collision with another particle or a quantum of radiation (*see* **photoionization**). This may occur as a result of the impact of *ionizing radiation or of thermal ionization and the reaction takes the form

$$A \rightarrow A^+ + e$$

Alternatively, ions can be formed by electron capture, i.e.

$$A + e \rightarrow A^-$$

ionization energy (IE) *See* **ionization potential.**

ionization gauge A vacuum gauge consisting of a three-electrode system inserted into the container in which the pressure is to be measured. Electrons from the cathode are attracted to the grid, which is positively biased. Some pass through the grid but do not reach the anode, as it is maintained at a negative potential. Some of these electrons do, however, collide with gas molecules, ionizing them and converting them to positive ions. These ions are attracted to the anode; the resulting anode current can be used as a measure of the number of gas molecules present. Pressure as low as 10^{-6} pascal can be measured in this way.

ionization potential (IP) Symbol *I*. The minimum energy required to remove an electron from a specified atom or molecule to such a distance that there is no electrostatic interaction between ion and electron. Originally defined as the minimum potential through which an electron would have to fall to ionize an atom, the ionization potential was measured in volts. It is now, however, defined as the energy to effect an ionization and is conveniently measured in electronvolts (although this is not an SI unit) or joules per mole. The synonymous term *ionization energy* (*IE*) is often used.

The energy to remove the least strongly bound electron is the *first ionization potential.* Second, third, and higher ionization potentials can also be measured, although there is some ambiguity in terminology. Thus, in chemistry the second ionization potential is often taken to be the minimum energy required to remove an electron from the singly charged ion; the second IP of lithium would be the energy for the process

$Li^+ \rightarrow Li^{2+} + e$

In physics, the second ionization potential is the energy required to remove an electron from the next to highest energy level in the neutral atom or molecule; e.g.

$Li \rightarrow Li^{*+} + e$,

where Li^{*+} is an excited singly charged ion produced by removing an electron from the K-shell.

ionizing radiation Radiation of sufficiently high energy to cause *ionization in the medium through which it passes. It may consist of a stream of high-energy particles (e.g. electrons, protons, alpha-particles) or short-wavelength electromagnetic radiation (ultraviolet, X-rays, gamma-rays). This type of radiation can cause extensive damage to the molecular structure of a substance either as a result of the direct transfer of energy to its atoms or molecules or as a result of the secondary electrons released by ionization. In biological tissue the effect of ionizing radiation can be very serious, usually as a consequence of the ejection of an electron from a water molecule and the oxidizing or reducing effects of the resulting highly reactive species:

$2H_2O \rightarrow e^- + H_2O + H_2O^+$

$H_2O^* \rightarrow .OH + .H$

$H_2O^+ + H_2O \rightarrow .OH + H_3O^+$

where the dot before a radical indicates an unpaired electron and an asterisk denotes an excited species.

ion-microprobe analysis A technique for analysing the surface composition of solids. The sample is bombarded with a narrow beam (as small as 2 μm diameter) of high-energy ions. Ions ejected from the surface by sputtering are detected by mass spectrometry. The technique allows quantitative analysis of both chemical and isotopic composition for concentrations as low as a few parts per million.

ionophore A relatively small hydrophobic molecule that facilitates the transport of ions across lipid membranes. Most ionophores are produced by microorganisms. There are two types of ionophore: *channel formers,* which combine to form a channel in the membrane through which ions can flow; and *mobile ion carriers,* which transport ions across a membrane by forming a complex with the ion.

ion pair A pair of oppositely charged ions produced as a result of a single ionization; e.g.

$HCl \rightarrow H^+ + Cl^-$.

Sometimes a positive ion and an electron are referred to as an ion pair, as in

$A \rightarrow A^+ + e^-$.

ion pump A type of *vacuum pump that can reduce the pressure in a container to about 1 nanopascal by passing a beam of electrons through the residual gas. The gas is ionized and the positive ions formed are attracted to a cathode within the container where they remain trapped. The pump is only useful at very low pressures, i.e. below about 1 micropascal. The pump has a limited capacity because the absorbed ions eventually saturate the surface of the cathode. A more effective pump can be made by simultaneously producing a film of metal by ion impact (sputtering), so that fresh surface is continuously produced. The device is then known as a *sputter-ion pump*.

IP *See* **ionization potential**.

IR *See* **infrared radiation**.

iridium Symbol Ir. A silvery metallic *transition element (*see also* **platinum metals**); a.n. 77; r.a.m. 192.20; r.d. 22.42; m.p. 2410°C; b.p. 4130°C. It occurs with platinum and is mainly used in alloys with platinum and osmium. The element forms a range of iridium(III) and iridium(IV) complexes. It was discovered in 1804 by Smithson Tennant (1761–1815).

iron Symbol Fe. A silvery malleable and ductile metallic *transition element; a.n. 26; r.a.m. 55.847; r.d. 7.87; m.p. 1535°C; b.p. 2750°C. The main sources are the ores *haematite (Fe_2O_3), *magnetite (Fe_3O_4), limonite ($FeO(OH)_nH_2O$), ilmenite ($FeTiO_3$), siderite ($FeCO_3$), and pyrite (FeS_2). The metal is smelted in a *blast furnace to give impure *pig iron, which is further processed to give *cast iron, *wrought iron, and various types of *steel. The pure element has three crystal forms: *alpha-iron*, stable below 906°C with a body-centred-cubic structure; *gamma-iron*, stable between 906°C and 1403°C with a nonmagnetic face-centred-cubic structure; and *delta-iron*, which is the body-centred-cubic form above 1403°C. Alpha-iron is ferromagnetic up to its Curie point (768°C). The element has nine isotopes (mass numbers 52–60), and is the fourth most abundant in the earth's crust. It is required as a trace element (*see* **essential element**) by living organisms. Iron is quite reactive, being oxidized by moist air, displacing hydrogen from dilute acids, and combining with nonmetallic elements. It forms ionic salts and numerous complexes with the metal in the +2 or +3 oxidation states. Iron(VI) also exists in the ferrate ion FeO_4^{2-}, and the element also forms complexes in which its oxidation number is zero (e.g. $Fe(CO)_5$).

iron(II) chloride A green-yellow deliquescent compound, $FeCl_2$; hexagonal;

r.d. 3.16; m.p. 670°C. It also exists in hydrated forms: $FeCl_2.2H_2O$ (green monoclinic; r.d. 2.36) and $FeCl_2.4H_2O$ (blue-green monoclinic deliquescent; r.d. 1.93). Anhydrous iron(II) chloride can be made by passing a stream of dry hydrogen chloride over the heated metal; the hydrated forms can be made using dilute hydrochloric acid or by recrystallizing with water. It is converted into iron(III) chloride by the action of chlorine.

iron(III) chloride A black-brown solid, $FeCl_3$; hexagonal; r.d. 2.9; m.p. 306°C; decomposes at 315°C. It also exists as the hexahydrate $FeCl_3.6H_2O$, a brown-yellow deliquescent crystalline substance (m.p. 37°C; b.p. 280–285°C). Iron(III) chloride is prepared by passing dry chlorine over iron wire or steel wool. The reaction proceeds with incandescence when started and iron(III) chloride sublimes as almost black iridescent scales. The compound is rapidly hydrolysed in moist air. In solution it is partly hydrolysed; hydrolysis can be suppressed by the addition of hydrochloric acid. The compound dissolves in many organic solvents, forming solutions of low electrical conductivity: in ethanol, ethoxyethane, and pyridine the molecular weight corresponds to $FeCl_3$ but is higher in other solvents corresponding to Fe_2Cl_6. The vapour is also dimerized. In many ways the compound resembles aluminium chloride, which it may replace in Friedel–Crafts reactions.

iron(II) oxide A black solid, FeO; cubic; r.d. 5.7; m.p. 1420°C. It can be obtained by heating iron(II) oxalate; the carbon monoxide formed produces a reducing atmosphere thus preventing oxidation to iron(III) oxide. The compound has the sodium chloride structure, indicating its ionic nature, but the crystal lattice is deficient in iron(II) ions and it is nonstoichiometric. Iron(II) oxide dissolves readily in dilute acids.

iron(III) oxide A red-brown to black insoluble solid, Fe_2O_3; trigonal; r.d. 5.24; m.p. 1565°C. There is also a hydrated form, $Fe_2O_3.xH_2O$, which is a red-brown powder; r.d. 2.44–3.60. (*See* **rusting**.)

Iron(III) oxide occurs naturally as *haematite and can be prepared by heating iron(III) hydroxide or iron(II) sulphate. It is readily reduced on heating in a stream of carbon monoxide or hydrogen.

iron pyrites *See* pyrite.

iron(II) sulphate An off-white solid, $FeSO_4.H_2O$; monoclinic; r.d. 2.970. There is also a heptahydrate, $FeSO_4.7H_2O$; blue-green monoclinic; r.d. 1.898; m.p. 64°C. The heptahydrate is the best known iron(II) salt and is sometimes called *green vitriol* or *copperas*. It is obtained by the action of dilute sulphuric acid on iron in a reducing atmosphere. The anhydrous compound is very hygroscopic. It decomposes at red heat to give iron(III) oxide, sulphur trioxide, and sulphur dioxide. A solution of iron(II) sulphate is gradually oxidized on exposure to air, a basic iron(III) sulphate being deposited.

iron(III) sulphate A yellow hygroscopic compound, $Fe_2(SO_4)_3$; rhombic; r.d.

3.097; decomposes above 480°C. It is obtained by heating an aqueous acidified solution of iron(II) sulphate with hydrogen peroxide:

$$2FeSO_4 + H_2SO_4 + H_2O_2 \rightarrow Fe_2(SO_4)_3 + 2H_2O$$

On crystallizing, the hydrate $Fe_2(SO_4)_3.9H_2O$ is formed. The acid sulphate $Fe_2(SO_4)_3.H_2SO_4.8H_2O$ is deposited from solutions containing a sufficient excess of sulphuric acid.

irreducible representation A representation of a symmetry operation of a group, which cannot be expressed in terms of a representation of lower dimension. When the representation of the group is in matrix form (i.e. a set of matrices that multiply in the same way as the elements of the group), the matrix representation cannot be put into block-diagonal form by constructing a linear combination of the basis functions. The importance of irreducible representations in *quantum mechanics is that the energy levels of the system are labelled by the irreducible representations of the symmetry group of the system, thus enabling *selection rules to be deduced. In contrast to an irreducible representation, a *reducible representation* can be expressed in terms of a representation of lower dimension, with a reducible matrix representation that can be put into block diagonal form by constructing a linear combination of the basis functions.

irreversibility The property of a system that precludes a change to the system from being a *reversible process. The paradox that although the equations describing the bodies in a system, such as Newton's laws of motion, Maxwell's equation, or Schrödinger's equation are invariant under time reversal, events involving systems made up from large numbers of these bodies are not reversible. The process of scrambling an egg is an example. The resolution of this paradox requires the concept of *entropy using *statistical mechanics. Irreversibility occurs in the transition from an ordered arrangement to a disordered arrangement, which is a natural trend, since changes in a closed system occur in the direction of increasing entropy.

irreversible process *See* **irreversibility; reversible process.**

irreversible reaction *See* **chemical reaction.**

IR spectroscopy *See* **infrared spectroscopy.**

isentropic process Any process that takes place without a change of *entropy. The quantity of heat transferred, δQ, in a reversible process is proportional to the change in entropy, δS, i.e. $\delta Q = T\delta S$, where T is the thermodynamic temperature. Therefore, a reversible *adiabatic process is isentropic, i.e. when δQ equals zero, δS also equals zero.

Ising model A model for magnetic systems in which atomic *spins have to be aligned either parallel or antiparallel to a given direction. The Ising model was introduced, and solved in the case of one dimension, by E. Ising in 1925. Ising found that in one dimension, in the absence of an external magnetic field, there is no spontaneous magnetization at any temperature above absolute zero. The study of *phase transitions in the Ising model in dimensions greater than one has been very important to the general understanding of phase transitions. In two dimensions, the Ising model was first solved exactly by L. Onsager in 1944. In three dimensions, approximation techniques, frequently involving renormalization have to be used.

iso- Prefix denoting that a compound is an *isomer, e.g. isopentane ($CH_3CH(CH_3)C_2H_5$, 2-methylbutane) is an isomer of pentane.

isobar **1.** A curve on a graph indicating readings taken at constant pressure. **2.** One of two or more nuclides that have the same number of nucleons but different *atomic numbers. Radium-88, actinium-89, and thorium-90 are isobars as each has a *nucleon number of 228.

isocyanate *See* **cyanic acid.**

isocyanic acid *See* **cyanic acid.**

isocyanide *See* **isonitrile.**

isocyanide test A test for primary amines by reaction with an alcoholic solution of potassium hydroxide and trichloromethane.

$$RNH_2 + 3KOH + CHCl_3 \rightarrow RNC + 3KCl + 3H_2O$$

The isocyanide RNC is recognized by its unpleasant smell. This reaction of primary amines is called the *carbylamine reaction.*

isoelectric point The point at which a substance (such as a colloid or protein) has zero net electric charge. Usually such substances are positively or negatively charged, depending on whether hydrogen ions or hydroxyl ions are predominantly absorbed. At the isoelectric point the net charge on the substance is zero, as positive and negative ions are absorbed equally. The substance has its minimum conductivity at its isoelectric point and therefore coagulates best at this point. In the case of hydrophilic substances, in which the surrounding water prevents coagulation, the isoelectric point is at the minimum of stability. The isoelectric point is characterized by the value of the pH at that point. Above the isoelectric pH level the substance acts as a base and below this level it acts as an acid. For example, at the isoelectric point the pH of gelatin is 4.7. Proteins precipitate most readily at their isoelectric points; this property can be utilized to separate mixtures of proteins or amino acids.

isoelectronic Denoting different molecules that have the same number of electrons. For example N_2 and CO are isoelectronic. The energy level diagrams of isoelectronic molecules are therefore similar.

isoenzyme *See* **isozyme.**

isoleucine *See* **amino acid.**

isomerism The existence of chemical compounds (*isomers*) that have the

1-chloropropane 2-chloropropane

structural isomers in which the functional group has different positions

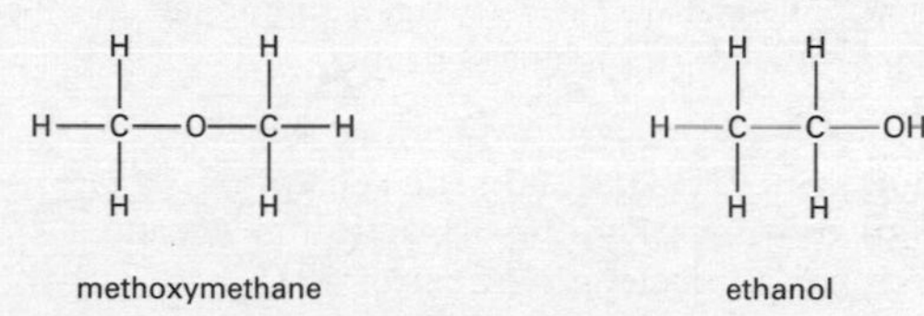

methoxymethane ethanol

structural isomers in which the functional groups are different

trans-but-2-ene *cis*-but-2-ene

cis–trans isomers in which the groups are distributed on a double bond

cis–trans isomers in a square-planar complex

keto form enol form

keto–enol tautomerism

Isomerism

same molecular formulae but different molecular structures or different arrangements of atoms in space. In *structural* (or *constitutional*) *isomerism* the molecules have different molecular structures: i.e. they may be different types of compound or they may simply differ in the position of the functional group in the molecule. Structural isomers generally have different physical and chemical properties. In *stereoisomerism*, the isomers

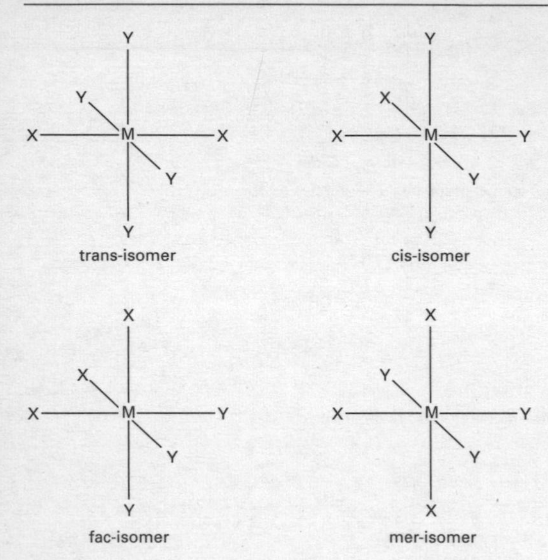

Isomers of octahedral complexes

have the same formula and functional groups, but differ in the arrangement of groups in space. Optical isomerism is one form of this (*see* **optical activity**). Another type is *cis–trans isomerism* (formerly *geometrical isomerism*), in which the isomers have different positions of groups with respect to a double bond or ring or central atom (see illustration). Octahedral complexes can display cis-trans isomerism if they have formulae of the type MX_2Y_4. Octahedral complexes with formulae of the type MX_3Y_3 can display a different type of isomerism. If the three X ligands are in a plane that includes the metal atom and the three Y ligands are in a different plane at right angles, then the structure is a *mer-isomer* (meridional). If the three X ligands are all on one face of the octahedron and the three Y ligands are on an opposite face, then it is a *fac-isomer* (facial). *See also* **E–Z convention**; **ambidentate**.

isomers *See* **isomerism**.

isometric 1. ***(in crystallography)*** Denoting a system in which the axes are perpendicular to each other, as in cubic crystals. **2.** Denoting a line on a graph illustrating the way in which temperature and pressure are interrelated at constant volume.

isomorphism The existence of two or more substances (*isomorphs*) that have the same crystal structure, so that they form *solid solutions.

isonitrile (isocyanide; carbylamine) An organic compound containing the group –NC, in which the bonding is to the nitrogen atom.

iso-octane *See* **octane**; **octane number**.

isopleth A vertical line in a liquid–vapour phase diagram consisting of a line of constant composition of the whole system as the pressure is changed. The word isopleth comes from the Greek for 'equal abundance'. *See also* **tie line**.

isopoly compound *See* **cluster compound**.

isopolymorphism The property of a substance with more than one crystalline structure that is isomorphous with the crystalline structures of another substance (*see* **polymorphism**). For example, antimony(III) oxide, Sb_2O_3, has both rhombic and octahedral structures and these are isomorphous with the similar structures of arsenic(III) oxide, As_2O_3; both oxides therefore exhibit isopolymorphism.

isoprene A colourless liquid diene, $CH_2{:}C(CH_3)CH{:}CH_2$. The systematic name is *2-methylbuta-1,3-diene*. It is the structural unit in *terpenes and natural *rubber, and is used in making synthetic rubbers.

isotactic polymer *See* **polymer**.

isotherm 1. A line on a map or chart joining points or places of equal temperature. **2.** A curve on a graph representing readings taken at constant temperature (e.g. the relationship between the pressure and volume of a gas at constant temperature).

isothermal process Any process that takes place at constant temperature. In such a process heat is, if necessary, supplied or removed from the system at just the right rate to maintain constant temperature. *Compare* **adiabatic process**.

isotonic Describing solutions that have the same osmotic pressure.

isotope One of two or more atoms of the same element that have the same number of protons in their nucleus but different numbers of neutrons. Hydrogen (1 proton, no neutrons), deuterium (1 proton, 1 neutron), and tritium (1 proton, 2 neutrons) are isotopes of hydrogen. Most elements in nature consist of a mixture of isotopes.

isotope separation The separation of the *isotopes of an element from each other on the basis of slight differences in their physical properties. For laboratory quantities the most suitable device is often the mass spectrometer. On a larger scale the methods used include gaseous diffusion (widely used for separating isotopes of uranium in the form of the gas uranium hexafluoride), distillation (formerly used to produce heavy water), electrolysis (requiring cheap electrical power), thermal diffusion (formerly used to separate uranium isotopes, but now considered uneconomic), centrifuging, and laser methods (involving the excitation of one isotope and its subsequent separation by electromagnetic means).

isotopic isomers *See* **isotopomers**.

isotopic number **(neutron excess)** The difference between the number of neutrons in an isotope and the number of protons.

isotopologues Substances that differ only in isotopic composition, e.g. CH_3OH, CH_2DOH, CHD_2OH, CD_3OH, CD_3OD.

isotopomers **(isotopic isomers)** Molecules that have the same numbers of each type of isotopic atom but in different arrangements in the molecule. For example, CH_2DOH and CH_3OD are isotopomers.

isotropic Denoting a medium whose physical properties are independent of direction. *Compare* **anisotropic**.

isozyme **(isoenzyme)** One of several forms of an enzyme in an individual or population that catalyse the same reaction but differ from each other in such properties as substrate affinity and maximum rates of enzyme–substrate reaction (*see* **Michaelis–Menten curve**).

J

Jablonski diagram A diagram that represents the electronic energy levels (and their relative positions) of a molecule. A Jablonski diagram enables radiative transitions between energy levels in molecules, as well as such phenomena as internal conversion, to be shown. The vertical axis measures energy, with the electronic energy levels in their lowest vibrational states being short horizontal lines. The vibrational energy levels of a particular electronic state are drawn as shaded regions above the horizontal line for that state. In a Jablonski diagram the horizontal location of the electronic state is not related to the internuclear distance between the atoms in these states.

jade A hard semiprecious stone consisting either of jadeite or nephrite. *Jadeite*, the most valued of the two, is a sodium aluminium pyroxene, $NaAlSi_2O_6$. It is prized for its intense translucent green colour but white, green and white, brown, and orange varieties also occur. The only important source of jadeite is in the Mogaung region of upper Burma. *Nephrite* is one of the amphibole group of rock-forming minerals. It occurs in a variety of colours, including green, yellow, white, and black. Sources include the Siberia, Turkistan, New Zealand, Alaska, China, and W USA.

jadeite *See* jade.

Jahn–Teller effect If a likely structure of a nonlinear molecule or ion would have degenerate orbitals (i.e. two molecular orbitals with the same energy levels) the actual structure of the molecule or ion is distorted so as to split the energy levels ('raise' the degeneracy). The effect is observed in inorganic complexes. For example, the ion $[Cu(H_2O)_6]^{2+}$ is octahedral and the six ligands might be expected to occupy equidistant positions at the corners of a regular octahedron. In fact, the octahedron is distorted, with four ligands in a square and two opposite ligands further away. If the 'original' structure has a centre of symmetry, the distorted structure must also have a centre of symmetry. The effect was predicted theoretically by H. A. Jahn (1907–) and Edward Teller (1908–) in 1937.

jasper An impure variety of *chalcedony. It is associated with iron ores and as a result contains iron oxide impurities that give the mineral its characteristic red or reddish-brown colour. Jasper is used as a gemstone.

jet A variety of *coal that can be cut and polished and is used for jewellery, ornaments, etc.

jeweller's rouge Red powdered haematite, iron(III) oxide, Fe_2O_3. It is a mild abrasive used in metal cleaners and polishes.

j-j coupling A type of *coupling occurring between electrons in atoms and nucleons in nuclei, in which the energies associated with the spin–orbit interactions are much higher than the energies associated with electrostatic repulsion. *Multiplets of many-electron atoms having a large atomic number are characterized by j-j coupling. The multiplets of many atoms and nuclei are intermediate between j-j coupling and *Russell–Saunders coupling, a state called *intermediate coupling*.

Joliot-Curie, Irène (1897–1956) French physicist, daughter of Marie and Pierre *Curie, who was educated by her mother and her scientist associates. In 1921 she began work at the Radium Institute, becoming director in 1946. In 1926 she married *Frédéric Joliot* (1900–58). They shared the 1935 Nobel Prize for chemistry for their discovery of artificial radioactivity the previous year.

joliotium *See* **transactinide elements**.

joule Symbol J. The *SI unit of work and energy equal to the work done when the point of application of a force of one newton moves, in the direction of the force, a distance of one metre. 1 joule = 10^7 ergs = 0.2388 calorie. It is named after James Prescott *Joule.

Joule, James Prescott (1818–89) British physicist. In 1840 he discovered the relationship between electric current passing through a wire, its resistance, and the amount of heat produced. In 1849 he gave an account of the kinetic theory of gases, and a year later announced his best-known finding, the mechanical equivalent of heat. Later, with William Thomson (Lord *Kelvin), he discovered the *Joule–Thomson effect.

Joule's law The *internal energy of a given mass of gas is independent of its volume and pressure, being a function of temperature alone. This law, which was formulated by James Prescott Joule, applies only to *ideal gases (for which it provides a definition of thermodynamic temperature) as in a real gas intermolecular forces would cause changes in the internal energy should a change of volume occur. *See also* **Joule–Thomson effect**.

Joule–Thomson effect (Joule–Kelvin effect) The change in temperature that occurs when a gas expands through a porous plug into a region of lower pressure. For most real gases the temperature falls under these circumstances as the gas has to do internal work in overcoming the intermolecular forces to enable the expansion to take place. This is a deviation from *Joule's law. There is usually also a deviation from *Boyle's law, which can cause either a rise or a fall in temperature since any increase in the product of pressure and volume is a measure of external work done. At a given pressure, there is a particular temperature, called the *inversion temperature* of the gas, at which the rise in temperature from the Boyle's law deviation is balanced by the fall from the Joule's law deviation. There is then no temperature change. Above the inversion temperature the gas is heated by expansion, below it, it is cooled. The effect was discovered by James Prescott Joule working in collaboration with William Thomson (later Lord Kelvin).

kainite A naturally occurring double salt of magnesium sulphate and potassium chloride, $MgSO_4.KCl.3H_2O$.

kalinite A mineral form of *aluminium potassium sulphate ($Al_2(SO_4)_3.K_2SO_4.24H_2O$).

kaolin **(china clay)** A soft white clay that is composed chiefly of the mineral kaolinite (*see* **clay minerals**). It is formed during the weathering and hydrothermal alteration of other clays or feldspar. Kaolin is mined in the UK, France, the Czech Republic, and USA. Besides its vital importance in the ceramics industry it is also used extensively as a filler in the manufacture of rubber, paper, paint, and textiles and as a constituent of medicines.

katharometer An instrument for comparing the thermal conductivities of two gases by comparing the rate of loss of heat from two heating coils surrounded by the gases. The instrument can be used to detect the presence of a small amount of an impurity in air and is also used as a detector in gas chromatography.

Kekulé, Friedrich August von Stradonitz (1829–96) German chemist, who became professor at Ghent (1858) and later at Bonn (1867). He studied the structures of organic molecules and was the first to recognize that carbon atoms form stable chains. Kekulé is best remembered for his proposed structure for *benzene in 1865, which he correctly interpreted as having a symmetrical ring of six carbon atoms.

Kekulé structure A proposed structure of *benzene in which the molecule has a hexagonal ring of carbon atoms linked by alternating double and single bonds. Kekulé structures contribute to the resonance hybrid of benzene. The structure was suggested in 1865 by Friedrich August *Kekulé.

kelvin Symbol K. The *SI unit of thermodynamic *temperature equal to the fraction 1/273.16 of the thermodynamic temperature of the *triple point of water. The magnitude of the kelvin is equal to that of the degree Celsius (centigrade), but a temperature expressed in degrees celsius is numerically equal to the temperature in kelvins less 273.15 (i.e. °C = K – 273.15). The *absolute zero of temperature has a temperature of 0 K (–273.15°C). The former name *degree kelvin* (symbol °K) became obsolete by international agreement in 1967. The unit is named after Lord Kelvin.

Kelvin, Lord (William Thomson; 1824–1907) British physicist, born in Belfast, who became professor of natural philosophy at Glasgow University in 1846. He carried out important experimental work on electromagnetism, inventing the mirror galvanometer and contributing to the development of telegraphy. He also worked with James *Joule on the *Joule–Thomson (or Joule–Kelvin) effect. His main theoretical work was in *thermodynamics, in which he stressed the importance of the conservation of energy. He also introduced the concept of absolute zero and the Kelvin temperature scale based on it; the unit of thermodynamic temperature is named after him. In 1896 he was created Baron Kelvin of Largs.

keratin Any of a group of fibrous *proteins occurring in hair, feathers, hooves, and horns. Keratins have coiled polypeptide chains that combine to form supercoils of several polypeptides linked by disulphide bonds between adjacent cysteine amino acids.

kerosine *See* **petroleum**.

Kerr effect The ability of certain substances to refract differently light waves whose vibrations are in two directions (*see* **double refraction**) when the substance is placed in an electric field. The effect, discovered in 1875 by John Kerr (1824–1907), is caused by the fact that certain molecules have electric *dipoles, which tend to be orientated by the applied field; the normal random motions of the molecules tends to destroy this orientation and the balance is struck by the relative magnitudes of the field strength, the temperature, and the magnitudes of the dipole moments.

The Kerr effect is observed in a *Kerr cell*, which consists of a glass cell containing the liquid or gaseous substance; two capacitor plates are inserted into the cell and light is passed through it at right angles to the electric field. There are two principal indexes of refraction: n_o (the ordinary index) and n_e (the extraordinary index). The difference in the velocity of propagation in the cell causes a phase difference, δ, between the two waves formed from a beam of monochromatic light, wavelength λ, such that

$$\delta = (n_o - n_e)x/\lambda,$$

where x is the length of the light path in the cell. Kerr also showed empirically that the ratio

$$(n_o - n_e)\lambda = BE^2,$$

where E is the field strength and B is a constant, called the *Kerr constant*, which is characteristic of the substance and approximately inversely proportional to the thermodynamic temperature.

The *Kerr shutter* consists of a Kerr cell filled with a liquid, such as nitrobenzene, placed between two crossed polarizers; the electric field is arranged to be perpendicular to the axis of the light beam and at 45° to the axis of the polarizers. In the absence of a field there is no optical path through the device. When the field is switched on the nitrobenzene becomes doubly refracting and a path opens between the crossed polarizers.

ketals Organic compounds, similar to *acetals, formed by addition of an alcohol to a ketone. If one molecule of ketone (RR′CO) reacts with one molecule of alcohol R″OH, then a *hemiketal* is formed. The rings of ketose sugars are hemiketals. Further reaction produces a full ketal ($RR'C(OR'')_2$).

ketene **(keten)** **1.** The compound $CH_2{=}C{=}O$ (*ethenone*). It can be prepared by pyrolysis of propanone. The compound reacts readily with compounds containing hydroxyl or amino groups and is used as an acetylating agent. **2.** Any of a class of compounds of the type $R_1R_2{=}C{=}O$, where R_1 and R_2 are organic groups. Ketenes are reactive compounds and are often generated in a reaction medium for organic synthesis.

keto–enol tautomerism A form of tautomerism in which a compound containing a $-CH_2-CO-$ group (the *keto form* of the molecule) is in equilibrium with one containing the $-CH{=}C(OH)-$ group (the *enol*). It occurs by migration of a hydrogen atom between a carbon atom and the oxygen on an adjacent carbon. *See* **isomerism.**

keto form *See* **keto–enol tautomerism.**

ketohexose *See* **monosaccharide.**

ketol An organic compound that has both an alcohol ($-CH_2OH$) and a keto ($-CO$) group. Ketols are made by a condensation reaction between two ketones or by the oxidation of dihydric alcohols (glycols).

ketone body Any of three compounds, acetoacetic acid (3-oxobutanoic acid, CH_3COCH_2COOH), β-hydroxybutyric acid (3-hydroxybutanoic acid, $CH_3CH(OH)CH_2COOH$), and acetone or (propanone, CH_3COCH_3), produced by the liver as a result of the metabolism of body fat deposits. Ketone bodies are normally used as energy sources by peripheral tissues.

ketones Organic compounds that contain the carbonyl group ($>C{=}O$) linked to two hydrocarbon groups. The *ketone group* is a carbonyl group with two single bonds to other carbon atoms. In systematic chemical nomenclature, ketone names end with the suffix *-one*. Examples are propanone (acetone), CH_3COCH_3, and butanone (methyl ethyl ketone), $CH_3COC_2H_5$. Ketones can be made by oxidizing secondary alcohols to convert the C–OH group to C=O. Certain ketones form addition compounds with sodium hydrogensulphate(IV) (sodium hydrogensulphite). They also form addition compounds with hydrogen cyanide to give *cyanohydrins and with alcohols to give *ketals. They undergo condensation reactions to yield *oximes, *hydrazones, phenylhydrazones, and *semicarbazones. These are reactions that they share with aldehydes. Unlike aldehydes, they do not affect Fehling's solution or Tollens reagent and do not easily oxidize. Strong oxidizing agents produce a mixture of carboxylic acids; butanone, for example, gives ethanoic and propanoic acids.

ketopentose *See* **monosaccharide.**

ketose *See* **monosaccharide.**

kieselguhr **(diatomaceous earth; diatomite)** A soft fine-grained deposit consisting of the siliceous skeletal remains of diatoms, formed in lakes and ponds. Kieselguhr is used as an absorbent, filtering material, filler, and insulator.

kieserite A mineral form of *magnesium sulphate monohydrate, $MgSO_4.H_2O$.

kilo- Symbol k. A prefix used in the metric system to denote 1000 times. For example, 1000 volts = 1 kilovolt (kV).

kilogram Symbol kg. The *SI unit of mass defined as a mass equal to that of the international platinum–iridium prototype kept by the International Bureau of Weights and Measures at Sèvres, near Paris.

kimberlite A rare igneous rock that often contains diamonds. It occurs as narrow pipe intrusions but is often altered and fragmented. It consists of olivine and phlogopite mica, usually with calcite, serpentine, and other minerals. The chief occurrences of kimberlite are in South Africa, especially at Kimberley (after which the rock is named), and in the Yakutia area of Siberia.

kinematic viscosity Symbol ν. The ratio of the *viscosity of a liquid to its density. The SI unit is $m^2\ s^{-1}$.

kinetic effect A chemical effect that depends on reaction rate rather than on thermodynamics. For example, diamond is thermodynamically less stable than graphite; its apparent stability depends on the vanishingly slow rate at which it is converted. *Overpotential in electrolytic cells is another example of a kinetic effect. *Kinetic isotope effects* are changes in reaction rates produced by isotope substitution. For example, if the slow step in a chemical reaction is the breaking of a C–H bond, the rate for the deuterated compound would be slightly lower because of the lower vibrational frequency of the C–D bond. Such effects are used in investigating the mechanisms of chemical reactions.

kinetic energy *See* **energy**.

kinetic isotope effect *See* **kinetic effect**.

kinetics The branch of physical chemistry concerned with measuring and studying the rates of chemical reactions. The main aim of chemical kinetics is to determine the mechanism of reactions by studying the rate under different conditions (temperature, pressure, etc.). *See also* **activated-complex theory**; **Arrhenius equation**.

kinetic theory A theory, largely the work of Count Rumford, James Prescott *Joule, and James Clerk *Maxwell, that explains the physical properties of matter in terms of the motions of its constituent particles. In a gas, for example, the pressure is due to the incessant impacts of the gas molecules on the walls of the container. If it is assumed that the molecules occupy negligible space, exert negligible forces on each other except during collisions, are perfectly elastic, and make only brief collisions with each other, it can be shown that the pressure p exerted by one mole of gas containing n molecules each of mass m in a container of volume V, will be given by:

$$p = nm\overline{c}^2/3V,$$

where $\bar{c}^2$ is the mean square speed of the molecules. As according to the *gas laws for one mole of gas: $pV = RT$, where T is the thermodynamic temperature, and R is the molar *gas constant, it follows that:

$$RT = nm\bar{c}^2/3$$

Thus, the thermodynamic temperature of a gas is proportional to the mean square speed of its molecules. As the average kinetic *energy of translation of the molecules is $m\bar{c}^2/2$, the temperature is given by:

$$T = (m\bar{c}^2/2)(2n/3R)$$

The number of molecules in one mole of any gas is the *Avogadro constant, N_A; therefore in this equation $n = N_A$. The ratio R/N_A is a constant called the *Boltzmann constant (k). The average kinetic energy of translation of the molecules of one mole of any gas is therefore $3kT/2$. For monatomic gases this is proportional to the *internal energy (U) of the gas, i.e.

$$U = N_A 3kT/2$$

and as $k = R/N_A$,

$$U = 3RT/2$$

For diatomic and polyatomic gases the rotational and vibrational energies also have to be taken into account (*see* **degrees of freedom**).

In liquids, according to the kinetic theory, the atoms and molecules still move around at random, the temperature being proportional to their average kinetic energy. However, they are sufficiently close to each other for the attractive forces between molecules to be important. A molecule that approaches the surface will experience a resultant force tending to keep it within the liquid. It is, therefore, only some of the fastest moving molecules that escape; as a result the average kinetic energy of those that fail to escape is reduced. In this way evaporation from the surface of a liquid causes its temperature to fall.

In a crystalline solid the atoms, ions, and molecules are able only to vibrate about the fixed positions of a *crystal lattice; the attractive forces are so strong at this range that no free movement is possible.

Kipp's apparatus A laboratory apparatus for making a gas by the reaction of a solid with a liquid (e.g. the reaction of hydrochloric acid with iron sulphide to give hydrogen sulphide). It consists of three interconnected glass globes arranged vertically, with the solid in the middle globe. The upper and lower globes are connected by a tube and contain the liquid. The middle globe has a tube with a tap for drawing off gas. When the tap is closed, pressure of gas forces the liquid down in the bottom reservoir and up into the top, and reaction does not occur. When the tap is opened, the release in pressure allows the liquid to rise into the middle globe, where it reacts with the solid. It is named after Petrus Kipp (1808–64).

Kirchhoff, Gustav Robert (1824–87) German physicist, who in 1850 became a professor at Breslau and four years later joined Robert *Bunsen at Heidelberg. In 1845, while still a student, he formulated Kirchhoff's laws concerning electric circuits. With Bunsen he worked on spectroscopy, a technique that led them to discover the elements caesium (1860) and rubidium (1861).

Kjeldahl's method A method for measuring the percentage of nitrogen in an organic compound. The compound is boiled with concentrated sulphuric acid and copper(II) sulphate catalyst to convert any nitrogen to ammonium sulphate. Alkali is added and the mixture heated to distil off ammonia. This is passed into a standard acid solution, and the amount of ammonia can then be found by estimating the amount of unreacted acid by titration. The amount of nitrogen in the original specimen can then be calculated. The method was developed by the Danish chemist Johan Kjeldahl (1849–1900).

knocking The metallic sound produced by a spark-ignition petrol engine under certain conditions. It is caused by rapid combustion of the unburnt explosive mixture in the combustion chambers ahead of the flame front. As the flame travels from the sparking plug towards the piston it compresses and heats the unburnt gases ahead of it. If the flame front moves fast enough, normal combustion occurs and the explosive mixture is ignited progressively by the flame. If it moves too slowly, ignition of the last part of the unburnt gas can occur very rapidly before the flame reaches it, producing a shock wave that travels back and forth across the combustion chamber. The result is overheating, possible damage to the plugs, an undesirable noise, and loss of power (probably due to preignition caused by overheated plugs). Knocking can be avoided by an engine design that increases turbulence in the combustion chamber and thereby increases flame speed. It also can be avoided by reducing the compression ratio, but this involves loss of efficiency. The most effective method is to use high-octane fuel (*see* **octane number**), which has a longer self-ignition delay than low-octane fuels. This can be achieved by the addition of an *antiknock agent*, such as lead(IV) tetraethyl, to the fuel, which retards the combustion chain reactions. However, lead-free petrol is now preferred to petrol containing lead tetraethyl owing to environmental dangers arising from lead in the atmosphere. In the USA the addition of lead compounds is now forbidden. New formulae for petrol are designed to raise the octane number without polluting the atmosphere. These new formulae include increasing the content of aromatics and oxygenates (oxygen-containing compounds, such as alcohols). However, it is claimed that the presence in the atmosphere of incompletely burnt aromatics constitutes a cancer risk.

Knoevenagel reaction A reaction in which an aldehyde RCHO reacts with malonic acid ($CH_2(COOH)_2$) with subsequent loss of CO_2 to give an unsaturated carboxylic acid RCH=CHCOOH. Thus, the chain length is increased by 2. The reaction is base-catalysed; usually pyridine is used. The reaction is named after Emil Knoevenagel (1865–1921).

knot theory A branch of mathematics used to classify knots and entanglements. Knot theory has applications to the study of the properties of polymers and the statistical mechanics of certain models of phase transitions.

Knudsen flow *See* **molecular flow.**

Koopmans' theorem The principle that the ionization energy of a molecule is equal to the orbital energy of the ejected electron. It is the basis of the interpretation of spectra in *photoelectron spectroscopy. Koopmans' theorem is an approximation in that it ignores any reorganization of electrons in the ion formed.

Kohlrausch equation An equation that describes the molar conductivities of strong electrolytes at low concentration, i.e. $\Lambda_m = \Lambda^{\ominus}{}_m - Kc^{1/2}$, where Λ_m is the molar conductivity, $\Lambda^{\ominus}{}_m$ is the *limiting molar conductivity* (the molar conductivity in the limit of zero concentration when the ions do not interact with each other), K is a coefficient related to the stoichiometry of the electrolyte, and c is the concentration of the electrolyte. It is possible to express $\Lambda^{\circ}{}_m$ as a sum of the contribution of each of its ions. The Kohlrausch equation was first stated in the 19th century by the German chemist Friedrich Kohlrausch (1840–1910) as the result of a considerable amount of experimental work. With its characteristic $c^{1/2}$ dependence, the equation is explained quantitatively by the existence of an ionic atmosphere round the ion as analysed by the *Debye–Hückel–Onsager theory.

Kohlrausch's law If a salt is dissolved in water, the conductivity of the (dilute) solution is the sum of two values – one depending on the positive ions and the other on the negative ions. The law, which depends on the independent migration of ions, was deduced experimentally by Friedrich Kohlrausch.

Kolbe's method A method of making alkanes by electrolysing a solution of a carboxylic acid salt. For a salt Na^+RCOO^-, the carboxylate ions lose electrons at the cathode to give radicals:

$$RCOO^- - e \rightarrow RCOO\cdot$$

These decompose to give alkyl radicals

$$RCOO\cdot \rightarrow R\cdot + CO_2$$

Two alkyl radicals couple to give an alkane

$$R\cdot + R\cdot \rightarrow RR$$

The method can only be used for hydrocarbons with an even number of carbon atoms, although mixtures of two salts can be electrolysed to give a mixture of three products. The method was discovered by the German chemist Herman Kolbe (1818–84), who electrolysed pentanoic acid (C_4H_9COOH) in 1849 and obtained a hydrocarbon, which he assumed was the substance 'butyl' C_4H_9 (actually octane, C_8H_{18}).

Kovar A tradename for an alloy of iron, cobalt, and nickel with an expansivity similar to that of glass. It is therefore used in making glass-to-metal seals, especially in circumstances in which a temperature variation can be expected.

Kramers theorem The energy levels of a system, such as an atom that contains an odd number of *spin-½ particles (e.g. electrons), are at least double *degenerate in the absence of an external magnetic field. This degeneracy, known as *Kramers degeneracy*, is a consequence of time reversal invariance. Kramers theorem was stated by the Dutch physicist Hendrick Anton Kramers (1894–1952) in 1930. Kramers degeneracy is removed by placing the system in an external magnetic field. The theorem holds even in the presence of crystal fields (*see* **crystal-field theory**) and *spin–orbit coupling.

Krebs, Sir Hans Adolf (1900–81) German-born British biochemist, who emigrated to Britain in 1933, working at Sheffield University before moving to Oxford in 1954. Krebs is best known for the *Krebs cycle, the basis of which he discovered in 1937. Details were later added by Fritz Lipmann (1899–1986), with whom Krebs shared the 1953 Nobel Prize for physiology or medicine.

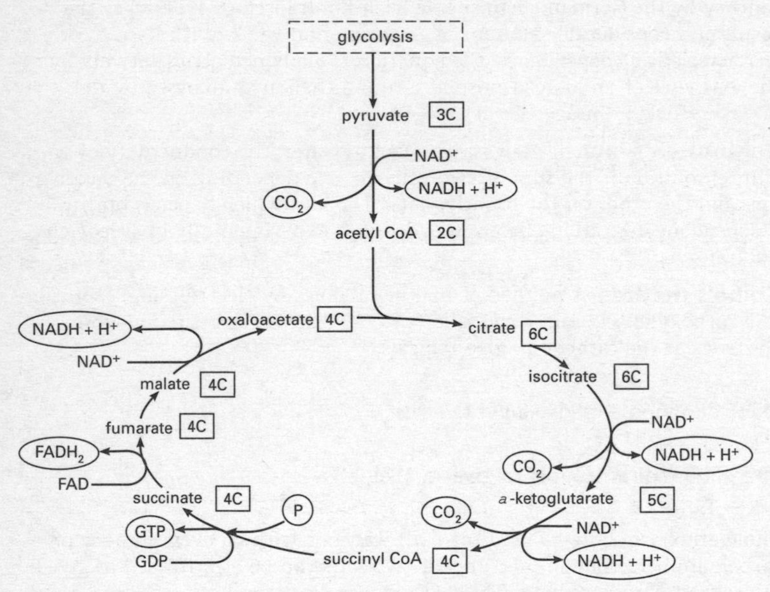

Krebs cycle

Krebs cycle (citric acid cycle; tricarboxylic acid cycle; TCA cycle) A cyclical series of biochemical reactions that is fundamental to the metabolism of aerobic organisms, i.e. animals, plants, and many microorganisms (see illustration). The enzymes of the Krebs cycle are located in the mitochondria and are in close association with the components of the *electron transport chain. The two-carbon acetyl coenzyme A (acetyl CoA) reacts with the four-carbon oxaloacetate to form the six-carbon citrate. In a series of seven reactions, this is reconverted to oxaloacetate and produces two molecules of carbon dioxide. Most importantly, the cycle generates one molecule of guanosine triphosphate (GTP – equivalent to 1 ATP) and reduces three molecules of the coenzyme *NAD to NADH and one molecule of the coenzyme *FAD to $FADH_2$. NADH and $FADH_2$ are then oxidized by the electron transport chain to generate three and two molecules of ATP respectively. This gives a net yield of 12 molecules of ATP per molecule of acetyl CoA.

Acetyl CoA can be derived from carbohydrates (via *glycolysis), fats, or certain amino acids. (Other amino acids may enter the cycle at different stages.) Thus the Krebs cycle is the central 'crossroads' in the complex system of metabolic pathways and is involved not only in degradation and energy production but also in the synthesis of biomolecules. It is named after its principal discoverer, Sir Hans Adolf *Krebs.

Kroll process A process for producing certain metals by reducing the chloride with magnesium metal, e.g.

$TiCl_4 + 2Mg \rightarrow Ti + 2MgCl_2$.

It is named after William Kroll, who devised the process in 1940.

krypton Symbol Kr. A colourless gaseous element belonging to group 0 (the *noble gases) of the periodic table; a.n. 36; r.a.m. 83.80; d. 3.73 $g\,m^{-3}$; m.p. −156.6°C; b.p. −152.3°C. Krypton occurs in air (0.0001% by volume) from which it can be extracted by fractional distillation of liquid air. Usually, the element is not isolated but is used with other inert gases in fluorescent lamps, etc. The element has five natural isotopes (mass numbers 78, 80, 82, 83, 84) and there are five radioactive isotopes (76, 77, 79, 81, 85). Krypton-85 (half-life 10.76 years) is produced in fission reactors and it has been suggested that an equilibrium amount will eventually occur in the atmosphere. The element is practically inert and forms very few compounds (certain fluorides, such as KrF_2, have been reported).

Kupfer nickel A naturally occurring form of nickel arsenide, NiAs; an important ore of nickel.

kurchatovium *See* **transactinide elements**.

labelling The process of replacing a stable atom in a compound with a radioisotope of the same element to enable its path through a biological or mechanical system to be traced by the radiation it emits. In some cases a different stable isotope is used and the path is detected by means of a mass spectrometer. A compound containing either a radioactive or stable isotope is called a *labelled compound* and the atom used is a *label*. If a hydrogen atom in each molecule of the compound has been replaced by a tritium atom, the compound is called a *tritiated compound*. A radioactive labelled compound will behave chemically and physically in the same way as an otherwise identical stable compound, and its presence can easily be detected using a Geiger counter. This process of *radioactive tracing* is widely used in chemistry, biology, medicine, and engineering. For example, it can be used to follow the course of the reaction of a carboxylic acid with an alcohol to give an ester, e.g.

$$CH_3COOH + C_2H_5OH \rightarrow C_2H_5COOCH_3 + H_2O$$

To determine whether the noncarbonyl oxygen in the ester comes from the acid or the alcohol, the reaction is performed with the labelled compound $CH_3CO^{18}OH$, in which the oxygen in the hydroxyl group of the acid has been 'labelled' by using the ^{18}O isotope. It is then found that the water product is $H_2{}^{18}O$; i.e. the oxygen in the ester comes from the alcohol, not the acid.

labile Describing a chemical compound in which certain atoms or groups can easily be replaced by other atoms or groups. The term is applied to coordination complexes in which ligands can easily be replaced by other ligands in an equilibrium reaction.

amino acid $\xrightarrow{-H_2O}$ γ lactam (γ butyrolactam) ⇌ γ lactim

Lactam formation

lactams Organic compounds containing a ring of atoms in which the group –NH.CO.– forms part of the ring. Lactams can be formed by reaction of an $-NH_2$ group in one part of a molecule with a –COOH group in the other to give a cyclic amide (see illustration). They can exist in an alternative tautomeric form, the *lactim* form, in which the hydrogen atom on the nitrogen has migrated to the oxygen of the carbonyl to give –N=C(OH)–. The pyrimidine base uracil is an example of a lactam.

lactate A salt or ester of lactic acid (i.e. a 2-hydroxypropanoate).

lactic acid (2-hydroxypropanoic acid) A clear odourless hygroscopic syrupy liquid, $CH_3CH(OH)COOH$, with a sour taste; r.d. 1.206; m.p. 18°C; b.p. 122°C. It is prepared by the hydrolysis of ethanal cyanohydrin or the oxidation of propan-1,2-diol using dilute nitric acid. Lactic acid is manufactured by the fermentation of lactose (from milk) and used in the dyeing and tanning industries. It is an alpha hydroxy *carboxylic acid. *See also* **optical activity**.

Lactic acid is produced from pyruvic acid in active muscle tissue when oxygen is limited and subsequently removed for conversion to glucose by the liver. During strenuous exercise it may build up in the muscles, causing cramplike pains. It is also produced by fermentation in certain bacteria and is characteristic of sour milk.

lactims *See* **lactams**.

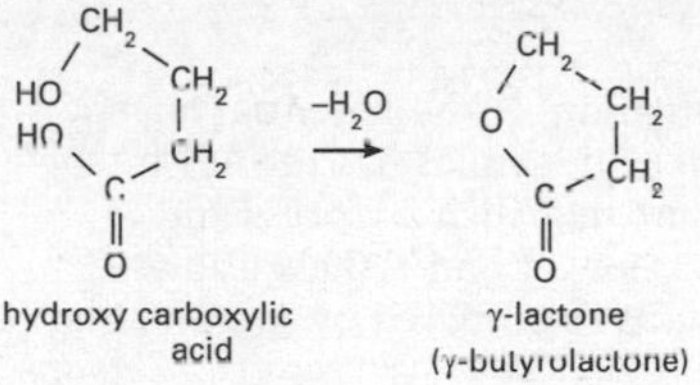

hydroxy carboxylic acid

γ-lactone (γ-butyrolactone)

Lactone formation

lactones Organic compounds containing a ring of atoms in which the group –CO.O– forms part of the ring. Lactones can be formed (or regarded as formed) by reaction of an –OH group in one part of a molecule with a –COOH group in the other to give a cyclic ester (see illustration). This type of reaction occurs with γ-hydroxy carboxylic acids such as the compound $CH_2(OH)CH_2CH_2COOH$ (in which the hydroxyl group is on the third carbon from the carboxyl group). The resulting γ-lactone has a five-membered ring. Similarly, δ-lactones have six-membered rings. β-lactones, with a four-membered ring, are not produced directly from β-hydroxy acids, but can be synthesized by other means.

lactose (milk sugar) A sugar comprising one glucose molecule linked to a galactose molecule. Lactose is manufactured by the mammary gland and occurs only in milk. For example, cows' milk contains about 4.7% lactose. It is less sweet than sucrose (cane sugar).

Ladenburg benzene An (erroneous) structure for *benzene proposed by Albert Ladenburg (1842–1911), in which the six carbon atoms were arranged at the corners of a triangular prism and linked by single bonds to each other and to the six hydrogen atoms.

laevo form *See* **optical activity**.

laevorotatory Designating a chemical compound that rotates the plane of plane-polarized light to the left (anticlockwise for someone facing the oncoming radiation). *See* **optical activity**.

laevulose *See* **fructose**.

Lagrange multipliers (undetermined multipliers) Parameters, usually denoted λ, introduced to assist in finding the maximum or minimum value of a function f of several variables $x_1,x_2,...x_n$, subject to some constraint that connects the variables. An important application of the method of Lagrange multipliers is its use in the derivation of the Boltzmann distribution in statistical mechanics, in which one of the Lagrange multipliers is $-1/kT$, where k is the *Boltzmann constant and T is the thermodynamic temperature. Lagrange multipliers were introduced by the Italian-born French mathematician Joseph-Louis Lagrange (1736–1813).

Lagrangian Symbol L. A function used to define a dynamical system in terms of functions of coordinates, velocities, and times given by:

$$L = T - V$$

where T is the kinetic energy of the system and V is the potential energy of the system. The Lagrangian formulation of dynamics has the advantage that it does not deal with many vector quantities, such as forces and accelerations, but only with two scalar functions, T and V. This leads to great simplifications. *Lagrangian dynamics* was formulated by Joseph-Louis Lagrange.

LAH Lithium aluminium hydride; *see* **lithium tetrahydroaluminate(III)**.

lake A pigment made by combining an organic dyestuff with an inorganic compound (usually an oxide, hydroxide, or salt). Absorption of the organic compound on the inorganic substrate yields a coloured complex, as in the combination of a dyestuff with a *mordant. Lakes are used in paints and printing inks.

lambda point *See* **superfluidity**.

Lamb-dip spectroscopy A spectroscopic technique enabling the centres of absorption lines to be determined very precisely by making use of the Doppler shift associated with very rapidly moving molecules. An intense monochromatic beam of radiofrequency electromagnetic radiation is passed through a sample of a gas with the frequency being slightly higher than that of maximum absorption. Only certain molecules moving at a certain specific speed can absorb radiation. The beam is then reflected back through the sample so that radiation is absorbed by molecules moving at exactly this same speed, except that they are moving away from the mirror. If the radiation is exactly at the absorption peak, only molecules moving perpendicular to the line of the beam (which hence have no Doppler shift) absorb both in the initial path and the reflected path of the radiation. Since molecules being excited in the initial path leave fewer molecules to be excited in the return path this causes a less intense absorption to be observed. As a result a dip appears in the curve, thus enabling the absorption peak to be found very accurately. Lamb-dip spectroscopy is named after Willis Eugene Lamb (1913–).

Lamb shift A small energy difference between two levels ($^2S_{1/2}$ and $^2P_{1/2}$) in the *hydrogen spectrum. The shift results from the quantum interaction between the atomic electron and the electromagnetic radiation. It was first explained by Willis Eugene Lamb.

lamellar solids Solid substances in which the crystal structure has distinct layers (i.e. has a layer lattice). The *micas are an example of this type of compound. *Intercalation compounds are lamellar compounds formed by interposition of atoms, ions, etc., between the layers of an existing element or compound. For example, graphite is a lamellar solid. With strong oxidizing agents (e.g. a mixture of concentrated sulphuric and nitric acids) it forms a nonstoichiometric 'graphitic oxide', which is an intercalation compound having oxygen atoms between the layers of carbon atoms. Substances of this type are called *graphitic compounds.*

lamp black A finely divided (microcrystalline) form of carbon made by burning organic compounds in insufficient oxygen. It is used as a black pigment and filler.

Landau levels The energy levels found by *quantum mechanics of free electrons in a uniform magnetic field. These energy levels, named after the Soviet physicist Lev Davidovich Landau (1908–68), who analysed the problem in 1930, have discrete values, which are integer multiples of heB/m, where h is the Planck constant, e is the charge of the electron, m is the mass of the electron, and B is the magnetic flux density. Each Landau level is a highly *degenerate level, with each level being filled by $2eB/h$, where the factor 2 is due to the spin of the electron.

Landé interval rule A rule used in interpreting atomic spectra stating that if the *spin–orbit coupling is weak in a given multiplet, the energy differences between two successive J levels (where J is the total resultant angular momentum of the coupled electrons) are proportional to the larger of the two values of J. The rule was stated by the German-born US physicist Alfred Landé (1888–1975) in 1923. It can be deduced from the quantum theory of angular momentum. In addition to assuming *Russell–Saunders coupling, the Landé interval rule assumes that the interactions between spin magnetic moments can be ignored, an assumption that is not correct for very light atoms, such as helium. Thus the Landé interval rule is best obeyed by atoms with medium atomic numbers.

Langevin equation A type of random equation of motion (*see* **stochastic process**) used to study *Brownian movement. The Langevin equation can be written in the form $\dot{v} = \xi v + A(t)$, where v is the velocity of a particle of mass m immersed in a fluid and $\dot{v}$ is the acceleration of the particle; ξv is a frictional force resulting from the viscosity of the fluid, with ξ being a constant friction coefficient, and $A(t)$ is a random force describing the average effect of the Brownian motion. The Langevin equation is named after the French physicist Paul Langevin (1872–1946). It is necessary to use statistical methods and the theory of probability to solve the Langevin equation.

Langmuir adsorption isotherm An equation used to describe the amount of gas adsorbed on a plane surface, as a function of the pressure of the gas in equilibrium with the surface. The Langmuir adsorption isotherm can be written:

$$\theta = bp/(1 + bp),$$

where θ is the fraction of the surface covered by the adsorbate, p is the pressure of the gas, and b is a constant called the *adsorption coefficient*, which is the equilibrium constant for the process of adsorption. The Langmuir adsorption isotherm was derived by the US chemist Irving Langmuir (1881–1957), using the *kinetic theory of gases and making the assumptions that:
(1) the adsorption consists entirely of a monolayer at the surface;
(2) there is no interaction between molecules on different sites and each site can hold only one adsorbed molecule;
(3) the heat of adsorption does not depend on the number of sites and is equal for all sites.
The Langmuir adsorption isotherm is of limited application since for real surfaces the energy is not the same for all sites and interactions between adsorbed molecules cannot be ignored.

Langmuir–Blodgett film A film of molecules on a surface that can contain multiple layers of film. A Langmuir–Blodgett film with multiple layers can be made by dipping a plate into a liquid so that it is covered by a monolayer and then repeating the process. This process, called the *Langmuir–Blodgett technique*, enables a multilayer to be built up, one monolayer at a time. Langmuir–Blodgett films have many potential practical applications, including insulation for optical and semiconductor devices and selective membranes in biotechnology.

Langmuir–Hinshelwood mechanism A possible mechanism for a bimolecular process catalyzed at a solid surface. It is assumed that two molecules are adsorbed on adjacent sites and that a reaction takes place via an activated complex on the surface to yield the products of the reaction. This mechanism was suggested by the US chemist Irving Langmuir (1881–1957) in 1921 and developed by the British chemist Sir Cyril Hinshelwood (1897–1967) in 1926. Some bimolecular reactions at surfaces are in agreement with the predictions of this model. *See also* **Langmuir–Rideal mechanism**.

Langmuir isotherm *See* **Langmuir adsorption isotherm**.

Langmuir–Rideal mechanism A possible mechanism for a bimolecular process catalysed at a solid surface. It is envisaged that one of the molecules is adsorbed and then reacts with a second molecule that is not adsorbed. This mechanism was suggested by the US chemist Irving Langmuir (1881–1957) in 1921 and developed by the British chemist Sir Eric Rideal in 1939. The Langmuir–Rideal mechanism is not common but certain reactions involving hydrogen atoms probably occur by this mechanism. *See also* **Langmuir–Hinshelwood mechanism**.

lanolin An emulsion of purified wool fat in water, containing cholesterol and certain terpene alcohols and esters. It is used in cosmetics.

lansfordite A mineral form of *magnesium carbonate pentahydrate, $MgCO_3.5H_2O$.

lanthanides *See* **lanthanoids**.

lanthanoid contraction *See* **lanthanoids**.

lanthanoids (lanthanides; lanthanons; rare-earth elements) A series of elements in the *periodic table, generally considered to range in proton number from cerium (58) to lutetium (71) inclusive. The lanthanoids all have two outer *s*-electrons (a $6s^2$ configuration), follow lanthanum, and are classified together because an increasing proton number corresponds to increase in number of $4f$ electrons. In fact, the $4f$ and $5d$ levels are close in energy and the filling is not smooth. The outer electron configurations are as follows:

57 lanthanum (La) $5d^16s^2$
58 cerium (Ce) $4f5d^16s^2$ (or $4f^26s^2$)
59 praseodymium (Pr) $4f^36s^2$
60 neodymium (Nd) $4f^46s^2$
61 promethium (Pm) $4f^56s^2$
62 samarium (Sm) $4f^66s^2$
63 europium (Eu) $4f^76s^2$
64 gadolinium (Gd) $4f^75d^16s^2$
65 terbium (Tb) $4f^96s^2$
66 dysprosium (Dy) $4f^{10}6s^2$
67 holmium (Ho) $4f^{11}6s^2$
68 erbium (Er) $4f^{12}6s^2$
69 thulium (Tm) $4f^{13}6s^2$
70 ytterbium (Yb) $4f^{14}6s^2$
71 lutetium (Lu) $4f^{14}5d^16s^2$

Note that lanthanum itself does not have a $4f$ electron but it is generally classified with the lanthanoids because of its chemical similarities, as are yttrium (Yt) and scandium (Sc). Scandium, yttrium, and lanthanum are *d*-block elements; the lanthanoids and *actinoids make up the *f*-block.

The lanthanoids are sometimes simply called the *rare earths*, although strictly the 'earths' are their oxides. Nor are they particularly rare: they occur widely, usually together. All are silvery very reactive metals. The *f*-electrons do not penetrate to the outer part of the atom and there is no *f*-orbital participation in bonding (unlike the *d*-orbitals of the main *transition elements) and the elements form few coordination compounds. The main compounds contain M^{3+} ions. Cerium also has the highly oxidizing Ce^{4+} state and europium and ytterbium have a M^{2+} state.

The 4*f* orbitals in the atoms are not very effective in shielding the outer electrons from the nuclear charge. In going across the series the increasing nuclear charge causes a contraction in the radius of the M^{3+} ion – from 0.1061 nm in lanthanum to 0.0848 nm in lutetium. This effect, the *lanthanoid contraction* (or *lanthanide contraction*), accounts for the similarity between the transition elements zirconium and hafnium.

lanthanons *See* **lanthanoids**.

lanthanum Symbol La. A silvery metallic element belonging to group 3 (formerly IIIA) of the periodic table and often considered to be one of the *lanthanoids; a.n. 57; r.a.m. 138.91; r.d. 6.146 (20°C); m.p. 921°C; b.p. 3457°C. Its principal ore is bastnasite, from which it is separated by an ion-exchange process. There are two natural isotopes, lanthanum–139 (stable) and lanthanum–138 (half-life 10^{10}–10^{15} years). The metal, being pyrophoric, is used in alloys for lighter flints and the oxide is used in some optical glasses. The largest use of lanthanum, however, is as a catalyst in cracking crude oil. Its chemistry resembles that of the lanthanoids. The element was discovered by Carl Mosander (1797–1858) in 1839.

lapis lazuli A blue rock that is widely used as a semiprecious stone and for ornamental purposes. It is composed chiefly of the deep blue mineral *lazurite* embedded in a matrix of white calcite and usually also contains small specks of pyrite. It occurs in only a few places in crystalline limestones as a contact metamorphic mineral. The chief source is Afghanistan; lapis lazuli also occurs near Lake Baikal in Siberia and in Chile. It was formerly used to make the artists' pigment ultramarine.

Laporte selection rule A selection rule in atomic spectra stating that spectral lines associated with electric-dipole radiation must arise from transitions between states of opposite parity. The Laporte selection rule was discovered by O. Laporte in 1924 and was explained by the application of group theory to the *quantum mechanics of atoms. In the case of magnetic-dipole and quadrupole radiation the selection rule for spectral lines is the opposite of the Laporte rule, i.e. transitions are only allowed between states of the same parity in these cases.

Larmor precession A precession of the motion of charged particles in a magnetic field. It was first deduced in 1897 by Sir Joseph Larmor (1857–1942). Applied to the orbital motion of an electron around the nucleus of an atom in a magnetic field of flux density B, the frequency of precession is given by $eB/4\pi mv\mu$, where e and m are the electronic charge and mass respectively, μ is the permeability, and v is the velocity of the electron. This is known as the *Larmor frequency*.

laser (light *a*mplification by *s*timulated *e*mission of *r*adiation) A light amplifier (also called an *optical maser*) usually used to produce monochromatic coherent radiation in the infrared, visible, and ultraviolet regions of the *electromagnetic spectrum. Lasers that operate in the X-ray region of the spectrum are also being developed.

Nonlaser light sources emit radiation in all directions as a result of the spontaneous emission of photons by thermally excited solids (filament lamps) or electronically excited atoms, ions, or molecules (fluorescent lamps, etc.). The emission accompanies the spontaneous return of the excited species to the *ground state and occurs randomly, i.e. the radiation is not coherent. In a laser, the atoms, ions, or molecules are first 'pumped' to an excited state and then stimulated to emit photons by collision of a photon of the same energy. This is called *stimulated emission*. In order to use it, it is first necessary to create a condition in the amplifying medium, called *population inversion*, in which the majority of the relevant entities are excited. Random emission from one entity can then trigger coherent emission from the others that it passes. In this way amplification is achieved.

The laser amplifier is converted to an oscillator by enclosing the amplifying medium within a resonator. Radiation then introduced along the axis of the resonator is reflected back and forth along its path by a mirror at one end and by a partially transmitting mirror at the other end. Between the mirrors the waves are amplified by stimulated emission. The radiation emerges through the semitransparent mirror at one end as a powerful coherent monochromatic parallel beam of light. The emitted beam is uniquely parallel because waves that do not bounce back and forth between the mirrors quickly escape through the sides of the oscillating medium without amplification.

Some lasers are solid, others are liquid or gas devices. Population inversion can be achieved by *optical pumping* with flashlights or with other lasers. It can also be achieved by such methods as chemical reactions and discharges in gases.

Lasers have found many uses since their invention in 1960. In chemistry, their main use has been in the study of photochemical reactions, in the spectroscopic investigation of molecules, and in *femtochemistry. *See also* **dye laser; four-level laser; Pockels cell.**

laser spectroscopy Spectroscopy that makes use of lasers. The beams of coherent monochromatic radiation produced by lasers have several significant advantages compared with other spectroscopic techniques, particularly in those that employ the *Raman effect.

Lassaigne's test A method of testing for the presence of a halogen, nitrogen, or sulphur in an organic compound. A sample is heated in a test tube with a pellet of sodium. The hot tube is dropped into pure water and the fragments ground up in a mortar. The presence of a halogen (now in the form of a sodium halide) is detected by precipitation with silver nitrate solution. Nitrogen is revealed by the formation of a precipitate of Prussian blue on heating part of the solution with iron(II) sulphate solution containing hydrochloric acid and a trace of iron(III) ions. Lead ethanoate or sodium nitroprusside gives a precipitate with any sulphur present.

latent heat Symbol *L*. The quantity of heat absorbed or released when a substance changes its physical phase at constant temperature (e.g. from solid to liquid at the melting point or from liquid to gas at the boiling point). For example, the latent heat of vaporization is the energy a substance absorbs from its surroundings in order to overcome the attractive forces between its molecules as it changes from a liquid to a gas and in order to do work against the external atmosphere as it expands. In thermodynamic terms the latent heat is the *enthalpy of evaporation (ΔH), i.e. $L = \Delta H = \Delta U + p\Delta V$, where ΔU is the change in the internal energy, p is the pressure, and ΔV is the change in volume.

The *specific latent heat* (symbol l) is the heat absorbed or released per unit mass of a substance in the course of its isothermal change of phase. The *molar latent heat* is the heat absorbed or released per unit amount of substance during an isothermal change of state.

latex Natural rubber as it is obtained from a rubber tree or any stable suspension in water of a similar synthetic polymer. Synthetic latexes are used to make articles from rubber or plastics by such techniques as dipping (rubber gloves), spreading (waterproof cloth), and electrodeposition (plastic-coated metal). They are also employed in paints and adhesives.

lattice The regular arrangement of atoms, ions, or molecules in a crystalline solid. *See* **crystal lattice**.

lattice energy A measure of the stability of a *crystal lattice, given by the energy that would be released per mole if atoms, ions, or molecules of the crystal were brought together from infinite distances apart to form the lattice. *See* **Born–Haber cycle**.

lattice vibrations The periodic vibrations of the atoms, ions, or molecules in a *crystal lattice about their mean positions. On heating, the amplitude of the vibrations increases until they are so energetic that the lattice breaks down. The temperature at which this happens is the melting point of the solid and the substance becomes a liquid. On cooling, the amplitude of the vibrations diminishes. At *absolute zero a residual vibration persists, associated with the *zero-point energy of the substance. The increase in the electrical resistance of a conductor is due to increased scattering of the free conduction electrons by the vibrating lattice particles.

laughing gas *See* **dinitrogen oxide**.

lauric acid *See* **dodecanoic acid**.

Lavoisier, Antoine Laurent (1743–1794) French chemist, who collected taxes for the government in Paris. In the 1770s he discovered oxygen and nitrogen in air and demolished the *phlogiston theory of combustion by demonstrating the role of oxygen in the process. In 1783 he made water by burning hydrogen in oxygen (*see* **Cavendish, Henry**). He also devised a rational nomenclature for chemical compounds. In 1794 he was tried by the Jacobins as an opponent of the Revolution (because of his tax-gathering), found guilty, and guillotined.

law of chemical equilibrium *See* **equilibrium constant.**

law of conservation of energy *See* **conservation law.**

law of conservation of mass *See* **conservation law.**

law of constant composition *See* **chemical combination.**

law of definite proportions *See* **chemical combination.**

law of mass action *See* **mass action.**

law of multiple proportions *See* **chemical combination.**

law of octaves **(Newlands' law)** An attempt at classifying elements made by John Newlands (1837–98) in 1863. He arranged 56 elements in order of increasing atomic mass in groups of eight, pointing out that each element resembled the element eight places from it in the list. He drew an analogy with the notes of a musical scale. *Newlands' octaves* were groups of similar elements distinguished in this way: e.g. oxygen and sulphur; nitrogen and phosphorus; and fluorine, chlorine, bromine, and iodine. In some cases it was necessary to put two elements in the same position. The proposal was rejected at the time. *See* **periodic table.**

law of reciprocal proportions *See* **chemical combination.**

lawrencium Symbol Lr. A radioactive metallic transuranic element belonging to the *actinoids; a.n. 103; mass number of the first discovered isotope 257 (half-life 8 seconds). A number of very short-lived isotopes have now been synthesized. The element was identified by Albert Ghiorso and associates in 1961. It was named after E. O. Lawrence (1901–58).

laws of chemical combination *See* **chemical combination.**

laws, theories, and hypotheses In science, a law is a descriptive principle of nature that holds in all circumstances covered by the wording of the law. There are no loopholes in the laws of nature and any exceptional event that did not comply with the law would require the existing law to be discarded or would have to be described as a miracle. Eponymous laws are named after their discoverers (e.g. *Boyle's law); some laws, however, are known by their subject matter (e.g. the law of conservation of mass), while other laws use both the name of the discoverer and the subject matter to describe them (e.g. Newton's law of gravitation).

A description of nature that encompasses more than one law but has not achieved the uncontrovertible status of a law is sometimes called a *theory*. Theories are often both eponymous and descriptive of the subject matter (e.g. Einstein's theory of relativity and Darwin's theory of evolution).

A *hypothesis* is a theory or law that retains the suggestion that it may not be universally true. However, some hypotheses about which no doubt still lingers have remained hypotheses (e.g. Avogadro's hypothesis), for no clear reason. Clearly there is a degree of overlap between the three concepts.

layer lattice A crystal structure in which the atoms are chemically bonded in plane layers, with relatively weak forces between atoms in adjacent layers. Graphite and micas are examples of substances having layer lattices (i.e. they are *lamellar solids).

lazurite *See* **lapis lazuli**.

LCAO (linear combination of atomic orbitals) A molecular *orbital formed by the linear combination of atomic orbitals. The LCAO approximation arises because with an electron that is very close to a nucleus, the potential energy of the electron is dominated by the interaction between the electron and the nucleus. Thus, very close to the nucleus of an atom, *A*, in a molecule, the *wave function of the molecule is very similar to the wave function of the atom *A*. The LCAO approximation shows the increase in electron density associated with chemical bonding. The LCAO method takes account of the symmetry of the molecule using symmetry-adapted linear combinations (SALC).

LCP *See* **liquid-crystal polymer**.

L-dopa *See* **dopa**.

L–D process *See* **basic-oxygen process**.

leaching Extraction of soluble components of a solid mixture by percolating a solvent through it.

lead Symbol Pb. A heavy dull grey soft ductile metallic element belonging to *group 14 (formerly IVB) of the periodic table; a.n. 82; r.a.m. 207.19; r.d. 11.35; m.p. 327.5°C; b.p. 1740°C. The main ore is the sulphide galena (PbS); other minor sources include anglesite ($PbSO_4$), cerussite ($PbCO_3$), and litharge (PbO). The metal is extracted by roasting the ore to give the oxide, followed by reduction with carbon. Silver is also recovered from the ores. Lead has a variety of uses including building construction, lead-plate accumulators, bullets, and shot, and is a constituent of such alloys as solder, pewter, bearing metals, type metals, and fusible alloys. Chemically, it forms compounds with the +2 and +4 oxidation states, the lead(II) state being the more stable.

lead(II) acetate *See* **lead(II) ethanoate**.

lead–acid accumulator An accumulator in which the electrodes are made of lead and the electrolyte consists of dilute sulphuric acid. The electrodes are usually cast from a lead alloy containing 7–12% of antimony (to give increased hardness and corrosion resistance) and a small amount of tin (for better casting properties). The electrodes are coated with a paste of lead(II) oxide (PbO) and finely divided lead; after insertion into the

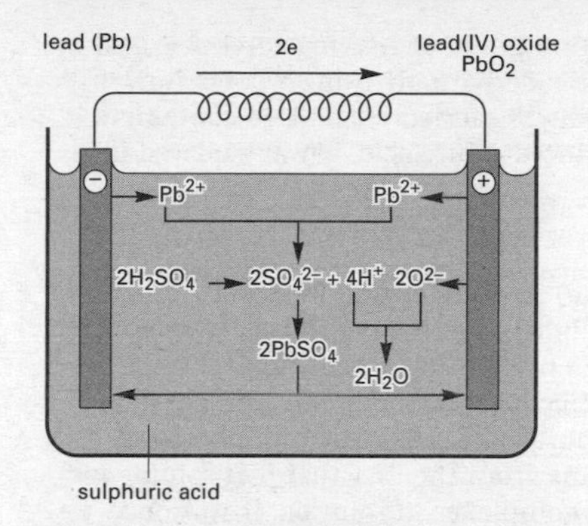

Lead–acid accumulator: reactions during discharge

electrolyte a 'forming' current is passed through the cell to convert the PbO on the negative plate into a sponge of finely divided lead. On the positive plate the PbO is converted to lead(IV) oxide (PbO_2). The equation for the overall reaction during discharge is:

$$PbO_2 + 2H_2SO_4 + Pb \rightarrow 2PbSO_4 + 2H_2O$$

The reaction is reversed during charging. Each cell gives an e.m.f. of about 2 volts and in motor vehicles a 12-volt battery of six cells is usually used. The lead–acid battery produces 80–120 kJ per kilogram. *Compare* **nickel–iron accumulator**.

lead(II) carbonate A white solid, $PbCO_3$, insoluble in water; rhombic; r.d. 6.6. It occurs as the mineral ***cerussite**, which is isomorphous with aragonite and may be prepared in the laboratory by the addition of cold ammonium carbonate solution to a cold solution of a lead(II) salt (acetate or nitrate). It decomposes at 315°C to lead(II) oxide and carbon dioxide.

lead(II) carbonate hydroxide (white lead; basic lead carbonate) A powder, $2PbCO_3.Pb(OH)_2$, insoluble in water, slightly soluble in aqueous carbonate solutions; r.d. 6.14; decomposes at 400°C. Lead(II) carbonate hydroxide occurs as the mineral *hydroxycerussite* (of variable composition). It was previously manufactured from lead in processes using spent tanning bark or horse manure, which released carbon dioxide. It is currently made by electrolysis of mixed solutions (e.g. ammonium nitrate, nitric acid, sulphuric acid, and acetic acid) using lead anodes. For the highest grade product the lead must be exceptionally pure (known in the trade as 'corroding lead') as small amounts of metallic impurity impart grey or pink discolorations. The material was used widely in paints, both for art work and for commerce, but it has the disadvantage of reacting with hydrogen sulphide in industrial atmospheres and producing black lead sulphide. The poisonous nature of lead compounds has also contributed to the declining importance of this material.

lead-chamber process An obsolete method of making sulphuric acid by the catalytic oxidation of sulphur dioxide with air using a potassium nitrate catalyst in water. The process was carried out in lead containers (which was expensive) and only produced dilute acid. It was replaced in 1876 by the *contact process.

lead dioxide *See* **lead(IV) oxide.**

lead(II) ethanoate **(lead(II) acetate)** A white crystalline solid, $Pb(CH_3COO)_2$, soluble in water and slightly soluble in ethanol. It exists as the anhydrous compound (r.d. 3.25; m.p. 280°C), as a trihydrate, $Pb(CH_3COO)_2.3H_2O$ (monoclinic; r.d. 2.55; loses water at 75°C), and as a decahydrate, $Pb(CH_3COO)_2.10H_2O$ (rhombic; r.d. 1.69). The common form is the trihydrate. Its chief interest stems from the fact that it is soluble in water and it also forms a variety of complexes in solution. It was once known as *sugar of lead* because of its sweet taste.

lead(IV) ethanoate **(lead tetra-acetate)** A colourless solid, $Pb(CH_3COO)_4$, which decomposes in water and is soluble in pure ethanoic acid; monoclinic; r.d. 2.228; m.p. 175°C. It may be prepared by dissolving dilead(II) lead(IV) oxide in warm ethanoic acid. In solution it behaves essentially as a covalent compound (no measurable conductivity) in contrast to the lead(II) salt, which is a weak electrolyte.

lead(IV) hydride *See* **plumbane.**

lead monoxide *See* **lead(II) oxide.**

lead(II) oxide **(lead monoxide)** A solid yellow compound, PbO, which is insoluble in water; m.p. 886°C. It exists in two crystalline forms: *litharge* (tetrahedral; r.d. 9.53) and *massicot* (rhombic; r.d. 8.0). It can be prepared by heating the nitrate, and is manufactured by heating molten lead in air. If the temperature used is lower than the melting point of the oxide, the product is massicot; above this, litharge is formed. Variations in the temperature and in the rate of cooling give rise to crystal vacancies and red, orange, and brown forms of litharge can be produced. The oxide is amphoteric, dissolving in acids to give lead(II) salts and in alkalis to give *plumbates.

lead(IV) oxide **(lead dioxide)** A dark brown or black solid with a rutile lattice, PbO_2, which is insoluble in water and slightly soluble in concentrated sulphuric and nitric acids; r.d. 9.375; decomposes at 290°C. Lead(IV) oxide may be prepared by the oxidation of lead(II) oxide by heating with alkaline chlorates or nitrates, or by anodic oxidation of lead(II) solutions. It is an oxidizing agent and readily reverts to the lead(II) oxidation state, as illustrated by its conversion to Pb_3O_4 and PbO on heating. It reacts with hydrochloric acid to evolve chlorine. Lead(IV) oxide has been used in the manufacture of safety matches and was widely used until the mid-1970s as an adsorbent for sulphur dioxide in pollution monitoring.

lead(II) sulphate A white crystalline solid, $PbSO_4$, which is virtually insoluble in water and soluble in solutions of ammonium salts; r.d. 6.2; m.p. 1170°C. It occurs as the mineral *anglesite*; it may be prepared in the laboratory by adding any solution containing sulphate ions to solutions of lead(II) ethanoate. The material known as *basic lead(II) sulphate* may be made by shaking together lead(II) sulphate and lead(II) hydroxide in water. This material has been used in white paint in preference to lead(II) carbonate hydroxide, as it is not so susceptible to discoloration through reaction with hydrogen sulphide. The toxicity of lead compounds has led to a decline in the use of these compounds.

lead(II) sulphide A black crystalline solid, PbS, which is insoluble in water; r.d. 7.5; m.p. 1114°C. It occurs naturally as the metallic-looking mineral *galena (the principal ore of lead). It may be prepared in the laboratory by the reaction of hydrogen sulphide with soluble lead(II) salts. Lead(II) sulphide has been used as an electrical rectifier.

lead tetra-acetate *See* **lead(IV) ethanoate.**

lead(IV) tetraethyl **(tetraethyl lead)** A colourless liquid, $Pb(C_2H_5)_4$, insoluble in water, soluble in benzene, ethanol, ether, and petroleum; r.d. 1.659; m.p. −137°C; b.p. 200°C. It may be prepared by the reaction of hydrogen and ethene with lead but a more convenient laboratory and industrial method is the reaction of a sodium–lead alloy with chloroethane. A more recent industrial process is the electrolysis of ethylmagnesium chloride (the Grignard reagent) using a lead anode and slowly running additional chloroethane onto the cathode. Lead tetraethyl is used in fuel for internal-combustion engines (along with 1,2-dibromoethane) to increase the *octane number and reduce knocking. The use of lead(IV) tetraethyl in petrol results in the emission of hazardous lead compounds into the atmosphere. Pressure from environmental groups has encouraged a reduction in the use of lead(IV) tetraethyl and an increasing use of lead-free petrol.

Leblanc process An obsolete process for manufacturing sodium carbonate. The raw materials were sodium chloride, sulphuric acid, coke, and limestone (calcium carbonate), and the process involved two stages. First the sodium chloride was heated with sulphuric acid to give sodium sulphate:

$$2NaCl(s) + H_2SO_4(l) \rightarrow Na_2SO_4(s) + 2HCl(g)$$

The sodium sulphate was then heated with coke and limestone:

$$Na_2SO_4 + 2C + CaCO_3 \rightarrow Na_2CO_3 + CaS + 2CO_2$$

Calcium sulphide was a by-product, the sodium carbonate being extracted by crystallization. The process, invented in 1783 by the French chemist Nicolas Leblanc (1742–1806), was the first for producing sodium carbonate synthetically (earlier methods were from wood ash and other vegetable sources). By the end of the 19th century it had been largely replaced by the *Solvay process.

lechatelierite A mineral form of *silicon(IV) oxide, SiO_2.

Le Chatelier's principle If a system is in equilibrium, any change imposed on the system tends to shift the equilibrium to nullify the effect of the applied change. The principle, which is a consequence of the law of conservation of energy, was first stated in 1888 by Henri Le Chatelier (1850–1936). It is applied to chemical equilibria. For example, in the gas reaction

$$2SO_2 + O_2 \rightleftharpoons 2SO_3$$

an increase in pressure on the reaction mixture displaces the equilibrium to the right, since this reduces the total number of molecules present and thus decreases the pressure. The standard enthalpy change for the forward reaction is negative (i.e. the reaction is exothermic). Thus, an increase in temperature displaces the equilibrium to the left since this tends to reduce the temperature. The *equilibrium constant thus falls with increasing temperature.

Leclanché cell A primary *voltaic cell consisting of a carbon rod (the anode) and a zinc rod (the cathode) dipping into an electrolyte of a 10–20% solution of ammonium chloride. *Polarization is prevented by using a mixture of manganese dioxide mixed with crushed carbon, held in contact with the anode by means of a porous bag or pot; this reacts with the hydrogen produced. This wet form of the cell, devised in 1867 by Georges Leclanché (1839–82), has an e.m.f. of about 1.5 volts. The *dry cell based on it is widely used in torches, radios, and calculators.

lectin Any of a group of proteins, derived from plants, that can bind to specific oligosaccharides on the surface of cells, causing the cells to clump together. Lectins can be used to identify mutant cells in cell cultures and to determine blood groups as they can cause the agglutination of red blood cells. Lectins are found in seeds of legumes and in other tissues, in which they are thought to act as a toxin.

LEED (low-energy electron diffraction) A technique used to study the structure of crystal surfaces and processes taking place on these surfaces. The surface is bombarded with a monochromatic electron beam 10^{-4} to 10^{-3} m in diameter, with energies between 6 and 600 V. The electrons are diffracted by the surface atoms and then collected on a fluorescent screen. Both the surface structure and changes that occur after chemisorption and surface reactions can be investigated in this way. It is necessary for the surface to be carefully cleaned and kept at ultrahigh vacuum pressure. Although many surfaces are altered by the electron beam and therefore cannot be studied using this method, there are enough surfaces and surface processes that can be studied using LEED to make it a very useful technique. Difficulties in interpreting LEED patterns arise as multiple-scattering theory, rather than single-scattering theory (as in X-ray or neutron scattering), is required. *See also* **electron diffraction**.

Lennard–Jones potential A potential used to give an approximate description of the potential energy interaction, V, of molecules as a function of intermolecular distance r. The general form of the Lennard–Jones potential is

$$V = C_n/r^n - C_6/r^6,$$

where C_n and C_6 are coefficients that depend on the specific molecules and n is greater than 6 so that at small separations the repulsion term dominates the interaction, the r^{-6} term being attractive. The value $n = 12$ is frequently chosen. In this case the Lennard–Jones potential is given by:

$$V = 4W[(r_0/r)^{12} - (r_0/r)^6],$$

where W is the depth of the potential well and r_0 is the separation at which $V = 0$. The minimum value of the well occurs at the separation $r_e = 2^{1/6}r_0$. The representation of the repulsive part of the interaction by a $1/r^{12}$ term is not realistic; a much more realistic term is the exponential term, $\exp(-r/r_0)$, as it is closer to the exponential decay of the wave functions and thus of their overlap, which describes the repulsion.

leucine *See* **amino acid.**

leuco form *See* **dyes.**

lever rule A rule enabling the relative amounts of two phases a and b, which are in equilibrium, to be found by a construction in a phase diagram. (For example, a can be gas and b can be liquid.) The distances l_a and l_b along the horizontal *tie line of the phase diagram are measured. The lever rule states that $n_a l_a = n_b l_b$, where n_a is the amount of phase a and n_b is the amount of phase b. The rule takes its name from the similar form of the rule, $m_a l_a = m_b l_b$, relating the moments of two masses m_a and m_b about a pivot in a lever.

Lewis, Gilbert Newton (1875–1946) US physical chemist who spent most of his career at Berkeley, California. His ideas on chemical bonding were extremely influential, and he introduced the idea of a stable octet of electrons and of a covalent bond being a shared pair of electrons. He also introduced the concept of Lewis acids and bases (*see* **acid**).

Lewis acid and base *See* **acid.**

Liebermann's reaction A method of testing for phenols. A small sample of the test substance and a crystal of sodium nitrite are dissolved in warm sulphuric acid. The solution is then poured into excess aqueous alkali, when the formation of a blue-green colour indicates the presence of a phenol.

Liebig, Justus von (1803–73) German organic chemist who worked at Gessen near Frankfurt. Liebig was the first to recognize that two different chemical compounds can have the same formula. He also developed a method of analysing organic compounds by burning them and weighing the carbon dioxide and water produced. With his students, he analysed many compounds and was extremely influential in the development of organic chemistry.

Liebig condenser A laboratory condenser having a straight glass tube surrounded by a coaxial glass jacket through which cooling water is passed. The device is named after Justus von *Liebig.

ligand An ion or molecule that donates a pair of electrons to a metal atom or ion in forming a coordination *complex. Molecules that function as ligands are acting as Lewis bases (*see* **acid**). For example, in the complex hexaquocopper(II) ion $[Cu(H_2O)_6]^{2+}$ six water molecules coordinate to a central Cu^{2+} ion. In the tetrachloroplatinate(II) ion $[PtCl_4]^{2-}$, four Cl^- ions are coordinated to a central Pt^{2+} ion. A feature of such ligands is that they have lone pairs of electrons, which they donate to empty metal orbitals. A certain class of ligands also have empty *p*- or *d*-orbitals in addition to their lone pair of electrons and can produce complexes in which the metal has low oxidation state. A double bond is formed between the metal and the ligand: a sigma bond by donation of the lone pair from ligand to metal, and a pi bond by *back donation* of electrons on the metal to empty *d*-orbitals on the ligand. Carbon monoxide is the most important such ligand, forming metal carbonyls (e.g. $Ni(CO)_4$).

The examples given above are examples of *monodentate* ligands (literally: 'having one tooth'), in which there is only one point on each ligand at which coordination can occur. Some ligands are *polydentate*; i.e. they have two or more possible coordination points. For instance, 1,2-diaminoethane, $H_2NC_2H_4NH_2$, is a *bidentate* ligand, having two coordination points. Certain polydentate ligands can form *chelates.

ligand-field theory An extension of *crystal-field theory describing the properties of compounds of transition-metal ions or rare-earth ions in which covalent bonding between the surrounding molecules (*see* **ligand**) and the transition-metal ions is taken into account. This may involve using valence-bond theory or molecular-orbital theory. Ligand-field theory was developed extensively in the 1930s. As with crystal-field theory, ligand-field theory indicates that energy levels of the transition-metal ions are split by the surrounding ligands, as determined by *group theory. The theory has been very successful in explaining the optical, spectroscopic, and magnetic properties of the compounds of transition-metal and rare-earth ions.

ligase Any of a class of enzymes that catalyse the formation of covalent bonds using the energy released by the cleavage of ATP. Ligases are important in the synthesis and repair of many biological molecules, including DNA, and are used in genetic engineering to insert foreign DNA into cloning vectors.

light-dependent reaction *See* **photosynthesis**.

light-independent reaction *See* **photosynthesis**.

lignin A complex organic polymer that is deposited within the cellulose of plant cell walls during secondary thickening. Lignification makes the walls woody and therefore rigid.

lignite *See* **coal**.

lime *See* **calcium oxide.**

limestone A sedimentary rock that is composed largely of carbonate minerals, especially carbonates of calcium and magnesium. *Calcite and *aragonite are the chief minerals; *dolomite is also present in the dolomitic limestones. There are many varieties of limestones but most are deposited in shallow water. *Organic limestones* (e.g. *chalk) are formed from the calcareous skeletons of organisms; *precipitated limestones* include oolite, which is composed of ooliths – spherical bodies formed by the precipitation of carbonate around a nucleus; and *clastic limestones* are derived from fragments of pre-existing calcareous rocks.

limewater A saturated solution of *calcium hydroxide in water. When carbon dioxide gas is bubbled through limewater, a 'milky' precipitate of calcium carbonate is formed:

$$Ca(OH)_2(aq) + CO_2(g) \rightarrow CaCO_3(s) + H_2O(l)$$

If the carbon dioxide continues to be bubbled through, the calcium carbonate eventually redissolves to form a clear solution of calcium hydrogencarbonate:

$$CaCO_3(s) + CO_2(g) + H_2O(g) \rightarrow Ca(HCO_3)_2(aq)$$

If cold limewater is used the original calcium carbonate precipitated has a calcite structure; hot limewater yields an aragonite structure.

limit cycle *See* **attractor.**

limonite A generic term for a group of hydrous iron oxides, mostly amorphous. *Goethite and *haematite are important constituents, together with colloidal silica, clays, and manganese oxides. Limonite is formed by direct precipitation from marine or fresh water in shallow seas, lagoons, and bogs (thus it is often called *bog iron ore*) and by oxidation of iron-rich minerals. It is used as an ore of iron and as a pigment.

Lindemann–Hinshelwood mechanism A mechanism for unimolecular chemical reactions put forward by the British physicist Frederick Lindermann (1886–1957) in 1921 and examined in more detail by the British chemist Sir Cyril Hinshelwood (1897–1967) in 1927. The mechanism postulates that a molecule of A becomes excited by colliding with another molecule of A, and that having been excited there is a possibility that it undergoes unimolecular decay. If the process of unimolecular decay is sufficiently slow, the reaction has a first-order rate law, in agreement with experiment. The Lindemann–Hinshelwood mechanism predicts that if the concentration of A is reduced, the reaction kinetics become second order. This change from first to second order agrees with experiment qualitatively, although it does not do so quantitatively. The mechanism fails quantitatively because the molecule has to be excited in a specific way for a reaction to take place. The *RRK and *RRKM theories improve on this deficiency of the Lindemann–Hinshelwood mechanism.

linear combination of atomic orbitals *See* **LCAO.**

linear molecule A molecule in which the atoms are in a straight line, as in carbon dioxide, O=C=O. Linear molecules have only two rotational degrees of freedom.

linear rotor *See* **moment of inertia**.

line spectrum *See* **spectrum**.

linoleic acid A liquid polyunsaturated *fatty acid with two double bonds, $CH_3(CH_2)_4CH{:}CHCH_2CH{:}CH(CH_2)_7COOH$. Linoleic acid is abundant in plant fats and oils, e.g. linseed oil, groundnut oil, and soya-bean oil. It is an *essential fatty acid.

linolenic acid A liquid polyunsaturated *fatty acid with three double bonds in its structure: $CH_3CH_2CH{:}CHCH_2CH{:}CHCH_2CH{:}CH(CH_2)_7COOH$. It occurs in certain plant oils, e.g. linseed and soya-bean oil, and in algae. It is one of the *essential fatty acids.

linseed oil A pale yellow oil pressed from flax seed. It contains a mixture of glycerides of fatty acids, including linoleic acid and linolenic acid. It is a *drying oil, used in oil paints, varnishes, linoleum, etc.

Linz–Donawitz process *See* **basic-oxygen process**.

lipase An enzyme secreted by the pancreas and the glands of the small intestine of vertebrates that catalyses the breakdown of fats into fatty acids and glycerol.

lipid Any of a diverse group of organic compounds, occurring in living organisms, that are insoluble in water but soluble in organic solvents, such as chloroform, benzene, etc. Lipids are broadly classified into two categories: *complex lipids*, which are esters of long-chain fatty acids and include the *glycerides (which constitute the *fats and *oils of animals and plants), *glycolipids, *phospholipids, and *waxes; and *simple lipids*, which do not contain fatty acids and include the *steroids and *terpenes.

Lipids have a variety of functions in living organisms. Fats and oils are a convenient and concentrated means of storing food energy in plants and animals. Phospholipids and *sterols, such as cholesterol, are major components of cell membranes (*see* **lipid bilayer**). Waxes provide vital waterproofing for body surfaces. Terpenes include vitamins A, E, and K, and phytol (a component of chlorophyll) and occur in essential oils, such as menthol and camphor. Steroids include the adrenal hormones, sex hormones, and bile acids.

Lipids can combine with proteins to form *lipoproteins*, e.g. in cell membranes. In bacterial cell walls, lipids may associate with polysaccharides to form *lipopolysaccharides*.

lipid bilayer The arrangement of lipid molecules in biological membranes, which takes the form of a double sheet. Each lipid molecule comprises a hydrophilic 'head' (having a high affinity for water) and a hydrophobic 'tail' (having a low affinity for water). In the lipid bilayer the molecules are aligned so that their hydrophilic heads face outwards, forming the outer and inner surfaces of the membrane, while the hydrophobic tails face inwards, away from the external aqueous environment.

lipoic acid A vitamin of the *vitamin B complex. It is one of the *coenzymes involved in the decarboxylation of pyruvate by the enzyme pyruvate dehydrogenase. Good sources of lipoic acid include liver and yeast.

lipolysis The breakdown of storage lipids in living organisms. Most long-term energy reserves are in the form of triglycerides in fats and oils. When these are needed, e.g. during starvation, lipase enzymes convert the triglycerides into glycerol and the component fatty acids. These are then transported to tissues and oxidized to provide energy.

lipoprotein *See* **lipid.**

lipowitz alloy A low-melting (70–74°C) alloy of bismuth (50%), lead (27%), tin (13%), and cadmium (10%).

liquation The separation of mixtures of solids by heating to a temperature at which lower-melting components liquefy.

liquefaction of gases The conversion of a gaseous substance into a liquid. This is usually achieved by one of four methods or by a combination of two of them:
(1) by vapour compression, provided that the substance is below its *critical temperature;
(2) by refrigeration at constant pressure, typically by cooling it with a colder fluid in a countercurrent heat exchanger;
(3) by making it perform work adiabatically against the atmosphere in a reversible cycle;
(4) by the *Joule–Thomson effect.

Large quantities of liquefied gases are now used commercially, especially *liquefied petroleum gas and liquefied natural gas.

liquefied natural gas **(LNG)** *See* **liquefied petroleum gas.**

liquefied petroleum gas **(LPG)** Various petroleum gases, principally propane and butane, stored as a liquid under pressure. It is used as an engine fuel and has the advantage of causing very little cylinder-head deposits.

Liquefied natural gas (*LNG*) is a similar product and consists mainly of methane. However, it cannot be liquefied simply by pressure as it has a low critical temperature of 190 K and must therefore be cooled to below this temperature before it will liquefy. Once liquefied it has to be stored in well-insulated containers. It provides a convenient form in which to ship natural gas in bulk from oil wells or gas-only wells to users. It is also used as an engine fuel.

liquid A phase of matter between that of a crystalline solid and a *gas. In a liquid, the large-scale three-dimensional atomic (or ionic or molecular) regularity of the solid is absent but, on the other hand, so is the total disorganization of the gas. Although liquids have been studied for many years there is still no comprehensive theory of the liquid state. It is clear, however, from diffraction studies that there is a short-range structural regularity extending over several molecular diameters. These bundles of ordered atoms, molecules, or ions move about in relation to each other, enabling liquids to have almost fixed volumes, which adopt the shape of their containers.

liquid crystal A substance that flows like a liquid but has some order in its arrangement of molecules. *Nematic crystals* have long molecules all aligned in the same direction, but otherwise randomly arranged. *Cholesteric* and *smectic* liquid crystals also have aligned molecules, which are arranged in distinct layers. In cholesteric crystals, the axes of the molecules are parallel to the plane of the layers; in smectic crystals they are perpendicular.

liquid-crystal polymer A polymer with a liquid-crystal structure, this being the most thermodynamically stable. Liquid-crystal polymers contain long rigid chains and combine strength with lightness. They are, however, difficult to produce commercially.

***l*-isomer** *See* **optical activity**.

L-isomer *See* **absolute configuration**.

litharge *See* **lead(II) oxide**.

lithia *See* **lithium oxide**.

lithium Symbol Li. A soft silvery metal, the first member of group 1 (formerly IA) of the periodic table (*see* **alkali metals**); a.n. 3; r.a.m. 6.939; r.d. 0.534; m.p. 180.54°C; b.p. 1347°C. It is a rare element found in spodumene ($LiAlSi_2O_6$), petalite ($LiAlSi_4O_{10}$), the mica lepidolite, and certain brines. It is usually extracted by treatment with sulphuric acid to give the sulphate, which is converted to the chloride. This is mixed with a small amount of potassium chloride, melted, and electrolysed. The stable isotopes are lithium-6 and lithium-7. Lithium-5 and lithium-8 are short-lived radioisotopes. The metal is used to remove oxygen in metallurgy and as a constituent of some Al and Mg alloys. It is also used in batteries and is a potential tritium source for fusion research. Lithium salts are used in psychomedicine. The element reacts with oxygen and water; on heating it also reacts with nitrogen and hydrogen. Its chemistry differs somewhat from that of the other group 1 elements because of the small size of the Li^+ ion.

lithium aluminium hydride *See* **lithium tetrahydroaluminate(III)**.

lithium carbonate A white solid, Li_2CO_3; r.d. 2.11; m.p. 723°C; decomposes above 1310°C. It is produced commercially by treating the ore with sulphuric acid at 250°C and leaching the product to give a solution of lithium sulphate. The carbonate is then obtained by precipitation with sodium carbonate solution. Lithium carbonate is used in the prevention and treatment of manic-depressive disorders. It is also used industrially in ceramic glazes.

lithium deuteride *See* **lithium hydride.**

lithium hydride A white solid, LiH; cubic; r.d. 0.82; m.p. 680°C; decomposes at about 850°C. It is produced by direct combination of the elements at temperatures above 500°C. The bonding in lithium hydride is believed to be largely ionic; i.e. Li^+H^- as supported by the fact that hydrogen is released from the anode on electrolysis of the molten salt. The compound reacts violently and exothermically with water to yield hydrogen and lithium hydroxide. It is used as a reducing agent to prepare other hydrides and the ^2H isotopic compound, *lithium deuteride*, is particularly valuable for deuterating a range of organic compounds. Lithium hydride has also been used as a shielding material for thermal neutrons.

lithium hydrogencarbonate A compound, $LiHCO_3$, formed by the reaction of carbon dioxide with aqueous lithium carbonate and known only in solution. It has found medicinal uses similar to those of lithium carbonate and is sometimes included in proprietary mineral waters.

lithium hydroxide A white crystalline solid, LiOH, soluble in water, slightly soluble in ethanol and insoluble in ether. It is known as the monohydrate (monoclinic; r.d. 1.51) and in the anhydrous form (tetragonal, r.d. 1.46; m.p. 450°C; decomposes at 924°C). The compound is made by reacting lime with lithium salts or lithium ores. Lithium hydroxide is basic but has a closer resemblance to group 2 hydroxides than to the other group 1 hydroxides (an example of the first member of a periodic group having atypical properties).

lithium oxide **(lithia)** A white crystalline compound, Li_2O; cubic; r.d. 2.01; m.p. 1700°C. It can be obtained from a number of lithium ores; the main uses are in lubricating greases, ceramics, glass and refractories, and as a flux in brazing and welding.

lithium sulphate A white or colourless crystalline material, Li_2SO_4, soluble in water and insoluble in ethanol. It forms a monohydrate (monoclinic; r.d. 1.88) and an anhydrous form, which exists in α- (monoclinic), β- (hexagonal) and γ- (cubic) forms; r.d. 2.23. The compound is prepared by the reaction of the hydroxide or carbonate with sulphuric acid. It is not isomorphous with other group 1 sulphates and does not form alums.

lithium tetrahydroaluminate(III) **(lithium aluminium hydride; LAH)** A white or light grey powder, $LiAlH_4$; r.d. 0.917; decomposes at 125°C. It is prepared by the reaction of excess lithium hydride with aluminium chloride. The compound is soluble in ethoxyethane, reacts violently with water to release hydrogen, and is widely used as a powerful reducing agent in organic chemistry. It should always be treated as a serious fire risk in storage.

litmus A water-soluble dye extracted from certain lichens. It turns red under acid conditions and blue under alkaline conditions, the colour change occurring over the pH range 4.5–8.3 (at 25°C). It is not suitable for titrations because of the wide range over which the colour changes, but is used as a rough *indicator of acidity or alkalinity, both in solution and as litmus paper (absorbent paper soaked in litmus solution).

litre Symbol l. A unit of volume in the metric system regarded as a special name for the cubic decimetre. It was formerly defined as the volume of 1 kilogram of pure water at 4°C at standard pressure, which is equivalent to 1.000 028 dm^3.

lixiviation The separation of mixtures by dissolving soluble constituents in water.

LNG *See* **liquefied petroleum gas.**

localization The confinement of electrons to a particular atom in a molecule or to a particular chemical bond.

localized bond A *chemical bond in which the electrons forming the bond remain between (or close to) the linked atoms. *Compare* **delocalization.**

lock-and-key mechanism A mechanism proposed in 1890 by Emil *Fischer to explain binding between the active site of an enzyme and a substrate molecule. The active site was thought to have a fixed structure (the lock), which exactly matched the structure of a specific substrate (the key). Thus the enzyme and substrate interact to form an *enzyme–substrate complex. The substrate is converted to products that no longer fit the active site and are therefore released, liberating the enzyme. Recent observations made by X-ray diffraction studies have shown that the active site of an enzyme is more flexible than the lock-and-key theory would suggest.

lodestone *See* **magnetite.**

logarithmic scale 1. A scale of measurement in which an increase or decrease of one unit represents a tenfold increase or decrease in the quantity measured. Decibels and pH measurements are common examples of logarithmic scales of measurement. **2.** A scale on the axis of a graph in which an increase of one unit represents a tenfold increase in the variable quantity. If a curve $y = x^n$ is plotted on graph paper with logarithmic scales on both axes, the result is a straight line of slope n, i.e. $\log y = n \log x$, which enables n to be determined.

London formula A formula giving an expression for the induced-dipole–induced-dipole interaction between molecules (called the *dispersion interaction* or *London interaction*). The London formula for the interaction energy V is given by $V = -C/r^6$, where $C = \frac{2}{3}\alpha'_1\alpha'_2 I_1 I_2/(I_1 + I_2)$. Here α'_1 and α'_2 are the polarizability volumes of molecule 1 and 2 respectively, I_1 and I_2 are the ionization energies of molecules 1 and 2 respectively, and r is the distance between the molecules. The London formula is named after Fritz London (1900–54), who derived it. The interaction described by the London formula is usually the dominant term in intermolecular forces (unless hydrogen bonds are present).

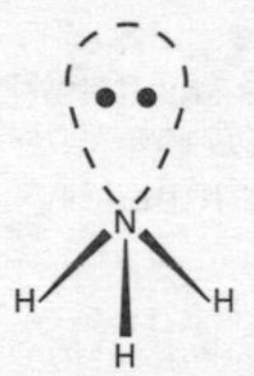

Lone pair of electrons in ammonia

lone pair A pair of electrons having opposite spin in an orbital of an atom. For instance, in ammonia the nitrogen atom has five electrons, three of which are used in forming single bonds with hydrogen atoms. The other two occupy a filled atomic orbital and constitute a lone pair (see illustration). The orbital containing these electrons is equivalent to a single bond (sigma orbital) in spatial orientation, accounting for the pyramidal shape of the molecule. In the water molecule, there are two lone pairs on the oxygen atom. In considering the shapes of molecules, repulsions between bonds and lone pairs can be taken into account:

lone pair–lone pair > lone pair–bond > bond–bond.

long period *See* **periodic table.**

Lorentz–Lorenz equation A relation between the *polarizability α of a molecule and the refractive index n of a substance made up of molecules with this polarizability. The Lorentz–Lorenz equation can be written in the form $\alpha = (3/4\pi N)\,[(n^2-1/(n^2+2)]$, where N is the number of molecules per unit volume. The equation provides a link between a microscopic quantity (the polarizability) and a macroscopic quantity (the refractive index). It was derived using macroscopic electrostatics in 1880 by Hendrik Lorentz (1853–1928) and independently by the Danish physicist Ludwig Valentin Lorenz also in 1880. *Compare* **Clausius–Mossotti equation.**

Loschmidt's constant (Loschmidt number) The number of particles per unit volume of an *ideal gas at STP. It has the value $2.686\,763(23) \times 10^{25}\ m^{-3}$ and was first worked out by Joseph Loschmidt (1821–95).

Lotka–Volterra mechanism A simple chemical reaction mechanism proposed as a possible mechanism of *oscillating reactions. The process involves a conversion of a reactant R into a product P. The reactant flows into the reaction chamber at a constant rate and the product is removed at a constant rate, i.e. the reaction is in a steady state (but not in chemical equilibrium). The mechanism involves three steps:

$$R + X \rightarrow 2X$$

X + Y → 2Y

Y → P

The first two steps involve *autocatalysis: the first step is catalysed by the reactant X and the second by the reactant Y. The kinetics of such a reaction can be calculated numerically, showing that the concentrations of both X and Y increase and decrease periodically with time. This results from the autocatalytic action. Initially, the concentration of X is small, but, as it increases, there is a rapid increase in the rate of the first reaction because of the autocatalytic action of X. As the concentration of X builds up, the rate of the second reaction also increases. Initially, the concentration of Y is low but there is a sudden surge in the rate of step 2, resulting from the autocatalytic action of Y. This lowers the concentration of X and slows down step 1, so the concentration of X falls. Less X is now available for the second step and the concentration of Y also starts to fall. With this fall in the amount of Y, less X is removed, and the first reaction again begins to increase. These processes are repeated, leading to repeated rises and falls in the concentrations of both X and Y. The cycles are not in phase, peaks in the concentration of Y occurring later than peaks in X.

In fact, known oscillating chemical reactions have different mechanisms to the above, but the scheme illustrates how oscillation may occur. This type of process is found in fields other than chemistry; they were investigated by the Italian mathematician Vito Volterra (1860–1940) in models of biological systems (e.g. predator–prey relationships).

low-energy electron diffraction *See* **LEED**.

lowering of vapour pressure A reduction in the saturated vapour pressure of a pure liquid when a solute is introduced. If the solute is a solid of low vapour pressure, the decrease in vapour pressure of the liquid is proportional to the concentration of particles of solute; i.e. to the number of dissolved molecules or ions per unit volume. To a first approximation, it does not depend on the nature of the particles. *See* **colligative properties; Raoult's law**.

lowest unoccupied molecular orbital (LUMO) The orbital in a molecule that has the lowest unoccupied energy level at the absolute zero of temperature. The lowest unoccupied molecular orbital and the highest occupied molecular orbital (HOMO) are the two *frontier orbitals of the molecule.

Lowry–Brønsted theory *See* **acid**.

LSD *See* **lysergic acid diethylamide**.

L-series *See* **absolute configuration**.

lubrication The use of a substance to prevent contact between solid surfaces in relative motion in order to reduce friction, wear, overheating, and rusting. Liquid hydrocarbons (oils), either derived from petroleum or made synthetically, are the most widely used lubricants as they are relatively inexpensive, are good coolants, provide the appropriate range of viscosities, and are thermally stable. Additives include polymeric substances that maintain the desired viscosity as the temperature increases, antioxidants that prevent the formation of a sludge, and alkaline-earth phenates that neutralize acids and reduce wear.

At high temperatures, solid lubricants, such as graphite or molybdenum disulphide, are often used. Semifluid lubricants (greases) are used to provide a seal against moisture and dirt and to remain attached to vertical surfaces. They are made by adding gelling agents, such as metallic soaps, to liquid lubricants.

luciferase *See* **bioluminescence**.

luciferin *See* **bioluminescence**.

lumen Symbol lm. The SI unit of luminous flux equal to the flux emitted by a uniform point source of 1 candela in a solid angle of 1 steradian.

luminescence The emission of light by a substance for any reason other than a rise in its temperature. In general, atoms of substances emit *photons of electromagnetic energy when they return to the *ground state after having been in an excited state (*see* **excitation**). The causes of the excitation are various. If the exciting cause is a photon, the process is called *photoluminescence*; if it is an electron it is called *electroluminescence*. *Chemiluminescence* is luminescence resulting from a chemical reaction (such as the slow oxidation of phosphorus); *bioluminescence is the luminescence produced by a living organism (such as a firefly). If the luminescence persists significantly after the exciting cause is removed it is called *phosphorescence*; if it does not it is called *fluorescence*. This distinction is arbitrary; in some definitions a persistence of more than 10 nanoseconds (10^{-8} s) is treated as phosphorescence.

LUMO *See* **lowest unoccupied molecular orbital**.

lutetium Symbol Lu. A silvery metallic element belonging to the *lanthanoids; a.n. 71; r.a.m. 174.97; r.d. 9.8404 (20°C); m.p. 1663°C; b.p. 3402°C. Lutetium is the least abundant of the elements and the little quantities that are available have been obtained by processing other metals. There are two natural isotopes, lutetium–175 (stable) and lutetium–176 (half-life 2.2×10^{10} years). The element is used as a catalyst. It was first identified by Gerges Urbain (1872–1938) in 1907.

lux Symbol lx. The SI unit of illuminance equal to the illumination produced by a luminous flux of 1 lumen distributed uniformly over an area of 1 square metre.

lyate ion The ion formed by removing a hydron from a molecule of a solvent. In water, for example, the hydroxide ion (OH^-) is the lyate ion.

lye *See* **potassium hydroxide**.

Lyman series *See* **hydrogen spectrum**.

lyonium ion The ion formed by adding a hydron (H^+) to a solvent molecule. For example, in ethanol, $C_2H_5OH_2^+$ is the lyonium ion.

lyophilic Having an affinity for a solvent ('solvent-loving'; if the solvent is water the term *hydrophilic* is used). *See* **colloids**.

lyophobic Lacking any affinity for a solvent ('solvent-hating'; if the solvent is water the term *hydrophobic* is used). *See* **colloids**.

lyotropic mesomorph An arrangement taken by micelles formed from surfactant molecules in concentrated solutions. It consists of long cylinders in a fairly close-packed hexagonal arrangement. Lyotropic mesomorphs are sometimes called liquid crystalline phases for micelles.

lysergic acid diethylamide (LSD) A chemical derivative of lysergic acid that has potent hallucinogenic properties (*see* **hallucinogen**). It occurs in the cereal fungus ergot and was first synthesized in 1943.

lysine *See* **amino acid**.

macromolecular crystal A crystalline solid in which the atoms are all linked together by covalent bonds. Carbon (in diamond), boron nitride, and silicon carbide are examples of substances that have macromolecular crystals. In effect, the crystal is a large molecule (hence the alternative description *giant-molecular*), which accounts for the hardness and high melting point of such materials.

macromolecule A very large molecule. Natural and synthetic polymers have macromolecules, as do such substances as haemoglobin. *See also* **colloids**.

macroscopic Designating a size scale very much larger than that of atoms and molecules. Macroscopic objects and systems are described by classical physics although *quantum mechanics can have macroscopic consequences. *Compare* **mesoscopic**; **microscopic**.

Madelung constant A constant arising in calculations of the cohesion of ionic crystals. The electrostatic interaction per ion pair, U, is given by $U(r) = -\alpha e^2/r$, where α is the Madelung constant and e^2/r is the Coulomb interaction between the ions, with r being the lattice constant. The value of α depends on the type of lattice. For the sodium chloride lattice, α has a value of about 1.75. A more realistic calculation of cohesion is obtained if short-range repulsions with an inverse power law are included, i.e.

$$U(r) = \alpha e^2/r - C/r^n,$$

where C and n are constants. The value of α can be used in calculations to determine C and n. It was first calculated by Erwin Madelung in 1918.

Magic acid *See* **superacid**.

magic-angle spinning A technique used in solid-state *nuclear magnetic resonance (NMR) for making the line widths smaller. In magic-angle spinning, both the dipole–dipole interaction and the chemical shift anisotropy have the angular dependence $1-3\cos^2\theta$, where θ is the angle between the principal axis of the molecule and the applied magnetic field. The 'magic angle' is the angle θ that satisfies $1-3\cos^2\theta = 0$ and is given by $\theta = 54.74°$. In magic-angle spinning the material is spun very rapidly at the magic angle to the applied magnetic field so that the dipole–dipole interactions and chemical shift anisotropies average to zero. It is necessary for the frequency of spinning to be at least as large as the width of the spectrum. This technique has been extensively used, with the spinning between 4 and 5 kHz.

Magnadur Tradename for a ceramic material used to make permanent magnets. It consists of sintered iron oxide and barium oxide.

Magnalium Tradename for an aluminium-based alloy of high reflectivity for light and ultraviolet radiation that contains 1–2% of copper and between 5% and 30% of magnesium. Strong and light, these alloys also sometimes contain other elements, such as tin, lead, and nickel.

magnesia *See* **magnesium oxide**.

magnesite A white, colourless, or grey mineral form of *magnesium carbonate, $MgCO_3$, crystallizing in the trigonal system. It is formed as a replacement mineral of magnesium-rich rocks when carbon dioxide is available. Magnesite is mined both as an ore for magnesium and as a source of magnesium carbonate. It occurs in Austria, USA, Greece, Norway, India, Australia, and South Africa.

magnesium Symbol Mg. A silvery metallic element belonging to group 2 (formerly IIA) of the periodic table (*see* **alkaline-earth metals**); a.n. 12; r.a.m. 24.305; r.d. 1.74; m.p. 648.8°C; b.p. 1090°C. The element is found in a number of minerals, including magnesite ($MgCO_3$), dolomite ($MgCO_3.CaCO_3$), and carnallite ($MgCl_2.KCl.6H_2O$). It is also present in sea water, and it is an *essential element for living organisms. Extraction is by electrolysis of the fused chloride. The element is used in a number of light alloys (e.g. for aircraft). Chemically, it is very reactive. In air it forms a protective oxide coating but when ignited it burns with an intense white flame. It also reacts with the halogens, sulphur, and nitrogen. Magnesium was first isolated by Bussy in 1828.

magnesium bicarbonate *See* **magnesium hydrogencarbonate**.

magnesium carbonate A white compound, $MgCO_3$, existing in anhydrous and hydrated forms. The anhydrous material (trigonal; r.d. 2.96) is found in the mineral *magnesite. There is also a trihydrate, $MgCO_3.3H_2O$ (rhombic; r.d. 1.85), which occurs naturally as *nesquehonite*, and a pentahydrate, $MgCO_3.5H_2O$ (monoclinic; r.d. 1.73), which occurs as *lansfordite*. Magnesium carbonate also occurs in the mixed salt *dolomite ($CaCO_3.MgCO_3$) and as *basic magnesium carbonate* in the two minerals *artinite* ($MgCO_3.Mg(OH)_2.3H_2O$) and *hydromagnesite* ($3MgCO_3.Mg(OH)_2.3H_2O$). The anhydrous salt can be formed by heating magnesium oxide in a stream of carbon dioxide:

$$MgO(s) + CO_2(g) \rightarrow MgCO_3(s)$$

Above 350°C, the reverse reaction predominates and the carbonate decomposes. Magnesium carbonate is used in making magnesium oxide and is a drying agent (e.g. in table salt). It is also used as a medical antacid and laxative (the basic carbonate is used) and is a component of certain inks and glasses.

magnesium chloride A white solid compound, $MgCl_2$. The anhydrous salt (hexagonal; r.d. 2.32; m.p. 714°C; b.p. 1412°C) can be prepared by the direct combination of dry chlorine with magnesium:

$$Mg(s) + Cl_2(g) \rightarrow MgCl_2(s)$$

The compound also occurs naturally as a constituent of carnallite ($KCl.MgCl_2$). It is a deliquescent compound that commonly forms the hexahydrate, $MgCl_2.6H_2O$ (monoclinic; r.d. 1.57). When heated, this hydrolyses to give magnesium oxide and hydrogen chloride gas. The fused chloride is electrolysed to produce magnesium and it is also used for fireproofing wood, in magnesia cements and artificial leather, and as a laxative.

magnesium hydrogencarbonate **(magnesium bicarbonate)** A compound, $Mg(HCO_3)_2$, that is stable only in solution. It is formed by the action of carbon dioxide on a suspension of magnesium carbonate in water:

$$MgCO_3(s) + CO_2(g) + H_2O(l) \rightarrow Mg(HCO_3)_2(aq)$$

On heating, this process is reversed. Magnesium hydrogencarbonate is one of the compounds responsible for temporary *hardness in water.

magnesium hydroxide A white solid compound, $Mg(OH)_2$; trigonal; r.d. 2.36; decomposes at 350°C. Magnesium hydroxide occurs naturally as the mineral *brucite* and can be prepared by reacting magnesium sulphate or chloride with sodium hydroxide solution. It is used in the refining of sugar and in the processing of uranium. Medicinally it is important as an antacid (*milk of magnesia*) and as a laxative.

magnesium oxide **(magnesia)** A white compound, MgO; cubic; r.d. 3.58; m.p. 2800°C. It occurs naturally as the mineral *periclase* and is prepared commercially by thermally decomposing the mineral *magnesite:

$$MgCO_3(s) \rightarrow MgO(s) + CO_2(g)$$

It has a wide range of uses, including reflective coatings on optical instruments and aircraft windscreens and in semiconductors. Its high melting point makes it useful as a refractory lining in metal and glass furnaces.

magnesium peroxide A white solid, MgO_2. It decomposes at 100°C to release oxygen and also releases oxygen on reaction with water:

$$2MgO_2(s) + 2H_2O \rightarrow 2Mg(OH)_2 + O_2$$

The compound is prepared by reacting sodium peroxide with magnesium sulphate solution and is used as a bleach for cotton and silk.

magnesium sulphate A white soluble compound, $MgSO_4$, existing as the anhydrous compound (rhombic; r.d. 2.66; decomposes at 1124°C) and in hydrated crystalline forms. The monohydrate $MgSO_4.H_2O$ (monoclinic; r.d. 2.45) occurs naturally as the mineral *kieserite*. The commonest hydrate is the heptahydrate, $MgSO_4.7H_2O$ (rhombic; r.d. 1.68), which is called *Epsom salt(s)*, and occurs naturally as the mineral *epsomite*. This is a white powder with a bitter saline taste, which loses $6H_2O$ at 150°C and $7H_2O$ at 200°C. It is used in sizing and fireproofing cotton and silk, in tanning leather, and in the manufacture of fertilizers, explosives, and matches. In medicine, it is used as a laxative. It is also used in veterinary medicine for treatment of local inflammations and infected wounds.

magnetic moment The ratio between the maximum torque (T_{max}) exerted on a magnet, current-carrying coil, or moving charge situated in a magnetic field and the strength of that field. It is thus a measure of the strength of a magnet or current-carrying coil. In the Sommerfeld approach this quantity (also called *electromagnetic moment* or *magnetic area moment*) is the ratio T_{max}/B. In the Kennelly approach the quantity (also called *magnetic dipole moment*) is T_{max}/H.

In the case of a magnet placed in a magnetic field of field strength H, the maximum torque T_{max} occurs when the axis of the magnet is perpendicular to the field. In the case of a coil of N turns and area A carrying a current I, the magnetic moment can be shown to be $m = T/B = NIA$ or $m = T/H = \mu NIA$. Magnetic moments are measured in $A\,m^2$.

An orbital electron has an orbital magnetic moment IA, where I is the equivalent current as the electron moves round its orbit. It is given by $I = q\omega/2\pi$, where q is the electronic charge and ω is its angular velocity. The orbital magnetic moment is therefore $IA = q\omega A/2\pi$, where A is the orbital area. If the electron is spinning there is also a spin magnetic moment (*see* **spin**); atomic nuclei also have magnetic moments.

magnetic quantum number *See* **atom**.

magnetism A group of phenomena associated with magnetic fields. Whenever an electric current flows a magnetic field is produced; as the orbital motion and the *spin of atomic electrons are equivalent to tiny current loops, individual atoms create magnetic fields around them, when their orbital electrons have a net *magnetic moment as a result of their angular momentum. The magnetic moment of an atom is the vector sum of the magnetic moments of the orbital motions and the spins of all the electrons in the atom. The macroscopic magnetic properties of a substance arise from the magnetic moments of its component atoms and molecules. Different materials have different characteristics in an applied magnetic field; there are four main types of magnetic behaviour:

(a) In *diamagnetism* the magnetization is in the opposite direction to that of the applied field, i.e. the susceptibility is negative. Although all substances are diamagnetic, it is a weak form of magnetism and may be masked by other, stronger, forms. It results from changes induced in the orbits of electrons in the atoms of a substance by the applied field, the direction of the change opposing the applied flux. There is thus a weak negative susceptibility (of the order of $-10^{-8}\ m^3\,mol^{-1}$) and a relative permeability of slightly less than one.

(b) In *paramagnetism* the atoms or molecules of the substance have net orbital or spin magnetic moments that are capable of being aligned in the direction of the applied field. They therefore have a positive (but small) susceptibility and a relative permeability slightly in excess of one. Paramagnetism occurs in all atoms and molecules with unpaired electrons; e.g. free atoms, free radicals, and compounds of transition metals containing ions with unfilled electron shells. It also occurs in metals as a result of the magnetic moments associated with the spins of the conducting electrons.

(c) In *ferromagnetic* substances, within a certain temperature range, there are net atomic magnetic moments, which line up in such a way that magnetization persists after the removal of the applied field. Below a certain temperature, called the *Curie point* (or Curie temperature) an increasing magnetic field applied to a ferromagnetic substance will cause increasing magnetization to a high value, called the *saturation magnetization*. This is because a ferromagnetic substance consists of small (1–0.1 mm across) magnetized regions called *domains*. The total magnetic moment of a sample of the substance is the vector sum of the magnetic moments of the component domains. Within each domain the individual atomic magnetic moments are spontaneously aligned by *exchange forces*, related to whether or not the atomic electron spins are parallel or antiparallel. However, in an unmagnetized piece of ferromagnetic material the magnetic moments of the domains themselves are not aligned; when an external field is applied those domains that are aligned with the field increase in size at the expense of the others. In a very strong field all the domains are lined up in the direction of the field and provide the high observed magnetization. Iron, nickel, cobalt, and their alloys are ferromagnetic. Above the Curie point, ferromagnetic materials become paramagnetic.

(d) Some metals, alloys, and transition-element salts exhibit another form of magnetism called *antiferromagnetism*. This occurs below a certain temperature, called the *Néel temperature*, when an ordered array of atomic magnetic moments spontaneously forms in which alternate moments have opposite directions. There is therefore no net resultant magnetic moment in the absence of an applied field. In manganese fluoride, for example, this antiparallel arrangement occurs below a Néel temperature of 72 K. Below this temperature the spontaneous ordering opposes the normal tendency of the magnetic moments to align with the applied field. Above the Néel temperature the substance is paramagnetic.

A special form of antiferromagnetism is *ferrimagnetism*, a type of magnetism exhibited by the *ferrites. In these materials the magnetic moments of adjacent ions are antiparallel and of unequal strength, or the number of magnetic moments in one direction is greater than those in the opposite direction. By suitable choice of rare-earth ions in the ferrite lattices it is possible to design ferrimagnetic substances with specific magnetizations for use in electronic components.

magnetite A black mineral form of iron oxide crystallizing in the cubic system. It is a mixed iron(II)-iron(III) oxide, Fe_3O_4, and is one of the major ores of iron. It is strongly magnetic and some varieties, known as *lodestone*, are natural magnets; these were used as compasses in the ancient world. Magnetite is widely distributed and occurs as an accessory mineral in almost all igneous and metamorphic rocks. The largest deposits of the mineral occur in N Sweden.

magnetochemistry The branch of physical chemistry concerned with measuring and investigating the magnetic properties of compounds. It is used particularly for studying transition-metal complexes, many of which are paramagnetic because they have unpaired electrons. Measurement of the magnetic susceptibility allows the magnetic moment of the metal atom to be calculated, and this gives information about the bonding in the complex.

magnetomechanical ratio *See* **gyromagnetic ratio.**

magneton A unit for measuring magnetic moments of nuclear, atomic, or molecular magnets. The *Bohr magneton* μ_B has the value of the classical magnetic moment of an electron, given by

$$\mu_B = eh/4\pi m_e = 9.274 \times 10^{-24}\ \text{A m}^2,$$

where e and m_e are the charge and mass of the electron and h is the Planck constant. The *nuclear magneton*, μ_N is obtained by replacing the mass of the electron by the mass of the proton and is therefore given by

$$\mu_N = \mu_B.m_e/m_p = 5.05 \times 10^{-27}\ \text{A m}^2.$$

Main-Smith–Stoner rule An empirical rule in the theory of atomic structure stating that for a principal quantum number n the number of electronic quantum states that can have the orbital quantum number l is $2(2l+1)$. This rule describes the subshells of atoms. It was put forward on the basis of chemical evidence by J. D. Main-Smith and independently on the basis of magnetic and spectroscopic evidence by Edmund Stoner in 1924. The Main-Smith–Stoner rule is a consequence of the *Pauli exclusion principle. The rule was one of the key developments that led to the enunciation of the Pauli exclusion principle in 1925.

malachite A secondary mineral form of copper carbonate–hydroxide, $CuCO_3.Cu(OH)_2$. It is bright green and crystallizes in the monoclinic system but usually occurs as aggregates of fibres or in massive form. It is generally found with *azurite in association with the more important copper ores and is itself mined as an ore of copper (e.g. in the Congo). It is also used as an ornamental stone and as a gemstone.

maleic acid *See* **butenedioic acid.**

maleic anhydride A colourless solid, $C_4H_2O_3$, m.p. 53°C, the anhydride of *cis*-butenedioic acid (maleic acid). It is a cyclic compound with a ring containing four carbon atoms and one oxygen atom, made by the catalytic oxidation of benzene or its derivatives at high temperatures. It is used mainly in the manufacture of alkyd and polyester resins and copolymers.

malic acid (2-hydroxybutanedioic acid) A crystalline solid, $HOOCCH(OH)CH_2COOH$. L-malic acid occurs in living organisms as an intermediate metabolite in the *Krebs cycle and also (in certain plants) in photosynthesis. It is found especially in the juice of unripe fruits, e.g. green apples.

malonic acid *See* **propanedioic acid.**

malt The product of the hydrolysis of starch by β-amylase that occurs during the germination of barley in brewing. *See also* **maltose**.

maltose **(malt sugar)** A sugar consisting of two linked glucose molecules that results from the action of the enzyme amylase on starch. Maltose occurs in barley seeds following germination and drying, which is the basis of the malting process used in the manufacture of beer and malt whisky.

malt sugar *See* **maltose.**

mancude (*ma*ximum *n*on-*cu*mulative *d*oubl*e*) Describing an organic compound that contains the maximum possible number of noncumulative double bonds, as in the *annulenes.

manganate(VI) A salt containing the ion MnO_4^{2-}. Manganate(VI) ions are dark green; they are produced by manganate(VII) ions in basic solution.

manganate(VII) **(permanganate)** A salt containing the ion MnO_4^-. Manganate(VII) ions are dark purple and strong oxidizing agents.

manganese Symbol Mn. A grey brittle metallic *transition element, a.n. 25; r.a.m. 54.94; r.d. 7.2; m.p. 1244°C; b.p. 1962°C. The main sources are pyrolusite (MnO_2) and rhodochrosite ($MnCO_3$). The metal can be extracted by reduction of the oxide using magnesium (*Kroll process) or aluminium (*Goldschmidt process). Often the ore is mixed with iron ore and reduced in an electric furnace to produce ferromanganese for use in alloy steels. The element is fairly electropositive; it combines with oxygen, nitrogen, and other nonmetals when heated (but not with hydrogen). Salts of manganese contain the element in the +2 and +3 oxidation states. Manganese(II) salts are the more stable. It also forms compounds in higher oxidation states, such as manganese(IV) oxide and manganate(VI) and manganate(VII) salts. The element was discovered in 1774 by Karl *Scheele.

manganese(IV) oxide **(manganese dioxide)** A black oxide made by heating manganese(II) nitrate. The compound also occurs naturally as *pyrolusite*. It is a strong oxidizing agent, used as a depolarizing agent in voltaic cells.

manganic compounds Compounds of manganese in its +3 oxidation state; e.g. manganic oxide is manganese(III) oxide, Mn_2O_3.

manganin A copper alloy containing 13–18% of manganese and 1–4% of nickel. It has a high electrical resistance, which is relatively insensitive to temperature changes. It is therefore suitable for use in resistance wire.

manganous compounds Compounds of manganese in its +2 oxidation state; e.g. manganous oxide is manganese(II) oxide, MnO.

mannan *See* **mannose.**

mannitol A polyhydric alcohol, $CH_2OH(CHOH)_4CH_2OH$, derived from mannose or fructose. It is the main soluble sugar in fungi and an important carbohydrate reserve in brown algae. Mannitol is used as a sweetener in certain foodstuffs and as a diuretic to relieve fluid retention.

mannose A *monosaccharide, $C_6H_{12}O_6$, stereoisomeric with glucose, that occurs naturally only in polymerized forms called *mannans*. These are found in plants, fungi, and bacteria, serving as food energy stores.

manometer A device for measuring pressure differences, usually by the difference in height of two liquid columns. The simplest type is the U-tube manometer, which consists of a glass tube bent into the shape of a U. If a pressure to be measured is fed to one side of the U-tube and the other is open to the atmosphere, the difference in level of the liquid in the two limbs gives a measure of the unknown pressure.

many-body problem The problem that it is very difficult to obtain exact solutions to systems involving interactions between more than two bodies – using either classical mechanics or quantum mechanics. To understand the physics of many-body systems it is necessary to make use of approximation techniques or model systems. For some problems, such as the *three-body problem* in classical mechanics, it is possible to obtain qualitative information about the system. If there are a great many bodies interacting, such as the molecules in a gas, the problem can be analysed using the techniques of *statistical mechanics.

marble A metamorphic rock composed of recrystallized *calcite or *dolomite. Pure marbles are white but such impurities as silica or clay minerals result in variations of colour. Marble is extensively used for building purposes and ornamental use; the pure white marble from Carrara in Italy is especially prized by sculptors. The term is applied commercially to any limestone or dolomite that can be cut and polished.

Markoffian process (Markov process) A random process (*see* **stochastic process**) in which the rate of change of a time-dependent quantity $\partial a(t)/\partial t$ depends on the instantaneous value of the quantity $a(t)$, where t is the time, but not on its previous history. If a random process can be assumed to be a Markov process, an analysis of the process is greatly simplified enabling useful equations in *nonequilibrium statistical mechanics and disordered solids to be derived. Problems involving Markov processes are solved using statistical methods and the theory of probability. Markov processes are named after the Russian mathematician Andrei Andreevich Markov (1856–1922).

Markovnikoff's rule When an acid HA adds to an alkene, a mixture of products can be formed if the alkene is not symmetrical. For instance, the reaction between $C_2H_5CH{:}CH_2$ and HCl can give $C_2H_5CH_2CH_2Cl$ or $C_2H_5CHClCH_3$. In general, a mixture of products occurs in which one predominates over the other. In 1870, Vladimir Markovnikoff (1837–1904) proposed the rule that the main product would be the one in which the hydrogen atom adds to the carbon having the larger number of hydrogen atoms (the latter product above). This occurs when the mechanism is *electrophilic addition, in which the first step is addition of H^+. The electron-releasing effect of the alkyl group (C_2H_5) distorts the electron-distribution in the double bond, making the carbon atom furthest from the alkyl group negative. This is the atom attacked by H^+ giving the carbonium ion $C_2H_5C^+HCH_3$, which further reacts with the negative ion Cl^-.

In some circumstances *anti-Markovnikoff* behaviour occurs, in which the opposite effect is found. This happens when the mechanism involves free radicals and is common in addition of hydrogen bromide when peroxides are present.

marsh gas Methane formed by rotting vegetation in marshes.

Marsh's test A chemical test for arsenic in which hydrochloric acid and zinc are added to the sample, arsine being produced by the nascent hydrogen generated. Gas from the sample is led through a heated glass tube and, if arsine is present, it decomposes to give a brown deposit of arsenic metal. The arsenic is distinguished from antimony (which gives a similar result) by the fact that antimony does not dissolve in sodium chlorate(I) (hypochlorite). The test was devised in 1836 by the British chemist James Marsh (1789–1846).

martensite A solid solution of carbon in alpha-iron (*see* **iron**) formed when *steel is cooled too rapidly for pearlite to form from austenite. It is responsible for the hardness of quenched steel.

mascagnite A mineral form of *ammonium sulphate, $(NH_4)_2SO_4$.

maser (*m*icrowave *a*mplification by *s*timulated *e*mission of *r*adiation) A device for amplifying or generating microwaves by means of stimulated emission (*see* **laser**).

mass A measure of a body's inertia, i.e. its resistance to acceleration. According to Newton's laws of motion, if two unequal masses, m_1 and m_2, are allowed to collide, in the absence of any other forces both will experience the same force of collision. If the two bodies acquire accelerations a_1 and a_2 as a result of the collision, then $m_1a_1 = m_2a_2$. This equation enables two masses to be compared. If one of the masses is regarded as a standard of mass, the mass of all other masses can be measured in terms of this standard. The body used for this purpose is a 1-kg cylinder of platinum–iridium alloy, called the international standard of mass.

mass action The law of mass action states that the rate at which a chemical reaction takes place at a given temperature is proportional to the product of the *active masses* of the reactants. The active mass of a reactant is taken to be its molar concentration. For example, for a reaction

$$xA + yB \rightarrow \text{products}$$

the rate is given by

$$R = k[A]^x[B]^y$$

where k is the *rate constant. The principle was introduced by C. M.

Guldberg and P. Waage in 1863. It is strictly correct only for ideal gases. In real cases *activities can be used. *See also* **equilibrium constant**.

mass concentration *See* **concentration**.

massicot *See* **lead(II) oxide**.

mass number *See* **nucleon number**.

mass spectroscopy A technique used to determine relative atomic masses and the relative abundance of isotopes, and for chemical analysis and the study of ion reactions. In a *mass spectrometer* a sample (usually gaseous) is ionized and the positive ions produced are accelerated into a high-vacuum region containing electric and magnetic fields. These fields deflect and focus the ions onto a detector. The fields can be varied in a controlled way so that ions of different types can impinge on the detector. A *mass spectrum* is thus obtained consisting of a series of peaks of variable intensity to which mass/charge (m/e) values can be assigned. The original ions are usually produced by electron impact, although ion impact, photoionization, and field ionization are also used. For organic molecules, the mass spectrum consists of a series of peaks, one corresponding to the parent ion and the others to fragment ions produced by the ionization process. Different molecules can be identified by their characteristic pattern of lines. Analysis of mixtures can be done by gas chromatography–mass spectroscopy (*see* **gas chromatography**).

masurium A former name for *technetium.

matrix (*pl.* **matrices**) **1.** ***(in chemistry)*** A continuous solid phase in which particles (atoms, ions, etc.) are embedded. Unstable species, such as free radicals, can be trapped in an unreactive substrate, such as solid argon, and studied by spectroscopy. The species under investigation are separated by the matrix, hence the term *matrix isolation* for this technique. **2.** ***(in geology)*** The fine-grained material of rock in which the coarser-grained material is embedded. **3.** ***(in mathematics)*** A set of quantities in a rectangular array, used in certain mathematical operations. The array is usually enclosed in large parentheses or in square brackets.

matrix mechanics A formulation of *quantum mechanics using matrices (*see* **matrix**) to represent states and operators. Matrix mechanics was the first formulation of quantum mechanics to be stated (by Werner Heisenberg in 1925) and was developed by Heisenberg and Max Born (1882–1970) and the German physicist Pascual Jordan (1902–80). It was shown by Erwin Schrödinger in 1926 to be equivalent to the *wave mechanics formulation of quantum mechanics.

Maxwell, James Clerk (1831–79) British physicist, born in Edinburgh, who held academic posts at Aberdeen, London, and Cambridge. In the 1860s he was one of the founders of the *kinetic theory of gases, but his best-known work was a mathematical analysis of electromagnetic radiation, published in 1864.

Maxwell–Boltzmann distribution A law describing the distribution of speeds among the molecules of a gas. In a system consisting of N molecules that are independent of each other except that they exchange energy on collision, it is clearly impossible to say what velocity any particular molecule will have. However, statistical statements regarding certain functions of the molecules were worked out by James Clerk *Maxwell and Ludwig *Boltzmann. One form of their law states that $n = N\exp(-E/RT)$, where n is the number of molecules with energy in excess of E, T is the thermodynamic temperature, and R is the *gas constant.

Maxwell's demon An imaginary creature that is able to open and shut a partition dividing two volumes of a gas in a container, when the two volumes are initially at the same temperature. The partition operated by the demon is only opened to allow fast molecules through. Such a process would make the volume of gas containing the fast molecules hotter than it was at the start; the volume of gas remaining would accordingly become cooler. This process would be a violation of the second law of *thermodynamics and therefore cannot occur. Maxwell's demon was invented by James Clerk Maxwell in a letter written in 1867 to show that the second law of thermodynamics has its origins in *statistical mechanics, although the name was suggested by Sir William Thomson (subsequently Lord Kelvin).

Maxwell's thermodynamic equations Equations in thermodynamics for a given mass of a homogeneous system, relating the entropy (S), pressure (p), volume (V), and thermodynamic temperature (T). The four equations are:

$$(\partial T/\partial V)_S = -(\partial p/\partial S)_V;$$
$$(\partial T/\partial p)_S = -(\partial V/\partial S)_p;$$
$$(\partial V/\partial T)_p = -(\partial S/\partial p)_T;$$
$$(\partial S/\partial V)_T = -(\partial p/\partial T)_V.$$

Mayer *f*-function A quantity that occurs in the calculation of virial coefficients; it is defined by $f = \exp(-V_2/kT) - 1$, where V_2 is the two-body interaction potential energy, k is the *Boltzmann constant, and T is the thermodynamic temperature. It is related to the second virial coefficient B by:

$$B = (-N_A/V)\int f\,\mathrm{d}r_1\mathrm{d}r_2,$$

where N_A is the Avogadro number and V is the volume of the system. This equation simplified to:

$$B = -2\pi N_A\int_0^\infty fr^2\mathrm{d}r$$

in the case of closed-shell atoms and octahedral and tetrahedral molecules. When particles are so far apart that the interaction $V \to 0$, then $f \to 0$ also, but when the particles are so close together that the interaction $V \to \infty$, then $f \to -1$. This enables strong repulsive interactions between particles to be analysed in terms of f but not of V. The function is named after the US physicist Joseph Mayer.

McLeod gauge A vacuum pressure gauge, devised by Herbert McLeod (1841–1923), in which a relatively large volume of a low-pressure gas is compressed to a small volume in a glass apparatus. The volume is reduced to an extent that causes the pressure to rise sufficiently to support a column of fluid high enough to read. This simple device, which relies on *Boyle's law, is suitable for measuring pressures in the range 10^3 to 10^{-3} pascal.

McMillan–Mayer theory A theory of solutions of nonelectrolytes developed by the US scientists W. G. McMillan and J. E. Mayer in 1945. The theory shows that there is a one-to-one correspondence between the equations describing a nonideal gas and those describing dilute solutions of nonelectrolytes. In particular, they showed that there is a correspondence between the pressure of the gas and the osmotic pressure of the solution. This enables an expansion for solutions to be written, which is analogous to the virial expansion of nonideal gases with analogues of the virial coefficients. These coefficients can be calculated with the analogue of potential being the potential of mean force of N solute molecules in the pure solvent. The McMillan–Mayer theory can also be extended to distribution functions.

mean free path The average distance travelled between collisions by the molecules in a gas, the electrons in a metallic crystal, the neutrons in a moderator, etc. According to the *kinetic theory the mean free path between elastic collisions of gas molecules of diameter d (assuming they are rigid spheres) is $1/\sqrt{2}n\pi d^2$, where n is the number of molecules per unit volume in the gas. As n is proportional to the pressure of the gas, the mean free path is inversely proportional to the pressure.

mean free time The average time that elapses between the collisions of the molecules in a gas, the electrons in a crystal, the neutrons in a moderator, etc. *See* **mean free path**.

mechanism The way in which a particular chemical reaction occurs, described in terms of the steps involved. For example, the hydrolysis of an alkyl chloride proceeds by the S_N1 mechanism (*see* **nucleophilic substitution**).

medium frequency (MF) A radio frequency in the range 0.3–3 megahertz; i.e. having a wavelength in the range 100–1000 metres.

mega- Symbol M. A prefix used in the metric system to denote one million times. For example, 10^6 volts = 1 megavolt (MV).

meitnerium Symbol Mt. A radioactive *transactinide element; a.n. 109. It was first made in 1982 by Peter Armbruster and a team in Darmstadt, Germany, by bombarding bismuth-209 nuclei with iron-58 nuclei. Only a few atoms have ever been detected.

Meitner, Lise (1878–1968) Austrian-born Swedish physicist and radiochemist who worked in Berlin with Otto *Hahn. Together they discovered protactinium. Meitner and Hahn worked together on neutron bombardment of uranium. In the 1930s, she escaped from Austria to Sweden to avoid Nazi persecution. In Stockholm, along with her nephew Otto Frisch (1904–79), she formulated the theory of nuclear fission.

melamine A white crystalline compound, $C_3N_6H_6$. Melamine is a cyclic compound having a six-membered ring of alternating C and N atoms, with three NH_2 groups. It can be copolymerized with methanal to give thermosetting *melamine resins*, which are used particularly for laminated coatings.

mellitic acid (benzenehexacarboxylic acid) A colourless crystalline compound, $C_6(COOH)_6$, m.p. 288°C. Its molecules consist of a benzene ring in which all six hydrogen atoms have been substituted by carboxyl (–COOH) groups. It occurs naturally in some lignite beds as honeystone (the aluminium salt), and is made by oxidizing charcoal with concentrated nitric acid. It decomposes on heating to form pyromellitic anhydride, used in making epoxy resins. Condensation products of mellitic acid are employed in making a wide range of dyes.

melting point (m.p.) The temperature at which a solid changes into a liquid. A pure substance under standard conditions of pressure (usually 1 atmosphere) has a single reproducible melting point. If heat is gradually and uniformly supplied to a solid the consequent rise in temperature stops at the melting point until the fusion process is complete.

Mendeleev, Dmitri Ivanovich (1834–1907) Russian chemist, who became professor of chemistry at St Petersburg in 1866. His most famous work, published in 1869, was the compilation of the *periodic table of the elements, based on the periodic law.

Mendeleev's law *See* **periodic law**.

mendelevium Symbol Md. A radioactive metallic transuranic element belonging to the *actinoids; a.n. 101; mass number of the first discovered nuclide 256 (half-life 1.3 hours). Several short-lived isotopes have now been synthesized. The element was first identified by Albert Ghiorso, Glenn Seaborg (1912–99), and associates in 1955.

Mendius reaction A reaction in which an organic nitrile is reduced by nascent hydrogen (e.g. from sodium in ethanol) to a primary amine:

$$RCN + 2H_2 \rightarrow RCH_2NH_2$$

menthol A white crystalline terpene alcohol, $C_{10}H_{19}OH$; r.d. 0.89; m.p. 42°C; b.p. 103–104°C. It has a minty taste and is found in certain essential oils (e.g. peppermint) and used as a flavouring.

mercaptans *See* **thiols**.

mercapto group *See* **thiols**.

mercuric compounds Compounds of mercury in its +2 oxidation state; e.g. mercuric chloride is mercury(II) chloride, $HgCl_2$.

mercurous compounds Compounds of mercury in its +1 oxidation state; e.g. mercury(I) chloride is mercurous chloride, HgCl.

mercury Symbol Hg. A heavy silvery liquid metallic element belonging to the *zinc group; a.n. 80; r.a.m. 200.59; r.d. 13.55; m.p. −38.87°C; b.p. 356.58°C. The main ore is the sulphide cinnabar (HgS), which can be decomposed to the elements. Mercury is used in thermometers, barometers, and other scientific apparatus, and in dental amalgams. The element is less reactive than zinc and cadmium and will not displace hydrogen from acids. It is also unusual in forming mercury(I) compounds containing the Hg_2^{2+} ion, as well as mercury(II) compounds containing Hg^{2+} ions. It also forms a number of complexes and organomercury compounds.

mercury cell A primary *voltaic cell consisting of a zinc anode and a cathode of mercury(II) oxide (HgO) mixed with graphite. The electrolyte is potassium hydroxide (KOH) saturated with zinc oxide, the overall reaction being:

$$Zn + HgO \rightarrow ZnO + Hg$$

The e.m.f. is 1.35 volts and the cell will deliver about 0.3 ampere-hour per cm^3.

mercury(I) chloride A white salt, Hg_2Cl_2; r.d. 7.15; sublimes at 400°C. It is made by heating mercury(II) chloride with mercury and is used in calomel cells (so called because the salt was formerly called *calomel*) and as a fungicide.

mercury(II) chloride A white salt, $HgCl_2$; r.d. 5.4; m.p. 276°C; b.p. 302°C. It is made by reacting mercury with chlorine and used in making other mercury compounds.

mercury(II) fulminate A grey crystalline solid, $Hg(CNO)_2.\frac{1}{2}H_2O$, made by the action of nitric acid on mercury and treating the solution formed with ethanol. It is used as a detonator for cartridges and can be handled safely only under cold water.

mercury(II) oxide A yellow or red oxide of mercury, HgO. The red form is made by heating mercury in oxygen at 350°C; the yellow form, which differs from the red in particle size, is precipitated when sodium hydroxide solution is added to a solution of mercury(II) nitrate. Both forms decompose to the elements at high temperature. The black precipitate formed when sodium hydroxide is added to mercury(I) nitrate solution is sometimes referred to as mercury(I) oxide (Hg_2O) but is probably a mixture of HgO and free mercury.

mercury(II) sulphide A red or black compound, HgS, occurring naturally as the minerals cinnabar (red) and metacinnabar (black). It can be obtained as a black precipitate by bubbling hydrogen sulphide through a solution of mercury(II) nitrate. The red form is obtained by sublimation. The compound is also called *vermilion* (used as a pigment).

mer-isomer *See* **isomerism.**

meso-isomer *See* **optical activity.**

mesomerism **(mesomeric effect)** A former name for *resonance in molecules. *See also* **electronic effects.**

mesomorph *See* **lyotropic mesomorph.**

mesoscopic Designating a size scale intermediate between those of the *microscopic and the *macroscopic states. Mesoscopic objects and systems require quantum mechanics to describe them.

meta- 1. Prefix designating a benzene compound in which two substituents are in the 1,3 positions on the benzene ring. The abbreviation *m*- is used; for example, *m*-xylene is 1,3-dimethylbenzene. *Compare* **ortho-**; **para-**. **2.** Prefix designating a lower oxo acid, e.g. metaphosphoric acid. *Compare* **ortho-**.

metabolic pathway *See* **metabolism.**

metabolism The sum of the chemical reactions that occur within living organisms. The various compounds that take part in or are formed by these reactions are called *metabolites*. In animals many metabolites are obtained by the digestion of food, whereas in plants only the basic starting materials (carbon dioxide, water, and minerals) are externally derived. The synthesis (*anabolism) and breakdown (*catabolism) of most compounds occurs by a number of reaction steps, the reaction sequence being termed a *metabolic pathway*. Some pathways (e.g. *glycolysis) are linear; others (e.g. the *Krebs cycle) are cyclic.

metabolite *See* **metabolism.**

metaboric acid *See* **boric acid.**

metal Any of a class of chemical elements that are typically lustrous solids that are good conductors of heat and electricity. Not all metals have all these properties (e.g. mercury is a liquid). In chemistry, metals fall into two distinct types. Those of the *s*- and *p*-blocks (e.g. sodium and aluminium) are generally soft silvery reactive elements. They tend to form positive ions and so are described as electropositive. This is contrasted with typical nonmetallic behaviour of forming negative ions. The *transition elements (e.g. iron and copper) are harder substances and generally less reactive. They form coordination complexes. All metals have oxides that are basic, although some, such as aluminium, have *amphoteric properties.

metaldehyde A solid compound, $C_4O_4H_4(CH_3)_4$, formed by polymerization of ethanal (acetaldehyde) in dilute acid solutions below 0°C. The compound, a tetramer of ethanal, is used in slug pellets and as a fuel for portable stoves.

metal fatigue A cumulative effect causing a metal to fail after repeated applications of stress, none of which exceeds the ultimate tensile strength. The *fatigue strength* (or *fatigue limit*) is the stress that will cause failure after a specified number (usually 10^7) of cycles. The number of cycles required to produce failure decreases as the level of stress or strain increases. Other factors, such as corrosion, also reduce the fatigue life.

metallic bond A chemical bond of the type holding together the atoms in a solid metal or alloy. In such solids, the atoms are considered to be ionized, with the positive ions occupying lattice positions. The valence electrons are able to move freely (or almost freely) through the lattice, forming an 'electron gas'. The bonding force is electrostatic attraction between the positive metal ions and the electrons. The existence of free electrons accounts for the good electrical and thermal conductivities of metals. *See also* **energy bands**.

metallic crystal A crystalline solid in which the atoms are held together by *metallic bonds. Metallic crystals are found in some *interstitial compounds as well as in metals and alloys.

metallized dye *See* **dyes**.

metallocene *See* **sandwich compound**.

metallography The microscopic study of the structure of metals and their alloys. Both optical microscopes and electron microscopes are used in this work.

metalloid (semimetal) Any of a class of chemical elements intermediate in properties between metals and nonmetals. The classification is not clear cut, but typical metalloids are boron, silicon, germanium, arsenic, and tellurium. They are electrical semiconductors and their oxides are amphoteric.

metallurgy The branch of applied science concerned with the production of metals from their ores, the purification of metals, the manufacture of alloys, and the use and performance of metals in engineering practice. *Process metallurgy* is concerned with the extraction and production of metals, while *physical metallurgy* concerns the mechanical behaviour of metals.

metamict state The amorphous state of a substance that has lost its crystalline structure as a result of the radioactivity of uranium or thorium. *Metamict minerals* are minerals whose structure has been disrupted by this process. The metamictization is caused by alpha-particles and the recoil nuclei from radioactive disintegration.

metaphosphoric acid *See* **phosphoric(V) acid**.

metaplumbate *See* **plumbate**.

metastable state A condition of a system in which it has a precarious stability that can easily be disturbed. It is unlike a state of stable equilibrium in that a minor disturbance will cause a system in a metastable state to fall to a lower energy level. A book lying on a table is in a state of stable equilibrium; a thin book standing on edge is in metastable equilibrium. Supercooled water is also in a metastable state. It is liquid below 0°C; a grain of dust or ice introduced into it will cause it to freeze. An excited state of an atom or nucleus that has an appreciable lifetime is also metastable.

metastannate *See* **stannate.**

metathesis *See* **double decomposition.**

methacrylate A salt or ester of methacrylic acid (2-methylpropenoic acid).

methacrylate resins *Acrylic resins obtained by polymerizing 2-methylpropenoic acid or its esters.

methacrylic acid *See* **2-methylpropenoic acid.**

methanal (formaldehyde) A colourless gas, HCHO; r.d. 0.815 (at −20°C); m.p. −92°C; b.p. −21°C. It is the simplest *aldehyde, made by the catalytic oxidation of methanol (500°C; silver catalyst) by air. It forms two polymers: *methanal trimer and polymethanal. *See also* **formalin.**

methanal trimer A cyclic trimer of methanal, $C_3O_3H_6$, obtained by distillation of an acidic solution of methanal. It has a six-membered ring of alternating –O– and $-CH_2-$ groups.

methanation A method of manufacturing methane from carbon monoxide or dioxide by high-pressure catalytic hydrogenation. It is often used to improve the calorific value of town gas.

methane A colourless odourless gas, CH_4; m.p. −182.5°C; b.p. −164°C. Methane is the simplest hydrocarbon, being the first member of the *alkane series. It is the main constituent of natural gas (~99%) and as such is an important raw material for producing other organic compounds. It can be converted into methanol by catalytic oxidation.

methanide *See* **carbide.**

methanoate (formate) A salt or ester of methanoic acid.

methanoic acid (formic acid) A colourless pungent liquid, HCOOH; r.d. 1.2; m.p. 8°C; b.p. 101°C. It can be made by the action of concentrated sulphuric acid on the sodium salt (sodium methanoate), and occurs naturally in ants and stinging nettles. Methanoic acid is the simplest of the *carboxylic acids.

methanol (methyl alcohol) A colourless liquid, CH_3OH; r.d. 0.79; m.p. −93.9°C; b.p. 64.96°C. It is made by catalytic oxidation of methane (from natural gas) using air. Methanol is used as a solvent (*see* **methylated spirits**) and as a raw material for making methanal (mainly for urea–formaldehyde resins). It was formerly made by the dry distillation of wood (hence the name *wood alcohol*).

methionine *See* amino acid.

methoxy group The organic group CH_3O-.

methyl acetate *See* methyl ethanoate.

methyl alcohol *See* methanol.

methylamine A colourless flammable gas, CH_3NH_2; m.p. −93.5°C; b.p. −6.3°C. It can be made by a catalytic reaction between methanol and ammonia and is used in the manufacture of other organic chemicals.

methylated spirits A mixture consisting mainly of ethanol with added methanol (~9.5%), pyridine (~0.5%), and blue dye. The additives are included to make the ethanol undrinkable so that it can be sold without excise duty for use as a solvent and a fuel (for small spirit stoves).

methylation A chemical reaction in which a methyl group (CH_3-) is introduced in a molecule. A particular example is the replacement of a hydrogen atom by a methyl group, as in a *Friedel–Crafts reaction.

methylbenzene **(toluene)** A colourless liquid, $CH_3C_6H_5$; r.d. 0.9; m.p. −95°C; b.p. 111°C. Methylbenzene is derived from benzene by replacement of a hydrogen atom by a methyl group. It can be obtained from coal tar or made from methylcyclohexane (extracted from crude oil) by catalytic dehydrogenation. Its main uses are as a solvent and as a raw material for producing TNT.

methyl bromide *See* bromomethane.

2-methylbuta-1,3-diene *See* isoprene.

methyl chloride *See* chloromethane.

methyl cyanide *See* ethanenitrile.

methylene The highly reactive *carbene, $:CH_2$. The divalent CH_2 group in a compound is the *methylene group.*

methylene chloride *See* dichloromethane.

methyl ethanoate **(methyl acetate)** A colourless volatile fragrant liquid, CH_3COOCH_3; r.d. 0.92; m.p. −98°C; b.p. 54°C. A typical *ester, it can be made from methanol and methanoic acid and is used mainly as a solvent.

methyl ethyl ketone *See* butanone.

methyl group **(methyl radical)** The organic group CH_3-.

methylidyne *See* carbyne.

methyl methacrylate An ester of methacrylic acid (2-methylpropenoic acid), $CH_2:C(CH_3)COOCH_3$, used in making *methacrylate resins.

methyl orange An organic dye used as an acid–base *indicator. It changes from red below pH 3.1 to yellow above pH 4.4 (at 25°C) and is used for titrations involving weak bases.

methylphenols **(cresols)** Organic compounds having a methyl group and a hydroxyl group bound directly to a benzene ring. There are three isomeric methylphenols with the formula $CH_3C_6H_4OH$, differing in the relative positions of the methyl and hydroxyl groups. A mixture of the three can be obtained by distilling coal tar and is used as a germicide and antiseptic.

2-methylpropenoic acid **(methacrylic acid)** A white crystalline unsaturated carboxylic acid, $CH_2{:}C(CH_3)COOH$, used in making *methacrylate resins.

methyl red An organic dye similar in structure and use to methyl orange. It changes from red below pH 4.4 to yellow above pH 6.0 (at 25°C).

methyl violet A violet dye used as a chemical indicator and as a biological stain. It is also the colouring matter in methylated spirits. It is a mixture of compounds of rosaniline, made by oxidizing dimethylphenylamine with copper(II) chloride.

metol *See* **aminophenol**.

metre Symbol m. The SI unit of length, being the length of the path travelled by light in vacuum during a time interval of $1/(2.99792458 \times 10^8)$ second. This definition, adopted by the General Conference on Weights and Measures in October, 1983, replaced the 1967 definition based on the krypton lamp, i.e. 1 650 763.73 wavelengths in a vacuum of the radiation corresponding to the transition between the levels $2p^{10}$ and $5d^5$ of the nuclide krypton–86. This definition (in 1958) replaced the older definition of a metre based on a platinum–iridium bar of standard length. When the *metric system was introduced in 1791 in France, the metre was intended to be one ten-millionth of the earth's meridian quadrant passing through Paris. However, the original geodetic surveys proved the impractibility of such a standard and the original platinum metre bar, the *mètre des archives*, was constructed in 1793.

metric system A decimal system of units originally devised by a committee of the French Academy, which included J. L. Lagrange and P. S. Laplace, in 1791. It was based on the *metre, the gram defined in terms of the mass of a cubic centimetre of water, and the second. This centimetre-gram-second system (*see* **c.g.s. units**) later gave way for scientific work to the metre-kilogram-second system (*see* **m.k.s. units**) on which *SI units are based.

Meyer, Viktor (1848–97) German chemist who worked at Zürich and later at Heidelberg on a wide range of topics. He was the first to prepare oximes and the sulphur compound thiophene. He is known for his method of measuring relative molecular mass by determining vapour density (*see* **Viktor Meyer's method**). Meyer also worked on stereochemistry and was the first to identify steric hindrance in chemical reactions.

mica Any of a group of silicate minerals with a layered structure. Micas are composed of linked SiO_4 tetrahedra with cations and hydroxyl groupings between the layers. The general formula is $X_2Y_{4-6}Z_8O_{20}(OH,F)_4$, where X = K,Na,Ca; Y = Al,Mg,Fe,Li; and Z = Si,Al. The three main mica minerals are: *muscovite, $K_2Al_4(Si_6Al_2O_{20})(OH,F)_4$; *biotite, $K_2(Mg,Fe^{2+})_{6-4}(Fe^{3+},Al,Ti)_{0-2}(Si_{6-5}Al_{2-3}O_{20})(OH,F)_4$; lepidolite, $K_2(Li,Al)_{5-6}(Si_{6-7}Al_{2-1}O_{20})(OH,F)_4$. Micas have perfect basal cleavage and the thin cleavage flakes are flexible and elastic. Flakes of mica are used as electrical insulators and as the dielectric in capacitors.

micelle An aggregate of molecules in a *colloid. For example, when soap or other *detergents dissolve in water they do so as micelles – small clusters of molecules in which the nonpolar hydrocarbon groups are in the centre and the hydrophilic polar groups are on the outside solvated by the water molecules.

Michaelis–Menten curve A graph that shows the relationship between the concentration of a substrate and the rate of the corresponding enzyme-controlled reaction. The curve only applies to enzyme reactions involving a single substrate. It was devised by Leonor Michaelis (1875–1949) and Maud Menten (1879–1960). The graph can be used to calculate the *Michaelis constant* (K_m), which is the concentration of a substrate required in order for an enzyme to act at half of its maximum velocity (V_{max}). The Michaelis constant is a measure of the affinity of an enzyme for a substrate. A low value corresponds to a high affinity, and vice versa. *See also* **enzyme kinetics**.

micro- Symbol μ. A prefix used in the metric system to denote one millionth. For example, 10^{-6} metre = 1 micrometre (μm).

microbalance A sensitive *balance capable of weighing masses of the order 10^{-6} to 10^{-9} kg.

microscopic Designating a size scale comparable to the subatomic particles, atoms, and molecules. Microscopic objects and systems are described by *quantum mechanics. *Compare* **macroscopic**; **mesoscopic**.

microscopic reversibility The principle that in a reversible reaction the mechanism in one direction is the exact reverse of the mechanism in the other direction. *See also* **detailed balance**.

microwaves Electromagnetic waves with wavelengths in the range 10^{-3} to 0.03 m.

microwave spectroscopy A sensitive technique for chemical analysis and the determination of molecular structure (bond lengths, bond angles, and dipole moments), and also relative atomic masses. It is based on the principle that microwave radiation (*see* **microwaves**) causes changes in the rotational energy levels of molecules and absorption consequently occurs at characteristic frequencies. In a microwave spectrometer a microwave source, usually a klystron valve, produces a beam that is passed through a gaseous sample. The beam then impinges on the detector, usually a crystal detector, and the signal (wavelength against intensity) is displayed, either as a printed plot or on an oscilloscope. As microwaves are absorbed by air the instrument is evacuated.

migration 1. The movement of a group, atom, or double bond from one part of a molecule to another. **2.** The movement of ions under the influence of an electric field.

milk of magnesia *See* **magnesium hydroxide.**

milk sugar *See* **lactose.**

Miller indices A set of three numbers that characterize a face of a crystal. The French mineralogist René Just Haüy (1743–1822) proposed the *law of rational intercepts*, which states that there is always a set of axes, known as *crystal axes*, that allows a crystal face to be characterized in terms of intercepts of the face with these axes. The reciprocals of these intercepts are small rational numbers. When the fractions are cleared there is a set of three integers. These integers are known as the *Miller indices* of the crystal face after the British mineralogist William Hallowes Miller (1810–80), who pointed out that crystal faces could be characterized by these indices. If a plane is parallel to one of the crystal axes then its intercept is at infinity and hence its reciprocal is 0. If a face cuts a crystal axis on the negative side of the origin then the intercept, and hence its reciprocal, i.e. the Miller index for that axis, are negative. This is indicated by a bar over the Miller index. For example, the Miller indices for the eight faces of an octahedron are (III), $(\bar{\mathrm{I}}\mathrm{II})$, $(\mathrm{I}\bar{\mathrm{I}}\mathrm{I})$, $(\mathrm{II}\bar{\mathrm{I}})$, $(\bar{\mathrm{I}}\bar{\mathrm{I}}\mathrm{I})$, $(\mathrm{I}\bar{\mathrm{I}}\bar{\mathrm{I}})$, $(\bar{\mathrm{I}}\mathrm{I}\bar{\mathrm{I}})$ and $(\bar{\mathrm{I}}\bar{\mathrm{I}}\bar{\mathrm{I}})$.

milli- Symbol m. A prefix used in the metric system to denote one thousandth. For example, 0.001 volt = 1 millivolt (mV).

Millon's reagent A solution of mercury(II) nitrate and nitrous acid used to test for proteins. The sample is added to the reagent and heated for two minutes at 95°C; the formation of a red precipitate indicates the presence of protein in the sample. The reagent is named after French chemist Auguste Millon (1812–67).

mineral A naturally occurring substance that has a characteristic chemical composition and, in general, a crystalline structure. The term is also often applied generally to organic substances that are obtained by mining (e.g. coal, petroleum, and natural gas) but strictly speaking these are not minerals, being complex mixtures without definite chemical formulas. Rocks are composed of mixtures of minerals. Minerals may be identified by the properties of their crystal system, hardness (measured on the Mohs' scale), relative density, lustre, colour, cleavage, and fracture. Many names of minerals end in *-ite*.

mineral acid A common inorganic acid, such as hydrochloric acid, sulphuric acid, or nitric acid.

mirabilite A mineral form of *sodium sulphate, $Na_2SO_4.10H_2O$.

misch metal An alloy of cerium (50%), lanthanum (25%), neodymium (18%), praseodymium (5%), and other rare earths. It is used alloyed with iron (up to 30%) in lighter flints, and in small quantities to improve the malleability of iron. It is also added to copper alloys to make them harder, to aluminium alloys to make them stronger, to magnesium alloys to reduce creep, and to nickel alloys to reduce oxidation.

Mitscherlich's law (law of isomorphism) Substances that have the same crystal structure have similar chemical formulae. The law can be used to determine the formula of an unknown compound if it is isomorphous with a compound of known formula. It is named after Eilhard Mitscherlich (1794–1863).

mixture A system of two or more distinct chemical substances. *Homogeneous mixtures* are those in which the atoms or molecules are interspersed, as in a mixture of gases or in a solution. *Heterogeneous mixtures* have distinguishable phases, e.g. a mixture of iron filings and sulphur. In a mixture there is no redistribution of valence electrons, and the components retain their individual chemical properties. Unlike compounds, mixtures can be separated by physical means (distillation, crystallization, etc.).

m.k.s. units A *metric system of units devised by A. Giorgi (and sometimes known as *Giorgi units*) in 1901. It is based on the metre, kilogram, and second and grew from the earlier *c.g.s. units. The electrical unit chosen to augment these three basic units was the ampere and the permeability of space (magnetic constant) was taken as 10^{-7} Hm^{-1}. To simplify electromagnetic calculations the magnetic constant was later changed to $4\pi \times 10^{-7}$ Hm^{-1} to give the *rationalized MKSA system*. This system, with some modifications, formed the basis of *SI units, now used in all scientific work.

mmHg A unit of pressure equal to that exerted under standard gravity by a height of one millimetre of mercury, or 133.322 pascals.

mobility (of an ion) Symbol u. The terminal speed of an ion in an electric field divided by the field strength.

mode The pattern of motion in a vibrating body. If the body has several component particles, such as a molecule consisting of several atoms, the modes of vibration are the different types of molecular vibrations possible.

moiety A characteristic part of a molecule. For example, in the ester $C_2H_5COOCH_3$, the OCH_3 can be regarded as the alcohol (methanol) moiety.

molal concentration *See* **concentration**.

molality *See* **concentration**.

molar 1. Denoting that an extensive physical property is being expressed per *amount of substance, usually per mole. For example, the molar heat capacity of a compound is the heat capacity of that compound per unit amount of substance; in SI units it would be expressed in J K^{-1} mol^{-1}. **2.** Having a concentration of one mole per dm^3.

molar conductivity Symbol Λ. The conductivity of that volume of an electrolyte that contains one mole of solution between electrodes placed one metre apart.

molar heat capacity *See* **heat capacity.**

molarity *See* **concentration.**

molar volume (molecular volume) The volume occupied by a substance per unit amount of substance.

mole Symbol mol. The SI unit of *amount of substance. It is equal to the amount of substance that contains as many elementary units as there are atoms in 0.012 kg of carbon–12. The elementary units may be atoms, molecules, ions, radicals, electrons, etc., and must be specified. 1 mole of a compound has a mass equal to its *relative molecular mass expressed in grams.

molecular beam A beam of atoms, ions, or molecules at low pressure, in which all the particles are travelling in the same direction and there are few collisions between them. They are formed by allowing a gas or vapour to pass through an aperture into an enclosure, which acts as a collimator by containing several additional apertures and vacuum pumps to remove any particles that do not pass through the apertures. Molecular beams are used in studies of surfaces and chemical reactions and in spectroscopy.

molecular chaperone Any of a group of proteins in living cells that assist newly synthesized or denatured proteins to fold into their functional three-dimensional structures. The chaperones bind to the protein and prevent improper interactions within the polypeptide chain, so that it assumes the correct folded orientation. This process requires energy in the form of ATP.

molecular distillation Distillation in high vacuum (about 0.1 pascal) with the condensing surface so close to the surface of the evaporating liquid that the molecules of the liquid travel to the condensing surface without collisions. This technique enables very much lower temperatures to be used than are used with distillation at atmospheric pressure and therefore heat-sensitive substances can be distilled. Oxidation of the distillate is also eliminated as there is no oxygen present.

molecular flow (Knudsen flow) The flow of a gas through a pipe in which the mean free path of gas molecules is large compared to the dimensions of the pipe. This occurs at low pressures; because most collisions are with the walls of the pipe rather than other gas molecules, the flow characteristics depend on the relative molecular mass of the gas rather than its viscosity. The effect was studied by M. H. C. Knudsen (1871–1949).

molecular formula *See* **formula**.

molecularity The number of molecules involved in forming the activated complex in a step of a chemical reaction. Reactions are said to be *unimolecular*, *bimolecular*, or *trimolecular* according to whether 1, 2, or 3 molecules are involved.

molecular modelling The use of computer software to produce simulations of molecular structures. Various chemical drawing programs exist to allow a graphic representation of chemical formulas in two dimensions. More sophisticated three-dimensional modelling programs also exist. The information about the molecule is stored in a data file giving the numbers and types of atoms present and the coordinates of these atoms. The software converts this into an on-screen three-dimensional view of the molecular structure in some specified format (e.g. ball-and-stick, wireframe, etc.). The structure can be rotated on the screen and, depending on the software, calculations may be done on the molecule.

molecular orbital *See* **orbital**.

molecular-orbital theory A method of computational chemistry in which the electrons are not assigned to individual bonds between atoms, but are treated as moving under the influence of the nuclei in the whole molecule. The molecule has a set of molecular orbitals (*see* **orbital**). The usual technique is to obtain molecular orbitals by a linear combination of atomic orbitals (LCAO). *See also* **density-function theory**; **valence-bond theory**.

molecular recognition The way in which a molecule may have a highly specific response to another molecule or atom. It is a feature of *host–guest chemistry.

molecular sieve Porous crystalline substances, especially aluminosilicates (*see* **zeolite**), that can be dehydrated with little change in crystal structure. As they form regularly spaced cavities, they provide a high surface area for the adsorption of smaller molecules.

The general formula of these substances is $M_nO.Al_2O_3.xSiO_2.yH_2O$, where M is a metal ion and n is twice the reciprocal of its valency. Molecular sieves are used as drying agents and in the separation and purification of fluids. They can also be loaded with chemical substances, which remain separated from any reaction that is taking place around them, until they are released by heating or by displacement with a more strongly adsorbed substance. They can thus be used as cation exchange mediums and as catalysts and catalyst supports. They are also used as the stationary phase in certain types of *chromatography (*molecular-sieve chromatography*).

molecular symmetry The set of symmetry operations (rotations, reflections, etc.) that can be applied to a molecule. This set forms the *point group of the molecule. For an isolated molecule the *Schoenflies system rather than the *Hermann–Mauguin system is used to denote its symmetry. Molecular symmetry is analysed systematically using *group theory. It is possible to make definite statements about certain properties of molecules, such as whether they can have an electric dipole moment or exhibit optical activity on the basis of symmetry.

molecular volume *See* **molar volume.**

molecular weight *See* **relative molecular mass.**

molecule One of the fundamental units forming a chemical compound; the smallest part of a chemical compound that can take part in a chemical reaction. In most covalent compounds, molecules consist of groups of atoms held together by covalent or coordinate bonds. Covalent substances that form *macromolecular crystals have no discrete molecules (in a sense, the whole crystal is a molecule). Similarly, ionic compounds do not have single molecules, being collections of oppositely charged ions.

mole fraction Symbol X. A measure of the amount of a component in a mixture. The mole fraction of component A is given by $X_A = n_A/N$, where n_A is the amount of substance of A (for a given entity) and N is the total amount of substance of the mixture (for the same entity).

Molisch's test *See* **alpha-naphthol test.**

molybdenum Symbol Mo. A silvery hard metallic *transition element; a.n. 42; r.a.m. 95.94; r.d. 10.22; m.p. 2617°C; b.p. 4612°C. It is found in molybdenite (MoS_2), the metal being extracted by roasting to give the oxide, followed by reduction with hydrogen. The element is used in alloy steels. Molybdenum(IV) sulphide (MoS_2) is used as a lubricant. Chemically, it is unreactive, being unaffected by most acids. It oxidizes at high temperatures and can be dissolved in molten alkali to give a range of molybdates and polymolybdates. Molybdenum was discovered in 1778 by Karl *Scheele.

moment of inertia Symbol I. A quantity associated with a body that is rotating about an axis. If a body consists of i particles of mass m_i a perpendicular distance r_i from an axis of rotation, the moment of inertia I of the body about that axis of rotation is given by $I = \sum m_i r_i^2$. The moment of inertia of a body is an important quantity because it is the analogue for rotational motion of mass for linear motion. In a molecule all rotational motion can be analysed using three perpendicular axes of rotation. Each of these has a moment of inertia associated with it. To a first approximation, a molecule can be regarded as a *rigid rotor*, i.e. a body that is not distorted by its rotation. There are four types of rigid rotor:

In a *spherical top* all three moments of inertia are equal (e.g. SF_6).

In a *symmetric top* two of the moments of inertia are equal (e.g. NH_3).

In an *asymmetric top* all three moments of inertia are different (e.g. H_2O).

In a *linear rotor* the moment of inertia about the axis of the molecule is zero (e.g. HCl, CO_2).

The type of rotor depends on the symmetry of the molecule. If a molecule has cubic or icosahedral symmetry it is a spherical top. If it has a threefold or higher axis of symmetry it is a symmetric top. If it does not have a threefold or higher axis of symmetry it is an asymmetric top. All linear molecules (and hence all diatomic molecules) are linear rotors. In reality, molecules are not rigid rotors because of the centrifugal forces associated with the rotation. The effect of such forces can be taken into account by adding a correction term to the rigid rotor model.

monatomic molecule A 'molecule' consisting of only one atom (e.g. Ar or He), distinguished from diatomic and polyatomic molecules.

Mond process A method of obtaining pure nickel by heating the impure metal in a stream of carbon monoxide at 50–60°C. Volatile nickel carbonyl ($Ni(CO)_4$) is formed, and this can be decomposed at higher temperatures (180°C) to give pure nickel. The method was invented by the German–British chemist Ludwig Mond (1839–1909).

Monel metal An alloy of nickel (60–70%), copper (25–35%), and small quantities of iron, manganese, silicon, and carbon. It is used to make acid-resisting equipment in the chemical industry.

monobasic acid An *acid that has only one acidic hydrogen atom in its molecules. Hydrochloric (HCl) and nitric (HNO_3) acids are common examples.

monoclinic *See* **crystal system**.

monoethanolamine *See* **ethanolamine**.

monoglyceride *See* **glyceride**.

monohydrate A crystalline compound having one mole of water per mole of compound.

monomer A molecule (or compound) that joins with others in forming a dimer, trimer, or polymer.

monosaccharide (simple sugar) A carbohydrate that cannot be split into smaller units by the action of dilute acids. Monosaccharides are classified according to the number of carbon atoms they possess: *trioses* have three carbon atoms; *tetroses*, four; *pentoses*, five; *hexoses*, six; etc. Each of these is further divided into *aldoses* and *ketoses*, depending on whether the molecule contains an aldehyde group (–CHO) or a ketone group (–CO–). For example glucose, having six carbon atoms and an aldehyde group, is an *aldohexose* whereas fructose is a *ketohexose*. These aldehyde and ketone groups confer reducing properties on monosaccharides: they can be oxidized to yield sugar acids. They also react with phosphoric acid to produce phosphate esters (e.g. in *ATP), which are important in cell metabolism. Monosaccharides can exist as either straight-chain or ring-shaped molecules. They also exhibit *optical activity, giving rise to both dextrorotatory and laevorotatory forms.

monosodium glutamate (MSG) A white solid, $C_8H_8NNaO_4.H_2O$, used extensively as a flavour enhancer, especially in convenience foods. It is a salt of glutamic acid (an *amino acid), from which it is prepared. It can cause an allergic reaction in some susceptible people who consume it.

monotropy *See* **allotropy**.

monovalent (univalent) Having a valency of one.

Monte Carlo method A numerical method for solving mathematical and physical problems that involves the random sampling of numbers. Applications of Monte Carlo methods include calculations in the theory of liquids, phase transitions, and quantum mechanical systems. The name comes from the casino at Monte Carlo and alludes to the use of random numbers in this technique.

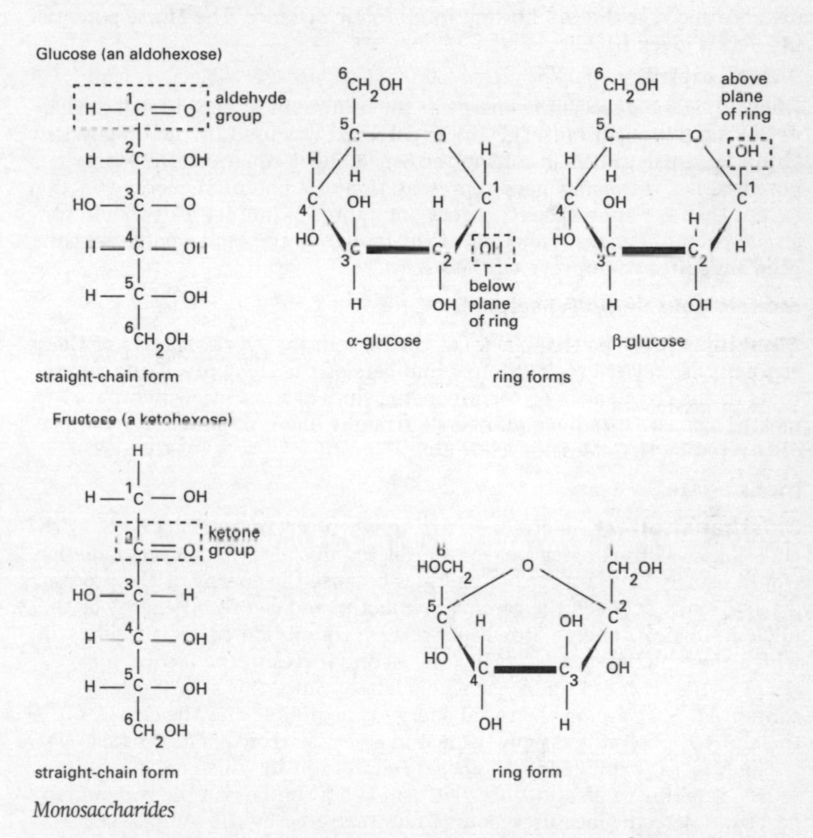

Monosaccharides

montmorillonite A clay mineral of variable composition. It is a hydrated aluminosilicate with a large capacity for exchanging cations (*see* **ion exchange**) and is the principal constituent of bentonite and fuller's earth.

One type of montmorillonite readily absorbs water, and another type swells in water to form a *gel.

mordant A substance used in certain dyeing processes. Mordants are often inorganic oxides or salts, which are absorbed on the fabric. The dyestuff then forms a coloured complex with the mordant, the colour depending on the mordant used as well as the dyestuff. *See also* **lake**.

morphine An alkaloid present in opium. It is an analgesic and narcotic, used medically for the relief of severe pain.

Morse potential The potential energy of a diatomic molecule as a function of the difference $(r - r_e)$, where r is the variable interatomic distance and r_e is the equilibrium interatomic distance. The Morse potential $U(r - r_e)$ is given by

$$D_e\{1 - \exp[-\beta(-r - r_e)]\}^2,$$

where D_e is the dissociation energy at the minimum of the curve (i.e. when $r = r_e$) and β is a constant. The Morse potential was used by the US physicist Philip M. Morse in 1929 in solving the Schrödinger equation. The Morse potential is a reasonably good representation of a potential-energy function except that as r approaches 0, U does not approach infinity as it should for a true potential energy function. Modifications of the Morse potential have been suggested to improve on this aspect.

mosaic gold *See* **tin(IV) sulphide**.

Moseley's law The frequencies of the lines in the *X-ray spectra of the elements are related to the atomic numbers of the elements. If the square roots of the frequencies of corresponding lines of a set of elements are plotted against the atomic numbers a straight line is obtained. The law was discovered by H. G. Moseley (1887–1915).

moss agate *See* **agate**.

Mössbauer effect An effect occurring when certain nuclides decay with emission of gamma radiation. For an isolated nucleus, the gamma radiation would usually have a spread of energies because the energy of the process is partitioned between the gamma-ray photon and the recoil energy of the nucleus. In 1957 Rudolph Mössbauer (1929–) found that in certain solids, in which the emitting nucleus is held by strong forces in the lattice, the recoil energy is taken up by the whole lattice. Since this may typically contain 10^{10}–10^{20} atoms, the recoil energy is negligible and the energy of the emitted photon is sharply defined in a very narrow energy spread.

The effect is exploited in *Mössbauer spectroscopy* in which a gamma-ray source is mounted on a moving platform and a similar sample is mounted nearby. A detector measures gamma rays scattered by the sample. The source is moved slowly towards the sample at a varying speed, so as to continuously change the frequency of the emitted gamma radiation by the Doppler effect. A sharp decrease in the signal from the detector at a particular speed (i.e. frequency) indicates resonance absorption in the sample nuclei. The effect is used to investigate nuclear energy levels. In chemistry, Mössbauer spectroscopy can also give information about the bonding and structure of compounds because *chemical shifts* in the resonance energy are produced by the presence of surrounding atoms.

m.p. *See* **melting point**.

MSG *See* **monosodium glutamate**.

mucopolysaccharide *See* **glycosaminoglycan**.

multicentre bond A bond formed between three, and sometimes more, atoms that contains only a single pair of electrons. The structure of *boranes can be explained by considering them to be *electron-deficient compounds containing multicentre bonds.

multidecker sandwich *See* **sandwich compound**.

multiple bond A bond between two atoms that contains more than one pair of electrons. Such bonds primarily involve sigma bonding with secondary contribution from pi bonding (or, sometimes, delta bonding). *See* **orbital**.

multiple proportions *See* **chemical combination**.

multiplet 1. A spectral line formed by more than two (*see* **doublet**) closely spaced lines. **2.** A group of elementary particles that are identical in all respects except that of electric charge.

multiplicity A quantity used in atomic *spectra to describe the energy levels of many-electron atoms characterized by *Russell–Saunders coupling given by $2S + 1$, where S is the total electron *spin quantum number. The multiplicity of an energy level is indicated by a left superscript to the value of L, where L is the resultant electron *orbital angular momentum of the individual electron orbital angular momenta l.

multipole interactions Interactions between arrays of point charges (*multipoles*) associated with potential energies that depend on distance. Multipole interactions are an important feature of intermolecular forces. An n-pole is a set of n charges having an n-pole moment but not a lower moment. A *monopole* is a single charge and the *monopole moment* is the charge. A *dipole* consists of two opposite charges and thus has no overall charge (and hence no monopole moment). Higher multipoles are the *quadrupole and the *octupole. The interaction potential energy between multipoles falls off with distance increasingly rapidly as the order of the multipoles increases. If a 2^m-pole interacts with a 2^n-pole, the interaction potential energy, V, varies with distance r as $V = c/(r^{m+n-1})$, where c is a constant. The rapid fall-off with distance, as the order increases, is due to the set of charges appearing to tend towards neutrality (as seen from outside). The multipole interactions with the largest interactions are: ion–ion, ion–dipole, dipole–dipole, and dispersion interactions.

Mumetal The original trade name for a ferromagnetic alloy, containing

78% nickel, 17% iron, and 5% copper, that had a high permeability and a low coercive force. More modern versions also contain chromium and molybdenum. These alloys are used in some transformer cores and for shielding various devices from external magnetic fields.

Muntz metal A form of *brass containing 60% copper, 39% zinc, and small amounts of lead and iron. Stronger than alpha-brass, it is used for hot forgings, brazing rods, and large nuts and bolts. It is named after G. F. Muntz (1794–1857).

muscovite **(white mica; potash mica)** A mineral form of potassium aluminosilicate, $K_2Al_4(Si_6Al_2)O_{20}(OH,F)_4$; one of the most important members of the *mica group of minerals. It is chemically complex and has a sheetlike crystal structure (*see* **intercalation compound**). It is usually silvery-grey in colour, sometimes tinted with green, brown, or pink. Muscovite is a common constituent of certain granites and pegmatites. It is also common in metamorphic and sedimentary rocks. It is widely used in industry, for example in the manufacture of electrical equipment and as a filler in roofing materials, wallpapers, and paint.

mustard gas A highly poisonous gas, $(ClCH_2CH_2)_2S$; dichlorodiethyl sulphide. It is made from ethene and disulphur dichloride (S_2Cl_2), and used as a war gas.

mutarotation Change of optical activity with time as a result of spontaneous chemical reaction.

myoglobin A globular protein occurring widely in muscle tissue as an oxygen carrier. It comprises a single polypeptide chain and a *haem group, which reversibly binds a molecule of oxygen. This is only relinquished at relatively low external oxygen concentrations, e.g. during strenuous exercise when muscle oxygen demand outpaces supply from the blood. Myoglobin thus acts as an emergency oxygen store.

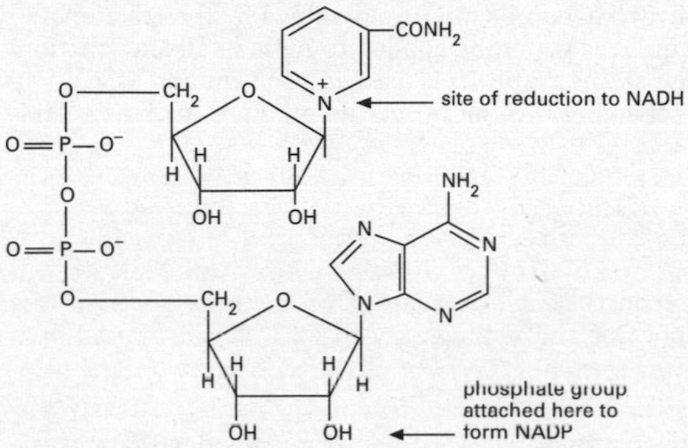

NAD

NAD (nicotinamide adenine dinucleotide) A *coenzyme, derived from the B vitamin *nicotinic acid, that participates in many biological dehydrogenation reactions. NAD is characteristically loosely bound to the enzymes concerned. It normally carries a positive charge and can accept one hydrogen atom and two electrons to become the reduced form, *NADH*. NADH is generated during the oxidation of food; it then gives up its two electrons (and single proton) to the electron transport chain, thereby reverting to NAD^+ and generating three molecules of ATP per molecule of NADH.

NADP (*nicotinamide adenine dinucleotide phosphate*) differs from NAD only in possessing an additional phosphate group. It functions in the same way as NAD although anabolic reactions (*see* **anabolism**) generally use NADPH (reduced NADP) as a hydrogen donor rather than NADH. Enzymes tend to be specific for either NAD or NADP as coenzyme.

nano- Symbol n. A prefix used in the metric system to denote 10^{-9}. For example, 10^{-9} second = 1 nanosecond (ns).

nanotechnology The development and use of devices that have a size of only a few nanometres. Research has been carried out into very small components, which depend on electronic effects and may involve movement of a countable number of electrons in their action. Such devices would act much faster than larger components. Considerable interest has been shown in the production of structures on a molecular level by suitable sequences of chemical reactions. It is also possible to manipulate individual atoms on surfaces using a variant of the *atomic force microscope.

nanotube *See* **buckminsterfullerene**.

napalm A substance used in incendiary bombs and flame throwers, made by forming a gel of petrol with aluminium soaps (aluminium salts of long-chain carboxylic acids, such as palmitic acid).

naphtha Any liquid hydrocarbon or mixture obtained by the fractional distillation of petroleum. It is generally applied to higher *alkane fractions with nine or ten carbon atoms. Naphtha is used as a solvent and as a starting material for *cracking into more volatile products, such as petrol.

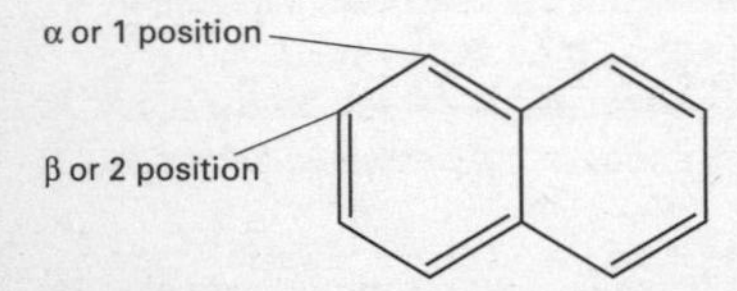

Naphthalene

naphthalene A white volatile solid, $C_{10}H_8$ (see formula); r.d. 1.025; m.p. 80.55°C; b.p. 218°C. Naphthalene is an aromatic hydrocarbon with an odour of mothballs and is obtained from crude oil. It is a raw material for making certain synthetic resins.

naphthols Two phenols derived from naphthalene with the formula $C_{10}H_7OH$, differing in the position of the –OH group. The most important is naphthalen-2-ol (β-naphthol), with the –OH in the 2-position. It is a white solid (r.d. 1.28; m.p. 123–124°C; b.p. 295°C) used in rubber as an antioxidant. Naphthalen-2-ol will couple with diazonium salts at the 1-position to form red *azo compounds, a reaction used in testing for the presence of primary amines (by making the diazonium salt and adding naphthalen-2-ol).

naphthyl group The group $C_{10}H_7$– obtained by removing a hydrogen atom from naphthalene. There are two forms depending on whether the hydrogen is removed from the 1- or 2-position.

nascent hydrogen A reactive form of hydrogen generated *in situ* in the reaction mixture (e.g. by the action of acid on zinc). Nascent hydrogen can reduce elements and compounds that do not readily react with 'normal' hydrogen. It was once thought that the hydrogen was present as atoms, but this is not the case. Probably hydrogen molecules are formed in an excited state and react before they revert to the ground state.

natron A mineral form of hydrated sodium carbonate, $Na_2CO_3.H_2O$.

Natta process *See* **Ziegler process**.

natural abundance *See* **abundance**.

natural gas A naturally occurring mixture of gaseous hydrocarbons that is found in porous sedimentary rocks in the earth's crust, usually in association with *petroleum deposits. It consists chiefly of methane (about 85%), ethane (up to about 10%), propane (about 3%), and butane. Carbon dioxide, nitrogen, oxygen, hydrogen sulphide, and sometimes helium may also be present. Natural gas, like petroleum, originates in the decomposition of organic matter. It is widely used as a fuel and also to produce carbon black and some organic chemicals. Natural gas occurs on every continent, the major reserves occurring in the USA, Russia, Kazakhstan, Turkmenistan, Ukraine, Algeria, Canada, and the Middle East. *See also* **liquefied petroleum gas**.

Néel temperature The temperature above which an antiferromagnetic substance becomes paramagnetic (*see* **magnetism**). The susceptibility increases with temperature, reaching a maximum at the Néel temperature, after which it abruptly declines. The phenomenon was discovered around 1930 by L. E. F. Néel (1904–).

neighbouring-group participation An effect in an organic chemical reaction in which the reactive centre interacts with a lone pair or with electrons in other bonds in the molecule that are not conjugated with the centre. This may affect the rate or the stereochemistry of the products.

nematic crystal *See* **liquid crystal**.

neodymium Symbol Nd. A soft silvery metallic element belonging to the *lanthanoids; a.n. 60; r.a.m. 144.24; r.d. 7.004 (20°); m.p. 1021°C; b.p. 3068°C. It occurs in bastnasite and monazite, from which it is recovered by an ion-exchange process. There are seven naturally occurring isotopes, all of which are stable, except neodymium-144, which is slightly radioactive (half-life 10^{10}–10^{15} years). Seven artificial radioisotopes have been produced. The metal is used to colour glass violet-purple and to make it dichroic. It is also used in misch metal (18% neodymium) and in neodymium–iron–boron alloys for magnets. It was discovered by Carl von Welsbach (1856–1929) in 1885.

neon Symbol Ne. A colourless gaseous element belonging to group 18 (formerly group 0) of the periodic table (the *noble gases); a.n. 10; r.a.m. 20.179; d. 0.9 g dm^{-3}; m.p. −248.67°C; b.p. −246.05°C. Neon occurs in air (0.0018% by volume) and is obtained by fractional distillation of liquid air. It is used in discharge tubes and neon lamps, in which it has a characteristic red glow. It forms hardly any compounds (neon fluorides have been reported). The element was discovered in 1898 by Sir William Ramsey and M. W. Travers.

neoprene A synthetic rubber made by polymerizing the compound 2-chlorobuta-1,2-diene. Neoprene is often used in place of natural rubber in applications requiring resistance to chemical attack.

nephrite *See* **jade**.

neptunium Symbol Np. A radioactive metallic transuranic element belonging to the *actinoids; a.n. 93; r.a.m. 237.0482. The most stable isotope, neptunium-237, has a half-life of 2.2×10^6 years and is produced in small quantities as a by-product by nuclear reactors. Other isotopes have mass numbers 229–236 and 238–241. The only other relatively long-lived isotope is neptunium–236 (half-life 5×10^3 years). The element was first produced by Edwin McMillan (1907–91) and Philip Abelson (1913–) in 1940.

neptunium series *See* **radioactive series.**

Nernst, (Hermann) Walther (1864–1941) German physical chemist noted for his work on electrochemistry and chemical thermodynamics. He is remembered as the discoverer of the third law of thermodynamics. Nernst was awarded the 1920 Nobel prize for chemistry.

Nernst–Einstein equation An equation relating the limiting molar conductivity $\Lambda_m^{\ominus}$ (*see* **Kohlrausch's law**) to the ionic diffusion coefficients, devised by Nernst and Albert *Einstein. The equation is

$$\Lambda_m^{\ominus} = (F^2/RT)(v_+z_+^2D_+ + v_-z_-^2D_-),$$

where F is the Faraday constant, R is the gas constant, T is the thermodynamic temperature, v_+ and v_- are the number of cations and anions per formula unit of electrolyte, z_+ and z_- are the valences of the ions, and D_+ and D_- are the diffusion coefficients of the ions. An application of the Nernst–Einstein equation is to calculate the ionic diffusion coefficients from experimental determinations of conductivity.

Nernst equation An equation that relates the *electrode potential E of an electrode that is in contact with an ionic solution to the ionic concentration c. The equation is:

$$E = E^{\ominus} - RTzF\ln c,$$

where $E^{\ominus}$ is the standard electrode potential, R is the gas constant, T is the absolute temperature, z is the valence of the ion, and F is the Faraday constant. The equation was derived by Walther *Nernst in 1889. It is fundamental to the thermodynamics of electrochemical cells.

Nernst heat theorem A statement of the third law of *thermodynamics in a restricted form given by Walther Nernst: if a chemical change takes place between pure crystalline solids at *absolute zero there is no change of entropy.

nesquehonite A mineral form of *magnesium carbonate trihydrate, $MgCO_3.3H_2O$.

Nessler's reagent A solution of mercury(II) iodide (HgI_2) in potassium iodide and potassium hydroxide named after Julius Nessler (1827–1905). It is used in testing for ammonia, with which it forms a brown coloration or precipitate.

neutral Describing a compound or solution that is neither acidic nor basic. A neutral solution is one that contains equal numbers of both protonated and deprotonated forms of the solvent.

neutralization The process in which an acid reacts with a base to form a salt and water.

neutron A neutral hadron that is stable in the atomic nucleus but decays into a proton, an electron, and an antineutrino with a mean life of 12 minutes outside the nucleus. Its rest mass is slightly greater than that of the proton, being $1.674\,9286(10) \times 10^{-27}$ kg. Neutrons occur in all atomic nuclei except normal hydrogen. The neutron was first reported in 1932 by James *Chadwick.

neutron diffraction The scattering of neutrons by atoms in solids, liquids, or gases. This process has given rise to a technique, analogous to *X-ray diffraction techniques, using a flux of thermal neutrons from a nuclear reactor to study solid-state phenomena. Thermal neutrons have average kinetic energies of about 0.025 eV (4×10^{-21} J) giving them an equivalent wavelength of about 0.1 nanometre, which is suitable for the study of interatomic interference. There are two types of interaction in the scattering of neutrons by atoms: one is the interaction between the neutrons and the atomic nucleus, the other is the interaction between the *magnetic moments of the neutrons and the spin and orbital magnetic moments of the atoms. The latter interaction has provided valuable information on antiferromagnetic and ferrimagnetic materials (*see* **magnetism**). Interaction with the atomic nucleus gives diffraction patterns that complement those from X-rays. X-rays, which interact with the extranuclear electrons, are not suitable for investigating light elements (e.g. hydrogen), whereas neutrons do give diffraction patterns from such atoms because they interact with nuclei.

neutron number Symbol N. The number of neutrons in an atomic nucleus of a particular nuclide. It is equal to the difference between the *nucleon number and the *atomic number.

Newlands' law *See* **law of octaves**.

Newman projection A type of *projection formula in which a molecule is viewed along a bond. *See* **conformation**.

newton Symbol N. The *SI unit of force, being the force required to give a mass of one kilogram an acceleration of $1\ \mathrm{m\,s^{-2}}$. It is named after Sir Isaac Newton (1642–1727).

Newtonian fluid A fluid in which the velocity gradient is directly proportional to the shear stress. If two flat plates of area A are separated by a layer of fluid of thickness d and move relative to each other at a velocity v, then the rate of shear is v/d and the shear stress is F/A, where F is the force applied to each. For a Newtonian fluid $F/A = \mu v/d$, where μ is the constant of proportionality and is called the *Newtonian viscosity*. Many liquids are Newtonian fluids over a wide range of temperatures and pressures. However, some are not; these are called *non-Newtonian fluids*. In such fluids there is a departure from the simple Newtonian relationships. For example, in some liquids the viscosity increases as the velocity gradient increases, i.e. the faster the liquid moves the more viscous it becomes. Such liquids are said to be *dilatant* and the phenomenon they exhibit is called *dilatancy*. It occurs in some pastes and suspensions. More common, however, is the opposite effect in which the viscosity depends not only on the velocity gradient but also on the time for which it has been applied. These liquids are said to exhibit *thixotropy*. The faster a *thixotropic liquid* moves the less viscous it becomes. This property is used in nondrip paints (which are more viscous on the brush than on the wall) and in lubricating oils (which become thinner when the parts they are lubricating start to move). Another example is the non-Newtonian flow of macromolecules in solution or in polymer melts. In this case the shearing force F is not parallel to the shear planes and the linear relationship does not apply. In general, the many types of non-Newtonian fluid are somewhat complicated and no theory has been developed to accommodate them fully.

NHOMO Next-to-highest occupied molecular orbital. *See* **subjacent orbitals**.

niacin *See* **nicotinic acid**.

Nichrome Tradename for a group of nickel–chromium alloys used for wire in heating elements as they possess good resistance to oxidation and have a high resistivity. Typical is Nichrome V containing 80% nickel and 19.5% chromium, the balance consisting of manganese, silicon, and carbon.

nickel Symbol Ni. A malleable ductile silvery metallic *transition element; a.n. 28; r.a.m. 58.70; r.d. 8.9; m.p. 1450°C; b.p. 2732°C. It is found in the minerals pentlandite (NiS), pyrrhoite ((Fe,Ni)S), and garnierite ($(Ni,Mg)_6(OH)_6Si_4O_{11}.H_2O$). Nickel is also present in certain iron meteorites (up to 20%). The metal is extracted by roasting the ore to give the oxide, followed by reduction with carbon monoxide and purification by the *Mond process. Alternatively electolysis is used. Nickel metal is used in special steels, in Invar, and, being ferromagnetic, in magnetic alloys, such as *Mumetal. It is also an effective catalyst, particularly for hydrogenation reactions (*see also* **Raney nickel**). The main compounds are formed with nickel in the +2 oxidation state; the +3 state also exists (e.g. the black oxide, Ni_2O_3). Nickel was discovered by Axel Cronstedt (1722–65) in 1751.

nickel–cadmium cell *See* **nickel–iron accumulator**.

nickel carbonyl A colourless volatile liquid, $Ni(CO)_4$; m.p. −25°C; b.p. 43°C. It is formed by direct combination of nickel metal with carbon monoxide at 50–60°C. The reaction is reversed at higher temperatures, and the reactions are the basis of the *Mond process for purifying nickel. The nickel in the compound has an oxidation state of zero, and the compound is a typical example of a complex with pi-bonding *ligands, in which filled *d*-orbitals on the nickel overlap with empty *p*-orbitals on the carbon.

nickelic compounds Compounds of nickel in its +3 oxidation state; e.g. nickelic oxide is nickel(III) oxide (Ni_2O_3).

nickel–iron accumulator **(Edison cell; NIFE cell)** A *secondary cell devised by Thomas Edison (1847–1931) having a positive plate of nickel oxide and a negative plate of iron both immersed in an electrolyte of potassium hydroxide. The reaction on discharge is

$$2NiOOH.H_2O + Fe \rightarrow 2Ni(OH)_2 + Fe(OH)_2,$$

the reverse occurring during charging. Each cell gives an e.m.f. of about 1.2 volts and produces about 100 kJ per kilogram during each discharge. The *nickel–cadmium cell* is a similar device with a negative cadmium electrode. It is often used as a *dry cell. *Compare* **lead–acid accumulator**.

nickelous compounds Compounds of nickel in its +2 oxidation state; e.g. nickelous oxide is nickel(II) oxide (NiO).

nickel(II) oxide A green powder, NiO; r.d. 6.6. It can be made by heating nickel(II) nitrate or carbonate with air excluded.

nickel(III) oxide **(nickel peroxide; nickel sesquioxide)** A black or grey powder, Ni_2O_3; r.d. 4.8. It is made by heating nickel(II) oxide in air and used in *nickel–iron accumulators.

nickel silver *See* **German silver**.

Nicol prism A device for producing plane-polarized light (*see* **polarizer**). It consists of two pieces of calcite cut with a 68° angle and stuck together with Canada balsam. The extraordinary ray (*see* **double refraction**) passes through the prism while the ordinary ray suffers total internal reflection at the interface between the two crystals, as the refractive index of the calcite is 1.66 for the ordinary ray and that of the Canada balsam is 1.53. Modifications of the prism using different shapes and cements are used for special purposes. It was devised in 1828 by William Nicol (1768–1851).

nicotinamide *See* **nicotinic acid**.

nicotinamide adenine dinucleotide *See* NAD.

nicotinamide adenine dinucleotide phosphate **(NADP)** *See* NAD.

nicotine A colourless poisonous *alkaloid present in tobacco. It is used as an insecticide.

nicotinic acid **(niacin)** A vitamin of the *vitamin B complex. It can be manufactured by plants and animals from the amino acid tryptophan. The amide derivative, *nicotinamide*, is a component of the coenzymes *NAD and NADP. These take part in many metabolic reactions as hydrogen acceptors. Deficiency of nicotinic acid causes the disease pellagra in humans. Good sources are liver and groundnut and sunflower meals.

nido-structure *See* **borane**.

nielsbohrium *See* **transactinide elements**.

NIFE cell *See* **nickel–iron accumulator**.

ninhydrin A brown crystalline substance, $C_9H_4O_3.H_2O$ (see formula). It reacts with amino acids to give a blue colour. Ninhydrin is commonly used in chromatography to analyse the amino-acid content of proteins.

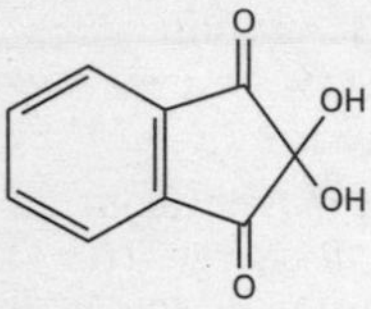

Ninhydrin

niobium Symbol Nb. A soft ductile grey-blue metallic transition element; a.n. 41; r.a.m. 92.91; r.d. 8.57; m.p. 2468°C; b.p. 4742°C. It occurs in several minerals, including niobite ($Fe(NbO_3)_2$), and is extracted by several methods including reduction of the complex fluoride K_2NbF_7 using sodium. It is used in special steels and in welded joints (to increase strength). Niobium–zirconium alloys are used in superconductors. Chemically, the element combines with the halogens and oxidizes in air at 200°C. It forms a number of compounds and complexes with the metal in oxidation states 2, 3, or 5. The element was discovered by Charles Hatchett (*c.* 1765–1847) in 1801 and first isolated by Christian Blomstrand (1826–97) in 1864. Formerly, it was called *columbium.*

nitrate A salt or ester of nitric acid.

nitration A type of chemical reaction in which a nitro group ($-NO_2$) is added to or substituted in a molecule. Nitration can be carried out by a mixture of concentrated nitric and sulphuric acids (*see* **nitrating mixture**).

nitrating mixture A mixture of concentrated sulphuric and nitric acids, used to introduce a nitro group ($-NO_2$) into an organic compound. Its action depends on the presence of the nitronium ion, NO_2^+. It is mainly used to introduce groups into the molecules of *aromatic compounds (the nitro group can subsequently be converted into or replaced by others) and to make commercial *nitro compounds, such as the explosives cellulose trinitrate (nitrocellulose), glyceryl trinitrate (nitroglycerine), trinitrotoluene (TNT), and picric acid (trinitrophenol).

nitre (saltpetre) Commercial *potassium nitrate; the name was formerly applied to natural crustlike efflorescences, occurring in some arid regions.

nitre cake *See* **sodium hydrogensulphate**.

nitrene A species of the type HN: or RN:, analogous to a *carbene. The nitrogen atom has four nonbonding electrons, two of which are a lone pair. The other two may have parallel spins (a triplet state) or antiparallel spins (a singlet state).

nitric acid A colourless corrosive poisonous liquid, HNO_3; r.d. 1.50; m.p. −42°C; b.p. 83°C. Nitric acid may be prepared in the laboratory by the distillation of a mixture of an alkali-metal nitrate and concentrated sulphuric acid. The industrial production is by the oxidation of ammonia to nitrogen monoxide, the oxidation of this to nitrogen dioxide, and the reaction of nitrogen dioxide with water to form nitric acid and nitrogen monoxide (which is recycled). The first reaction (NH_3 to NO) is catalysed by platinum or platinum/rhodium in the form of fine wire gauze. The oxidation of NO and the absorption of NO_2 to form the product are noncatalytic and proceed with high yields but both reactions are second-order and slow. Increases in pressure reduce the selectivity of the reaction and therefore rather large gas absorption towers are required. In practice the absorbing acid is refrigerated to around 2°C and a commercial 'concentrated nitric acid' at about 67% is produced.

Nitric acid is a strong acid (highly dissociated in aqueous solution) and dilute solutions behave much like other mineral acids. Concentrated nitric acid is a strong oxidizing agent. Most metals dissolve to form nitrates but with the evolution of nitrogen oxides. Concentrated nitric acid also reacts with several nonmetals to give the oxo acid or oxide. Nitric acid is generally stored in dark brown bottles because of the photolytic decomposition to dinitrogen tetroxide. *See also* **nitration**.

nitric oxide *See* **nitrogen monoxide**.

nitrides Compounds of nitrogen with a more electropositive element. Boron nitride is a covalent compound having macromolecular crystals. Certain electropositive elements, such as lithium, magnesium, and calcium, react directly with nitrogen to form ionic nitrides containing the N^{3-} ion. Transition elements form a range of interstitial nitrides (e.g. Mn_4N, W_2N), which can be produced by heating the metal in ammonia.

nitriding The process of hardening the surface of steel by producing a layer of iron nitride. One technique is to heat the metal in ammonia gas. Another is to dip the hot metal in a bath of molten sodium cyanide.

nitrification A chemical process in which nitrogen (mostly in the form of ammonia) in plant and animal wastes and dead remains is oxidized at first to nitrites and then to nitrates. These reactions are effected mainly by the bacteria *Nitrosomonas* and *Nitrobacter* respectively. Unlike ammonia, nitrates are readily taken up by plant roots; nitrification is therefore a crucial part of the *nitrogen cycle. Nitrogen-containing compounds are often applied to soils deficient in this element, as fertilizer. *Compare* **denitrification**.

nitrile rubber A copolymer of buta-1,3-diene and propenonitrile. It is a commercially important synthetic rubber because of its resistance to oil and many solvents.

nitriles **(cyanides)** Organic compounds containing the group –CN bound to an organic group. Nitriles are made by reaction between potassium cyanide and haloalkanes in alcoholic solution, e.g.

$$KCN + CH_3Cl \rightarrow CH_3CN + KCl$$

An alternative method is dehydration of amides

$$CH_3CONH_2 - H_2O \rightarrow CH_3CN$$

They can be hydrolysed to amides and carboxylic acids and can be reduced to amines.

nitrite A salt or ester of nitrous acid. The salts contain the dioxonitrate (III) ion, NO_2^-, which has a bond angle of 115°.

nitroalkane (nitroparaffin) A type of organic compound of general formula $C_nH_{2n+1}NO_2$. The nitroalkanes are colourless pleasant-smelling liquids made by treating a haloalkane with silver nitrate. They can be reduced to amines by the action of tin and hydrochloric acid. Lower nitroalkanes such as nitromethane (CH_3NO_2, b.p. 100°C) are used as high-performance fuels (for internal combustion engines and rockets), as chemical intermediates, and as polar solvents.

nitrobenzene A yellow oily liquid, $C_6H_5NO_2$; r.d. 1.2; m.p. 6°C; b.p. 211°C. It is made by the *nitration of benzene using a mixture of nitric and sulphuric acids.

nitrocellulose *See* **cellulose nitrate**.

nitro compounds Organic compounds containing the group $-NO_2$ (the *nitro group*) bound to a carbon atom. Nitro compounds are made by *nitration reactions. They can be reduced to aromatic amines (e.g. nitrobenzene can be reduced to phenylamine). *See also* **explosive**.

nitrogen Symbol N. A colourless gaseous element belonging to *group 15 (formerly VB) of the periodic table; a.n. 7; r.a.m. 14.0067; d. 1.2506 g dm^{-3}; m.p. −209.86°C; b.p. −195.8°C. It occurs in air (about 78% by volume) and is an essential constituent of proteins and nucleic acids in living organisms (*see* **nitrogen cycle**). Nitrogen is obtained for industrial purposes by fractional distillation of liquid air. Pure nitrogen can be obtained in the laboratory by heating a metal azide. There are two natural isotopes: nitrogen–14 and nitrogen–15 (about 3%). The element is used in the *Haber process for making ammonia and is also used to provide an inert atmosphere in welding and metallurgy. The gas is diatomic and relatively inert – it reacts with hydrogen at high temperatures and with oxygen in electric discharges. It also forms *nitrides with certain metals. Nitrogen was discovered in 1772 by Daniel Rutherford (1749–1819).

nitrogen cycle One of the major cycles of chemical elements in the environment. Nitrates in the soil are taken up by plant roots and may then pass along food chains into animals. Decomposing bacteria convert nitrogen-containing compounds (especially ammonia) in plant and animal wastes and dead remains back into nitrates, which are released into the soil and can again be taken up by plants (*see* **nitrification**). Though nitrogen is essential to all forms of life, the huge amount present in the atmosphere is not directly available to most organisms (*compare* **carbon cycle**). It can, however, be assimilated by some specialized bacteria (*see* **nitrogen fixation**) and is thus made available to other organisms indirectly. Lightning flashes also make some nitrogen available to plants by causing the combination of atmospheric nitrogen and oxygen to form oxides of nitrogen, which enter

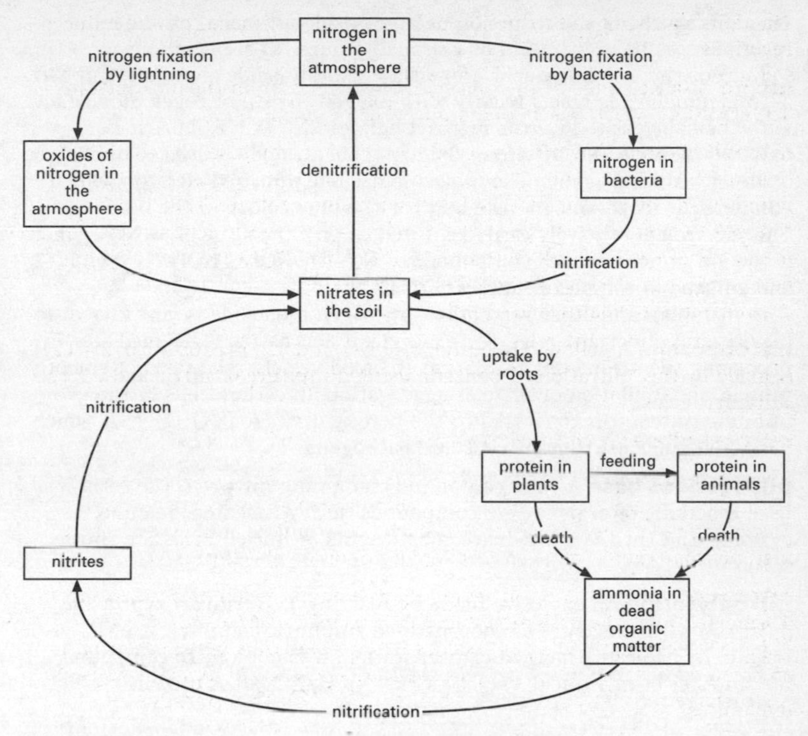

The nitrogen cycle

the soil and form nitrates. Some nitrogen is returned from the soil to the atmosphere by denitrifying bacteria (*see* **denitrification**).

nitrogen dioxide *See* **dinitrogen tetroxide**.

nitrogen fixation A chemical process in which atmospheric nitrogen is assimilated into organic compounds in living organisms and hence into the *nitrogen cycle. The ability to fix nitrogen is limited to certain bacteria (e.g. *Azotobacter*, *Anabaena*). *Rhizobium* bacteria are able to fix nitrogen in association with cells in the roots of leguminous plants, such as peas and beans, in which they form characteristic root nodules; cultivation of legumes is therefore one way of increasing soil nitrogen. Various chemical processes are used to fix atmospheric nitrogen in the manufacture of *fertilizers. These include the *Birkeland–Eyde process, the cyanamide process (*see* **calcium dicarbide**), and the *Haber process.

nitrogen monoxide (nitric oxide) A colourless gas, NO; m.p. −163.6°C; b.p. −151.8°C. It is soluble in water, ethanol, and ether. In the liquid state nitrogen monoxide is blue in colour (r.d. 1.26). It is formed in many reactions involving the reduction of nitric acid, but more convenient reactions for the preparation of reasonably pure NO are reactions of sodium nitrite, sulphuric acid, and either sodium iodide or iron(II) sulphate. Nitrogen monoxide reacts readily with oxygen to give nitrogen dioxide and with the halogens to give the nitrosyl halides XNO (X = F,Cl,Br). It is oxidized to nitric acid by strong oxidizing agents and reduced to dinitrogen oxide by reducing agents. The molecule has one unpaired electron, which accounts for its paramagnetism and for the blue colour in the liquid state. This electron is relatively easily removed to give the *nitrosyl ion* NO^+, which is the ion present in such compounds as $NOClO_4$, $NOBF_4$, $NOFeCl_4$, $(NO)_2PtCl_6$ and a ligand in complexes, such as $Co(CO)_3NO$.

In mammals and other vertebrates, nitrogen monoxide is now known to play several important roles. For example, it acts as a gaseous mediator in producing such responses as dilation of blood vessels, relaxation of smooth muscle, and inhibition of platelet aggregation. In certain cells of the immune system it is converted to the peroxynitrite ion ($^-O–O–N=O$), which has activity against tumour cells and pathogens.

nitrogenous base A basic compound containing nitrogen. The term is used especially of organic ring compounds, such as adenine, guanine, cytosine, and thymine, which are constituents of nucleic acids. *See* **amine salts**.

nitroglycerine An explosive made by reacting 1,2,3-trihydroxypropane (glycerol) with a mixture of concentrated sulphuric and nitric acids. Despite its name and method of preparation, it is not a nitro compound, but an ester of nitric acid, $CH_2(NO_3)CH(NO_3)CH_2(NO_3)$. It is used in dynamites.

nitro group *See* **nitro compounds**.

nitronium ion *See* **nitryl ion**.

nitroparaffin *See* **nitroalkane**.

nitrosamines A group of carcinogenic compounds with the general formula RR′NNO, where R and R′ are side groups with a variety of possible structures. Nitrosamines, which are a component of cigarette smoke, cause cancer in a number of organs, particularly in the liver, kidneys, and lungs. An example of a nitrosamine is dimethylnitrosamine, which has two methyl side groups (CH_3–).

nitrosyl ion The ion NO^+. *See* **nitrogen monoxide**.

nitrous acid A weak acid, HNO_2, known only in solution and in the gas phase. It is prepared by the action of acids upon nitrites, preferably using a combination that removes the salt as an insoluble precipitate (e.g. $Ba(NO_2)_2$ and H_2SO_4). The solutions are unstable and decompose on heating to give nitric acid and nitrogen monoxide. Nitrous acid can function both as an oxidizing agent (forms NO) with I^- and Fe^{2+}, or as a reducing agent (forms NO_3^-) with, for example, Cu^{2+}; the latter is most common. It is widely used (prepared *in situ*) for the preparation of diazonium compounds in organic chemistry. The full systematic name is *dioxonitric(III) acid*.

nitrous oxide *See* **dinitrogen oxide.**

nitryl ion **(nitronium ion)** The ion NO_2^+, found in mixtures of nitric acid and sulphuric acid and solutions of nitrogen oxides in nitric acid. Nitryl salts, such as $NO_2^+ClO_4^-$, can be isolated but are extremely reactive. Nitryl ions generated *in situ* are used for *nitration in organic chemistry.

NMR *See* **nuclear magnetic resonance.**

nobelium Symbol No. A radioactive metallic transuranic element belonging to the *actinoids; a.n. 102; mass number of most stable element 254 (half-life 55 seconds). Seven isotopes are known. The element was first identified with certainty by Albert Ghiorso and Glenn Seaborg (1912–99) in 1966.

noble gases **(inert gases; rare gases; group 18 elements)** A group of monatomic gaseous elements forming group 18 (formerly group 0) of the *periodic table: helium (He), neon (Ne), argon (Ar), krypton (Kr), xenon (Xe), and radon (Rn). The electron configuration of helium is $1s^2$. The configurations of the others terminate in ns^2np^6 and all inner shells are fully occupied. The elements thus represent the termination of a period and have closed-shell configuration and associated high ionization energies (He 2370 to Rn 1040 kJ mol^{-1}) and lack of chemical reactivity. Being monatomic the noble gases are spherically symmetrical and have very weak interatomic interactions and consequent low enthalpies of vaporization. The behaviour of the lighter members approaches that of an ideal gas at normal temperatures; with the heavier members increasing polarizability and dispersion forces lead to easier liquefaction under pressure. Four types of 'compound' have been described for the noble gases but of these only one can be correctly described as compounds in the normal sense. One type consists of such species as HHe^+, He_2^+, Ar_2^+, $HeLi^+$, which form under highly energetic conditions, such as those in arcs and sparks. They are short-lived and only detected spectroscopically. A second group of materials described as inert-gas–metal compounds do not have defined compositions and are simply noble gases adsorbed onto the surface of dispersed metal. The third type, previously described as 'hydrates' are in fact clathrate compounds with the noble gas molecule trapped in a water lattice. True compounds of the noble gases were first described in 1962 and several fluorides, oxyfluorides, fluoroplatinates, and fluoroantimonates of *xenon are known. A few krypton fluorides and a radon fluoride are also known although the short half-life of radon and its intense alpha activity restrict the availability of information. Apart from argon, the noble gases are present in the atmosphere at only trace levels. Helium may be found along with natural gas (up to 7%), arising from the radioactive decay of heavier elements (via alpha particles).

noble metal A metal characterized by it lack of chemical reactivity, particularly to acids and atmospheric corrosion. Examples include gold, palladium, platinum, and rhodium.

no-bond resonance *See* **hyperconjugation**.

NOE *See* **nuclear Overhauser effect**.

nonahydrate A crystalline compound that has nine moles of water per mole of compound.

nonbenzenoid aromatics Aromatic compounds that have rings other than benzene rings. Examples are the cyclopentadienyl anion, $C_5H_5^-$, and the tropylium cation, $C_7H_7^+$. *See also* **annulenes**.

noncompetitive inhibition *See* **inhibition**.

nonequilibrium statistical mechanics The statistical mechanics of systems not in thermal equilibrium. One of the main purposes of nonequilibrium statistical mechanics is to calculate *transport coefficients and inverse transport coefficients, such as conductivity and viscosity, from first principles and to provide a basis for transport theory. The nonequilibrium systems easiest to understand are those near thermal equilibrium. For systems far from equilibrium, such possibilities as *chaos and self-organization can arise due to nonlinearity.

nonequilibrium thermodynamics The thermodynamics of systems not in thermal equilibrium. The nonequilibrium systems easiest to understand are those near thermal equilibrium; these systems are described by the *Onsager relations. For systems far from equilibrium, more complicated patterns, such as *chaos and *self-organization, can arise due to nonlinearity. Which behaviour is observed depends on the value of certain parameters in the system. The transition from one type of behaviour to another as the parameters are altered occurs at *bifurcations.

nonferrous metal Any metal other than iron or any alloy that does not contain iron. In commercial terms this usually means aluminium, copper, lead, nickel, tin, zinc, or their alloys.

nonmetal An element that is not a *metal. Nonmetals can either be insulators or semiconductors. At low temperatures nonmetals are poor conductors of both electricity and heat as few free electrons move through the material. If the conduction band is near to the valence band (*see* **energy bands**) it is possible for nonmetals to conduct electricity at high temperatures but, in contrast to metals, the conductivity increases with increasing temperature. Nonmetals are electronegative elements, such as carbon, nitrogen, oxygen, phosphorus, sulphur, and the halogens. They form compounds that contain negative ions or covalent bonds. Their oxides are either neutral or acidic.

nonpolar compound A compound that has covalent molecules with no permanent dipole moment. Examples of nonpolar compounds are methane and benzene.

nonpolar solvent *See* solvent.

nonreducing sugar A sugar that cannot donate electrons to other molecules and therefore cannot act as a reducing agent. Sucrose is the most common nonreducing sugar. The linkage between the glucose and fructose units in sucrose, which involves aldehyde and ketone groups, is responsible for the inability of sucrose to act as a *reducing sugar.

nonrelativistic quantum theory *See* quantum theory.

nonrenewable energy sources *See* renewable energy sources.

nonstoichiometric compound **(Berthollide compound)** A chemical compound in which the elements do not combine in simple ratios. For example, rutile (titanium(IV) oxide) is often deficient in oxygen, typically having a formula $TiO_{1.8}$.

noradrenaline **(norepinephrine)** A hormone produced by the adrenal glands and also secreted from nerve endings in the sympathetic nervous system as a chemical transmitter of nerve impulses. Many of its general actions are similar to those of *adrenaline, but it is more concerned with maintaining normal body activity than with preparing the body for emergencies.

Nordhausen sulphuric acid *See* disulphuric(VI) acid.

norepinephrine *See* noradrenaline.

normal Having a concentration of one gram equivalent per dm^3.

normal mode of vibration A basic vibration of a polyatomic molecule. All vibrational motion of a polyatomic molecule can be treated as a superposition of the normal modes of vibration of the molecule. If N is the number of atoms in a molecule, the number of modes of vibration is $3N-5$ for a linear molecule and $3N-6$ for a nonlinear molecule. Each of these vibrational modes has a characteristic frequency, although it is possible for some of them to be degenerate. For example, in a linear triatomic molecule there are four normal modes of vibration since $3N-5=4$ for $N=3$. These vibrational modes are: (a) symmetric stretching (breathing) vibrations; (b) antisymmetric stretching vibrations; and (c) two bending vibrations, which are degenerate.

N.T.P. *See* s.t.p.

n-type semiconductor *See* semiconductor.

nuclear magnetic resonance **(NMR)** The absorption of electromagnetic radiation at a suitable precise frequency by a nucleus with a nonzero magnetic moment in an external magnetic field. The phenomenon occurs if the nucleus has nonzero *spin, in which case it behaves as a small magnet. In an external magnetic field, the nucleus's magnetic moment vector precesses about the field direction but only certain orientations are allowed by quantum rules. Thus, for hydrogen (spin of ½) there are two possible states in the presence of a field, each with a slightly different energy. Nuclear magnetic resonance is the absorption of radiation at a photon energy equal to the difference between these levels, causing a transition from a lower to a higher energy state. For practical purposes, the difference in energy levels is small and the radiation is in the radiofrequency region of the electromagnetic spectrum. It depends on the field strength.

NMR can be used for the accurate determination of nuclear moments. It can also be used in a sensitive form of magnetometer to measure magnetic fields. In medicine, *magnetic resonance imaging* (*MRI*) has been developed, in which images of tissue are produced by magnetic-resonance techniques.

The main application of NMR is as a technique for chemical analysis and structure determination, known as *NMR spectroscopy*. It depends on the fact that the electrons in a molecule shield the nucleus to some extent from the field, causing different atoms to absorb at slightly different frequencies (or at slightly different fields for a fixed frequency). Such effects are known as *chemical shifts*. There are two methods of NMR spectroscopy. In *continuous wave* (CW) NMR, the sample is subjected to a strong field, which can be varied in a controlled way over a small region. It is irradiated with radiation at a fixed frequency, and a detector monitors the field at the sample. As the field changes, absorption corresponding to transitions occurs at certain values, and this causes oscillations in the field, which induce a signal in the detector. *Fourier transform* (FT) NMR uses a fixed magnetic field and the sample is subjected to a high-intensity pulse of radiation covering a range of frequencies. The signal produced is analysed mathematically to give the NMR spectrum. The most common nucleus studied is ^{1}H. For instance, an NMR spectrum of ethanol (CH_3CH_2OH) has three peaks in the ratio 3:2:1, corresponding to the three different hydrogen-atom environments. The peaks also have a fine structure caused by interaction between spins in the molecule. Other nuclei can also be used for NMR spectroscopy (e.g. ^{13}C, ^{14}N, ^{19}F) although these generally have lower magnetic moment and natural abundance than hydrogen. *See also* **electron-spin resonance**.

nuclear magneton *See* **magneton**.

nuclear Overhauser effect **(NOE)** An effect in *nuclear magnetic resonance (NMR) used to increase the intensities of resonance lines. The large population differences between the states given by the Boltzmann distribution gives rise to strong intensities of resonance lines. In the nuclear Overhauser effect, spin relaxation is used to transfer population difference from one type of nucleus to another type of nucleus so that the intensities of the resonance lines of the second type of nucleus are increased. An example of such an enhancement, which is widely used, is between protons and ^{13}C, in which case it is possible to attain enhancement of about three times. In the NOE proton, irradiation is used to produce the enhancement of the ^{13}C line. Another use of the NOE is to determine the distance between protons.

nucleon A *proton or a *neutron.

nucleon number (mass number) Symbol A. The number of *nucleons in an atomic nucleus of a particular nuclide.

nucleophile An ion or molecule that can donate electrons. Nucleophiles are often oxidizing agents and Lewis bases. They are either negative ions (e.g. Cl^-) or molecules that have electron pairs (e.g. NH_3). In organic reactions they tend to attack positively charged parts of a molecule. *Compare* **electrophile**.

nucleophilic addition A type of addition reaction in which the first step is attachment of a *nucleophile to a positive (electron-deficient) part of the molecule. *Aldehydes and *ketones undergo reactions of this type because of polarization of the carbonyl group (carbon positive).

nucleophilic substitution A type of substitution reaction in which a *nucleophile displaces another group or atom from a compound. For example, in

$$CR_3Cl + OH^- \rightarrow CR_3OH + Cl^-$$

the nucleophile is the OH^- ion. There are two possible mechanisms of nucleophilic substitution. In *S_N1 reactions*, a positive carbonium ion is first formed:

$$CR_3Cl \rightarrow CR_3^+ + Cl^-$$

This then reacts with the nucleophile

$$CR_3^+ + OH^- \rightarrow CR_3OH$$

The CR_3^+ ion is planar and the OH^- ion can attack from either side. Consequently, if the original molecule is optically active (the three R groups are different) then a racemic mixture of products results.

The alternative mechanism, the *S_N2 reaction*, is a concerted reaction in which the nucleophile approaches from the side of the R groups as the other group (Cl in the example) leaves. In this case the configuration of the molecule is inverted. If the original molecule is optically active, the product has the opposite activity, an effect known as *Walden inversion*. The notations S_N1 and S_N2 refer to the kinetics of the reactions. In the S_N1 mechanism, the slow step is the first one, which is unimolecular (and first order in CR_3Cl). In the S_N2 reaction, the process is bimolecular (and second order overall).

nucleoside An organic compound consisting of a nitrogen-containing *purine or *pyrimidine base linked to a sugar (ribose or deoxyribose). An example is *adenosine. *Compare* **nucleotide**.

nucleosynthesis The synthesis of chemical elements by nuclear processes. There are several ways in which nucleosynthesis can take place. *Primordial nucleosynthesis* took place very soon after the big bang, when the universe was extremely hot. This process was responsible for the cosmic abundances observed for light elements, such as helium. Explosive nucleosynthesis can also occur during the explosion of a supernova.

However, *stellar nucleosynthesis*, which takes place in the centre of stars at very high temperatures, is now the principal form of nucleosynthesis. The exact process occurring in stellar nucleosynthesis depends on the temperature, density, and chemical composition of the star. The synthesis of helium from protons and of carbon from helium can both occur in stellar nucleosynthesis.

nucleotide An organic compound consisting of a nitrogen-containing *purine or *pyrimidine base linked to a sugar (ribose or deoxyribose) and a phosphate group. *DNA and *RNA are made up of long chains of nucleotides (i.e. *polynucleotides*). *Compare* **nucleoside**.

nucleus The central core of an atom that contains most of its mass. It is positively charged and consists of one or more nucleons (protons or neutrons). The positive charge of the nucleus is determined by the number of protons it contains (*see* **atomic number**) and in the neutral atom this is balanced by an equal number of electrons, which move around the nucleus. The simplest nucleus is the hydrogen nucleus, consisting of one proton only. All other nuclei also contain one or more neutrons. The neutrons contribute to the atomic mass (*see* **nucleon number**) but not to the nuclear charge. The most massive nucleus that occurs in nature is uranium-238, containing 92 protons and 146 neutrons. The symbol used for this *nuclide is $^{238}_{92}U$, the upper figure being the nucleon number and the lower figure the atomic number. In all nuclei the nucleon number (A) is equal to the sum of the atomic number (Z) and the neutron number (N), i.e. $A = Z + N$.

nuclide A type of atom as characterized by its *atomic number and its *neutron number. An *isotope refers to a series of different atoms that have the same atomic number but different neutron numbers (e.g. uranium-238 and uranium-235 are isotopes of uranium), whereas a nuclide refers only to a particular nuclear species (e.g. the nuclides uranium-235 and plutonium-239 are fissile). The term is also used for the type of nucleus.

nylon Any of various synthetic polyamide fibres having a protein-like structure formed by the condensation between an amino group of one molecule and a carboxylic acid group of another. There are three main nylon fibres, nylon 6, nylon 6,6, and nylon 6,10. Nylon 6, for example Enkalon and Celon, is formed by the self-condensation of 6-aminohexanoic acid. Nylon 6,6, for example Bri nylon, is made by polycondensation of hexanedioic acid (adipic acid) and 1,6-diaminohexane (hexamethylenediamine) having an average formula weight between 12 000 and 15 000. Nylon 6,10 is prepared by polymerizing decanedioic acid and 1,6-diaminohexane.

occlusion 1. The trapping of small pockets of liquid in a crystal during crystallization. **2.** The absorption of a gas by a solid such that atoms or molecules of the gas occupy interstitial positions in the solid lattice. Palladium, for example, can occlude hydrogen.

ochre A yellow or red mineral form of iron(III) oxide, Fe_2O_3, used as a pigment.

octadecanoate *See* **stearate.**

octadecanoic acid *See* **stearic acid.**

octadecenoic acid A straight-chain unsaturated fatty acid with the formula $C_{17}H_{33}COOH$. *Cis-octadec-9-enoic acid* (*see* **oleic acid**) has the formula $CH_3(CH_2)_7CH{:}CH(CH_2)_7COOH$. The glycerides of this acid are found in many natural fats and oils.

octahedral *See* **complex.**

octahydrate A crystalline hydrate that has eight moles of water per mole of compound.

octane A straight-chain liquid *alkane, C_8H_{18}; r.d. 0.7; m.p. −56.79°C; b.p. 125.66°C. It is present in petroleum. The compound is isomeric with 2,2,4-trimethylpentane, $(CH_3)_3CCH_2CH(CH_3)_2$, *iso-octane*). *See* **octane number**.

octane number A number that provides a measure of the ability of a fuel to resist knocking when it is burnt in a spark-ignition engine. It is the percentage by volume of iso-octane (C_8H_{18}; 2,2,4-trimethylpentane) in a blend with normal heptane (C_7H_{16}) that matches the knocking behaviour of the fuel being tested in a single cylinder four-stroke engine of standard design. *Compare* **cetane number**.

octanoic acid (caprylic acid) A colourless liquid straight-chain saturated *carboxylic acid, $CH_3(CH_2)_6COOH$; b.p. 239.3°C.

octavalent Having a valency of eight.

octave *See* **law of octaves.**

octet A stable group of eight electrons in the outer shell of an atom (as in an atom of a noble gas).

octupole A set of eight point charges that has zero net charge and does not have either a dipole moment or a quadrupole moment. An example of an octupole is a methane molecule (CH_4). Octupole interactions are much smaller than quadrupole interactions and very much smaller than dipole interactions.

ohm Symbol Ω. The derived *SI unit of electrical resistance, being the resistance between two points on a conductor when a constant potential difference of one volt, applied between these points, produces a current of one ampere in the conductor. The former *international ohm* (sometimes called the 'mercury ohm') was defined in terms of the resistance of a column of mercury. The unit is named after Georg Ohm (1787–1854).

oil Any of various viscous liquids that are generally immiscible with water. Natural plant and animal oils are either volatile mixtures of terpenes and simple esters (e.g. *essential oils) or are *glycerides of fatty acids. Mineral oils are mixtures of hydrocarbons (e.g. *petroleum).

oil of vitriol *See* **sulphuric acid.**

oil of wintergreen Methyl salicylate (methyl 2-hydroxybenzoate, $C_8H_8O_3$), a colourless aromatic liquid ester, b.p. 223°C. It occurs in the essential oils of some plants, and is manufactured from salicylic acid. It is easily absorbed through the skin and used in medicine for treating muscular and sciatic pain. Because of its attractive smell it is also used in perfumes and food flavourings.

oil sand **(tar sand; bituminous sand)** A sandstone or porous carbonate rock that is impregnated with hydrocarbons. The largest deposit of oil sand occurs in Alberta, Canada (the Athabasca tar sands); there are also deposits in the Orinoco Basin of Venezuela, Russia, USA, Madagascar, Albania, Trinidad, and Romania.

oil shale A fine-grained carbonaceous sedimentary rock from which oil can be extracted. The rock contains organic matter – *kerogen* – which decomposes to yield oil when heated. Deposits of oil shale occur on every continent, the largest known reserves occurring in Colorado, Utah, and Wyoming in the USA. Commercial production of oil from oil shale is generally considered to be uneconomic unless the price of petroleum rises above the recovery costs for oil from oil shale. However, threats of declining conventional oil resources have resulted in considerable interest and developments in recovery techniques.

oleaginous Producing or containing oil or lipids. Oleaginous microorganisms, which normally contain 20–25% oil, are of interest in biotechnology as alternative sources of conventional oils or as possible sources for novel oils. The majority of the oils produced by oleaginous eukaryotic microorganisms are similar to plant oils. One possibility under consideration is the production of oils and fats from waste material, to be used in animal feed.

oleate A salt or ester of *oleic acid.

olefines *See* **alkenes.**

oleic acid An unsaturated *fatty acid with one double bond, $CH_3(CH_2)_7CH{:}CH(CH_2)_7COOH$; r.d. 0.9; m.p. 13°C. Oleic acid is one of the most abundant constituent fatty acids of animal and plant fats, occurring in butterfat, lard, tallow, groundnut oil, soya-bean oil, etc. Its systematic chemical name is *cis-octadec-9-enoic acid.*

oleum *See* **disulphuric(VI) acid.**

oligonucleotide A short polymer of *nucleotides.

oligopeptide *See* **peptide.**

oligosaccharide A carbohydrate (a type of *sugar) whose molecules contain a chain of up to 20 united monosaccharides. Oligosaccharides are formed as intermediates during the digestion of *polysaccharides, such as cellulose and starch.

olivine An important group of rock-forming silicate minerals crystallizing in the orthorhombic system. Olivine conforms to the general formula $(Mg,Fe)_2SiO_4$ and comprises a complete series from pure magnesium silicate (forsterite, Mg_2SiO_4) to pure iron silicate (fayalite, Fe_2SiO_4). It is green, brown-green, or yellow-green in colour. A gem variety of olivine is *peridot.*

one-pot synthesis A method of synthesizing organic compounds in which the materials used are mixed together in a single vessel and allowed to react, rather than conducting the reaction in a sequence of separate stages.

onium ion An ion formed by adding a proton to a neutral molecule, e.g. the hydroxonium ion (H_3O^+) or the ammonium ion (NH_4^+).

Onsager relations In a system that is not at equilibrium, various changes are occurring. For example, there may be a flow of energy from one part of the system to another and, at the same time, a flow of mass (diffusion). Flows of this type are coupled, i.e. each depends on the other. Equations exist of the type

$$J_1 = L_{11}X_1 + L_{12}X_2$$

$$J_2 = L_{21}X_1 + L_{22}X_2$$

Here J_1 is the flow of energy and J_2 the flow of matter. X_1 is the 'force' producing energy flow and X_2 that producing matter flow. L_{11} is the coefficient for thermal conductance and L_{22} the coefficient for diffusion. The coefficients L_{12} and L_{21} represent coupling of the flows with each other. Equations of this type can be generalized to any number of flows and are known as the *phenomenological relations.* In Onsager's theory the coupling coefficients are equal, i.e. $L_{12} = L_{21}$, etc. These are known as *reciprocal relations.* It follows that:

$$(\partial J_1/\partial X_2)_{X_1} = (\partial J_2/\partial X_1)_{X_2}.$$

The theory was developed by the Swedish physicist Lars Onsager in 1931.

opal A hydrous amorphous form of silica. Many varieties of opal occur, some being prized as gemstones. Common opal is usually milk white but the presence of impurities may colour it yellow, green, or red. Precious opals, which are used as gemstones, display the property of *opalescence* – a characteristic internal play of colours resulting from the interference of light rays within the stone. Black opal has a black background against which the colours are displayed. The chief sources of precious opals are Australia and Mexico. Geyserite is a variety deposited by geysers or hot springs. Another variety, diatomite, is made up of the skeletons of diatoms.

open chain *See* chain.

open-hearth process A traditional but now obsolete method for manufacturing steel by heating together scrap, pig iron, etc., in a refractory-lined shallow open furnace heated by burning producer gas in air. It has been replaced by the *basic-oxygen process.

operator A mathematical entity that performs a specific operation on a function to transform the function into another function. For example, the square root sign $\sqrt{}$ and the differentiation symbol d/dx are operators.

opiate One of a group of drugs derived from *opium*, an extract of the poppy plant *Papaver somniferum* that depresses brain function (a *narcotic* action). Opiates include *morphine and its synthetic derivatives, such as *heroin and codeine. They are used in medicine chiefly to relieve pain, but the use of morphine and heroin is strictly controlled since they can cause drug dependence and tolerance.

opioid Any one of a group of substances that produce pharmacological and physiological effects similar to those of morphine. Opioids are not necessarily structurally similar to morphine, although a subgroup of opioids, the *opiates, are morphine-derived compounds.

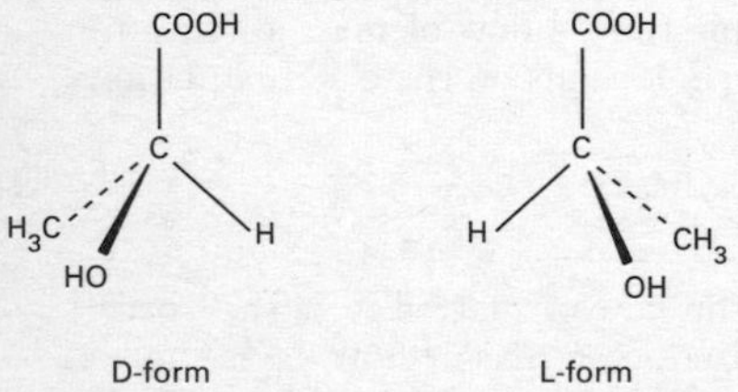

D-form L-form

Isomers of lactic acid

optical activity The ability of certain substances to rotate the plane of plane-polarized light as it passes through a crystal, liquid, or solution. It occurs when the molecules of the substance are asymmetric, so that they can exist in two different structural forms each being a mirror image of the other (*see* **chirality element**). The two forms are *optical isomers* or *enantiomers*. The existence of such forms is also known as *enantiomorphism* (the mirror images being *enantiomorphs*). One form will rotate the light in one direction and the other will rotate it by an equal amount in the other. The two possible forms are described as *dextrorotatory or *laevorotatory according to the direction of rotation. Prefixes are used to designate the isomer: (+)- (dextrorotatory), (−)- (laevorotatory), and (±)- (racemic mixture) are now preferred to, and increasingly used for, the former *d*-, *l*-, and *dl*-, respectively. An equimolar mixture of the two forms is not optically active. It is called a *racemic mixture* (or *racemate*). In addition, certain molecules can have a *meso form* in which one part of the molecule is a mirror image of the other. Such molecules are not optically active.

Molecules that show optical activity have no plane of symmetry. The commonest case of this is in organic compounds in which a carbon atom is linked to four different groups. An atom of this type is said to be a *chiral centre*. Asymmetric molecules showing optical activity can also occur in inorganic compounds. For example, an octahedral complex in which the central ion coordinates to six different ligands would be optically active. Many naturally occurring compounds show optical isomerism and usually only one isomer occurs naturally. For instance, glucose is found in the dextrorotatory form. The other isomer, (−)- or *l*-glucose, can be synthesized in the laboratory, but cannot be synthesized by living organisms. *See also* **absolute configuration**.

optical brightener *See* **brighteners**.

optical glass Glass used in the manufacture of lenses, prisms, and other optical parts. It must be homogeneous and free from bubbles and strain. Optical *crown glass* may contain potassium or barium in place of the sodium of ordinary crown glass and has a refractive index in the range 1.51 to 1.54. *Flint glass* contains lead oxide and has a refractive index between 1.58 and 1.72. Higher refractive indexes are obtained by adding lanthanoid oxides to glasses; these are now known as lanthanum crowns and flints.

optical isomers *See* **optical activity**.

optical maser *See* **laser**.

optical pumping *See* **laser**.

optical rotary dispersion (ORD) The effect in which the amount of rotation of plane-polarized light by an optically active compound depends on the wavelength. A graph of rotation against wavelength has a characteristic shape showing peaks or troughs.

optical rotation Rotation of plane-polarized light. *See* **optical activity**.

optoacoustic spectroscopy A spectroscopic technique in which electromagnetic radiation is absorbed by materials that generate sound waves. This technique has been used particularly in gases. The principle underlying optoacoustic spectroscopy is that the absorbed electromagnetic radiation is converted into motion, which is associated with the production of sound waves. *See also* **photoacoustic spectroscopy**.

orbit The path of an electron as it travels round the nucleus of an atom. *See* **orbital**.

orbital A region in which an electron may be found in an atom or molecule. In the original *Bohr theory of the atom the electrons were assumed to move around the nucleus in circular orbits, but further advances in quantum mechanics led to the view that it is not possible to give a definite path for an electron. According to *wave mechanics, the electron has a certain probability of being in a given element of space. Thus for a hydrogen atom the electron can be anywhere from close to the nucleus to out in space but the maximum probability in spherical shells of equal thickness occurs in a spherical shell around the nucleus with a radius

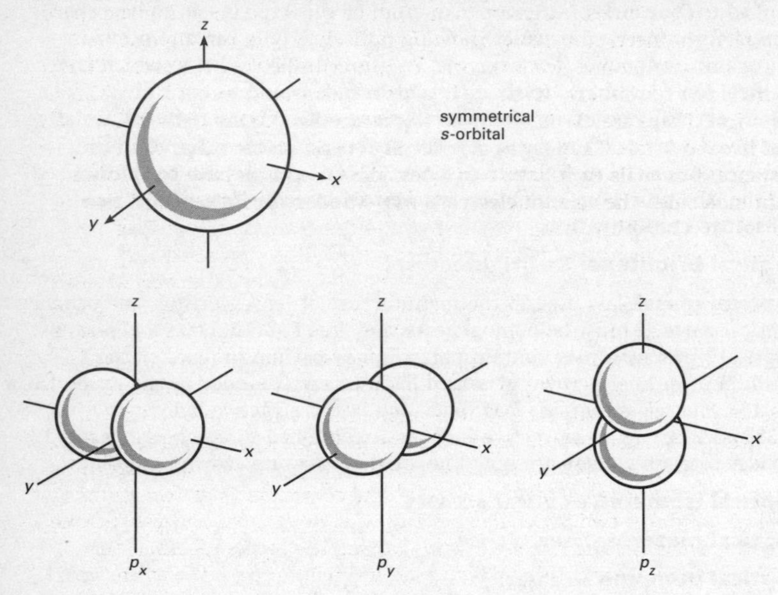

Atomic orbitals

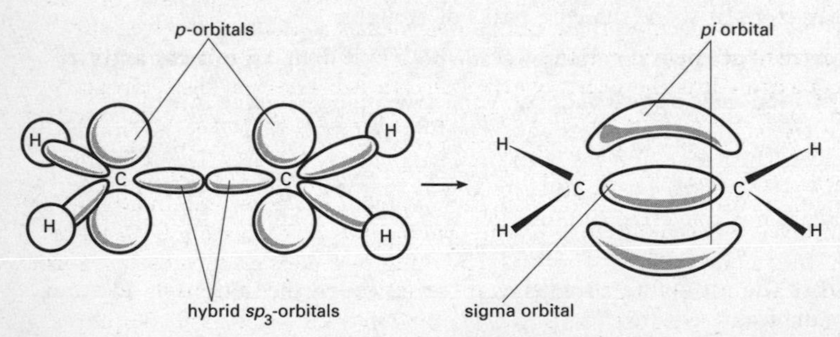

Molecular orbitals: formation of the double bond in ethene

Orbitals

equal to the Bohr radius of the atom. The probabilities of finding an electron in different regions can be obtained by solving the Schrödinger wave equation to give the wave function ψ, and the probability of location per unit volume is then proportional to $|\psi|^2$. Thus the idea of electrons in fixed orbits has been replaced by that of a probability distribution around the nucleus – an *atomic orbital* (see illustration). Alternatively, the orbital can be thought of as an electric charge distribution (averaged over time). In representing orbitals it is convenient to take a surface enclosing the space in which the electron is likely to be found with a high probability.

The possible atomic orbitals correspond to subshells of the atom. Thus there is one *s*-orbital for each shell (orbital quantum number $l = 0$). This is spherical. There are three *p*-orbitals (corresponding to the three values of l) and five *d*-orbitals. The shapes of orbitals depend on the value of l. For instance, *p*-orbitals each have two lobes; most *d*-orbitals have four lobes.

In molecules, the valence electrons move under the influence of two nuclei (in a bond involving two atoms) and there are corresponding *molecular orbitals* for electrons (see illustration). It is convenient in considering these to regard them as formed by overlap of atomic orbitals. In a hydrogen molecule the *s*-orbitals on the two atoms overlap and form a molecular orbital between the two nuclei. This is an example of a *sigma orbital*. In a double bond, as in ethene, one bond is produced by overlap along the line of axes to form a sigma orbital. The other is produced by sideways overlap of the lobes of the *p*-orbitals (see illustration). The resulting molecular orbital has two parts, one on each side of the sigma orbital – this is a *pi orbital*. It is also possible for a *delta orbital* to form by lateral overlap of two *d*-orbitals. In fact, the combination of two atomic orbitals produces two molecular orbitals with different energies. The one of lower energy is the *bonding orbital*, holding the atoms together; the other is the *antibonding orbital*, which would tend to push the atoms apart. In the case of valence electrons, only the lower (bonding) orbital is filled.

In considering the formation of molecular orbitals it is often useful to think in terms of *hybrid* atomic orbitals. For instance, carbon has in its outer shell one *s*-orbital and three *p*-orbitals. In forming methane (or other tetrahedral molecules) these can be regarded as combining to give four equivalent sp^3 hybrid orbitals, each with a lobe directed to a corner of a tetrahedron. It is these that overlap with the *s*-orbitals on the hydrogen atoms. In ethene, two *p*-orbitals combine with the *s*-orbital to give three sp^2 hybrids with lobes in a plane pointing to the corners of an equilateral triangle. These form the sigma orbitals in the C–H and C–C bonds. The remaining *p*-orbitals (one on each carbon) form the pi orbital. In ethyne, sp^2 hybridization occurs to give two hybrid orbitals on each atom with lobes pointing along the axis. The two remaining *p*-orbitals on each carbon form two pi orbitals. Hybrid atomic orbitals can also involve *d*-orbitals. For instance, square-planar complexes use sp^2d hybrids; octahedral complexes use sp^3d^2.

orbital quantum number *See* atom.

ORD *See* **optical rotary dispersion**.

order In the expression for the rate of a chemical reaction, the sum of the powers of the concentrations is the overall order of the reaction. For instance, in a reaction

$A + B \rightarrow C$

the rate equation may have the form

$R = k[A][B]^2$

This reaction would be described as *first order* in A and *second order* in B. The overall order is three. The order of a reaction depends on the mechanism and it is possible for the rate to be independent of concentration (*zero order*) or for the order to be a fraction. *See also* **molecularity**; **pseudo order**.

ore A naturally occurring mineral from which a metal and certain other elements (e.g. phosphorus) can be extracted, usually on a commercial basis. Metals may be present in ores in the native form, but more commonly they occur combined as oxides, sulphides, sulphates, silicates, etc.

ore dressing *See* **beneficiation**.

oregonator A type of chemical reaction mechanism that causes an *oscillating reaction. It is the type of mechanism responsible for the *B-Z reaction, and involves five steps of the form:

$A + Y \rightarrow X$

$X + Y \rightarrow C$

$A + X \rightarrow 2X + Z$

$2X \rightarrow D$

$Z \rightarrow Y$

Autocatalysis occurs as in the *Lotka–Volterra mechanism and the *brusselator. The mechanism was named after Oregon in America, where the research group that discovered it is based.

organic chemistry The branch of chemistry concerned with compounds of carbon.

organo- Prefix used before the name of an element to indicate compounds of the elements containing organic groups (with the element bound to carbon atoms). For example, lead(IV) tetraethyl is an organolead compound.

organometallic compound A compound in which a metal atom or ion is bound to an organic group. Organometallic compounds may have single metal–carbon bonds, as in the aluminium alkyls (e.g. $Al(CH_3)_3$). In some cases, the bonding is to the pi electrons of a double bond, as in complexes formed between platinum and ethene, or to the pi electrons of a ring, as in *ferrocene.

organophosphorus compound A compound that has at least one carbon–phosphorus bond. Organophosphorus compounds are used as pesticides (as in malathion and parathion) and as catalysts and solvents. Phosphate esters, sometimes included in this category, are employed for fire-proofing textiles.

Orgel diagram A diagram showing how the energy levels of a transition-metal atom split when it is placed in a ligand field. The vertical axis shows the energy and the horizontal axis shows the strength of the ligand field, with zero ligand field strength at the centre of the horizontal axis. That the splitting for d^n is the same as d^{n+5} and the opposite of d^{10-n} is readily seen on an Orgel diagram, both for octahedral and tetrahedral fields. The spectroscopic, optical, and magnetic properties of complexes of transition metals are made clear in such diagrams. Orgel diagrams are named after Leslie Orgel, who developed them in 1955.

origin of elements The nuclear processes that give rise to chemical elements. There is not one single process that can account for all the elements. The abundance of the chemical elements is determined not just by the stability of the nuclei of the atoms but also how readily the nuclear processes leading to the existence of these atoms occur. Most of the helium in the universe was produced by fusion in the early universe when the temperature and the pressure were very high. Most of the elements between helium and iron were made in nuclear fusion reactions inside stars. Since iron is at the bottom of the energy *valley of stability*, energy needs to be put into a nucleus heavier than iron for a fusion reaction to occur. Inside stars some heavy elements are built up by the *s-process*, where s stands for slow, in which high-energy neutrons are absorbed by a nucleus, with the resulting nucleus undergoing beta decay to produce a nucleus with a higher atomic number. Some heavy elements are produced by the *r-process*, where r stands for rapid, which occurs in supernova explosions when a great deal of gravitational energy is released when a large star that has exhausted its nuclear fuel collapses to a neutron star, with the release of a very large number of neutrons.

ornithine (Orn) An *amino acid, $H_2N(CH_2)_3CH(NH_2)COOH$, that is not a constituent of proteins but is important in living organisms as an intermediate in the reactions of the *urea cycle and in arginine synthesis.

ornithine cycle *See* **urea cycle**.

orpiment A natural yellow mineral form of arsenic(III) sulphide, As_2S_3. The name is also used for the synthetic compound, which is used as a pigment.

ortho- 1. Prefix indicating that a benzene compound has two substituted groups in the 1,2 positions (i.e. on adjacent carbon atoms). The abbreviation *o*- is used; for example *o*-dichlorobenzene is 1,2-dichlorobenzene. *Compare* **meta-**; **para-**. **2.** Prefix formerly used to indicate the most hydrated form of an acid. For example, phosphoric(V) acid, H_3PO_4, was called orthophosphoric acid to distinguish it from the lower metaphosphoric acid,

HPO_3 (which is actually $(HPO_3)_n$). **3.** Prefix denoting the form of diatomic molecules in which nuclei have parallel spins, e.g. orthohydrogen (*see* **hydrogen**). *Compare* **para-**.

orthoboric acid *See* **boric acid.**

orthoclase *See* **feldspars.**

orthohelium A form of helium once thought to exist as of one of two species of the element. Because the spectrum of atomic helium consists of transitions between singlet states (including transitions from the ground state) and transitions between triplet states but not transitions between singlet and triplet states, early spectroscopists thought that two species of helium existed. The other form was called *parahelium.*

orthohydrogen *See* **hydrogen.**

orthophosphoric acid *See* **phosphoric(V) acid.**

orthoplumbate *See* **plumbate.**

orthorhombic *See* **crystal system.**

orthosilicate *See* **silicate.**

orthostannate *See* **stannate.**

oscillating reaction (clock reaction) A type of chemical reaction in which the concentrations of the products and reactants change periodically, either with time or with position in the reacting medium. Thus, the concentration of a component may increase with time to a maximum, decrease to a minimum, then increase again, and so on, continuing the oscillation over a period of time. Systems are also known in which spirals and other patterns spread through the reacting medium, demonstrating a periodic spatial variation. Oscillating chemical reactions have certain features in common. They all occur under conditions far from chemical equilibrium and all involve *autocatalysis, i.e. a product of a reaction step acts as a catalyst for that step. This autocatalysis drives the oscillation by a process of positive feedback. Moreover, oscillating chemical reactions are associated with the phenomenon known as *bistability*. In this, a reaction may be in a steady-state condition, with reactants flowing into a reaction zone while products are flowing out of it. Under these conditions, the concentrations in the reaction zone may not change with time, although the reaction is not in a state of chemical equilibrium. Bistable systems have two possible stable steady states. Interaction with an additional substance in the reaction medium causes the system to oscillate between the states as the concentrations change. Oscillating chemical reactions are thought to occur in a number of biochemical processes. For example, they occur in glycolysis, in which ATP is produced by enzyme-catalysed reactions. They are also known to regulate the rhythm of the heartbeat. Most have highly complex reaction mechanisms. *See also* **Lotka–Volterra mechanism**; **brusselator**; **oregonator**; **chaotic reaction**.

osmiridium A hard white naturally occurring alloy consisting principally of osmium (17–48%) and iridium (49%). It also contains small quantities of platinum, rhodium, and ruthenium. It is used for making small items subject to wear, e.g. electrical contacts or the tips of pen nibs.

osmium Symbol Os. A hard blue-white metallic *transition element; a.n. 76; r.a.m. 190.2; r.d. 22.57; m.p. 3045°C; b.p. 5027°C. It is found associated with platinum and is used in certain alloys with platinum and iridium (*see* **osmiridium**). Osmium forms a number of complexes in a range of oxidation states. It was discovered by Smithson Tennant (1761–1815) in 1804.

osmium(IV) oxide **(osmium tetroxide)** A yellow solid, OsO_4, made by heating osmium in air. It is used as an oxidizing agent in organic chemistry, as a catalyst, and as a fixative in electron microscopy.

osmometer *See* **osmosis**.

osmosis The passage of a solvent through a *semipermeable membrane* separating two solutions of different concentrations. A semipermeable membrane is one through which the molecules of a solvent can pass but the molecules of most solutes cannot. There is a thermodynamic tendency for solutions separated by such a membrane to become equal in concentration, the water (or other solvent) flowing from the weaker to the stronger solution. Osmosis will stop when the two solutions reach equal concentration, and can also be stopped by applying a pressure to the liquid on the stronger-solution side of the membrane. The pressure required to stop the flow from a pure solvent into a solution is a characteristic of the solution, and is called the *osmotic pressure* (symbol Π). Osmotic pressure depends only on the concentration of particles in the solution, not on their nature (i.e. it is a *colligative property). For a solution of n moles in volume V at thermodynamic temperature T, the osmotic pressure is given by $\Pi V = nRT$, where R is the gas constant. Osmotic-pressure measurements are used in finding the relative molecular masses of compounds, particularly macromolecules. A device used to measure osmotic pressure is called an *osmometer*.

osmotic pressure *See* **osmosis**.

Ostwald, Friedrich Wilhelm (1853–1932) German physical chemist. In 1887 he went to Leipzig where he worked on a wide range of topics including hydrolysis, viscosity, ionization, and catalysis. Ostwald was highly influential in the development of physical chemistry as a subject. He was awarded the 1909 Nobel Prize for chemistry.

Ostwald ripening A process used in crystal growth in which a mixture of large and small crystals are both in contact with a solvent. The large crystals grow and the small crystals disappear. This occurs because there is a higher energy associated with the smaller crystals. When they dissolve the heat associated with this higher energy is released, enabling recrystallization to occur on the large crystals. Ostwald ripening is used in such applications as photography, requiring crystals with specific properties.

Ostwald's dilution law An expression for the degree of dissociation of a weak electrolyte. For example, if a weak acid dissociates in water

$$HA \rightleftharpoons H^+ + A^-$$

the dissociation constant K_a is given by

$$K_a = \alpha^2 n/(1 - \alpha)V$$

where α is the degree of dissociation, n the initial amount of substance (before dissociation), and V the volume. If α is small compared with 1, then $\alpha^2 = KV/n$; i.e. the degree of dissociation is proportional to the square root of the dilution. The law was first put forward by Wilhelm Ostwald to account for electrical conductivities of electrolyte solutions.

outer-sphere mechanism *See* **electron-transfer reaction**.

overpotential (overvoltage) A potential that must be applied in an electrolytic cell in addition to the theoretical potential required to liberate a given substance at an electrode. The value depends on the electrode material and on the current density. It is a kinetic effect occurring because of the significant activation energy for electron transfer at the electrodes, and is particularly important for the liberation of such gases as hydrogen and oxygen. For example, in the electrolysis of a solution of zinc ions, hydrogen ($E^{\ominus}$ = 0.00 V) would be expected to be liberated at the cathode in preference to zinc ($E^{\ominus}$ = −0.76 V). In fact, the high overpotential of hydrogen on zinc (about 1 V under suitable conditions) means that zinc can be deposited instead. The relation between the current and the overpotential is given by the *Butler–Volmer equation.

overtones *See* **harmonic**.

oxalate A salt or ester of *oxalic acid.

oxalic acid (ethanedioic acid) A crystalline solid, $(COOH)_2$, that is slightly soluble in water. Oxalic acid is strongly acidic and very poisonous. It occurs in certain plants, e.g. sorrel and the leaf blades of rhubarb.

oxaloacetic acid A compound, $HO_2CCH_2COCO_2H$, that plays an integral role in the *Krebs cycle. The anion, oxaloacetate, reacts with the acetyl group from acetyl coenzyme A to form citrate.

oxfuel A liquid fuel containing added alcohols or ethers to act as an additional source of oxygen during combustion of the fuel. It has been claimed that such additives help to lower the concentration of carbon monoxide in engine emissions.

oxidant *See* **oxidizing agent**.

oxidation *See* **oxidation–reduction**.

oxidation number (oxidation state) *See* **oxidation–reduction**.

oxidation–reduction (redox) Originally, *oxidation* was simply regarded as a chemical reaction with oxygen. The reverse process – loss of oxygen – was called *reduction*. Reaction with hydrogen also came to be regarded as reduction. Later, a more general idea of oxidation and reduction was developed in which oxidation was loss of electrons and reduction was gain of electrons. This wider definition covered the original one. For example, in the reaction

$4Na(s) + O_2(g) \rightarrow 2Na_2O(s)$

the sodium atoms lose electrons to give Na^+ ions and are oxidized. At the same time, the oxygen atoms gain electrons and are reduced. These definitions of oxidation and reduction also apply to reactions that do not involve oxygen. For instance in

$2Na(s) + Cl_2(g) \rightarrow 2NaCl(s)$

the sodium is oxidized and the chlorine reduced. Oxidation and reduction also occurs at the electrodes in *cells.

This definition of oxidation and reduction applies only to reactions in which electron transfer occurs – i.e. to reactions involving ions. It can be extended to reactions between covalent compounds by using the concept of *oxidation number* (or *state*). This is a measure of the electron control that an atom has in a compound compared to the atom in the pure element. An oxidation number consists of two parts:
(1) Its sign, which indicates whether the control has increased (negative) or decreased (positive).
(2) Its value, which gives the number of electrons over which control has changed.

The change of electron control may be complete (in ionic compounds) or partial (in covalent compounds). For example, in SO_2 the sulphur has an oxidation number +4, having gained partial control over 4 electrons compared to sulphur atoms in pure sulphur. The oxygen has an oxidation number −2, each oxygen having lost partial control over 2 electrons compared to oxygen atoms in gaseous oxygen. Oxidation is a reaction involving an increase in oxidation number and reduction involves a decrease. Thus in

$2H_2 + O_2 \rightarrow 2H_2O$

the hydrogen in water is +1 and the oxygen −2. The hydrogen is oxidized and the oxygen is reduced.

The oxidation number is used in naming inorganic compounds. Thus in H_2SO_4, sulphuric(VI) acid, the sulphur has an oxidation number of +6. Compounds that tend to undergo reduction readily are *oxidizing agents; those that undergo oxidation are *reducing agents.

oxidation state *See* **oxidation–reduction**.

oxidative deamination A reaction involved in the catabolism of amino acids that assists their excretion from the body. An example of an oxidative deamination is the conversion of glutamate to α-ketoglutarate, a reaction catalysed by the enzyme glutamate dehydrogenase.

oxidative decarboxylation The reaction in the *Krebs cycle in which oxygen, derived from two water molecules, is used to oxidize two carbon atoms to two molecules of carbon dioxide. The two carbon atoms result from the decarboxylation reactions that occur during the Krebs cycle as the six-carbon compound citrate is converted to the four-carbon compound oxaloacetate.

oxides Binary compounds formed between elements and oxygen. Oxides of nonmetals are covalent compounds having simple molecules (e.g. CO, CO_2, SO_2) or giant molecular lattices (e.g. SiO_2). They are typically acidic or neutral. Oxides of metals are ionic, containing the O^{2-} ion. They are generally basic or *amphoteric. Various other types of ionic oxide exist (*see* **ozonides**; **peroxides**; **superoxides**).

oxidizing acid An acid that can act as a strong oxidizing agent as well as an acid. Nitric acid is a common example. It is able to attack metals, such as copper, that are below hydrogen in the electromotive series, by oxidizing the metal:

$$2HNO_3 + Cu \rightarrow CuO + H_2O + 2NO_2$$

This is followed by reaction between the acid and the oxide:

$$2HNO_3 + CuO \rightarrow Cu(NO_3)_2 + H_2O$$

oxidizing agent **(oxidant)** A substance that brings about oxidation in other substances. It achieves this by being itself reduced. Oxidizing agents contain atoms with high oxidation numbers; that is the atoms have suffered electron loss. In oxidizing other substances these atoms gain electrons.

$$R{-}C(=O){-}R' + H_2N{-}O{-}H \xrightarrow{-H_2O} R{-}C(=N{-}O{-}H){-}R'$$

ketone hydroxylamine oxime

Formation of an oxime from a ketone. The same reaction occurs with an aldehyde (R′ = H)

oximes Compounds containing the group C:NOH, formed by reaction of an aldehyde or ketone with hydroxylamine (H_2NOH) (see illustration). Ethanal (CH_3CHO), for example, forms the oxime $CH_3CH{:}NOH$.

oxo- Prefix indicating the presence of oxygen in a chemical compound.

oxoacid An acid in which the acidic hydrogen is part of a hydroxyl group bound to an atom that is bound to an oxo group (=O). Sulphuric acid is an example. *Compare* **hydroxoacid**.

3-oxobutanoic acid **(acetoacetic acid)** A colourless syrupy liquid, CH_3COCH_2COOH. It is an unstable compound, decomposing into propanone and carbon dioxide. The acid can be prepared from its ester, *ethyl 3-oxobutanoate.

oxonium ion An ion of the type R_3O^+, in which R indicates hydrogen or an organic group especially the ion H_3O^+, which is formed when *acids dissociate in water. This is also called the *hydroxonium ion* or the *hydronium ion*.

oxo process An industrial process for making aldehydes by reaction between alkanes, carbon monoxide, and hydrogen (cobalt catalyst using high pressure and temperature).

oxyacetylene burner A welding or cutting torch that burns a mixture of oxygen and acetylene (ethyne) in a specially designed jet. The flame temperature of about 3300°C enables all ferrous metals to be welded. For cutting, the point at which the steel is to be cut is preheated with the oxyacetylene flame and a powerful jet of oxygen is then directed onto the steel. The oxygen reacts with the hot steel to form iron oxide and the heat of this reaction melts more iron, which is blown away by the force of the jet.

oxygen Symbol O. A colourless odourless gaseous element belonging to *group 16 (formerly VIB) of the periodic table; a.n. 8; r.a.m. 15.9994; d. 1.429 g dm^{-3}; m.p. −218.4°C; b.p. −183°C. It is the most abundant element in the earth's crust (49.2% by weight) and is present in the atmosphere (28% by volume). Atmospheric oxygen is of vital importance for all organisms that carry out aerobic respiration. For industrial purposes it is obtained by fractional distillation of liquid air. It is used in metallurgical processes, in high-temperature flames (e.g. for welding), and in breathing apparatus. The common form is diatomic (*dioxygen*, O_2); there is also a reactive allotrope *ozone (O_3). Chemically, oxygen reacts with most other elements forming *oxides. The element was discovered by Joseph Priestley in 1774.

oxygenates Oxygen-containing organic compounds, such as ethanol and acetone, present in motor fuels.

oxyhaemoglobin *See* **haemoglobin**.

ozonation The formation of *ozone (O_3) in the earth's atmosphere. In the upper atmosphere (stratosphere) about 20–50 km above the surface of the earth, oxygen molecules (O_2) dissociate into their constituent atoms under the influence of ultraviolet light of short wavelength (below about 240 nm). These atoms combine with oxygen molecules to form ozone (*see* **ozone layer**). Ozone is also formed in the lower atmosphere from nitrogen oxides and other pollutants by photochemical reactions (*see* **photochemical smog**).

ozone (trioxygen) A colourless gas, O_3, soluble in cold water and in alkalis; m.p. −192.7°C; b.p. −111.9°C. Liquid ozone is dark blue in colour and is diamagnetic (dioxygen, O_2, is paramagnetic). The gas is made by passing oxygen through a silent electric discharge and is usually used in mixtures with oxygen. It is produced in the stratosphere by the action of high-energy ultraviolet radiation on oxygen and its presence there acts as a screen for ultraviolet radiation (*see* **ozone layer**). Ozone is also one of the greenhouse gases (*see* **greenhouse effect**). It is a powerful oxidizing agent and is used to form ozonides by reaction with alkenes and subsequently by hydrolysis to carbonyl compounds.

ozone hole *See* **ozone layer**.

ozone layer (ozonosphere) A layer of the *earth's atmosphere in which most of the atmosphere's ozone is concentrated. It occurs 15–50 km above the earth's surface and is virtually synonymous with the stratosphere. In this layer most of the sun's ultraviolet radiation is absorbed by the ozone molecules, causing a rise in the temperature of the stratosphere and preventing vertical mixing so that the stratosphere forms a stable layer. By absorbing most of the solar ultraviolet radiation the ozone layer protects living organisms on earth. The fact that the ozone layer is thinnest at the equator is believed to account for the high equatorial incidence of skin cancer as a result of exposure to unabsorbed solar ultraviolet radiation. In the 1980s it was found that depletion of the ozone layer was occurring over both the poles, creating *ozone holes*. This is thought to have been caused by a series of complex photochemical reactions involving nitrogen oxides produced from aircraft and, more seriously, *chlorofluorocarbons (CFCs) and *halons. CFCs rise to the stratosphere, where they react with ultraviolet light to release chlorine atoms; these atoms, which are highly reactive, catalyse the destruction of ozone. Use of CFCs is now much reduced in an effort to reverse this human-induced damage to the ozone layer.

ozonides 1. A group of compounds formed by reaction of ozone with alkali metal hydroxides and formally containing the ion O_3^-. **2.** Unstable compounds formed by the addition of ozone to the C=C double bond in alkenes. *See* **ozonolysis**.

ozonolysis A reaction of alkenes with ozone to form an ozonide. It was once used to investigate the structure of alkenes by hydrolysing the ozonide to give aldehydes or ketones. For instance

$$R_2C{:}CHR' \rightarrow R_2CO + R'CHO$$

These could be identified, and the structure of the original alkene determined.

PAGE (polyacrylamide gel electrophoresis) A type of *electrophoresis used to determine the size and composition of proteins. Proteins are placed on a matrix of polyacrylamide gel and an electric field is applied. The protein molecules migrate towards the positive pole, the smaller molecules moving at a faster rate through the pores of the gel. The proteins are then detected by applying a stain.

PAH Polyaromatic hydrocarbon. *See* **particulate matter**.

palladium Symbol Pd. A soft white ductile *transition element (*see also* **platinum metals**); a.n. 46; r.a.m. 106.4; r.d. 12.02; m.p. 1552°C; b.p. 3140±1°C. It occurs in some copper and nickel ores and is used in jewellery and as a catalyst for hydrogenation reactions. Chemically, it does not react with oxygen at normal temperatures. It dissolves slowly in hydrochloric acid. Palladium is capable of occluding 900 times its own volume of hydrogen. It forms few simple salts, most compounds being complexes of palladium(II) with some palladium(IV). It was discovered by William Woolaston (1766–1828) in 1803.

palmitate (hexadecanoate) A salt or ester of palmitic acid.

palmitic acid (hexadecanoic acid) A 16-carbon saturated fatty acid, $CH_3(CH_2)_{14}COOH$; r.d. 0.85; m.p. 63°C; b.p. 390°C. Glycerides of palmitic acid occur widely in plant and animal oils and fats.

pantothenic acid A vitamin of the *vitamin B complex. It is a constituent of coenzyme A, which performs a crucial role in the oxidation of fats, carbohydrates, and certain amino acids. Deficiency rarely occurs because the vitamin occurs in many foods, especially cereal grains, peas, egg yolk, liver, and yeast.

papain A protein-digesting enzyme occurring in the fruit of the West Indian papaya tree (*Carica papaya*). It is used as a digestant and in the manufacture of meat tenderizers.

paper chromatography A technique for analysing mixtures by *chromatography, in which the stationary phase is absorbent paper. A spot of the mixture to be investigated is placed near one edge of the paper and the sheet is suspended vertically in a solvent, which rises through the paper by capillary action carrying the components with it. The components move at different rates, partly because they absorb to different extents on the cellulose and partly because of partition between the solvent and the moisture in the paper. The paper is removed and dried, and the different components form a line of spots along the paper. Colourless substances are detected by using ultraviolet radiation or by spraying with a substance that reacts to give a coloured spot (e.g. ninhydrin gives a blue coloration with amino acids). The components can be identified by the distance they move in a given time.

para- 1. Prefix designating a benzene compound in which two substituents are in the 1,4 positions, i.e. directly opposite each other, on the benzene ring. The abbreviation *p*- is used; for example, *p*-xylene is 1,4-dimethylbenzene. *Compare* **ortho-**; **meta-**. **2.** Prefix denoting the form of diatomic molecules in which the nuclei have opposite spins, e.g. parahydrogen (*see* **hydrogen**). *Compare* **ortho-**.

paraffin *See* **petroleum**.

paraffins *See* **alkanes**.

paraffin wax *See* **petroleum**.

paraformaldehyde *See* **methanal**.

parahelium *See* **orthohelium**.

parahydrogen *See* **hydrogen**.

paraldehyde *See* **ethanal**.

parallel spins Neighbouring spinning electrons in which the *spins, and hence the magnetic moments, of the electrons are aligned in the same direction.

paramagnetism *See* **magnetism**.

Paraquat Tradename for an organic herbicide (*see* **dipyridyl**) used to control broadleaved weeds and grasses. It is poisonous to humans, having toxic effects on the liver, lungs, and kidneys if ingested. Paraquat is not easily broken down and can persist in the environment adsorbed to soil particles. *See also* **pesticide**.

parent *See* **daughter**.

partially permeable membrane A membrane that is permeable to the small molecules of water and certain solutes but does not allow the passage of large solute molecules. This term is preferred to semipermeable membrane when describing biological membranes. *See* **osmosis**.

partial pressure *See* **Dalton's law**.

particle in a box A system in quantum mechanics used to illustrate important features of quantum mechanics, such as quantization of energy levels and the existence of *zero-point energy. A particle with a mass m is allowed to move between two walls having the coordinates $x = 0$ and $x = L$. The potential energy of the particle is taken to be zero between the walls and infinite outside the walls. The time-independent *Schrödinger equation is exactly soluble for this problem, with the energies E_n being given by $n^2h^2/8mL^2$, $n = 1,2,...$, and the wave functions ψ_n being given by $\psi_n = (2/L)^{1/2} \sin(n\pi x/L)$, where n are the quantum numbers that label the energy levels and h is the *Planck constant. The particle in a box can be used as a rough model for delocalized electrons in a molecule or metal. The problem of a particle in a box can also be solved in two and three dimensions, enabling relations between symmetry and degenerate energy levels to be seen.

particulate matter **(PM)** Matter present as small particles. Particulate matter, also known as *particulates*, is produced as an airborne pollutant in many processes. It may be inorganic, e.g. silicate or carbon, or organic, e.g. polyaromatic hydrocarbons (*PAHs*). Particulates are classified by size; e.g. PM10, with particles less than 10 μm and PM2.5, with particles less than 2.5 μm.

parting agent *See* **release agent.**

partition If a substance is in contact with two different phases then, in general, it will have a different affinity for each phase. Part of the substance will be absorbed or dissolved by one and part by the other, the relative amounts depending on the relative affinities. The substance is said to be *partitioned* between the two phases. For example, if two immiscible liquids are taken and a third compound is shaken up with them, then an equilibrium is reached in which the concentration in one solvent differs from that in the other. The ratio of the concentrations is the *partition coefficient* of the system. The *partition law* states that this ratio is a constant for given liquids.

partition coefficient *See* **partition.**

partition function The quantity Z defined by

$$Z = \sum \exp(-E_i/kT),$$

where the sum is taken over all states i of the system. E_i is the energy of the ith state, k is the Boltzmann constant, and T is the thermodynamic temperature. Z is a quantity of fundamental importance in equilibrium statistical mechanics. For a system in which there are nontrivial interactions, it is very difficult to calculate the partition function exactly. For such systems it is necessary to use approximation techniques. The partition function links results at the atomic level to thermodynamics, since Z is related to the Helmholtz free energy F by $F = kT\ln Z$.

pascal The *SI unit of pressure equal to one newton per square metre. It is named after the French mathematician Blaise Pascal (1623–62).

Paschen–Back effect An effect on atomic line spectra that occurs when the atoms are placed in a strong magnetic field. Spectral lines that give the anomalous *Zeeman effect when the atoms are placed in a weaker magnetic field have a splitting pattern in a very strong magnetic field. The Paschen–Back effect is named after the German physicists Louis Carl Heinrich Friedrich Paschen (1865–1947) and Ernest E. A. Back (1881–1959), who discovered it in 1912. In the quantum theory of atoms the Paschen–Back effect is explained by the fact that the energies of precession of the electron's orbital angular momentum l and the spin angular momentum s about the direction of the magnetic field H are greater than the energies of coupling between l and s. In the Paschen–Back effect the orbital magnetic moment and the spin magnetic moment precess independently about the direction of H.

Paschen series *See* **hydrogen spectrum**.

passive Describing a solid that has reacted with another substance to form a protective layer, so that further reaction stops. The solid is said to have been 'rendered passive'. For example, aluminium reacts spontaneously with oxygen in air to form a thin layer of *aluminium oxide, which prevents further oxidation. Similarly, pure iron forms a protective oxide layer with concentrated nitric acid and is not dissolved further.

Pasteur, Louis (1822–95) French chemist and microbiologist, who held appointments in Strasbourg (1849–54) and Lille (1854–57), before returning to Paris to the Ecole Normale and the Sorbonne. From 1888 to his death he was director of the Pasteur Institute. In 1848 he discovered *optical activity, in 1860 relating it to molecular structure. In 1856 he began work on fermentation, and by 1862 was able to disprove the existence of spontaneous generation. He introduced pasteurization (originally for wine) in 1863. He went on to study disease and developed vaccines against cholera (1880), anthrax (1882), and rabies (1885).

Patterson function A function, denoted $P(x,y,z)$, used in the analysis of the results of *X-ray crystallography. $P(x,y,z)$ is defined by

$$P(x,y,z) = \sum_{hkl} |F_{hkl}| \cos 2\pi(hx + ky + lz),$$

where h, k, and l are the Miller indices of the crystal. This function is plotted as a contour map with maxima of the contours at vector distances from the origin, i.e. vectors from the origin to the point (x,y,z), where the maximum occurs, corresponding to vector distances between pairs of maxima in electron density. Thus, the Patterson function enables interatomic vectors, which give the lengths and directions between atomic centres, to be determined. This technique, introduced by A. L. Patterson in 1934, is called *Patterson synthesis* and is most useful if the unit cell contains heavy atoms.

Patterson synthesis *See* **Patterson function**.

Pauli, Wolfgang Ernst (1900–58) Austrian-born Swiss physicist. After studying with Niels *Bohr and Max *Born, he taught at Heidelberg and, finally Zurich. His formulation in 1924 of the *Pauli exclusion principle explained the electronic make-up of atoms. For this work he was awarded the 1945 Nobel Prize for physics. In 1931 he predicted the existence of the neutrino.

Pauli exclusion principle The principle that no two electrons in an atom can have all four quantum numbers the same. It was first formulated in 1925 by Wolfgang *Pauli and more generally applies to the quantum states of all elementary particles with half-integral spin.

Pauling, Linus Carl (1901–94) US chemist. After spending two years in Europe, he became a professor at the Californian Institute of Technology. His original work was on chemical bonding; in the mid-1930s he turned to the structure of proteins, for which he was awarded the 1954 Nobel Prize for chemistry. He was also an active campaigner against nuclear weapons and in 1962 was awarded the Nobel Peace Prize.

Pauling's rules A set of rules that enables the structures of many complex ionic crystals to be understood. These rules involve the sizes and electrical charges of the ions and were proposed by Linus *Pauling in 1929. They work well for mainly ionic crystals, such as silicates, but less well for crystals, such as sulphides, in which the bonding is mainly covalent. Pauling put forward his rules on the basis of empirical observations and calculations of crystal energies. A number of extensions and modifications have been proposed.

***p*-block elements** The block of elements in the periodic table consisting of the main groups 13 (B to Tl), 14 (C to Pb), 15 (N to Bi), 16 (O to Po), 17 (F to At) and 18 (He to Rn). The outer electronic configurations of these elements all have the form ns^2np^x where $x = 1$ to 6. Members at the top and on the right of the *p*-block are nonmetals (C, N, P, O, F, S, Cl, Br, I, At). Those on the left and at the bottom are metals (Al, Ga, In, Tl, Sn, Pb, Sb, Bi, Po). Between the two, from the top left to bottom right, lie an ill-defined group of metalloid elements (B, Si, Ge, As, Te).

PCB *See* **polychlorinated biphenyl.**

peacock ore *See* **bornite.**

pearl ash *See* **potassium carbonate.**

pearlite *See* **steel.**

penicillin An antibiotic derived from the mould *Penicillium notatum*; specifically it is known as *penicillin G* and belongs to a class of similar substances called penicillins. They produce their effects by disrupting synthesis of the bacterial cell wall, and are used to treat a variety of infections caused by bacteria.

Penning ionization A photochemical process that produces a positively charged ion. For atoms A and B, the process of Penning ionization is written:

$$A^* + B \rightarrow A + B^+ + e^-,$$

where A* denotes that the atom A has absorbed a photon, thus acquiring enough energy for the process to take place, and e^- is an electron. An example of Penning ionization is the ionization of mercury by argon:

$$Ar^* + Hg \rightarrow Ar + Hg^+ + e^-,$$

which occurs because the energy of the metastable state of argon is higher than the ionization energy of mercury. The process of Penning ionization was discovered by F. M. Penning in 1927.

Penrose tiling A method of tiling a two-dimensional plane using tiles with fivefold symmetry. Since this type of symmetry is not allowed in crystallography, two types of tile are needed, which are called 'fat' and 'skinny'. Adjacent tiles have to obey certain matching rules. Penrose tiling, named after the British mathematician and physicist Sir Roger Penrose (1931–), who put forward this idea in 1974, is the two-dimensional analogue of a *quasicrystal.

pentahydrate A crystalline hydrate that has five moles of water per mole of compound.

pentane A straight-chain alkane hydrocarbon, C_5H_{12}; r.d. 0.63; m.p. −129.7°C; b.p. 36.1°C. It is obtained by distillation of petroleum.

pentanoic acid **(valeric acid)** A colourless liquid *carboxylic acid, $CH_3(CH_2)_3COOH$; r.d. 0.9; m.p. −34°C; b.p. 186.05°C. It is used in the perfume industry.

pentavalent **(quinquevalent)** Having a valency of five.

pentlandite A mineral consisting of a mixed iron–nickel sulphide, $(Fe,Ni)_9S_8$, crystallizing in the cubic system; the chief ore of nickel. It is yellowish-bronze in colour with a metallic lustre. The chief occurrence of the mineral is at Sudbury in Ontario, Canada.

pentose A sugar that has five carbon atoms per molecule. *See* **monosaccharide**.

pentose phosphate pathway **(pentose shunt)** A series of biochemical reactions that results in the conversion of glucose 6-phosphate to ribose 5-phosphate and generates NADPH, which provides reducing power for other metabolic reactions, such as synthesis of fatty acids. Ribose 5-phosphate and its derivatives are components of such molecules as ATP, coenzyme A, NAD, FAD, DNA, and RNA. In plants the pentose phosphate pathway also plays a role in the synthesis of sugars from carbon dioxide. In animals the pathway occurs at various sites, including the liver and adipose tissue.

pentyl group **(pentyl radical)** The organic group $CH_3CH_2CH_2CH_2CH_2$–, derived from pentane.

pepsin An enzyme that catalyses the breakdown of proteins to polypeptides in the vertebrate stomach. It is secreted as an inactive precursor, *pepsinogen*.

peptide Any of a group of organic compounds comprising two or more amino acids linked by *peptide bonds*. These bonds are formed by the reaction between adjacent carboxyl (–COOH) and amino (–NH_2) groups with the elimination of water (see illustration). *Dipeptides* contain two amino acids, *tripeptides* three, and so on. *Polypeptides contain more than ten and usually 100–300. Naturally occurring *oligopeptides* (of less than ten amino acids) include the tripeptide glutathione and the pituitary hormones

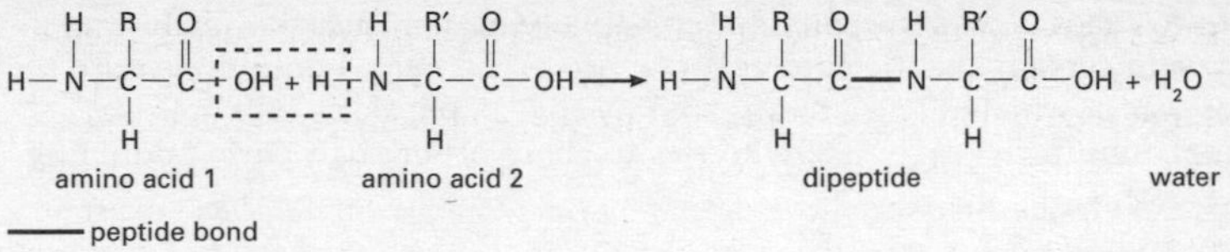

$H_2N{-}CHR{-}CO{-}OH + H{-}NH{-}CHR'{-}COOH \rightarrow H_2N{-}CHR{-}CO{-}NH{-}CHR'{-}COOH + H_2O$

amino acid 1 amino acid 2 dipeptide water

—— peptide bond

Formation of a peptide bond

vasopressin and oxytocin, which are octapeptides. Peptides also result from protein breakdown, e.g. during digestion.

peptide mapping (peptide fingerprinting) The technique of forming two-dimensional patterns of peptides (on paper or gel) by partial hydrolysis of a protein followed by electrophoresis and chromatography. The peptide pattern (or *fingerprint*) produced is characteristic for a particular protein and the technique can be used to separate a mixture of peptides.

peptidoglycan A macromolecule that is a component of the cell wall of bacteria; it is not found in eukaryotes. Consisting of chains of amino sugars (N-acetylglucosamine and N-acetylmuramic acid) linked to a tripeptide (of alanine, glutamic acid, and lysine or diaminopimelic acid), it confers strength and shape to the cell wall.

per- Prefix indicating that a chemical compound contains an excess of an element, e.g. a peroxide.

perchlorate *See* **chlorates.**

perchloric acid *See* **chloric(VII) acid.**

perdisulphuric acid *See* **peroxosulphuric(VI) acid.**

perfect gas *See* **ideal gas; gas.**

perfect solution *See* **Raoult's law.**

pericyclic reaction A type of concerted chemical reaction that proceeds through a cyclic conjugated transition state. Pericyclic reactions include *cheletropic reactions, some *cycloadditions, and *electrocyclic reactions.

peridot *See* **olivine.**

period 1. The time taken for one complete cycle of an oscillating system or wave. **2.** *See* **periodic table.**

period doubling A mechanism for describing the transition to *chaos in certain dynamical systems. If the force on a body produces a regular orbit with a specific *period a sudden increase in the force can suddenly double the period of the orbit and the motion becomes more complex. The original simple motion is called a *one-cycle*, while the more complicated motion after the period doubling is called a *two-cycle*. The process of period doubling can continue until a motion called an *n-cycle* is produced. As *n* increases to infinity the motion becomes non-periodic. The period-doubling route to chaos occurs in many systems involving nonlinearity, including lasers and certain chaotic chemical reactions. The period-doubling route to chaos was postulated and investigated by the US physicist Mitchell Feigenbaum in the early 1980s. Routes to chaos other than period doubling also exist.

periodic acid *See* **iodic(VII) acid**.

periodic law The principle that the physical and chemical properties of elements are a periodic function of their proton number. The concept was first proposed in 1869 by Dimitri *Mendeleev, using relative atomic mass rather than proton number, as a culmination of efforts to rationalize chemical properties by Johann Döbereiner (1817), John Newlands (1863), and Lothar Meyer (1864). One of the major successes of the periodic law was its ability to predict chemical and physical properties of undiscovered elements and unknown compounds that were later confirmed experimentally. *See* **periodic table**.

periodic table A table of elements arranged in order of increasing proton number to show the similarities of chemical elements with related electronic configurations. (The original form was proposed by Dimitri Mendeleev in 1869 using relative atomic masses.) In the modern *short form*, the *lanthanoids and *actinoids are not shown. The elements fall into vertical columns, known as *groups*. Going down a group, the atoms of the elements all have the same outer shell structure, but an increasing number of inner shells. Traditionally, the alkali metals were shown on the left of the table and the groups were numbered IA to VIIA, IB to VIIB, and 0 (for the noble gases). All the elements in the middle of the table are classified as *transition elements and the nontransition elements are regarded as *main-group* elements. Because of confusion in the past regarding the numbering of groups and the designations of subgroups, modern practice is to number the groups across the table from 1 to 18 (see Appendix). Horizontal rows in the table are *periods*. The first three are called *short periods*; the next four (which include transition elements) are *long periods*. Within a period, the atoms of all the elements have the same number of shells, but with a steadily increasing number of electrons in the outer shell. The periodic table can also be divided into four *blocks* depending on the type of shell being filled: the **s*-block, the **p*-block, the **d*-block, and the **f*-block.

There are certain general features of chemical behaviour shown in the periodic table. In moving down a group, there is an increase in metallic character because of the increased size of the atom. In going across a period, there is a change from metallic (electropositive) behaviour to nonmetallic (electronegative) because of the increasing number of electrons in the outer shell. Consequently, metallic elements tend to be those on the left and towards the bottom of the table; nonmetallic elements are towards the top and the right.

There is also a significant difference between the elements of the second short period (lithium to fluorine) and the other elements in their respective groups. This is because the atoms in the second period are smaller and their valence electrons are shielded by a small $1s^2$ inner shell. Atoms in the other periods have inner *s*- and *p*-electrons shielding the outer electrons from the nucleus. Moreover, those in the second period only have *s*- and *p*-orbitals available for bonding. Heavier atoms can also promote electrons to vacant *d*-orbitals in their outer shell and use these for bonding. *See also* **diagonal relationship; inert-pair effect.**

periplanar *See* **torsion angle.**

Perkin, Sir William Henry (1838–1907) British chemist, who while still a student accidentally produced mauvine, the first aniline dye and the first dyestuff to be synthesized. Perkin built a factory to produce it, and made a fortune.

Permalloys A group of alloys of high magnetic permeability consisting of iron and nickel (usually 40–80%) often with small amounts of other elements (e.g. 3–5% molybdenum, copper, chromium, or tungsten). They are used in thin foils in electronic transformers, for magnetic shielding, and in computer memories.

permanent gas A gas, such as oxygen or nitrogen, that was formerly thought to be impossible to liquefy. A permanent gas is now regarded as one that cannot be liquefied by pressure alone at normal temperatures (i.e. a gas that has a critical temperature below room temperature).

permanent hardness *See* **hardness of water.**

permanganate *See* **manganate(VII).**

permittivity Symbol ε. The ratio of the electric displacement in a medium to the intensity of the electric field producing it. It is important for electrical insulators used as dielectrics.

If two charges Q_1 and Q_2 are separated by a distance r in a vacuum, the force F between the charges is given by:

$$F = Q_1Q_2/r^2 4\pi\varepsilon_0$$

In this statement of Coulomb's law using *SI units, ε_0 is called the absolute permittivity of free space, which is now known as the *electric constant*. It has the value 8.854×10^{-12} F m^{-1}.

If the medium between the charges is anything other than a vacuum the equation becomes:

$$F = Q_1Q_2/r^2 4\pi\varepsilon$$

and the force between the charges is reduced. ε is the *absolute permittivity* of the new medium. The *relative permittivity* (ε_r) of a medium, formerly called the *dielectric constant*, is given by $\varepsilon_r = \varepsilon/\varepsilon_0$.

permonosulphuric(VI) acid *See* **peroxosulphuric(VI) acid.**

Permutit Tradename for a *zeolite used for water softening.

peroxides A group of inorganic compounds that contain the O_2^{2-} ion.

They are notionally derived from hydrogen peroxide, H_2O_2, but these ions do not exist in aqueous solution due to extremely rapid hydrolysis to OH^-.

peroxodisulphuric acid *See* **peroxosulphuric(VI) acid**.

peroxomonosulphuric(VI) acid *See* **peroxosulphuric(VI) acid**.

peroxosulphuric(VI) acid The term commonly refers to *peroxomonosulphuric(VI) acid*, H_2SO_5, which is also called *permonosulphuric(VI) acid* and *Caro's acid*. It is a crystalline compound made by the action of hydrogen peroxide on concentrated sulphuric acid. It decomposes in water and the crystals decompose, with melting, above 45°C. The compound *peroxodisulphuric acid*, $H_2S_2O_8$, also exists (formerly called *perdisulphuric acid*). It is made by the high-current electrolysis of sulphate solutions. It decomposes at 65°C (with melting) and is hydrolysed in water to give the mono acid and sulphuric acid. Both peroxo acids are very powerful oxidizing agents. *See also* **sulphuric acid** (for structural formulas).

persistent Describing a pesticide or other pollutant that is not readily broken down and can persist for long periods, causing damage in the environment. For example, the herbicides *Paraquat and *DDT can persist in the soil for many years after their application.

Perspex Tradename for a form of *polymethylmethacrylate.

perturbation theory A method used in calculations in both classical physics (e.g. planetary orbits) and quantum mechanics (e.g. atomic structure), in which the system is divided into a part that is exactly calculable and a small term, which prevents the whole system from being exactly calculable. The technique of perturbation theory enables the effects of the small term to be calculated by an infinite series (which in general is an asymptotic series). Each term in the series is a 'correction term' to the solutions of the exactly calculable system. In classical physics, perturbation theory can be used for calculating planetary orbits. In quantum mechanics, it can be used to calculate the energy levels in molecules.

PES *See* **photoelectron spectroscopy**.

PESM *See* **photoelectron spectromicroscopy**.

pesticide Any chemical compound used to kill pests that destroy agricultural production or are in some way harmful to humans. Pesticides include *herbicides* (such as 2,4-*D and *Paraquat), which kill unwanted plants or weeds; *insecticides* (such as pyrethrum), which kill insect pests; *fungicides*, which kill fungi; and *rodenticides* (such as *warfarin), which kill rodents. The problems associated with pesticides are that they are very often nonspecific and may therefore be toxic to organisms that are not pests; they may also be nonbiodegradable, so that they persist in the environment and may accumulate in living organisms (*see* **bioaccumulation**). Organophosphorus insecticides, such as malathion and parathion, are biodegradable but can also damage the respiratory and nervous systems in humans as well as killing useful insects, such as bees.

They act by inhibiting the action of the enzyme cholinesterase. Organochlorine insecticides, such as dieldrin, aldrin, and *DDT, are very persistent and not easily biodegradable.

peta- Symbol P. A prefix used in the metric system to denote one thousand million million times. For example, 10^{15} metres = 1 petametre (Pm).

petrochemicals Organic chemicals obtained from petroleum or natural gas.

petroleum A naturally occurring oil that consists chiefly of hydrocarbons with some other elements, such as sulphur, oxygen, and nitrogen. In its unrefined form petroleum is known as *crude oil* (sometimes *rock oil*). Petroleum is believed to have been formed from the remains of living organisms that were deposited, together with rock particles and biochemical and chemical precipitates, in shallow depressions, chiefly in marine conditions. Under burial and compaction the organic matter went through a series of processes before being transformed into petroleum, which migrated from the source rock to become trapped in large underground reservoirs beneath a layer of impermeable rock. The petroleum often floats above a layer of water and is held under pressure beneath a layer of *natural gas.

Petroleum reservoirs are discovered through geological exploration: commercially important oil reserves are detected by exploratory narrow-bore drilling. The major known reserves of petroleum are in Saudi Arabia, Russia, China, Kuwait, Iran, Iraq, Mexico, USA, United Arab Emirates, Libya, and Venezuela. The oil is actually obtained by the sinking of an oil well. Before it can be used it is separated by fractional distillation in oil refineries. The main fractions obtained are:
(1) *Refinery gas* A mixture of methane, ethane, butane, and propane used as a fuel and for making other organic chemicals.
(2) *Gasoline* A mixture of hydrocarbons containing 5 to 8 carbon atoms, boiling in the range 40–180°C. It is used for motor fuels and for making other chemicals.
(3) *Kerosine* (or *paraffin oil*) A mixture of hydrocarbons having 11 or 12 carbon atoms, boiling in the range 160–250°C. Kerosine is a fuel for jet aircraft and for oil-fired domestic heating. It is also cracked to produce smaller hydrocarbons for use in motor fuels.
(4) *Diesel oil* (or *gas oil*) A mixture of hydrocarbons having 13 to 25 carbon atoms, boiling in the range 220–350°C. It is a fuel for diesel engines.

The residue is a mixture of higher hydrocarbons. The liquid components are obtained by vacuum distillation and used in lubricating oils. The solid components (*paraffin wax*) are obtained by solvent extraction. The final residue is a black tar containing free carbon (*asphalt* or *bitumen*).

petroleum ether A colourless volatile flammable mixture of hydrocarbons (not an ether), mainly pentane and hexane. It boils in the range 30–70°C and is used as a solvent.

pewter An alloy of lead and tin. It usually contains 63% tin; pewter tankards and food containers should have less than 35% of lead so that the lead remains in solid solution with the tin in the presence of weak acids in the food and drink. Copper is sometimes added to increase ductility and antimony is added if a hard alloy is required.

Pfund series *See* **hydrogen spectrum**.

PGA *See* **phosphoglyceric acid**.

pH *See* **pH scale**.

phase A homogeneous part of a heterogeneous system that is separated from other parts by a distinguishable boundary. A mixture of ice and water is a two-phase system. A solution of salt in water is a single-phase system.

phase diagram A graph showing the relationship between solid, liquid, and gaseous *phases over a range of conditions (e.g. temperature and pressure). *See* **steel** (illustration).

phase rule For any system at equilibrium, the relationship $P + F = C + 2$ holds, where P is the number of distinct phases, C the number of components, and F the number of degrees of freedom of the system. The relationship derived by Josiah Willard *Gibbs in 1876, is often called the *Gibbs phase rule.*

phase space For a system with n degrees of freedom, the $2n$-dimensional space with coordinates $(q_1, q_2, ..., q_n, p_1, p_2, ..., p_n)$, where the qs describe the degrees of freedom of the system and the ps are the corresponding momenta. Each point represents a state of the system. In a gas of N point particles, each particle has three positional coordinates and three corresponding momentum coordinates, so that the phase space has $6N$-dimensions. If the particles have internal degrees of freedom, such as the vibrations and rotations of molecules, then these must be included in the phase space, which is consequently of higher dimension than that for point particles. As the system changes with time the representative points trace out a curve in phase space known as a *trajectory*. *See also* **attractor**; **configuration space**; **statistical mechanics**.

phase transition A change in a feature that characterizes a system. Examples of phase transitions are changes from solid to liquid, liquid to gas, and the reverse changes. Other examples of phase transitions include the transition from a paramagnet to a ferromagnet (*see* **magnetism**) and the transition from a normally conducting metal to a superconductor. Phase transitions can occur by altering such variables as temperature and pressure.

Phase transitions can be classified by their *order*. If there is non-zero *latent heat at the transition it is said to be a *first-order transition*. If the latent heat is zero it is said to be a *second-order transition*.

Some models describing phase transitions, particularly in low-dimensional systems, are amenable to exact mathematical solutions.

Techniques for investigating transitions may include the feature of *universality*, in which very different physical systems behave in the same way near a phase transition.

phenol **(carbolic acid)** A white crystalline solid, C_6H_5OH; r.d. 1.1; m.p. 43°C; b.p. 182°C. It is made by the *cumene process or by the *Raschig process. Phenol reacts readily to form substituted derivatives. It is an important industrial chemical used in making Nylon, phenolic and epoxy resins, dyestuffs, explosives, and pharmaceuticals. *See also* **phenols**.

phenolphthalein A dye used as an acid-base *indicator. It is colourless below pH 8 and red above pH 9.6. It is used in titrations involving weak acids and strong bases. It is also used as a laxative.

phenols Organic compounds that contain a hydroxyl group (–OH) bound directly to a carbon atom in a benzene ring. Unlike normal alcohols, phenols are acidic because of the influence of the aromatic ring. Thus, phenol itself (C_6H_5OH) ionizes in water:

$$C_6H_5OH \rightarrow C_6H_5O^- + H^+$$

Phenols are made by fusing a sulphonic acid salt with sodium hydroxide to form the sodium salt of the phenol. The free phenol is liberated by adding sulphuric acid.

phenomenological relations *See* **Onsager relations**.

phenoxy resins Thermoplastic materials made by condensation of phenols, used mainly for packaging.

phenylalanine *See* **amino acid**.

phenylamine **(aniline; aminobenzene)** A colourless oily liquid aromatic *amine, $C_6H_5NH_2$, with an 'earthy' smell; r.d. 1.0217; m.p. −6.3°C; b.p. 184.1°C. The compound turns brown on exposure to sunlight. It is basic, forming the *phenylammonium* (or *anilinium*) *ion*, $C_6H_5NH_3^+$, with strong acids. It is manufactured by the reduction of nitrobenzene or by the addition of ammonia to chlorobenzene using a copper(II) salt catalyst at 200°C and 55 atm. The compound is used extensively in the rubber industry and in the manufacture of drugs and dyes.

phenylammonium ion The ion $C_6H_5NH_3^+$, derived from *phenylamine.

***N*-phenylethanamide** *See* **acetanilide**.

phenylenediamine *See* **diaminobenzene**.

phenylethene **(styrene)** A liquid hydrocarbon, $C_6H_5CH{:}CH_2$; r.d. 0.9; m.p. −31°C; b.p. 145°C. It can be made by dehydrogenating ethylbenzene and is used in making polystyrene.

phenyl group The organic group C_6H_5–, present in benzene.

phenylhydrazine A toxic colourless dense liquid, $C_6H_5NHNH_2$, b.p. 240°C, which turns brown on exposure to air. It is a powerful reducing agent, made from *diazonium salts of benzene. It is used to identify aldehydes and ketones, with which it forms condensation products called *hydrazones. With glucose and similar sugars it forms osazones. For such tests, the nitro derivative 2,4-dinitrophenylhydrazine (DNP) is often preferred as this generally forms crystalline derivatives that can be identified by their melting points. Phenylhydrazine is also used to make dyes and derivatives of *indole.

phenylhydrazones *See* **hydrazones.**

phenylmethanol (benzyl alcohol) A liquid aromatic alcohol, $C_6H_5CH_2OH$; r.d. 1.04; m.p. −15.3°C; b.p. 205.4°C. It is used mainly as a solvent.

phenylmethylamine *See* **benzylamine.**

phenyl methyl ketone (acetophenone) A colourless, crystalline ketone with an odour of bitter almonds, $C_6H_5COCH_3$, m.p. 20°C. It is made by treating benzene with ethanoyl chloride in the presence of aluminium chloride. It is used as a solvent and in making perfumes.

3-phenylpropenoic acid *See* **cinnamic acid.**

Phillips process A process for making high-density polyethene by polymerizing ethene at high pressure (30 atmospheres) and 150°C. The catalyst is chromium(III) oxide supported on silica and alumina.

phlogiston theory A former theory of combustion in which all flammable objects were supposed to contain a substance called *phlogiston*, which was released when the object burned. The existence of this hypothetical substance was proposed in 1669 by Johann Becher, who called it 'combustible earth' (*terra pinguis*: literally 'fat earth'). For example, according to Becher, the conversion of wood to ashes by burning was explained on the assumption that the original wood consisted of ash and *terra pinguis*, which was released on burning. In the early 18th century Georg Stahl renamed the substance *phlogiston* (from the Greek for 'burned') and extended the theory to include the calcination (and corrosion) of metals. Thus, metals were thought to be composed of *calx* (a powdery residue) and phlogiston; when a metal was heated, phlogiston was set free and the calx remained. The process could be reversed by heating the metal over charcoal (a substance believed to be rich in phlogiston, because combustion almost totally consumed it). The calx would absorb the phlogiston released by the burning charcoal and become metallic again.

The theory was finally demolished by Antoine Lavoisier, who showed by careful experiments with reactions in closed containers that there was no *absolute* gain in mass – the gain in mass of the substance was matched by a corresponding loss in mass of the air used in combustion. After experiments with Priestley's dephlogisticated air, Lavoisier realized that this gas, which he named oxygen, was taken up to form a calx (now called an oxide). The role of oxygen in the new theory was almost exactly the opposite of phlogiston's role in the old. In combustion and corrosion phlogiston was released; in the modern theory, oxygen is taken up to form an oxide.

phonochemistry The use of high-frequency sound (ultrasound, with a frequency greater than about 20 kHz) to induce or accelerate certain types of chemical reaction.

phosgene *See* **carbonyl chloride**.

phosphagen A compound found in animal tissues that provides a reserve of chemical energy in the form of high-energy phosphate bonds. The most common phosphagens are creatine phosphate, occurring in vertebrate muscle and nerves, and arginine phosphate, found in most invertebrates. During tissue activity (e.g. muscle contraction) phosphagens give up their phosphate groups, thereby generating *ATP from ADP. The phosphagens are then reformed when ATP is available.

phosphates Salts based formally on phosphorus(V) oxoacids and in particular salts of *phosphoric(V) acid, H_3PO_4. A large number of polymeric phosphates also exist, containing P–O–P bridges. These are formed by heating the free acid and its salts under a variety of conditions; as well as linear polyphosphates, cyclic polyphosphates and cross-linked polyphosphates or ultraphosphates are known.

phosphatide *See* **phospholipid**.

phosphide A binary compound of phosphorus with a more electropositive element. Phosphides show a wide range of properties. Alkali and alkaline earth metals form ionic phosphides, such as Na_3P and Ca_3P_2, which are readily hydrolysed by water. The other transition-metal phosphides are inert metallic-looking solids with high melting points and electrical conductivities.

phosphine A colourless highly toxic gas, PH_3; m.p. 133°C; b.p. −87.7°C; slightly soluble in water. Phosphine may be prepared by reacting water or dilute acids with calcium phosphide or by reaction between yellow phosphorus and concentrated alkali. Solutions of phosphine are neutral but phosphine does react with some acids to give phosphonium salts containing PH_4^+ ions, analogous to the ammonium ions. Phosphine prepared in the laboratory is usually contaminated with diphosphine and is spontaneously flammable but the pure compound is not so. Phosphine can function as a ligand in binding to transition-metal ions. Dilute gas mixtures of very pure phosphine and the rare gases are used for doping semiconductors.

phosphinic acid (hypophosphorus acid) A white crystalline solid, H_3PO_2; r.d. 1.493; m.p. 26.5°C; decomposes above 130°C. It is soluble in water, ethanol, and ethoxyethane. Salts of phosphinic acid may be prepared by boiling white phosphorus with the hydroxides of group I or group II metals. The free acid is made by the oxidation of phosphine with iodine. It is a weak monobasic acid in which it is the –O–H group that is ionized to give the ion $H_2PO_2^-$. The acid and its salts are readily oxidized to the orthophosphate and consequently are good reducing agents.

phosphite *See* **phosphonic acid**.

phosphodiester bond The covalent bond that links a phosphate group and a sugar group, by means of an oxygen bridge, in the sugar–phosphate backbone of a nucleic acid molecule. *See* **DNA**.

phosphoglyceric acid (PGA; 3-phosphoglycerate) *See* **glycerate 3-phosphate**.

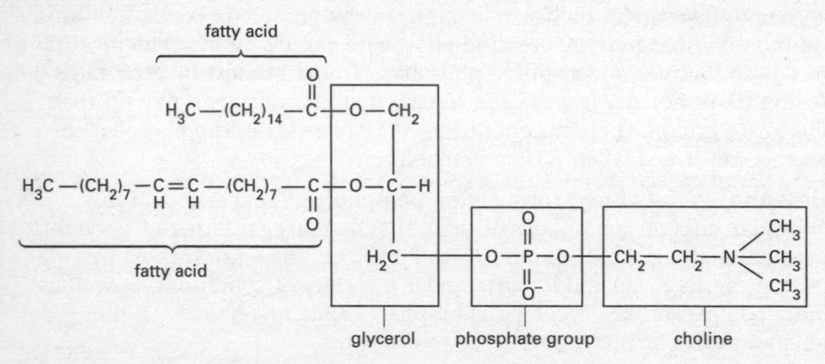

Structure of lecithin (phosphatidylcholine), a phosphoglyceride

phospholipid (phosphatide) One of a group of lipids having both a phosphate group and one or more fatty acids. *Glycerophospholipids* (or *phosphoglycerides*) are based on *glycerol; the three hydroxyl groups are esterified with two fatty acids and a phosphate group, which may itself be bound to one of a variety of simple organic groups (see illustration). *Sphingolipids* are based on the alcohol sphingosine and contain only one fatty acid linked to an amino group. With their hydrophilic polar phosphate groups and long hydrophobic hydrocarbon 'tails', phospholipids readily form membrane-like structures in water. They are a major component of cell membranes.

phosphonate *See* **phosphonic acid**.

phosphonic acid (phosphorous acid; orthophosphorous acid) A colourless to pale-yellow deliquescent crystalline solid, H_3PO_3; r.d. 1.65; m.p. 73.6°C; decomposes at 200°C; very soluble in water and soluble in alcohol. Phosphonic acid may be crystallized from the solution obtained by adding ice-cold water to phosphorus(III) oxide or phosphorus trichloride. The structure of this material is unusual in that it contains one direct P–H bond and is more correctly written $(HO)_2HPO$. The acid is dibasic, giving rise to the ions $H_2PO_3^-$ and HPO_3^{2-} (*phosphonates*; formerly *phosphites*), and has moderate reducing properties. On heating it gives phosphine and phosphoric(V) acid.

phosphonium ion The ion PH_4^+, or the corresponding organic derivatives of the type R_3PH^+, RPH_3^+. The phosphonium ion PH_4^+ is formally analogous to the ammonium ion NH_4^+ but PH_3 has a much lower proton affinity than

NH_3 and reaction of PH_3 with acids is necessary for the production of phosphonium salts.

phosphor A substance that is capable of *luminescence (including phosphorescence). Phosphors that release their energy after a short delay of between 10^{-10} and 10^{-4} second are sometimes called *scintillators.*

phosphor bronze An alloy of copper containing 4% to 10% of tin and 0.05% to 1% of phosphorus as a deoxidizing agent. It is used particularly for marine purposes and where it is exposed to heavy wear, as in gear wheels. *See also* **bronze.**

phosphorescence *See* **luminescence.**

phosphoric(V) acid (orthophosphoric acid) A white rhombic solid, H_3PO_4; r.d. 1.834; m.p. 42.35°C; loses water at 213°C; very soluble in water and soluble in ethanol. Phosphoric(V) acid is very deliquescent and is generally supplied as a concentrated aqueous solution. It is the most commercially important derivative of phosphorus, accounting for over 90% of the phosphate rock mined. It is manufactured by two methods; the *wet process,* in which the product contains some of the impurities originally present in the rock and applications are largely in the fertilizer industry, and the *thermal process,* which produces a much purer product suitable for the foodstuffs and detergent industries. In the wet process the phosphate rock, $Ca_3(PO_4)_2$, is treated with sulphuric acid and the calcium sulphate removed either as gypsum or the hemihydrate. In the thermal process, molten phosphorus is sprayed and burned in a mixture of air and steam. Phosphoric(V) acid is a weak tribasic acid, which is best visualized as $(HO)_3PO$. Its full systematic name is *tetraoxo-phosphoric(V) acid.* It gives rise to three series of salts containing *phosphate(V) ions* based on the anions $[(HO)_2PO_2]^-$, $[(HO)PO_3]^{2-}$, and PO_4^{3-}. These salts are acidic, neutral, and alkaline in character respectively and phosphate ions often feature in buffer systems. There is also a wide range of higher acids and acid anions in which there is some P–O–P chain formation. The simplest of these is *pyrophosphoric acid* (technically *heptaoxodiphosphoric(V) acid),* $H_4P_2O_7$, produced by heating phosphoric(V) acid (solid) and phosphorus(III) chloride oxide. *Metaphosphoric acid* is a glassy polymeric solid $(HPO_2)_x$.

phosphorous acid *See* **phosphonic acid.**

phosphorus Symbol P. A nonmetallic element belonging to *group 15 (formerly VB) of the periodic table; a.n. 15; r.a.m. 30.9738; r.d. 1.82 (white), 2.34 (red); m.p. 44.1°C (α-white); b.p. 280°C (α-white). It occurs in various phosphate rocks, from which it is extracted by heating with carbon (coke) and silicon(IV) oxide in an electric furnace (1500°C). Calcium silicate and carbon monoxide are also produced. Phosphorus has a number of allotropic forms. The α-white form consists of P_4 tetrahedra (there is also a β-white form stable below −77°C). If α-white phosphorus is dissolved in lead and heated at 500°C a violet form is obtained. Red phosphorus, which is a combination of violet and white phosphorus, is obtained by heating α-white phosphorus at 250°C with air excluded. There is also a black allotrope, which has a graphite-like structure, made by heating white phosphorus at 300°C with a mercury catalyst. The element is highly reactive. It forms metal *phosphides and covalently bonded phosphorus(III) and phosphorus(V) compounds. Phosphorus is an *essential element for living organisms. It is an important constituent of tissues (especially bones and teeth) and of cells, being required for the formation of nucleic acids and energy-carrying molecules (e.g. ATP) and also involved in various metabolic reactions. The element was discovered by Hennig Brand (*c.* 1630–92) in 1669.

phosphorus(III) bromide **(phosphorus tribromide)** A colourless fuming liquid, PBr_3; r.d. 2.85; m.p. −40°C; b.p. 173°C. It is prepared by passing bromine vapour over phosphorus but avoiding an excess, which would lead to the phosphorus(V) bromide. Like the other phosphorus(III) halides, PBr_3 is pyramidal in the gas phase. In the liquid phase the P–Br bonds are labile; for example, PBr_3 will react with PCl_3 to give a mixture of products in which the halogen atoms have been redistributed. Phosphorus(III) bromide is rapidly hydrolysed by water to give phosphonic acid and hydrogen bromide. It reacts readily with many organic hydroxyl groups and is used as a reagent for introducing bromine atoms into organic molecules.

phosphorus(V) bromide **(phosphorus pentabromide)** A yellow readily sublimable solid, PBr_5, which decomposes below 100°C and is soluble in benzene and carbon tetrachloride (tetrachloromethane). It may be prepared by the reaction of phosphorus(III) bromide with bromine or the direct reaction of phosphorus with excess bromine. It is very readily hydrolysed to give hydrogen bromide and phosphoric(V) acid. An interesting feature of this material is that in the solid state it has the structure $[PBr_4]^+Br^-$. It is used in organic chemistry as a brominating agent.

phosphorus(III) chloride **(phosphorus trichloride)** A colourless fuming liquid, PCl_3; r.d. 1.57; m.p. −112°C; b.p. 75.5°C. It is soluble in ether and in carbon tetrachloride but reacts with water and with ethanol. It may be prepared by passing chlorine over excess phosphorus (excess chlorine contaminates the product with phosphorus(V) chloride). The molecule is pyramidal in the gas phase and possesses weak electron-pair donor properties. It is hydrolysed violently by water to phosphonic acid and hydrogen chloride. Phosphorus(III) chloride is an important starting point for the synthesis of a variety of inorganic and organic derivatives of phosphorus.

phosphorus(V) chloride **(phosphorus pentachloride)** A yellow-white rhombic solid, PCl_5, which fumes in air; r.d. 4.65; m.p. 166.8°C (under pressure); sublimes at 160–162°C. It is decomposed by water to give hydrogen chloride and phosphoric(V) acid. It is soluble in organic solvents. The compound may be prepared by the reaction of chlorine with phosphorus(III) chloride. Phosphorus(V) chloride is structurally interesting in that in the gas phase it has the expected trigonal bipyramidal form but in the solid phase it consists of the ions $[PCl_4]^+[PCl_6]^-$. The same ions are detected when phosphorus(V) chloride is dissolved in polar solvents. It is used in organic chemisty as a chlorinating agent.

phosphorus(III) chloride oxide **(phosphorus oxychloride; phosphoryl chloride)** A colourless fuming liquid, $POCl_3$; r.d. 1.67; m.p. 2°C; b.p. 105.3°C. It may be prepared by the reaction of phosphorus(III) chloride with oxygen or by the reaction of phosphorus(V) oxide with phosphorus(V) chloride. Its reactions are very similar to those of phosphorus(III) chloride. Hydrolysis with water gives phosphoric(V) acid. Phosphorus(III) chloride oxide has a distorted tetrahedral shape and can act as a donor towards metal ions, thus giving rise to a series of complexes.

phosphorus cycle The cycling of phosphorus between the biotic and abiotic components of the environment. Inorganic phosphates (PO_4^{3-}, HPO_4^{2-}, or $H_2PO_4^-$) are absorbed by plants from the soil and bodies of water and eventually pass into animals through food chains. Within living organisms phosphates are built up into nucleic acids and other organic molecules. When plants and animals die, phosphates are released and returned to the abiotic environment through the action of bacteria. On a geological time scale, phosphates in aquatic environments eventually become incorporated into and form part of rocks; through a gradual process of erosion, these phosphates are returned to the soil, seas, rivers, and lakes. Phosphorus-containing rocks are mined for the manufacture of fertilizers, which provide an additional supply of inorganic phosphate to the abiotic environment.

phosphorus(III) oxide **(phosphorus trioxide)** A white or colourless waxy solid, P_4O_6; r.d. 2.13; m.p. 23.8°C; b.p. 173.8°C. It is soluble in ether, chloroform, and benzene but reacts with cold water to give phosphonic acid, H_3PO_3, and with hot water to give phosphine and phosphoric(V) acid. The compound is formed when phosphorus is burned in an oxygen-deficient atmosphere (about 50% yield). As it is difficult to separate from white phosphorus by distillation, the mixture is irradiated with ultraviolet radiation to convert excess white phosphorus into the red form, after which the oxide can be separated by dissolution in organic solvents. Although called a trioxide for historical reasons, phosphorus(III) oxide consists of P_4O_6 molecules of tetrahedral symmetry in which each phosphorus atom is linked to the three others by an oxygen bridge. The chemistry is very complex. Above 210°C it decomposes into red phosphorus and polymeric oxides. It reacts with chlorine and bromine to give oxohalides and with alkalis to give phosphonates (*see* **phosphonic acid**).

phosphorus(V) oxide **(phosphorus pentoxide; phosphoric anhydride)** A white powdery and extremely deliquescent solid, P_4O_{10}; r.d. 2.39; m.p. 580°C (under pressure); sublimes at 300°C. It reacts violently with water to give phosphoric(V) acid. It is prepared by burning elemental phosphorus in a plentiful supply of oxygen, then purified by sublimation. The hexagonal crystalline form consists of P_4O_{10} molecular units; these have the phosphorus atoms arranged tetrahedrally, each P atom linked to three others by oxygen bridges and having in addition one terminal oxygen atom. The compound is used as a drying agent and as a dehydrating agent; for example, amides are converted into nitrites and sulphuric acid is converted to sulphur trioxide.

phosphorus oxychloride *See* **phosphorus(III) chloride oxide.**

phosphorus pentabromide *See* **phosphorus(V) bromide.**

phosphorus pentachloride *See* **phosphorus(V) chloride.**

phosphorus tribromide *See* **phosphorus(III) bromide.**

phosphorus trichloride *See* **phosphorus(III) chloride.**

phosphorus trioxide *See* **phosphorus(III) oxide.**

phosphoryl chloride *See* **phosphorus(III) chloride oxide.**

photoacoustic spectroscopy A spectroscopic technique in which spectra of opaque materials, such as powders, are obtained by exposing the material to light that is modulated at acoustic frequencies. This produces a photoacoustic signal, the magnitude of which can be related to the ultraviolet or infrared absorption coefficient of the material. *See also* **optoacoustic spectroscopy.**

photochemical reaction A chemical reaction caused by light or ultraviolet radiation. The incident photons are absorbed by reactant molecules to give excited molecules or free radicals, which undergo further reaction.

photochemical smog A noxious smog produced by the reaction of nitrogen oxides with hydrocarbons in the presence of ultraviolet light from the sun. The reaction is very complex and one of the products is ozone.

photochemistry The branch of chemistry concerned with *photochemical reactions.

photochromism A change of colour occurring in certain substances when exposed to light. Photochromic materials are used in sunglasses that darken in bright sunlight.

photoconductive effect *See* **photoelectric effect.**

photoelectric effect The liberation of electrons from a substance exposed to electromagnetic radiation. The number of such electrons (*photoelectrons*) emitted depends on the intensity of the radiation. The kinetic energy of the electrons emitted depends on the frequency of the radiation. The effect is a quantum process in which the radiation is regarded as a stream of *photons, each having an energy hf, where h is the Planck constant and f is the frequency of the radiation. A photon can only eject an electron if the photon energy exceeds the *work function, ϕ, of the solid, i.e. if $hf_0 = \phi$ an electron will be ejected; f_0 is the minimum frequency (or *threshold frequency*) at which ejection will occur. For many solids the photoelectric effect occurs at ultraviolet frequencies or above, but for some materials (having low work functions) it occurs with light. The maximum kinetic energy, E_m, of the photoelectron is given by *Einstein's equation: $E_m = hf - \phi$.

Apart from the liberation of electrons from atoms, other phenomena are also referred to as photoelectric effects. These are the *photoconductive effect* and the *photovoltaic effect*. In the photoconductive effect, an increase in the electrical conductivity of a semiconductor is caused by radiation as a result of the excitation of additional free charge carriers by the incident photons. *Photoconductive cells*, using such photosensitive materials as cadmium sulphide, are widely used as radiation detectors and light switches (e.g. to switch on street lighting).

In the photovoltaic effect, an e.m.f. is produced between two layers of different materials as a result of irradiation. The effect is made use of in *photovoltaic cells*, most of which consist of *p–n* semiconductor junctions. When photons are absorbed near a *p–n* junction new free charge carriers are produced (as in photoconductivity); however, in the photovoltaic effect the electric field in the junction region causes the new charge carriers to move, creating a flow of current in an external circuit without the need for a battery.

photoelectron An electron emitted from a substance by irradiation as a result of the *photoelectric effect or *photoionization.

photoelectron spectromicroscopy (PESM) A technique for investigating the composition of surfaces, based on ionization of atoms at the surfaces. Ionizing radiation (such as X-rays or ultraviolet radiation) is used to eject electrons from atoms in the surface. The ejected electrons are then focused, thus enabling an image of the surface to be constructed.

photoelectron spectroscopy (PES) A technique for determining the *ionization potentials of molecules. In ultraviolet photoelectron spectroscopy (*UPS*) the sample is a gas or vapour irradiated with a narrow beam of ultraviolet radiation (usually from a helium source at 58.4 nm, 21.21 eV photon energy). The photoelectrons produced in accordance with *Einstein's equation are passed through a slit into a vacuum region, where they are deflected by magnetic or electrostatic fields to give an energy spectrum. The photoelectron spectrum obtained has peaks corresponding to the ionization potentials of the molecule (and hence the orbital energies). The technique also gives information on the vibrational energy levels of the ions formed. X-ray photoelectron spectroscopy (*XPS*), also known as *ESCA* (electron spectroscopy for chemical analysis), is a similar analytical technique in which a beam of X-rays is used. In this case, the electrons ejected are from the inner shells of the atoms. Peaks in the electron spectrum for a particular element show characteristic chemical shifts, which depend on the presence of other atoms in the molecule. *See also* **Koopmans' theorem**.

photoemission The process in which electrons are emitted by a substance as a result of irradiation. *See* **photoelectric effect**; **photoionization**.

photography The process of forming a permanent record of an image on specially treated film or paper. In normal black-and-white photography a camera is used to expose a film or plate to a focused image of the scene for a specified time. The film or plate is coated with an emulsion containing silver salts and the exposure to light causes the silver salts to break down into silver atoms; where the light is bright dark areas of silver are formed on the film after development (by a mild reducing agent) and fixing. The negative so formed is printed, either by a contact process or by projection. In either case light passing through the negative film falls on a sheet of paper also coated with emulsion. Where the negative is dark, less light passes through and the resulting positive is light in this area, corresponding with a light area in the original scene. As photographic emulsions are sensitive to ultraviolet and X-rays, they are widely used in studies involving these forms of electromagnetic radiation. *See also* **colour photography**.

photoionization The *ionization of an atom or molecule as a result of irradiation by electromagnetic radiation. For a photoionization to occur the incident photon of the radiation must have an energy in excess of the *ionization potential of the species being irradiated. The ejected photoelectron will have an energy, E, given by $E = hf - I$, where h is the Planck constant, f is the frequency of the incident radiation, and I is the ionization potential of the irradiated species.

photoluminescence *See* **luminescence**.

photolysis A chemical reaction produced by exposure to light or ultraviolet radiation. Photolytic reactions often involve free radicals, the first step being homolytic fission of a chemical bond. (*See* **flash photolysis**.) The photolysis of water, using energy from sunlight absorbed by chlorophyll, produces gaseous oxygen, electrons, and hydrogen ions and is a key reaction in *photosynthesis.

photon A particle with zero rest mass consisting of a *quantum of electromagnetic radiation. The photon may also be regarded as a unit of energy equal to hf, where h is the *Planck constant and f is the frequency of the radiation in hertz. Photons travel at the speed of light. They are required to explain the photoelectric effect and other phenomena that require light to have particle character.

photosensitive substance 1. Any substance that when exposed to electromagnetic radiation produces a photoconductive, photoelectric, or photovoltaic effect. **2.** Any substance, such as the emulsion of a photographic film, in which electromagnetic radiation produces a chemical change.

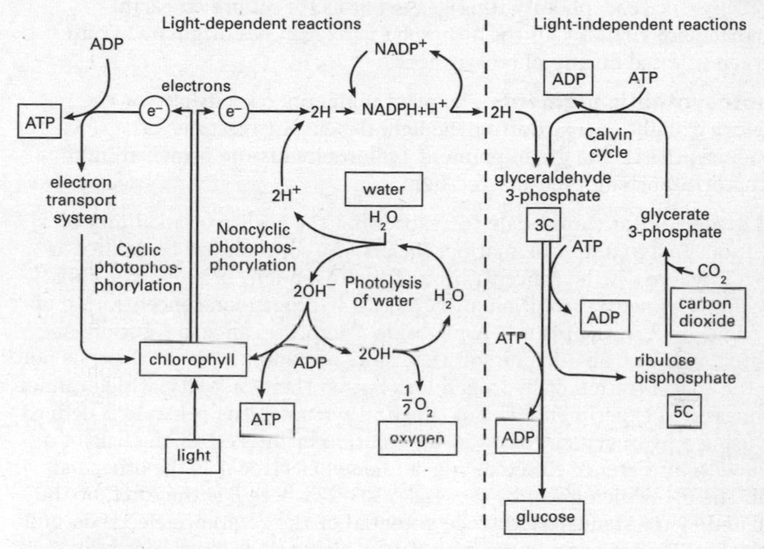

Photosynthesis

photosynthesis The chemical process by which green plants synthesize organic compounds from carbon dioxide and water in the presence of sunlight. It occurs in the chloroplasts (most of which are in the leaves) and there are two principal types of reactions. In the *light-dependent reactions*, which require the presence of light, energy from sunlight is absorbed by *photosynthetic pigments (chiefly the green pigment *chlorophyll) and used to bring about the photolysis of water:

$$H_2O \rightarrow 2H^+ + 2e^- + \tfrac{1}{2}O_2.$$

The electrons released by this reaction pass along a series of electron carriers (*see* **electron transport chain**); as they do so they lose their energy, which is used to convert ADP to ATP in the process of photophosphorylation. The electrons and protons produced by the photolysis of water are used to reduce NADP:

$$2H^+ + 2e^- + NADP^+ \rightarrow NADPH + H^+.$$

The ATP and NADPH produced during the light-dependent reactions provide energy and reducing power, respectively, for the ensuing *light-independent reactions* (formerly called the 'dark reaction'), which nevertheless cannot be sustained without the ATP generated by the light-dependent reactions. During these reactions carbon dioxide is reduced to carbohydrate in a metabolic pathway known as the *Calvin cycle.

Photosynthesis can be summarized by the equation:

$CO_2 + 2H_2O \rightarrow [CH_2O] + H_2O + O_2$.

Since virtually all other forms of life are directly or indirectly dependent on plants for food, photosynthesis is the basis for all life on earth. Furthermore virtually all the atmospheric oxygen has originated from oxygen released during photosynthesis.

photosynthetic pigments The plant pigments responsible for the capture of light energy during the light-dependent reactions of *photosynthesis. The green pigment *chlorophyll is the principal light receptor, absorbing blue and red light.

pH scale A logarithmic scale for expressing the acidity or alkalinity of a solution. To a first approximation, the pH of a solution can be defined as $-\log_{10}c$, where c is the concentration of hydrogen ions in moles per cubic decimetre. A neutral solution at 25°C has a hydrogen-ion concentration of 10^{-7} mol dm^{-3}, so the pH is 7. A pH below 7 indicates an acid solution; one above 7 indicates an alkaline solution. More accurately, the pH depends not on the concentration of hydrogen ions but on their *activity, which cannot be measured experimentally. For practical purposes, the pH scale is defined by using a hydrogen electrode in the solution of interest as one half of a cell, with a reference electrode (e.g. a calomel electrode) as the other half cell. The pH is then given by $(E - E_R)F/2.303RT$, where E is the e.m.f. of the cell and E_R the standard electrode potential of the reference electrode, and F the Faraday constant. In practice, a glass electrode is more convenient than a hydrogen electrode.

pH stands for 'potential of hydrogen'. The scale was introduced by Søren Sørensen (1868–1939) in 1909.

phthalic acid A colourless crystalline dicarboxylic acid, $C_6H_4(COOH)_2$; r.d. 1.6; m.p. 207°C. The two –COOH groups are substituted on adjacent carbon atoms of the ring, the technical name being *benzene-1,2-dicarboxylic acid*. The acid is made from *phthalic anhydride* (benzene-1,2-dicarboxylic anhydride, $C_8H_4O_3$), which is made by the catalytic oxidation of naphthalene. The anhydride is used in making plasticizers and polyester resins.

phthalic anhydride *See* **phthalic acid.**

phthalic ester An ester of phthalic acid. Phthalic esters are made by reacting phthalic anhydride with an alcohol in the presence of sulphuric acid. They are high boiling-point liquids used as plasticizers and in the manufacture of polymers and alkyd resins. The dimethyl ester is used as an insect repellent.

physical chemistry The branch of chemistry concerned with the effect of chemical structure on physical properties. It includes chemical thermodynamics and electrochemistry.

physisorption *See* **adsorption.**

pi-adduct An adduct that is formed by donation of an electron pair between a pi orbital and a sigma or pi orbital.

piano stool *See* **sandwich compound**.

pi bond *See* **orbital**.

pico- Symbol p. A prefix used in the metric system to denote 10^{-12}. For example, 10^{-12} farad = 1 picofarad (pF).

picrate A salt or ester of picric acid.

picric acid **(2,4,6-trinitrophenol)** A yellow highly explosive nitro compound, $C_6H_2(NO_2)_3$; r.d. 1.8; m.p. 122°C.

pi electron An electron in a pi orbital. *See* **orbital**.

pig iron The impure form of iron produced by a blast furnace, which is cast into pigs (blocks) for converting at a later date into cast iron, steel, etc. The composition depends on the ores used, the smelting procedure, and the use to which the pigs will later be put.

pi orbital *See* **orbital**.

pipette A graduated tube used for transferring measured volumes of liquid or, sometimes, gases.

Pirani gauge An instrument used to measure low pressures (1–10^{-4} torr; 100 0.01 Pa). It consists of an electrically heated filament, which is exposed to the gas whose pressure is to be measured. The extent to which heat is conducted away from the filament depends on the gas pressure, which thus controls its equilibrium temperature. Since the resistance of the filament is dependent on its temperature, the pressure is related to the resistance of the filament. The filament is arranged to be part of a Wheatstone bridge circuit and the pressure is read from a microammeter calibrated in pressure units. As the effect depends on the thermal conductivity of the gas, the calibration has to be made each time the pressure of a different gas is measured.

pirsonnite A mineral consisting of a hydrated mixed carbonate of sodium and calcium, $Na_2CO_3.CaCO_3.2H_2O$.

pitch A black or dark-brown residue resulting from the distillation of coal tar, wood tar, or petroleum (bitumen). The term is also sometimes used for the naturally occurring petroleum residue (asphalt). Pitch is used as a binding agent (e.g. in road tars), for waterproofing (e.g. in roofing felts), and as a fuel.

pitchblende *See* **uraninite**.

p*K* value A measure of the strength of an acid on a logarithmic scale. The pK value is given by $\log_{10}(1/K_a)$, where K_a is the acid dissociation constant. pK values are often used to compare the strengths of acids.

Planck, Max Karl Ernst Ludwig (1858–1947) German physicist, who became a professor at Berlin University in 1892. Here he formulated the *quantum theory, which had its basis in a paper of 1900. One of the most important scientific discoveries of the century, this work earned him the 1918 Nobel Prize for physics.

Planck constant Symbol *h*. The fundamental constant equal to the ratio of the energy of a quantum of energy to its frequency. It has the value $6.626\,0755(40) \times 10^{-34}$ J s. It is named after Max Planck. In quantum-mechanical calculations the *rationalized Planck constant* (or *Dirac constant*) $\hbar = h/2\pi = 1.054\,589 \times 10^{-34}$ J s is frequently used.

plane-polarized light *See* **polarization of light**.

plaster of Paris The hemihydrate of *calcium sulphate, $2CaSO_4.H_2O$, prepared by heating the mineral gypsum. When ground to a fine powder and mixed with water, plaster of Paris sets hard, forming interlocking crystals of gypsum. The setting results in an increase in volume and so the plaster fits tightly into a mould. It is used in pottery making, as a cast for setting broken bones, and as a constituent of the plaster used in the building industry.

plasticizer A substance added to a synthetic resin to make it flexible. *See* plastics.

plastics Materials that can be shaped by applying heat or pressure. Most plastics are made from polymeric synthetic *resins, although a few are based on natural substances (e.g. cellulose derivatives or shellac). They fall into two main classes. *Thermoplastic materials* can be repeatedly softened by heating and hardened again on cooling. *Thermosetting materials* are initially soft, but change irreversibly to a hard rigid form on heating. Plastics contain the synthetic resin mixed with such additives as pigments, plasticizers (to improve flexibility), antioxidants and other stabilizers, and fillers. See Chronology.

plastocyanin A blue copper-containing protein that is found in chloroplasts and acts as an electron carrier molecule in the light-dependent reactions of *photosynthesis. Plastocyanin consists of amino acid groups in association with a copper molecule which gives this compound a blue colour.

plastoquinone A quinone, found in chloroplasts, that functions as one of the carrier molecules of the electron transport chain in the light-dependent reactions of *photosynthesis.

platinum Symbol Pt. A silvery white metallic *transition element (*see also* **platinum metals**); a.n. 78; r.a.m. 195.09; r.d. 21.45; m.p. 1772°C; b.p. 3827±100°C. It occurs in some nickel and copper ores and is also found native in some deposits. The main source is the anode sludge obtained in copper–nickel refining. The element is used in jewellery, laboratory apparatus (e.g. thermocouples, electrodes, etc.), electrical contacts, and in certain alloys (e.g. with iridium or rhodium). It is also a hydrogenation

PLASTICS

1851	Scottish chemist Charles Macintosh (1766–1843) makes ebonite (from rubber).
1855	British chemist Alexander Parkes (1813–90) patents Parkesine, a plastic made from nitrocellulose, methanol, and wood pulp; it is later called 'celluloid'.
1860	British chemist Charles Williams (1829–1910) prepares isoprene (synthetic rubber).
1868	US printer John Hyatt (1837–1920) develops commercial process for making celluloid.
1884	French chemist Hilaire de Chardonnet (1839–1924) develops process for making rayon.
1892	British chemists Edward Bevan (1856–1921) and Charles Cross (1855–1935) develop the viscose process for making rayon.
1899	British chemist Frederick Kipping (1863–1949) discovers silicone plastics.
1901	German chemists Krische and Spitteler make formaldehyde–casein plastic (Galalith).
1905	Belgian-born US chemist Leo Baekland (1863–1944) invents Bakelite.
1912	Swiss chemist Jacques Brandenberger produces Cellophane (viscose cellulose film).
1913	US Formica Insulation company markets plastic laminate made from formaldehyde resins.
1918	Hans John prepares urea–formaldehyde resin.
1926	German chemist Hermann Staudinger (1881–1965) discovers the polymeric nature of plastics.
1930	US chemist Waldo Semon develops PVC (polyvinyl chloride).
1930	Canadian chemist William Chalmers discovers polymethylmethacrylate (Perspex and Plexiglass)
1930	German chemists at IG Farbenindustrie produce polystyrene.
1931	Wallace Carothers invents nylon.
1938	US chemist Roy Plunkett produces polytetrafluoroethene (PTFE).
1939	British company ICI develops commercial process for making polyethene.
1941	British chemists John Whinfield (1901–66) and J. Dickson develop Terylene (Dacron).
1941	German company IG Farbenindustrie produces polyurethane.
1943	US Dow Corning company produces silicone plastics.
1947	British chemists produce acrylic fibres.
1953	German chemist Karl Ziegler (1896–1973) discovers catalyst for making high-density polyethene.
1954	Italian chemist Giulio Natta (1903–79) develops industrial process for making high-density polyethene (using Ziegler catalyst).
1989	Italian company Ferruzzi produces biodegradable plastic (based on starch).

catalyst. The element does not oxidize nor dissolve in hydrochloric acid. Most of its compounds are platinum(II) or platinum(IV) complexes.

platinum black Black finely divided platinum metal produced by vacuum evaporation and used as an absorbent and a catalyst.

platinum metals The three members of the second and third transition series immediately proceeding silver and gold: ruthenium (Ru), rhodium (Rh), and palladium (Pd); and osmium (Os), iridium (Ir), and platinum (Pt). These elements, together with iron, cobalt, and nickel, were formerly classed as group VIII of the periodic table. The platinum-group metals are relatively hard and resistant to corrosion and are used in jewellery and in some industrial applications (e.g. electrical contacts). They have certain chemical similarities that justify classifying them together. All are resistant to chemical attack. In solution they form a vast range of complex ions. They also form coordination compounds with carbon monoxide and other pi-bonding ligands. A number of complexes can be made in which a hydrogen atom is linked directly to the metal. The metals and their organic compounds have considerable catalytic activity. *See also* **transition elements.**

pleochroic Denoting a crystal that appears to be of different colours, depending on the direction from which it is viewed. It is caused by polarization of light as it passes through an anisotropic medium.

plumbago *See* **carbon.**

plumbane (lead(IV) hydride) An extremely unstable gas, PbH_4, said to be formed by the action of acids on magnesium–lead alloys. It was first reported in 1924, although doubts have since been expressed about the existence of the compound. It demonstrates the declining stability of the hydrides in group 14. More stable organic derivatives are known; e.g. trimethyl plumbane, $(CH_3)_3PbH$.

plumbate A compound formed by reaction of lead oxides (or hydroxides) with alkali. The oxides of lead are amphoteric (weakly acidic) and react to give plumbate ions. With the lead(IV) oxide, reaction with molten alkali gives the plumbate(IV) ion

$$PbO_2 + 2OH^- \rightarrow PbO_3^{2-} + H_2O$$

In fact, various ions are present in which the lead is bound to hydroxide groups, the principal one being the hexahydroxoplumbate(IV) ion $Pb(OH)_6^{2-}$. This is the negative ion present in crystalline 'trihydrates' of the type $K_2PbO_3.3H_2O$. Lead(II) oxide gives the trihydroxoplumbate(II) ion in alkaline solutions

$$PbO(s) + OH^-(aq) + H_2O(l) \rightarrow Pb(OH)_3^{2-}(aq)$$

Plumbate(IV) compounds were formerly referred to as *orthoplumbates* (PbO_4^{4-}) or *metaplumbates* (PbO_3^{2-}). Plumbate(II) compounds were called *plumbites.*

plumbic compounds Compounds of lead in its higher (+4) oxidation state; e.g. plumbic oxide is lead(IV) oxide, PbO_2.

plumbite *See* **plumbate.**

plumbous compounds Compounds of lead in its lower (+2) oxidation state; e.g. plumbous oxide is lead(II) oxide, PbO.

plutonium Symbol Pu. A dense silvery radioactive metallic transuranic element belonging to the *actinoids; a.n. 94; mass number of most stable isotope 244 (half-life 7.6×10^7 years); r.d. 19.84; m.p. 641°C; b.p. 3232°C. Thirteen isotopes are known, by far the most important being plutonium–239 (half-life 2.44×10^4 years), which undergoes nuclear fission with slow neutrons and is therefore a vital power source for nuclear weapons and some nuclear reactors. About 20 tonnes of plutonium are produced annually by the world's nuclear reactors. The element was first produced by Seaborg, McMillan, Kennedy, and Wahl in 1940.

PM *See* **particulate matter.**

Pockels cell An electro-optical device used to produce population inversion in lasers and a pulse of radiation. In a Pockels cell, crystals of ammonium dihydrogenphosphate are used to convert plane-polarized light to circularly polarized light when a potential difference is applied. A Pockels cell can be made part of a laser cavity. When this happens, a change of polarization occurs when light is reflected from a mirror and the light that is polarized in one plane is converted into light polarized in the perpendicular plane. The effect is that the reflected light does not stimulate emission. When the Pockels cell is turned off the polarization does not occur and the energy in the cavity can be released in the form of a pulse of stimulated radiation. *See also* **laser.**

point group A group of symmetry elements that leave a point unchanged, used in classifying molecules. *Compare* **space group.**

poise A *c.g.s. unit of viscosity equal to the tangential force in dynes per square centimetre required to maintain a difference in velocity of one centimetre per second between two parallel planes of a fluid separated by one centimetre. 1 poise is equal to 10^{-1} N s m^{-2}.

poison 1. Any substance that is injurious to the health of a living organism. **2.** A substance that prevents the activity of a catalyst. **3.** A substance that absorbs neutrons in a nuclear reactor and therefore slows down the reaction. It may be added intentionally for this purpose or may be formed as a fission product and need to be periodically removed.

polar compound A compound that is either ionic (e.g. sodium chloride) or that has molecules with a large permanent dipole moment (e.g. water).

polarimeter (polariscope) An instrument used to determine the angle through which the plane of polarization of plane-polarized light is rotated on passing through an optically active substance. Essentially, a polarimeter consists of a light source, a *polarizer* (e.g. a sheet of Polaroid) for producing plane-polarized light, a transparent cell containing the sample, and an *analyser*. The analyser is a polarizing material that can be rotated. Light from the source is plane-polarized by the polarizer and passes through the sample, then through the analyser into the eye or onto a light-detector. The angle of polarization is determined by rotating the analyser until the maximum transmission of light occurs. The angle of rotation is read off a scale. Simple portable polarimeters are used for estimating the concentrations of sugar solutions in confectionary manufacture.

polariscope (polarimeter) A device used to study optically active substances (*see* **optical activity**). The simplest type of instrument consists of a light source, collimator, polarizer, and analyser. The specimen is placed between polarizer and analyser, so that any rotation of the plane of polarization of the light can be assessed by turning the analyser.

polarizability Symbol α. A measure of the response of a molecule to an external electric field. When a molecule is placed in an external electric field, the displacement of electric charge induces a dipole in the molecule. If the electric field strength is denoted $\boldsymbol{E}$ and the electrical dipole moment induced by this electric field $\boldsymbol{p}$, the polarizability α is defined by $\boldsymbol{p} = \alpha\boldsymbol{E}$. To calculate the polarizability from first principles it is necessary to use the quantum mechanics of molecules. However, if regarded as a parameter, the polarizability α provides a link between microscopic and macroscopic theories as in the *Clausius–Mossotti equation and the *Lorentz–Lorenz equation.

polarization 1. *See* **polarization of light**. **2.** The formation of products of the chemical reaction in a *voltaic cell in the vicinity of the electrodes resulting in increased resistance to current flow and, frequently, to a reduction in the e.m.f. of the cell. *See also* **depolarization**. **3.** The partial separation of electric charges in an insulator subjected to an electric field. **4.** The separation of charge in a polar *chemical bond.

polarization of light The process of confining the vibrations of the electric vector of light waves to one direction. In unpolarized light the electric field vibrates in all directions perpendicular to the direction of propagation. After reflection or transmission through certain substances (*see* **Polaroid**) the electric field is confined to one direction and the radiation is said to be *plane-polarized light*. The plane of plane-polarized light can be rotated when it passes through certain substances (*see* **optical activity**).

In *circularly polarized light*, the tip of the electric vector describes a circular helix about the direction of propagation with a frequency equal to the frequency of the light. The magnitude of the vector remains constant. In *elliptically polarized light*, the vector also rotates about the direction of propagation but the amplitude changes; a projection of the vector on a plane at right angles to the direction of propagation describes an ellipse. Circularly and elliptically polarized light are produced using a retardation plate.

polarizer A device used to plane-polarize light (*see* **polarization of light**). *Nicol prisms or *Polaroid can be used as polarizers. If a polarizer is placed in front of a source of unpolarized light, the transmitted light is plane-polarized in a specific direction. As the human eye is unable to detect that light is polarized, it is necessary to use an *analyser to detect the direction of polarization. *Crossing* a polarizer and analyser causes extinction of the light, i.e. if the plane of polarization of the polarizer and the plane of the analyser are perpendicular, no light is transmitted when the polarizer and analyser are combined. Both a polarizer and an analyser are components of a *polarimeter.

polar molecule A molecule that has a dipole moment; i.e. one in which there is some separation of charge in the *chemical bonds, so that one part of the molecule has a positive charge and the other a negative charge.

polarography An analytical technique having an electrochemical basis. A *dropping-mercury electrode* is used as the cathode along with a large nonpolarizable anode, and a dilute solution of the sample. The dropping-mercury electrode consists of a narrow tube through which mercury is slowly passed into the solution so as to form small drops at the end of the tube, which fall away. In this way the cathode can have a low surface area and be kept clean. A variable potential is applied to the cell and a plot of current against potential (a *polarogram*) made. As each chemical species is reduced at the cathode (in order of their electrode potentials) a step-wise increase in current is obtained. The height of each step is proportional to the concentration of the component. The technique is useful for detecting trace amounts of metals and for the investigation of solvated complexes.

Polaroid A doubly refracting material that plane-polarizes unpolarized light passed through it. It consists of a plastic sheet strained in a manner that makes it birefringent by aligning its molecules. Sunglasses incorporating a Polaroid material absorb light that is vibrating horizontally – produced by reflection from horizontal surfaces – and thus reduce glare.

polar solvent *See* **solvent**.

pollutant Any substance, produced and released into the environment as a result of human activities, that has damaging effects on living organisms. Pollutants may be toxic substances (e.g. *pesticides) or natural constituents of the atmosphere (e.g. carbon dioxide) that are present in excessive amounts. *See* **pollution**.

pollution An undesirable change in the physical, chemical, or biological characteristics of the natural environment, brought about by man's activities. It may be harmful to human or nonhuman life. Pollution may affect the soil, rivers, seas, or the atmosphere. There are two main classes of *pollutants: those that are *biodegradable* (e.g. sewage), i.e. can be rendered harmless by natural processes and need therefore cause no permanent harm if adequately dispersed or treated; and those that are *nonbiodegradable* (e.g. *heavy metals (such as lead) in industrial effluents and *DDT and other chlorinated hydrocarbons used as pesticides), which eventually accumulate in the environment and may be concentrated in food chains. Other forms of pollution in the environment include noise (e.g. from jet aircraft, traffic, and industrial processes) and thermal pollution (e.g. the release of excessive waste heat into lakes or rivers causing harm to wildlife). Recent pollution problems include the disposal of radioactive waste; *acid rain; *photochemical smog; increasing levels of human waste; high levels of carbon dioxide and other greenhouse gases in the atmosphere (*see* **greenhouse effect**); damage to the *ozone layer by nitrogen oxides, *chlorofluorocarbons (CFCs), and *halons; and pollution of inland waters by agricultural *fertilizers and sewage effluent, causing eutrophication. Attempts to contain or prevent pollution include strict regulations concerning factory emissions, the use of smokeless fuels, the banning of certain pesticides, the increasing use of lead-free petrol, restrictions on the use of chlorofluorocarbons, and the introduction, in some countries, of catalytic converters to cut pollutants in car exhausts.

polonium Symbol Po. A rare radioactive metallic element of group 16 (formerly VIB) of the periodic table; a.n. 84; r.a.m. 210; r.d. 9.32; m.p. 254°C; b.p. 962°C. The element occurs in uranium ores to an extent of about 100 micrograms per 1000 kilograms. It has over 30 isotopes, more than any other element. The longest-lived isotope is polonium–209 (half-life 103 years). Polonium has attracted attention as a possible heat source for spacecraft as the energy released as it decays is 1.4×10^5 J kg^{-1} s^{-1}. It was discovered by Marie Curie in 1898 in a sample of pitchblende.

poly- Prefix indicating a polymer, e.g. polyethene. Sometimes brackets are used in polymer names to indicate the repeated unit, e.g. poly(ethene).

polyacrylamide A polymer formed from 2-propenamide (CH_2:CHCONH_2). Polyacrylamide gels are made using a cross-linking agent to form three-dimensional matrices. They are used in gel electrophoresis (*see* **PAGE**).

polyacrylamide gel electrophoresis *See* **PAGE**.

polyamide A type of condensation polymer produced by the interaction of an amino group of one molecule and a carboxylic acid group of another molecule to give a protein-like structure. The polyamide chains are linked together by hydrogen bonding.

polyatomic molecule A molecule formed from several atoms (e.g. pyridine, C_5H_5N, or dinitrogen tetroxide, N_2O_4), as distinguished from diatomic and monatomic molecules.

polybasic acid An acid with more than one replaceable hydrogen atom. Examples include the dibasic sulphuric acid (H_2SO_4) and tribasic phosphoric(V) (orthophosphoric) acid (H_3PO_4). Replacement of all the hydrogens by metal atoms forms normal salts. If not all the hydrogens are replaced, *acid salts are formed.

polychlorinated biphenyl (PCB) Any of a number of derivatives of biphenyl ($C_6H_5C_6H_5$) in which some of the hydrogen atoms on the benzene rings have been replaced by chlorine atoms. Polychlorinated biphenyls are used in the manufacture of certain polymers as electrical insulators. They are highly toxic and are suspected to be carcinogenic; their increasing use has caused concern because they have been shown to accumulate in the food chain.

polychloroethene **(PVC; polyvinyl chloride)** A tough white solid material, which softens with the application of a plasticizer, manufactured from chloroethene by heating in an inert solvent using benzoyl peroxide as an initiator, or by the free-radical mechanism initiated by heating chloroethene in water with potassium persulphate or hydrogen peroxide. The polymer is used in a variety of ways, being easy to colour and resistant to fire, chemicals, and weather.

polycyclic Denoting a compound that has two or more rings in its molecules. Polycyclic compounds may contain single rings (as in phenylbenzene, $C_6H_5.C_6H_5$) or fused rings (as in naphthalene, $C_{10}H_8$).

polydioxoboric(III) acid *See* **boric acid.**

polyene An unsaturated hydrocarbon that contains two or more double carbon–carbon bonds in its molecule.

polyester A condensation polymer formed by the interaction of polyhydric alcohols and polybasic acids. Linear polyesters are saturated thermoplastics and linked by dipole–dipole attraction as the carbonyl groups are polarized. They are extensively used as fibres (e.g. *Terylene*). Unsaturated polyesters readily copolymerize to give thermosetting products. They are used in the manufacture of glass-fibre products. *See also* **alkyd resin.**

polyethene **(polyethylene; polythene)** A flexible waxy translucent polyalkene thermoplastic made in a variety of ways producing a polymer of varying characteristics. In the ICI process, ethene containing a trace of oxygen is subjected to a pressure in excess of 1500 atmospheres and a temperature of 200°C. Low-density polyethene (r.d. 0.92) has a formula weight between 50 000 and 300 000, softening at a temperature around 110°C, while the high-density polythene (r.d. 0.945–0.96) has a formula weight up to 3 000 000, softening around 130°C. The low-density polymer is less crystalline, being more atactic. Polyethene is used as an insulator; it is acid resistant and is easily moulded and blown. *See* **Phillips process**; **Ziegler process.**

polyethylene *See* **polyethene.**

polyhydric alcohol **(polyol)** An *alcohol that has several hydroxyl groups per molecule.

polymer A substance having large molecules consisting of repeated units (the monomers). See Feature (pp 452–453).

Polymers are substances that have *macromolecules composed of many repeating molecular structural units (known as 'mers'). A large number of naturally occurring substances are polymers, including *rubber and many substances based on glucose, such as the polysaccharides *cellulose and *starch (in plants) and *glycogen (in animals). *Proteins, nucleic acids, and inorganic macromolecular substances, such as rock-forming *silicates, can all be considered as polymers.

Synthetic polymers

One of the unique features of the chemistry of carbon is its ability to form long chains of atoms. This property is the basis of an important area of industrial chemistry concerned with the manufacture of polymeric organic materials with a variety of properties and uses (see **plastics**). The molecules in these materials are essentially long chains of atoms of various lengths. In some polymers, cross-linkage occurs between the chains.

Synthetic polymers are formed by chemical reactions in which individual molecules (*monomers*) join together to form larger units (*see* **polymerization**). Two types of polymer, *homopolymers* and *heteropolymers*, can be distinguished

Homopolymers

These are polymers formed from a single monomer. An example is *polyethene (polyethylene), which is made by polymerization of ethene ($CH_2{:}CH_2$). Typically such polymers are formed by *addition reactions involving unsaturated molecules. Other similar examples are *polypropene (polypropylene), polystyrene, and *polytetrafluoroethene (PTFE). Homopolymers may also be made by *condensation reactions (as in the case of *polyurethane).

$$H_2C{=}CH_2 \longrightarrow -CH_2-CH_2-CH_2-CH_2 \cdots CH_2-CH_2-$$

Addition polymerization of ethene to form polyethene: a homopolymer

$$H_2N-(CH_2)_6-NH_2 + HOOC-(CH_2)_4-COOH$$

1,6-diaminohexane hexanedioic acid

$$\downarrow$$

$$-NH-(CH_2)_6-NH-CO-(CH_2)_4-CO-NH-(CH_2)_6-NH- \cdots + nH_2O$$

Condensation polymerization to form nylon: a heteropolymer

Heteropolymers

These are also known as *copolymers*. They are made from two (or more) different monomers, which usually undergo a condensation reaction with the elimination of a simple molecule, such as ammonia or water. A typical example is the condensation of 1,6-diaminohexane (hexamethylenediamine) with hexanedioic acid (adipic acid) to form nylon 6,6. Here the reaction occurs between the amine groups on the diaminohexane and the carboxyl groups on the hexanedioic acid, with elimination of water molecules (see diagram opposite).

The properties of a polymeric plastic can most easily be modified if it is a copolymer of two or more different monomers. A well-known example is ABS (acrylonitrile–butadiene–styrene) copolymer, commonly used for the body shells of computers and other electronic apparatus. Its properties can be preselected by varying the proportions of the component monomers.

Stereospecific polymers

In both normal polyethene and nylon the polymer molecules take the form of long chains of various lengths with no regular arrangement of the subunits. Such polymers are said to be *atactic*. If the constituent subunits repeat along the chain in a regular way, a stereospecific polymer may result. Such polymers have crystalline properties in that there is a repeating pattern along the polymer chain. The polymer may be *isotactic*, with a particular group always along the same side of the main chain, or *syndiotactic*, with the group alternating from side to side of the chain. Stereospecific polymerization can be performed by use of certain catalytic agents (*see* **Ziegler process**).

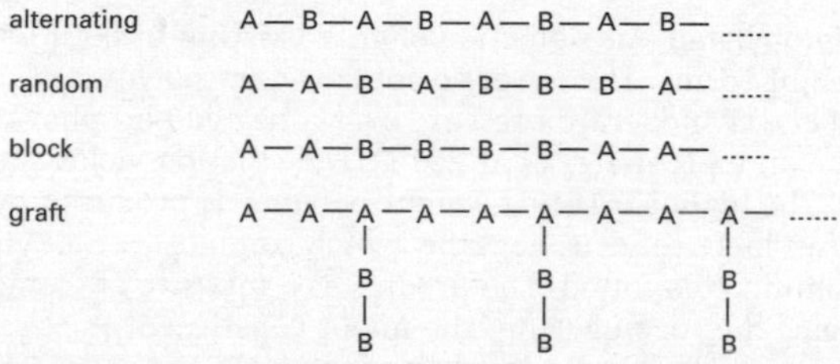

alternating A—B—A—B—A—B—A—B— ......

random A—A—B—A—B—B—B—A— ......

block A—A—B—B—B—B—A—A— ......

graft A—A—A—A—A—A—A—A—A— ...... (with B—B side chains on the third, sixth and ninth A)

Types of copolymer depending on the arrangement of the monomers A and B

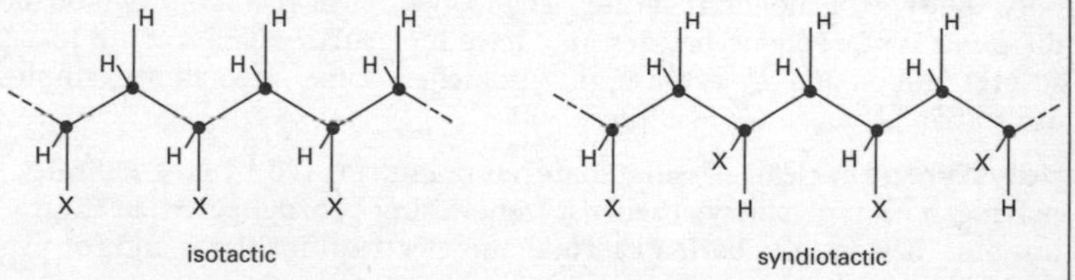

isotactic

syndiotactic

Types of stereospecific polymer

polymerization A chemical reaction in which molecules join together to form a polymer. If the reaction is an addition reaction, the process is *addition polymerization*; condensation reactions cause *condensation polymerization*, in which a small molecule is eliminated during the reaction. Polymers consisting of a single monomer are *homopolymers*; those formed from two different monomers are *copolymers*.

polymethanal A solid polymer of methanal, formed by evaporation of an aqueous solution of methanal.

polymethylmethacrylate A clear thermoplastic acrylic material made by polymerizing methyl methacrylate. The technical name is *poly(methyl 2-methylpropenoate)*. It is used in such materials as *Perspex*.

polymorphism The existence of chemical substances in two (*dimorphism*) or more physical forms. *See* **allotropy**.

polyol *See* **polyhydric alcohol**.

polypeptide A *peptide comprising ten or more amino acids. Polypeptides that constitute proteins usually contain 100–300 amino acids. Shorter ones include certain antibiotics, e.g. gramicidin, and some hormones, e.g. ACTH, which has 39 amino acids. The properties of a polypeptide are determined by the type and sequence of its constituent amino acids.

polypropene (polypropylene) An isotactic polymer existing in both low and high formula-weight forms. The lower-formula-weight polymer is made by passing propene at moderate pressure over a heated phosphoric acid catalyst spread on an inert material at 200°C. The reaction yields the trimer and tetramer. The higher-formula-weight polymer is produced by passing propene into an inert solvent, heptane, which contains a trialkyl aluminium and a titanium compound. The product is a mixture of isotactic and atactic polypropene, the former being the major constituent. Polypropene is used as a thermoplastic moulding material.

polypropylene *See* **polypropene**.

polysaccharide Any of a group of carbohydrates comprising long chains of monosaccharide (simple-sugar) molecules. *Homopolysaccharides* consist of only one type of monosaccharide; *heteropolysaccharides* contain two or more different types. Polysaccharides may have molecular weights of up to several million and are often highly branched. Some important examples are starch, glycogen, and cellulose.

polystyrene A clear glasslike material manufactured by free-radical polymerization of phenylethene (styrene) using benzoyl peroxide as an initiator. It is used as both a thermal and electrical insulator and for packing and decorative purposes.

polysulphides *See* **sulphides**.

polytetrafluoroethene (PTFE) A thermosetting plastic with a high softening point (327°C) prepared by the polymerization of tetrafluoroethene under pressure (45–50 atmospheres). The reaction requires an initiator, ammonium peroxosulphate. The polymer has a low coefficient of friction and its 'anti-stick' properties are probably due to its helical structure with the fluorine atoms on the surface of an inner ring of carbon atoms. It is used for coating cooking utensils and nonlubricated bearings.

polythene *See* **polyethene.**

polythionate A salt of a polythionic acid.

polythionic acids Oxo acids of sulphur with the general formula $HO.SO_2.S_n.SO_2.OH$, where $n = 0–4$. *See also* **sulphuric acid.**

polyurethane A polymer containing the urethane group –NH.CO.O–, prepared by reacting di-isocyanates with appropriate diols or triols. A wide range of polyurethanes can be made, and they are used in adhesives, durable paints and varnishes, plastics, and rubbers. Addition of water to the polyurethane plastics turns them into foams.

polyvinylacetate (PVA) A thermoplastic polymer used in adhesives and coatings. It is made by polymerizing vinyl acetate (CH_2:$CHCOOCH_3$).

polyvinyl alcohol A water-soluble polymer made from polyvinylacetate (PVA) by hydrolysis with sodium hydroxide. It is used for making adhesives that are miscible with water, as a size for textiles, and for making synthetic fibres.

polyvinyl chloride *See* **polychloroethene.**

polyyne An unsaturated hydrocarbon that contains two or more triple carbon–carbon bonds in its molecule.

population inversion *See* **laser.**

porphyrin Any of a group of organic pigments characterized by the possession of a cyclic group of four linked nitrogen-containing rings (a

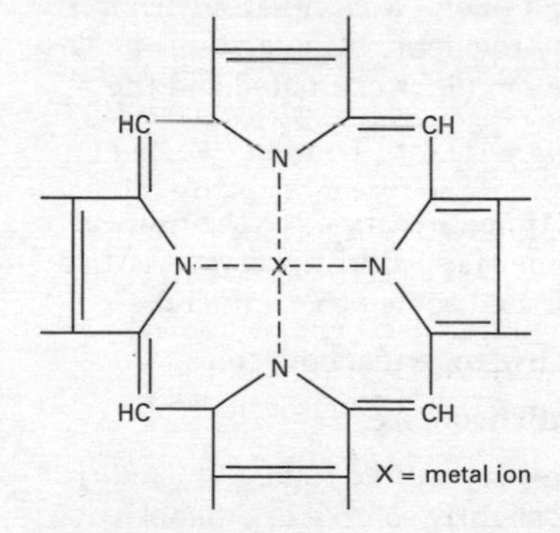

Generalized structure of a porphyrin

tetrapyrrole nucleus), the nitrogen atoms of which are often coordinated to metal ions. Porphyrins differ in the nature of their side-chain groups. They include the *chlorophylls, which contain magnesium; and *haem, which contains iron and forms the *prosthetic group of haemoglobin, myoglobin, and the cytochromes.

positronium *See* **exotic atom.**

potash Any of a number of potassium compounds, such as the carbonate or the hydroxide.

potash alum *See* **aluminium potassium sulphate; alums.**

potash mica *See* **muscovite.**

potassamide *See* **potassium monoxide.**

potassium Symbol K. A soft silvery metallic element belonging to group 1 (formerly IA) of the periodic table (*see* **alkali metals**); a.n. 19; r.a.m. 39.098; r.d. 0.86; m.p. 63.7°C; b.p. 774°C. The element occurs in seawater and in a number of minerals, such as sylvite (KCl), carnallite ($KCl.MgCl_2.6H_2O$), and kainite ($MgSO_4.KCl.3H_2O$). It is obtained by electrolysis. The metal has few uses but potassium salts are used for a wide range of applications. Potassium is an *essential element for living organisms. The potassium ion, K^+, is the most abundant cation in plant tissues, being absorbed through the roots and being used in such processes as protein synthesis. In animals the passage of potassium and sodium ions across the nerve-cell membrane is responsible for the changes of electrical potential that accompany the transmission of impulses. Chemically, it is highly reactive, resembling sodium in its behaviour and compounds. It also forms an orange-coloured superoxide, KO_2, which contains the O_2^- ion. Potassium was discovered by Sir Humphry *Davy in 1807.

potassium–argon dating A *dating technique for certain rocks that depends on the decay of the radioisotope potassium–40 to argon–40, a process with a half-life of about 1.27×10^{10} years. It assumes that all the argon–40 formed in the potassium-bearing mineral accumulates within it and that all the argon present is formed by the decay of potassium–40. The mass of argon–40 and potassium–40 in the sample is estimated and the sample is then dated from the equation:

$$^{40}Ar = 0.1102\ ^{40}K(e^{\lambda t} - 1),$$

where λ is the decay constant and t is the time in years since the mineral cooled to about 300°C, when the ^{40}Ar became trapped in the crystal lattice. The method is effective for micas, feldspar, and some other minerals.

potassium bicarbonate *See* **potassium hydrogencarbonate.**

potassium bichromate *See* **potassium dichromate.**

potassium bromide A white or colourless crystalline solid, KBr, slightly hygroscopic and soluble in water and very slightly soluble in ethanol; cubic; r.d. 2.75; m.p. 734°C; b.p. 1435°C. Potassium bromide may be prepared by the action of bromine on hot potassium hydroxide solution or by the action of iron(III) bromide or hydrogen bromide on potassium carbonate solution. It is used widely in the photographic industry and is also used as a sedative. Because of its range of transparency to infrared radiation, KBr is used both as a matrix for solid samples and as a prism material in infrared spectroscopy.

potassium carbonate **(pearl ash; potash)** A translucent (granular) or white (powder) deliquescent solid known in the anhydrous and hydrated forms. K_2CO_3 (monoclinic; r.d. 2.4; m.p. 891°C) decomposes without boiling. $2K_2CO_3.3H_2O$ (monoclinic; r.d. 2.04) dehydrates to $K_2CO_3.H_2O$ above 100°C and to K_2CO_3 above 130°C. It is prepared by the Engel–Precht process in which potassium chloride and magnesium oxide react with carbon dioxide to give the compound *Engel's salt*, $MgCO_3.KHCO_3.4H_2O$. This is decomposed in solution to give the hydrogencarbonate, which can then be calcined to K_2CO_3. Potassium carbonate is soluble in water (insoluble in alcohol) with significant hydrolysis to produce basic solutions. Industrial uses include glasses and glazes, the manufacture of soft soaps, and in dyeing and wool finishing. It is used in the laboratory as a drying agent.

potassium chlorate A colourless crystalline compound, $KClO_3$, which is soluble in water and moderately soluble in ethanol; monoclinic; r.d. 2.32; m.p. 356°C; decomposes above 400°C giving off oxygen. The industrial route to potassium chlorate involves the fractional crystallization of a solution of potassium chloride and sodium chlorate but it may also be prepared by electrolysis of hot concentrated solutions of potassium chloride. It is a powerful oxidizing agent finding applications in weedkillers and disinfectants and, because of its ability to produce oxygen, it is used in explosives, pyrotechnics, and matches.

potassium chloride A white crystalline solid, KCl, which is soluble in water and very slightly soluble in ethanol; cubic; r.d. 1.98; m.p. 772°C; sublimes at 1500°C. Potassium chloride occurs naturally as the mineral *sylvite* (KCl) and as *carnallite* ($KCl.MgCl_2.6H_2O$); it is produced industrially by fractional crystallization of these deposits or of solutions from lake brines. It has the interesting property of being more soluble than sodium chloride in hot water but less soluble in cold. It is used as a fertilizer, in photography, and as a source of other potassium salts, such as the chlorate and the hydroxide. It has low toxicity.

potassium chromate A bright yellow crystalline solid, K_2CrO_4, soluble in water and insoluble in alcohol; rhombic; r.d. 2.73; m.p. 968.3°C; decomposes without boiling. It is produced industrially by roasting powdered chromite ore with potassium hydroxide and limestone and leaching the resulting cinder with hot potassium sulphate solution. Potassium chromate is used in leather finishing, as a textile mordant, and in enamels and pigments. In the laboratory it is used as an analytical reagent and as an indicator. Like other chromium(III) compounds it is toxic when ingested or inhaled.

potassium chromium sulphate **(chrome alum)** A violet or ruby-red crystalline solid, $K_2SO_4.Cr_2(SO_4)_3.24H_2O$, that is soluble in water and insoluble in ethanol; cubic or octahedral; r.d. 1.826; m.p. 89°C; loses $10H_2O$ at 100°C, $12H_2O$ at 400°C. Six water molecules surround each of the chromium(III) ions and the remaining ones are hydrogen bonded to the sulphate ions. Like all alums, the compound may be prepared by mixing equimolar quantities of the constituent sulphates. *See* **alums**.

potassium cyanide **(cyanide)** A white crystalline or granular deliquescent solid, KCN, soluble in water and in ethanol and having a faint characteristic odour of almonds (due to hydrolysis forming hydrogen cyanide at the surface); cubic; r.d. 1.52; m.p. 634°C. It is prepared industrially by the absorption of hydrogen cyanide in potassium hydroxide. The compound is used in the extraction of silver and gold, in some metal-finishing processes and electroplating, as an insecticide and fumigant (source of HCN), and in the preparation of cyanogen derivatives. In the laboratory it is used in analysis, as a reducing agent, and as a stabilizing *ligand for low oxidation states. The salt itself is highly toxic and aqueous solutions of potassium cyanide are strongly hydrolysed to give rise to the slow release of equally toxic hydrogen cyanide gas.

potassium dichromate **(potassium bichromate)** An orange-red crystalline solid, $K_2Cr_2O_7$, soluble in water and insoluble in alcohol; monoclinic or triclinic; r.d. 2.68; monoclinic changes to triclinic at 241.6°C; m.p. 396°C; decomposes above 500°C. It is prepared by acidification of crude potassium chromate solution (the addition of a base to solutions of potassium dichromate reverses this process). The compound is used industrially as an oxidizing agent in the chemical industry and in dyestuffs manufacture, in electroplating, pyrotechnics, glass manufacture, glues, tanning, photography and lithography, and in ceramic products. Laboratory uses include application as an analytical reagent and as an oxidizng agent. Potassium dichromate is toxic and considered a fire risk on account of its oxidizing properties.

potassium dioxide *See* **potassium superoxide**.

potassium hydride A white or greyish white crystalline solid, KH; r.d. 1.43–1.47. It is prepared by passing hydrogen over heated potassium and marketed as a light grey powder dispersed in oil. The solid decomposes on heating and in contact with moisture and is an excellent reducing agent. Potassium hydride is a fire hazard because it produces hydrogen on reaction with water.

potassium hydrogencarbonate **(potassium bicarbonate)** A white crystalline solid, $KHCO_3$, soluble in water and insoluble in ethanol; r.d. 2.17; decomposes about 120°C. It occurs naturally as *calcinite* and is prepared by passing carbon dioxide into saturated potassium carbonate solution. It is used in baking, soft-drinks manufacture, and in CO_2 fire extinguishers. Because of its buffering capacity, it is added to some detergents and also used as a laboratory reagent.

potassium hydrogentartrate **(cream of tartar)** A white crystalline acid salt, $HOOC(CHOH)_2COOK$. It is obtained from deposits on wine vats (argol) and used in baking powders.

potassium hydroxide **(caustic potash; lye)** A white deliquescent solid, KOH, often sold as pellets, flakes, or sticks, soluble in water and in ethanol and very slightly soluble in ether; rhombic; r.d. 2.044; m.p. 360.4°C; b.p. 1320°C. It is prepared industrially by the electrolysis of concentrated potassium chloride solution but it can also be made by heating potassium carbonate or sulphate with slaked lime, $Ca(OH)_2$. It closely resembles sodium hydroxide but is more soluble and is therefore preferred as an absorber for carbon dioxide and sulphur dioxide. It is also used in the manufacture of soft soap, other potassium salts, and in Ni–Fe and alkaline storage cells. Potassium hydroxide is extremely corrosive to body tissues and especially damaging to the eyes.

potassium iodate A white crystalline solid, KIO_3, soluble in water and insoluble in ethanol; monoclinic; r.d. 3.9; m.p. 560°C. It may be prepared by the reaction of iodine with hot concentrated potassium hydroxide or by careful electrolysis of potassium iodide solution. It is an oxidizing agent and is used as an analytical reagent. Some potassium iodate is used as a food additive.

potassium iodide A white crystalline solid, KI, with a strong bitter taste, soluble in water, ethanol, and acetone; cubic; r.d. 3.13; m.p. 681°C; b.p. 1330°C. It may be prepared by the reaction of iodine with hot potassium hydroxide solution followed by separation from the iodate (which is also formed) by fractional crystallization. In solution it has the interesting property of dissolving iodine to form the triiodide ion I_3^-, which is brown. Potassium iodide is widely used as an analytical reagent, in photography, and also as an additive to table salt to prevent goitre and other disorders due to iodine deficiency.

potassium manganate(VII) **(potassium permanganate)** A compound, $KMnO_4$, forming purple crystals with a metallic sheen, soluble in water (intense purple solution), acetone, and methanol, but decomposed by ethanol; r.d. 2.70; decomposition begins slightly above 100°C and is complete at 240°C. The compound is prepared by fusing manganese(IV) oxide with potassium hydroxide to form the manganate and electrolysing the manganate solution using iron electrodes at about 60°C. An alternative route employs production of sodium manganate by a similar fusion process, oxidation with chlorine and sulphuric acid, then treatment with potassium chloride to crystallize the required product.

Potassium manganate(VII) is widely used as an oxidizing agent and as a disinfectant in a variety of applications, and as an analytical reagent.

potassium monoxide A grey crystalline solid, K_2O; cubic; r.d. 2.32; decomposition occurs at 350°C. It may be prepared by the oxidation of potassium metal with potassium nitrate. It reacts with ethanol to form potassium ethoxide (KOC_2H_5), and with liquid ammonia to form potassium hydroxide and *potassamide* (KNH_2).

potassium nitrate **(saltpetre)** A colourless rhombohedral or trigonal solid, KNO_3, soluble in water, insoluble in alcohol; r.d. 2.109; transition to trigonal form at 129°C; m.p. 334°C; decomposes at 400°C. It occurs naturally as *nitre* and may be prepared by the reaction of sodium nitrate with potassium chloride followed by fractional crystallization. It is a powerful oxidizing agent (releases oxygen on heating) and is used in gunpowder and fertilizers.

potassium nitrite A white or slightly yellow deliquescent solid, KNO_2, soluble in water and insoluble in ethanol; r.d. 1.91; m.p. 440°C; may explode at 600°C. Potassium nitrite is prepared by the reduction of potassium nitrate. It reacts with cold dilute mineral acids to give nitrous acid and is also able to behave as a reducing agent (if oxidized to the nitrate) or as an oxidizing agent (if reduced to nitrogen). It is used in organic synthesis because of its part in diazotization, and in detecting the presence of the amino groups in organic compounds.

potassium permanganate *See* **potassium manganate(VII)**.

potassium sulphate A white crystalline powder, K_2SO_4, soluble in water and insoluble in ethanol; rhombic or hexagonal; r.d. 2.66; m.p. 1069°C. It occurs naturally as *schönite* (Strassfurt deposits) and in lake brines, from which it is separated by fractional crystallization. It has also been produced by the *Hargreaves process*, which involves the oxidation of potassium chloride with sulphuric acid. In the laboratory it may be obtained by the reaction of either potassium hydroxide or potassium carbonate with sulphuric acid. Potassium sulphate is used in cements, in glass manufacture, as a food additive, and as a fertilizer (source of K^+) for chloride-sensitive plants, such as tobacco and citrus.

potassium sulphide A yellow-red or brown-red deliquescent solid, K_2S, which is soluble in water and in ethanol but insoluble in diethyl ether; cubic; r.d. 1.80; m.p. 840°C. It is made industrially by reducing potassium sulphate with carbon at high temperatures in the absence of air. In the laboratory it may be prepared by the reaction of hydrogen sulphide with potassium hydroxide. The pentahydrate is obtained on crystallization. Solutions are strongly alkaline due to hydrolysis. It is used as an analytical reagent and as a depilatory. Potassium sulphide is generally regarded as a hazardous chemical with a fire risk; dusts of K_2S have been known to explode.

potassium sulphite A white crystalline solid, K_2SO_3, soluble in water and very sparingly soluble in ethanol; r.d. 1.51; decomposes on heating. It is a reducing agent and is used as such in photography and in the food and brewing industries, where it prevents oxidation.

potassium superoxide **(potassium dioxide)** A yellow paramagnetic solid, KO_2, produced by burning potassium in an excess of oxygen; it is very soluble (by reaction) in water, soluble in ethanol, and slightly soluble in diethyl ether; m.p. 380°C. When treated with cold water or dilute mineral acids, hydrogen peroxide is obtained. The compound is a powerful oxidizing agent and on strong heating releases oxygen with the formation of the monoxide, K_2O.

potential barrier A region containing a maximum of potential that prevents a particle on one side of it from passing to the other side. According to classical theory a particle must possess energy in excess of the height of the potential barrier to pass it. However, in quantum theory there is a finite probability that a particle with less energy will pass through the barrier (*see* **tunnel effect**). A potential barrier surrounds the atomic nucleus and is important in nuclear physics; a similar but much lower barrier exists at the interface between semiconductors and metals and between differently doped semiconductors. These barriers are important in the design of electronic devices.

potential energy *See* energy.

potential-energy curve A graph of the potential energy of electrons in a diatomic molecule, in which the potential energy of an electronic state is plotted vertically and the interatomic distance is plotted horizontally, with the minimum of the curve being the average internuclear distance. The *Morse potential gives a good analytic description of a potential energy curve. The dissociation energy and certain quantities of interest in vibrational spectroscopy can be found from the potential-energy curve.

potential-energy surface A multidimensional surface used in the theory of electronic states of polyatomic molecules and chemical reactions. A *potential-energy curve is the simplest type of potential energy surface. Each internuclear distance in the molecule or reacting system is represented by one dimension, as is the potential energy. This means that if a nonlinear molecule has N atoms its potential energy surface is a $(3N-6)$-dimensional surface in a $(3N-5)$-dimensional space. In the case of a linear molecule the potential-energy surface is a $(3N-5)$-dimensional space. If the surface has a minimum then that electronic state is stable. It is possible for there to be more than one minimum. In potential energy surfaces for chemical reactions the reactants and products are frequently 'valleys' connected by a region of higher energy, which is associated with the *activation energy of the reaction.

potentiometric titration A titration in which the end point is found by measuring the potential on an electrode immersed in the reaction mixture.

powder metallurgy A process in which powdered metals or alloys are pressed into a variety of shapes at high temperatures. The process started with the pressing of powdered tungsten into incandescent lamp filaments in the first decade of this century and is now widely used for making self-lubricating bearings and cemented tungsten carbide cutting tools.

The powders are produced by atomization of molten metals, chemical decomposition of a compound of the metal, or crushing and grinding of the metal or alloy. The parts are pressed into moulds at pressures ranging from 140×10^6 Pa to 830×10^6 Pa after which they are heated in a controlled atmosphere to bond the particles together (*see* **sintering**).

praseodymium Symbol Pr. A soft silvery metallic element belonging to the *lanthanoids; a.n. 59; r.a.m. 140.91; r.d. 6.773; m.p. 931°C; b.p. 3512°C. It occurs in bastnasite and monazite, from which it is recovered by an ion-exchange process. The only naturally occurring isotope is praseodymium–141, which is not radioactive; however, fourteen radioisotopes have been produced. It is used in mischmetal, a rare-earth alloy containing 5% praseodymium, for use in lighter flints. Another rare-earth mixture containing 30% praseodymium is used as a catalyst in cracking crude oil. The element was discovered by C. A. von Welsbach in 1885.

precessional motion A form of motion that occurs when a torque is applied to a rotating body in such a way that it tends to change the direction of its axis of rotation. It arises because the resultant of the angular velocity of rotation and the increment of angular velocity produced by the torque is an angular velocity about a new direction; this commonly changes the axis of the applied torque and leads to sustained rotation of the original axis of rotation.

A spinning top, the axis of which is not exactly vertical, has a torque acting on it as a result of gravity. Instead of falling over, the top precesses about a vertical line through the pivot. Precessional effects occur in atoms in magnetic fields.

precipitate A suspension of small solid particles produced in a liquid by chemical reaction.

precipitation **1.** All liquid and solid forms of water that are deposited from the atmosphere; it includes rain, drizzle, snow, hail, dew, and hoar frost. **2.** The formation of a precipitate.

precursor A compound that leads to another compound in a series of chemical reactions.

pressure The force acting normally on unit area of a surface or the ratio of force to area. It is measured in *pascals in SI units. *Absolute pressure* is pressure measured on a gauge that reads zero at zero pressure rather than at atmospheric pressure. *Gauge pressure* is measured on a gauge that reads zero at atmospheric pressure.

pressure gauge Any device used to measure *pressure. Three basic types are in use: the liquid-column gauge (e.g. the mercury barometer and the manometer), the expanding-element gauge (e.g. the Bourdon gauge and the aneroid barometer), and the electrical transducer. In the last category the strain gauge is an example. Capacitor pressure gauges also come into this category. In these devices, the pressure to be measured displaces one plate of a capacitor and thus alters its capacitance.

Priestley, Joseph (1733–1804) British chemist, who in 1755 became a Presbyterian minister. In Leeds, in 1767, he experimented with carbon dioxide ('fixed air') from a nearby brewery; with it he invented soda water. He moved to a ministry in Birmingham in 1780, and in 1791 his revolutionary views caused a mob to burn his house, as a result of which he emigrated to the USA in 1794. In the early 1770s he experimented with combustion and produced the gases hydrogen chloride, sulphur dioxide, and dinitrogen oxide (nitrous oxide). In 1774 he isolated oxygen (*see also* **Lavoisier, Antoine**).

primary alcohol *See* **alcohols**.

primary amine *See* **amines**.

primary cell A *voltaic cell in which the chemical reaction producing the e.m.f. is not satisfactorily reversible and the cell cannot therefore be recharged by the application of a charging current. *See* **Daniell cell**; **Leclanché cell**; **Weston cell**; **mercury cell**. *Compare* **secondary cell**.

principal quantum number *See* **atom**.

prochiral Denoting an organic molecule lacking a chiral centre but containing one or more carbon atoms attached to two identical ligands and two other different ligands (C_{aabc}). The carbon atom is said to be a *prochiral centre*. The name prochiral is used because if *a* is replaced by a ligand *d*, different from either *a*, *b*, or *c*, there is a chiral centre in the molecule C_{abcd}.

producer gas **(air gas)** A mixture of carbon monoxide and nitrogen made by passing air over very hot carbon. Usually some steam is added to the air and the mixture contains hydrogen. The gas is used as a fuel in some industrial processes.

product *See* **chemical reaction**.

progesterone A hormone, produced primarily by the corpus luteum of the ovary but also by the placenta, that prepares the inner lining of the uterus for implantation of a fertilized egg cell. If implantation fails, the corpus luteum degenerates and progesterone production ceases accordingly. If implantation occurs, the corpus luteum continues to secrete progesterone, under the influence of luteinizing hormone and prolactin, for several months of pregnancy, by which time the placenta has taken over this function. During pregnancy, progesterone maintains the constitution of the uterus and prevents further release of eggs from the ovary. Small amounts of progesterone are produced by the testes. *See also* **progestogen**.

progestogen One of a group of naturally occurring or synthetic hormones that maintain the normal course of pregnancy. The best known is *progesterone. In high doses progestogens inhibit secretion of luteinizing hormone, thereby preventing ovulation, and alter the consistency of mucus in the vagina so that conception tends not to occur. They are therefore used as major constituents of oral contraceptives.

projection formula A convention for representing the three-dimensional structure of a molecule in two dimensions. *See* **Fischer projection**; **Newman projection**; **sawhorse projection**.

prolactin (lactogenic hormone; luteotrophic hormone; luteotrophin) A hormone produced by the anterior pituitary gland. In mammals it stimulates the mammary glands to produce milk and the corpus luteum of the ovary to secrete the hormone *progesterone.

proline *See* **amino acid**.

promethium Symbol Pm. A soft silvery metallic element belonging to the *lanthanoids; a.n. 61; r.a.m. 145; r.d. 7.26 (20°C); m.p. 1080°C; b.p. 2460°C. The only naturally occurring isotope, promethium–147, has a half-life of only 2.52 years. Eighteen other radioisotopes have been produced, but they have very short half-lives. The only known source of the element is nuclear-waste material. Promethium–147 is of interest as a beta-decay power source but the promethium–146 and –148, which emit penetrating gamma radiation, must first be removed. It was discovered by J. A. Marinsky, L. E. Glendenin, and C. D. Coryell in 1947.

promoter A substance added to a catalyst to increase its activity.

proof A measure of the amount of alcohol (ethanol) in drinks. *Proof spirit* contains 49.28% ethanol by weight (about 57% by volume). Degrees of proof express the percentage of proof spirit present, so 70° proof spirit contains 0.7 × 57% alcohol.

1,2-propadiene (allene) A colourless gas, CH_2CCH_2; r.d. 1.79; m.p. −136°C; b.p. −34.5°C. Propadiene may be prepared from 1,3-dibromopropane ($CH_2BrCHCH_2Br$) by the action of zinc dust. *See also* **allenes**.

propanal (propionaldehyde) A colourless liquid *aldehyde, C_2H_5CHO; m.p. −81°C; b.p. 48.8°C.

propane A colourless gaseous hydrocarbon, C_3H_8; m.p. −190°C; b.p. −42°C. It is the third member of the *alkane series and is obtained from petroleum. Its main use is as bottled gas for fuel.

propanedioic acid (malonic acid) A white crystalline dicarboxylic acid, $HOOCCH_2COOH$; m.p. 132°C. It decomposes above its melting point to ethanoic acid. Propanedioic acid is used in the synthesis of other dicarboxylic acids.

propanoic acid (propionic acid) A colourless liquid *carboxylic acid, CH_3CH_2COOH; r.d. 0.99; m.p. −20.8°C; b.p. 141°C. It is used to make calcium propanate – an additive in bread.

propanol Either of two *alcohols with the formula C_3H_7OH. Propan-1-ol is $CH_3CH_2CH_2OH$ and propan-2-ol is $CH_3CH(OH)CH_3$. Both are colourless volatile liquids. Propan-2-ol is used in making propanone (acetone).

propanone (acetone) A colourless flammable volatile compound,

CH_3COCH_3; r.d. 0.79; m.p. −95.4°C; b.p. 56.2°C. The simplest *ketone, propanone is miscible with water. It is made by oxidation of propan-2-ol (*see* **propanol**) or is obtained as a by-product in the manufacture of phenol from cumene; it is used as a solvent and as a raw material for making plastics.

propenal **(acrolein)** A colourless pungent liquid unsaturated aldehyde, CH_2:CHCHO; r.d. 0.84; m.p. −87°C; b.p. 53°C. It is made from propene and can be polymerized to give acrylate resins.

propene **(propylene)** A colourless gaseous hydrocarbon, CH_3CH:CH_2; m.p. −185.25°C; b.p. −47.4°C. It is an *alkene obtained from petroleum by cracking alkanes. Its main use is in the manufacture of polypropene.

propenoate **(acrylate)** A salt or ester of *propenoic acid.

propenoic acid **(acrylic acid)** An unsaturated liquid *carboxylic acid, CH_2:CHCOOH; m.p. 13°C; b.p. 141.6°C. It readily polymerizes and it is used in the manufacture of *acrylic resins.

propenol **(allyl alcohol)** A pungent-smelling colourless unsaturated liquid alcohol, CH_2=$CHCH_2OH$; b.p. 97°C. It is found in wood alcohol and can be prepared in the laboratory by reducing propenal or by heating glycerol with oxalic acid. Industrially, it is made by the high-temperature catalytic reaction between propenal and 2-propanol or by the hydrolysis of 2-chloropropene, which is made by chlorinating propene. Potassium permanganate oxidizes propenol to glycerol and silver oxide converts it to a mixture of propenoic acid and propenal.

propenonitrile **(acrylonitrile; vinyl cyanide)** A colourless liquid, H_2C.CHCN, r.d. 0.81; m.p. −83.5°C. It is an unsaturated nitrile, made from propene and used to make acrylic resins.

propenyl group **(allyl group)** The organic group H_2C=$CHCH_2$–.

propionaldehyde *See* **propanal**.

propylene *See* **propene**.

propyl group The organic group $CH_3CH_2CH_2$–.

prostaglandin Any of a group of organic compounds derived from *essential fatty acids and causing a range of physiological effects in animals. Prostaglandins have been detected in most body tissues. They act at very low concentrations to cause the contraction of smooth muscle; natural and synthetic prostaglandins are used to induce abortion or labour in humans and domestic animals. Two prostaglandin derivatives have antagonistic effects on blood circulation: *thromboxane* A_2 causes blood clotting while *prostacyclin* causes blood vessels to dilate. Inflammation in allergic reactions and other diseases is also thought to involve prostaglandins.

prosthetic group A tightly bound nonpeptide inorganic or organic component of a protein. Prosthetic groups may be lipids, carbohydrates, metal ions, phosphate groups, etc. Some *coenzymes are more correctly regarded as prosthetic groups.

protactinium Symbol Pa. A radioactive metallic element belonging to the *actinoids; a.n. 91; r.a.m. 231.036; r.d. 15.37 (calculated); m.p. <1600°C (estimated). The most stable isotope, protactinium–231, has a half-life of 3.43×10^4 years; at least ten other radioisotopes are known. Protactinium–231 occurs in all uranium ores as it is derived from uranium–235. Protactinium has no practical applications; it was discovered by Lise Meitner and Otto Hahn in 1917.

protamine Any of a group of proteins of relatively low molecular weight found in association with the chromosomal *DNA of vertebrate sperm cells. They contain a single polypeptide chain comprising about 67% arginine. Protamines are thought to protect and support the chromosomes.

protease (peptidase; proteinase; proteolytic enzyme) Any enzyme that catalyses the hydrolysis of proteins into smaller *peptide fractions and amino acids, a process known as *proteolysis*. Examples are *pepsin and *trypsin. Several proteases, acting sequentially, are normally required for the complete digestion of a protein to its constituent amino acids.

protein Any of a large group of organic compounds found in all living organisms. Proteins comprise carbon, hydrogen, oxygen, and nitrogen and most also contain sulphur; molecular weights range from 6000 to several million. Protein molecules consist of one or several long chains (*polypeptides) of *amino acids linked in a characteristic sequence. This sequence is called the *primary structure* of the protein. These polypeptides may undergo coiling or pleating, the nature and extent of which is described as the *secondary structure*. The three-dimensional shape of the coiled or pleated polypeptides is called the *tertiary structure*. *Quaternary structure* specifies the structural relationship of the component polypeptides.

Proteins may be broadly classified into globular proteins and fibrous proteins. *Globular proteins* have compact rounded molecules and are usually water-soluble. Of prime importance are the *enzymes, proteins that catalyse biochemical reactions. Other globular proteins include the antibodies, which combine with foreign substances in the body; the carrier proteins, such as *haemoglobin; the storage proteins (e.g. casein in milk and albumin in egg white), and certain hormones (e.g. insulin). *Fibrous proteins* are generally insoluble in water and consist of long coiled strands or flat sheets, which confer strength and elasticity. In this category are keratin and collagen. Actin and myosin are the principal fibrous proteins of muscle, the interaction of which brings about muscle contraction. Blood clotting involves the fibrous protein called fibrin.

When heated over 50°C or subjected to strong acids or alkalis, proteins lose their specific tertiary structure and may form insoluble coagulates (e.g. egg white). This usually inactivates their biological properties.

protein engineering The techniques used to alter the structure of proteins (especially enzymes) in order to improve their use to humans. This involves artificially modifying the DNA sequences that encode them so that, for example, new amino acids are inserted into existing proteins. Synthesized lengths of novel DNA can be used to produce new proteins by cells or other systems containing the necessary factors for transcription and translation. Alternatively, new proteins can be synthesized by *solid state synthesis*, in which polypeptide chains are assembled under the control of chemicals. One end of the chain is anchored to a solid support and the chemicals selectively determine which amino acids are added to the free end. The appropriate chemicals can be renewed during the process; when synthesized, the polypeptide is removed and purified. Protein engineering is used to synthesize enzymes (so-called 'designer enzymes') used in biotechnology. The three-dimensional tertiary structure of proteins is crucially important for their function, and this can be investigated using computer-aided modelling.

protein synthesis The process by which living cells manufacture proteins from their constituent amino acids, in accordance with the genetic information carried in the DNA of the chromosomes. This information is encoded in messenger *RNA, which is transcribed from DNA in the nucleus of the cell: the sequence of amino acids in a particular protein is determined by the sequence of nucleotides in messenger RNA. At the ribosomes the information carried by messenger RNA is translated into the sequence of amino acids of the protein in the process of translation.

proteolysis The enzymic splitting of proteins. *See* **protease**.

proteolytic enzyme *See* **protease**.

proton An elementary particle that is stable, bears a positive charge equal in magnitude to that of the *electron, and has a mass of $1.672\,614 \times 10^{-27}$ kg, which is 1836.12 times that of the electron. The proton is a hydrogen ion and occurs in all atomic nuclei. *See also* **hydron**.

protonic acid An *acid that forms positive hydrogen ions (or, strictly, oxonium ions) in aqueous solution. The term is used to distinguish 'traditional' acids from Lewis acids or from Lowry–Brønsted acids in nonaqueous solvents.

proton number *See* **atomic number**.

Prout's hypothesis The hypothesis put forward by the British chemist William Prout (1785–1850) in 1815 that all atomic weights are integer multiples of the atomic weight of hydrogen and hence that all atoms are made out of hydrogen. Subsequent work on atomic weights in the 19th century showed that this hypothesis is incorrect (with chlorine having an atomic weight of 35.5 being a glaring example of this). The understanding of atomic structure that emerged in the 20th century, with atomic number being the number of protons in an atom and non-integer atomic weights being due to mixtures of isotopes, has vindicated the spirit of Prout's hypothesis.

prussic acid *See* **hydrogen cyanide**.

pseudoaromatic **(antiaromatic)** A compound that has a ring of atoms containing alternating double and single bonds, yet does not have the characteristic properties of *aromatic compounds. Such compounds do not obey the Hückel rule. Cyclooctatetraene (C_8H_8), for instance, has a ring of eight carbon atoms with conjugated double bonds, but the ring is not planar and the compound acts like an alkene, undergoing addition reactions.

pseudo-axial *See* **ring conformations**.

pseudo-equatorial *See* **ring conformations**.

pseudohalogens A group of compounds, including cyanogen $(CN)_2$ and thiocyanogen $(SCN)_2$, that have some resemblance to the halogens. Thus, they form hydrogen acids (HCN and HSCN) and ionic salts containing such ions as CN^- and SCN^-.

pseudo order An order of a chemical reaction that appears to be less than the true order because of the experimental conditions used. Pseudo orders occur when one reactant is present in large excess. For example, a reaction of substance A undergoing hydrolysis may appear to be proportional only to [A] because the amount of water present is so large.

PTFE *See* **polytetrafluoroethene**.

ptyalin An enzyme that digests carbohydrates (*see* **amylase**). It is present in mammalian saliva and is responsible for the initial stages of starch digestion.

p-type semiconductor See **semiconductor**.

pullulan A water-soluble polysaccharide composed of glucose units that are polymerized in such a way as to make it viscous and impermeable to oxygen. Pullulan is used in adhesives, food packaging, and moulded articles. It is derived from the fungus *Aureobasidium pullulans*.

pumice A porous volcanic rock that is light and full of cavities due to expanding gases that were liberated from solution in the lava while it solidified. Pumice is often light enough to float on water. It is usually acid (siliceous) in composition, and is used as an abrasive and for polishing.

pump A device that imparts energy to a fluid in order to move it from one place or level to another or to raise its pressure (*compare* **vacuum pump**). *Centrifugal pumps* and turbines have rotating impellers, which increase the velocity of the fluid, part of the energy so acquired by the fluid then being converted to pressure energy. Displacement pumps act directly on the fluid, forcing it to flow against a pressure. They include piston, plunger, gear, screw, and cam pumps.

Purine

purine An organic nitrogenous base (see formula), sparingly soluble in water, that gives rise to a group of biologically important derivatives, notably *adenine and *guanine, which occur in nucleotides and nucleic acids (DNA and RNA).

PVA *See* **polyvinylacetate.**

PVC *See* **polychloroethene.**

pyranose A *sugar having a six-membered ring containing five carbon atoms and one oxygen atom.

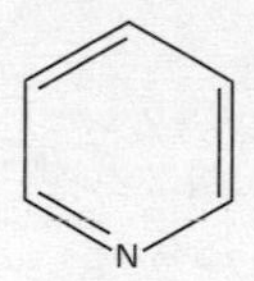

Pyridine

pyridine A colourless liquid with a strong unpleasant smell, C_5H_5N (see formula); r.d. 0.98; m.p. −42°C; b.p. 115°C. Pyridine is an aromatic heterocyclic compound present in coal tar. It is used in making other organic chemicals.

pyridoxine *See* **vitamin B complex.**

Pyrimidine

pyrimidine An organic nitrogenous base (see formula), sparingly soluble in water, that gives rise to a group of biologically important derivatives, notably *uracil, *thymine, and *cytosine, which occur in *nucleotides and nucleic acids (DNA and RNA).

pyrite (iron pyrites) A mineral form of iron(II) sulphide, FeS_2. Superficially it resembles gold in appearance, hence it is also known as *fool's gold*, but it is harder and more brittle than gold (which may be cut with a knife). Pyrite crystallizes in the cubic system, is brass yellow in colour, has a metallic lustre, and a hardness of 6–6.5 on the Mohs' scale. It is the most common and widespread of the sulphide minerals and is used as a source of sulphur for the production of sulphuric acid. Sources include the Rio Tinto mines in Spain.

pyro- Prefix denoting an oxo acid that could be obtained from a lower acid by dehydration of two molecules. For example, pyrosulphuric acid is $H_2S_2O_7$ (i.e. $2H_2SO_4$ minus H_2O).

pyroboric acid *See* **boric acid.**

pyroelectricity The property of certain crystals, such as tourmaline, of acquiring opposite electrical charges on opposite faces when heated. In tourmaline a rise in temperature of 1 K at room temperature produces a polarization of some 10^{-5} C m^{-2}.

pyrogallol 1,2,3-trihydroxybenzene, $C_6H_3(OH)_3$, a white crystalline solid, m.p. 132°C. Alkaline solutions turn dark brown on exposure to air through reaction with oxygen. It is a powerful reducing agent, employed in photographic developers. It is also used in volumetric gas analysis as an absorber of oxygen.

pyrolusite See **manganese(IV) oxide.**

pyrolysis Chemical decomposition occurring as a result of high temperature.

pyrometric cones *See* **Seger cones.**

pyrometry The measurement of high temperatures from the amount of radiation emitted, using a *pyrometer*. Modern *narrow-band* or *spectral* pyrometers use infrared-sensitive photoelectric cells behind filters that exclude visible light. In the *optical pyrometer* (or disappearing filament pyrometer) the image of the incandescent source is focused in the plane of a tungsten filament that is heated electrically. A variable resistor is used to adjust the current through the filament until it blends into the image of the source, when viewed through a red filter and an eyepiece. The temperature is then read from a calibrated ammeter or a calibrated dial on the variable resistor. In the *total-radiation pyrometer* radiation emitted by the source is focused by a concave mirror onto a blackened foil to which a thermopile is attached. From the e.m.f. produced by the thermopile the temperature of the source can be calculated.

pyrones Compounds containing a six-membered ring system having an

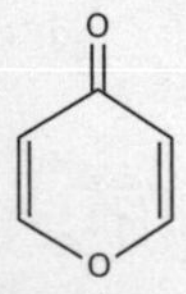

Pyrones

oxygen hetero atom and a carbonyl group attached to the ring. There are two forms depending on whether the carbonyl is in the 1 or 3 position. The pyrone ring system occurs in many naturally occurring compounds.

pyrophoric Igniting spontaneously in air. *Pyrophoric alloys* are alloys that give sparks when struck. *See* **misch metal**.

pyrophosphoric acid *See* **phosphoric(V) acid**.

pyrosilicate *See* **silicate**.

pyrosulphuric acid *See* **disulphuric(VI) acid**.

pyroxenes A group of ferromagnesian rock-forming silicate minerals. They are common in basic igneous rocks but may also be developed by metamorphic processes in gneisses, schists, and marbles. Pyroxenes have a complex crystal chemistry; they are composed of continuous chains of silicon and oxygen atoms linked by a variety of other elements. They are related to the *amphiboles, from which they differ in cleavage angles. The general formula is $X_{1-p}Y_{1+p}Z_2O_6$, where X = Ca,Na; Y = Mg,Fe^{2+},Mn,Li,Al,Fe^{3+},Ti; and Z = Si,Al.

Orthorhombic pyroxenes (*orthopyroxenes*), $(Mg,Fe)_2Si_2O_6$, vary in composition between the end-members enstatite ($Mg_2Si_2O_6$) and orthoferrosilite ($Fe_2Si_2O_6$). Monoclinic pyroxenes (*clinopyroxenes*), the larger group, include:

diopside, $CaMgSi_2O_6$;
hedenbergite, $CaFe^{2+}Si_2O_6$;
johannsenite, $CaMnSi_2O_6$;
augite, $(Ca,Mg,Fe,Ti,Al)_2(Si,Al)_2O_6$;
aegirine, $NaFe^{3+}Si_2O_6$;
jadeite (*see* **jade**);
pigeonite $(Mg,Fe^{2+},Ca)(Mg,Fe^{2+})Si_2O_6$.

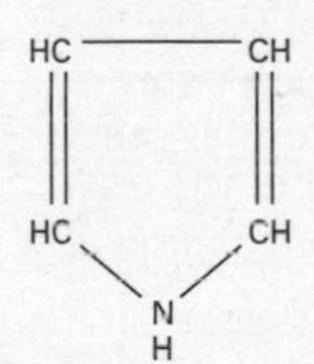

Pyrrole

pyrrole An organic nitrogen-containing compound (see formula) that forms part of the structure of *porphyrins.

pyruvic acid (2-oxopropanoic acid) A colourless liquid organic acid, $CH_3COCOOH$. Pyruvate is an important intermediate compound in metabolism, being produced during *glycolysis and converted to acetyl coenzyme A, required for the *Krebs cycle. Under anaerobic conditions pyruvate is converted to lactate or ethanol.

QSAR *See* **quantitative structure–activity relationship**.

QSMR *See* **quantitative structure–metabolism relationship**.

quadrivalent Having a valency of four.

quadrupole A set of four point charges that has zero net charge and dipole moment. An example of a quadrupole is a carbon dioxide (CO_2) molecule. Quadrupole interactions are much smaller than dipole interactions. Transitions involving quadrupole moments are much weaker than transitions involving dipole moments, but can allow transitions forbidden in dipole moment transitions.

qualitative analysis *See* **analysis**.

quantitative analysis *See* **analysis**.

quantitative structure–activity relationship (QSAR) A statistical algorithm that quantitatively defines the relationship between the chemical structure of a drug and its effect on an organism. QSAR studies are often used to predict the activity or toxicity of new drugs. Similar methods can be used to predict the metabolism of new drugs (*quantitative structure–metabolism relationships*).

quantitative structure–metabolism relationship (QSMR) *See* **quantitative structure–activity relationship**.

quantum (*pl.* **quanta**) The minimum amount by which certain properties, such as energy or angular momentum, of a system can change. Such properties do not, therefore, vary continuously, but in integral multiples of the relevant quantum, and are described as *quantized*. This concept forms the basis of the *quantum theory. In waves and fields the quantum can be regarded as an excitation, giving a particle-like interpretation to the wave or field. Thus, the quantum of the electromagnetic field is the *photon and the graviton is the quantum of the gravitational field.

quantum chaos The *quantum mechanics of systems for which the corresponding classical system can exhibit *chaos. This subject was initiated by Einstein in 1917, who showed that the quantization conditions associated with the *Bohr theory need to be modified for systems that show chaos in classical mechanics. The subject of quantum chaos is an active field of research in which many basic issues still require clarification. It appears that systems exhibiting chaos in classical mechanics do not necessarily exhibit chaos in quantum mechanics.

quantum chemistry The application of quantum mechanics to chemistry. Quantum chemistry mostly used empirical methods at first but as computing power has developed it has been increasingly based on the first principles of quantum mechanics.

quantum entanglement A phenomenon in quantum mechanics in which a particle or system does not have a definite state but exists as an intermediate form of two 'entangled' states. One of these states is realized when a 'measurement' is made.

quantum jump A change in a system (e.g. an atom or molecule) from one quantum state to another.

quantum mechanics A system of mechanics that was developed from *quantum theory and is used to explain the properties of atoms and molecules. Using the energy *quantum as a starting point it incorporates Heisenberg's *uncertainty principle and the *de Broglie wavelength to establish the wave–particle duality on which the Schrödinger equation is based. This form of quantum mechanics is called *wave mechanics*. An alternative but equivalent formalism, *matrix mechanics, is based on mathematical operators. *See also* **computational chemistry**.

quantum number *See* **atom**; **spin**.

quantum simulation The mathematical modelling of systems of large numbers of atoms or molecules by computer studies of relatively small clusters. It is possible to obtain information about solids and liquids in this way and to study surface properties and reactions.

quantum state The state of a quantized system as described by its quantum numbers. For instance, the state of a hydrogen *atom is described by the four quantum numbers n, l, m, m_s. In the ground state they have values 1, 0, 0, and ½ respectively.

quantum statistics A statistical description of a system of particles that obeys the rules of *quantum mechanics rather than classical mechanics. In quantum statistics, energy states are considered to be quantized. If the particles are treated as indistinguishable, *Bose–Einstein statistics* apply if any number of particles can occupy a given quantum state. Such particles are called *bosons*. All known bosons have an angular momentum nh, where n is zero or an integer and h is the Planck constant. For identical bosons the *wave function is always symmetric. If only one particle may occupy each quantum state, *Fermi–Dirac statistics* apply and the particles are called *fermions*. All known fermions have a total angular momentum $(n + ½)h/2\pi$ and any wave function that involves identical fermions is always antisymmetric.

quantum theory The theory devised by Max *Planck in 1900 to account for the emission of the black-body radiation from hot bodies. According to this theory energy is emitted in quanta (*see* **quantum**), each of which has an energy equal to $h\nu$, where h is the *Planck constant and ν is the frequency of the radiation. This theory led to the modern theory of the interaction between matter and radiation known as *quantum mechanics, which generalizes and replaces classical mechanics and Maxwell's electromagnetic theory. In *nonrelativistic quantum theory* particles are assumed to be neither created nor destroyed, to move slowly relative to the speed of light, and to have a mass that does not change with velocity. These assumptions apply to atomic and molecular phenomena and to some aspects of nuclear physics. *Relativistic quantum theory* applies to particles that have zero rest mass or travel at or near the speed of light.

quantum theory of radiation The theory that describes the emission and absorption of electromagnetic radiation. Since it is a quantum theory the processes of emission and absorption are described in terms of the creation and disappearance of photons, with photons being created in emission processes and disappearing in absorption. *Absorption* occurs when electromagnetic radiation that impinges on a quantum mechanical system, such as an atom or molecules, induces a transition from the ground state of the system to an excited state by virtue of a photon of the radiation being absorbed. Similarly, emission of a photon occurs when an atom returns to the ground state from an excited state, i.e. *spontaneous emission occurs. It is also possible for an atom in an excited state to emit a photon by *induced emission, a process that is used in *lasers.

quantum yield In a photochemical reaction, the number of events occurring per photon absorbed. It is given by the number of moles of product formed (or reactant consumed) divided by the number of moles of photons absorbed.

quartz The most abundant and common mineral, consisting of crystalline silica (silicon dioxide, SiO_2), crystallizing in the trigonal system. It has a hardness of 7 on the Mohs' scale. Well-formed crystals of quartz are six-sided prisms terminating in six-sided pyramids. Quartz is ordinarily colourless and transparent, in which form it is known as *rock crystal*. Coloured varieties, a number of which are used as gemstones, include *amethyst, citrine quartz (yellow), rose quartz (pink), milk quartz (white), smoky quartz (grey-brown), *chalcedony, *agate, and *jasper. Quartz occurs in many rocks, especially igneous rocks such as granite and quartzite (of which it is the chief constituent), metamorphic rocks such as gneisses and schists, and sedimentary rocks such as sandstone and limestone. The mineral is piezoelectric and is used in oscillators. It is also used in optical instruments and in glass, glaze, and abrasives.

quasicrystal A solid structure in which there is (1) long-range incommensurate translational order (*see* **incommensurate lattice**) and (2) long-range orientational order with a point group, which is not allowed in *crystallography. Condition (1) is called *quasiperiodicity*. In two dimensions, the fivefold symmetry of a pentagon is an example of a point symmetry, which is not allowed in crystallography, but for which quasicrystals exist. In three dimensions the symmetry of an icosahedron is not allowed in crystallography, but quasicrystals with this symmetry exist (e.g. AlMn). Diffraction patterns for quasicrystals have Bragg peaks, with the density of

Bragg peaks in each plane being higher than would be expected for a perfect periodic crystal.

quasiperiodicity *See* **quasicrystal**.

quaternary ammonium compounds *See* **amine salts**.

quenching 1. ***(in metallurgy)*** The rapid cooling of a metal by immersing it in a bath of liquid in order to improve its properties. Steels are quenched to make them harder but some nonferrous metals are quenched for other reasons (copper, for example, is made softer by quenching). **2.** ***(in physics)*** The process of inhibiting a continuous discharge in a *Geiger counter so that the incidence of further ionizing radiation can cause a new discharge. This is achieved by introducing a quenching vapour, such as methane mixed with argon or with neon, into the tube.

quicklime *See* **calcium oxide**.

quinhydrone electrode A *half cell consisting of a platinum electrode in an equimolar solution of quinone (cyclohexadiene-1,4-dione) and hydroquinone (benzene-1,4-diol). It depends on the oxidation–reduction reaction

$$C_6H_4(OH)_2 \rightleftharpoons C_6H_4O_2 + 2H^+ 2e.$$

quinine A white solid, $C_{20}H_{24}N_2O_2.3H_2O$, m.p. 57°C. It is a poisonous alkaloid occurring in the bark of the South American cinchona tree, although it is now usually produced synthetically. It forms salts and is toxic to the malarial parasite, and so quinine and its salts are used to treat malaria; in small doses it may be prescribed for colds and influenza. In dilute solutions it has a pleasant astringent taste and is added to some types of tonic water.

quinol *See* **benzene-1,4-diol**.

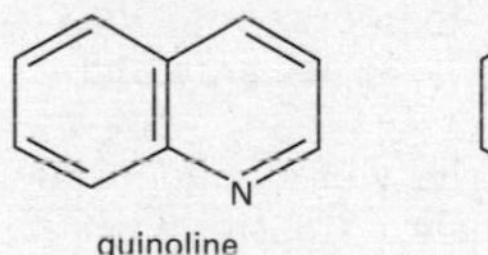

Isomers of quinoline

quinoline A hygroscopic unpleasant-smelling colourless oily liquid, C_9H_7N; b.p. 240°C. Its molecules consist of a benzene ring fused to a pyridine ring. It occurs in coal tar and bone oil, and is made from phenylamine and nitrobenzene. Quinoline is a basic compound, forming salts with mineral acids and forming quaternary ammonium compounds with haloalkanes. It is used for making medicines and dyes.

quinone 1. *See* **cyclohexadiene-1,4-dione**. **2.** Any similar compound containing C=O groups in an unsaturated ring.

racemate *See* **racemic mixture.**

racemic mixture (racemate) A mixture of equal quantities of the *d*- and *l*-forms of an optically active compound. Racemic mixtures are denoted by the prefix *dl*- (e.g. *dl*-lactic acid). A racemic mixture shows no *optical activity.

racemization A chemical reaction in which an optically active compound is converted into a *racemic mixture.

rad *See* **radiation units.**

radial distribution function The average number of atoms found at a distance *r* from a central atom. In the case of a crystal, the radial distribution function is a series of regular spikes, corresponding to the atomic positions in the lattice. Calculating the radial distribution function is of fundamental importance in the theory of liquids.

radiation 1. Energy travelling in the form of electromagnetic waves or photons. **2.** A stream of particles, especially alpha- or beta-particles from a radioactive source or neutrons from a nuclear reactor.

radiationless decay Decay of an atom or molecule from an excited state to a lower energy state without the emission of electromagnetic radiation. A common example of a radiationless process is the *Auger effect, in which an electron rather than a photon is emitted as a result of decay.

radiation units Units of measurement used to express the activity of a radionuclide and the dose of ionizing radiation. The units *curie*, *roentgen*, *rad*, and *rem* are not coherent with SI units but their temporary use with SI units has been approved while the derived SI units *becquerel*, *gray*, and *sievert* become familiar.

The becquerel (Bq), the SI unit of activity, is the activity of a radionuclide decaying at a rate, on average, of one spontaneous nuclear transition per second. Thus 1 Bq = 1 s^{-1}. The former unit, the curie (Ci), is equal to 3.7×10^{10} Bq. The curie was originally chosen to approximate the activity of 1 gram of radium–226.

The gray (Gy), the SI unit of absorbed dose, is the absorbed dose when the energy per unit mass imparted to matter by ionizing radiation is 1 joule per kilogram. The former unit, the rad (rd), is equal to 10^{-2} Gy.

The sievert (Sv), the SI unit of dose equivalent, is the dose equivalent when the absorbed dose of ionizing radiation multiplied by the stipulated dimensionless factors is 1 J kg^{-1}. As different types of radiation cause different effects in biological tissue a weighted absorbed dose, called the *dose equivalent*, is used in which the absorbed dose is modified by multiplying it by dimensionless factors stipulated by the International

Commission on Radiological Protection. The former unit of dose equivalent, the rem (originally an acronym for *r*oentgen *e*quivalent *m*an), is equal to 10^{-2} Sv.

In SI units, exposure to ionizing radiation is expressed in coulombs per kilogram, the quantity of X- or gamma-radiation that produces ion pairs carrying 1 coulomb of charge of either sign in 1 kilogram of pure dry air. The former unit, the roentgen (R), is equal to 2.58×10^{-4} C kg^{-1}.

radiative capture *See* **capture.**

radical A group of atoms, either in a compound or existing alone. *See* **free radical; functional group.**

radical ion A radical that has a positive or negative charge. An example is the benzene radical cation $C_6H_6^{\cdot +}$. In radical ions, the odd electron and the charge are usually (but not necessarily) on the same atom.

radioactive age The age of an archaeological or geological specimen as determined by a process that depends on a radioactive decay. *See* **carbon dating; fission-track dating; potassium–argon dating; rubidium–strontium dating; uranium–lead dating.**

radioactive dating *See* **radiometric dating.**

radioactive isotope *See* **radioisotope.**

radioactive nuclide *See* **radionuclide.**

radioactive series A series of radioactive nuclides in which each member of the series is formed by the decay of the nuclide before it. The series ends with a stable nuclide. Three radioactive series occur naturally, those headed by thorium–232 (*thorium series*), uranium–235 (*actinium series*), and uranium–238 (*uranium series*). All three series end with an isotope of lead. The *neptunium series* starts with the artificial isotope plutonium–241, which decays to neptunium–237, and ends with bismuth–209.

radioactive tracing *See* **labelling.**

radioactivity The spontaneous disintegration of certain atomic nuclei accompanied by the emission of alpha-particles (helium nuclei), beta-particles (electrons or positrons), or gamma radiation (short-wavelength electromagnetic waves).

Natural radioactivity is the result of the spontaneous disintegration of naturally occurring radioisotopes. Many radioisotopes can be arranged in three *radioactive series. The rate of disintegration is uninfluenced by chemical changes or any normal changes in their environment. However, radioactivity can be induced in many nuclides by bombarding them with neutrons or other particles. *See also* **decay; ionizing radiation; radiation units.**

radiocarbon dating *See* **carbon dating.**

radiochemistry The branch of chemistry concerned with radioactive compounds and with ionization. It includes the study of compounds of radioactive elements and the preparation and use of compounds containing radioactive atoms. *See* **labelling**; **radiolysis**.

radiogenic Resulting from radioactive decay.

radioimmunoassay (RIA) A sensitive quantitative method for detecting trace amounts of a biomolecule, based on its capacity to displace a radioactively labelled form of the molecule from combination with its antibody.

radioisotope (radioactive isotope) An isotope of an element that is radioactive. *See* **labelling**.

radiolysis The use of ionizing radiation to produce chemical reactions. The radiation used includes alpha particles, electrons, neutrons, X-rays, and gamma rays from radioactive materials or from accelerators. Energy transfer produces ions and excited species, which undergo further reaction. A particular feature of radiolysis is the formation of short-lived solvated electrons in water and other polar solvents.

radiometric dating (radioactive dating) *See* **dating techniques**; **radioactive age**.

radionuclide (radioactive nuclide) A *nuclide that is radioactive.

radium Symbol Ra. A radioactive metallic element belonging to *group 2 (formerly IIA) of the periodic table; a.n. 88; r.a.m. 226.0254; r.d. ~5; m.p. 700°C; b.p. 1140°C. It occurs in uranium ores (e.g. pitchblende). The most stable isotope is radium–226 (half-life 1602 years), which decays to radon. It is used as a radioactive source in research and, to some extent, in radiotherapy. The element was isolated from pitchblende in 1898 by Marie and Pierre Curie.

radon Symbol Rn. A colourless radioactive gaseous element belonging to group 18 of the periodic table (the *noble gases); a.n. 86; r.a.m. 222; d. 9.73 g dm^{-3}; m.p. −71°C; b.p. −61.8°C. At least 20 isotopes are known, the most stable being radon–222 (half-life 3.8 days). It is formed by decay of radium–226 and undergoes alpha decay. It is used in radiotherapy. Radon occurs naturally, particularly in areas underlain by granite, where it is thought to be a health hazard. As a noble gas, radon is practically inert, although a few compounds, e.g. radon fluoride, can be made. It was first isolated by William Ramsey and Robert Whytlaw-Gray (1877–1958) in 1908.

raffinate A liquid purified by solvent extraction.

raffinose A white solid carbohydrate, $C_{18}H_{32}O_{16}$, m.p. 80°C. It is a trisaccharide (a type of *sugar) consisting of fructose, galactose and glucose. It occurs naturally in sugar-beet and cotton-seed residues.

r.a.m. *See* **relative atomic mass**.

Raman effect A type of scattering of electromagnetic radiation in which light suffers a change in frequency and a change in phase as it passes through a material medium. The intensity of Raman scattering is about one-thousandth of that in Rayleigh scattering in liquids; for this reason it was not discovered until 1928. However, it was not until the development of the laser that the effect was put to use.

In *Raman spectroscopy* light from a laser is passed through a substance and the scattering is analysed spectroscopically. The new frequencies in the *Raman spectrum* of monochromatic light scattered by a substance are characteristic of the substance. Both inelastic and superelastic scattering occurs. The technique is used as a means of determining molecular structure and as a tool in chemical analysis. The effect was discovered by the Indian scientist Sir C. V. Raman (1888–1970).

Ramsay, Sir William (1852–1916) British chemist, born in Glasgow. After working under Robert *Bunsen, he returned to Glasgow before taking up professorships at Bristol (1880–87) and London (1887–1912). In the early 1890s he worked with Lord *Rayleigh on the gases in air and in 1894 they discovered argon. In 1898, with Morris Travers (1872–1961), he discovered neon, krypton, and xenon. Six years later he discovered the last of the noble gases, radon. He was awarded the Nobel Prize for chemistry in 1904, the year in which Rayleigh received the physics prize.

random alloy *See* **disordered solid.**

random walk The problem of determining the distance from a starting position made by a walker, who can either move forward (toward $+x$) or backwards (toward $-x$) with the choice being made randomly (e.g. by tossing a coin). The progress of the walker is characterized by the net distance D_N travelled in N steps. After N steps the root mean square value D_{rms}, which is the average distance away from the starting position, is given by $D_{rms} = \sqrt{N}$. Physical applications of the random walk include diffusion and the related problem of Brownian motion as well as problems involving the structures of polymers and disordered solids.

Raney nickel A spongy form of nickel made by the action of sodium hydroxide on a nickel–aluminium alloy. The sodium hydroxide dissolves the aluminium leaving a highly active form of nickel with a large surface area. The material is a black pyrophoric powder saturated with hydrogen. It is an extremely efficient catalyst, especially for hydrogenation reactions at room temperature. It was discovered in 1927 by the American chemist M. Raney (1885–1966).

ranksite A mineral consisting of a mixed sodium carbonate, sodium sulphate, and potassium chloride, $2Na_2CO_3.9Na_2SO_4.KCl$.

Raoult's law The partial vapour pressure of a solvent is proportional to its mole fraction. If p is the vapour pressure of the solvent (with a substance dissolved in it) and X the mole fraction of solvent (number of moles of solvent divided by total number of moles) then $p = p_0X$, where p_0 is the vapour pressure of the pure solvent. A solution that obeys Raoult's law is said to be an *ideal solution*. In general the law holds only for dilute solutions, although some mixtures of liquids obey it over a whole range of concentrations. Such solutions are *perfect solutions* and occur when the intermolecular forces between molecules of the pure substances are similar to the forces between molecules of one and molecules of the other. Deviations in Raoult's law for mixtures of liquids cause the formation of *azeotropes. The law was discovered by the French chemist François Raoult (1830–1901).

rapeseed methyl ester *See* **biofuel.**

rare-earth elements *See* **lanthanoids.**

rarefaction A reduction in the pressure of a fluid and therefore of its density.

rare gases *See* **noble gases.**

Raschig process An industrial process for making chlorobenzene (and phenol) by a gas-phase reaction between benzene vapour, hydrogen chloride, and oxygen (air) at 230°C:

$$2C_6H_6 + 2HCl + O_2 \rightarrow 2H_2O + 2C_6H_5Cl$$

The catalyst is copper(II) chloride. The chlorobenzene is mainly used for making phenol by the reaction

$$C_6H_5Cl + H_2O \rightarrow HCl + C_6H_5OH$$

This reaction proceeds at 430°C with a silicon catalyst. The process was invented by the German chemist Fritz Raschig (1863–1928).

Raschig synthesis *See* **hydrazine.**

rate constant (velocity constant) Symbol k. The constant in an expression for the rate of a chemical reaction in terms of concentrations (or activities). For instance, in a simple unimolecular reaction $A \rightarrow B$, the rate is proportional to the concentration of A, i.e. rate = $k[A]$, where k is the rate constant, which depends on the temperature. The equation is the *rate equation* of the reaction, and its form depends on the reaction mechanism.

rate-determining step The slowest step in a chemical reaction that involves a number of steps. In such reactions, there is often a single step that is appreciably slower than the other steps, and the rate of this determines the overall rate of the reaction.

rationalized Planck constant *See* **Planck constant.**

rationalized units A system of units in which the defining equations have been made to conform to the geometry of the system in a logical way. Thus equations that involve circular symmetry contain the factor 2π, while those involving spherical symmetry contain the factor 4π. *SI units are rationalized; c.g.s. units are unrationalized.

Rayleigh, Lord (John William Strutt; 1842–1919) British physicist, who built a private laboratory after working at Cambridge University. His work in this laboratory included the discovery of *Rayleigh scattering of electromagnetic radiation. He also worked in acoustics, electricity, and optics, as well as collaborating with William *Ramsay on the discovery of argon. He was awarded the 1904 Nobel Prize for physics.

Rayleigh scattering Scattering of electromagnetic radiation by molecules in which the frequency of the scattered radiation is unchanged. This type of scattering was analysed by Lord *Rayleigh in his papers in the late 19th century, which showed that the blue colour of the sky is a result of this type of light scattering, with molecules of the atmosphere of the earth scattering light from the sun.

rayon A textile made from cellulose. There are two types, both made from wood pulp. In the viscose process, the pulp is dissolved in carbon disulphide and sodium hydroxide to give a thick brown liquid containing cellulose xanthate. The liquid is then forced through fine nozzles into acid, where the xanthate is decomposed and a cellulose filament is produced. The product is *viscose rayon*. In the acetate process cellulose acetate is made and dissolved in a solvent. The solution is forced through nozzles into air, where the solvent quickly evaporates leaving a filament of *acetate rayon*.

RBS *See* **Rutherford backscattering spectrometry**.

reactant *See* **chemical reaction**.

reaction *See* **chemical reaction**.

reactive dye *See* **dyes**.

reagent A substance reacting with another substance. Laboratory reagents are compounds, such as sulphuric acid, hydrochloric acid, sodium hydroxide, etc., used in chemical analysis or experiments.

realgar A red mineral form of arsenic(II) sulphide, As_2S_2.

real gas A gas that does not have the properties assigned to an *ideal gas. Its molecules have a finite size and there are forces between them (*see* **equation of state**).

rearrangement A type of chemical reaction in which the atoms in a molecule rearrange to form a new molecule.

reciprocal lattice A lattice for a crystal that can be defined from the lattice in real space. If the primitive translation vectors of the lattice in real space are denoted by $\boldsymbol{a}, \boldsymbol{b}, \boldsymbol{c}$ then the primitive translations $\boldsymbol{a}', \boldsymbol{b}', \boldsymbol{c}'$ in the reciprocal lattice are defined by $\boldsymbol{a}' = \boldsymbol{b} \times \boldsymbol{c}[\boldsymbol{abc}]$, $\boldsymbol{b}' = \boldsymbol{c} \times \boldsymbol{a}[\boldsymbol{abc}]$, $\boldsymbol{c}' = \boldsymbol{a} \times \boldsymbol{b}[\boldsymbol{abc}]$, where $[\boldsymbol{abc}]$ denotes the *scalar triple product* $\boldsymbol{a}.(\boldsymbol{b} \times \boldsymbol{c})$. The reciprocal lattice is a fundamental concept in the theory of *X-ray diffraction and energy bands with a diffraction pattern being much more closely related to the reciprocal lattice than the real-space lattice.

reciprocal proportions *See* **chemical combination**.

recombination process The process in which a neutral atom or molecule is formed by the combination of a positive ion and a negative ion or electron; i.e. a process of the type:

$A^+ + B^- \rightarrow AB$

or

$A^+ + e^- \rightarrow A$

In recombination, the neutral species formed is usually in an excited state, from which it can decay with emission of light or other electromagnetic radiation.

recrystallization A process of repeated crystallization in order to purify a substance or to obtain more regular crystals of a purified substance.

rectification The process of purifying a liquid by *distillation. *See* **fractional distillation**.

rectified spirit A constant-boiling mixture of *ethanol (95.6°) and water; it is obtained by distillation.

recycling 1. The recovery and processing of materials after they have been used, which enables them to be reused. For example, used paper, cans, and glass can be broken down into their constituents, which form the raw materials for the manufacture of new products. **2.** The continual movement of *essential elements between the biotic (living) and abiotic (nonliving) components of the environment. *See* **carbon cycle**; **nitrogen cycle**; **phosphorus cycle**; **sulphur cycle**.

red lead *See* **dilead(II) lead(IV) oxide**.

redox *See* **oxidation–reduction**.

reducing agent (reductant) A substance that brings about reduction in other substances. It achieves this by being itself oxidized. Reducing agents contain atoms with low oxidation numbers; that is the atoms have gained electrons. In reducing other substances, these atoms lose electrons.

reducing sugar A monosaccharide or disaccharide sugar that can donate electrons to other molecules and can therefore act as a reducing agent. The possession of a free ketone (–CO–) or aldehyde (–CHO) group enables most monosaccharides and disaccharides to act as reducing sugars. Reducing sugars can be detected by *Benedict's test. *Compare* **nonreducing sugar**.

reductant *See* **reducing agent**.

reduction *See* **oxidation–reduction**.

refinery gas *See* **petroleum**.

refining The process of purifying substances or extracting substances from mixtures.

refluxing A laboratory technique in which a liquid is boiled in a container attached to a condenser (*reflux condenser*), so that the liquid continuously flows back into the container. It is used for carrying out reactions over long periods in organic synthesis.

reforming The conversion of straight-chain alkanes into branched-chain alkanes by *cracking or by catalytic reaction. It is used in petroleum refining to produce hydrocarbons suitable for use in gasoline. Benzene is also manufactured from alkane hydrocarbons by catalytic reforming. *Steam reforming* is a process used to convert methane (from natural gas) into a mixture of carbon monoxide and hydrogen, which is used to synthesize organic chemicals. The reaction

$$CH_4 + H_2O \rightarrow CO + 3H_2$$

occurs at about 900°C using a nickel catalyst.

refractory 1. Having a high melting point. Metal oxides, carbides, and silicides tend to be refractory, and are extensively used for lining furnaces. **2.** A refractory material.

Regnault's method A technique for measuring gas density by evacuating and weighing a glass bulb of known volume, admitting gas at known pressure, and reweighing. The determination must be carried out at constant known temperature and the result corrected to standard temperature and pressure. The method is named after the French chemist Henri Victor Regnault (1810–78).

relative atomic mass (atomic weight; r.a.m.) Symbol A_r. The ratio of the average mass per atom of the naturally occurring form of an element to 1/12 of the mass of a carbon–12 atom.

relative density (r.d.) The ratio of the *density of a substance to the density of some reference substance. For liquids or solids it is the ratio of the density (usually at 20°C) to the density of water (at its maximum density). This quantity was formerly called *specific gravity*. Sometimes relative densities of gases are used; for example, relative to dry air, both gases being at s.t.p.

relative molecular mass (molecular weight) Symbol M_r. The ratio of the average mass per molecule of the naturally occurring form of an element or compound to 1/12 of the mass of a carbon–12 atom. It is equal to the sum of the relative atomic masses of all the atoms that comprise a molecule.

relative permittivity *See* permittivity.

relativistic quantum mechanics Quantum mechanics that is in accord with special relativity theory. The main equation of relativistic quantum mechanics is the *Dirac equation. It is necessary to use relativistic quantum mechanics to describe the electronic properties of heavy atoms and all the *fine structure of atomic spectra. The colour of solid gold and mercury existing as a liquid are both due to relativistic effects in quantum mechanics.

relativistic quantum theory *See* **quantum theory**.

relaxation The return to the equilibrium state of a system after it has experienced a sudden change due to an external influence. The time for relaxation to take place is called the *relaxation time*. An example is the average time that a system remains in the higher energy state before it falls to the lower energy state in *nuclear magnetic resonance (NMR). In general, any process involving decay is assumed to have exponential decay, with the relaxation time being the time it takes for the variable to fall from its initial value to 1/e of its initial value. Another example of relaxation time is the time needed for a gas to return to the Maxwell distribution of velocities, after it has been suddenly disturbed from that state.

release agent A substance that is applied to surfaces to prevent them from sticking together. Release agents are used in many industrial processes, including the manufacture of food, glass, paper, and plastics. They include polyethene, polytetrafluoroethene (PTFE), polyvinyl alcohol (PVA), silicones, and waxes, as well as stearates and other glycerides. They are also known as *abherents* or *parting agents*.

rem *See* **radiation units**.

renaturation The reconstruction of a protein or nucleic acid that has been denatured such that the molecule resumes its original function. Some proteins can be renatured by reversing the conditions (of temperature, pH, etc.) that brought about denaturation.

renewable energy sources Sources of energy that do not use up the earth's finite mineral resources. *Nonrenewable energy sources* are *fossil fuels and fission fuels. Various renewable energy sources are being used or investigated, including geothermal energy, hydroelectric power, nuclear fusion, solar energy, tides, wind power, and wave power.

rennin An enzyme secreted by cells lining the stomach in mammals that is responsible for clotting milk. It acts on a soluble milk protein (*caseinogen*), which it converts to the insoluble form casein. This ensures that milk remains in the stomach long enough to be acted on by protein-digesting enzymes.

reptation A motion describing the dynamics of a polymer in a highly entangled state, such as a network. Regarding the entangled state as a set of chains between crosslinks it is possible to regard the chain as being in a 'tube', with the tube being formed by topological constraints. The chain is longer than the tube so that the 'slack' of the chain moves through the tube, which causes the tube itself to change with time. This motion was called *reptation* (from the Latin *reptare*, to creep), by the French physicist P. G. de Gennes, who postulated it in 1971. Many experiments indicate that reptation dominates the dynamics of polymer chains when they are entangled.

resin A synthetic or naturally occurring *polymer. Synthetic resins are used in making *plastics. Natural resins are acidic chemicals secreted by many trees (especially conifers) into ducts or canals. They are found either as brittle glassy substances or dissolved in essential oils. Their functions are probably similar to those of gums and mucilages.

resolution The process of separating a racemic mixture into its optically active constituents. In some cases the crystals of the two forms have a different appearance, and the separation can be done by hand. In general, however, physical methods (distillation, crystallization, etc.) cannot be used because the optical isomers have identical physical properties. The most common technique is to react the mixture with a compound that is itself optically active, and then separate the two. For instance, a racemic mixture of *l*-A and *d*-A reacted with *l*-B, gives two compounds AB that are not optical isomers but diastereoisomers and can be separated and reconverted into the pure *l*-A and *d*-A. Biological techniques using bacteria that convert one form but not the other can also be used.

resonance The representation of the structure of a molecule by two or more conventional formulae. For example, the formula of methanal can be represented by a covalent structure $H_2C{=}O$, in which there is a double bond in the carbonyl group. It is known that in such compounds the oxygen has some negative charge and the carbon some positive charge. The true bonding in the molecule is somewhere between $H_2C{=}O$ and the ionic compound $H_2C^+O^-$. It is said to be a *resonance hybrid* of the two, indicated by

$$H_2C{=}O \leftrightarrow H_2C^+O^-$$

The two possible structures are called *canonical forms*, and they need not contribute equally to the actual form. Note that the double-headed arrow does not imply that the two forms are in equilibrium.

resonance effect *See* **electronic effects**.

resonance ionization spectroscopy **(RIS)** A spectroscopic technique in which single atoms in a gas are detected using a laser to ionize that atom. A sample containing the atoms to be excited is subjected to light from a laser, tuned so that only that type of atom is excited by the light. If the frequency of light at which the atom is excited is ν, the atoms in the excited state can be ionized if the ionization potential of the atom is less than 2ν. In contrast to other techniques of ionization, this type of ionization only occurs for atoms that are 'in tune' with the frequency of light. Because RIS is very selective in determining which atom is ionized for a given frequency it has many applications in chemistry.

resorcinol *See* **1,3-dihydroxybenzene**.

retinol *See* **vitamin A**.

retort 1. A laboratory apparatus consisting of a glass bulb with a long neck. **2.** A vessel used for reaction or distillation in industrial chemical processes.

retrosynthetic analysis A systematic approach to organic synthesis in which the target molecule is considered and thought of as broken up into smaller parts at certain strategic bonds according to certain rules. These parts are similarly treated until simpler compounds are reached, with known methods of synthesis. Logical analysis of methods of chemical synthesis can be done by computer programs. The technique, which was developed by the American chemist Elias James Corey (1928–), has been very productive in enabling the synthesis of many complex natural products.

reverberatory furnace A metallurgical furnace in which the charge to be heated is kept separate from the fuel. It consists of a shallow hearth on which the charge is heated by flames that pass over it and by radiation reflected onto it from a low roof.

reverse osmosis A method of obtaining pure water from water containing a salt, as in *desalination. Pure water and the salt water are separated by a semipermeable membrane and the pressure of the salt water is raised above the osmotic pressure, causing water from the brine to pass through the membrane into the pure water. This process requires a pressure of some 25 atmospheres, which makes it difficult to apply on a large scale.

reversible process Any process in which the variables that define the state of the system can be made to change in such a way that they pass through the same values in the reverse order when the process is reversed. It is also a condition of a reversible process that any exchanges of energy, work, or matter with the surroundings should be reversed in direction and order when the process is reversed. Any process that does not comply with these conditions when it is reversed is said to be an *irreversible process*. All natural processes are irreversible, although some processes can be made to approach closely to a reversible process.

R_F value ***(in chromatography)*** The distance travelled by a given component divided by the distance travelled by the solvent front. For a given system at a known temperature, it is a characteristic of the component and can be used to identify components.

rhe A unit of fluidity equal to the reciprocal of the *poise.

rhenium Symbol Re. A silvery-white metallic *transition element; a.n. 75; r.a.m. 186.2; r.d. 20.53; m.p. 3180°C; b.p. 5627 (estimated)°C. The element is obtained as a by-product in refining molybdenum, and is used in certain alloys (e.g. rhenium–molybdenum alloys are superconducting). The element forms a number of complexes with oxidation states in the range 1–7. It was discovered by Walter Noddack (1893–1960) and Ida Tacke (1896–) in 1925.

rheology The study of the deformation and flow of matter.

rheopexy The process by which certain thixotropic substances set more rapidly when they are stirred, shaken, or tapped. Gypsum in water is such a *rheopectic substance.*

rhodinol *See* **aminophenol.**

rhodium Symbol Rh. A silvery-white metallic *transition element; a.n. 45; r.a.m. 102.9; r.d. 12.4; m.p. 1966°C; b.p. 3727°C. It occurs with platinum and is used in certain platinum alloys (e.g. for thermocouples) and in plating jewellery and optical reflectors. Chemically, it is not attacked by acids (dissolves only slowly in aqua regia) and reacts with nonmetals (e.g. oxygen and chlorine) at red heat. Its main oxidation state is +3 although it also forms complexes in the +4 state. The element was discovered in 1803 by William Wollaston (1766–1828).

RIA *See* **radioimmunoassay.**

riboflavin *See* **vitamin B complex.**

ribonucleic acid *See* **RNA.**

ribose A *monosaccharide, $C_5H_{10}O_5$, rarely occurring free in nature but important as a component of *RNA (ribonucleic acid). Its derivative *deoxyribose*, $C_5H_{10}O_4$, is equally important as a constituent of *DNA (deoxyribonucleic acid), which carries the genetic code in chromosomes.

ribulose A ketopentose sugar (*see* **monosaccharide**), $C_5H_{11}O_5$, that is involved in carbon dioxide fixation in photosynthesis as a component of *ribulose bisphosphate.

ribulose bisphosphate (RuBP) A five-carbon sugar that is combined with carbon dioxide to form two three-carbon intermediates in the first stage of the light-dependent reactions of *photosynthesis (*see* **Calvin cycle**). The enzyme that mediates the carboxylation of ribulose bisphosphate is *ribulose bisphosphate carboxylase.*

Rice–Herzfeld mechanism A mechanism enabling complex chain reactions to give simple rate laws in chemical kinetics. An example of a Rice–Herzfeld mechanism occurs with the pyrolysis of acetaldehyde, the mechanism consisting of initiation, propagation (in two steps), and termination. This leads to the experimental result that the overall reaction is three-halves order in CH_3CHO. To ascertain that such a mechanism is correct, either the steady-state approximation is used or the differential equations for the reaction rates are solved numerically. It is possible, as in this example, for the Rice–Herzfeld mechanism to give a correct description of the main part of the reaction but not to take account of reactions that produce by-products. The Rice–Herzfeld mechanism is named after F. O. Rice and K. F. Herzfeld, who put forward the scheme in 1934.

Rice–Ramsperger–Kassel theory (RRK theory) A statistical theory used to describe unimolecular reactions. The RRK theory, put forward by O. K. Rice, H. C. Ramsperger, and independently by L. S. Kassel in the late 1920s, assumes that a molecule is a system of loosely coupled oscillators.

Regarding the coupling of the oscillators as loose enables calculations to be made on the statistical distribution of energy freely flowing between the vibrational modes. In RRK theory the rate constant for the decomposition of a molecule increases with the energy of the molecule. The RRK theory has been applied with success to some chemical reactions but is limited in its predictive power. The unsatisfactory features of the RRK theory are improved by the *RRKM theory (Rice–Ramsperger–Kassel–Marcus theory), in which the individual vibrational modes of the system are taken into account.

ring A closed chain of atoms in a molecule. In compounds, such as naphthalene, in which two rings share a common side, the rings are *fused rings*. *Ring closures* are chemical reactions in which one part of a chain reacts with another to form a ring, as in the formation of *lactams and *lactones.

ring conformations Shapes that can be taken by nonplanar rings and can be interconverted by formal rotations about single bonds. In cyclohexane, for example, the most stable conformation is the *chair* conformation in which the 2, 3, 5, and 6 atoms lie in a plane and the 1 and 4 carbon atoms lie on opposite sides of the plane. In this form, there is no ring strain. In the *boat* conformation, the 1 and 4 carbon atoms lie on the same side of the plane. Other conformers are the *twist* conformation and the *half chair* conformation (see diagram). The energy difference between the chair conformation and the (least stable) half chair is 10.8 kcal/mol and the forms interconvert quickly at normal temperatures (about 99% of molecules are in the chair form).

Other rings also have different conformations. In a five-membered ring, such as that of cyclopentane, the *twist* conformation has three adjacent carbon atoms in a plane with one of the carbons above this plane and the other below it. The less stable *envelope* conformation has four atoms in a plane with one atom out of the plane. In an eight-membered ring, such as that of cyclooctane, the *tub* conformation has four atoms corresponding to a pair of diametrically opposite bonds in one plane, with the other four atoms on the same side of this plane. It is analogous to the boat form of cyclohexane. The *crown* conformation has atoms alternately above and below the average plane of the molecule.

Various names are given to positions of groups attached to atoms in different ring conformations. In the chair conformation of cyclohexane, an *axial* bond is one that makes a large angle with the average plane of the ring. An equatorial bond is one that makes a small angle. In a cyclopentane ring, the terms *pseudo-axial* and *pseudo-equatorial* are used. In the boat conformation of cyclohexane the bond roughly parallel to the plane of four atoms is the *bowsprit*, the other bond being the *flagpole*.

RIS *See* **resonance ionization spectroscopy.**

R-isomer *See* **absolute configuration.**

RME *See* **biofuel.**

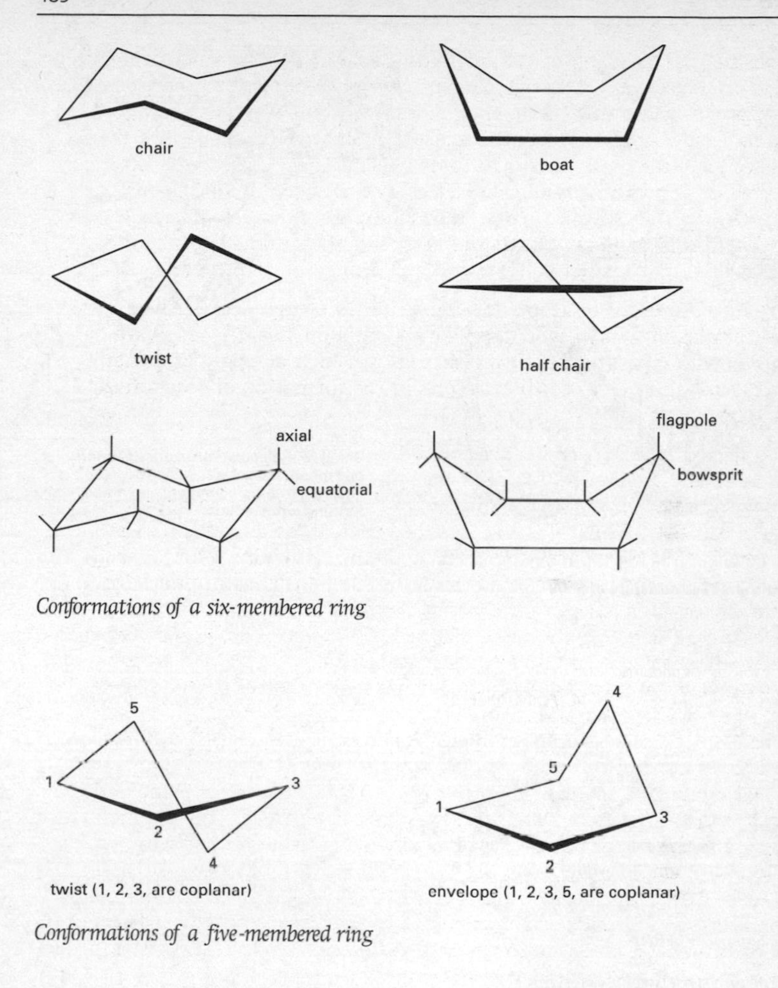

Conformations of a six-membered ring

Conformations of a five-membered ring

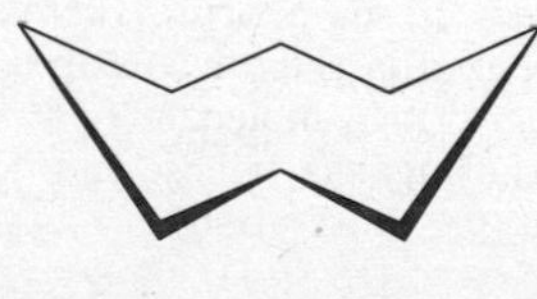

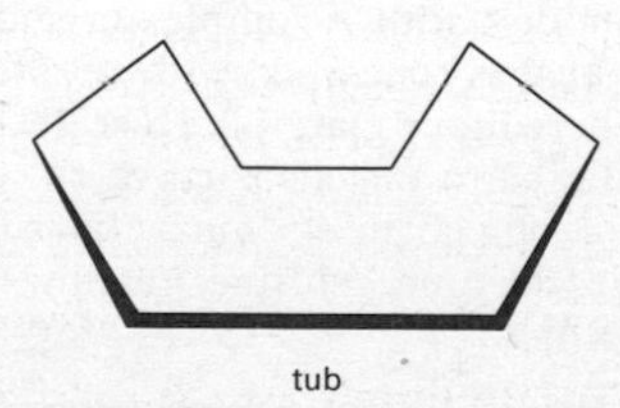

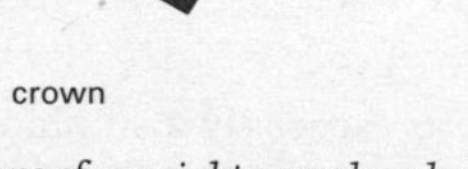

Conformations of an eight-membered ring

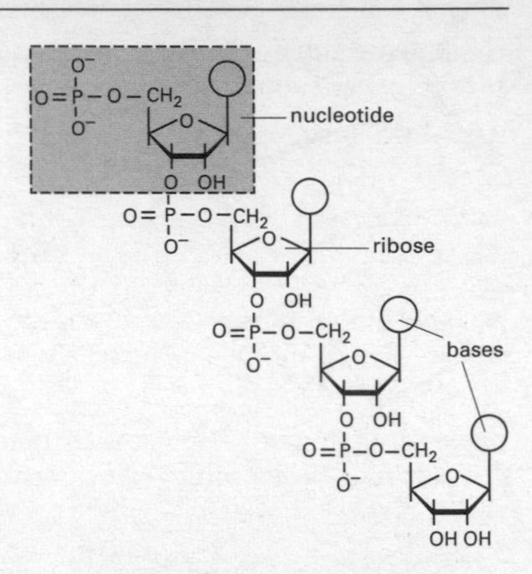

Detail of molecular structure of sugar–phosphate backbone. Each ribose unit is attached to a phosphate group and a base, forming a nucleotide.

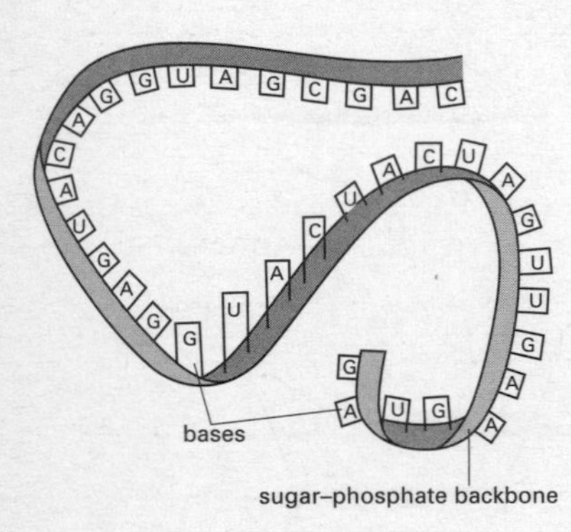

Single-stranded structure of RNA

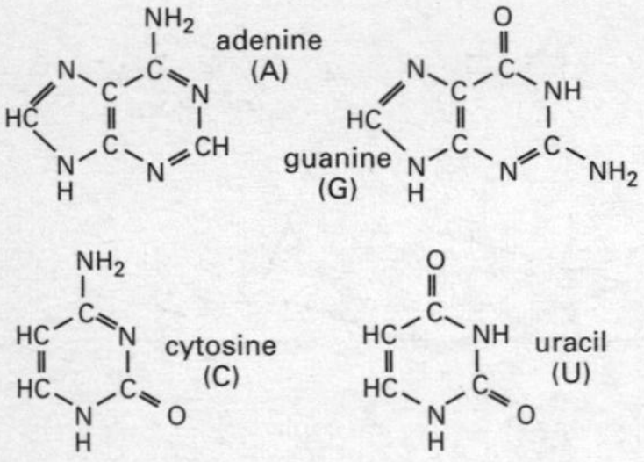

adenine (A)

guanine (G)

cytosine (C)

uracil (U)

The four bases of RNA

Molecular structure of RNA

RNA (ribonucleic acid) A complex organic compound (a nucleic acid) in living cells that is concerned with *protein synthesis. In some viruses, RNA is also the hereditary material. Most RNA is synthesized in the nucleus and then distributed to various parts of the cytoplasm. An RNA molecule consists of a long chain of *nucleotides in which the sugar is *ribose and the bases are adenine, cytosine, guanine, and uracil (see illustration). *Compare* **DNA**.

roasting The heating of a finely ground ore, especially a sulphide, in air prior to *smelting. The roasting process expels moisture, chemically combined water, and volatile matter; in the case of sulphides, the sulphur is expelled as sulphur dioxide and the ore is converted into an oxide.

Rochelle salt Potassium sodium tartrate tetrahydrate, $KNaC_4H_4O_6.4H_2O$. A colourless crystalline salt used for its piezoelectric properties.

Rochon prism An optical device consisting of two quartz prisms; the first, cut parallel to the optic axis, receives the light; the second, with the optic axis at right angles, transmits the ordinary ray without deviation but the extraordinary ray is deflected and can be absorbed by a screen. The device can be used to produce plane-polarized light and it can also be used with ultraviolet radiation.

rock An aggregate of mineral particles that makes up part of the earth's crust. It may be consolidated or unconsolidated (e.g. sand, gravel, mud, shells, coral, and clay).

rock crystal *See* **quartz**.

rocking-chair cell *See* **intercalation cell**.

rock salt *See* **halite**.

roentgen The former unit of dose equivalent (*see* **radiation units**). It is named after the discoverer of X-rays, W. K. Roentgen (1845–1923).

roentgenium Symbol Rg. A radioactive transactinide; a.n. 111. It was made by fusion of ^{209}Bi with ^{64}Ni. Only a few atoms have been detected.

root-mean-square value (RMS value) **1.** ***(in statistics)*** A typical value of a number (n) of values of a quantity ($x_1,x_2,x_3...$) equal to the square root of the sum of the squares of the values divided by n, i.e.

$$\text{RMS value} = \sqrt{[(x_1^2 + x_2^2 + x_3^2...)/n]}$$

2. ***(in physics)*** A typical value of a continuously varying quantity, such as an alternating electric current, obtained similarly from many samples taken at regular time intervals during a cycle. Theoretically this can be shown to be the *effective value*, i.e. the value of the equivalent direct current that would produce the same power dissipation. For a sinusoidal current this is equal to $I_m/\sqrt{2}$, where I_m is the maximum value of the current.

Rose's metal An alloy of low melting point (about 100°C) consisting of 50% bismuth, 25–28% lead, and 22–25% tin.

rosin A hard natural resin obtained from pine tree oil and the wastes from processing wood pulp. It may be colourless, yellow, brown, or black. It is used as a flotation agent, solder flux, sizing compound, and in lacquers and plasticizers. It is also used to provide 'grip' to violinists' bows (when it may be called *colophony*) and dancers' and boxers' shoes.

rotamer One of a set of *conformational isomers that differ from each other by restricted rotation about one or more single bonds.

rotational spectroscopy The spectroscopic study of the rotational motion of molecules. Rotational spectroscopy gives information about interatomic distances. The transitions between different rotational energy levels in molecules correspond to the microwave and far-infrared regions of the electromagnetic spectrum. It is only possible for there to be transitions between rotational energy levels in pure rotational spectra if the molecule has a permanent dipole moment. In the near-infrared region rotational transitions are superimposed on vibrational transitions, resulting in a *vibrational–rotational* spectrum. This type of spectrum is considerably more complicated than a purely rotational spectrum. *See* **vibrational spectroscopy**.

rotation group A group formed by the set of all rotations about a point. The rotation group is associated with the angular momentum of a system and is important in the theory of the rotational motion of molecules. The group representations of the rotation group are closely associated with the quantum theory of angular momentum.

RRK theory *See* **Rice–Ramsperger–Kassel theory**.

RRKM theory (Rice–Ramsperger–Kassel–Marcus theory) A theory of unimolecular chemical reactions in which the *Rice–Ramsperger–Kassel theory was improved by the US chemists O. K. Rice and R. A. Marcus in 1950; it was further improved subsequently in several papers by Marcus and associates by taking into account the individual vibrational frequencies, rotations, and zero-point energies of the energized species and activated complexes taking part. The RRKM theory is very successful in explaining the results of experiments for unimolecular reactions for a variety of reactions.

R–S convention *See* **absolute configuration**.

rubber A polymeric substance obtained from the sap of the tree *Hevea brasiliensis*. Crude natural rubber is obtained by coagulating and drying the sap (latex), and is then modified by *vulcanization and compounding with fillers. It is a polymer of *isoprene containing the unit $-CH_2C(CH_3):CHCH_2-$. Various synthetic rubbers can also be made. *See* **neoprene**; **nitrile rubber**; **silicones**.

rubidium Symbol Rb. A soft silvery-white metallic element belonging to *group 1 (formerly IA) of the periodic table; a.n. 37; r.a.m. 85.47; r.d. 1.53; m.p. 38.89°C; b.p. 688°C. It is found in a number of minerals (e.g. lepidolite) and in certain brines. The metal is obtained by electrolysis of molten rubidium chloride. The naturally occurring isotope rubidium–87 is radioactive (*see* **rubidium–strontium dating**). The metal is highly reactive, with properties similar to those of other group 1 elements, igniting spontaneously in air. It was discovered spectroscopically by Robert *Bunsen and Gustav *Kirchhoff in 1861.

rubidium–strontium dating A method of dating geological specimens based on the decay of the radioisotope rubidium–87 into the stable isotope strontium–87. Natural rubidium contains 27.85% of rubidium–87, which has a half-life of 4.7×10^{10} years. The ratio $^{87}Rb/^{87}Sr$ in a specimen gives an estimate of its age (up to several thousand million years).

ruby The transparent red variety of the mineral *corundum, the colour being due to the presence of traces of chromium. It is a valuable gemstone, more precious than diamonds. The finest rubies are obtained from Mogok in Burma, where they occur in metamorphic limestones; Sri Lanka and Thailand are the only other important sources. Rubies have been produced synthetically by the Verneuil flame-fusion process. Industrial rubies are used in lasers, watches, and other precision instruments.

Russell–Saunders coupling **(L–S coupling)** A type of coupling in systems involving many *fermions. These systems include electrons in atoms and nucleons in nuclei, in which the energies associated with electrostatic repulsion are much greater than the energies associated with *spin–orbit coupling. *Multiplets of many-electron atoms with a low atomic number are characterized by Russell–Saunders coupling. It is named after the US physicists Henry Norris Russell (1877–1957) and Frederick Saunders (1875–1963) who postulated this type of coupling to explain the spectra of many-electron atoms with low atomic number in 1925. The multiplets of heavy atoms and nuclei are better described by *j-j coupling or *intermediate coupling*, i.e. a coupling in which the energies of electrostatic repulsion and spin–orbit coupling are similar in size.

rusting Corrosion of iron (or steel) to form a hydrated iron(III) oxide $Fe_2O_3.xH_2O$. Rusting occurs only in the presence of both water and oxygen. It is an electrochemical process in which different parts of the iron surface act as electrodes in a cell reaction. At the anode, iron atoms dissolve as Fe^{2+} ions:

ruthenium Symbol Ru. A hard white metallic *transition element; a.n. 44; r.a.m. 101.07; r.d. 12.3; m.p. 2310°C; b.p. 3900°C. It is found associated with platinum and is used as a catalyst and in certain platinum alloys. Chemically, it dissolves in fused alkalis but is not attacked by acids. It reacts with oxygen and halogens at high temperatures. It also forms complexes with a range of oxidation states. The element was isolated by K. K. Klaus in 1844.

Rutherford, Ernest, Lord (1871–1937) New Zealand-born British physicist, who worked under Sir J. J. *Thomson at Cambridge University (1895–98). He then took up a professorship at McGill University, Canada, and collaborated with Frederick *Soddy in studying radioactivity. In 1899 he discovered *alpha particles and beta particles, followed by the discovery of *gamma radiation the following year. In 1905, with Soddy, he announced that radioactive *decay involves a series of transformations. He moved to Manchester University in 1907 and there, with Hans Geiger and E. Marsden, devised the alpha-particle scattering experiment that led in 1911 to the discovery of the atomic nucleus. After moving to Cambridge in 1919 he achieved the artificial splitting of atoms. In 1908 he was awarded the Nobel Prize for chemistry.

Rutherford backscattering spectrometry (RBS) A technique for analysing samples of material by irradiation with a beam of alpha particles and measurement of the energies of the alpha particles after they have been scattered by the sample. This enables the elements present and their amounts to be determined because the energy of a scattered alpha particle depends on the mass of the nucleus with which it collides.

rutherfordium Symbol Rf. A radioactive *transactinide element; a.n. 104. It was first reported in 1964 at Dubna, near Moscow, and in 1969 it was detected by A. Ghiorso and a team at Berkeley, California. It can be made by bombarding californium-249 nuclei with carbon-12 nuclei.

Rutherford model The model of an atom put forward by Ernest *Rutherford in 1911 on the basis of experiments on the scattering of alpha particles. The model consisted of a very dense positively charged nucleus, with electrons orbiting round the nucleus. This model presented a serious difficulty for the classical theory of electricity and magnetism, which predicts that the electron should spiral into the nucleus in a fraction of a second, radiating electromagnetic energy while doing so. This difficulty led to the development of the *Bohr theory in 1913 and was definitively solved by the development of quantum mechanics and its application to atomic structure in the mid-1920s.

$Fe(s) \rightarrow Fe^{2+}(aq) + 2e$

At the cathode, hydroxide ions are formed:

$$O_2(aq) + 2H_2O(l) + 4e \rightarrow 4OH^-(aq)$$

The $Fe(OH)_2$ in solution is oxidized to Fe_2O_3. Rusting is accelerated by impurities in the iron and by the presence of acids or other electrolytes in the water.

rutile A mineral form of titanium(IV) oxide, TiO_2.

Rydberg constant Symbol R. A constant that occurs in the formulae for atomic spectra and is related to the binding energy between an electron and a nucleon. It is connected to other constants by the relationship $R = \mu_0{}^2me^4c^3/8h^3$, where μ_0 is the magnetic constant, m and e are the mass and charge of an electron, c is the speed of light, and h is the *Planck constant. It has the value $1.097 \times 10^7\ m^{-1}$. It is named after the Swedish physicist Johannes Robert Rydberg (1854–1919), who developed a formula for the spectrum of hydrogen.

Rydberg spectrum An absorption spectrum of a gas in the ultraviolet region, consisting of a series of lines that become closer together towards shorter wavelengths, merging into a continuous absorption region. The absorption lines correspond to electronic transitions to successively higher energy levels. The onset of the continuum corresponds to photoionization of the atom or molecule, and can thus be used to determine the ionization potential.

S

Sabatier–Senderens process A method of organic synthesis employing hydrogenation and a heated nickel catalyst. It is employed commercially for hydrogenating unsaturated vegetable oils to make margarine. It is named after Paul Sabatier (1854–1941) and Jean-Baptiste Senderens (1856–1937).

saccharide *See* sugar.

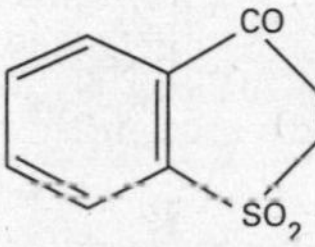

Saccharin

saccharin A white crystalline solid, $C_7H_5NO_3S$, m.p. 224°C. It is made from a compound of toluene, derived from petroleum or coal tar. It is a well-known artificial sweetener, being some 500 times as sweet as sugar (sucrose), and is usually marketed as its sodium salt. Because of an association with cancer in laboratory animals, its use is restricted in some countries.

saccharose *See* sucrose.

Sachse reaction A reaction of methane at high temperature to produce ethyne:

$$2CH_4 \rightarrow C_2H_2 + 3H_2$$

The reaction occurs at about 1500°C, the high temperature being obtained by burning part of the methane in air.

Sackur–Tetrode equation An equation for the entropy of a perfect monatomic gas. The entropy S is given by:

$$S = nR\ln(e^{5/2}V/nN_A\Lambda^3),$$

where $\Lambda = h/(2\pi mkT)^{1/2}$, where n is the amount of the gas, R is the gas constant, e is the base of natural logarithms, V is the volume of the system, N_A is the Avogadro constant, h is the Planck constant, m is the mass of each atom, k is the Boltzmann constant, and T is the thermodynamic temperature. To calculate the *molar entropy* of the gas both sides are divided by n. The Sackur–Tetrode equation can be used to show that the entropy change ΔS, when a perfect gas expands isothermally from V_i to V_f, is given by:

$$\Delta S = nR\ln(aV_f) - nR\ln(aV_i) = nR\ln(V_f/V_i),$$

where *aV* is the quantity inside the logarithm bracket in the Sackur–Tetrode equation.

sacrificial protection (cathodic protection) The protection of iron or steel against corrosion (*see* **rusting**) by using a more reactive metal. A common form is galvanizing (*see* **galvanized iron**), in which the iron surface is coated with a layer of zinc. Even if the zinc layer is scratched, the iron does not rust because zinc ions are formed in solution in preference to iron ions. Pieces of magnesium alloy are similarly used in protecting pipelines, etc.

sal ammoniac *See* **ammonium chloride**.

SALC (symmetry-adapted linear combinations) A linear combination of atomic orbitals (LCAO), which is a 'building block' of the LCAO orbitals making up a molecular *orbital. The SALC are constructed by group theory appropriate for the symmetry group of the molecule. SALC are used in the construction of molecular orbitals.

salicylic acid (1-hydroxybenzoic acid) A naturally occurring carboxylic acid, HOC_6H_4COOH, found in certain plants; r.d. 1.44; m.p. 159°C; sublimes at 211°C. It is used in making aspirin and in the foodstuffs and dyestuffs industries.

saline Describing a chemical compound that is a salt, or a solution containing a salt.

salinometer An instrument for measuring the salinity of a solution. There are two main types: one is a type of hydrometer to measure density; the other is an apparatus for measuring the electrical conductivity of the solution.

sal soda Anhydrous *sodium carbonate, Na_2CO_3.

salt A compound formed by reaction of an acid with a base, in which the hydrogen of the acid has been replaced by metal or other positive ions. Typically, salts are crystalline ionic compounds such as Na^+Cl^- and $NH_4^+NO_3^-$. Covalent metal compounds, such as $TiCl_4$, are also often regarded as salts.

salt bridge An electrical connection made between two half cells. It usually consists of a glass U-tube filled with agar jelly containing a salt, such as potassium chloride. A strip of filter paper soaked in the salt solution can also be used.

salt cake Industrial *sodium sulphate.

salting in *See* **salting out**.

salting out The effect in which the solubility of a substance in a certain solvent is reduced by the presence of a second solute dissolved in the solvent. For example, certain substances dissolved in water can be precipitated (or evolved as a gas) by addition of an ionic salt. The substance is more soluble in pure water than in the salt solution. The opposite effect involving an increase in solubility may occur. This is known as *salting in.*

saltpetre *See* **nitre**.

samarium Symbol Sm. A soft silvery metallic element belonging to the *lanthanoids; a.n. 62; r.a.m. 150.35; r.d. 7.52 (20°C); m.p. 1077°C; b.p. 1791°C. It occurs in monazite and bastnatite. There are seven naturally occurring isotopes, all of which are stable except samarium-147, which is weakly radioactive (half-life 2.5×10^{11} years). The metal is used in special alloys for making nuclear-reactor parts as it is a neutron absorber. Samarium oxide (Sm_2O_3) is used in small quantities in special optical glasses. The largest use of the element is in the ferromagnetic alloy $SmCo_5$, which produces permanent magnets five times stronger than any other material. The element was discovered by François Lecoq de Boisbaudran in 1879.

sand Particles of rock with diameters in the range 0.06–2.00 mm. Most sands are composed chiefly of particles of quartz, which are derived from the weathering of quartz-bearing rocks.

Sandmeyer reaction A reaction of diazonium salts used to prepare chloro- or bromo-substituted aromatic compounds. The method is to diazotize an aromatic amide at low temperature and add an equimolar solution of the halogen acid and copper(I) halide. A complex of the diazonium salt and copper halide forms, which decomposes when the temperature is raised. The copper halide acts as a catalyst in the reaction of the halide ions from the acid, for example

$$C_6H_5N_2^+(aq) + Cl^-(aq) + CuCl(aq) \rightarrow C_6H_5Cl(l) + N_2(g) + CuCl(aq)$$

The reaction was discovered in 1884 by the German chemist Traugott Sandmeyer (1854–1922). *See also* **Gattermann reaction**.

sandwich compound A transition-metal complex in which a metal atom or ion is 'sandwiched' between two rings of atoms. *Ferrocene was the first such compound to be prepared, having two parallel cyclopentadienyl rings with an iron ion between them. In such compounds (also known as *metallocenes*) the metal coordinates to the pi electron system of the ring, rather than to individual atoms. A wide variety of these compounds are known, having five-, six-, seven-, or eight-membered rings and involving such metals as Cr, Mn, Co, Ni, and Fe. Other similar compounds are known. A *multidecker sandwich* has three or more parallel rings with metal atoms between them. In a *bent sandwich*, the rings are not parallel. A *half sandwich* (or *piano stool*) has one ring, with single ligands on the other side of the metal.

Sanger's reagent 2,4-dinitrofluorobenzene, $C_6H_3F(NO_2)_2$, used to identify the end *amino acid in a protein chain. It is named after Frederick Sanger (1918–).

saponification The reaction of esters with alkalis to give alcohols and salts of carboxylic acids:

$$RCOOR' + OH^- \rightarrow RCOO^- + R'OH$$

See **esterification**; **soap**.

saponin A type of toxic *glycoside that forms a frothy colloidal solution on shaking with water. Saponins occur in many plants (such as horse chestnut). They break down red blood cells and have been used for poisoning fish. On hydrolysis they yield a variety of sugars.

sapphire Any of the gem varieties of *corundum except ruby, especially the blue variety, but other colours of sapphire include yellow, brown, green, pink, orange, and purple. Sapphires are obtained from igneous and metamorphic rocks and from alluvial deposits. The chief sources are Sri Lanka, Kashmir, Burma, Thailand, East Africa, the USA, and Australia. Sapphires are used as gemstones and in record-player styluses and some types of laser. They are synthesized by the Verneuil flame-fusion process.

saturable laser *See* **dye laser**.

saturated 1. (of a compound) Consisting of molecules that have only single bonds (i.e. no double or triple bonds). Saturated compounds can undergo substitution reactions but not addition reactions. *Compare* **unsaturated**. **2.** (of a solution) Containing the maximum equilibrium amount of solute at a given temperature. In a saturated solution the dissolved substance is in equilibrium with undissolved substance; i.e. the rate at which solute particles leave the solution is exactly balanced by the rate at which they dissolve. A solution containing less than the equilibrium amount is said to be *unsaturated*. One containing more than the equilibrium amount is *supersaturated*. Supersaturated solutions can be made by slowly cooling a saturated solution. Such solutions are metastable; if a small crystal seed is added the excess solute crystallizes out of solution. **3.** (of a vapour) *See* **vapour pressure**.

saturation *See* **supersaturation**.

saturation spectroscopy A spectroscopic technique using lasers to locate absorption maxima with great precision. In saturation spectroscopy the laser beam is split into an intense saturating beam and a less intense beam that pass through the cavity containing the sample in almost opposite directions. The saturating beam sometimes excites molecules, which are shifted to its frequency by the Doppler effect. The other beam gives a modulated signal at the detector if it is interacting with the same Doppler-shifted molecules. These molecules are not moving parallel to the beams and have an extremely small Doppler shift, thus providing very high resolution. *See also* **Lamb-dip spectroscopy**.

sawhorse projection A type of *projection formula in which a three-dimensional view is drawn. *See* **conformation**.

***s*-block elements** The elements of the first two groups of the *periodic table; i.e. groups 1 (Li, Na, K, Rb, Cs, Fr) and 2 (Be, Mg, Ca, Sr, Ba, Ra). The outer electronic configurations of these elements all have inert-gas structures plus outer ns^1 (group 1) or ns^2 (group 2) electrons. The term thus excludes elements with incomplete inner *d*-levels (transition metals) or with incomplete inner *f*-levels (lanthanoids and actinoids) even though these often have outer ns^2 or occasionally ns^1 configurations. Typically, the *s*-block elements are reactive metals forming stable ionic compounds containing M^+ or M^{2+} ions. *See* **alkali metals; alkaline-earth metals.**

sc Synclinal. *See* **torsion angle.**

scandium Symbol Sc. A rare soft silvery metallic element belonging to group 3 (formerly IIIA) of the periodic table; a.n. 21; r.a.m. 44.956; r.d. 2.989 (alpha form), 3.19 (beta form); m.p. 1541°C; b.p. 2831°C. Scandium often occurs in *lanthanoid ores, from which it can be separated on account of the greater solubility of its thiocyanate in ether. The only natural isotope, which is not radioactive, is scandium-45, and there are nine radioactive isotopes, all with relatively short half-lives. Because of the metal's high reactivity and high cost no substantial uses have been found for either the metal or its compounds. Predicted in 1869 by Dmitri *Mendeleev, and then called *ekaboron*, the oxide (called *scandia*) was isolated by Lars Nilson (1840–99) in 1879.

scanning electron microscope *See* **electron microscope.**

scanning tunnelling microscope (STM) A type of electron microscope that uses the quantum-mechanical *tunnel effect to study atomic structures and observe single atoms. A fine-pointed conducting tip near a surface of a sample causes electrons to tunnel from the surface to the tip. Its occurrence depends on the electron density of the surface and the distance between the tip and the surface itself. The electric current produced is kept constant by moving the tip up or down as moves across the surface.

scavenger A reagent that removes a trace component from a system or that removes a reactive intermediate from a reaction.

SCF *See* **self-consistent field.**

Scheele, Karl Wilhelm (1742–86) Swedish chemist, who became an apothecary and in 1775 set up his own pharmacy at Köping. He made many chemical discoveries. In 1772 he prepared oxygen (*see also* **Lavoisier, Antoine**; **Priestley, Joseph**) and in 1774 he isolated chlorine. He also discovered manganese, glycerol, hydrocyanic (prussic) acid, citric acid, and many other substances.

scheelite A mineral form of calcium tungstate, $CaWO_4$, used as an ore of tungsten. It occurs in contact metamorphosed deposits and vein deposits as colourless or white tetragonal crystals.

Schiff's base A compound formed by a condensation reaction between an aromatic amine and an aldehyde or ketone, for example

$$RNH_2 + R'CHO \rightarrow RN{:}CHR' + H_2O$$

The compounds are often crystalline and are used in organic chemistry for characterizing aromatic amines (by preparing the Schiff's base and measuring the melting point). They are named after the German chemist Hugo Schiff (1834–1915).

Schiff's reagent A reagent, devised by Hugo Schiff, used for testing for aldehydes and ketones; it consists of a solution of fuchsin dye that has been decolorized by sulphur dioxide. Aliphatic aldehydes restore the pink immediately, whereas aromatic ketones have no effect on the reagent. Aromatic aldehydes and aliphatic ketones restore the colour slowly.

Schoenflies system A system for categorizing symmetries of molecules. C_n groups contain only an *n*-fold rotation axis. C_{nv} groups, in addition to the *n*-fold rotation axis, have a mirror plane that contains the axis of rotation (and mirror planes associated with the existence of the *n*-fold axis). C_{nh} groups, in addition to the *n*-fold rotation axis, have a mirror plane perpendicular to the axis. S_n groups have an *n*-fold rotation–reflection axis. D_n groups have an *n*-fold rotation axis and a two-fold axis perpendicular to the *n*-fold axis (and two-fold axes associated with the existence of the *n*-fold axis). D_{nh} groups have all the symmetry operations of D_n and also a mirror plane perpendicular to the *n*-fold axis. D_{nd} groups contain all the symmetry operations of D_n and also mirror planes that contain the *n*-fold axis and bisect the angles between the two-fold axes. In the Schoenflies notation *C* stands for 'cyclic', *S* stands for 'spiegel' (mirror), and *D* stands for 'dihedral'. The subscripts *h*, *v*, and *d* stand for horizontal, vertical, and diagonal respectively, where these words refer to the position of the mirror planes with respect to the *n*-fold axis (considered to be vertical). In addition to the *noncubic groups* referred to so far, there are *cubic groups*, which have several rotation axes with the same value of *n*. These are the *tetrahedral groups* T, T_h, and T_d, the *octahedral groups* O and O_h, and the *icosahedral group* I. The Schoenflies system is commonly used for isolated molecules, while the *Hermann–Mauguin system is commonly used in crystallography.

schönite A mineral form of potassium sulphate, K_2SO_4.

Schottky defect *See* **crystal defect.**

Schrödinger, Erwin (1887–1961) Austrian physicist, who became professor of physics at Berlin University in 1927. He left for Oxford to escape the Nazis in 1933, returned to Graz in Austria in 1936, and then left again in 1938 for Dublin's Institute of Advanced Studies. He finally returned to Austria in 1956. He is best known for the development of *wave mechanics and the *Schrödinger equation, work that earned him a share of the 1933 Nobel Prize for physics with Paul *Dirac.

Schrödinger equation An equation used in wave mechanics (*see* **quantum mechanics**) for the wave function of a particle. The time-independent Schrödinger equation is:

$$\nabla^2\psi + 8\pi^2 m(E - U)\psi/h^2 = 0,$$

where ψ is the wave function, ∇^2 the Laplace operator, h the Planck constant, m the particle's mass, E its total energy, and U its potential energy. It can also be written as:

$$H\psi = E\psi,$$

where H is the *Hamiltonian operator. It was devised by Erwin Schrödinger, who was mainly responsible for wave mechanics.

Schweizer's reagent A solution made by dissolving copper(II) hydroxide in concentrated ammonia solution. It has a deep blue colour and is used as a solvent for cellulose in the cuprammonium process for making rayon. When the cellulose solution is forced through spinnarets into an acid bath, fibres of cellulose are reformed.

scrubber *See* **absorption tower**.

seaborgium Symbol Sg. A radioactive *transactinide element; a.n. 106. It was first detected in 1974 by Albert Ghiorso and a team in California. It can be produced by bombarding californium-249 nuclei with oxygen-18 nuclei. It is named after the US physicist Glenn Seaborg (1912–99).

secondary alcohol *See* **alcohols**.

secondary amine *See* **amines**.

secondary cell A *voltaic cell in which the chemical reaction producing the e.m.f. is reversible and the cell can therefore be charged by passing a current through it. *See* **accumulator; intercalation cell**. *Compare* **primary cell**.

secondary emission The emission of electrons from a surface as a result of the impact of other charged particles, especially as a result of bombardment with (primary) electrons. As the number of secondary electrons can exceed the number of primary electrons, the process is important in photomultipliers. *See also* **Auger effect**.

secondary-ion mass spectrometry (SIMS) A technique for analysing the chemical structure of a solid by bombarding it with ions. A sample of the solid to be analysed is bombarded with ions, referred to as *primary ions*, having energies between 5 and 20 keV. A number of different types of entity are detached from the surface of the solid, including neutral atoms and molecules, electrons, photons, and negative and positive ions. The negative and positive ions given off, called the *secondary ions*, can be identified using mass spectrometry. Using this technique solid samples can be characterized with great accuracy.

second-order reaction *See* **order**.

sedimentation The settling of the solid particles through a liquid either to produce a concentrated slurry from a dilute suspension or to clarify a liquid containing solid particles. Usually this relies on the force of gravity, but if the particles are too small or the difference in density between the solid and liquid phases is too small, a *centrifuge may be used. In the simplest case the rate of sedimentation is determined by Stokes's law, but in practice the predicted rate is rarely reached. Measurement of the rate of sedimentation in an *ultracentrifuge can be used to estimate the size of macromolecules.

seed A crystal used to induce other crystals to form from a gas, liquid, or solution.

Seger cones **(pyrometric cones)** A series of cones used to indicate the temperature inside a furnace or kiln. The cones are made from different mixtures of clay, limestone, feldspars, etc., and each one softens at a different temperature. The drooping of the vertex is an indication that the known softening temperature has been reached and thus the furnace temperature can be estimated.

selection rules Rules that determine which transitions between different energy levels are possible in a system, such as an elementary particle, nucleus, atom, molecule, or crystal, described by quantum mechanics. Transitions cannot take place between any two energy levels. Group theory, associated with the symmetry of the system, determines which transitions, called *allowed transitions, can take place and which transitions, called *forbidden transitions, cannot take place. Selection rules determined in this way are very useful in analysing the spectra of quantum-mechanical systems.

selenides Binary compounds of selenium with other more electropositive elements. Selenides of nonmetals are covalent (e.g. H_2Se). Most metal selenides can be prepared by direct combination of the elements. Some are well-defined ionic compounds (containing Se^{2-}), while others are nonstoichiometric interstitial compounds (e.g. Pd_4Se, $PdSe_2$).

selenium Symbol Se. A metalloid element belonging to group 16 (formerly VIB) of the periodic table; a.n. 34; r.a.m. 78.96; r.d. 4.81 (grey); m.p. 217°C (grey); b.p. 684.9°C. There are a number of allotropic forms, including grey, red, and black selenium. It occurs in sulphide ores of other metals and is obtained as a by-product (e.g. from the anode sludge in electrolytic refining). The element is a semiconductor; the grey allotrope is light-sensitive and is used in photocells, xerography, and similar applications. Chemically, it resembles sulphur, and forms compounds with selenium in the +2, +4, and +6 oxidation states. Selenium was discovered in 1817 by Jöns Berzelius.

selenium cell Either of two types of photoelectric cell; one type relies on the photoconductive effect, the other on the photovoltaic effect (*see* **photoelectric effect**). In the photoconductive selenium cell an external e.m.f. must be applied; as the selenium changes its resistance on exposure to light, the current produced is a measure of the light energy falling on the selenium. In the photovoltaic selenium cell, the e.m.f. is generated within the cell. In this type of cell, a thin film of vitreous or metallic selenium is applied to a metal surface, a transparent film of another metal, usually gold or platinum, being placed over the selenium. Both types of cell are used as light meters in photography.

self-assembly *See* **self-organization.**

self-consistent field (SCF) A concept used to find approximate solutions to the many-body problem in *quantum mechanics. The procedure starts with an approximate solution for a particle moving in a single-particle potential, which derives from its average interaction with all the other particles. This average interaction is determined by the *wave functions of all the other particles. The equation describing this average interaction is solved and the improved solution obtained is used in the calculation of the interaction term. This procedure is repeated for wave functions until the wave functions and associated energies are not significantly changed in the cycle, self-consistency having been attained. In atomic theory the *Hartree–Fock procedure makes use of self-consistent fields.

self-organization The spontaneous order arising in a system when certain parameters of the system reach critical values. Self-organization occurs in many systems in physics, chemistry, and biology. It can occur when a system is driven far from thermal *equilibrium. Since a self-organizing system is open to its environment, the second law of *thermodynamics is not violated by the formation of an ordered phase, as entropy can be transferred to the environment. Self-organization is related to the concepts of broken symmetry, complexity, nonlinearity, and *nonequilibrium statistical mechanics. Many systems that undergo transitions to self-organization can also undergo transitions to *chaos. In chemistry, *self-assembly* is one of the features of *supramolecular chemistry.

Seliwanoff's test A biochemical test to identify the presence of ketonic sugars, such as fructose, in solution. It was devised by the Russian chemist F. F. Seliwanoff. A few drops of the reagent, consisting of resorcinol crystals dissolved in equal amounts of water and hydrochloric acid, are heated with the test solution and the formation of a red precipitate indicates a positive result.

semicarbazones Organic compounds containing the unsaturated group $=C:N.NH.CO.NH_2$. They are formed when aldehydes or ketones react with a semicarbazide ($H_2N.NH.CO.NH_2$). Semicarbazones are crystalline compounds with relatively high melting points. They are used to identify aldehydes and ketones in quantitative analysis: the semicarbazone derivative is made and identified by its melting point. Semicarbazones are also used in separating ketones from reaction mixtures: the derivative is crystallized out and hydrolysed to give the ketone.

semiclassical approximation An approximation technique used to calculate quantities in quantum mechanics. This technique is called the semiclassical approximation because the *wave function is written as an asymptotic series with ascending powers of the Planck constant, *h*, with the first term being purely classical. It is also known as the *Wentzel–Kramers–Brillouin (WKB) approximation*, named after Gregor Wentzel (1898–1978), Hendrik Anton Kramers (1894–1952), and Léon Brillouin (1889–1969), who invented it independently in 1926. The semiclassical approximation is particularly successful for calculations involving the *tunnel effect, such as field emission, and radioactive decay producing alpha particles.

semiconductor A crystalline solid with an electrical conductivity (typically 10^5–10^{-7} siemens per metre) intermediate between that of a conductor (up to 10^9 S m^{-1}) and an insulator (as low as 10^{-15} S m^{-1}). Semiconducting properties are a feature of *metalloid elements, such as silicon and germanium. As the atoms in a crystalline solid are close together, the orbitals of their electrons overlap and their individual energy levels are spread out into energy bands. Conduction occurs in semiconductors as the result of a net movement, under the influence of an electric field, of electrons in the conduction band and empty states, called *holes*, in the valence band. A hole behaves as if it was an electron with a positive charge. Electrons and holes are known as the *charge carriers* in a semiconductor. The type of charge carrier that predominates in a particular region or material is called the *majority carrier* and that with the lower concentration is the *minority carrier*. An *intrinsic semiconductor* is one in which the concentration of charge carriers is a characteristic of the material itself; electrons jump to the conduction band from the valence band as a result of thermal excitation, each electron that makes the jump leaving behind a hole in the valence band. Therefore, in an intrinsic semiconductor the charge carriers are equally divided between electrons and holes. In *extrinsic semiconductors* the type of conduction that predominates depends on the number and valence of the impurity atoms present. Germanium and silicon atoms have a valence of four. If impurity atoms with a valence of five, such as arsenic, antimony, or phosphorus, are added to the lattice, there will be an extra electron per atom available for conduction, i.e. one that is not required to pair with the four valence electrons of the germanium or silicon. Thus extrinsic semiconductors doped with atoms of valence five give rise to crystals with electrons as majority carriers, the so-called *n-type conductors*. Similarly, if the impurity atoms have a valence of three, such as boron, aluminium, indium, or gallium, one hole per atom is created by an unsatisfied bond. The majority carriers are therefore holes, i.e. *p-type conductors*.

semiconductor laser A type of *laser in which semiconductors provide the excitation. The laser action results from electrons in the conduction band (*see* **energy bands**) being stimulated to recombine with holes in the valence band. When this occurs the electrons give up the energy corresponding to the band gap. Materials, such as gallium arsenide, are suitable for this purpose. A junction between p-type and n-type semiconductors can be used with the light passing along the plane of the junction. Mirrors for the laser action are provided by the ends of the crystals. Semiconductor lasers can be as small as 1 mm in length.

semi-empirical calculations A technique for calculating atomic and molecular quantities in which the values of integrals are found using quantities derived from experiment (such as ionization energies obtained spectroscopically). Semi-empirical calculations were formerly widely used, particularly for large molecules involving extensive computation to calculate all the integrals using *ab-initio calculations. As computing power has increased, the properties of large molecules can be calculated by ab-initio calculations, therefore the use of semi-empirical calculations has steadily declined, first for small molecules and more recently for large molecules.

semimetal *See* metalloid.

semipermeable membrane A membrane that is permeable to molecules of the solvent but not the solute in *osmosis. Semipermeable membranes can be made by supporting a film of material (e.g. cellulose) on a wire gauze or porous pot.

semipolar bond *See* chemical bond.

Semtex A nitrogen-based stable odourless plastic *explosive.

septivalent **(heptavalent)** Having a valency of seven.

sequence rule *See* CIP system.

sequestration The process of forming coordination complexes of an ion in solution. Sequestration often involves the formation of chelate complexes, and is used to prevent the chemical effect of an ion without removing it from the solution (e.g. the sequestration of Ca^{2+} ions in water softening). It is also a way of supplying ions in a protected form, e.g. sequestered iron solutions for plants in regions having alkaline soil.

serine *See* amino acid.

serpentine Any of a group of hydrous magnesium silicate minerals with the general composition $Mg_3Si_2O_5(OH)_4$. Serpentine is monoclinic and occurs in two main forms: *chrysotile*, which is fibrous and the chief source of *asbestos; and *antigorite*, which occurs as platy masses. It is generally green or white with a mottled appearance, sometimes resembling a snakeskin – hence the name. It is formed through the metamorphic alteration of ultrabasic rocks rich in olivine, pyroxene, and amphibole. *Serpentinite* is a rock consisting mainly of serpentine; it is used as an ornamental stone.

sesqui- Prefix indicating a ratio of 2:3 in a chemical compound. For example, a sesquioxide has the formula M_2O_3.

sesquiterpene *See* terpenes.

SEXAFS **(surface-extended X-ray absorption fine-structure spectroscopy)** A technique that makes use of the oscillations in X-ray absorbance found in the high-frequency side of the absorption edge to investigate the structure of surfaces. The physical principle behind SEXAFS is quantum-mechanical interference between the wave function of a photoelectron and the wave function associated with scattering by surrounding atoms. In the case of constructive interference, the photoelectron has a higher probability of appearing, while in the case of destructive interference, the photoelectron has a lower probability of appearing. The technique therefore provides information about the surroundings of an atom. Analysis of results from SEXAFS has given a great deal of information about the structure of surfaces and how they respond to adsorption.

sexivalent **(hexavalent)** Having a valency of six.

shale A form of *clay that occurs in thin layers. Shales are very common sedimentary rocks. *See also* **oil shale**.

shell *See* **atom**.

shellac A hard resin produced as a secretion by a plant parasite, the south-east Asian lac insect *Lacifer lacca*. It is used in sealing wax, varnish (French polish), and electrical insulators.

sherardizing The process of coating iron or steel with a zinc corrosion-resistant layer by heating it in contact with zinc dust to a temperature slightly below the melting point of zinc. At a temperature of about 371°C the two metals amalgamate to form internal layers of zinc–iron alloys and an external layer of pure zinc. The process was invented by Sherard Cowper-Coles (d. 1935).

shock wave A very narrow region of high pressure and temperature formed in a fluid when the fluid flows supersonically over a stationary object or a projectile flying supersonically passes through a stationary fluid. A shock wave may also be generated by violent disturbances in a fluid, such as a lightning stroke or a bomb blast. Shock waves are generated experimentally to excite molecules for spectroscopic investigations.

short period *See* **periodic table**.

sial The rocks that form the earth's continental crust. These are granite rock types rich in *si*lica (SiO_2) and *al*uminium (Al), hence the name. *Compare* sima.

side chain *See* **chain**.

side reaction A chemical reaction that occurs at the same time as a main reaction but to a lesser extent, thus leading to other products mixed with the main products.

siderite A brown or grey-green mineral form of iron(II) carbonate, $FeCO_3$, often with magnesium and manganese substituting for the iron. It occurs in sedimentary deposits or in hydrothermal veins and is an important iron ore. It is found in England, Greenland, Spain, N Africa, and the USA.

siemens Symbol S. The SI unit of electrical conductance equal to the conductance of a circuit or element that has a resistance of 1 ohm. 1 S = 10^{-1} Ω. The unit was formerly called the mho or reciprocal ohm. It is named after Ernst Werner von Siemens (1816–92).

sievert The SI unit of dose equivalent (*see* **radiation units**). It is named after the Swedish physicist Rolf Sievert (1896–1966).

sigma bond *See* **orbital**.

sigma electron An electron in a sigma orbital. *See* **orbital**.

silane (silicane) 1. A colourless gas, SiH_4, which is insoluble in water; d. 1.44 g dm^{-3}; r.d. 0.68 (liquid); m.p. −185°C; b.p. −112°C. Silane is produced by reduction of silicon with lithium tetrahydridoaluminate(III). It is also formed by the reaction of magnesium silicide (Mg_2Si) with acids, although other silicon hydrides are also produced at the same time. Silane itself is stable in the absence of air but is spontaneously flammable, even at low temperatures. It is a reducing agent and has been used for the removal of corrosion in inaccessible plants (e.g. pipes in nuclear reactors). **2.** (*or* **silicon hydride**) Any of a class of compounds of silicon and hydrogen. They have the general formula Si_nH_{2n+2}. The first three in the series are silane itself (SiH_4), *disilane* (Si_2H_6), and *trisilane* (Si_3H_8). The compounds are analogous to the alkanes but are much less stable and only the lower members of the series can be prepared in any quantity (up to Si_6H_{14}). No silicon hydrides containing double or triple bonds exist (i.e. there are no analogues of the alkenes and alkynes).

silica *See* **silicon(IV) oxide**.

silica gel A rigid gel made by coagulating a sol of sodium silicate and heating to drive off water. It is used as a support for catalysts and also as a drying agent because it readily absorbs moisture from the air. The gel itself is colourless but, when used in desiccators, etc., a blue cobalt salt is added. As moisture is taken up, the salt turns pink, indicating that the gel needs to be regenerated (by heating).

silicane *See* **silane**.

silicate Any of a group of substances containing negative ions composed of silicon and oxygen. The silicates are a very extensive group and natural silicates form the major component of most rocks (*see* **silicate minerals**). The basic structural unit is the tetrahedral SiO_4 group. This may occur as a simple discrete SiO_4^{4-} anion as in the *orthosilicates*, e.g. *phenacite* (Be_2SiO_4) and *willemite* (Zn_2SiO_4). Many larger silicate species are also found (see illustration). These are composed of SiO_4 tetrahedra linked by sharing oxygen atoms as in the *pyrosilicates*, $Si_2O_7^{6-}$, e.g. $Sc_2Si_2O_7$. The linking can extend to such forms as benitoite, $BaTiSi_3O_9$, or alternatively infinite chain anions, which are single strand (*pyroxenes) or double strand (*amphiboles). Spodumene, $LiAl(SiO_3)_2$, is a pyroxene and the asbestos minerals are amphiboles. Large two-dimensional sheets are also possible, as in the various *micas (see illustration), and the linking can extend to full three-dimensional framework structures, often with substituted trivalent atoms in the lattice. The *zeolites are examples of this.

silicate minerals A group of rock-forming minerals that make up the bulk of the earth's outer crust (about 90%) and constitute one-third of all minerals. All silicate minerals are based on a fundamental structural unit – the SiO_4 tetrahedron (*see* **silicate**). They consist of a metal (e.g. calcium, magnesium, aluminium) combined with silicon and oxygen. The silicate minerals are classified on a structural basis according to how the tetrahedra are linked together. The six groups are: nesosilicates (e.g. olivine and *garnet); sorosilicates (e.g. hemimorphite); cyclosilicates (e.g. axinite, *beryl, and *tourmaline); inosilicates (e.g. *amphiboles and *pyroxenes); phyllosilicates (e.g. *micas, *clay minerals, and *talc); and tektosilicates (e.g. *feldspars and *feldspathoids). Many silicate minerals are of economic importance.

silicide A compound of silicon with a more electropositive element. The silicides are structurally similar to the interstitial carbides but the range encountered is more diverse. They react with mineral acids to form a range of *silanes.

silicon Symbol Si. A metalloid element belonging to *group 14 (formerly IVB) of the periodic table; a.n. 14; r.a.m. 28.086; r.d. 2.33; m.p. 1410°C; b.p. 2355°C. Silicon is the second most abundant element in the earth's crust (25.7% by weight) occurring in various forms of silicon(IV) oxide (e.g. *quartz) and in *silicate minerals. The element is extracted by reducing the oxide with carbon in an electric furnace and is used extensively for its semiconductor properties. It has a diamond-like crystal structure; an amorphous form also exists. Chemically, silicon is less reactive than carbon. The element combines with oxygen at red heat and is also dissolved by molten alkali. There is a large number of organosilicon compounds (e.g. *siloxanes) although silicon does not form the range of silicon–hydrogen compounds and derivatives that carbon does (*see* **silane**). The element was identified by Antoine *Lavoisier in 1787 and first isolated in 1823 by Jöns *Berzelius.

silicon carbide **(carborundum)** A black solid compound, SiC, insoluble in water and soluble in molten alkali; r.d. 3.217; m.p. *c.* 2700°C. Silicon carbide is made by heating silicon(IV) oxide with carbon in an electric furnace (depending on the grade required sand and coke may be used). It is extremely hard and is widely used as an abrasive. The solid exists in both zinc blende and wurtzite structures.

silicon dioxide *See* **silicon(IV) oxide**.

silicones Polymeric compounds containing chains of silicon atoms alternating with oxygen atoms, with the silicon atoms linked to organic groups. A variety of silicone materials exist, including oils, waxes, and rubbers. They tend to be more resistant to temperature and chemical attack than their carbon analogues.

SiO_4^{4-} as in Be_2SiO_4 (phenacite)

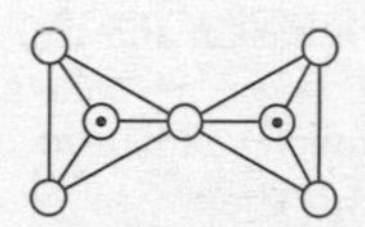

$Si_2O_5^{2-}$ as in $Sc_2Si_2O_7$ (thortveitite)

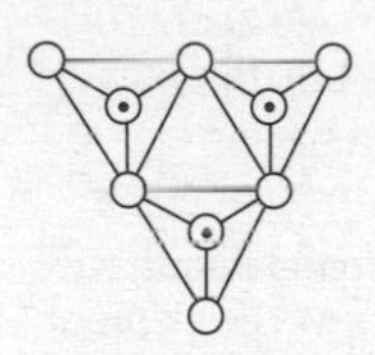

$Si_3O_9^{6-}$ as in $BaTiSi_3O_9$ (bentonite)

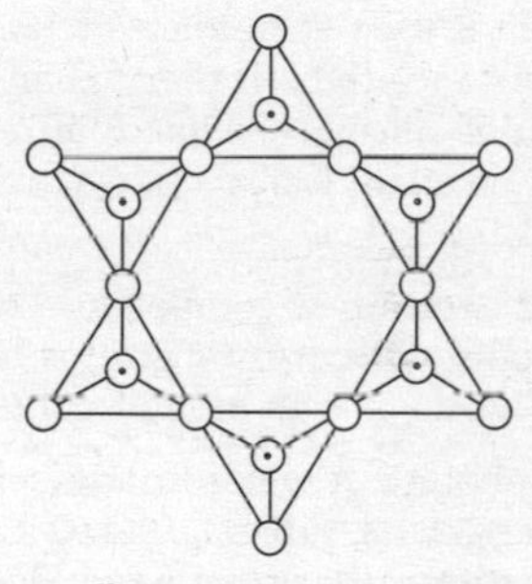

$Si_6O_{18}^{12-}$ as in $Be_3Al_2Si_6O_{18}$ (beryl)

Structure of some discrete silicate ions

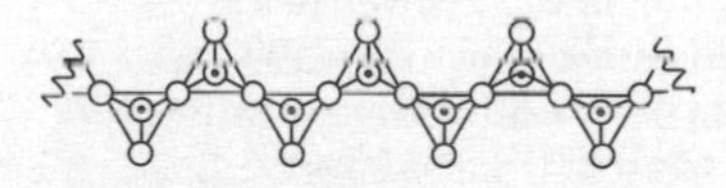

single chain : pyroxenes

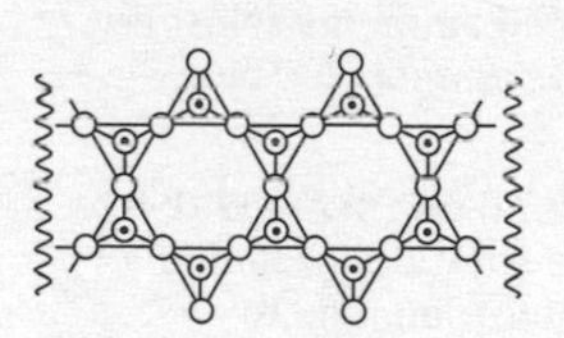

double chain : amphiboles

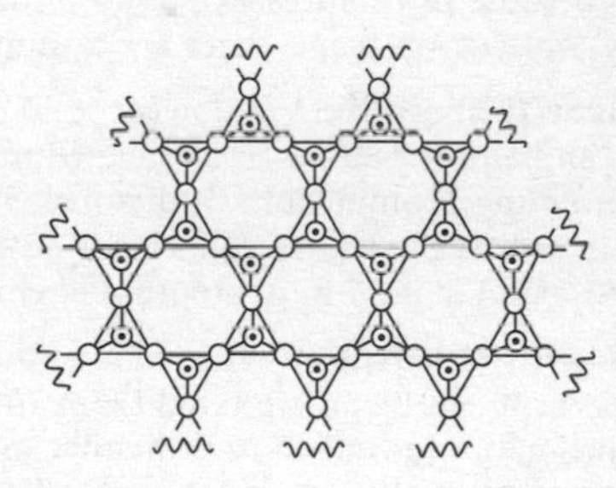

sheet : micas

Structure of some polymeric silicate ions

silicon hydride *See* silane.

silicon(IV) oxide **(silicon dioxide; silica)** A colourless or white vitreous solid, SiO_2, insoluble in water and soluble (by reaction) in hydrofluoric acid and in strong alkali; m.p. 1713°C; b.p. 2230°C. The following forms occur naturally: *cristobalite* (cubic or tetragonal crystals; r.d. 2.32); *tridymite* (rhombic; r.d. 2.26); *quartz (hexagonal; r.d. 2.63–2.66); *lechatelierite* (r.d. 2.19). Quartz has two modifications: α-quartz below 575°C and β-quartz above 575°C; above 870°C β-quartz is slowly transformed to tridymite and above 1470°C this is slowly converted to cristobalite. Various forms of silicon(IV) oxide occur widely in the earth's crust; yellow sand for example is quartz with iron(III) oxide impurities and flint is essentially amorphous silica. The gemstones amethyst, opal, and rock crystal are also forms of quartz.

Silica is an important commercial material in the form of *silica brick*, a highly refractive furnace lining, which is also resistant to abrasion and to corrosion. Silicon(IV) oxide is also the basis of both clear and opaque silica glass, which is used on account of its transparency to ultraviolet radiation and its resistance to both thermal and mechanical shock. A certain proportion of silicon(IV) oxide is also used in ordinary glass and in some glazes and enamels. It also finds many applications as a drying agent in the form of *silica gel.

siloxanes A group of compounds containing silicon atoms bound to oxygen atoms, with organic groups linked to the silicon atoms, e.g. $R_3SiOSiR_3$, where R is an organic group. *Silicones are polymers of siloxanes.

silver Symbol Ag. A white lustrous soft metallic *transition element; a.n. 47; r.a.m. 107.87; r.d. 10.5; m.p. 961.93°C; b.p. 2212°C. It occurs as the element and as the minerals argentite (Ag_2S) and horn silver (AgCl). It is also present in ores of lead and copper, and is extracted as a by-product of smelting and refining these metals. The element is used in jewellery, tableware, etc., and silver compounds are used in photography. Chemically, silver is less reactive than copper. A dark silver sulphide forms when silver tarnishes in air because of the presence of sulphur compounds. Silver(I) ionic salts exist (e.g. $AgNO_3$, AgCl) and there are a number of silver(II) complexes.

silver(I) bromide A yellowish solid compound, AgBr; r.d. 6.5; m.p. 432°C. It can be precipitated from silver(I) nitrate solution by adding a solution containing bromide ions. It dissolves in concentrated ammonia solutions (but, unlike the chloride, does not dissolve in dilute ammonia). The compound is used in photographic emulsions.

silver(I) chloride A white solid compound, AgCl; r.d. 5.6; m.p. 455°C; b.p. 1550°C. It can be precipitated from silver(I) nitrate solution by adding a solution of chloride ions. It dissolves in ammonia solution (due to formation of the complex ion $[Ag(NH_3)_2]^+$). The compound is used in photographic emulsions.

silver(I) iodide A yellow solid compound, AgI; r.d. 6.01; m.p. 558°C; b.p. 1506°C. It can be precipitated from silver(I) nitrate solutions by adding a solution of iodide ions. Unlike the chloride and bromide, it does not dissolve in ammonia solutions.

silver-mirror test *See* **Tollens reagent**.

silver(I) nitrate A colourless solid, $AgNO_3$; r.d. 4.3; m.p. 212°C. It is an important silver salt because it is water-soluble. It is used in photography. In the laboratory, it is used as a test for chloride, bromide, and iodide ions and in volumetric analysis of chlorides using an *adsorption indicator.

silver(I) oxide A brown slightly water-soluble amorphous powder, Ag_2O; r.d. 7.14. It can be made by adding sodium hydroxide solution to silver(I) nitrate solution. Silver(I) oxide is strongly basic and is also an oxidizing agent. It is used in certain reactions in preparative organic chemistry; for example, moist silver(I) oxide converts haloalkanes into alcohols; dry silver oxide converts haloalkanes into ethers. The compound decomposes to the elements at 300°C and can be reduced by hydrogen to silver. With ozone it gives the oxide AgO (which is diamagnetic and probably $Ag^{I}Ag^{III}O_2$).

sima The rocks that form the earth's oceanic crust and underlie the upper crust. These are basaltic rock types rich in *si*lica (SiO_2) and *ma*gnesium (Mg), hence the name. The sima is denser and more plastic than the *sial that forms the continental crust.

SIMS *See* **secondary-ion mass spectrometry.**

single bond *See* **chemical bond.**

singlet An atomic state in which two spin angular momenta of electrons cancel each other, resulting in zero net spin. A singlet state usually has a higher energy than a *triplet state because of the effect of correlations of spin on the Coulomb interactions between electrons (as in *Hund's rules). This can lead to substantial energy differences between triplet and singlet states.

sintered glass Porous glass made by sintering powdered glass, used for filtration of precipitates in gravimetric analysis.

sintering The process of heating and compacting a powdered material at a temperature below its melting point in order to weld the particles together into a single rigid shape. Materials commonly sintered include metals and alloys, glass, and ceramic oxides. Sintered magnetic materials, cooled in a magnetic field, make especially retentive permanent magnets.

S-isomer *See* **absolute configuration.**

SI units Système International d'Unités: the international system of units now recommended for all scientific purposes. A coherent and rationalized system of units derived from the *m.k.s. units, SI units have now replaced *c.g.s. units and *Imperial units. The system has seven *base units* and two *dimensionless units* (formerly called *supplementary units*), all other units being derived from these nine units. There are 18 derived units with special names. Each unit has an agreed symbol (a capital letter or an initial capital letter if it is named after a scientist, otherwise the symbol consists of one or two lower-case letters). Decimal multiples of the units are indicated by a set of prefixes; whenever possible a prefix representing 10 raised to a power that is a multiple of three should be used. See Appendix.

skew *See* **torsion angle.**

slag Material produced during the *smelting or refining of metals by reaction of the flux with impurities (e.g. calcium silicate formed by reaction of calcium oxide flux with silicon dioxide impurities). The liquid slag can be separated from the liquid metal because it floats on the surface. *See also* **basic slag.**

slaked lime *See* **calcium hydroxide.**

SLUMO Second-lowest unoccupied molecular orbital. *See* **subjacent orbitals.**

slurry A paste consisting of a suspension of a solid in a liquid.

smectic *See* **liquid crystal.**

smelting The process of separating a metal from its ore by heating the ore to a high temperature in a suitable furnace in the presence of a reducing agent, such as carbon, and a fluxing agent, such as limestone. Iron ore is smelted in this way so that the metal melts and, being denser than the molten *slag, sinks below the slag, enabling it to be removed from the furnace separately.

smoke A fine suspension of solid particles in a gas.

S_N1 reaction *See* **nucleophilic substitution.**

S_N2 reaction *See* **nucleophilic substitution.**

SNG Substitute (or synthetic) natural gas; a mixture of gaseous hydrocarbons produced from coal, petroleum, etc., and suitable for use as a fuel. Before the discovery of natural gas *coal gas was widely used as a domestic and industrial fuel. This gave way to natural gas in the early part of this century in the US and other countries where natural gas was plentiful. The replacement of coal gas occurred somewhat later in the UK and other parts of Europe. More recently, interest has developed in ways of manufacturing hydrocarbon gas fuels. The main sources are coal and the naphtha fraction of petroleum. In the case of coal three methods have been used: (1) pyrolysis – i.e. more efficient forms of destructive distillation, often with further hydrogenation of the hydrocarbon products; (2) heating the coal with hydrogen and catalysts to give hydrocarbons – a process known as *hydroliquefaction* (*see also* **Bergius process**); (3) producing carbon monoxide and hydrogen and obtaining hydrocarbons by the *Fischer–Tropsch process. SNG from naptha is made by steam *reforming.

soap A substance made by boiling animal fats with sodium hydroxide. The reaction involves the hydrolysis of *glyceride esters of fatty acids to glycerol and sodium salts of the acids present (mainly the stearate, oleate, and palmitate), giving a soft semisolid with *detergent action. Potassium hydroxide gives a more liquid product (*soft soap*). By extension, other metal salts of long-chain fatty acids are also called soaps. *See also* **saponification.**

soda Any of a number of sodium compounds, such as caustic soda (NaOH) or, especially, washing soda ($Na_2CO_3.10H_2O$).

soda ash Anhydrous *sodium carbonate, Na_2CO_3.

soda lime A mixed hydroxide of sodium and calcium made by slaking lime with caustic soda solution (to give NaOH + $Ca(OH)_2$) and recovering greyish white granules by evaporation. The material is produced largely for industrial adsorption of carbon dioxide and water, but also finds some applications in pollution and effluent control. It is also used as a laboratory drying agent.

sodamide *See* **sodium amide**.

Soddy, Frederick (1877–1956) British chemist, who worked with Ernest *Rutherford in Canada and William *Ramsay in London before finally settling in Oxford in 1919. His announcement in 1913 of the existence of *isotopes won him the 1921 Nobel Prize for physics.

sodium Symbol Na. A soft silvery reactive element belonging to group 1 (formerly IA) of the periodic table (*see* **alkali metals**); a.n. 11; r.a.m. 22.9898; r.d. 0.97; m.p. 97.8°C; b.p. 882–889°C. Sodium occurs as the chloride in sea water and in the mineral halite. It is extracted by electrolysis in a Downs cell. The metal is used as a reducing agent in certain reactions and liquid sodium is also a coolant in nuclear reactors. Chemically, it is highly reactive, oxidizing in air and reacting violently with water (it is kept under oil). It dissolves in liquid ammonia to form blue solutions containing solvated electrons. Sodium is a major *essential element required by living organisms. The element was first isolated by Humphry Davy in 1807.

sodium acetate *See* **sodium ethanoate**.

sodium aluminate A white solid, $NaAlO_2$ or $Na_2Al_2O_4$, which is insoluble in ethanol and soluble in water giving strongly alkaline solutions; m.p. 1800°C. It is manufactured by heating bauxite with sodium carbonate and extracting the residue with water, or it may be prepared in the laboratory by adding excess aluminium to hot concentrated sodium hydroxide. In solution the ion $Al(OH)_4^-$ predominates. Sodium aluminate is used as a mordant, in the production of zeolites, in effluent treatment, in glass manufacture, and in cleansing compounds.

sodium amide **(sodamide)** A white crystalline powder, $NaNH_2$, which decomposes in water and in warm ethanol, and has an odour of ammonia; m.p. 210°C; b.p. 400°C. It is produced by passing dry ammonia over metallic sodium at 350°C. It reacts with red-hot carbon to give sodium cyanide and with nitrogen(I) oxide to give sodium azide.

sodium azide A white or colourless crystalline solid, NaN_3, soluble in water and slightly soluble in alcohol; hexagonal; r.d. 1.846; decomposes on heating. It is made by the action of nitrogen(I) oxide on hot sodamide ($NaNH_2$) and is used as an organic reagent and in the manufacture of detonators.

sodium benzenecarboxylate (sodium benzoate) An either colourless crystalline or white amorphous powder, C_6H_5COONa, soluble in water and slightly soluble in ethanol. It is made by the reaction of sodium hydroxide with benzoic acid and is used in the dyestuffs industry and as a food preservative. It was formerly used as an antiseptic.

sodium benzoate *See* **sodium benzenecarboxylate.**

sodium bicarbonate *See* **sodium hydrogencarbonate.**

sodium bisulphate *See* **sodium hydrogensulphate.**

sodium bisulphite *See* **sodium hydrogensulphite.**

sodium bromide A white crystalline solid, NaBr, known chiefly as the dihydrate (monoclinic; r.d. 2.17), and as the anhydrous salt (cubic; r.d. 3.20; m.p. 747°C; b.p. 1390°C). The dihydrate loses water at about 52°C and is very slightly soluble in alcohol. Sodium bromide is prepared by the reaction of bromine on hot sodium hydroxide solution or of hydrogen bromide on sodium carbonate solution. It is used in photographic processing and in analytical chemistry.

sodium carbonate Anhydrous sodium carbonate (*soda ash, sal soda*) is a white powder, which cakes and aggregates on exposure to air due to the formation of hydrates. The monohydrate, $Na_2CO_3.H_2O$, is a white crystalline material, which is soluble in water and insoluble in alcohol; r.d. 2.532; loses water at 109°C; m.p. 851°C.

The decahydrate, $Na_2CO_3.10H_2O$ (*washing soda*), is a translucent efflorescent crystalline solid; r.d. 1.44; loses water at 32–34°C to give the monohydrate; m.p. 851°C. Sodium carbonate may be manufactured by the *Solvay process or by suitable crystallization procedures from any one of a number of natural deposits, such as:

trona ($Na_2CO_3.NaHCO_3.2H_2O$),
natron ($Na_2CO_3.10H_2O$),
ranksite ($2Na_2CO_3.9Na_2SO_4.KCl$),
pirsonnite ($Na_2CO_3.CaCO_3.2H_2O$),
gaylussite ($Na_2CO_3.CaCO_3.5H_2O$).

The method of extraction is very sensitive to the relative energy costs and transport costs in the region involved. Sodium carbonate is used in photography, in cleaning, in pH control of water, in textile treatment, glasses and glazes, and as a food additive and volumetric reagent. *See also* **sodium sesquicarbonate.**

sodium chlorate(V) A white crystalline solid, $NaClO_3$; cubic; r.d. 2.49; m.p. 250°C. It decomposes above its melting point to give oxygen and sodium chloride. The compound is soluble in water and in ethanol and is prepared by the reaction of chlorine on hot concentrated sodium hydroxide. Sodium chlorate is a powerful oxidizing agent and is used in the manufacture of matches and soft explosives, in calico printing, and as a garden weedkiller.

sodium chloride (common salt) A colourless crystalline solid, NaCl, soluble in water and very slightly soluble in ethanol; cubic; r.d. 2.17; m.p. 801°C; b.p. 1413°C. It occurs as the mineral *halite (rock salt) and in natural brines and sea water. It has the interesting property of a solubility in water that changes very little with temperature. It is used industrially as the starting point for a range of sodium-based products (e.g. Solvay process for Na_2CO_3, Castner–Kellner process for NaOH), and is known universally as a preservative and seasoner of foods. Sodium chloride has a key role in biological systems in maintaining electrolyte balances.

sodium cyanide A white or colourless crystalline solid, NaCN, deliquescent, soluble in water and in liquid ammonia, and slightly soluble in ethanol; cubic; m.p. 564°C; b.p. 1496°C. Sodium cyanide is now made by absorbing hydrogen cyanide in sodium hydroxide or sodium carbonate solution. The compound is extremely poisonous because it reacts with the iron in haemoglobin in the blood, so preventing oxygen reaching the tissues of the body. It is used in the extraction of precious metals and in electroplating industries. Aqueous solutions are alkaline due to salt hydrolysis.

sodium dichromate A red crystalline solid, $Na_2Cr_2O_7.2H_2O$, soluble in water and insoluble in ethanol. It is usually known as the dihydrate (r.d. 2.52), which starts to lose water above 100°C; the compound decomposes above 400°C. It is made by melting chrome iron ore with lime and soda ash and acidification of the chromate thus formed. Sodium dichromate is cheaper than the corresponding potassium compound but has the disadvantage of being hygroscopic. It is used as a mordant in dyeing, as an oxidizing agent in organic chemistry, and in analytical chemistry.

sodium dihydrogenorthophosphate *See* **sodium dihydrogenphosphate(V).**

sodium dihydrogenphosphate(V) **(sodium dihydrogenorthophosphate)** A colourless crystalline solid, NaH_2PO_4, which is soluble in water and insoluble in alcohol, known as the monohydrate (r.d. 2.04) and the dihydrate (r.d. 1.91). The dihydrate loses one water molecule at 60°C and the second molecule of water at 100°C, followed by decomposition at 204°C. The compound may be prepared by treating sodium carbonate with an equimolar quantity of phosphoric acid or by neutralizing phosphoric acid with sodium hydroxide. It is used in the preparation of sodium phosphate (Na_3PO_4), in baking powders, as a food additive, and as a constituent of buffering systems. Both sodium dihydrogenphosphate and trisodium phosphate enriched in ^{32}P have been used to study phosphate participation in metabolic processes.

sodium dioxide *See* **sodium superoxide.**

sodium ethanoate **(sodium acetate)** A colourless crystalline compound, CH_3COONa, which is known as the anhydrous salt (r.d. 1.52; m.p. 324°C) or the trihydrate (r.d. 1.45; loses water at 58°C). Both forms are soluble in water and in ethoxyethane, and slightly soluble in ethanol. The compound may be prepared by the reaction of ethanoic acid (acetic acid) with sodium carbonate or with sodium hydroxide. Because it is a salt of a strong base and a weak acid, sodium ethanoate is used in buffers for pH control in many laboratory applications, in foodstuffs, and in electroplating. It is also used in dyeing, soaps, pharmaceuticals, and in photography.

sodium fluoride A crystalline compound, NaF, soluble in water and very slightly soluble in ethanol; cubic; r.d. 2.56; m.p. 993°C; b.p. 1695°C. It occurs naturally as villiaumite and may be prepared by the reaction of sodium hydroxide or of sodium carbonate with hydrogen fluoride. The reaction of sodium fluoride with concentrated sulphuric acid may be used as a source of hydrogen fluoride. The compound is used in ceramic enamels and as a preservative agent for fermentation. It is highly toxic but in very dilute solution (less than 1 part per million) it is used in the fluoridation of water for the prevention of tooth decay on account of its ability to replace OH groups with F groups in the material of dental enamel.

sodium formate *See* **sodium methanoate**.

sodium hexafluoraluminate A colourless monoclinic solid, Na_3AlF_6, very slightly soluble in water; r.d. 2.9; m.p. 1000°C. It changes to a cubic form at 580°C. The compound occurs naturally as the mineral *cryolite but a considerable amount is manufactured by the reaction of aluminium fluoride wth alumina and sodium hydroxide or directly with sodium aluminate. Its most important use is in the manufacture of aluminium in the *Hall–Heroult cell. It is also used in the manufacture of enamels, opaque glasses, and ceramic glazes.

sodium hydride A white crystalline solid, NaH; cubic; r.d. 0.92; decomposes above 300°C (slow); completely decomposed at 800°C. Sodium hydride is prepared by the reaction of pure dry hydrogen with sodium at 350°C. Electrolysis of sodium hydride in molten LiCl/KCl leads to the evolution of hydrogen; this is taken as evidence for the ionic nature of NaH and the presence of the hydride ion (H^-). It reacts violently with water to give sodium hydroxide and hydrogen, with halogens to give the halide and appropriate hydrogen halide, and ignites spontaneously with oxygen at 230°C. It is a powerful reducing agent with several laboratory applications.

sodium hydrogencarbonate (bicarbonate of soda; sodium bicarbonate) A white crystalline solid, $NaHCO_3$, soluble in water and slightly soluble in ethanol; monoclinic; r.d. 2.159; loses carbon dioxide above 270°C. It is manufactured in the *Solvay process and may be prepared in the laboratory by passing carbon dioxide through sodium carbonate or sodium hydroxide solution. Sodium hydrogencarbonate reacts with acids to give carbon dioxide and, as it does not have strongly corrosive or strongly basic properties itself, it is employed in bulk for the treatment of acid spillage and in medicinal applications as an antacid. Sodium hydrogencarbonate is also used in baking powders (and is known as *baking soda*), dry-powder fire extinguishers, and in the textiles, tanning, paper, and ceramics industries. The hydrogencarbonate ion has an important biological role as an intermediate between atmospheric CO_2/H_2CO_3 and the carbonate ion CO_3^{2-}. For water-living organisms this is the most important and in some cases the only source of carbon.

sodium hydrogensulphate **(sodium bisulphate)** A colourless solid, $NaHSO_4$, known in anhydrous and monohydrate forms. The anhydrous solid is triclinic (r.d. 2.435; m.p. >315°C). The monohydrate is monoclinic and deliquescent (r.d. 2.103; m.p. 59°C). Both forms are soluble in water and slightly soluble in alcohol. Sodium hydrogensulphate was originally made by the reaction between sodium nitrate and sulphuric acid, hence its old name of *nitre cake*. It may be manufactured by the reaction of sodium hydroxide with sulphuric acid, or by heating equimolar proportions of sodium chloride and concentrated sulphuric acid. Solutions of sodium hydrogensulphate are acidic. On heating the compound decomposes (via $Na_2S_2O_7$) to give sulphur trioxide. It is used in paper making, glass making, and textile finishing.

sodium hydrogensulphite **(sodium bisulphite)** A white solid, $NaHSO_3$, which is very soluble in water (yellow in solution) and slightly soluble in ethanol; monoclinic; r.d. 1.48. It decomposes on heating to give sodium sulphate, sulphur dioxide, and sulphur. It is formed by saturating a solution of sodium carbonate with sulphur dioxide. The compound is used in the brewing industry and in the sterilization of wine casks. It is a general antiseptic and bleaching agent. *See also* **aldehydes**.

sodium hydroxide **(caustic soda)** A white transluscent deliquescent solid, NaOH, soluble in water and ethanol but insoluble in ether; r.d. 2.13; m.p. 318°C; b.p. 1390°C. Hydrates containing 7, 5, 4, 3.5, 3, 2, and 1 molecule of water are known.

Sodium hydroxide was formerly made by the treatment of sodium carbonate with lime but its main source today is from the electrolysis of brine using mercury cells or any of a variety of diaphragm cells. The principal product demanded from these cells is chlorine (for use in plastics) and sodium hydroxide is almost reduced to the status of a by-product. It is strongly alkaline and finds many applications in the chemical industry, particularly in the production of soaps and paper. It is also used to adsorb acidic gases, such as carbon dioxide and sulphur dioxide, and is used in the treatment of effluent for the removal of heavy metals (as hydroxides) and of acidity. Sodium hydroxide solutions are extremely corrosive to body tissue and are particularly hazardous to the eyes.

sodium iodide A white crystalline solid, NaI, very soluble in water and soluble in both ethanol and ethanoic acid. It is known in both the anhydrous form (cubic; r.d. 3.67; m.p. 661°C; b.p. 1304°C) and as the dihydrate (monoclinic; r.d. 2.45). It is prepared by the reaction of hydrogen iodide with sodium carbonate or sodium hydroxide in solution. Like potassium iodide, sodium iodide in aqueous solution dissolves iodine to form a brown solution containing the I_3^- ion. It finds applications in photography and is also used in medicine as an expectorant and in the administration of radioactive iodine for studies of thyroid function and for treatment of diseases of the thyroid.

sodium methanoate **(sodium formate)** A colourless deliquescent solid, HCOONa, soluble in water and slightly soluble in ethanol; monoclinic; r.d. 1.92; m.p. 253°C; decomposes on further heating. The monohydrate is also known. The compound may be produced by the reaction of carbon monoxide with solid sodium hydroxide at 200°C and 10 atmospheres pressure; in the laboratory it can be conveniently prepared by the reaction of methanoic acid and sodium hydroxide. Its uses are in the production of oxalic acid (ethanedioic acid) and methanoic acid and in the laboratory it is a convenient source of carbon monoxide.

sodium monoxide A whitish-grey deliquescent solid, Na_2O; r.d. 2.27; sublimes at 1275°C. It is manufactured by oxidation of the metal in a limited supply of oxygen and purified by sublimation. Reaction with water produces sodium hydroxide. Its commercial applications are similar to those of sodium hydroxide.

sodium nitrate **(Chile saltpetre)** A white solid, $NaNO_3$, soluble in water and in ethanol; trigonal; r.d. 2.261; m.p. 306°C; decomposes at 380°C. A rhombohedral form is also known. It is obtained from deposits of caliche or may be prepared by the reaction of nitric acid with sodium hydroxide or sodium carbonate. It was previously used for the manufacture of nitric acid by heating with concentrated sulphuric acid. Its main use is in nitrate fertilizers.

sodium nitrite A yellow hygroscopic crystalline compound, $NaNO_2$, soluble in water, slightly soluble in ether and in ethanol; rhombohedral; r.d. 2.17; m.p. 271°C; decomposes above 320°C. It is formed by the thermal decomposition of sodium nitrate and is used in the preparation of nitrous acid (reaction with cold dilute hydrochloric acid). Sodium nitrite is used in organic *diazotization and as a corrosion inhibitor.

sodium orthophosphate *See* **trisodium phosphate(V)**.

sodium peroxide A whitish solid (yellow when hot), Na_2O_2, soluble in ice-water and decomposed in warm water or alcohol; r.d. 2.80; decomposes at 460°C. A crystalline octahydrate (hexagonal) is obtained by crystallization from ice-water. The compound is formed by the combustion of sodium metal in excess oxygen. At normal temperatures it reacts with water to give sodium hydroxide and hydrogen peroxide. It is a powerful oxidizing agent reacting with iodine vapour to give the iodate and periodate, with carbon at 300°C to give the carbonate, and with nitrogen(II) oxide to give the nitrate. It is used as a bleaching agent in wool and yarn processing, in the refining of oils and fats, and in the production of wood pulp.

sodium sesquicarbonate A white crystalline hydrated double salt, $Na_2CO_3.NaHCO_3.2H_2O$, soluble in water but less alkaline than sodium carbonate; r.d. 2.12; decomposes on heating. It may be prepared by crystallizing equimolar quantities of the constituent materials; it also occurs naturally as *trona* and in Searles Lake brines. It is widely used as a detergent and soap builder and, because of its mild alkaline properties, as a water-softening agent and bath-salt base. *See also* **sodium carbonate**.

sodium sulphate A white crystalline compound, Na_2SO_4, usually known as the anhydrous compound (orthorhombic; r.d. 2.67; m.p. 888°C) or the decahydrate (monoclinic; r.d. 1.46; which loses water at 100°C). The decahydrate is known as *Glauber's salt*. A metastable heptahydrate ($Na_2SO_4.7H_2O$) also exists. All forms are soluble in water, dissolving to give a neutral solution. The compound occurs naturally as

mirabilite ($Na_2SO_4.10H_2O$),
threnardite (Na_2SO_4), and
glauberite ($Na_2SO_4.CaSO_4$).

Sodium sulphate may be produced industrially by the reaction of magnesium sulphate with sodium chloride in solution followed by crystallization, or by the reaction of concentrated sulphuric acid with solid sodium chloride. The latter method was used in the *Leblanc process for the production of alkali and has given the name *salt cake* to impure industrial sodium sulphate. Sodium sulphate is used in the manufacture of glass and soft glazes and in dyeing to promote an even finish. It also finds medicinal application as a purgative and in commercial aperient salts.

sodium sulphide A yellow-red solid, Na_2S, formed by the reduction of sodium sulphate with carbon (coke) at elevated temperatures. It is a corrosive and readily oxidized material of variable composition and usually contains polysulphides of the type Na_2S_2, Na_2S_3, and Na_2S_4, which cause the variety of colours. It is known in an anhydrous form (r.d. 1.85; m.p. 1180°C) and as a nonahydrate, $Na_2S.9H_2O$ (r.d. 1.43; decomposes at 920°C). Other hydrates of sodium sulphide have been reported. The compound is deliquescent, soluble in water with extensive hydrolysis, and slightly soluble in alcohol. It is used in wood pulping, dyestuffs manufacture, and metallurgy on account of its reducing properties. It has also been used for the production of sodium thiosulphate (for the photographic industry) and as a depilatory agent in leather preparation. It is a strong skin irritant.

sodium sulphite A white solid, Na_2SO_3, existing in an anhydrous form (r.d. 2.63) and as a heptahydrate (r.d. 1.59). Sodium sulphite is soluble in water and because it is readily oxidized it is widely used as a convenient reducing agent. It is prepared by reacting sulphur dioxide with either sodium carbonate or sodium hydroxide. Dilute mineral acids reverse this process and release sulphur dioxide. Sodium sulphite is used as a bleaching agent in textiles and in paper manufacture. Its use as an antioxidant in some canned foodstuffs gives rise to a slightly sulphurous smell immediately on opening, but its use is prohibited in meats or foods that contain vitamin B_1. Sodium sulphite solutions are occasionally used as biological preservatives.

sodium–sulphur cell A type of *secondary cell that has molten electrodes of sodium and sulphur separated by a solid electrolyte consisting of beta alumina (a crystalline form of aluminium oxide). When the cell is producing current, sodium ions flow through the alumina to the sulphur, where they form sodium polysulphide. Electrons from the sodium flow in the external circuit. The opposite process takes place during charging of the cell. Sodium–sulphur batteries have been considered for use in electric vehicles because of their high peak power levels and relatively low weight. However, some of the output has to be used to maintain the operating temperature (about 370°C) and the cost of sodium is high.

sodium superoxide **(sodium dioxide)** A whitish-yellow solid, NaO_2, formed by the reaction of sodium peroxide with excess oxygen at elevated temperatures and pressures. It reacts with water to form hydrogen peroxide and oxygen.

sodium thiosulphate **(hypo)** A colourless efflorescent solid, $Na_2S_2O_3$, soluble in water but insoluble in ethanol, commonly encountered as the pentahydrate (monoclinic; r.d. 1.73; m.p. 42°C), which loses water at 100°C to give the anhydrous form (r.d. 1.66). It is prepared by the reaction of sulphur dioxide with a suspension of sulphur in boiling sodium hydroxide solution. Aqueous solutions of sodium thiosulphate are readily oxidized in the presence of air to sodium tetrathionate and sodium sulphate. The reaction with dilute acids gives sulphur and sulphur dioxide. It is used in the photographic industry and in analytical chemistry.

soft soap *See* soap.

soft water *See* **hardness of water**.

sol A *colloid in which small solid particles are dispersed in a liquid continuous phase.

solder An alloy used to join metal surfaces. A *soft solder* melts at a temperature in the range 200–300°C and consists of a tin–lead alloy. The tin content varies between 80% for the lower end of the melting range and 31% for the higher end. *Hard solders* contain substantial quantities of silver in the alloy. *Brazing solders* are usually alloys of copper and zinc, which melt at over 800°C.

solid A state of matter in which there is a three-dimensional regularity of structure, resulting from the proximity of the component atoms, ions, or molecules and the strength of the forces between them. True solids are crystalline (*see also* **amorphous**). If a crystalline solid is heated, the kinetic energy of the components increases. At a specific temperature, called the *melting point*, the forces between the components become unable to contain them within the crystal structure. At this temperature, the lattice breaks down and the solid becomes a liquid.

solid solution A crystalline material that is a mixture of two or more components, with ions, atoms, or molecules of one component replacing some of the ions, atoms, or molecules of the other component in its normal crystal lattice. Solid solutions are found in certain alloys. For example, gold and copper form solid solutions in which some of the copper atoms in the lattice are replaced by gold atoms. In general, the gold atoms are distributed at random, and a range of gold–copper compositions is possible. At a certain composition, the gold and copper atoms can each form regular individual lattices (referred to as *superlattices*). Mixed crystals of double salts (such as alums) are also examples of solid solutions. Compounds can form solid solutions if they are isomorphous (*see* **isomorphism**).

solubility The quantity of solute that dissolves in a given quantity of solvent to form a saturated solution. Solubility is measured in kilograms per metre cubed, moles per kilogram of solvent, etc. The solubility of a substance in a given solvent depends on the temperature. Generally, for a solid in a liquid, solubility increases with temperature; for a gas, solubility decreases. *See also* **concentration**.

solubility product Symbol K_s. The product of the concentrations of ions in a saturated solution. For instance, if a compound A_xB_y is in equilibrium with its solution

$$A_xB_y(s) \rightleftharpoons xA^+(aq) + yB^-(aq)$$

the equilibrium constant is

$$K_c = [A^+]^x[B^-]^y/[A_xB_y]$$

Since the concentration of the undissolved solid can be put equal to 1, the solubility product is given by

$$K_s = [A^+]^x[B^-]^y$$

The expression is only true for sparingly soluble salts. If the product of ionic concentrations in a solution exceeds the solubility product, then precipitation occurs.

solute The substance dissolved in a solvent in forming a *solution.

solution A homogeneous mixture of a liquid (the *solvent) with a gas or solid (the *solute*). In a solution, the molecules of the solute are discrete and mixed with the molecules of solvent. There is usually some interaction between the solvent and solute molecules (*see* **solvation**). Two liquids that can mix on the molecular level are said to be *miscible*. In this case, the solvent is the major component and the solute the minor component. *See also* **solid solution**.

solvation The interaction of ions of a solute with the molecules of solvent. For instance, when sodium chloride is dissolved in water the sodium ions attract polar water molecules, with the negative oxygen atoms pointing towards the positive Na^+ ion. Solvation of transition-metal ions can also occur by formation of coordinate bonds, as in the hexaquocopper(II) ion $[Cu(H_2O)_6]^{2+}$. Solvation is the process that causes ionic solids to dissolve, because the energy released compensates for the energy necessary to break down the crystal lattice. It occurs only with polar solvents. Solvation in which the solvent is water is called *hydration*.

Solvay process (ammonia–soda process) An industrial method of making sodium carbonate from calcium carbonate and sodium chloride. The calcium carbonate is first heated to give calcium oxide and carbon dioxide, which is bubbled into a solution of sodium chloride in ammonia. Sodium hydrogencarbonate is precipitated:

$$H_2O + CO_2(g) + NaCl(aq) + NH_3(aq) \rightarrow NaHCO_3(s) + NH_4Cl(aq)$$

The sodium hydrogencarbonate is heated to give sodium carbonate and carbon dioxide. The ammonium chloride is heated with calcium oxide (from the first stage) to regenerate the ammonia. The process was patented in 1861 by the Belgian chemist Ernest Solvay (1838–1922).

solvent A liquid that dissolves another substance or substances to form a *solution. *Polar solvents* are compounds such as water and liquid ammonia, which have dipole moments and consequently high dielectric constants. These solvents are capable of dissolving ionic compounds or covalent compounds that ionize (*see* **solvation**). *Nonpolar solvents* are compounds such as ethoxyethane and benzene, which do not have permanent dipole moments. These do not dissolve ionic compounds but will dissolve nonpolar covalent compounds. Solvents can be further categorized according to their proton-donating and accepting properties. *Amphiprotic solvents* self-ionize and can therefore act both as proton donators and acceptors. A typical example is water:

$$2H_2O \rightleftharpoons H_3O^+ + OH^-$$

Aprotic solvents neither accept nor donate protons; tetrachloromethane (carbon tetrachloride) is an example.

solvent extraction The process of separating one constituent from a mixture by dissolving it in a solvent in which it is soluble but in which the other constituents of the mixture are not. The process is usually carried out in the liquid phase, in which case it is also known as *liquid–liquid extraction.* In liquid–liquid extraction, the solution containing the desired constituent must be immiscible with the rest of the mixture. The process is widely used in extracting oil from oil-bearing materials.

solvolysis A reaction between a compound and its solvent. *See* **hydrolysis**.

SOMO Singly occupied molecular orbital.

sonochemistry The study of chemical reactions in liquids subjected to high-intensity sound or ultrasound. This causes the formation, growth, and collapse of tiny bubbles within the liquid, generating localized centres of very high temperature and pressure, with extremely rapid cooling rates. Such conditions are suitable for studying novel reactions, decomposing polymers, and producing amorphous materials.

sorbitol A polyhydric alcohol, $CH_2OH(CHOH)_4CH_2OH$, derived from glucose; it is isomeric with *mannitol. It is found in rose hips and rowan berries and is manufactured by the catalytic reduction of glucose with hydrogen. Sorbitol is used as a sweetener (in diabetic foods) and in the manufacture of vitamin C and various cosmetics, foodstuffs, and medicines.

sorption *Absorption of a gas by a solid.

sorption pump A type of vacuum pump in which gas is removed from a system by absorption on a solid (e.g. activated charcoal or a zeolite) at low temperature.

sp Synperiplanar. *See* **torsion angle**.

space group A group of symmetry elements applying to a lattice. *Compare* **point group**.

species A chemical entity, such as a particular atom, ion, or molecule.

specific 1. Denoting that an extensive physical quantity so described is expressed per unit mass. For example, the *specific latent heat* of a body is its latent heat per unit mass. When the extensive physical quantity is denoted by a capital letter (e.g. L for latent heat), the specific quantity is denoted by the corresponding lower-case letter (e.g. l for specific latent heat). **2.** In some older physical quantities the adjective 'specific' was added for other reasons (e.g. specific gravity, specific resistance). These names are now no longer used.

specific activity *See* **activity**.

specific gravity *See* **relative density**; **specific**.

specific heat capacity *See* **heat capacity**.

spectrochemical series A series of ligands arranged in the order in which they cause splitting of the energy levels of *d*-orbitals in metal complexes (*see* **crystal-field theory**). The series for some common ligands has the form:

$CN^- > NO_2^- > NH_3 > C_5H_5N > H_2O > OH^- > F^- > Cl^- > Br^- > I^-$

spectrograph *See* **spectroscope**.

spectrometer Any of various instruments for producing a spectrum and measuring the wavelengths, energies, etc., involved. A simple type, for visible radiation, is a spectroscope equipped with a calibrated scale allowing wavelengths to be read off or calculated. In the X-ray to infrared region of the electromagnetic spectrum, the spectrum is produced by dispersing the radiation with a prism or diffraction grating (or crystal, in the case of hard X-rays). Some form of photoelectric detector is used, and the spectrum can be obtained as a graphical plot, which shows how the intensity of the radiation varies with wavelength. Such instruments are also called *spectrophotometers*. Spectrometers also exist for investigating the gamma-ray region and the microwave and radio-wave regions of the spectrum (*see* **electron-spin resonance**; **nuclear magnetic resonance**). Instruments for obtaining spectra of particle beams are also called spectrometers (*see* **spectrum**; **photoelectron spectroscopy**).

spectrophotometer *See* **spectrometer**.

spectroscope An optical instrument that produces a *spectrum for visual observation. The first such instrument was made by R. W. Bunsen; in its simplest form it consists of a hollow tube with a slit at one end by which the light enters and a collimating lens at the other end to produce a parallel beam, a prism to disperse the light, and a telescope for viewing the spectrum. In the *spectrograph*, the spectroscope is provided with a camera to record the spectrum.

For a broad range of spectroscopic work, from the ultraviolet to the infrared, a diffraction grating is used instead of a prism. *See also* **spectrometer**.

spectroscopy The study of methods of producing and analysing *spectra using *spectroscopes, *spectrometers, spectrographs, and spectrophotometers. The interpretations of the spectra so produced can be used for chemical analysis, examining atomic and molecular energy levels and molecular structures, and for determining the composition and motions of celestial bodies.

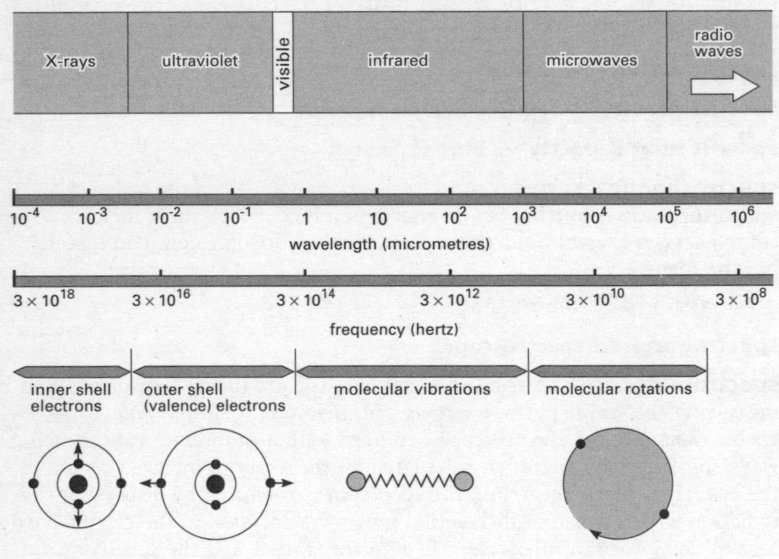

Sources of electromagnetic spectra

spectrum (*pl.* **spectra**) **1.** A distribution of entities or properties arrayed in order of increasing or decreasing magnitude. For example, a beam of ions passed through a mass spectrograph, in which they are deflected according to their charge-to-mass ratios, will have a range of masses called a *mass spectrum*. A *sound spectrum* is the distribution of energy over a range of frequencies of a particular source. **2.** A range of electromagnetic energies arrayed in order of increasing or decreasing wavelength or frequency (*see* **electromagnetic spectrum**). The *emission spectrum* of a body or substance is the characteristic range of radiations it emits when it is heated, bombarded by electron or ions, or absorbs photons. The *absorption spectrum* of a substance is produced by examining, through the substance and through a spectroscope, a continuous spectrum of radiation. The energies removed from the continuous spectrum by the absorbing medium show up as black lines or bands. With a substance capable of emitting a spectrum, these are in exactly the same positions in the spectrum as some of the lines and bands in the emission spectrum.

Emission and absorption spectra may show a *continuous spectrum*, a *line spectrum*, or a *band spectrum*. A continuous spectrum contains an unbroken sequence of frequencies over a relatively wide range; it is produced by incandescent solids, liquids, and compressed gases. Line spectra are discontinuous lines produced by excited atoms and ions as they fall back to a lower energy level. Band spectra (closely grouped bands of lines) are characteristic of molecular gases or chemical compounds. *See also* **spectroscopy**.

speculum An alloy of copper and tin formerly used in reflecting telescopes to make the main mirror as it could be cast, ground, and polished to make a highly reflective surface. It has now been largely replaced by silvered glass for this purpose.

sphalerite (zinc blende) A mineral form of zinc sulphide, ZnS, crystallizing in the cubic system; the principal ore of zinc. It is usually yellow-brown to brownish-black in colour and occurs, often with galena, in metasomatic deposits and also in hydrothermal veins and replacement deposits. Sphalerite is mined on every continent, the chief sources including the USA, Canada, Mexico, Russia, Australia, Peru, and Poland.

spherical top *See* **moment of inertia**.

sphingolipid *See* **phospholipid**.

spiegel (spiegeleisen) A form of *pig iron containing 15–30% of manganese and 4–5% of carbon. It is added to steel in a Bessemer converter as a deoxidizing agent and to raise the manganese content of steel.

spin (intrinsic angular momentum) Symbol *s*. The part of the total angular momentum of a particle, atom, nucleus, etc., that can continue to exist even when the particle is apparently at rest, i.e. when its translational motion is zero and therefore its orbital angular momentum is zero. A molecule, atom, or nucleus in a specified energy level, or a particular elementary particle, has a particular spin, just as it has a particular charge or mass. According to *quantum theory, this is quantized and is restricted to multiples of $h/2\pi$, where h is the *Planck constant. Spin is characterized by a quantum number s. For example, for an electron $s = \pm\frac{1}{2}$, implying a spin of $+ h/4\pi$ when it is spinning in one direction and $-h/4\pi$ when it is spinning in the other. Because of their spin, particles also have their own intrinsic magnetic moments and in a magnetic field the spin of the particles lines up at an angle to the direction of the field, precessing around this direction. *See also* **nuclear magnetic resonance**.

spinel A group of oxide minerals with the general formula $F^{2+}R_2{}^{3+}O_4$, where F^{2+} = Mg, Fe, Zn, Mn, or Ni and R^{3+} = Al, Fe, or Cr, crystallizing in the cubic system. The spinels are divided into three series: spinel ($MgAl_2O_4$), *magnetite, and *chromite. They occur in high-temperature igneous or metamorphic rocks.

spin glass An alloy of a small amount of a magnetic metal (0.1–10%) with a nonmagnetic metal, in which the atoms of the magnetic element are randomly distributed through the crystal lattice of the nonmagnetic element. Examples are AuFe and CuMn. Theories of the magnetic and other properties of spin glasses are complicated by the random distribution of the magnetic atoms.

spin–lattice relaxation A process in which electrons in a crystal return to the distribution for equilibrium statistical mechanics after some perturbation, such as magnetic resonance, has caused more electron spins to be in high-energy states. The excess energy of these high-energy states is taken up by vibrations of the lattice of the solid.

spinodal curve A curve that separates a metastable region from an unstable region in the coexistence region of a binary fluid. Above the spinodal curve the process of moving towards equilibrium occurs by droplet nucleation, while below the spinodal curve there are periodic modulations of the order parameter, which have a small amplitude at first (*see* **spinodal decomposition**). The spinodal curve is not a sharp boundary in real systems as a result of fluctuations.

spinodal decomposition The process of moving towards equilibrium in a part of a phase diagram in which the order parameter is conserved. Spinodal decomposition is observed in the quenching of binary mixtures. *See also* **spinodal curve**.

spin–orbit coupling An interaction between the orbital angular momentum and the spin angular momentum of an individual particle, such as an electron. For light atoms, spin–orbit coupling is small so that *multiplets of many-electron atoms are described by *Russell–Saunders coupling. For heavy atoms, spin–orbit coupling is large so that multiplets of many-electron atoms are described by *j-j coupling. For medium-sized atoms the sizes of the energies associated with spin–orbit coupling are comparable to the sizes of energies associated with electrostatic repulsion between the electrons, the multiplets in this case being described as having *intermediate coupling*. Spin–orbit coupling is large in many nuclei, particularly heavy nuclei.

spirits of salt A name formerly given to hydrogen chloride because this compound can be made by adding sulphuric acid to common salt (sodium chloride).

spontaneous combustion Combustion in which a substance produces sufficient heat within itself, usually by a slow oxidation process, for ignition to take place without the need for an external high-temperature energy source.

spontaneous emission The emission of a photon by an atom as it makes a transition from an excited state to the ground state. Spontaneous emission occurs independently of any external electromagnetic radiation; the transition is caused by interactions between atoms and vacuum fluctuations of the quantized electromagnetic field. The process of spontaneous emission, which cannot be described by nonrelativistic *quantum mechanics, as given by formulations such as the *Schrödinger equation, is responsible for the limited lifetime of an excited state of an atom before it emits a photon.

sputtering The process by which some of the atoms of an electrode (usually a cathode) are ejected as a result of bombardment by heavy positive ions. Although the process is generally unwanted, it can be used to produce a clean surface or to deposit a uniform film of a metal on an object in an evacuated enclosure.

squalene An intermediate compound formed in the synthesis of cholesterol; it is a hydrocarbon containing 30 carbon atoms. The immediate oxidation of squalene to squalene 2,3-epoxide is the last common step in the synthesis of *sterols in animals, plants, and fungi.

square-planar Describing a coordination compound in which four ligands positioned at the corners of a square coordinate to a metal ion at the centre of the square. *See* **complex.**

stabilization energy The amount by which the energy of a delocalized chemical structure is less than the theoretical energy of a structure with localized bonds. It is obtained by subtracting the experimental heat of formation of the compound (in $kJ\,mol^{-1}$) from that calculated on the basis of a classical structure with localized bonds.

stabilizer 1. A substance used to inhibit a chemical reaction, i.e. a negative catalyst. **2.** A substance used to prevent a colloid from coagulating.

stable equilibrium *See* **equilibrium.**

staggered conformation *See* **conformation.**

stainless steel A form of *steel containing at least 11–12% of chromium, a low percentage of carbon, and often some other elements, notably nickel and molybdenum. Stainless steel does not rust or stain and therefore has a wide variety of uses in industrial, chemical, and domestic environments. A particularly successful alloy is the steel known as 18–8, which contains 18% Cr, 8% Ni, and 0.08% C.

stalactites and stalagmites Accretions of calcium carbonate in limestone caves. Stalactites are tapering cones or pendants that hang down from the roofs of caves; stalagmites are upward projections from the cave floor and tend to be broader at their bases than stalactites. Both are formed from drips of water containing calcium hydrogencarbonate in solution and may take thousands of years to grow.

standard cell A *voltaic cell, such as a *Clark cell, or *Weston cell, used as a standard of e.m.f.

standard electrode An electrode (a half cell) used in measuring electrode potential. *See* **hydrogen half cell.**

standard electrode potential *See* **electrode potential.**

standard solution A solution of known concentration for use in volumetric analysis.

standard state A state of a system used as a reference value in thermodynamic measurements. Standard states involve a reference value of pressure (usually one atmosphere, 101.325 kPa) or concentration (usually 1 M). Thermodynamic functions are designated as 'standard' when they refer to changes in which reactants and products are all in their standard and their normal physical state. For example, the standard molar enthalpy of formation of water at 298 K is the enthalpy change for the reaction

$$H_2(g) + \frac{1}{2}O_2(g) \rightarrow H_2O(l)$$

$\Delta H^{\ominus}_{298} = -285.83\ \text{kJ mol}^{-1}$. Note the superscript $^{\ominus}$ is used to denote standard state and the temperature should be indicated.

standard temperature and pressure *See* **s.t.p.**

stannane *See* **tin(IV) hydride.**

stannate A compound formed by reaction of tin oxides (or hydroxides) with alkali. Tin oxides are amphoteric (weakly acidic) and react to give stannate ions. Tin(IV) oxide with molten alkali gives the stannate(IV) ion

$$SnO_2 + 2OH^- \rightarrow SnO_3^{2-} + H_2O$$

In fact, there are various ions present in which the tin is bound to hydroxide groups, the main one being the hexahydroxostannate(IV) ion, $Sn(OH)_6^{2-}$. This is the negative ion present in crystalline 'trihydrates' of the type $K_2Sn_2O_3.3H_2O$.
Tin(II) oxide gives the trihydroxostannate(II) ion in alkaline solutions

$$SnO(s) + OH^-(aq) + H_2O(l) \rightarrow Sn(OH)_3^-(aq)$$

Stannate(IV) compounds were formerly referred to as *orthostannates* (SnO_4^{4-}) or *metastannates* (SnO_3^{2-}). Stannate(II) compounds were called *stannites*.

stannic compounds Compounds of tin in its higher (+4) oxidation state; e.g. stannic chloride is tin(IV) chloride.

stannite *See* **stannate.**

stannous compounds Compounds of tin in its lower (+2) oxidation state; e.g. stannous chloride is tin(II) chloride.

starch A *polysaccharide consisting of various proportions of two glucose polymers, *amylose and *amylopectin. It occurs widely in plants, especially in roots, tubers, seeds, and fruits, as a carbohydrate storage product and energy source. Starch is therefore a major energy source for animals. When digested it ultimately yields glucose. Starch granules are insoluble in cold water but disrupt if heated to form a gelatinous solution. This gives an intense blue colour with iodine solutions and starch is used as an *indicator in certain titrations.

Stark effect The splitting of lines in the *spectra of atoms due to the presence of a strong electric field. It is named after the German physicist Johannes Stark (1874–1957), who discovered it in 1913. Like the normal *Zeeman effect, the Stark effect can be understood in terms of the classical electron theory of Lorentz. The Stark effect for hydrogen atoms was also described by the Bohr theory of the atom. In terms of quantum mechanics, the Stark effect is described by regarding the electric field as a perturbation on the quantum states and energy levels of an atom in the absence of an electric field. This application of perturbation theory was its first use in quantum mechanics.

Stark–Einstein law The law stating that in a photochemical process (such as a photochemical reaction) one photon is absorbed by each molecule causing the main photochemical process. In some circumstances, one molecule, having absorbed a photon, initiates a process involving several molecules. The Stark–Einstein law is named after Johannes Stark and Albert *Einstein.

state of matter One of the three physical states in which matter can exist, i.e. *solid, *liquid, or *gas. Plasma is sometimes regarded as the fourth state of matter.

stationary phase *See* **chromatography**.

stationary state A state of a system when it has an energy level permitted by *quantum mechanics. Transitions from one stationary state to another can occur by the emission or absorption of an appropriate quanta of energy (e.g. in the form of photons).

statistical mechanics The branch of physics in which statistical methods are applied to the microscopic constituents of a system in order to predict its macroscopic properties. The earliest application of this method was Boltzmann's attempt to explain the thermodynamic properties of gases on the basis of the statistical properties of large assemblies of molecules.

In classical statistical mechanics, each particle is regarded as occupying a point in *phase space*, i.e. to have an exact position and momentum at any particular instant. The probability that this point will occupy any small volume of the phase space is taken to be proportional to the volume. The Maxwell–Boltzmann law gives the most probable distribution of the particles in phase space.

With the advent of quantum theory, the exactness of these premises was disturbed (by the Heisenberg uncertainty principle). In the *quantum statistics that evolved as a result, the phase space is divided into cells, each having a volume h^f, where h is the Planck constant and f is the number of degrees of freedom of the particles. This new concept led to Bose–Einstein statistics, and for particles obeying the Pauli exclusion principle, to Fermi–Dirac statistics.

steam distillation A method of distilling liquids that are immiscible with water by bubbling steam through them. It depends on the fact that the vapour pressure (and hence the boiling point) of a mixture of two immiscible liquids is lower than the vapour pressure of either pure liquid.

steam point The temperature at which the maximum vapour pressure of water is equal to the standard atmospheric pressure (101 325 Pa). On the Celsius scale it has the value 100°C.

stearate (octadecanoate) A salt or ester of stearic acid.

stearic acid (octadecanoic acid) A solid saturated *fatty acid, $CH_3(CH_2)_{16}COOH$; r.d. 0.94; m.p. 71.5–72°C; b.p. 360°C (with decomposition). It occurs widely (as *glycerides) in animal and vegetable fats.

steel Any of a number of alloys consisting predominantly of iron with varying proportions of carbon (up to 1.7%) and, in some cases, small quantities of other elements (*alloy steels*), such as manganese, silicon, chromium, molybdenum, and nickel. Steels containing over 11–12% of chromium are known as *stainless steels.

Carbon steels exist in three stable crystalline phases: *ferrite* has a body-centred cubic crystal, *austenite* has a face-centred cubic crystal, and *cementite* has an orthorhombic crystal. *Pearlite* is a mixture of ferrite and cementite arranged in parallel plates. The phase diagram shows how the phases form at different temperatures and compositions.

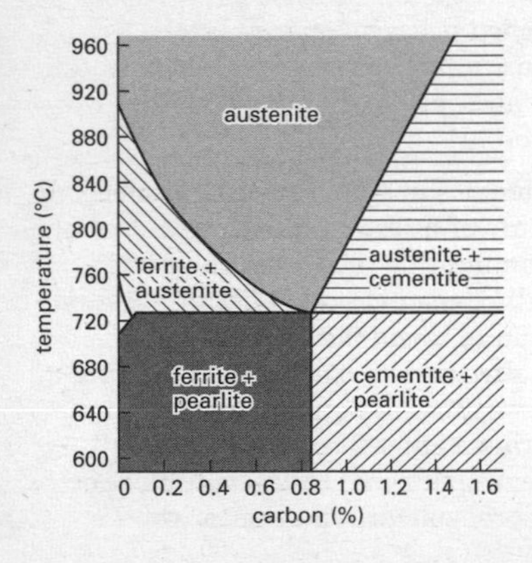

Phase diagram for steel

Steels are manufactured by the *basic-oxygen process (L–D process), which has largely replaced the *Bessemer process and the *open-hearth process, or in electrical furnaces.

step A single stage in a chemical reaction. For example, the addition of hydrogen chloride to ethene involves three steps:

$HCl \rightarrow H^+ + Cl^-$

$H^+ + C_2H_4 \rightarrow CH_3CH_2^+$

$CH_3CH_2^+ + Cl^- \rightarrow CH_3CH_2Cl$

steradian Symbol sr. The dimensionless (supplementary) *SI unit of solid angle equal to the solid angle that encloses a surface on a sphere equal to the square of the radius of the sphere.

stere A unit of volume equal to $1\,m^3$. It is not now used for scientific purposes.

stereochemistry The branch of chemistry concerned with the structure of molecules and the way the arrangement of atoms and groups affects the chemical properties.

stereoisomerism *See* **isomerism.**

stereoregular Describing a *polymer that has a regular pattern of side groups along its chain.

stereospecific Describing chemical reactions that give products with a particular arrangement of atoms in space. An example of a stereospecific reaction is the *Ziegler process for making polyethene.

steric effect An effect in which the rate or path of a chemical reaction depends on the size or arrangement of groups in a molecule.

steric hindrance An effect in which a chemical reaction is slowed down or prevented because large groups on a reactant molecule hinder the approach of another reactant molecule.

Stern–Gerlach experiment An experiment demonstrating the *space quantization* of rotating bodies, first conducted by the German scientists Otto Stern (1888–1969) and Walther Gerlach (1899–1979) in 1921. Their experiment involved passing a beam of silver atoms through an inhomogeneous magnetic field. The purpose of the experiment was to investigate the interaction between a charged rotating body and the field. A charged rotating body acts like a magnet. In classical mechanics the orientation of angular momentum can have any value so that the magnet associated with it can take any orientation. However, in quantum mechanics angular momentum is quantized, which means that the associated magnet can only lie in certain discrete orientations. As expected, sharp bands of atoms are observed, provided that the beam has low intensity. The low intensity reduces the blurring effects caused by collisions between atoms.

steroid nucleus

cholesterol (a sterol)

testosterone (an androgen)

Steroid structure

steroid Any of a group of lipids derived from saturated compound called cyclopentanoperhydrophenanthrene, which has a nucleus of four rings (see formula). Some of the most important steroid derivatives are the steroid alcohols, or sterols. Other steroids include the bile acids, which aid digestion of fats in the intestine; the sex hormones (androgens and oestrogens); and the corticosteroid hormones, produced by the adrenal cortex. *Vitamin D is also based on the steroid structure.

sterol Any of a group of *steroid-based alcohols having a hydrocarbon side-chain of 8–10 carbon atoms. Sterols exist either as free sterols or as esters of fatty acids. Animal sterols (*zoosterols*) include cholesterol and lanosterol. The major plant sterol (*phytosterol*) is beta-sitosterol, while fungal sterols (*mycosterols*) include ergosterol.

stimulated emission *See* **induced emission**; **laser**.

Stirling's approximation An approximation for the factorial, $n!$, of a number n. The precise form of Stirling's approximation is:

$$n! = (2\pi)^{1/2}n^{n+1/2}\exp(-n)$$

This approximation is valid for large values of n, being accurate for n greater than about 10. A simplified version of Stirling's expression is $\log_e n! = n\log_e n - n$, which is an approximation to the precise form, derived by dropping all those terms that do not increase at least as quickly as n. An important application of Stirling's approximation is in the derivation of the Boltzmann distribution (*see* **Boltzmann equation**). In this application n is very much greater than 10, enabling the simplified version of Stirling's approximation to be used.

STM *See* **scanning tunnelling microscope**.

stochastic process Any process in which there is a random element. Stochastic processes are important in *nonequilibrium statistical mechanics and *disordered solids. In a *time-dependent stochastic process*, a variable that changes with time does so in such a way that there is no correlation between different time intervals. An example of a stochastic process is *Brownian movement. Equations, such as the *Langevin equation and the *Fokker–Planck equation, that describe stochastic processes are called *stochastic equations*. It is necessary to use statistical methods and the theory of probability to analyse stochastic processes and their equations.

stoichiometric Describing chemical reactions in which the reactants combine in simple whole-number ratios.

stoichiometric coefficient *See* **chemical equation.**

stoichiometric compound A compound in which atoms are combined in exact whole-number ratios. *Compare* **nonstoichiometric compound.**

stoichiometric mixture A mixture of substances that can react to give products with no excess reactant.

stoichiometric sum *See* **chemical equation.**

stoichiometry The relative proportions in which elements form compounds or in which substances react.

stokes Symbol St. A c.g.s. unit of kinematic viscosity equal to the ratio of the viscosity of a fluid in poises to its density in grams per cubic centimetre. 1 stokes = 10^{-4} m^2 s^{-1}. It is named after the British mathematician and physicist Sir George Gabriel Stokes (1819–1903).

Stokes radiation The electromagnetic radiation that occurs in the *Raman effect when the frequency of the scattered radiation is lower than the frequency of the unscattered radiation, i.e. energy is transferred from the photon to the scattering molecule. The spectral Raman lines corresponding to this type of radiation are called *Stokes lines*, with the set of Stokes lines being called the *O-branch* of the Raman spectrum.

Similarly, *anti-Stokes radiation is the radiation when the frequency of the scattered electromagnetic radiation is higher than the frequency of the unscattered radiation, i.e. energy is transferred to the photon from the scattering molecule. The spectral Raman lines corresponding to this type of radiation are called *anti-Stokes lines*, with the set of anti-Stokes lines being called the *S-branch* of the Raman spectrum.

Stokes radiation is named after the British physicist Sir George Gabriel Stokes (1819–1903).

stopped-flow technique A technique for investigating fast reactions in solution. The reactant solutions are rapidly mixed and flow through a tube. The composition or properties of the mixture are monitored at some point in the tube (e.g. by photometry). The flow is suddenly stopped and the change of signal with time is used to elucidate the kinetics of the process. The stopped-flow technique is useful for reactions occurring in the millisecond time range.

s.t.p. Standard temperature and pressure, formerly known as *N.T.P.* (normal temperature and pressure). The standard conditions used as a basis for calculations involving quantities that vary with temperature and pressure. These conditions are used when comparing the properties of gases. They are 273.15 K (or 0°C) and 101 325 Pa (or 760.0 mmHg).

straight chain *See* **chain.**

strange attractor *See* **attractor.**

streaming potential A potential difference that occurs when a liquid under pressure is forced through a narrow opening or diaphragm. It can be measured between two electrodes at each end of a capillary tube made of the same material as the diaphragm.

strong acid An *acid that is completely dissociated in aqueous solution.

strontia *See* **strontium oxide**.

strontianite A mineral form of *strontium carbonate, $SrCO_3$.

strontium Symbol Sr. A soft yellowish metallic element belonging to group 2 (formerly IIA) of the periodic table (*see* **alkaline-earth metals**); a.n. 38; r.a.m. 87.62; r.d. 2.6; m.p. 769°C; b.p. 1384°C. The element is found in the minerals strontianite ($SrCO_3$) and celestine ($SrSO_4$). It can be obtained by roasting the ore to give the oxide, followed by reduction with aluminium (i.e. the *Goldschmidt process). The element, which is highly reactive, is used in certain alloys and as a vacuum getter. The isotope strontium–90 is present in radioactive fallout (half-life 28 years), and can be metabolized with calcium so that it collects in bone. Strontium was discovered by Martin Klaproth (1743–1817) and Thomas Hope (1766–1844) in 1798 and isolated by Humphry Davy in 1808.

strontium bicarbonate *See* **strontium hydrogencarbonate**.

strontium carbonate A white solid, $SrCO_3$; orthorhombic; r.d. 3.7; decomposes at 1340°C. It occurs naturally as the mineral *strontianite* and is prepared industrially by boiling celestine (strontium sulphate) with ammonium carbonate. It can also be prepared by passing carbon dioxide over strontium oxide or hydroxide or by passing the gas through a solution of strontium salt. It is a phosphor, used to coat the glass of cathode-ray screens, and is also used in the refining of sugar, as a slagging agent in certain metal furnaces, and to provide a red flame in fireworks.

strontium chloride A white compound, $SrCl_2$. The anhydrous salt (cubic; r.d. 3.05; m.p. 872°C; b.p. 1250°C) can be prepared by passing chlorine over heated strontium. It is deliquescent and readily forms the hexahydrate, $SrCl_2.6H_2O$ (r.d. 2.67). This can be made by neutralizing hydrochloric acid with strontium carbonate, oxide, or hydroxide. Strontium chloride is used for military flares.

strontium hydrogencarbonate **(strontium bicarbonate)** A compound, $Sr(HCO_3)_2$, which is stable only in solution. It is formed by the action of carbon dioxide on a suspension of strontium carbonate in water. On heating, this process is reversed.

strontium oxide **(strontia)** A white compound, SrO; r.d. 4.7; m.p. 2430°C, b.p. 3000°C. It can be prepared by the decomposition of heated strontium carbonate, hydroxide, or nitrate, and is used in the manufacture of other strontium salts, in pigments, soaps and greases, and as a drying agent.

strontium sulphate A white solid, $SrSO_4$; r.d. 3.96; m.p. 1605°C. It can be made by dissolving strontium oxide, hydroxide, or carbonate in sulphuric acid. It is used as a pigment in paints and ceramic glazes and to provide a red colour in fireworks.

structural formula *See* **formula**.

structural isomerism *See* **isomerism**.

structure factor A quantity denoted F_{hkl}, where h, k, and l are the Miller indices of the crystal, which occurs in *X-ray crystallography and other experiments involving scattering in crystals. F_{hkl} is defined by the equation:

$$F_{hkl} = \sum_{i} f_i \exp[2\pi i(hx_i + ky_i + lz_i)],$$

where the sum is over all atoms of the unit cell and f_i is the *scattering factor* for atom i defined by:

$$f_i = 4\pi\int_0^\infty (\rho \sin kr/kr) r^2 dr$$

Here k – $4\pi\sin\theta/\lambda$, where θ is the Bragg angle (*see* **Bragg's law**), λ is the wavelength of the X-rays, and ρ is the electron density distribution of the atom i. The structure factor is used in Patterson synthesis (*see* **Patterson function**).

strychnine A colourless poisonous crystalline alkaloid found in certain plants.

styrene *See* **phenylethene**.

subjacent orbitals The next-to-highest occupied molecular orbital (*NHOMO*) and the second-lowest unoccupied molecular orbital (*SLUMO*). In certain cases these subjacent orbitals are significant in *frontier-orbital theory.

sublimate A solid formed by sublimation.

sublimation A direct change of state from solid to gas.

submillimetre waves Electromagnetic radiation with wavelengths below one millimetre (and therefore frequencies greater than 300 gigahertz), extending to radiation of the far infrared. A source of submillimetre radiation is a medium pressure mercury lamp in quartz. Submillimetre waves can be detected by a *Golay cell.

subshell *See* **atom**.

substantive dye *See* **dyes**.

substantivity The affinity of a dye for its substrate.

substituent 1. An atom or group that replaces another in a substitution reaction. **2.** An atom or group regarded as having replaced a hydrogen atom in a chemical derivative. For example, dibromobenzene ($C_6H_4Br_2$) is a derivative of benzene with bromine substituents.

substitution reaction (displacement reaction) A reaction in which one atom or molecule is replaced by another atom or molecule. *See* **electrophilic substitution**; **nucleophilic substitution**.

substrate 1. The substance that is affected by the action of a catalyst; for example, the substance upon which an *enzyme acts in a biochemical reaction. **2.** The substance on which some other substance is adsorbed or in which it is absorbed. Examples include the material to which a dye is attached, the porous solid absorbing a gas, and the *matrix trapping isolated atoms, radicals, etc.

succinic acid *See* **butanedioic acid.**

sucrose (cane sugar; beet sugar; saccharose) A sugar comprising one molecule of glucose linked to a fructose molecule. It occurs widely in plants and is particularly abundant in sugar cane and sugar beet (15–20%), from which it is extracted and refined for table sugar. If heated to 200°C, sucrose becomes caramel.

sugar (saccharide) Any of a group of water-soluble *carbohydrates of relatively low molecular weight and typically having a sweet taste. The simple sugars are called *monosaccharides. More complex sugars comprise between two and ten monosaccharides linked together: *disaccharides contain two, trisaccharides three, and so on. The name is often used to refer specifically to *sucrose (table sugar).

sugar of lead *See* **lead(II) ethanoate.**

sulpha drugs *See* **sulphonamides.**

sulphamic acid A colourless crystalline solid, NH_2SO_2OH, which is extremely soluble in water and normally exists as the *zwitterion $H_3N^+.SO_3^-$. It is a strong acid, readily forming sulphamate salts. It is used in electroplating, hard-water scale removers, herbicides, and artificial sweeteners.

sulphanes Compounds of hydrogen and sulphur containing chains of sulphur atoms. They have the general formula H_2S_n. The simplest is hydrogen sulphide, H_2S; other members of the series are H_2S_2, H_2S_3, H_2S_4, etc. *See* **sulphides.**

sulphanilic acid (4-aminobenzene sulphonic acid) A colourless crystalline solid, $H_2NC_6H_4SO_2OH$, made by prolonged heating of *phenylamine (aniline) sulphate. It readily forms *diazo compounds and is used to make dyes and sulpha drugs.

sulphate A salt or ester of sulphuric(VI) acid. Organic sulphates have the formula R_2SO_4, where R is an organic group. Sulphate salts contain the ion SO_4^{2-}.

sulphides 1. Inorganic compounds of sulphur with more electropositive elements. Compounds of sulphur with nonmetals are covalent compounds, e.g. hydrogen sulphide (H_2S). Metals form ionic sulphides containing the S^{2-} ion; these are salts of hydrogen sulphide. *Polysulphides* can also be produced containing the polymeric ion S_x^{2-}. **2.** (*or* **thio ethers**) Organic compounds that contain the group –S– linked to two hydrocarbon groups. Organic sulphides are named from the linking groups, e.g. dimethyl sulphide (CH_3SCH_3), ethyl methyl sulphide ($C_2H_5SCH_3$). They are analogues of ethers in which the oxygen is replaced by sulphur (hence the alternative name) but are generally more reactive than ethers. Thus they react with halogen compounds to form *sulphonium compounds and can be oxidized to *sulphoxides.

sulphinate (dithionite; hyposulphite) A salt that contains the negative ion $S_2O_4^{2-}$, usually formed by the reduction of sulphites with excess SO_2. Solutions are not very stable and decompose to give thiosulphate and hydrogensulphite ions. The structure is $^-O_2S\text{-}SO_2^-$.

sulphinic acid (dithionous acid; hyposulphurous acid) An unstable acid, $H_2S_2O_4$, known in the form of its salts (sulphinates). *See also* **sulphuric acid.**

sulphite A salt or ester derived from sulphurous acid. The salts contain the trioxosulphate(IV) ion SO_3^{2-}. Sulphites generally have reducing properties.

sulphonamides Organic compounds containing the group $-SO_2.NH_2$. The sulphonamides are amides of sulphonic acids. Many have antibacterial action and are also known as *sulpha drugs*, including sulphadiazine, $NH_2C_6H_4SO_2NHC_4H_3N_2$, sulphathiazole, $NH_2C_6H_4SO_2NHC_3H_2NS$, and several others. They act by preventing bacteria from reproducing and are used to treat a variety of bacterial infections, especially of the gut and urinary system.

sulphonate A salt or ester of a sulphonic acid.

sulphonation A type of chemical reaction in which a $-SO_3H$ group is substituted on a benzene ring to form a *sulphonic acid. The reaction is carried out by refluxing with concentrated sulphuric(VI) acid for a long period. It can also occur with cold disulphuric(VI) acid ($H_2S_2O_7$). Sulphonation is an example of electrophilic substitution in which the electrophile is a sulphur trioxide molecule, SO_3.

sulphonic acids Organic compounds containing the $-SO_2.OH$ group. Sulphonic acids are formed by reaction of aromatic hydrocarbons with concentrated sulphuric acid. They are strong acids, ionizing completely in solution to form the sulphonate ion, $-SO_2.O^-$.

sulphonium compounds Compounds containing the ion R_3S^+ (sulphonium ion), where R is any organic group. Sulphonium compounds can be formed by reaction of organic sulphides with halogen compounds. For example, diethyl sulphide, $C_2H_5SC_2H_5$, reacts with chloromethane, CH_3Cl, to give diethylmethylsulphonium chloride, $(C_2H_5)_2.CH_3.S^+Cl^-$.

sulphoxides Organic compounds containing the group =S=O (*sulphoxide group*) linked to two other groups, e.g. dimethyl sulphoxide, $(CH_3)_2SO$.

sulphur Symbol S. A yellow nonmetallic element belonging to *group 16 (formerly VIB) of the periodic table; a.n. 16; r.a.m. 32.06; r.d. 2.07 (rhombic);

R—S—R sulphide (thio ether)

$R-S(=O)_2-OH$ sulphonic acid

$R_2S^{+}-R$ sulphonium ion

$R-S(=O)_2-O^-$ sulphonate ion

R—S—H thiol (mercaptan)

$R-S(=O)_2-NH_2$ sulphonamide

$R_2S=O$ sulphoxide

Examples of organic sulphur compounds

m.p. 112.8°C; b.p. 444.674°C. The element occurs in many sulphide and sulphate minerals and native sulphur is also found in Sicily and the USA (obtained by the *Frasch process). It can also be obtained from hydrogen sulphide by the *Claus process.

Sulphur has various allotropic forms. Below 95.6°C the stable crystal form is rhombic; above this temperature the element transforms into a triclinic form. These crystalline forms both contain cyclic S_8 molecules. At temperatures just above its melting point, molten sulphur is a yellow liquid containing S_8 rings (as in the solid form). At about 160°C, the sulphur atoms form chains and the liquid becomes more viscous and dark brown. If the molten sulphur is cooled quickly from this temperature (e.g. by pouring into cold water) a reddish-brown solid known as *plastic sulphur* is obtained. Above 200°C the viscosity decreases. Sulphur vapour contains a mixture of S_2, S_4, S_6, and S_8 molecules. *Flowers of sulphur* is a yellow powder obtained by subliming the vapour. It is used as a plant fungicide. The element is also used to produce sulphuric acid and other sulphur compounds.

Sulphur is an *essential element in living organisms, occurring in the amino acids cysteine and methionine and therefore in many proteins. It is also a constituent of various cell metabolites, e.g. coenzyme A. Sulphur is absorbed by plants from the soil as the sulphate ion (SO_4^{2-}). *See* **sulphur cycle**.

sulphur cycle The cycling of sulphur between the biotic (living) and abiotic (nonliving) components of the environment. Most of the sulphur in the abiotic environment is found in rocks, although a small amount is present in the atmosphere as sulphur dioxide (SO_2), produced by combustion of fossil fuels. Sulphate (SO_4^{2-}), derived from the weathering and oxidation of rocks, is taken up by plants and incorporated into sulphur-containing proteins. In this form sulphur is passed along food chains to animals. Decomposition of dead organic matter and faeces by anaerobic sulphate-reducing bacteria returns sulphur to the abiotic environment in the form of hydrogen sulphide (H_2S). Hydrogen sulphide can be converted back to sulphate or to elemental sulphur by the action of different groups of photosynthetic and sulphide-oxidizing bacteria. Elemental sulphur becomes incorporated into rocks.

sulphur dichloride *See* **disulphur dichloride**.

sulphur dichloride dioxide (sulphuryl chloride) A colourless liquid, SO_2Cl_2; r.d. 1.67; m.p. −54.1°C; b.p. 69°C. It decomposes in water but is soluble in benzene. The compound is formed by the action of chlorine on sulphur dioxide in the presence of an iron(III) chloride catalyst or sunlight. It is used as a chlorinating agent and a source of the related fluoride, SO_2F_2.

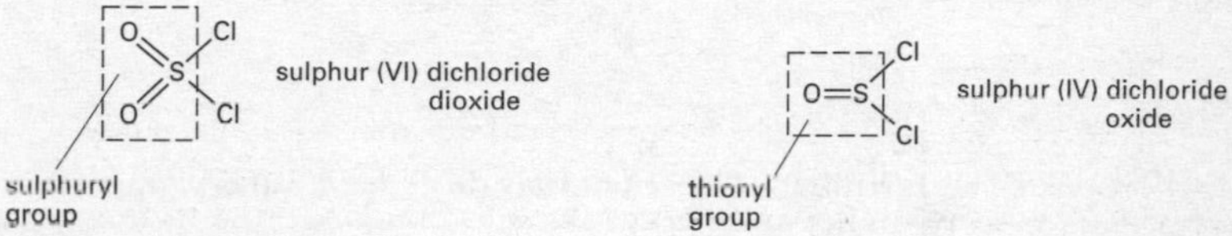

sulphur (VI) dichloride dioxide
sulphuryl group
sulphur (IV) dichloride oxide
thionyl group

Oxychlorides of sulphur

sulphur dichloride oxide (thionyl chloride) A colourless fuming liquid, $SOCl_2$; m.p. −105°C; b.p. 78.8°C. It hydrolyses rapidly in water but is soluble in benzene. It may be prepared by the direct action of sulphur on chlorine monoxide or, more commonly, by the reaction of phosphorus(V) chloride with sulphur dioxide. It is used as a chlorinating agent in synthetic organic chemistry (replacing –OH groups with Cl).

sulphur dioxide (sulphur(IV) oxide) A colourless liquid or pungent gas, SO_2, formed by sulphur burning in air; r.d. 1.43 (liquid); m.p. −72.7°C; b.p. −10°C. It can be made by heating iron sulphide (pyrites) in air. The compound is a reducing agent and is used in bleaching and as a fumigant and food preservative. Large quantities are also used in the *contact process for manufacturing sulphuric acid. It dissolves in water to give a mixture of sulphuric and sulphurous acids. *See also* **acid rain**.

sulphuretted hydrogen *See* **hydrogen sulphide**.

sulphuric acid (oil of vitriol) A colourless oily liquid, H_2SO_4; r.d. 1.84; m.p. 10.36°C; b.p. 338°C. The pure acid is rarely used; it is commonly available as a 96–98% solution (m.p. 3.0°C). The compound also forms a range of hydrates: $H_2SO_4.H_2O$ (m.p. 8.62°C); $H_2SO_4.2H_2O$ (m.p. −38/39°C); $H_2SO_4.6H_2O$ (m.p. −54°C); $H_2SO_4.8H_2O$ (m.p. −62°C). Its full systematic name is *tetraoxosulphuric(VI) acid.*

Until the 1930s, sulphuric acid was manufactured by the *lead-chamber process, but this has now been replaced by the *contact process (catalytic oxidation of sulphur dioxide). More sulphuric acid is made in the UK than any other chemical product; production levels (UK) are commonly 12 000 to 13 000 tonnes per day. It is extensively used in industry, the main

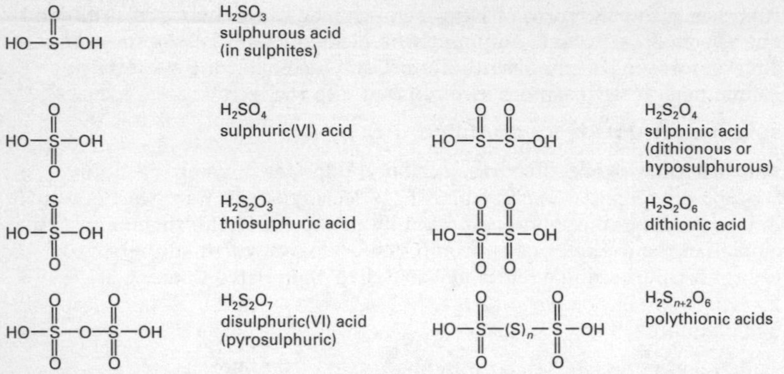

Structure	Formula / name	Structure	Formula / name
HO—S(=O)—OH	H_2SO_3 sulphurous acid (in sulphites)		
HO—S(=O)(=O)—OH	H_2SO_4 sulphuric(VI) acid	HO—S(=O)—S(=O)—OH	$H_2S_2O_4$ sulphinic acid (dithionous or hyposulphurous)
HO—S(=S)(=O)—OH	$H_2S_2O_3$ thiosulphuric acid	HO—S(=O)(=O)—S(=O)(=O)—OH	$H_2S_2O_6$ dithionic acid
HO—S(=O)(=O)—O—S(=O)(=O)—OH	$H_2S_2O_7$ disulphuric(VI) acid (pyrosulphuric)	HO—S(=O)(=O)—(S)$_n$—S(=O)(=O)—OH	$H_2S_{n+2}O_6$ polythionic acids

Structures of some oxo acids of sulphur

applications being fertilizers (32%), chemicals (16%), paints and pigments (15%), detergents (11%), and fibres (9%).

In concentrated sulphuric acid there is extensive hydrogen bonding and several competing equilibria, to give species such as H_3O^+, HSO_4^-, $H_3SO_4^+$, and $H_2S_2O_7$. Apart from being a powerful protonating agent (it protonates chlorides and nitrates producing hydrogen chloride and nitric acid), the compound is a moderately strong oxidizing agent. Thus, it will dissolve copper:

$$Cu(s) + H_2SO_4(l) \rightarrow CuO(s) + H_2O(l) + SO_2(g)$$

$$CuO(s) + H_2SO_4(l) \rightarrow CuSO_4(aq) + H_2O(l)$$

It is also a powerful dehydrating agent, capable of removing H_2O from many organic compounds (as in the production of acid *anhydrides). In dilute solution it is a strong dibasic acid forming two series of salts, the sulphates and the hydrogensulphates.

sulphuric(IV) acid *See* **sulphurous acid.**

sulphur monochloride *See* **disulphur dichloride.**

sulphurous acid **(sulphuric(IV) acid)** A weak dibasic acid, H_2SO_3, known in the form of its salts: the sulphites and hydrogensulphites. It is considered to be formed (along with sulphuric acid) when sulphur dioxide is dissolved in water. It is probable, however, that the molecule H_2SO_3 is not present and that the solution contains hydrated SO_2. It is a reducing agent. The systematic name is *trioxosulphuric(IV) acid*. *See also* **sulphuric acid.**

sulphur(IV) oxide *See* **sulphur dioxide.**

sulphur(VI) oxide *See* **sulphur trioxide.**

sulphur trioxide **(sulphur(VI) oxide)** A colourless fuming solid, SO_3, which has three crystalline modifications. In decreasing order of stability these are: α, r.d. 1.97; m.p. 16.83°C; b.p. 44.8°C; β, m.p. 16.24°C; sublimes at 50°C; r.d.

2.29; γ, m.p. 16.8°C; b.p. 44.8°C. All are polymeric, with linked SO_4 tetrahedra: the γ-form has an icelike structure and is obtained by rapid quenching of the vapour; the β-form has infinite helical chains; and the α-form has infinite chains with some cross-linking of the SO_4 tetrahedra. Even in the vapour, there are polymeric species, and not discrete sulphur trioxide molecules (hence the compound is more correctly called by its systematic name *sulphur(VI) oxide*).

Sulphur trioxide is prepared by the oxidation of sulphur dioxide with oxygen in the presence of a vanadium(V) oxide catalyst. It may be prepared in the laboratory by distilling a mixture of concentrated sulphuric acid and phosphorus(V) oxide. It reacts violently with water to give sulphuric(VI) acid and is an important intermediate in the preparation of sulphuric acid and oleum.

sulphuryl chloride *See* **sulphur dichloride dioxide.**

sulphuryl group The group $=SO_2$, as in *sulphur dichloride oxide.

sulphydryl group *See* **thiols.**

superacid An *acid that has a proton-donating ability equal to or greater than that of anhydrous sulphuric acid. A superacid is a particular type of Brønsted acid. Those that are much stronger than sulphuric acids can be made by adding certain pentafluorides and their derivatives to such acids as fluorosulphuric acid (HSO_3F) or hydrogen fluoride (HF). The pentafluorides, such as antimony pentafluoride (SbF_5), are very strong Lewis acids. The mixtures $HF\text{-}SbF_5$ and $HSO_3F\text{-}SbF_5$ are among the strongest acids known; their applications include the protonation of very weak bases in organic chemistry and the abstraction of hydrogen from saturated hydrocarbons to produce carbonium ions. An equimolar mixture of HSO_3F and SbF_5 is known by the tradename *Magic acid*. Very strong bases, such as lithium diisopropylamide, are sometimes known as *superbases*.

superbase *See* **superacid.**

superconductivity The property possessed by some substances below a certain temperature, the *transition point*, of zero resistance. Until recently, the known superconducting materials had very low transition points. However, synthetic organic conductors and certain metal oxide ceramics have now been produced that become superconducting at much higher temperatures. For example, much work has been done on ytterbium–barium–copper oxides, which have transition temperatures of about 100 K.

supercooling 1. The cooling of a liquid to below its freezing point without a change from the liquid to solid state taking place. In this metastable state the particles of the liquid lose energy but do not fall into place in the lattice of the solid crystal. If the liquid is seeded (*see* **seed**) crystallization usually takes place and the liquid returns to its normal freezing point. Crystallization can also be induced by the presence of particles of dust, by mechanical vibration, or by rough surfaces. **2.** The analogous cooling of a vapour to make it supersaturated.

superfluidity The exhibition by a fluid at extremely low temperatures, e.g. liquid helium at 2.186 K, of very high thermal conductivity and frictionless flow. The temperature at which superfluidity occurs is called the *lambda point*.

superheating The heating of a liquid to above its normal boiling point by increasing the pressure.

superheavy elements *See* **supertransuranics**.

superionic conductor An ionic solid in which the electrical conductivity due to the motion of ions is similar to that of a molten salt, i.e. the conductivity is much higher than is usually observed in ionic solids.

superlattice *See* **solid solution**.

supernatant liquid The clear liquid remaining when a precipitate has settled.

superoxides A group of inorganic compounds that contain the O_2^- ion. They are formed in significant quantities only for sodium, potassium, rubidium, and caesium. They are very powerful oxidizing agents and react vigorously with water to give oxygen gas and OH^- ions. The superoxide ion has an unpaired electron and is paramagnetic and coloured (orange).

superphosphate A commercial phosphate mixture consisting mainly of monocalcium phosphate. Single-superphosphate is made by treating phosphate rock with sulphuric acid; the product contains 16–20% 'available' P_2O_5:

$$Ca_{10}(PO_4)_6F_2 + 7H_2SO_4 = 3Ca(H_2PO_4)_2 + 7CaSO_4 + 2HF$$

Triple-superphosphate is made by using phosphoric(V) acid in place of sulphuric acid; the product contains 45–50% 'available' P_2O_5:

$$Ca_{10}(PO_4)_6F_2 + 14H_3PO_4 = 10Ca(H_2PO_4)_2 + 2HF$$

superplasticity The ability of some metals and alloys to stretch uniformly by several thousand percent at high temperatures, unlike normal alloys, which fail after being stretched 100% or less. Since 1962, when this property was discovered in an alloy of zinc and aluminium (22%), many alloys and ceramics have been shown to possess this property. For superplasticity to occur, the metal grain must be small and rounded and the alloy must have a slow rate of deformation.

superradiant Denoting a *laser in which the efficiency of the transition of stimulated emission is sufficiently large for the radiation to be produced by a single pulse of electromagnetic radiation without using mirrors. An example of a superradiant laser is the nitrogen laser based on the ultraviolet transition $C^3\Pi_u \rightarrow B^3\Pi_g$.

supersaturated solution *See* **saturated**.

supersaturation 1. The state of the atmosphere in which the relative humidity is over 100%. This occurs in pure air where no condensation nuclei are available. Supersaturation is usually prevented in the atmosphere by the abundance of condensation nuclei (e.g. dust, sea salt, and smoke particles). **2.** The state of any vapour whose pressure exceeds that at which condensation usually occurs (at the prevailing temperature).

supplementary units *See* **SI units**.

supramolecular chemistry A field of chemical research concerned with the formation and properties of large assemblies of molecules held together by intramolecular forces (hydrogen bonds, van der Waals' forces, etc.). One feature of supramolecular chemistry is that of *self-assembly* (*see* **self-organization**), in which the structure forms spontaneously as a consequence of the nature of the molecules. The molecular units are sometimes known as *synthons*. Another aspect is the study of very large molecules able to be used in complex chemical reactions in a fashion similar to, for example, the actions of the naturally occurring haemoglobin and nucleic acid molecules. Such molecules have great potential in such areas as medicine, electronics, and optics. The *helicate and *texaphyrin molecules and *dendrimers are typical examples of compounds of interest in this field. The field also includes *host–guest chemistry*, which is concerned with molecules specifically designed to accept other molecules. Examples include *crown ethers, *cryptands, and *calixarenes.

surface-extended X-ray absorption fine-structure spectroscopy *See* **SEXAFS**.

surface tension Symbol γ. The property of a liquid that makes it behave as if its surface is enclosed in an elastic skin. The property results from intermolecular forces: a molecule in the interior of a liquid experiences interactions from other molecules equally from all sides, whereas a molecule at the surface is only affected by molecules below it in the liquid. The surface tension is defined as the force acting over the surface per unit length of surface perpendicular to the force. It is measured in newtons per metre. It can equally be defined as the energy required to increase the surface area isothermally by one square metre, i.e. it can be measured in joules per metre squared (which is equivalent to N m^{-1}).

The property of surface tension is responsible for the formation of liquid drops, soap bubbles, and meniscuses, as well as the rise of liquids in a capillary tube (*capillarity*), the absorption of liquids by porous substances, and the ability of liquids to wet a surface.

surfactant **(surface active agent)** A substance, such as a *detergent, added to a liquid to increase its spreading or wetting properties by reducing its *surface tension.

suspension A mixture in which small solid or liquid particles are suspended in a liquid or gas.

sylvite **(sylvine)** A mineral form of *potassium chloride, KCl.

symmetric top *See* **moment of inertia**.

symmetry The property of an object that enables it to undergo certain manipulations, called *symmetry operations*, such that its new state is indistinguishable from its original state. Examples include inversion through a point, reflection through a plane, and rotation about an axis. The geometrical feature of the object with respect to which the symmetry operation is carried out is termed a *symmetry element*. In the above examples, a point through which inversion can be performed, a plane of reflection, and an axis of rotation are the symmetry elements. *See also* **molecular symmetry**.

symmetry-adapted linear combinations *See* **SALC**.

syn *See* **torsion angle**.

synclinal *See* **torsion angle**.

syndiotactic *See* **polymer**.

syneresis The spontaneous separation of the solid and liquid components of a gel on standing.

synperiplanar *See* **torsion angle**.

synthesis The formation of chemical compounds from more simple compounds.

synthesis gas *See* **Haber process**.

synthetic Describing a substance that has been made artificially; i.e. one that does not come from a natural source.

synthon *See* **supramolecular chemistry**.

Système International d'Unités *See* **SI units**.

Szilard–Chalmers effect An effect discovered by L. Szilard (1898–1964) and T. A. Chalmers in 1934; it has been used to separate radioactive products in a nuclear reaction involving the absorption of a neutron and the emission of gamma rays. If a material absorbs a neutron and subsequently emits a gamma ray, the emission of the gamma ray causes the nucleus to recoil. Frequently, the recoil energy is sufficient to break the chemical bond between the atom and the molecule of which it forms part. Thus, although the atom that has absorbed the neutron is an isotope of the original atom it is in a different form chemically, enabling separation to take place. For example, if an aqueous solution of sodium chlorate ($NaClO_3$) is subjected to bombardment by slow neutrons, the Cl^{37} is converted to Cl^{38}, with many of the Cl^{38} atoms breaking from the chlorate and moving into the solution in the form of chloride ions. This is an example of a 'hot atom' reaction. The Cl^{38} can be precipitated out using silver nitrate.

2,4,5-T 2,4,5-trichlorophenoxyacetic acid (2,4,5-trichlorophenoxyethanoic acid): a synthetic auxin formerly widely used as a herbicide and defoliant. It is now banned in many countries as it tends to become contaminated with the toxic chemical dioxin.

tactic polymer *See* polymer.

tactosol A colloidal sol containing nonspherical particles capable of orientating themselves. An example of a tactosol is given by an aged vanadium pentoxide sol, in which the particles are rod-shaped. When a tactosol is placed in a magnetic field its particles arrange themselves along the lines of force of the field.

Tafel plot The graph of the logarithm of the current density j against the overpotential η in electrochemistry in the *high overpotential limit*. This limit can occur in electrolysis either when the overpotential is large and positive, the electrode being the anode, or when the overpotential is large and negative, the electrode being the cathode. In the case of positive overpotential, the second exponential in the *Butler–Volmer equation can be ignored, as it is much smaller than the first exponential, and thus

$$j = j_0\exp[(1 - \alpha)f\eta]$$

In the case of a negative overpotential, the first exponential can be ignored, which gives $j = j_0\exp(-\alpha f\eta)$, so that $\ln(-j) = \ln j_0 - \alpha f\eta$. Thus, in the Tafel plot the value of α can be obtained from the slope and the value of j_0 can be obtained from the intercept at $\eta = 0$.

talc A white or pale-green mineral form of magnesium silicate, $Mg_3Si_4O_{10}(OH)_2$, crystallizing in the triclinic system. It forms as a secondary mineral by alteration of magnesium-rich olivines, pyroxenes, and amphiboles of ultrabasic rocks. It is soaplike to touch and very soft, having a hardness of 1 on the Mohs' scale. Massive fine-grained talc is known as *soapstone* or *steatite*. Talc in powdered form is used as a lubricant, as a filler in paper, paints, and rubber, and in cosmetics, ceramics, and French chalk. It occurs chiefly in the USA, Russia, France, and Japan.

tannic acid A yellowish complex organic compound present in certain plants. It is used in dyeing as a mordant.

tannin One of a group of complex organic chemicals commonly found in leaves, unripe fruits, and the bark of trees. Their function is uncertain though the unpleasant taste may discourage grazing animals. Some tannins have commercial uses, notably in the production of leather and ink.

tantalum Symbol Ta. A heavy blue-grey metallic *transition element; a.n. 73; r.a.m. 180.948; r.d. 16.63; m.p. 2996°C; b.p. 5427°C. It is found with niobium in the ore columbite–tantalite (Fe,Mn)(Ta,Nb)$_2$O$_6$. It is extracted by dissolving in hydrofluoric acid, separating the tantalum and niobium fluorides to give K_2TaF_7, and reduction of this with sodium. The element contains the stable isotope tantalum–181 and the long-lived radioactive isotope tantalum–180 (0.012%; half-life >10^7 years). There are several other short-lived isotopes. The element is used in certain alloys and in electronic components. Tantalum parts are also used in surgery because of the unreactive nature of the metal (e.g. in pins to join bones). Chemically, the metal forms a passive oxide layer in air. It forms complexes in the +2, +3, +4, and +5 oxidation states. Tantalum was identified in 1802 by Anders Ekeberg (1767–1813) and first isolated in 1820 by *Berzelius.

tar Any of various black semisolid mixtures of hydrocarbons and free carbon, produced by destructive distillation of *coal or by *petroleum refining.

tar sand *See* **oil sand.**

tartaric acid A crystalline naturally occurring carboxylic acid, $(CHOH)_2(COOH)_2$; r.d. 1.8; m.p. 171–174°C. It can be obtained from tartar (potassium hydrogen tartrate) deposits from wine vats, and is used in baking powders and as a foodstuffs additive. The compound is optically active (*see* **optical activity**). The systematic name is *2,3-dihydroxybutanedioic acid.*

tartrate A salt or ester of *tartaric acid.

tartrazine A food additive (E102) that gives foods a yellow colour. Tartrazine can cause a toxic response in the immune system and is banned in some countries.

tautomerism A type of *isomerism in which the two isomers (*tautomers*) are in equilibrium. *See* **keto–enol tautomerism.**

TCA cycle *See* **Krebs cycle.**

technetium Symbol Tc. A radioactive metallic *transition element; a.n. 43; m.p. 2172°C; b.p. 4877°C. The element can be detected in certain stars and is present in the fission products of uranium. It was first made by Carlo Perrier and Emilio Segré (1905–89) by bombarding molybdenum with deuterons to give technetium–97. The most stable isotope is technetium–99 (half-life 2.6×10^6 years); this is used to some extent in labelling for medical diagnosis. There are sixteen known isotopes. Chemically, the metal has properties intermediate between manganese and rhenium.

Teflon Tradename for a form of *polytetrafluoroethene.

tele-substitution A type of substitution reaction in which the entering group takes a position on an atom that is more than one atom away from the atom to which the leaving group is attached. *See also* **cine-substitution.**

tellurides Binary compounds of tellurium with other more electropositive elements. Compounds of tellurium with nonmetals are covalent (e.g. H_2Te).

Metal tellurides can be made by direct combination of the elements and are ionic (containing Te^{2-}) or nonstoichiometric interstitial compounds (e.g. Pd_4Te, $PdTe_2$).

tellurium Symbol Te. A silvery metalloid element belonging to *group 16 (formerly VIB) of the periodic table; a.n. 52; r.a.m. 127.60; r.d. 6.24 (crystalline); m.p. 449.5°C; b.p. 989.8°C. It occurs mainly as *tellurides in ores of gold, silver, copper, and nickel and it is obtained as a by-product in copper refining. There are eight natural isotopes and nine radioactive isotopes. The element is used in semiconductors and small amounts are added to certain steels. Tellurium is also added in small quantities to lead. Its chemistry is similar to that of sulphur. It was discovered by Franz Müller (1740–1825) in 1782.

temperature The property of a body or region of space that determines whether or not there will be a net flow of heat into it or out of it from a neighbouring body or region and in which direction (if any) the heat will flow. If there is no heat flow the bodies or regions are said to be in *thermodynamic equilibrium* and at the same temperature. If there is a flow of heat, the direction of the flow is from the body or region of higher temperature. Broadly, there are two methods of quantifying this property. The empirical method is to take two or more reproducible temperature-dependent events and assign *fixed points* on a scale of values to these events. For example, the Celsius temperature scale uses the freezing point and boiling point of water as the two fixed points, assigns the values 0 and 100 to them, respectively, and divides the scale between them into 100 degrees. This method is serviceable for many practical purposes (*see* **temperature scales**), but lacking a theoretical basis it is awkward to use in many scientific contexts. In the 19th century, Lord *Kelvin proposed a thermodynamic method to specify temperature, based on the measurement of the quantity of heat flowing between bodies at different temperatures. This concept relies on an absolute scale of temperature with an *absolute zero of temperature, at which no body can give up heat. He also used Sadi Carnot's concept of an ideal frictionless perfectly efficient heat engine (*see* **Carnot cycle**). This Carnot engine takes in a quantity of heat q_1 at a temperature T_1 and exhausts heat q_2 at T_2, so that $T_1/T_2 = q_1/q_2$. If T_2 has a value fixed by definition, a Carnot engine can be run between this fixed temperature and any unknown temperature T_1, enabling T_1 to be calculated by measuring the values of q_1 and q_2. This concept remains the basis for defining *thermodynamic temperature*, quite independently of the nature of the working substance. The unit in which thermodynamic temperature is expressed is the *kelvin. In practice thermodynamic temperatures cannot be measured directly; they are usually inferred from measurements with a gas thermometer containing a nearly ideal gas. This is possible because another aspect of thermodynamic temperature is its relationship to the *internal energy of a given amount of substance. This can be shown most simply in the case of an ideal monatomic gas, in which the internal energy per mole (U) is equal to the total kinetic energy of translation of the atoms in one mole of the gas (a monatomic gas has no rotational or vibrational energy). According to *kinetic theory, the thermodynamic temperature of such a gas is given by $T = 2U/3R$, where R is the universal *gas constant.

temperature-independent paramagnetism **(TIP)** Orbital paramagnetism that is independent of temperature. In some substances TIP can make the molecules paramagnetic, although all the electrons in the ground state are paired, if there are low-lying excited states to which electrons can readily move.

	T/K	t/°C
triple point of equilibrium hydrogen	13.81	–259.34
temperature of equilibrium hydrogen when its vapour pressure is 25/76 standard atmosphere	17.042	–256.108
b.p. of equilibrium hydrogen	20.28	–252.87
b.p. of neon	27.102	–246.048
triple point of oxygen	54.361	–218.789
b.p. of oxygen	90.188	–182.962
triple point of water	273.16	0.01
b.p. of water	373.15	100
f.p. of zinc	692.73	419.58
f.p. of silver	1235.08	961.93
f.p. of gold	1337.58	1064.43

Temperature scales

temperature scales A number of empirical scales of *temperature have been in use: the *Celsius scale is widely used for many purposes and in certain countries the *Fahrenheit scale is still used. These scales both rely on the use of *fixed points*, such as the freezing point and the boiling point of water, and the division of the *fundamental interval* between these two points into units of temperature (100 degrees in the case of the Celsius scale and 180 degrees in the Fahrenheit scale).

However, for scientific purposes the scale in use is the *International Practical Temperature Scale (1968)*, which is designed to conform as closely as possible to thermodynamic temperature and is expressed in the unit of thermodynamic temperature, the *kelvin. The eleven fixed points of the scale are given in the table, with the instruments specified for interpolating between them. Above the freezing point of gold, a radiation pyrometer is used, based on Planck's law of radiation. The scale is expected to be refined in the late 1980s.

tempering The process of increasing the toughness of an alloy, such as steel, by heating it to a predetermined temperature, maintaining it at this temperature for a predetermined time, and cooling it to room temperature at a predetermined rate. In steel, the purpose of the process is to heat the alloy to a temperature that will enable the excess carbide to precipitate out of the supersaturated solid solution of *martensite and then to cool the saturated solution fast enough to prevent further precipitation or grain growth. For this reason steel is quenched rapidly by dipping into cold water.

temporary hardness *See* **hardness of water**.

tera- Symbol T. A prefix used in the metric system to denote one million million times. For example, 10^{12} volts = 1 teravolt (TV).

terbium Symbol Tb. A silvery metallic element belonging to the *lanthanoids; a.n. 65; r.a.m. 158.92; r.d. 8.23 (20°C); m.p. 1356°C; b.p. 3123°C. It occurs in apatite and xenotime, from which it is obtained by an ion-exchange process. There is only one natural isotope, terbium–159, which is stable. Seventeen artificial isotopes have been identified. It is used as a dopant in semiconducting devices. It was discovered by Carl Mosander (1797–1858) in 1843.

terephthalic acid **(1,4-benzenedicarboxylic acid)** A colourless crystalline solid, $C_6H_4(COOH)_2$, m.p. 300°C. It is made by oxidizing *p*-xylene (1,4-dimethylebenzene) and used for making polyesters, such as Terylene.

term An electronic energy level in an atom. A term is characterized by a *term symbol*, which is given by a capital letter indicating the total orbital angular momentum L of the atom, with a left superscript that gives the value of $2S + 1$, where S is the total spin angular momentum of the atom. Analogously to the letters used for single electron angular momentum the letters S, P, D, F correspond to $L = 0, 1, 2, 3$ respectively. For example, a term for which $L = 1$ and $S = 1$ has the term symbol ^3P.

In the absence of *spin–orbit coupling the degeneracy of a term is $(2L + 1)(2S + 1)$. In the presence of spin–orbit coupling a term splits up into several closely spaced energy levels, with this set of closely spaced energy levels being called a *multiplet*. The *multiplicity* of a multiplet is given by $(2S + 1)$ since the number of atomic energy levels a term splits into because of spin–orbit coupling is $(2S + 1)$. This is the case because the total electronic angular momentum J of an atom can have the $(2S + 1)$ possible values: $L + S$, $L + S - 1$, etc. Each level in a multiplet has the value of J written as a right subscript to the term symbol. The degeneracy of a level in a multiplet is $2J + 1$. The names given to multiplets with $2S + 1 = 1, 2, 3, 4$ are *singlet*, *doublet*, *triplet*, *quartet*, respectively, to correspond to the number of atomic levels in the multiplet. For example, $^3P_1$ is the $J = 1$ level in a P triplet.

ternary compound A chemical compound containing three different elements.

terpenes A group of unsaturated hydrocarbons present in plants (*see* **essential oil**). Terpenes consist of isoprene units, $CH_2{:}C(CH_3)CH{:}CH_2$. Monoterpenes have two units, $C_{10}H_{16}$, sesquiterpenes three units, $C_{15}H_{24}$, diterpenes four units, $C_{20}H_{32}$, etc. *Terpenoids*, which are derivatives of terpenes, include abscisic acid and gibberellin (plant growth substances) and the *carotenoid and *chlorophyll pigments.

terpinene A cyclic *terpene, $C_{10}H_{16}$, used in perfumery and as a flavouring. It exists in three isomeric forms. α-Terpinene is an oily liquid that smells of lemons, b.p. 182°C. It occurs in the herbs cardamom, coriander and marjoram although it is usually extracted from the essential oils of other plants. β-Terpinene, b.p. 174°C, is generally synthesized from oil of savin. γ-Terpinene, b.p. 182°C, occurs in various plant oils, including those of coriander, cumin, lemon, and samphire. It can be made by the action of an alcoholic sulphuric acid solution on pinene (the major component of turpentine).

tertiary alcohol *See* **alcohols.**

tertiary amine *See* **amines.**

tervalent **(trivalent)** Having a valency of three.

Terylene Tradename for a type of *polyester used in synthetic fibres.

tesla Symbol T. The SI unit of magnetic flux density equal to one weber of magnetic flux per square metre, i.e. $1\ T = 1\ Wb\ m^{-2}$. It is named after Nikola Tesla (1870–1943), Croatian-born US electrical engineer.

tetrachloroethene A colourless nonflammable volatile liquid, $CCl_2{:}CCl_2$; r.d. 1.6; m.p. −22°C; b.p. 121°C. It is used as a solvent.

tetrachloromethane **(carbon tetrachloride)** A colourless volatile liquid with a characteristic odour, virtually insoluble in water but miscible with many organic liquids, such as ethanol and benzene; r.d. 1.586; m.p. −23°C; b.p. 76.54°C. It is made by the chlorination of methane (previously by chlorination of carbon disulphide). The compound is a good solvent for waxes, lacquers, and rubbers and the main industrial use is as a solvent, but safer substances (e.g. 1,1,1-trichloroethane) are increasingly being used. Moist carbon tetrachloride is partly decomposed to phosgene and hydrogen chloride and this provides a further restriction on its use.

tetraethyl lead *See* **lead(IV) tetraethyl.**

tetragonal *See* **crystal system.**

tetrahedral angle The angle between the bonds in a *tetrahedral compound (approximately 109° for a regular tetrahedron).

tetrahedral compound A compound in which four atoms or groups situated at the corners of a tetrahedron are linked (by covalent or coordinate bonds) to an atom at the centre of the tetrahedron. *See also* **complex.**

tetrahedron A polyhedron with four triangular faces. In a *regular tetrahedron* all four triangles are congruent equilateral triangles. It constitutes a regular triangular pyramid.

tetrahydrate A crystalline hydrate containing four moles of water per mole of compound.

tetrahydrofuran **(THF)** A colourless volatile liquid, C_4H_8O; r.d. 0.89; m.p. −65°C; b.p. 67°C. It is made by the acid hydrolysis of polysaccharides in oat husks, and is widely used as a solvent.

tetrahydroxomonoxodiboric(III) acid *See* **boric acid.**

tetraoxophosphoric(V) acid *See* **phosphoric(V) acid.**

tetraoxosulphuric(VI) acid *See* **sulphuric acid.**

tetravalent **(quadrivalent)** Having a valency of four.

texaphyrin A synthetic molecule similar to a porphyrin but containing five central nitrogen atoms rather than four, thus increasing the size of the central 'hole' and enabling larger cations, such as cadmium, to be bound stably. *See also* **supramolecular chemistry.**

thallium Symbol Tl. A greyish metallic element belonging to *group 13 (formerly IIIB) of the periodic table; a.n. 81; r.a.m. 204.39; r.d. 11.85 (20°C); m.p. 303.5°C; b.p. 1457±10°C. It occurs in zinc blende and some iron ores and is recovered in small quantities from lead and zinc concentrates. The naturally occurring isotopes are thallium-203 and thallium-205; eleven radioisotopes have been identified. It has few uses – experimental alloys for special purposes and some minor uses in electronics. The sulphate has been used as a rodenticide. Thallium(I) compounds resemble those of the alkali metals. Thallium(III) compounds are easily reduced to the thallium(I) state and are therefore strong oxidizing agents. The element was discovered by Sir William *Crookes in 1861.

theory *See* **laws, theories, and hypotheses.**

thermal analysis A technique for chemical analysis and the investigation of the products formed by heating a substance. In *differential thermal analysis* (*DTA*) a sample is heated, usually in an inert atmosphere, and a plot of weight against temperature made. In *differential scanning calorimetry* (*DSC*) heat is added to or removed from a sample electrically as the temperature is increased, thus allowing the enthalpy changes due to thermal decomposition to be studied.

thermal capacity *See* **heat capacity.**

thermal equilibrium *See* **equilibrium.**

thermite A stoichiometric powdered mixture of iron(III) oxide and aluminium for the reaction:

$$2Al + Fe_2O_3 \rightarrow Al_2O_3 + 2Fe$$

The reaction is highly exothermic and the increase in temperature is sufficient to melt the iron produced. It has been used for localized welding of steel objects (e.g. railway lines) in the *Thermit process.* Thermite is also used in incendiary bombs.

thermochemistry The branch of physical chemistry concerned with heats of chemical reaction, heats of formation of chemical compounds, etc.

thermodynamics The study of the laws that govern the conversion of energy from one form to another, the direction in which heat will flow, and the availability of energy to do work. It is based on the concept that in an isolated system anywhere in the universe there is a measurable quantity of energy called the *internal energy (U) of the system. This is the total kinetic and potential energy of the atoms and molecules of the system of all kinds that can be transferred directly as heat; it therefore excludes chemical and nuclear energy. The value of U can only be changed if the system ceases to be isolated. In these circumstances U can change by the transfer of mass to or from the system, the transfer of heat (Q) to or from the system, or by the work (W) being done on or by the system. For an adiabatic ($Q = 0$) system of constant mass, $\Delta U = W$. By convention, W is taken to be positive if work is done on the system and negative if work is done by the system. For nonadiabatic systems of constant mass, $\Delta U = Q + W$. This statement, which is equivalent to the law of conservation of energy, is known as the *first law of thermodynamics*.

All natural processes conform to this law, but not all processes conforming to it can occur in nature. Most natural processes are irreversible, i.e. they will only proceed in one direction (*see* **reversible process**). The direction that a natural process can take is the subject of the *second law of thermodynamics*, which can be stated in a variety of ways. R. Clausius (1822–88) stated the law in two ways: "heat cannot be transferred from one body to a second body at a higher temperature without producing some other effect" and "the entropy of a closed system increases with time". These statements introduce the thermodynamic concepts of *temperature (T) and *entropy (S), both of which are parameters determining the direction in which an irreversible process can go. The temperature of a body or system determines whether heat will flow into it or out of it; its entropy is a measure of the unavailability of its energy to do work. Thus T and S determine the relationship between Q and W in the statement of the first law. This is usually presented by stating the second law in the form $\Delta U = T\Delta S - W$.

The second law is concerned with changes in entropy (ΔS). The *third law of thermodynamics* provides an absolute scale of values for entropy by stating that for changes involving only perfect crystalline solids at *absolute zero, the change of the total entropy is zero. This law enables absolute values to be stated for entropies.

One other law is used in thermodynamics. Because it is fundamental to, and assumed by, the other laws of thermodynamics it is usually known as the *zeroth law of thermodynamics*. This states that if two bodies are each in thermal equilibrium with a third body, then all three bodies are in thermal equilibrium with each other. *See also* **enthalpy**; **free energy**.

thermodynamic temperature *See* **temperature**.

thermoluminescence *Luminescence produced in a solid when its temperature is raised. It arises when free electrons and holes, trapped in a solid as a result of exposure to ionizing radiation, unite and emit photons of light. The process is made use of in *thermoluminescent dating*, which assumes that the number of electrons and holes trapped in a sample of pottery is related to the length of time that has elapsed since the pottery was fired. By comparing the luminescence produced by heating a piece of pottery of unknown age with the luminescence produced by heating similar materials of known age, a fairly accurate estimate of the age of an object can be made.

thermoluminescent dating *See* **thermoluminescence**.

thermolysis (pyrolysis) The chemical decomposition of a substance by heat. It is an important process in chemical manufacture, such as the thermal *cracking of hydrocarbons in the petroleum industry.

thermometer An instrument used for measuring the *temperature of a substance. A number of techniques and forms are used in thermometers depending on such factors as the degree of accuracy required and the range of temperatures to be measured, but they all measure temperature by making use of some property of a substance that varies with temperature. For example, *liquid in glass thermometers* depend on the expansion of a liquid, usually mercury or alcohol coloured with dye. These consist of a liquid-filled glass bulb attached to a partially filled capillary tube. In the *bimetallic thermometer* the unequal expansion of two dissimilar metals that have been bonded together into a narrow strip and coiled is used to move a pointer round a dial. The *gas thermometer*, which is more accurate than the liquid-in-glass thermometer, measures the variation in the pressure of a gas kept at constant volume. The *resistance thermometer* is based on the change in resistance of conductors or semiconductors with temperature change. Platinum, nickel, and copper are the metals most commonly used in resistance thermometers.

thermoplastic *See* **plastics**.

thermosetting *See* **plastics**.

thermostat A device that controls the heating or cooling of a substance in order to maintain it at a constant temperature. It consists of a temperature-sensing instrument connected to a switching device. When the temperature reaches a predetermined level the sensor switches the heating or cooling source on or off according to a predetermined program. The sensing thermometer is often a bimetallic strip that triggers a simple electrical switch. Thermostats are used for space-heating controls, in water heaters and refrigerators, and to maintain the environment of a scientific experiment at a constant temperature.

THF *See* **tetrahydrofuran**.

thiamin(e) *See* **vitamin B complex**.

thiazole A heterocyclic compound containing a five-membered ring with

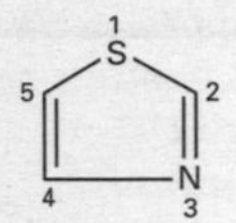

Thiazole

sulphur and nitrogen hetero atoms, C_3SNH_3. A range of thiazole dyes are manufactured containing this ring system.

thienyl ring *See* **thiophene.**

thin-layer chromatography A technique for the analysis of liquid mixtures using *chromatography. The stationary phase is a thin layer of an absorbing solid (e.g. alumina) prepared by spreading a slurry of the solid on a plate (usually glass) and drying it in an oven. A spot of the mixture to be analysed is placed near one edge and the plate is stood upright in a solvent. The solvent rises through the layer by capillary action carrying the components up the plate at different rates (depending on the extent to which they are absorbed by the solid). After a given time, the plate is dried and the location of spots noted. It is possible to identify constituents of the mixture by the distance moved in a given time. The technique needs careful control of the thickness of the layer and of the temperature. *See also* **R_F value.**

thiocyanate A salt or ester of thiocyanic acid.

thiocyanic acid An unstable gas, HSCN.

thio ethers *See* **sulphides.**

thiol group *See* **thiols.**

thiols (mercaptans; thio alcohols) Organic compounds that contain the group –SH (called the *thiol group*, *mercapto group*, or *sulphydryl group*). Thiols are analogues of alcohols in which the oxygen atom is replaced by a sulphur atom. They are named according to the parent hydrocarbon; e.g. ethane thiol (C_2H_5SH). A characteristic property is their strong disagreeable odour. For example the odour of garlic is produced by ethane thiol. Unlike alcohols they are acidic, reacting with alkalis and certain metals to form saltlike compounds. The older name, mercaptan, comes from their ability to react with ('seize') mercury.

thionyl chloride *See* **sulphur dichloride oxide.**

thionyl group The group =SO, as in *sulphur dichloride oxide.

thiophene A colourless liquid compound, C_4H_4S; m.p. –38°C; b.p. 84°C. The compound is present in commercial benzene. The ring system is also known as a *thienyl ring*.

thiosulphate A salt containing the ion $S_2O_3^{2-}$ formally derived from thiosulphuric acid. Thiosulphates readily decompose in acid solution to give elemental sulphur and hydrogensulphite (HSO_3^-) ions.

thiosulphuric acid An unstable acid, $H_2S_2O_3$, formed by the reaction of sulphur trioxide with hydrogen sulphide. *See also* **sulphuric acid**.

thiourea A white crystalline solid, $(NH_2)_2CS$; r.d. 1.4; m.p. 182°C. It is used as a fixer in photography.

thixotropy *See* **Newtonian fluid**.

Thomson, Sir Joseph John (1856–1940) British physicist, who became a professor at Cambridge University in 1884. He is best known for his work on cathode rays, which led to his discovery of the electron in 1897. He went on to study the conduction of electricity through gases, and it is for this work that he was awarded the Nobel Prize for physics in 1906. His son, *Sir George Paget Thomson* (1892–1975), discovered *electron diffraction, for which he shared the 1937 Nobel Prize for physics with Clinton J. Davisson (1881–1958), who independently made the same discovery.

Thomson, Sir William *See* **Kelvin, Lord**.

thoria *See* **thorium**.

thorium Symbol Th. A grey radioactive metallic element belonging to the *actinoids; a.n. 90; r.a.m. 232.038; r.d. 11.5–11.9 (17°C); m.p. 1740–1760°C; b.p. 4780–4800°C. It occurs in monazite sand in Brazil, India, and USA. The isotopes of thorium have mass numbers from 223 to 234 inclusive; the most stable isotope, thorium-232, has a half-life of 1.39×10^{10} years. It has an oxidation state of (+4) and its chemistry resembles that of the other actinoids. It can be used as a nuclear fuel for breeder reactors as thorium-232 captures slow neutrons to breed uranium-233. Thorium dioxide (*thoria*, ThO_2) is used on gas mantles and in special refractories. The element was discovered by J. J. *Berzelius in 1829.

thorium series *See* **radioactive series**.

three-body problem *See* **many-body problem**.

threnardite A mineral form of *sodium sulphate, Na_2SO_4.

threonine *See* **amino acid**.

thulium Symbol Tm. A soft grey metallic element belonging to the *lanthanoids; a.n. 69; r.a.m. 168.934; r.d. 9.321 (20°C); m.p. 1545°C; b.p. 1947°C. It occurs in apatite and xenotime. There is one natural isotope, thulium-169, and seventeen artificial isotopes. There are no uses for the element, which was discovered by Per Cleve (1840–1905) in 1879.

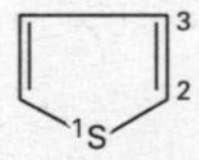

Thiophene

thymine A *pyrimidine derivative and one of the major component bases of *nucleotides and the nucleic acid *DNA.

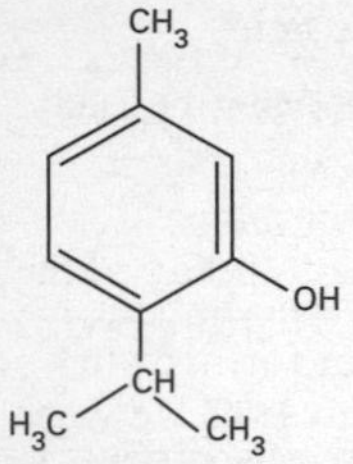

Thymol

thymol A pungent-smelling colourless crystalline compound, $C_{10}H_{14}O$; m.p. 51°C. It occurs in various essential oils, particularly oil of thyme, and can be made by using iron(III) chloride to oxidize piperitone (itself extracted from eucalyptus oil). Its antiseptic properties are exploited in gargles and mouthwashes.

tie line A horizontal line in a liquid–vapour phase diagram that is perpendicular to an *isopleth. The starting point is the first point at which the liquid can coexist with its vapour as the pressure is reduced; the end point is the point beyond which only vapour is present.

tin Symbol Sn. A silvery malleable metallic element belonging to *group 14 (formerly IVB) of the periodic table; a.n. 50; r.a.m. 118.69; r.d. 7.28; m.p. 231.88°C; b.p. 2260°C. It is found as tin(IV) oxide in ores, such as cassiterite, and is extracted by reduction with carbon. The metal (called *white tin*) has a powdery nonmetallic allotrope *grey tin*, into which it changes below 18°C. The formation of this allotrope is called *tin plague*; it can be reversed by heating to 100°C. The natural element has 21 isotopes (the largest number of any element); five radioactive isotopes are also known. The metal is used as a thin protective coating for steel plate and is a constituent of a number of alloys (e.g. phosphor bronze, gun metal, solder, Babbitt metal, and pewter). Chemically it is reactive. It combines directly with chlorine and oxygen and displaces hydrogen from dilute acids. It also dissolves in alkalis to form *stannates. There are two series of compounds with tin in the +2 and +4 oxidation states.

tin(II) chloride A white solid, $SnCl_2$, soluble in water and ethanol. It exists in the anhydrous form (rhombic; r.d. 3.95; m.p. 246°C; b.p. 652°C) and as a dihydrate, $SnCl_2.2H_2O$ (monoclinic; r.d. 2.71; m.p. 37.7°C). The compound is made by dissolving metallic tin in hydrochloric acid and is partially hydrolysed in solution.

$$Sn^{2+} + H_2O \rightleftharpoons SnOH^+ + H^+$$

Excess acid must be present to prevent the precipitation of basic salts. In the presence of additional chloride ions the pyramidal ion $[SnCl_3]^-$ is formed; in the gas phase the $SnCl_2$ molecule is bent. It is a reducing agent in acid solutions and oxidizes slowly in air:

$$Sn^{2+} \rightarrow Sn^{4+} + 2e$$

tin(IV) chloride A colourless fuming liquid, $SnCl_4$, hydrolysed in cold water, decomposed by hot water, and soluble in ethers; r.d. 2.226; m.p. −33°C; b.p. 114°C. Tin(IV) chloride is a covalent compound, which may be prepared directly from the elements. It dissolves sulphur, phosphorus, bromine, and iodine, and there is evidence for the presence of species such as $SnCl_2I_2$. In hydrochloric acid and in chloride solutions the coordination is extended from four to six by the formation of the $SnCl_6^{2-}$ ion.

tincture A solution with alcohol as the solvent (e.g. tincture of iodine).

tin(IV) hydride (stannane) A highly reactive and volatile gas (b.p. −53°), SnH_4, which decomposes on moderate heating (150°C). It is prepared by the reduction of tin chlorides using lithium tetrahydridoaluminate(III) and is used in the synthesis of some organo-tin compounds. The compound has reducing properties.

tin(IV) oxide (tin dioxide) A white solid, SnO_2, insoluble in water; tetrahedral; r.d. 6.95; m.p. 1127°C; sublimes between 1800°C and 1900°C. Tin(IV) oxide is trimorphic: the common form, which occurs naturally as the ore *cassiterite, has a rutile lattice but hexagonal and rhombic forms are also known. There are also two so-called dihydrates, $SnO_2.2H_2O$, known as α- and β-stannic acid. These are essentially tin hydroxides. Tin(IV) oxide is amphoteric, dissolving in molten alkalis to form *stannates; in the presence of sulphur, thiostannates are produced.

tin plague *See* tin.

tin(II) sulphide A grey-black cubic or monoclinic solid, SnS, virtually insoluble in water; r.d. 5.22; m.p. 882°C; b.p. 1230°C. It has a layer structure similar to that of black phosphorus. Its heat of formation is low and it can be made by heating the elements together. Above 265°C it slowly decomposes (disproportionates) to tin(IV) sulphide and tin metal. The compound reacts with hydrochloric acid to give tin(II) chloride and hydrogen sulphide.

tin(IV) sulphide (mosaic gold) A bronze or golden yellow crystalline compound, SnS_2, insoluble in water and in ethanol; hexagonal; r.d. 4.5; decomposes at 600°C. It is prepared by the reaction of hydrogen sulphide with a soluble tin(IV) salt or by the action of heat on thiostannic acid, H_2SnS_3. The golden-yellow form used for producing a gilded effect on wood is prepared by heating tin, sulphur, and ammonium chloride.

TIP *See* temperature-independent paramagnetism.

titania *See* titanium(IV) oxide.

titanic chloride *See* titanium(IV) chloride.

titanium Symbol Ti. A white metallic *transition element; a.n. 22; r.a.m. 47.9; r.d. 4.5; m.p. 1660±10°C; b.p. 3287°C. The main sources are rutile (TiO_2) and, to a lesser extent, ilmenite ($FeTiO_3$). The element also occurs in numerous other minerals. It is obtained by heating the oxide with carbon and chlorine to give $TiCl_4$, which is reduced by the *Kroll process. The main use is in a large number of strong light corrosion-resistant alloys for aircraft, ships, chemical plant, etc. The element forms a passive oxide coating in air. At higher temperatures it reacts with oxygen, nitrogen, chlorine, and other nonmetals. It dissolves in dilute acids. The main compounds are titanium(IV) salts and complexes; titanium(II) and titanium(III) compounds are also known. The element was first discovered by William Gregor (1761–1817) in 1789.

titanium(IV) chloride **(titanic chloride; titanium tetrachloride)** A colourless volatile liquid, $TiCl_4$, made by strongly heating titanium(IV) oxide and carbon in a stream of dry chorine gas. It fumes in moist air to produce titanium oxychlorides. It is used to produce other titanium compounds and pure titanium metal.

titanium dioxide *See* **titanium(IV) oxide**.

titanium(IV) oxide **(titania; titanium dioxide)** A white oxide, TiO_2, occurring naturally in various forms, particularly the mineral rutile. It is used as a white pigment and as a filler for plastics, rubber, etc.

titanium tetrachloride *See* **titanium(IV) chloride**.

titration A method of volumetric analysis in which a volume of one reagent (the *titrant*) is added to a known volume of another reagent slowly from a burette until an end point is reached (*see* **indicator**). The volume added before the end point is reached is noted. If one of the solutions has a known concentration, that of the other can be calculated.

TNT *See* **trinitrotoluene**.

tocopherol *See* **vitamin E**.

Tollens reagent A reagent used in testing for aldehydes. It is made by adding sodium hydroxide to silver nitrate to give silver(I) oxide, which is dissolved in aqueous ammonia (giving the complex ion $[Ag(NH_3)_2]^+$). The sample is warmed with the reagent in a test tube. Aldehydes reduce the complex Ag^+ ion to metallic silver, forming a bright silver mirror on the inside of the tube (hence the name *silver-mirror test*). Ketones give a negative result. It is named after Bernhard Tollens (1841–1918).

toluene *See* **methylbenzene**.

topaz A variably coloured aluminium silicate mineral, $Al_2(SiO_4)(OH,F)_2$, that forms orthorhombic crystals. It occurs chiefly in acid igneous rocks, such as granites and pegmatites. Topaz is valued as a gemstone because of its transparency, variety of colours (the wine-yellow variety being most highly prized), and great hardness (8 on the Mohs' scale). When heated, yellow or brownish topaz often becomes a rose-pink colour. The main sources of topaz are Brazil, Russia, and the USA.

torr A unit of pressure, used in high-vacuum technology, defined as 1 mmHg. 1 torr is equal to 133.322 pascals. The unit is named after Evangelista Torricelli (1609–47).

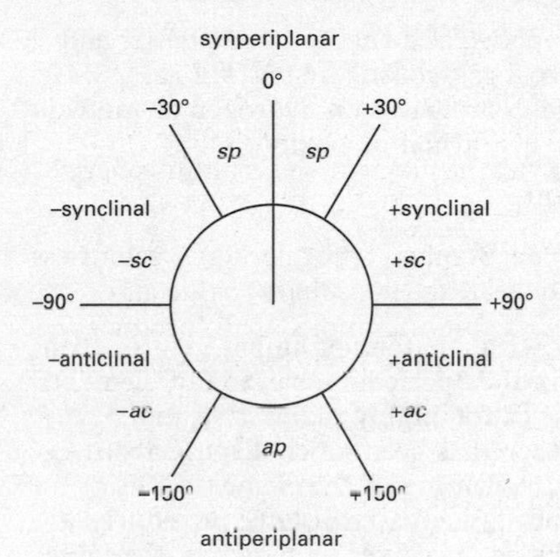

Torsion angle

torsion angle In a nonlinear chain of atoms A–B–C–D, the angle between the plane containing atoms ABC and the plane containing BCD. The torsion angle can have any value from 0° to 180°. If the chain is viewed along the line BC, the torsion angle is positive if the bond AB would have to be rotated in a clockwise sense (less than 180°) to eclipse (i.e. align with) the bond CD. If the rotation of AB has to be in an anticlockwise sense, the torsion angle would be negative.

There is a terminology relating to the steric arrangements of atoms based on the size of the torsion angle (*see also* **conformation**). The *syn* arrangement is one in which the size of the torsion angle is ±90°. The *anti* arrangement is one with a torsion angle between ±90° and 180°. Another distinction is between *clinal* arrangements (value between 30° and 150° or –30° and –150°) and *periplanar* arrangements (value between 0 and ±30° or ±150° and 180°). Combining the two gives four ranges of torsion angle in an arrangement:

synperiplanar (*sp*) 0° to ±30°;
synclinal (*sc*) 30° to 90° or –30° to –90°;
anticlinal (*ac*) 90° to 150° or –90° to –150°;
antiperiplanar (*ap*) ±150° to 180°.

The antiperiplanar conformation is also called a *trans* conformation; the synperiplanar conformation corresponds to a *cis* conformation. The synclinal conformation is also called a *gauche* or *skew* conformation.

total-radiation pyrometer *See* **pyrometry**.

tourmaline A group of minerals composed of complex cyclosilicates containing boron with the general formula $NaR_3^{2+}Al_6B_3Si_6O_{27}(H,F)_4$, where R = Fe^{2+}, Mg, or (Al + Li). The crystals are trigonal, elongated, and variably coloured, the two ends of the crystals often having different colours. Tourmaline is used as a gemstone and because of its double refraction and piezoelectric properties is also used in polarizers and some pressure gauges.

town gas Any manufactured gas supplied as a fuel gas to domestic and industrial users. Examples include *coal gas, substitute natural gas (*SNG), and *natural gas itself. Most town gases are based on hydrogen or methane, although coal gas also contains up to 8% carbon monoxide.

trace element *See* **essential element.**

tracing (radioactive tracing) *See* **labelling.**

trans *See* **isomerism; torsion angle.**

transactinide elements Elements with an atomic number greater than 103, i.e. elements above lawrencium in the *periodic table. So far, elements up to 116 have been detected. Because of the highly radioactive and transient nature of these elements, there has been much dispute about priority of discovery and, consequently, naming of the elements. The International Union of Pure and Applied Chemistry (IUPAC) introduced a set of systematic temporary names based on affixes, as shown in the table. All these element names end in -ium. So, for example, element 109 in this system is called un + nil + enn + ium, i.e. unnilennium, and given the symbol u+n+e, i.e. Une.

Affix	Number	Symbol
nil	0	n
un	1	u
bi	2	b
tri	3	t
quad	4	q
pent	5	p
hex	6	h
sept	7	s
oct	8	o
enn	9	e

One long-standing dispute was about the element 104 (rutherfordium), which has also been called kurchatovium (Ku). There have also been disputes between IUPAC and the American Chemical Union about names. In 1994 IUPAC suggested the following list:

mendelevium (Md, 101)
nobelium (No, 102)
lawrencium (Lr, 103)
dubnium (Db, 104)
joliotium (Jl, 105)

rutherfordium (Rf, 106)
bohrium (Bh, 107)
hahnium (Hn, 108)
meitnerium (Mt, 109)

The ACU favoured a different set of names:

mendelevium (Md, 101)
nobelium (No, 102)
lawrencium (Lr, 103)
rutherfordium (Rf, 104)
hahnium (Ha, 105)
seaborgium (Sg, 106)
nielsbohrium (Ns, 107)
hassium (Hs, 108)
meitnerium (Mt, 109)

A compromise list was adopted by IUPAC in 1997 and is generally accepted:

mendelevium (Md, 101)
nobelium (No, 102)
lawrencium (Lr, 103)
rutherfordium (Rf, 104)
dubnium (Db, 105)
seaborgium (Sg, 106)
bohrium (Bh, 107)
hassium (Hs, 108)
meitnerium (Mt, 109)

Element 110 was named as darmstadtium in 2003 and element 111 was named roentgenium in 2004. So far elements 112 (ununbium Uub), 113 (ununtrium, Uut), 114 (ununquadium, Uuq), 115 (ununpentium, Uup), and 116 (ununhexium, Uuh) are not officially named.

transamination A biochemical reaction in amino acid metabolism in which an amine group is transferred from an amino acid to a keto acid to form a new amino acid and keto acid. The coenzyme required for this reaction is pyridoxal phosphate.

trans effect An effect in the substitution of inorganic square-planar complexes, in which certain ligands in the original complex are able to direct the incoming ligand into the trans position. The order of ligands in decreasing trans-directing power is:
$CN^- > NO_2 > I^- > Br^- > Cl^- > NH_3 > H_2O$.

transfer coefficient Symbol α. A quantity used in the analysis of processes in electrochemistry; it is the fraction of the potential difference at the surface of an electrode that assists charge difference in one direction but discourages it in the other direction. Its value lies between 0 and 1, frequently being about ½. It is related to the slope of a graph of the logarithm of the current against the potential.

transformation The conversion of a compound into a particular product, irrespective of the reagents or mechanism involved.

transient species A short-lived intermediate in a chemical reaction.

transition A change of a system from one quantum state to another.

transition elements A set of elements in the *periodic table in which filling of electrons in an inner *d*- or *f*-level occurs. With increasing proton number, electrons fill atomic levels up to argon, which has the electron configuration $1s^22s^22p^63s^23p^6$. In this shell, there are 5 *d*-orbitals, which can each contain 2 electrons. However, at this point the subshell of lowest energy is not the 3*d* but the 4*s*. The next two elements, potassium and calcium, have the configurations $[Ar]4s^1$ and $[Ar]4s^2$ respectively. For the next element, scandium, the 3*d* level is of lower energy than the 4*p* level, and scandium has the configuration $[Ar]3d^14s^2$. This filling of the inner *d*-level continues up to zinc $[Ar]3d^{10}4s^2$, giving the first transition series. There is a further series of this type in the next period of the table: between yttrium ($[Kr]4d5s^2$) and cadmium ($[Kr]4d^{10}5s^2$). This is the second transition series. In the next period of the table the situation is rather more complicated. Lanthanum has the configuration $[Xe]5d^16s^2$. The level of lowest energy then becomes the 4*f* level and the next element, cerium, has the configuration $[Xe]4f^15d^16s^2$. There are 7 of these *f*-orbitals, each of which can contain 2 electrons, and filling of the *f*-levels continues up to lutetium ($[Xe]4f^{14}5d^16s^2$). Then the filling of the 5*d* levels continues from hafnium to mercury. The series of 14 elements from cerium to lutetium is a 'series within a series' called an *inner transition series*. This one is the *lanthanoid series. In the next period there is a similar inner transition series, the *actinoid series, from thorium to lawrencium. Then filling of the *d*-level continues from element 104 onwards.

In fact, the classification of chemical elements is valuable only in so far as it illustrates chemical behaviour, and it is conventional to use the term 'transition elements' in a more restricted sense. The elements in the inner transition series from cerium (58) to lutetium (71) are called the lanthanoids; those in the series from thorium (90) to lawrencium (103) are the actinoids. These two series together make up the *f*-block in the periodic table. It is also common to include scandium, yttrium, and lanthanum with the lanthanoids (because of chemical similarity) and to include actinium with the actinoids. Of the remaining transition elements, it is usual to speak of three *main transition series*: from titanium to copper; from zirconium to silver; and from hafnium to gold. All these elements have similar chemical properties that result from the presence of unfilled *d*-orbitals in the element or (in the case of copper, silver, and gold) in the ions. The elements from 104 to 109 and the undiscovered elements 110 and 111 make up a fourth transition series. The elements zinc, cadmium, and mercury have filled *d*-orbitals both in the elements and in compounds, and are usually regarded as nontransition elements forming group 12 of the periodic table.

The elements of the three main transition series are all typical metals (in the nonchemical sense), i.e. most are strong hard materials that are good conductors of heat and electricity and have high melting and boiling points. Chemically, their behaviour depends on the existence of unfilled *d*-orbitals. They exhibit variable valency, have coloured compounds, and form *coordination compounds. Many of their compounds are paramagnetic as a result of the presence of unpaired electrons. Many of them are good catalysts. They are less reactive than the *s*- and *p*-block metals.

transition point **(transition temperature)** **1.** The temperature at which one crystalline form of a substance changes to another form. **2.** The temperature at which a substance changes phase. **3.** The temperature at which a substance becomes superconducting. **4.** The temperature at which some other change, such as a change of magnetic properties, takes place.

transition state **(activated complex)** The association of atoms of highest energy formed during a chemical reaction. The transition state can be regarded as a short-lived intermediate that breaks down to give the products. For example, in a S_N2 substitution reaction, one atom or group approaches the molecule as the other leaves. The transition state is an intermediate state in which both attacking and leaving groups are partly bound to the molecule, e.g.

$$B + RA \rightarrow B\text{---}R\text{---}A \rightarrow BR + A$$

In the theory of reaction rates, the reactants are assumed to be in equilibrium with this activated complex, which decomposes to the products.

transmission electron microscope *See* **electron microscope.**

transmutation The transformation of one element into another by bombardment of nuclei with particles. For example, plutonium is obtained by the neutron bombardment of uranium.

transport coefficients Quantities that characterize transport in a system. Examples of transport coefficients include electrical and thermal conductivity. One of the main purposes of *nonequilibrium statistical mechanics is to calculate such coefficients from first principles. It is difficult to calculate transport coefficients exactly for noninteracting systems and it is therefore necessary to use approximation techniques. A transport coefficient gives a measure for flow in a system. An *inverse transport coefficient* gives a measure of resistance to flow in a system.

transport number Symbol *t*. The fraction of the total charge carried by a particular type of ion in the conduction of electricity through electrolytes.

transuranic elements Elements with an atomic number greater than 92, i.e. elements above uranium in the *periodic table. Most of these elements are unstable and have short half-lives. *See also* **transactinide elements.**

triacylglycerol *See* **triglyceride.**

triatomic molecule A molecule formed from three atoms (e.g. H_2O or CO_2).

triazine *See* **azine.**

tribology The study of friction, lubrication, and lubricants.

triboluminescence *Luminescence caused by friction; for example, some crystalline substances emit light when they are crushed as a result of static electric charges generated by the friction.

tribromomethane **(bromoform)** A colourless liquid *haloform, $CHBr_3$; r.d. 2.9; m.p. 8°C; b.p. 150°C.

tricarbon dioxide **(carbon suboxide)** A colourless gas, C_3O_2, with an unpleasant odour; r.d. 1.114 (liquid at 0°C); m.p. −111.3°C; b.p. 7°C. It is the acid anhydride of malonic acid, from which it can be prepared by dehydration using phosphorus(V) oxide. The molecule is linear (O:C:C:C:O).

tricarboxylic acid cycle *See* **Krebs cycle.**

trichloroethanal **(chloral)** A liquid aldehyde, CCl_3CHO; r.d. 1.51; m.p. −57.5°C; b.p. 97.8°C. It is made by chlorinating ethanal and used in making DDT. *See also* **2,2,2-trichloroethanediol.**

2,2,2-trichloroethanediol **(chloral hydrate)** A colourless crystalline solid, $CCl_3CH(OH)_2$; r.d. 1.91; m.p. 57°C; b.p. 96.3°C. It is made by the hydrolysis of trichloroethanal and is unusual in having two –OH groups on the same carbon atom. Gem diols of this type are usually unstable; in this case the compound is stabilized by the presence of the three Cl atoms. It is used as a sedative.

trichloroethene **(trichlorethylene)** A colourless liquid, CCl_2=CHCl, b.p. 87°C. It is toxic and nonflammable, with a smell resembling that of chloroform (trichloromethane). It is widely used as a solvent in dry cleaning and degreasing. It is also used to extract oils from nuts and fruit, as an anaesthetic, and as a fire extinguisher.

trichloromethane **(chloroform)** A colourless volatile sweet-smelling liquid *haloform, $CHCl_3$; r.d. 1.48; m.p. −63.5°C; b.p. 61.7°C. It can be made by chlorination of methane (followed by separation of the mixture of products) or by the haloform reaction. It is an effective anaesthetic but can cause liver damage and it has now been replaced by other halogenated hydrocarbons. Chloroform is used as a solvent and raw material for making other compounds.

triclinic *See* **crystal system.**

tridymite A mineral form of *silicon(IV) oxide, SiO_2.

triethanolamine *See* **ethanolamine.**

triglyceride **(triacylglycerol)** An ester of glycerol (propane-1,2,3-triol) in which all three hydroxyl groups are esterified with a fatty acid. Triglycerides are the major constituent of fats and oils and provide a concentrated food energy store in living organisms as well as cooking fats and oils, margarines, etc. Their physical and chemical properties depend on the nature of their constituent fatty acids. In *simple triglycerides* all three fatty acids are identical; in *mixed triglycerides* two or three different fatty acids are present.

trigonal bipyramid *See illustration at* **complex.**

trihydrate A crystalline hydrate that contains three moles of water per mole of compound.

trihydric alcohol *See* triol.

3,4,5-trihydroxybenzoic acid *See* **gallic acid.**

triiodomethane **(iodoform)** A yellow volatile solid sweet-smelling *haloform, CHI_3; r.d. 4.1; m.p. 115°C. It is made by the haloform reaction.

triiron tetroxide **(ferrosoferric oxide)** A black magnetic oxide, Fe_3O_4; r.d. 5.2. It is formed when iron is heated in steam and also occurs naturally as the mineral *magnetite. The oxide dissolves in acids to give a mixture of iron(II) and iron(III) salts.

trimethylaluminium **(aluminium trimethyl)** A colourless liquid, $Al(CH_3)_3$, which ignites in air and reacts with water to give aluminium hydroxide and methane, usually with extreme vigour; r.d. 0.752; m.p. 0°C; b.p. 130°C. Like other aluminium alkyls it may be prepared by reacting a Grignard reagent with aluminium trichloride. Aluminium alkyls are used in the *Ziegler process for the manufacture of high-density polyethene (polythene).

2,4,6-trinitrophenol *See* **picric acid.**

trinitrotoluene **(TNT)** A yellow highly explosive crystalline solid, $CH_3C_6H_2(NO_2)_3$; r.d. 1.65; m.p. 82°C. It is made by nitrating toluene (methylbenzene), the systematic name being 1-methyl-2,4,6-trinitrobenzene

triol **(trihydric alcohol)** An *alcohol containing three hydroxyl groups per molecule.

triose A sugar molecule that contains three carbon atoms. *See* **monosaccharide.**

trioxoboric(III) acid *See* **boric acid.**

trioxosulphuric(IV) acid *See* **sulphurous acid.**

trioxygen *See* ozone.

triphenylmethyl An unusual aromatic compound that can exist as an anion, cation or free radical, $(C_6H_5)_3C$. Treatment of triphenylcarbinol in alcoholic solution with mineral acids produces triphenylmethyl salts, containing the cation $(C_6H_5)_3C^+$, which crystallize as orange-red solids. The action of sodium on a solution of triphenylmethyl chloride in ether produces the yellow $(C_6H_5)_3C^-$ anion. The free radical form is made by treating triphenylmethyl chloride with mercury or zinc in the absence of oxygen.

triple bond *See* **chemical bond.**

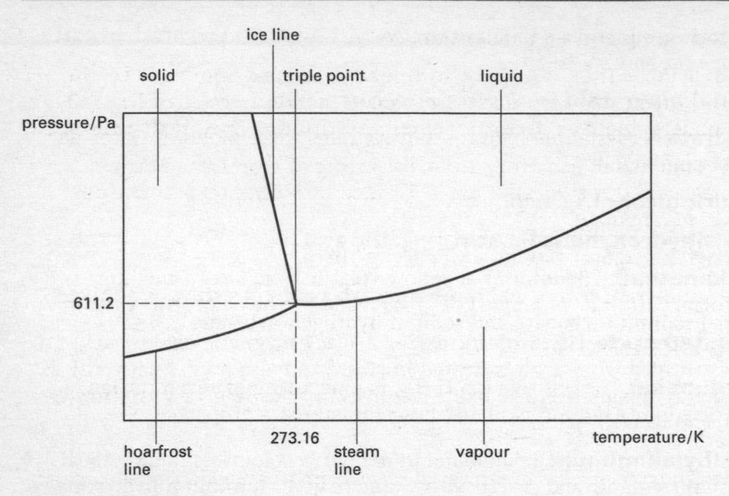

Triple point of water

triple point The temperature and pressure at which the vapour, liquid, and solid phases of a substance are in equilibrium. For water the triple point occurs at 273.16 K and 611.2 Pa (see illustration). This value forms the basis of the definition of the *kelvin and the thermodynamic *temperature scale.

triplet An atomic state in which two spin angular momenta of electrons combine together to give a total nonzero spin. A triplet state usually has a lower energy than a *singlet state because of the effect of correlations of spin on the Coulomb interactions between electrons (as in *Hund's rules). This can lead to large energy differences between triplet and singlet states.

trisilane *See* silane.

trisodium phosphate(V) **(sodium orthophosphate)** A colourless crystalline compound, Na_3PO_4, soluble in water and insoluble in ethanol. It is known both as the decahydrate (octagonal; r.d. 2.54) and the dodecahydrate (trigonal; r.d. 1.62) The dodecahydrate loses water at about 76°C and the decahydrate melts at 100°C. Trisodium phosphate may be prepared by boiling sodium carbonate with the stoichiometric amount of phosphoric acid and subsequently adding sodium hydroxide to the disodium salt thus formed. It is useful as an additive for high-pressure boiler feed water (for removal of calcium and magnesium as phosphates), in emulsifiers, as a water-softening agent, and as a component in detergents and cleaning agents. Sodium phosphate labelled with the radioactive isotope ^{32}P is used in the study of the role of phosphate in biological processes and is also used (intravenously) in the treatment of polycythaemia.

tritiated compound *See* **labelling**.

tritium Symbol T. An isotope of hydrogen with mass number 3; i.e. the nucleus contains 2 neutrons and 1 proton. It is radioactive (half-life 12.3 years), undergoing beta decay to helium–3. Tritium is used in *labelling.

triton A nucleus of a tritium atom, consisting of a proton and two neutrons bound together; the ion T^+ formed by ionization of a tritium atom. *See also* **hydron**.

trivalent (tervalent) Having a valency of three.

trona A mineral form of sodium sesquicarbonate, consisting of a mixed hydrated sodium carbonate and sodium hydrogencarbonate, $Na_2CO_3.NaHCO_3.2H_2O$.

tropylium ion The positive ion $C_7H_7^+$, having a ring of seven carbon atoms. The ion is symmetrical and has characteristic properties of *aromatic compounds.

trypsin An enzyme that digests proteins (*see* **protease**). It is secreted in an inactive form (*trypsinogen*) by the pancreas into the duodenum. There, trypsinogen is acted on by an enzyme (*enterokinase*) produced in the duodenum to yield trypsin. The active enzyme plays an important role in the digestion of proteins in the anterior portion of the small intestine. It also activates other proteases in the pancreatic juice.

trypsinogen *See* **trypsin**.

tryptophan *See* **amino acid**.

tub *See* **ring conformations**.

tunable laser *See* **dye laser**.

tungsten Symbol W. A white or grey metallic *transition element (formerly called *wolfram*); a.n. 74; r.a.m. 183.85; r.d. 19.3; m.p. 3410°C; b.p. 5660°C. It is found in a number of ores, including the oxides wolframite, $(Fe,Mn)WO_4$, and scheelite, $CaWO_4$. The ore is heated with concentrated sodium hydroxide solution to form a soluble *tungstate*. The oxide WO_3 is precipitated from this by adding acid, and is reduced to the metal using hydrogen. It is used in various alloys, especially high-speed steels (for cutting tools) and in lamp filaments. Tungsten forms a protective oxide in air and can be oxidized at high temperature. It does not dissolve in dilute acids. It forms compounds in which the oxidation state ranges from +2 to +6. The metal was first isolated by Juan d'Elhuyer and Fausto d'Elhuyer (1755–1833) in 1783.

tungsten carbide A black powder, WC, made by heating powdered tungsten metal with lamp black at 1600°C. It is extremely hard (9.5 on Mohs' scale) and is used in dies and cutting tools. A ditungsten carbide, W_2C, also exists.

tunnel effect An effect in which electrons are able to tunnel through a narrow potential barrier that would constitute a forbidden region if the electrons were treated as classical particles. That there is a finite probability of an electron tunnelling from one classically allowed region to another arises as a consequence of *quantum mechanics. The effect is made use of in the tunnel diode.

turpentine An oily liquid extracted from pine resin. It contains pinene, $C_{10}H_{16}$, and other terpenes and is mainly used as a solvent.

turquoise A mineral consisting of a hydrated phosphate of aluminium and copper, $CuAl_6(PO_4)_4(OH)_8.4H_2O)$, that is prized as a semiprecious stone. It crystallizes in the triclinic system and is generally blue in colour, the ‘robin’s egg’ blue variety being the most sought after. It usually occurs in veinlets and as masses and is formed by the action of surface waters on aluminium-rich rocks. The finest specimens are obtained from Iran.

twinning The growth of crystals in such a way that two distinct crystals form sharing a common plane of atoms.

twist *See* **ring conformations**.

Tyndall effect The scattering of light as it passes through a medium containing small particles. If a polychromatic beam of light is passed through a medium containing particles with diameters less than about one-twentieth of the wavlength of light, the scattered light appears blue. This accounts for the blue appearance of tobacco smoke. At higher particle diameters, the scattered light remains polychromatic. The effect is seen in suspensions and certain colloids. It is named after John Tyndall (1820–93).

tyrosine *See* **amino acid**.

ubiquinone **(coenzyme Q)** Any of a group of related quinone-derived compounds that serve as electron carriers in the *electron transport chain reactions of cellular respiration. Ubiquinone molecules have side chains of different lengths in different types of organisms but function in similar ways.

ultracentrifuge A high-speed centrifuge used to measure the rate of sedimentation of colloidal particles or to separate macromolecules, such as proteins or nucleic acids, from solutions. Ultracentrifuges are electrically driven and are capable of speeds up to 60 000 rpm.

ultrahigh frequency **(UHF)** A radio frequency in the range 3×10^9–0.3×10^9 Hz; i.e. having a wavelength in the range 10 cm to 1 m.

ultramicroscope A form of microscope that uses the Tyndall effect to reveal the presence of particles that cannot be seen with a normal optical microscope. Colloidal particles, smoke particles, etc., are suspended in a liquid or gas in a cell with a black background and illuminated by an intense cone of light that enters the cell from the side and has its apex in the field of view. The particles then produce diffraction-ring systems, appearing as bright specks on the dark background.

ultrasonics The study and use of pressure waves that have a frequency in excess of 20 000 Hz and are therefore inaudible to the human ear. Ultrasonics are used in medical diagnosis, particularly in conditions such as pregnancy, in which X-rays could have a harmful effect. Ultrasonic techniques are also used industrially to test for flaws in metals, to clean surfaces, to test the thickness of parts, and to form colloids.

ultraviolet radiation **(UV)** Electromagnetic radiation having wavelengths between that of violet light and long X-rays, i.e. between 400 nanometres and 4 nm. In the range 400–300 nm the radiation is known as the *near ultraviolet*. In the range 300–200 nm it is known as the *far ultraviolet*. Below 200 nm it is known as the *extreme ultraviolet* or the *vacuum ultraviolet*, as absorption by the oxygen in the air makes the use of evacuated apparatus essential. The sun is a strong emitter of UV radiation but only the near UV reaches the surface of the earth as the *ozone layer of the atmosphere absorbs all wavelengths below 290 nm. Ultraviolet radiation is classified in three ranges according to its effect on the skin. The ranges are:

UV-A (320–400 nm);
UV-B (290–320 nm);
UV-C (230–290 nm).

The longest-wavelength range, UV-A, is not harmful in normal doses and is used clinically in the treatment of certain skin complaints, such as psoriasis. It is also used to induce *vitamin D formation in patients that are allergic to vitamin D preparations. UV-B causes reddening of the skin followed by pigmentation (tanning). Excessive exposure can cause severe blistering. UV-C, with the shortest wavelengths, is particularly damaging. It is thought that short-wavelength ultraviolet radiation causes skin cancer and that the risk of contracting this has been increased by the depletion of the ozone layer.

Most UV radiation for practical use is produced by various types of mercury-vapour lamps. Ordinary glass absorbs UV radiation and therefore lenses and prisms for use in the UV are made from quartz.

ultraviolet–visible spectroscopy **(UV–visible spectroscopy)** A technique for chemical analysis and the determination of structure. It is based on the principle that electronic transitions in molecules occur in the visible and ultraviolet regions of the electromagnetic spectrum, and that a given transition occurs at a characteristic wavelength. The spectrometer has two sources, one of ultraviolet and the other of white visible light, which together cover the whole wavelength range of the instrument. If the whole wavelength range is used, the source is changed over at the appropriate point. The radiation from the source is split into two beams of equal intensity. One beam is passed through a dilute solution of the sample while the other is passed through the pure solvent and is used as a reference against which the first is compared after transmittance.

uncertainty principle **(Heisenberg uncertainty principle; principle of indeterminism)** The principle that it is not possible to know with unlimited accuracy both the position and momentum of a particle. This principle, discovered in 1927 by Werner *Heisenberg, is usually stated in the form: $\Delta x \Delta p_x \geq h/4\pi$, where Δx is the uncertainty in the *x*-coordinate of the particle, Δp_x is the uncertainty in the *x*-component of the particle's momentum, and h is the *Planck constant. An explanation of the uncertainty is that in order to locate a particle exactly, an observer must be able to bounce off it a photon of radiation; this act of location itself alters the position of the particle in an unpredictable way. To locate the position accurately, photons of short wavelength would have to be used. These would have associated large momenta and cause a large effect on the position. On the other hand, using long-wavelength photons would have less effect on the particle's position, but would be less accurate because of the longer wavelength. The principle has had a profound effect on scientific thought as it appears to upset the classical relationship between cause and effect at the atomic level.

undetermined multipliers *See* **Lagrange multipliers.**

ungerade *See* **gerade.**

uniaxial crystal A double-refracting crystal (*see* **double refraction**) having only one optic axis.

unimolecular reaction A chemical reaction or step involving only one molecule. An example is the decomposition of dinitrogen tetroxide:

$N_2O_4 \rightarrow 2NO_2$

Molecules colliding with other molecules acquire sufficient activation energy to react, and the activated complex only involves the atoms of a single molecule.

unit A specified measure of a physical quantity, such as length, mass, time, etc., specified multiples of which are used to express magnitudes of that physical quantity. For scientific purposes previous systems of units have now been replaced by *SI units.

unit cell The group of particles (atoms, ions, or molecules) in a crystal that is repeated in three dimensions in the *crystal lattice. *See also* **crystal system**.

univalent (monovalent) Having a valency of one.

universal constants *See* **fundamental constants**.

universal indicator A mixture of acid–base *indicators that changes colour (e.g. red-yellow-orange-green-blue) over a range of pH.

unnil- *See* **transactinide elements**.

unsaturated 1. (of a compound) Having double or triple bonds in its molecules. Unsaturated compounds can undergo addition reactions as well as substitution. *Compare* **saturated**. **2.** (of a solution) *See* **saturated**.

unstable equilibrium *See* **equilibrium**.

UPS Ultraviolet photoelectron spectroscopy. *See* **photoelectron spectroscopy**.

UPVC Unplasticized PVC: a tough hardwearing form of PVC used for window frames and similar applications.

uracil A *pyrimidine derivative and one of the major component bases of *nucleotides and the nucleic acid *RNA.

uraninite A mineral form of uranium(IV) oxide, containing minute amounts of radium, thorium, polonium, lead, and helium. When uraninite occurs in a massive form with a pitchy lustre it is known as *pitchblende*, the chief ore of uranium. Uraninite occurs in Saxony (Germany), Romania, Norway, the UK (Cornwall), E Africa (Congo), USA, and Canada (Great Bear Lake).

uranium Symbol U. A white radioactive metallic element belonging to the *actinoids; a.n. 92; r.a.m. 238.03; r.d. 19.05 (20°C); m.p. 1132±1°C; b.p. 3818°C. It occurs as *uraninite, from which the metal is extracted by an ion-exchange process. Three isotopes are found in nature: uranium–238 (99.28%), uranium–235 (0.71%), and uranium–234 (0.006%). As uranium–235 undergoes nuclear fission with slow neutrons it is the fuel used in nuclear reactors and nuclear weapons; uranium has therefore assumed enormous technical and political importance since their invention. It was discovered by Martin Klaproth (1747–1817) in 1789.

uranium(VI) fluoride **(uranium hexafluoride)** A volatile white solid, UF_6; r.d. 4.68; m.p. 64.5°C. It is used in the separation of uranium isotopes by gas diffusion.

uranium(IV) oxide A black solid, UO_2; r.d. 10.96; m.p. 2500°C. It occurs naturally as *uraninite and is used in nuclear reactors.

uranium hexafluoride *See* **uranium(VI) fluoride**.

uranium–lead dating A group of methods of dating certain rocks that depends on the decay of the radioisotope uranium–238 to lead–206 (half-life 4.5×10^9 years) or the decay of uranium–235 to lead–207 (half-life 7.1×10^8 years). One form of uranium–lead dating depends on measuring the ratio of the amount of helium trapped in the rock to the amount of uranium present (since the decay $^{238}U \rightarrow {}^{206}Pb$ releases eight alpha-particles). Another method of calculating the age of the rocks is to measure the ratio of radiogenic lead (^{206}Pb, ^{207}Pb, and ^{208}Pb) present to nonradiogenic lead (^{204}Pb). These methods give reliable results for ages of the order 10^7–10^9 years. *See* **dating techniques**.

uranium(IV) oxide A black solid, UO_2; r.d. 10.9; m.p. 3000°C. It occurs naturally as *uraninite and is used in nuclear reactors.

uranium series *See* **radioactive series**.

urea **(carbamide)** A white crystalline solid, $CO(NH_2)_2$; r.d. 1.3; m.p. 135°C. It is soluble in water but insoluble in certain organic solvents. Urea is the major end product of nitrogen excretion in mammals, being synthesized by the *urea cycle. Urea is synthesized industrially from ammonia and carbon dioxide for use in *urea–formaldehyde resins and pharmaceuticals, as a source of nonprotein nitrogen for ruminant livestock, and as a nitrogen fertilizer.

urea cycle **(ornithine cycle)** The series of biochemical reactions that converts ammonia to *urea during the excretion of metabolic nitrogen. Urea formation occurs in mammals and, to a lesser extent, in some other animals. The liver converts ammonia to the much less toxic urea, which is excreted in solution in urine.

urea–formaldehyde resins Synthetic resins made by copolymerizing urea with formaldehyde (methanal). They are used as adhesives or thermosetting plastics.

urethane resins **(polyurethanes)** Synthetic resins containing the repeating group –NH–CO–O–. There are numerous types made by copolymerizing isocyanate esters with polyhydric alcohols. They have a variety of uses in plastics, paints, and solid foams.

Urey, Harold Clayton (1894–1981) US physical chemist, who became a professor at the University of California in 1958. His best-known work was the discovery of *deuterium (heavy hydrogen) in 1932, for which he was awarded the 1939 Nobel Prize for physics.

uric acid The end product of purine breakdown in most primates, birds, terrestrial reptiles, and insects and also (except in primates) the major form in which metabolic nitrogen is excreted. Being fairly insoluble, uric acid can be expelled in solid form, which conserves valuable water in arid environments. The accumulation of uric acid in the synovial fluid of joints causes gout.

UV *See* **ultraviolet radiation.**

UV–visible spectroscopy *See* **ultraviolet–visible spectroscopy.**

vacancy *See* **crystal defect.**

vacuum A space in which there is a low pressure of gas, i.e. relatively few atoms or molecules. A *perfect vacuum* would contain no atoms or molecules, but this is unobtainable as all the materials that surround such a space have a finite *vapour pressure. In a *soft* (or *low*) *vacuum* the pressure is reduced to about 10^{-2} pascal, whereas a *hard* (or *high*) vacuum has a pressure of 10^{-2}–10^{-7} pascal. Below 10^{-7} pascal is known as an *ultrahigh vacuum. See also* **vacuum pump.**

vacuum distillation Distillation under reduced pressure. The depression in the boiling point of the substance distilled means that the temperature is lower, which may prevent the substance from decomposing.

vacuum pump A pump used to reduce the gas pressure in a container. The normal laboratory rotary oil-seal pump can maintain a pressure of 10^{-1} Pa. For pressures down to 10^{-7} Pa a *diffusion pump is required. *Ion pumps can achieve a pressure of 10^{-9} Pa and a *cryogenic pump combined with a diffusion pump can reach 10^{-13} Pa.

valence *See* **valency.**

valence band *See* **energy bands.**

valence-bond theory A method of computational chemistry in which the electrons are considered as belonging to definite bonds consisting of pairs of electrons associated with pairs of atoms in the molecule. The actual state of the molecule can be regarded as the result of a set of canonical forms (*see* **resonance**). *See also* **density-function theory; molecular-orbital theory.**

valence electron An electron in one of the outer shells of an atom that takes part in forming chemical bonds.

valency (valence) The combining power of an atom or radical, equal to the number of hydrogen atoms that the atom could combine with or displace in a chemical compound (hydrogen has a valency of 1). It is equal to the ionic charge in ionic compounds; for example, in Na_2S, sodium has a valency of 1 (Na^+) and sulphur a valency of 2 (S^{2-}). In covalent compounds it is equal to the number of bonds formed; in CO_2 oxygen has a valency of 2 and carbon has a valency of 4.

valine *See* **amino acid.**

vanadium Symbol V. A silvery-white metallic *transition element; a.n. 23; r.a.m. 50.94; r.d. 5.96; m.p. 1890°C; b.p. 3380°C. It occurs in a number of complex ores, including vanadinite ($Pb_5Cl(VO_4)_3$) and carnotite

$(K_2(ClO_2)_2(VO_4)_2)$. The pure metal can be obtained by reducing the oxide with calcium. The element is used in a large number of alloy steels. Chemically, it reacts with nonmetals at high temperatures but is not affected by hydrochloric acid or alkalis. It forms a range of complexes with oxidation states from +2 to +5. Vanadium was discovered in 1801 by Andrés del Rio (1764–1849), who allowed himself to be persuaded that what he had discovered was an impure form of chromium. The element was rediscovered and named by Nils Sefström (1787–1854) in 1880.

vanadium(V) oxide **(vanadium pentoxide)** A crystalline compound, V_2O_5, used extensively as a catalyst in industrial gas-phase oxidation processes.

vanadium pentoxide *See* **vanadium(V) oxide**.

van der Waals' equation *See* **equation of state**.

van der Waals' force An attractive force between atoms or molecules, named after Johannes van der Waals (1837–1923). The force accounts for the term a/V^2 in van der Waals' equation (*see* **equation of state**). These forces are much weaker than those arising from valence bonds and are inversely proportional to the seventh power of the distance between the atoms or molecules. They are the forces responsible for nonideal behaviour of gases and for the lattice energy of molecular crystals. There are three factors causing such forces: (1) dipole–dipole interaction, i.e. electrostatic attractions between two molecules with permanent dipole moments; (2) dipole-induced dipole interactions, in which the dipole of one molecule polarizes a neighbouring molecule; (3) dispersion forces arising because of small instantaneous dipoles in atoms.

van't Hoff, Jacobus *See* **Hoff, Jacobus van't**.

van't Hoff factor Symbol *i*. A factor appearing in equations for *colligative properties, equal to the ratio of the number of actual particles present to the number of undissociated particles.

van't Hoff's isochore An equation formulated by van't Hoff for the variation of equilibrium constant with temperature:

$$(\mathrm{d}\log_e K)/\mathrm{d}T = \Delta H/RT^2,$$

where K is the equilibrium constant, R is the gas constant, T is the thermodynamic temperature, and ΔH the enthalpy of the reaction.

vapour density The density of a gas or vapour relative to hydrogen, oxygen, or air. Taking hydrogen as the reference substance, the vapour density is the ratio of the mass of a particular volume of a gas to the mass of an equal volume of hydrogen under identical conditions of pressure and temperature. Taking the density of hydrogen as 1, this ratio is equal to half the relative molecular mass of the gas.

vapour pressure The pressure exerted by a vapour. All solids and liquids give off vapours, consisting of atoms or molecules of the substances that have evaporated from the condensed forms. These atoms or molecules exert a vapour pressure. If the substance is in an enclosed space, the vapour pressure will reach an equilibrium value that depends only on the nature of the substance and the temperature. This equilibrium value occurs when there is a dynamic equilibrium between the atoms or molecules escaping from the liquid or solid and those that strike the surface of the liquid or solid and return to it. The vapour is then said to be a *saturated vapour* and the pressure it exerts is the *saturated vapour pressure.*

variational principle In calculations in quantum mechanics the principle that if a trial *wave function is used to calculate the energy of a system, then the energy calculated cannot be lower than the actual energy of the ground state of the system. To use variational principles effectively in quantum mechanics a trial wave function contains parameters that can be varied, with these parameters being changed until the lowest possible energy is found.

verdigris A green patina of basic copper salts formed on copper. The composition of verdigris varies depending on the atmospheric conditions, but includes the basic carbonate $CuCO_3.Cu(OH)_2$, the basic sulphate $CuSO_4.Cu(OH)_2.H_2O$, and sometimes the basic chloride $CuCl_2.Cu(OH)_2$.

vermiculite *See* **clay minerals.**

very high frequency **(VHF)** A radio frequency in the range 3×10^8–0.3×10^8 Hz, i.e. having a wavelength in the range 1–10 m.

very low frequency **(VLF)** A radio frequency in the range 3×10^4–0.3×10^4 Hz, i.e. having a wavelength in the range 10–100 km.

vibrational relaxation A process in which a polyatomic molecule in an excited vibrational state returns to a lower vibrational state in the same electronic state by colliding with other molecules.

vibrational spectroscopy The spectroscopic investigation of the vibrational energy levels of molecules. In the infrared region of the electromagnetic spectrum vibrational transitions are accompanied by rotational transitions. Infrared spectra of molecules are series of *bands*, with each band being associated with a vibrational transition and every line in that band being associated with a rotational transition that accompanies the vibrational transition.

Some features of vibrational spectra can be analysed by regarding the vibrations as simple harmonic motion, but a realistic account of molecular vibrations requires that anharmonicity is taken into account.

A diatomic molecule can only have a vibrational–rotational spectrum if it has a permanent dipole moment. A polyatomic molecule can only have a vibrational–rotational spectrum if the normal modes of vibration cause the molecule to have an oscillating dipole moment.

vicinal **(vic)** Designating a molecule in which two atoms or groups are linked to adjacent atoms. For example, 1,2-dichloroethane (CH_2ClCH_2Cl) is a vicinal (or vic) dihalide and can be named *vic*-dichloroethane.

Viktor Meyer's method A method of measuring vapour density, devised by Viktor *Meyer. A weighed sample in a small tube is dropped into a heated bulb with a long neck. The sample vaporizes and displaces air, which is collected over water and the volume measured. The vapour density can then be calculated.

villiaumite A mineral form of sodium fluoride, NaF.

vinegar A dilute solution of *ethanoic acid (up to 6%), used as a flavouring and pickling medium. Natural vinegar is made by the continued fermentation of alcoholic liquors, usually by *Acetobacter* species, which oxidize ethanol to ethanoic acid. Vinegar is also made by diluting synthetic ethanoic acid.

vinyl acetate *See* **ethenyl ethanoate**.

vinylation The catalysed reaction between ethyne and a compound with an active hydrogen atom, such as alcohol, amine, or carboxylic acid. Addition takes place across the triple bond of ethyne to produce an ethenyl (vinyl) compound, containing the group $CH_2=CH-$.

vinyl chloride *See* **chloroethene**.

vinyl group The organic group $CH_2:CH-$.

virial coefficients *See* **virial equation**.

virial equation A gas law that attempts to account for the behaviour of real gases, as opposed to an ideal gas. It takes the form

$$pV = RT + Bp + Cp^2 + Dp^3 + ...,$$

where B, C, and D are known as *virial coefficients*.

viscose process *See* **rayon**.

viscosity A measure of the resistance to flow that a fluid offers when it is subjected to shear stress. For a Newtonian fluid, the force, F, needed to maintain a velocity gradient, dv/dx, between adjacent planes of a fluid of area A is given by: $F = \eta A(dv/dx)$, where η is a constant, called the coefficient of viscosity. In SI units it has the unit pascal second (in the c.g.s. system it was measured in *poise). Non-Newtonian fluids, such as clays, do not conform to this simple model. *See also* **kinematic viscosity**.

visible spectrum The *spectrum of electromagnetic radiations to which the human eye is sensitive.

vitamin One of a number of organic compounds required by living organisms in relatively small amounts to maintain normal health. There are some 14 generally recognized major vitamins: the water-soluble *vitamin B complex (containing 9) and *vitamin C and the fat-soluble *vitamin A, *vitamin D, *vitamin E, and *vitamin K. Most B vitamins and vitamin C occur in plants, animals, and microorganisms; they function typically as *coenzymes. Vitamins A, D, E, and K occur only in animals, especially vertebrates, and perform a variety of metabolic roles. Animals are unable to manufacture many vitamins themselves and must have adequate amounts in the diet. Foods may contain vitamin precursors (called *provitamins*) that are chemically changed to the actual vitamin on entering the body. Many vitamins are destroyed by light and heat, e.g. during cooking.

vitamin A (retinol) A fat-soluble vitamin that cannot be synthesized by mammals and other vertebrates and must be provided in the diet. Green plants contain precursors of the vitamin, notably carotenes, that are converted to vitamin A in the intestinal wall and liver. The aldehyde derivative of vitamin A, *retinal*, is a constituent of the visual pigment rhodopsin. Deficiency affects the eyes, causing night blindness, xerophthalmia, and eventually total blindness. The role of vitamin A in other aspects of metabolism is less clear.

vitamin B complex A group of water-soluble vitamins that characteristically serve as components of *coenzymes. Plants and many microorganisms can manufacture B vitamins but dietary sources are essential for most animals. Heat and light tend to destroy B vitamins.

Vitamin B_1 (*thiamin(e)*) is a precursor of the coenzyme thiamine pyrophosphate, which functions in carbohydrate metabolism. Deficiency leads to beriberi in humans and to polyneuritis in birds. Good sources include brewer's yeast, wheatgerm, beans, peas, and green vegetables.

Vitamin B_2 (*riboflavin*) occurs in green vegetables, yeast, liver, and milk. It is a constituent of the coenzymes *FAD and FMN, which have an important role in the metabolism of all major nutrients as well as in the oxidative phosphorylation reactions of the *electron transport chain. Deficiency of B_2 causes inflammation of the tongue and lips and mouth sores.

Vitamin B_6 (*pyridoxine*) is widely distributed in cereal grains, yeast, liver, milk, etc. It is a constituent of a coenzyme (pyridoxal phosphate) involved in amino acid metabolism. Deficiency causes retarded growth, dermatitis, convulsions, and other symptoms.

Vitamin B_{12} (*cyanocobalamin; cobalamin*) is manufactured only by microorganisms and natural sources are entirely of animal origin. Liver is especially rich in it. One form of B_{12} functions as a coenzyme in a number of reactions, including the oxidation of fatty acids and the synthesis of DNA. It also works in conjunction with *folic acid (another B vitamin) in the synthesis of the amino acid methionine and it is required for normal production of red blood cells. Vitamin B_{12} can only be absorbed from the gut in the presence of a glycoprotein called *intrinsic factor*; lack of this factor or deficiency of B_{12} results in pernicious anaemia.

Other vitamins in the B complex include *nicotinic acid, *pantothenic acid, *biotin, and *lipoic acid. *See also* **choline**.

vitamin C (ascorbic acid) A colourless crystalline water-soluble vitamin found especially in citrus fruits and green vegetables. Most organisms synthesize it from glucose but man and other primates and various other species must obtain it from their diet. It is required for the maintenance of healthy connective tissue; deficiency leads to scurvy. Vitamin C is readily destroyed by heat and light.

vitamin D A fat-soluble vitamin occurring in the form of two steroid derivatives: *vitamin* D_2 (*ergocalciferol*, or *calciferol*), found in yeast; and *vitamin* D_3 (*cholecalciferol*), which occurs in animals. Vitamin D_2 is formed from a steroid by the action of ultraviolet light and D_3 is produced by the action of sunlight on a cholesterol derivative in the skin. Fish-liver oils are the major dietary source. The active form of vitamin D is manufactured in response to the secretion of parathyroid hormone, which occurs when blood calcium levels are low. It causes increased uptake of calcium from the gut, which increases the supply of calcium for bone synthesis. Vitamin D deficiency causes rickets in growing animals and osteomalacia in mature animals. Both conditions are characterized by weak deformed bones.

vitamin E (tocopherol) A fat-soluble vitamin consisting of several closely related compounds, deficiency of which leads to a range of disorders in different species, including muscular dystrophy, liver damage, and infertility. Good sources are cereal grains and green vegetables. Vitamin E prevents the oxidation of unsaturated fatty acids in cell membranes, so maintaining their structure.

vitamin K A fat-soluble vitamin consisting of several related compounds that act as coenzymes in the synthesis of several proteins necessary for blood clotting. Deficiency of vitamin K, which leads to extensive bleeding, is rare because a form of the vitamin is manufactured by intestinal bacteria. Green vegetables and egg yolk are good sources.

vitreous Having a glasslike appearance or structure.

volt Symbol V. The SI unit of electric potential, potential difference, or e.m.f. defined as the difference of potential between two points on a conductor carrying a constant current of one ampere when the power dissipated between the points is one watt. It is named after Alessandro Volta.

Volta, Alessandro Giuseppe Antonio Anastasio (1745–1827) Italian physicist. In 1774 he began teaching in Como and in that year invented the electrophorus. He moved to Pavia University in 1778. In 1800 he made the *voltaic cell, thus providing the first practical source of electric current. The SI unit of potential difference is named after him.

voltaic cell (galvanic cell) A device that produces an e.m.f. as a result of chemical reactions that take place within it. These reactions occur at the surfaces of two electrodes, each of which dips into an electrolyte. The first voltaic cell, devised by Alessandro Volta, had electrodes of two different metals dipping into brine. *See* **primary cell; secondary cell.**

voltaic pile An early form of battery, devised by Alessandro Volta, consisting of a number of flat *voltaic cells joined in series. The liquid electrolyte was absorbed into paper or leather discs.

voltameter (coulometer) 1. An electrolytic cell formerly used to measure quantity of electric charge. The increase in mass (m) of the cathode of the cell as a result of the deposition on it of a metal from a solution of its salt enables the charge (Q) to be determined from the relationship $Q = m/z$, where z is the electrochemical equivalent of the metal. **2.** Any other type of electrolytic cell used for measurement.

volume Symbol V. The space occupied by a body or mass of fluid.

volumetric analysis A method of quantitative analysis using measurement of volumes. For gases, the main technique is in reacting or absorbing gases in graduated containers over mercury, and measuring the volume changes. For liquids, it involves *titrations.

VSEPR theory Valence-shell electron-pair repulsion theory; a theory for predicting the shapes of molecules. *See* **lone pair**.

vulcanite (ebonite) A hard black insulating material made by the vulcanization of rubber with a high proportion of sulphur (up to 30%).

vulcanization A process for hardening rubber by heating it with sulphur or sulphur compounds.

Wacker process A process for the manufacture of ethanal by the air oxidation of ethene. A mixture of air and ethene is bubbled through a solution containing palladium(II) chloride and copper(II) chloride. The Pd^{2+} ions form a complex with the ethene in which the ion is bound to the pi electrons in the C=C bond. This decreases the electron density in the bond, making it susceptible to nucleophilic attack by water molecules. The complex formed breaks down to ethanal and palladium metal. The Cu^{2+} ions oxidize the palladium back to Pd^{2+}, being reduced to Cu^{+} ions in the process. The air present oxidizes Cu^{+} back to Cu^{2+}. Thus the copper(II) and palladium(II) ions effectively act as catalysts in the process, which is now the main source of ethanal and, by further oxidation, ethanoic acid. It can also be applied to other alkenes. It is named after Alexander von Wacker (1846–1922).

WAHUHA A pulse sequence in *nuclear magnetic resonance (NMR) used to reduce linewidth. The WAHUHA sequence (named after its inventors Waugh, Huber, and Haberlen) effects an averaging procedure by twisting the magnetization vector in various directions.

Walden's rule An empirical rule suggested by P. Walden (1863–1957) concerning ions in solutions, stating that the product of the molar conductivity, Λ_m, and the viscosity, η, is approximately constant for the same ions in different solvents. Some justification for Walden's rule is provided by the proportional relationship between Λ_m and the diffusion coefficient, D; as D is inversely proportional to the viscosity, Λ_m is inversely proportional to η, which is in accordance with Walden's rule. However, different solvents hydrate the same ions differently, so that both the radius and the viscosity change when the solvent is changed. It is this fact that limits the validity of the rule.

warfarin 3-(alpha-acetonylbenzyl)-4-hydroxycoumarin: a synthetic anticoagulant used both therapeutically in clinical medicine and, in lethal doses, as a rodenticide (*see* **pesticide**).

washing soda *Sodium carbonate decahydrate, $Na_2CO_3.10H_2O$.

water A colourless liquid, H_2O; r.d. 1.000 (4°C); m.p. 0.000°C; b.p. 100.000°C. In the gas phase water consists of single H_2O molecules in which the H–O–H angle is 105°. The structure of liquid water is still controversial; hydrogen bonding of the type $H_2O...H–O–H$ imposes a high degree of structure and current models supported by X-ray scattering studies have short-range ordered regions, which are constantly disintegrating and re-forming. This ordering of the liquid state is sufficient to make the density of water at about 0°C higher than that of the relatively open-structured ice; the maximum density occurs at 3.98°C. This accounts for the well-known phenomenon of ice floating on water and the contraction of water below ice, a fact of enormous biological significance for all aquatic organisms.

Ice has nine distinct structural modifications of which ordinary ice, or ice I, has an open structure built of puckered six-membered rings in which each H_2O unit is tetrahedrally surrounded by four other H_2O units.

Because of its angular shape the water molecule has a permanent dipole moment and in addition it is strongly hydrogen bonded and has a high dielectric constant. These properties combine to make water a powerful solvent for both polar and ionic compounds. Species in solution are frequently strongly hydrated and in fact ions frequently written as, for example, Cu^{2+} are essentially $[Cu(H_2O)_6]^{2+}$. Crystalline *hydrates are also common for inorganic substances; polar organic compounds, particularly those with O–H and N–H bonds, also form hydrates.

Pure liquid water is very weakly dissociated into H_3O^+ and OH^- ions by self ionization:

$$H_2O \rightleftharpoons H^+ + OH^-$$

(*see* **ionic product**) and consequently any species that increases the concentration of the positive species, H_3O^+, is acidic and species increasing the concentration of the negative species, OH^-, are basic (*see* **acid**). The phenomena of ion transport in water and the division of materials into *hydrophilic* (water loving) and *hydrophobic* (water hating) substances are central features of almost all biological chemistry. A further property of water that is of fundamental importance to the whole planet is its strong absorption in the infrared range of the spectrum and its transparency to visible and near ultraviolet radiation. This allows solar radiation to reach the earth during hours of daylight but restricts rapid heat loss at night. Thus atmospheric water prevents violent diurnal oscillations in the earth's ambient temperature.

water gas A mixture of carbon monoxide and hydrogen produced by passing steam over hot carbon (coke):

$$H_2O(g) + C(s) \rightarrow CO(g) + H_2(g)$$

The reaction is strongly endothermic but the reaction can be used in conjunction with that for *producer gas for making fuel gas. The main use of water gas before World War II was in producing hydrogen for the *Haber process. Here the above reaction was combined with the *water-gas shift reaction* to increase the amount of hydrogen:

$$CO + H_2O \rightleftharpoons CO_2 + H_2$$

Most hydrogen for the Haber process is now made from natural gas by steam *reforming.

water glass A viscous colloidal solution of sodium silicates in water, used to make silica gel and as a size and preservative.

water of crystallization Water present in crystalline compounds in definite proportions. Many crystalline salts form hydrates containing 1, 2, 3, or more moles of water per mole of compound, and the water may be held in the crystal in various ways. Thus, the water molecules may simply occupy lattice positions in the crystal, or they may form bonds with the anions or the cations present. In the pentahydrate of copper sulphate ($CuSO_4.5H_2O$), for instance, each copper ion is coordinated to four water molecules through the lone pairs on the oxygen to form the *complex $[Cu(H_2O)_4]^{2+}$. Each sulphate ion has one water molecule held by hydrogen bonding. The difference between the two types of bonding is demonstrated by the fact that the pentahydrate converts to the monohydrate at 100°C and only becomes anhydrous above 250°C. *Water of constitution* is an obsolete term for water combined in a compound (as in a metal hydroxide $M(OH)_2$ regarded as a hydrated oxide $MO.H_2O$).

water softening *See* **hardness of water.**

watt Symbol W. The SI unit of power, defined as a power of one joule per second. In electrical contexts it is equal to the rate of energy transformation by an electric current of one ampere flowing through a conductor the ends of which are maintained at a potential difference of one volt. The unit is named after James Watt (1736–1819).

wave equation A partial differential equation of the form:

$$\nabla^2 u = (1/c^2)\partial^2 u/\partial t^2,$$

where

$$\nabla^2 = \partial^2/\partial x^2 + \partial^2/\partial y^2 + \partial^2/\partial z^2$$

is the *Laplace operator*. It represents the propagation of a wave, where u is the displacement and c the speed of propagation. *See also* **Schrödinger equation.**

wave function A function $\psi(x,y,z)$ appearing in the *Schrödinger equation in wave mechanics. The wave function is a mathematical expression involving the coordinates of a particle in space. If the Schrödinger equation can be solved for a particle in a given system (e.g. an electron in an atom) then, depending on the boundary conditions, the solution is a set of allowed wave functions (*eigenfunctions*) of the particle, each corresponding to an allowed energy level (*eigenvalue*). The physical significance of the wave function is that the square of its absolute value, $|\psi|^2$, at a point is proportional to the probability of finding the particle in a small element of volume, dxdydz, at that point. For an electron in an atom, this gives rise to the idea of atomic and molecular *orbitals.

wave mechanics A formulation of *quantum mechanics in which the dual wave–particle nature of such entities as electrons is described by the *Schrödinger equation. Schrödinger put forward this formulation of quantum mechanics in 1926 and in the same year showed that it was equivalent to matrix mechanics. Taking into account the *de Broglie wavelength, Schrödinger postulated a wave mechanics that bears the same relation to Newtonian mechanics as physical optics does to geometrical optics.

wave number Symbol *k*. The number of cycles of a wave in unit length. It is the reciprocal of the wavelength.

wave packet A superposition of waves with one predominant *wave number *k*, but with several other wave numbers near *k*. Wave packets are useful for the analysis of scattering in *quantum mechanics. Concentrated packets of waves can be used to describe localized particles of both matter and *photons. The Heisenberg *uncertainty principle can be derived from a wave-packet description of entities in quantum mechanics. The motion of a wave packet is in accord with the motion of the corresponding classical particle, if the potential energy change across the dimensions of the packet is very small. This proposition is known as *Ehrenfest's theorem*, named after the Dutch physicist Paul Ehrenfest (1880–1933), who proved it in 1927.

wave–particle duality The concept that waves carrying energy may have a corpuscular aspect and that particles may have a wave aspect; which of the two models is the more appropriate will depend on the properties the model is seeking to explain. For example, waves of electromagnetic radiation need to be visualized as particles, called *photons, to explain the photoelectric effect while electrons need to be thought of as de Broglie waves in *electron diffraction.

wave vector A vector $\mathbf{k}$ associated with a *wave number *k*. In the case of free electrons the wave vector $\mathbf{k}$ is related to the momentum $\mathbf{p}$ in *quantum mechanics by $\mathbf{p} = \hbar\mathbf{k}$, where $\hbar$ is the rationalized *Planck constant. In the case of *Bloch's theorem, the wave vector $\mathbf{k}$ can only have certain values and can be thought of as a quantum number associated with the translational symmetry of the crystal.

wax Any of various solid or semisolid substances. There are two main types. Mineral waxes are mixtures of hydrocarbons with high molecular weights. Paraffin wax, obtained from *petroleum, is an example. Waxes secreted by plants or animals are mainly esters of fatty acids and usually have a protective function.

weak acid An *acid that is only partially dissociated in aqueous solution.

weber Symbol Wb. The SI unit of magnetic flux equal to the flux that, linking a circuit of one turn, produces in it an e.m.f. of one volt as it is reduced to zero at a uniform rate in one second. It is named after Wilhelm Weber (1804–91).

Weissenberg technique A technique used to overcome the problem of overlapping reflections in the identification of the symmetry and the dimensions of a unit cell in *X-ray crystallography. In this technique, a screen is placed in front of the film allowing only one set of reflections to be exposed. The Weissenberg technique produces distorted photographs, but this can be overcome by having a coupling between the motions of the crystal and the film. Using the *precession camera technique* undistorted photographs can be obtained.

Weston cell **(cadmium cell)** A type of primary *voltaic cell devised by Edward Weston (1850–1936), which is used as a standard; it produces a constant e.m.f. of 1.0186 volts at 20°C. The cell is usually made in an H-shaped glass vessel with a mercury anode covered with a paste of cadmium sulphate and mercury(I) sulphate in one leg and a cadmium amalgam cathode covered with cadmium sulphate in the other leg. The electrolyte, which connects the two electrodes by means of the bar of the H, is a saturated solution of cadmium sulphate. In some cells sulphuric acid is added to prevent the hydrolysis of mercury sulphate.

white arsenic *See* **arsenic(III) oxide**.

white mica *See* **muscovite**.

white spirit A liquid mixture of hydrocarbons obtained from petroleum, used as a solvent for paint ('turpentine substitute').

Wigner–Seitz cell A polyhedron in a crystal that is bounded by planes formed by perpendicular bisectors of bonds between lattice sites. The Wigner–Seitz cell was used by Hungarian-born US physicist Eugene Wigner (1902–95) and Frederick Seitz in 1933 in the course of their analysis of the cohesion of metals. The concept has been used extensively in the theory of solids.

Wigner–Witmer rules A set of results found by applying group theory that states which molecular electronic states can exist, starting from the electronic states of the isolated atoms. The rules were stated for diatomic molecules by Hungarian-born US physicist Eugene Wigner (1902–95) and E. E. Witmer in 1928 and were subsequently extended to polyatomic molecules. The Wigner–Witmer rules are also known as *correlation rules* since they involve the correlation between atomic electronic states and molecular electronic states. They are useful in analysing the *electronic spectra of molecules.

Williamson's synthesis Either of two methods of producing ethers, both named after the British chemist Alexander Williamson (1824–1904).
1. The dehydration of alcohols using concentrated sulphuric acid. The overall reaction can be written

$$2ROH \rightarrow H_2O + ROR$$

The method is used for making ethoxyethane ($C_2H_5OC_2H_5$) from ethanol by heating at 140°C with excess of alcohol (excess acid at 170°C gives ethene). Although the steps in the reaction are all reversible, the ether is distilled off so the reaction can proceed to completion. This is *Williamson's continuous process*. In general, there are two possible mechanisms for this synthesis. In the first (favoured by primary alcohols), an alkylhydrogen sulphate is formed

$$ROH + H_2SO_4 \rightleftharpoons ROSO_3H + H_2O$$

This reacts with another alcohol molecule to give an oxonium ion

$$ROH + ROSO_3H \rightarrow ROHR^+$$

This loses a proton to give ROR.

The second mechanism (favoured by tertiary alcohols) is formation of a carbonium ion

$ROH + H^+ \rightarrow H_2O + R^+$

This is attacked by the lone pair on the other alcohol molecule

$R^+ + ROH \rightarrow ROHR^+$

and the oxonium ion formed again gives the product by loss of a proton.

The method can be used for making symmetric ethers (i.e. having both R groups the same). It can successfully be used for mixed ethers only when one alcohol is primary and the other tertiary (otherwise a mixture of the three possible products results).

2. A method of preparing ethers by reacting a haloalkane with an alkoxide. The reaction, discovered in 1850, is a nucleophilic substitution in which the negative alkoxide ion displaces a halide ion; for example:

$RI + {}^{-}OR' \rightarrow ROR' + I^-$

A mixture of the reagents is refluxed in ethanol. The method is particularly useful for preparing mixed ethers, although a possible side reaction under some conditions is an elimination to give an alcohol and an alkene.

Wiswesser line notation (WLN) A shorthand notation for chemical compounds that can be used instead of the usual notation for chemical compounds; it consists of symbols set out in a line. The symbols used are the upper-case letters of the alphabet (A–Z), the numerals (0–9), where 0 denotes zero, with three other symbols: the ampersand (&), the hyphen (-), and the oblique stroke (/), and a blank space. Atomic symbols with one letter, such as B and F, are unchanged. Frequently occurring elements with more than one letter and functional groups are also assigned one letter; for example, G stands for chlorine, Q for hydroxyl, and Z for NH_2. The numerals indicate the number of carbon atoms in an unbranched internally saturated alkyl chain. For example, CH_3 is denoted 1 and CH_3CH_2 is denoted 2. To establish the notation of a compound the characters for the fragments are given in an established order. For example, the notation for C_2H_5OH is Q2. Rules for structures with branched chains and fused rings are also given. WLN provides a short and unambiguous notation, which is very suitable for use with a computer.

witherite A mineral form of *barium carbonate, $BaCO_3$.

WLN *See* **Wiswesser line notation**.

Wöhler, Friedrich (1800–82) German physician and chemist, who became a professor of chemistry at Göttingen. In 1828 he made his best-known discovery, the synthesis of urea (an organic compound) from ammonium cyanate (an inorganic salt). This finally disproved the assertion that organic substances can be formed only in living things. Wöhler also isolated aluminium (1827), beryllium (1828), and yttrium (1828).

Wöhler's synthesis A synthesis of urea performed by Friedrich Wöhler in 1828. He discovered that urea ($CO(NH_2)_2$) was formed when a solution of ammonium isocyanate (NH_4NCO) was evaporated. At the time it was believed that organic substances such as urea could be made only by living organisms, and its production from an inorganic compound was a notable discovery. It is sometimes (erroneously) cited as ending the belief in vitalism.

wolfram *See* tungsten.

wolframite (iron manganese tungsten) A mineral consisting of a mixed iron–manganese tungstate, $(FeMn)WO_4$, crystallizing in the monoclinic system; the principal ore of tungsten. It commonly occurs as blackish or brownish tabular crystal groups. It is found chiefly in quartz veins associated with granitic rocks. China is the major producer of wolframite.

wood alcohol *See* methanol.

Wood's metal A low-melting (71°C) alloy of bismuth (50%), lead (25%), tin (12.5%), and cadmium (12.5%). It is used for fusible links in automatic sprinkler systems. The melting point can be changed by varying the composition. It is named after William Wood (1671–1730).

Woodward, Robert Burns (1917–79) US organic chemist who worked at Harvard. He is remembered for his work in organic synthesis, producing many organic compounds including quinine, cholesterol, cortisone, lysergic acid, strychnine, chlorophyll, and vitamin B_{12}. In 1965 he formulated the *Woodward–Hoffmann rules for certain types of addition reactions. Woodward was awarded the 1965 Nobel Prize for chemistry.

Woodward–Hoffmann rules Rules governing the formation of products during certain types of organic concerted reactions. The theory of such reactions was put forward in 1969 by Woodward and Roald Hoffmann (1937–), and is concerned with the way that orbitals of the reactants change continuously into orbitals of the products during reaction and with conservation of orbital symmetry during this process. It is sometimes known as *frontier-orbital theory.*

work function A quantity that determines the extent to which thermionic or photoelectric emission will occur according to the Richardson equation or Einstein's photoelectric equation. It is sometimes expressed as a potential difference (symbol ϕ) in volts and sometimes as the energy required to remove an electron (symbol W) in electronvolts or joules. The former has been called the *work function potential* and the latter the *work function energy.*

work hardening An increase in the hardness of metals as a result of working them cold. It causes a permanent distortion of the crystal structure and is particularly apparent with iron, copper, aluminium, etc., whereas with lead and zinc it does not occur as these metals are capable of recrystallizing at room temperature.

wrought iron A highly refined form of iron containing 1–3% of slag (mostly iron silicate), which is evenly distributed throughout the material in threads and fibres so that the product has a fibrous structure quite dissimilar to that of crystalline cast iron. Wrought iron rusts less readily than other forms of metallic iron and it welds and works more easily. It is used for chains, hooks, tubes, etc.

Wurtz reaction A reaction to prepare alkanes by reacting a haloalkane with sodium:

$$2RX + 2Na \rightarrow 2NaX + RR$$

The haloalkane is refluxed with sodium in dry ether. The method is named after the French chemist Charles-Adolphe Wurtz (1817–84). The analogous reaction using a haloalkane and a haloarene, for example:

$$C_6H_5Cl + CH_3Cl + 2Na \rightarrow 2NaCl + C_6H_5CH_3$$

is called the *Fittig reaction* after the German chemist Rudolph Fittig (1835–1910).

xanthates Salts or esters containing the group –SCS(OR), where R is an organic group. Cellulose xanthate is an intermediate in the manufacture of *rayon by the viscose process.

Xanthone

xanthone (dibenzo-4-pyrone) A colourless crystalline compound, $C_{13}H_8O_2$; m.p. 174°C. The ring system is present in *xanthone dyes.*

xenobiotic Any substance foreign to living systems. Xenobiotics include drugs, pesticides, and carcinogens. Detoxification of such substances occurs mainly in the liver.

xenon Symbol Xe. A colourless odourless gas belonging to group 18 of the periodic table (*see* **noble gases**); a.n. 54; r.a.m. 131.30; d. 5.887 g dm^{-3}; m.p. –111.9°C; b.p. –107.1°C. It is present in the atmosphere (0.00087%) from which it is extracted by distillation of liquid air. There are nine natural isotopes with mass numbers 124, 126, 128–132, 134, and 136. Seven radioactive isotopes are also known. The element is used in fluorescent lamps and bubble chambers. Liquid xenon in a supercritical state at high temperatures is used as a solvent for infrared spectroscopy and for chemical reactions. The compound $Xe^+PtF_6^-$ was the first noble-gas compound to be synthesized. Several other compounds of xenon are known, including XeF_2, XeF_4, $XeSiF_6$, XeO_2F_2, and XeO_3. Recently, compounds have been isolated that contain xenon–carbon bonds, such as $[C_6H_5Xe][B(C_6H_5)_3F]$ (pentafluorophenylxenon fluoroborate), which is stable under normal conditions. The element was discovered in 1898 by Ramsey and Travers.

XPS X-ray photoelectron spectroscopy. *See* **photoelectron spectroscopy**.

X-ray crystallography The use of *X-ray diffraction to determine the structure of crystals or molecules. The technique involves directing a beam of X-rays at a crystalline sample and recording the diffracted X-rays on a photographic plate. The diffraction pattern consists of a pattern of spots on the plate, and the crystal structure can be worked out from the positions and intensities of the diffraction spots. X-rays are diffracted by the electrons in the molecules and if molecular crystals of a compound are used, the electron density distribution in the molecule can be determined.

X-ray diffraction The diffraction of X-rays by a crystal. The wavelengths of X-rays are comparable in size to the distances between atoms in most crystals, and the repeated pattern of the crystal lattice acts like a diffraction grating for X-rays. Thus, a crystal of suitable type can be used to disperse X-rays in a spectrometer. X-ray diffraction is also the basis of X-ray crystallography.

X-ray fluorescence The emission of *X-rays from excited atoms produced by the impact of high-energy electrons, other particles, or a primary beam of other X-rays. The wavelengths of the fluorescent X-rays can be measured by an X-ray spectrometer as a means of chemical analysis. X-ray fluorescence is used in such techniques as *electron-probe microanalysis.

X-rays Electromagnetic radiation of shorter wavelength than ultraviolet radiation produced by bombardment of atoms by high-quantum-energy particles. The range of wavelengths is 10^{-11} m to 10^{-9} m. Atoms of all the elements emit a characteristic *X-ray spectrum* when they are bombarded by electrons. The X-ray photons are emitted when the incident electrons knock an inner orbital electron out of an atom. When this happens an outer electron falls into the inner shell to replace it, losing potential energy (ΔE) in doing so. The wavelength λ of the emitted photon will then be given by $\lambda = ch/\Delta E$, where c is the speed of light and h is the Planck constant.

X-rays can pass through many forms of matter and they are therefore used medically and industrially to examine internal structures. X-rays are produced for these purposes by an X-ray tube.

X-ray spectrum *See* **X-rays.**

xylenes *See* **dimethylbenzenes.**

xylenol (hydroxydimethylbenzene) Any of six isomeric solid aromatic compounds, $C_6H_3(CH_3)_2OH$. An impure mixture of isomers is a liquid made from coal tar and employed as a solvent. The pure substances resemble *phenol in their reactions. They are used to make thermosetting polymer resins, and the chloro- derivative of 1,2,5-xylenol is an industrial disinfectant.

yeasts A group of unicellular fungi of the class Hemiascomycetae and phylum Ascomycota. They occur as single cells or as groups or chains of cells; yeasts reproduce asexually by budding and sexually by producing ascospores. Yeasts of the genus *Saccharomyces* ferment sugars and are used in the baking and brewing industries.

ylide A chemical species derived from an onium ion by loss of a hydron. For example, the phosphorus ylide $(C_6H_5)_2P{=}CR_2$, derived from $(C_6H_5)_2PCHR_2^+$ by loss of H^+.

yocto- Symbol y. A prefix used in the metric system to indicate 10^{-24}. For example, 10^{-24} second = 1 yoctosecond (ys).

yotta- Symbol Y. A prefix used in the metric system to indicate 10^{24}. For example, 10^{24} metres = 1 yottametre (Ym).

ytterbium Symbol Yb. A silvery metallic element belonging to the *lanthanoids; a.n. 70; r.a.m. 173.04; r.d. 6.965 (20°C); m.p. 819°C; b.p. 1194°C. It occurs in gadolinite, monazite, and xenotime. There are seven natural isotopes and ten artificial isotopes are known. It is used in certain steels. The element was discovered by Jean de Marignac (1817–94) in 1878.

yttrium Symbol Y. A silvery-grey metallic element belonging to group 3 (formerly IIIA) of the periodic table; a.n. 39; r.a.m. 88.905; r.d. 4.469 (20°C); m.p. 1522°C; b.p. 3338°C. It occurs in uranium ores and in *lanthanoid ores, from which it can be extracted by an ion exchange process. The natural isotope is yttrium-89, and there are 14 known artificial isotopes. The metal is used in superconducting alloys and in alloys for strong permanent magnets (in both cases, with cobalt). The oxide (Y_2O_3) is used in colour-television phosphors, neodymium-doped lasers, and microwave components. Chemically it resembles the lanthanoids, forming ionic compounds containing Y^{3+} ions. The metal is stable in air below 400°C. It was discovered in 1828 by Friedrich Wöhler.

Zeeman effect The splitting of the lines in a spectrum when the source of the spectrum is exposed to a magnetic field. It was discovered in 1896 by Pieter Zeeman (1865–1943). In the *normal Zeeman effect* a single line is split into three if the field is perpendicular to the light path or two lines if the field is parallel to the light path. This effect can be explained by classical electromagnetic principles in terms of the speeding up and slowing down of orbital electrons in the source as a result of the applied field. The *anomalous Zeeman effect* is a complicated splitting of the lines into several closely spaced lines, so called because it does not agree with classical predictions. This effect is explained by quantum mechanics in terms of electron spin.

Zeisel reaction A method of determining the number of methoxy ($-OCH_3$) groups in an organic compound. The compound is heated wih excess hydriodic acid, forming an alcohol and iodomethane:

$$R{-}O{-}CH_3 + HI \rightarrow ROH + CH_3I$$

The iodomethane is distilled off and led into an alcoholic solution of silver nitrate, where it precipitates silver iodide. This is filtered and weighed, and the number of iodine atoms and hence methoxy groups can be calculated. The method was developed by S. Zeisel in 1886.

Zeise's salt A complex of platinum and ethene, $PtCl_3\,(CH_2CH_2)$, in which the Pt coordinates to the pi bond of the ethene. It was synthesized by W. C. Zeise in 1827.

zeolite A natural or synthetic hydrated aluminosilicate with an open three-dimensional crystal structure, in which water molecules are held in cavities in the lattice. The water can be driven off by heating and the zeolite can then absorb other molecules of suitable size. Zeolites are used for separating mixtures by selective absorption – for this reason they are often called *molecular sieves*. They are also used in sorption pumps for vacuum systems and certain types (e.g. *Permutit*) are used in ion-exchange (e.g. water-softening).

zepto- Symbol z. A prefix used in the metric system to indicate 10^{-21}. For example, 10^{-21} second = 1 zeptosecond (zs).

zero order *See* order.

zero-point energy The energy remaining in a substance at the *absolute zero of temperature (0 K). This is in accordance with quantum theory, in which a particle oscillating with simple harmonic motion does not have a stationary state of zero kinetic energy. Moreover, the *uncertainty principle does not allow such a particle to be at rest at exactly the centrepoint of its oscillations.

zeroth law of thermodynamics *See* thermodynamics.

zetta- Symbol Z. A prefix used in the metric system to indicate 10^{21}. For example, 10^{21} metres = 1 zettametre (Zm).

Ziegler process An industrial process for the manufacture of high-density polyethene using catalysts of titanium(IV) chloride ($TiCl_4$) and aluminium alkyls (e.g. triethylaluminium, $Al(C_2H_5)_3$). The process was introduced in 1953 by the German chemist Karl Ziegler (1898–1973). It allowed the manufacture of polythene at lower temperatures (about 60°C) and pressures (about 1 atm.) than used in the original process. Moreover, the polyethene produced had more straight-chain molecules, giving the product more rigidity and a higher melting point than the earlier low-density polyethene. The reaction involves the formation of a titanium alkyl in which the titanium can coordinate directly to the pi bond in ethene.

In 1954 the process was developed further by the Italian chemist Giulio Natta (1903–79), who extended the use of Ziegler's catalysts (and similar catalysts) to other alkenes. In particular he showed how to produce stereospecific polymers of propene.

zinc Symbol Zn. A blue-white metallic element; a.n. 30; r.a.m. 65.38; r.d. 7.1; m.p. 419.88°C; b.p. 907°C. It occurs in sphalerite (or zinc blende, ZnS), which is found associated with the lead sulphide, and in smithsonite ($ZnCO_3$). Ores are roasted to give the oxide and this is reduced with carbon (coke) at high temperature, the zinc vapour being condensed. Alternatively, the oxide is dissolved in sulphuric acid and the zinc obtained by electrolysis. There are five stable isotopes (mass numbers 64, 66, 67, 68, and 70) and six radioactive isotopes are known. The metal is used in galvanizing and in a number of alloys (brass, bronze, etc.). Chemically it is a reactive metal, combining with oxygen and other nonmetals and reacting with dilute acids to release hydrogen. It also dissolves in alkalis to give *zincates. Most of its compounds contain the Zn^{2+} ion.

zincate A salt formed in solution by dissolving zinc or zinc oxide in alkali.

The formula is often written ZnO_2^{2-} although in aqueous solution the ions present are probably complex ions in which the Zn^{2+} is coordinated to OH^- ions. ZnO_2^{2-} ions may exist in molten sodium zincate, but most solid 'zincates' are mixed oxides.

zinc blende A mineral form of *zinc sulphide, ZnS, the principal ore of zinc (*see* **sphalerite**). The *zinc-blende structure* is the crystal structure of this compound (and of other compounds). It has zinc atoms surrounded by four sulphur atoms at the corners of a tetrahedron. Each sulphur is similarly surrounded by four zinc atoms. The crystals belong to the cubic system.

zinc chloride A white crystalline compound, $ZnCl_2$. The anhydrous salt, which is deliquescent, can be made by the action of hydrogen chloride gas on hot zinc; r.d. 2.9; m.p. 283°C; b.p. 732°C. It has a relatively low melting point and sublimes easily, indicating that it is a molecular compound rather than ionic. Various hydrates also exist. Zinc chloride is used as a catalyst, dehydrating agent, and flux for hard solder. It was once known as *butter of zinc*.

zinc chloride cell *See* **dry cell**.

zinc group The group of elements in the periodic table consisting of zinc (Zn), cadmium (Cd), and mercury (Hg). *See* **group 2 elements**.

zincite A mineral form of *zinc oxide, ZnO.

zinc oxide A powder, white when cold and yellow when hot, ZnO; r.d. 5.606; m.p. 1975°C. It occurs naturally as a reddish orange ore *zincite*, and can also be made by oxidizing hot zinc in air. It is amphoteric, forming *zincates with bases. It is used as a pigment (*Chinese white*) and a mild antiseptic in zinc ointments. An archaic name is *philosopher's wool*.

zinc sulphate A white crystalline water-soluble compound made by heating zinc sulphide ore in air and dissolving out and recrystallizing the sulphate. The common form is the heptahydrate, $ZnSO_4.7H_2O$; r.d. 1.9. This loses water above 30°C to give the hexahydrate and more water is lost above 70°C to form the monohydrate. The anhydrous salt forms at 280°C and this decomposes above 500°C. The compound, which was formerly called *white vitriol*, is used as a mordant and as a styptic (to check bleeding).

zinc sulphide A yellow-white water-soluble solid, ZnS. It occurs naturally as *sphalerite (*see also* **zinc blende**) and wurtzite. The compound sublimes at 1180°C. It is used as a pigment and phosphor.

zirconia *See* **zirconium**.

zirconium Symbol Zr. A grey-white metallic *transition element; a.n. 40; r.a.m. 91.22; r.d. 6.49; m.p. 1852°C; b.p. 4377°C. It is found in zircon ($ZrSiO_4$; the main source) and in baddeleyite (ZnO_2). Extraction is by chlorination to give $ZrCl_4$ which is purified by solvent extraction and reduced with magnesium (Kroll process). There are five natural isotopes (mass numbers 90, 91, 92, 94, and 96) and six radioactive isotopes are known. The element is used in nuclear reactors (it is an effective neutron absorber) and in certain alloys. The metal forms a passive layer of oxide in air and burns at 500°C. Most of its compounds are complexes of zirconium(IV). *Zirconium(IV) oxide* (*zirconia*) is used as an electrolyte in fuel cells. The element was identified in 1789 by Klaproth and was first isolated by Berzelius in 1824.

zirconium(IV) oxide *See* zirconium.

***Z*-isomer** *See* E–Z convention.

zone refining A technique used to reduce the level of impurities in certain metals, alloys, semiconductors, and other materials. It is based on the observation that the solubility of an impurity may be different in the liquid and solid phases of a material. To take advantage of this observation, a narrow molten zone is moved along the length of a specimen of the material, with the result that the impurities are segregated at one end of the bar and the pure material at the other. In general, if the impurities lower the melting point of the material they are moved in the same direction as the molten zone moves, and vice versa.

zwitterion (ampholyte ion) An ion that has a positive and negative charge on the same group of atoms. Zwitterions can be formed from compounds that contain both acid groups and basic groups in their molecules. For example, aminoethanoic acid (the amino acid glycine) has the formula $H_2N.CH_2.COOH$. However, under neutral conditions, it exists in the different form of the zwitterion $^+H_3N.CH_2.COO^-$, which can be regarded as having been produced by an internal neutralization reaction (transfer of a proton from the carboxyl group to the amino group). Aminoethanoic acid, as a consequence, has some properties characteristic of ionic compounds; e.g. a high melting point and solubility in water. In acid solutions, the positive ion $^+H_3NCH_2COOH$ is formed. In basic solutions, the negative ion $H_2NCH_2COO^-$ predominates. The name comes from the German *zwei*, two.

Appendix 1. The Greek alphabet

Letters		Name	Letters		Name
Α	α	alpha	Ν	ν	nu
Β	β	beta	Ξ	ξ	xi
Γ	γ	gamma	Ο	ο	omicron
Δ	δ	delta	Π	π	pi
Ε	ϵ	epsilon	Ρ	ρ	rho
Ζ	ζ	zeta	Σ	σ	sigma
Η	η	eta	Τ	τ	tau
Θ	θ	theta	Υ	υ	upsilon
Ι	ι	iota	Φ	φ	phi
Κ	κ	kappa	Χ	χ	chi
Λ	λ	lambda	Ψ	ψ	psi
Μ	μ	mu	Ω	ω	omega

Appendix 2. Fundamental constants

Constant	Symbol	Value in SI units
acceleration of free fall	g	$9.806\ 65\ \mathrm{m\ s^{-2}}$
Avogadro constant	L, N_A	$6.022\ 1367(36) \times 10^{23}\ \mathrm{mol^{-1}}$
Boltzmann constant	$k = R/N_A$	$1.380\ 658(12) \times 10^{-23}\ \mathrm{J\ K^{-1}}$
electric constant	ε_0	$8.854\ 187\ 817 \times 10^{-12}\ \mathrm{F\ m^{-1}}$
electronic charge	e	$1.602\ 177\ 33(49) \times 10^{-19}\ \mathrm{C}$
electronic rest mass	m_e	$9.109\ 3897(54) \times 10^{-31}\ \mathrm{kg}$
Faraday constant	F	$9.648\ 5309(29) \times 10^{4}\ \mathrm{C\ mol^{-1}}$
gas constant	R	$8.314\ 510(70)\ \mathrm{J\ K^{-1}\ mol^{-1}}$
gravitational constant	G	$6.672\ 59(85) \times 10^{-11}\ \mathrm{m^3\ kg^{-1}\ s^{-2}}$
Loschmidt's constant	N_L	$2.686\ 763(23) \times 10^{25}\ \mathrm{m^{-3}}$
magnetic constant	μ_0	$4\pi \times 10^{-7}\ \mathrm{H\ m^{-1}}$
neutron rest mass	m_n	$1.674\ 9286(10) \times 10^{-27}\ \mathrm{kg}$
Planck constant	h	$6.626\ 0755(40) \times 10^{-34}\ \mathrm{J\ s}$
proton rest mass	m_p	$1.672\ 6231(10) \times 10^{-27}\ \mathrm{kg}$
speed of light	c	$2.997\ 924\ 58 \times 10^{8}\ \mathrm{m\ s^{-1}}$
Stefan–Boltzmann constant	σ	$5.670\ 51(19) \times 10^{-8}\ \mathrm{W\ m^{-2}\ K^{-4}}$

Appendix 3. SI units

TABLE 3.1 Base and dimensionless SI units

Physical quantity	Name	Symbol
length	metre	m
mass	kilogram	kg
time	second	s
electric current	ampere	A
thermodynamic temperature	kelvin	K
luminous intensity	candela	cd
amount of substance	mole	mol
*plane angle	radian	rad
*solid angle	steradian	sr

*dimensionless units

TABLE 3.2 Derived SI units with special names

Physical quantity	Name of SI unit	Symbol of SI unit
frequency	hertz	Hz
energy	joule	J
force	newton	N
power	watt	W
pressure	pascal	Pa
electric charge	coulomb	C
electric potential difference	volt	V
electric resistance	ohm	Ω
electric conductance	siemens	S
electric capacitance	farad	F
magnetic flux	weber	Wb
inductance	henry	H
magnetic flux density (magnetic induction)	tesla	T
luminous flux	lumen	lm
illuminance	lux	lx
absorbed dose	gray	Gy
activity	becquerel	Bq
dose equivalent	sievert	Sv

TABLE 3.3 Decimal multiples and submultiples to be used with SI units

Submultiple	Prefix	Symbol	Multiple	Prefix	Symbol
10^{-1}	deci	d	10	deca	da
10^{-2}	centi	c	10^{2}	hecto	h
10^{-3}	milli	m	10^{3}	kilo	k
10^{-6}	micro	μ	10^{6}	mega	M
10^{-9}	nano	n	10^{9}	giga	G
10^{-12}	pico	p	10^{12}	tera	T
10^{-15}	femto	f	10^{15}	peta	P
10^{-18}	atto	a	10^{18}	exa	E
10^{-21}	zepto	z	10^{21}	zetta	Z
10^{-24}	yocto	y	10^{24}	yotta	Y

TABLE 3.4 Conversion of units to SI units

From	To	Multiply by
in	m	2.54×10^{-2}
ft	m	0.3048
sq. in	m^2	6.4516×10^{-4}
sq. ft	m^2	9.2903×10^{-2}
cu. in	m^3	1.63871×10^{-5}
cu. ft	m^3	2.83168×10^{-2}
l(itre)	m^3	10^{-3}
gal(lon)	l(itre)	4.546 09
miles/hr	$m\ s^{-1}$	0.477 04
km/hr	$m\ s^{-1}$	0.277 78
lb	kg	0.453 592
$g\ cm^{-3}$	$kg\ m^{-3}$	10^{3}
lb/in^3	$kg\ m^{-3}$	$2.767\,99 \times 10^{4}$
dyne	N	10^{-5}
poundal	N	0.138 255
lbf	N	4.448 22
mmHg	Pa	133.322
atmosphere	Pa	$1.013\,25 \times 10^{5}$
hp	W	745.7
erg	J	10^{-7}
eV	J	$1.602\,10 \times 10^{-19}$
kW h	J	3.6×10^{6}
cal	J	4.1868

Appendix 4. The electromagnetic spectrum

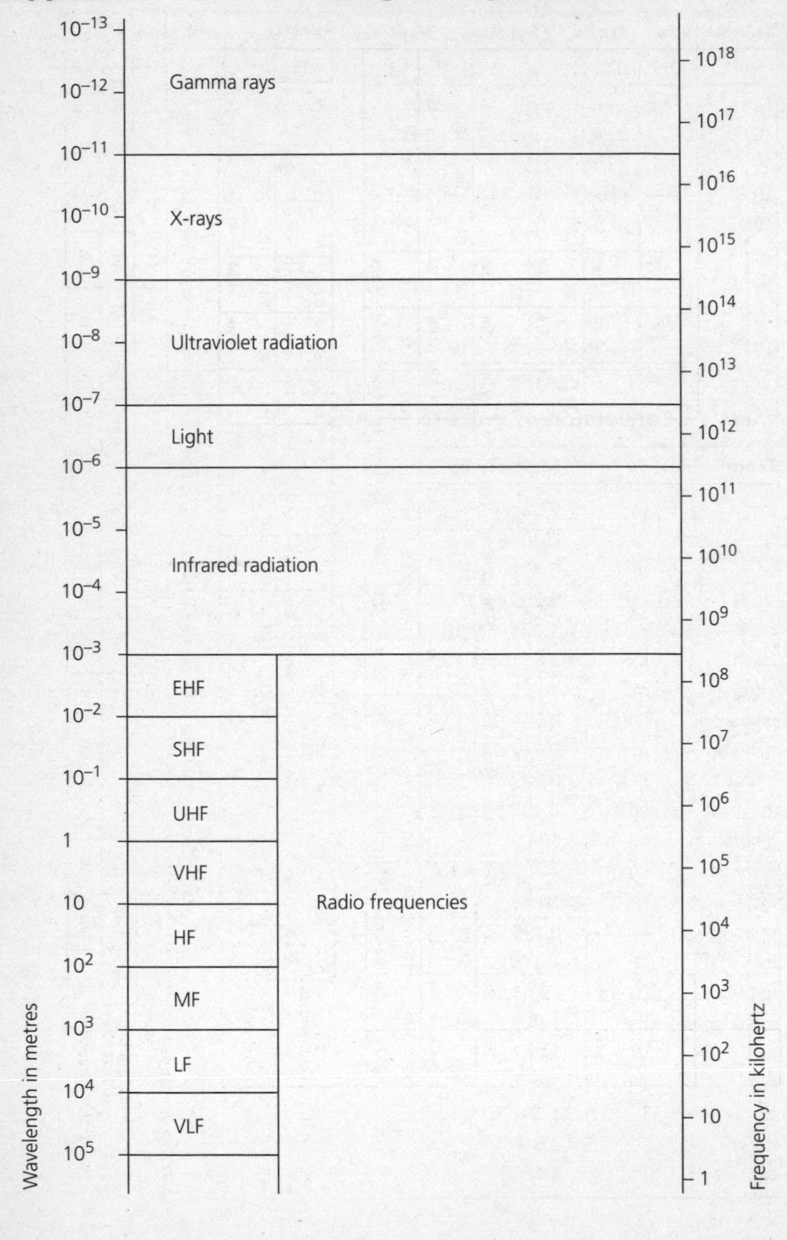

Appendix 5. The periodic table

Group	1	2	3	4	5	6	7	8	9	10	11	12	13	14	15	16	17	18	Period n
	1 **H**																	2 **He**	1
	3 **Li**	4 **Be**											5 **B**	6 **C**	7 **N**	8 **O**	9 **F**	10 **Ne**	2
	11 **Na**	12 **Mg**											13 **Al**	14 **Si**	15 **P**	16 **S**	17 **Cl**	18 **Ar**	3
	19 **K**	20 **Ca**	21 **Sc**	22 **Ti**	23 **V**	24 **Cr**	25 **Mn**	26 **Fe**	27 **Co**	28 **Ni**	29 **Cu**	30 **Zn**	31 **Gs**	32 **Ge**	33 **As**	34 **Se**	35 **Br**	36 **Kr**	4
	37 **Rb**	38 **Sr**	39 **Y**	40 **Zr**	41 **Nb**	42 **Mo**	43 **Tc**	44 **Ru**	45 **Rh**	46 **Pd**	47 **Ag**	48 **Cd**	49 **In**	50 **Sn**	51 **Sb**	52 **Te**	53 **I**	54 **Xe**	5
	55 **Cs**	56 **Ba**	57–71 **La–Lu**	72 **Hf**	73 **Ta**	74 **W**	75 **Re**	76 **Os**	77 **Ir**	78 **Pt**	79 **Au**	80 **Hg**	81 **Tl**	82 **Pb**	83 **Bi**	84 **Po**	85 **At**	86 **Rn**	6
	87 **Fr**	88 **Ra**	89–103 **Ac–Lr**	104 **Rf**	105 **Db**	106 **Sg**	107 **Bh**	108 **Hs**	109 **Mt**	110 **Ds**	111 **Rg**	112 **Uub**	113 **Uut**	114 **Uuq**	115 **Uup**	116 **Uuh**			7

																Period
Lanthanoids	57 **La**	58 **Ce**	59 **Pr**	60 **Nd**	61 **Pm**	62 **Sm**	63 **Eu**	64 **Gd**	65 **Tb**	66 **Dy**	67 **Ho**	68 **Er**	69 **Tm**	70 **Yb**	71 **Lu**	6
Actinoids	89 **Ac**	90 **Th**	91 **Pa**	92 **U**	93 **Np**	94 **Pu**	95 **Am**	96 **Cm**	97 **Bk**	98 **Cf**	99 **Fs**	100 **Fm**	101 **Md**	102 **No**	103 **Lr**	7

Correspondence of recommended group designations to other designations in recent use

IUPAC Recommendations 1990	1	2	3	4	5	6	7	8	9	10	11	12	13	14	15	16	17	18
Usual European Convention	IA	IIA	IIIA	IVA	VA	VIA	VIIA	VIII (or VIIIA)			IB	IIB	IIIB	IVB	VB	VIB	VIIB	0 (or VIIIB)
Usual US Convention	IA	IIA	IIIB	IVB	VB	VIB	VIIB	VIII			IB	IIB	IIIA	IVA	VA	VIA		VIIIA (or 0)

Appendix 6. The chemical elements

R.a.m. values with asterisk denote mass number of the most stable known isotope

Element	Symb	a.n.	r.a.m.
actinium	Ac	89	227*
aluminium	Al	13	26.98
americium	Am	95	243*
antimony	Sb	51	121.75
argon	Ar	18	39.948
arsenic	As	33	74.92
astatine	At	85	210*
barium	Ba	56	137.34
berkelium	Bk	97	247*
beryllium	Be	4	9.012
bismuth	Bi	83	208.98
bohrium	Bh	107	262*
boron	B	5	10.81
bromine	Br	35	79.909
cadmium	Cd	48	112.41
caesium	Cs	55	132.905
calcium	Ca	20	40.08
californium	Cf	98	251*
carbon	C	6	12.011
cerium	Ce	58	140.12
chlorine	Cl	17	35.453
chromium	Cr	24	52.00
cobalt	Co	27	58.933
copper	Cu	29	63.546
curium	Cm	96	247*
darmstadtium	Ds	110	271*
dubnium	Db	105	262*
dysprosium	Dy	66	162.50
einsteinium	Es	99	254*
erbium	Er	68	167.26
europium	Eu	63	151.96
fermium	Fm	100	257*
fluorine	F	9	18.9984
francium	Fr	87	223*
gadolinium	Gd	64	157.25
gallium	Ga	31	69.72
germanium	Ge	32	72.59
gold	Au	79	196.967
hafnium	Hf	72	178.49
hassium	Hs	108	265*
helium	He	2	4.0026
holmium	Ho	67	164.93
hydrogen	H	1	1.008
indium	In	49	114.82
iodine	I	53	126.9045
iridium	Ir	77	192.20
iron	Fe	26	55.847
krypton	Kr	36	83.80
lanthanum	La	57	138.91
lawrencium	Lr	103	256*
lead	Pb	82	207.19
lithium	Li	3	6.939
lutetium	Lu	71	174.97
magnesium	Mg	12	24.305
manganese	Mn	25	54.94
meitnerium	Mt	109	266*
mendelevium	Md	101	258*
mercury	Hg	80	200.59
molybdenum	Mo	42	95.94
neodymium	Nd	60	144.24
neon	Ne	10	20.179
neptunium	Np	93	237.0482
nickel	Ni	28	58.70
niobium	Nb	41	92.91
nitrogen	N	7	14.0067
nobelium	No	102	254*
osmium	Os	76	190.2
oxygen	O	8	15.9994
palladium	Pd	46	106.4
phosphorus	P	15	30.9738
platinum	Pt	78	195.09
plutonium	Pu	94	244*

Element	Symb	a.n.	r.a.m.
polonium	Po	84	210*
potassium	K	19	39.098
praseodymium	Pr	59	140.91
promethium	Pm	61	145
protactinium	Pa	91	231.036
radium	Ra	88	226.0254
radon	Rn	86	222*
rhenium	Re	75	186.2
rhodium	Rh	45	102.9
roentgenium	Rg	111	272*
rubidium	Rb	37	85.47
ruthenium	Ru	44	101.07
rutherfordium	Rf	104	261*
samarium	Sm	62	150.35
scandium	Sc	21	44.956
seaborgium	Sg	106	263*
selenium	Se	34	78.96
silicon	Si	14	28.086
silver	Ag	47	107.87
sodium	Na	11	22.9898
strontium	Sr	38	87.62
sulphur	S	16	32.06
tantalum	Ta	73	180.948
technetium	Tc	43	98*
tellurium	Te	52	127.60
terbium	Tb	65	158.92
thallium	Tl	81	204.39
thorium	Th	90	232.038
thulium	Tm	69	168.934
tin	Sn	50	118.69
titanium	Ti	22	47.9
tungsten	W	74	183.85
ununbium	Uub	112	285*
ununtrium	Uut	113	284*
ununquadium	Uuq	114	289*
ununpentium	Uup	115	288*
ununhexium	Uuh	116	292*
uranium	U	92	238.03
vanadium	V	23	50.94
xenon	Xe	54	131.30
ytterbium	Yb	70	173.04
yttrium	Y	39	88.905
zinc	Zn	30	65.38
zirconium	Zr	40	91.22

Appendix 7. Useful websites

AAAS www.sciencemag.org
Access to the *Science* magazine of the American Association for the Advancement of Science.

Alchemy http://levity.com/alchemy
An extensive collection of information on alchemy in all its facets. It provides over 200 complete alchemical texts, extensive bibliographical material, numerous articles, and introductory and general reference material on alchemy.

American Chemical Society www.chemistry.org/portal/a/c/s/1/home.html
The official website for the ACS. Packed with news, it also offers authoritative searches powered by Google.

Chemdex www.chemdex.org
An online directory of chemistry on the web established in 1993 and containing over 5000 links to further resources.

Chemistry and Industry www.chemind.org/CI/index.jsp
The website of the journal *Chemistry and Industry* offering daily news and selected current and past articles from the magazine.

Chemlab http://chemlab.pc.maricopa.edu/periodic/about.html
An interactive periodic table with a comprehensive database of element properties, which can be searched and collated in novel and useful ways.

CONFCHEM http://ched-ccce.org/confchem
Regular online conferences in chemistry education and research. Dates and topics are announced on this website, via the CONFCHEM list and other chemistry-related Internet sources, and in various print publications. CONFCHEM papers are available on this website and discussion takes place via the CONFCHEM email list.

IUPAC http://iupac.chemsoc.org/dhtml_home.html
The official home page of the International Union of Pure and Applied Chemistry.

IUPAC Nomenclature www.chem.qmul.ac.uk/iupac
A large amount of information on organic and biochemical nomenclature held at Queen Mary College, University of London.

Nature Magazine Online www.nature.com
An online weekly journal that offers news articles and features, complete reference works online, and information on the latest science research.

New Scientist www.newscientist.com
A popular news and archive site for all branches of science.

Royal Society of Chemistry www.chemsoc.org/timeline
An exploration of key events in the history of science with a particular emphasis on chemistry.

Scientific American www.sciam.com
A popular science news site containing selected recent articles.

Webelements www.webelements.com
A periodic table at the University of Sheffield linked to a very comprehensive database of the elements and their compounds.